Bright 2

ARCHITECTURAL ILLUMINATION AND LIGHT INSTALLATIONS

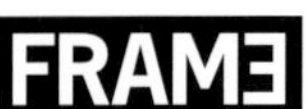

Contents

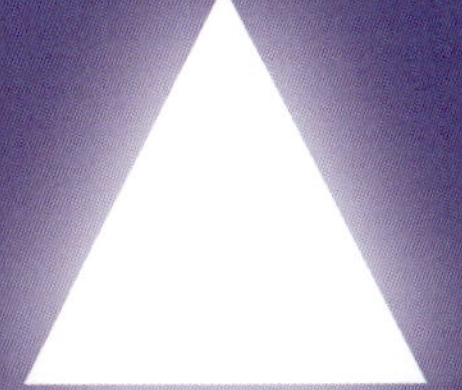

DYNAMIC

INTERACTIVE

STATIC

Dynamic

004

005

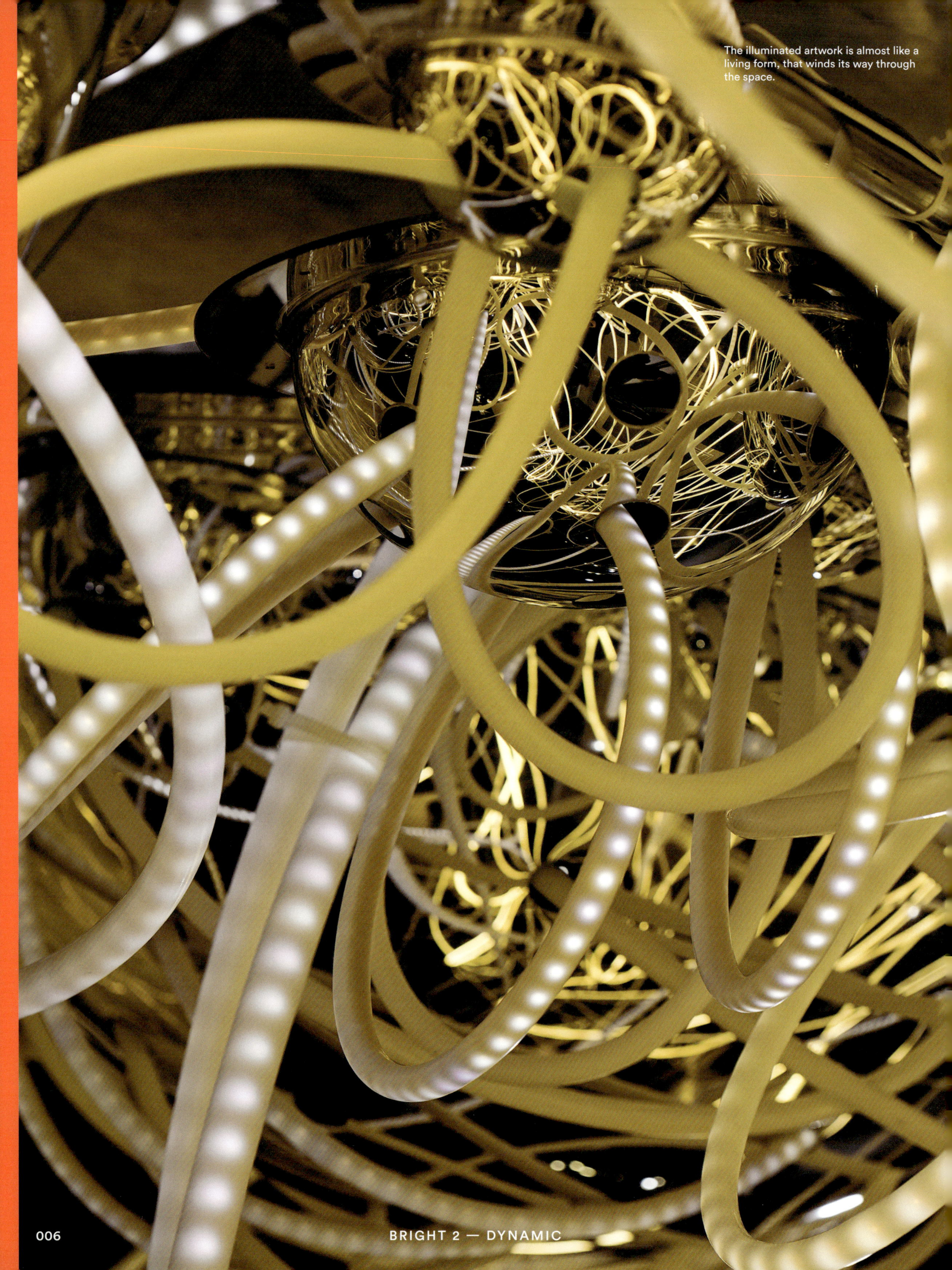

The illuminated artwork is almost like a living form, that winds its way through the space.

Breathless Maiden Lane

A cutting-edge, undulating and dynamic light installation by **Grimanesa Amorós Studio** in the heart of Lower Manhattan creates an immersive environment that responds to its surrounding architecture in the marble and granite atrium in New York's Financial District.

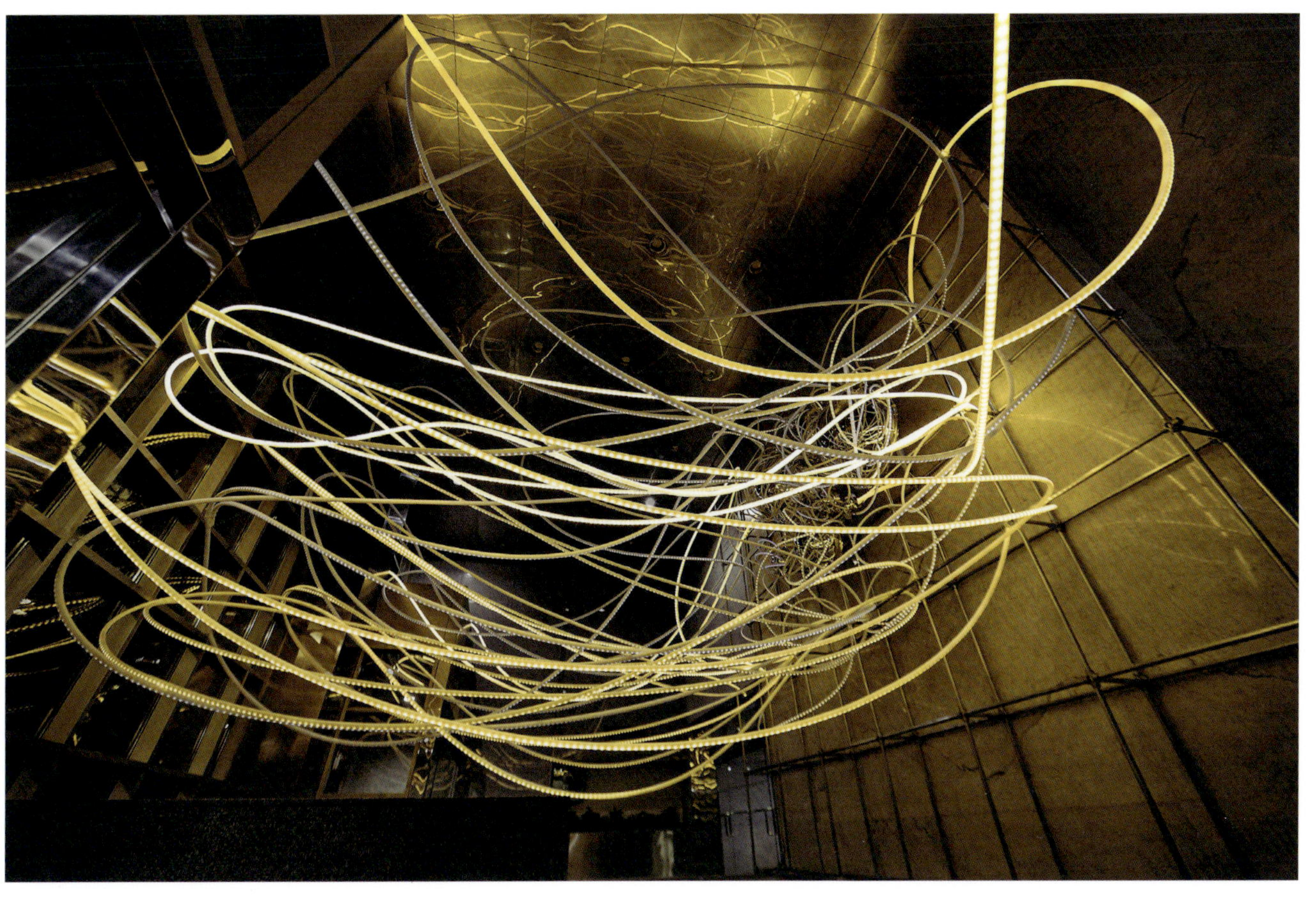

Photos Grimanesa Amorós Studio

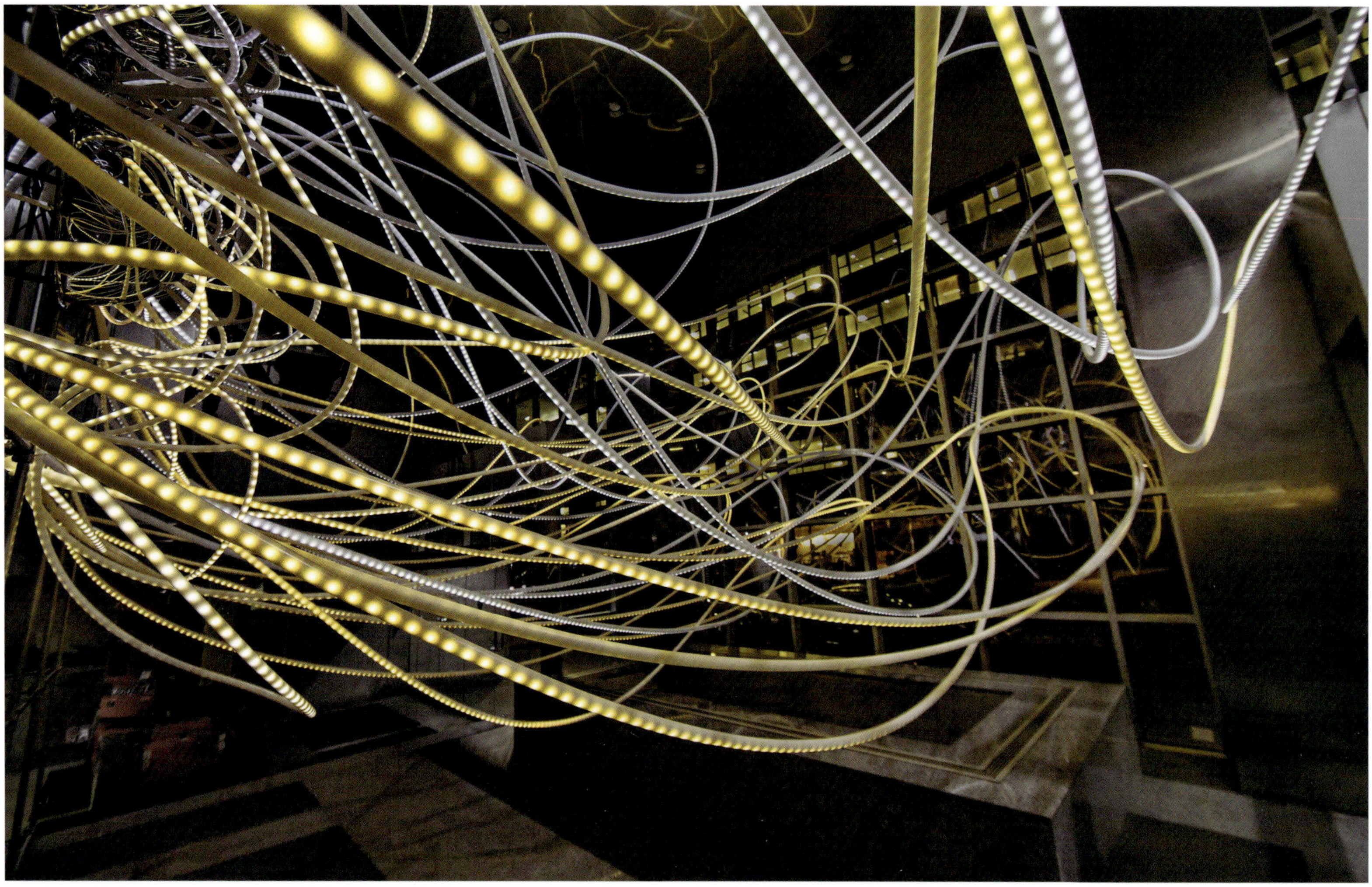

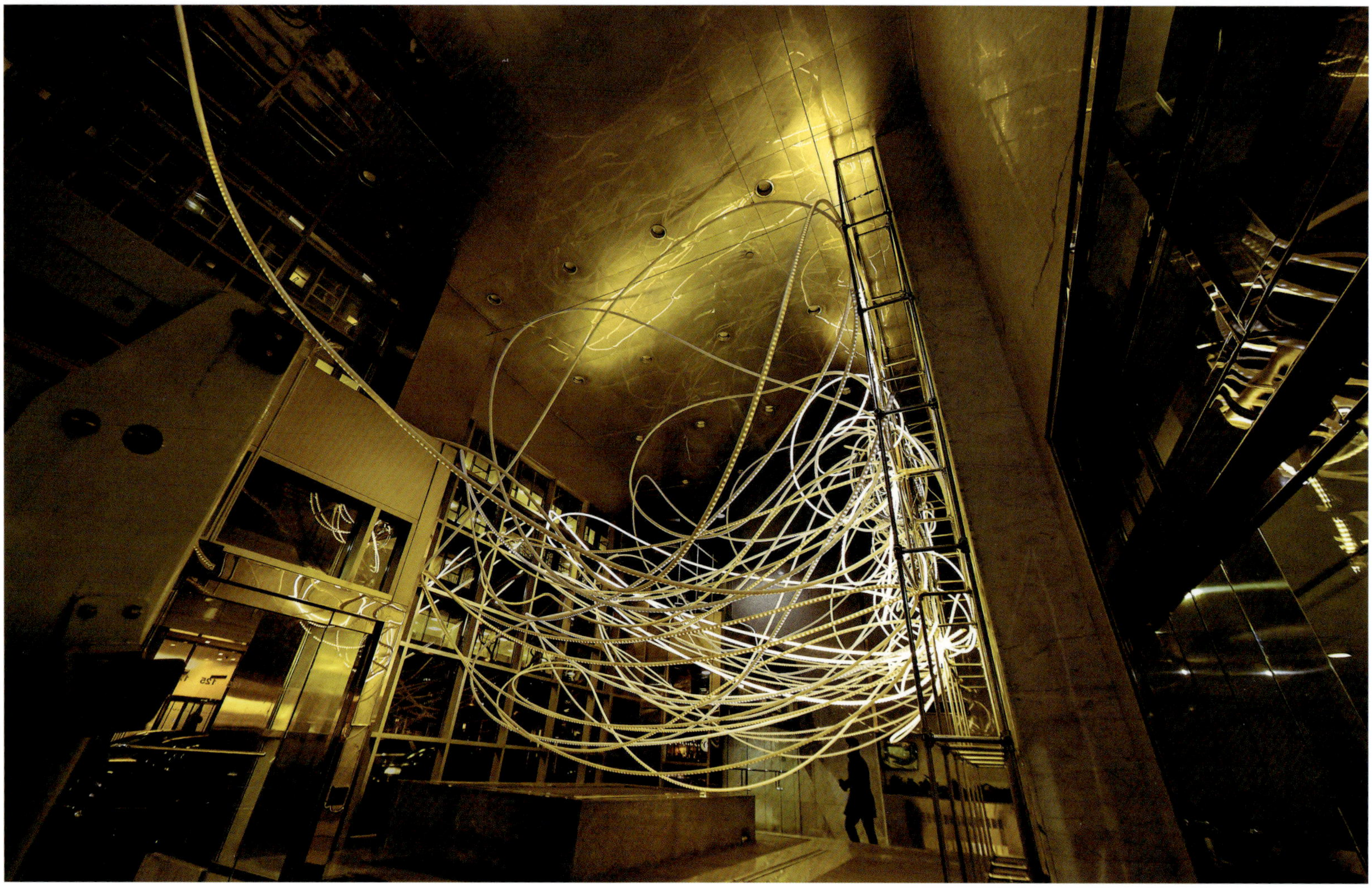

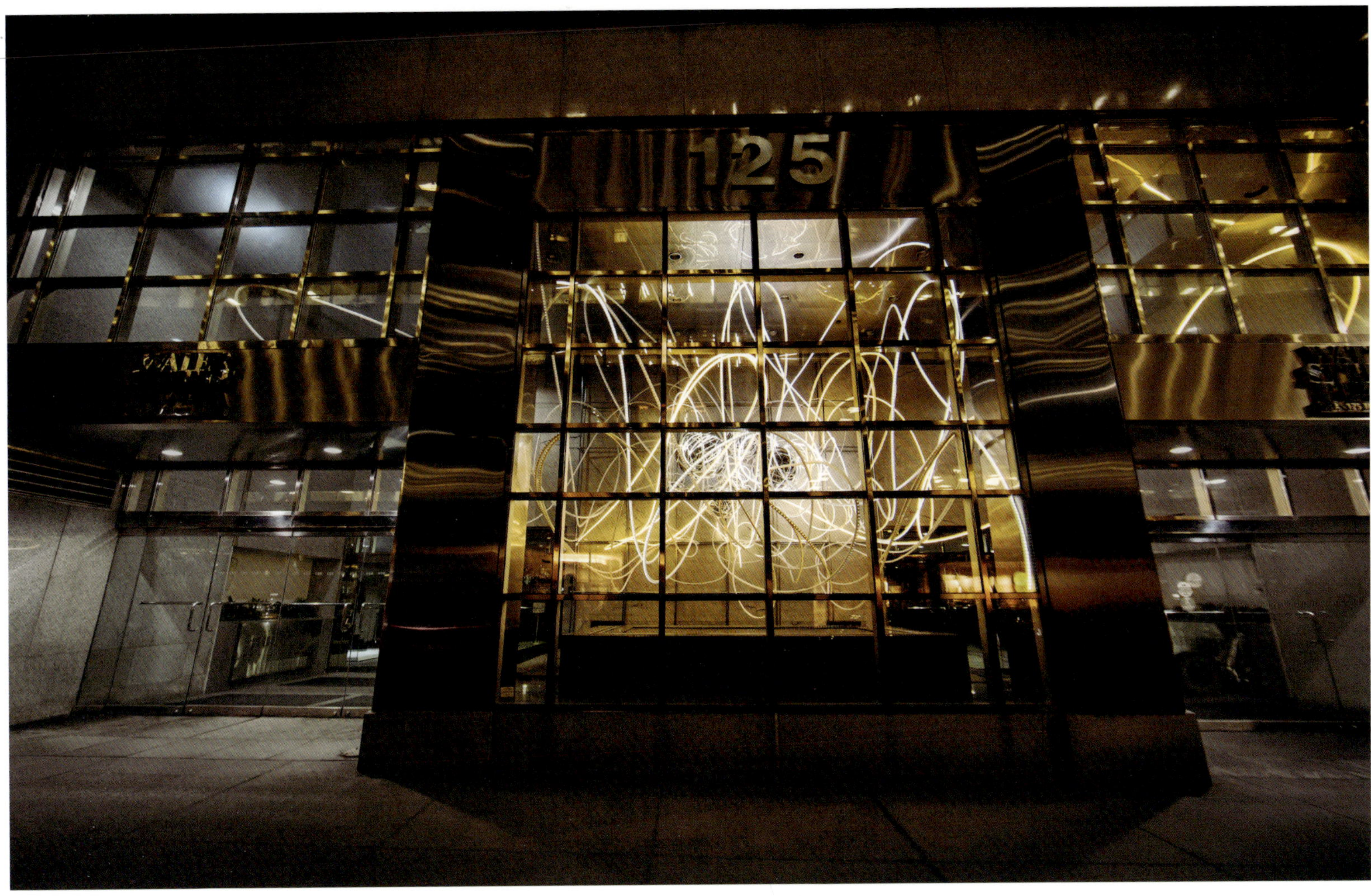

Illuminating the atrium with a dynamic pattern, the artwork work an ethereal, almost pulsating quality.

SUBTLE NUANCES IN THE LIGHTING SEQUENCE, THE CADENCE OF WHICH REVEALS ITSELF LIKE A MUSICAL SCORE

Designer
Grimanesa Amorós
Location
New York, United States
Client
Time Equities Incorporated
Collaborator(s)/ consultant
Theatre Consultants Collaborative
Manufacturers
Philips, Color Kinetics, OptiLED, Elemental LED
Date
February 2014

Breathless Maiden Lane by Grimanesa Amorós Studio is located in a unique exhibition space in New York's Financial District, right in the heart of Lower Manhattan. Grimanesa Amorós is a multidisciplinary artist whose practice incorporates elements from sculpture, video, sound art and cutting-edge technology to create site-responsive installations to engage architecture and create community. The sculpture enhances and reveals the atrium's architecture in which it is suspended: a glass, marble and granite space.

Utilising LED lights in combination with diffusive reflective material and her signature 'bubble' sculptures. Hovering above the ground as if weightless, the work appears to defy gravity – a structural grid, designed to echo the building's monumental windows, stands against the back wall of the atrium. This structure serves as the work's spine, supporting the bubble sculptures and graceful LED lines, which appear to stretch out and explore the architecture of the atrium in undulating loops. Some lines touch the window panes as if grasping for the street. The result is a marvelous tangle of coils, swirls and arcs. A dynamic pattern activates the LEDs, in four shades of white and a golden yellow, giving the work an ethereal quality. At night, reflections from Breathless Maiden Lane bounce off the high shine marble walls, stainless steel ceiling and windows to create an immersive environment of reflections on an endless feedback loop.

Amorós researches the history of many installation sites; however her process remains organic and instinctive. This intuitive relationship to technology is a distinctive feature of her studio's practice. Some elements must be planned and programmed but others, such as the exact placement of the lines of lights, come to artist as the installation is created on-site. In this sense, the technology does not determine but complements the aesthetics of her work. The glittering lights are undeniably spectacular, but it is the subtle nuances – the cadence of the custom lighting sequence revealing itself like a musical score – that compels continued and focused viewing. —

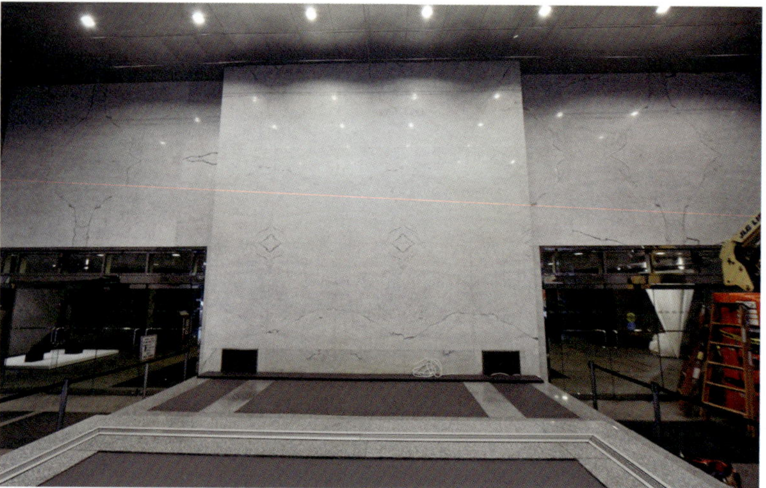

The framework was affixed to the back wall so that the installation could fill the atrium with an explosion of light.

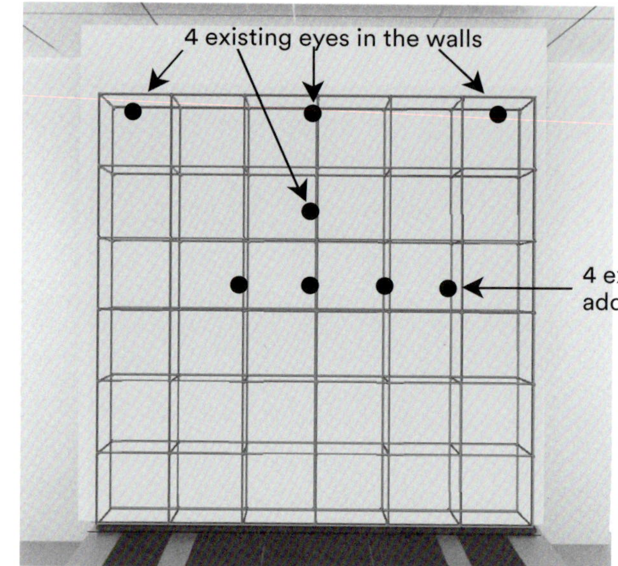

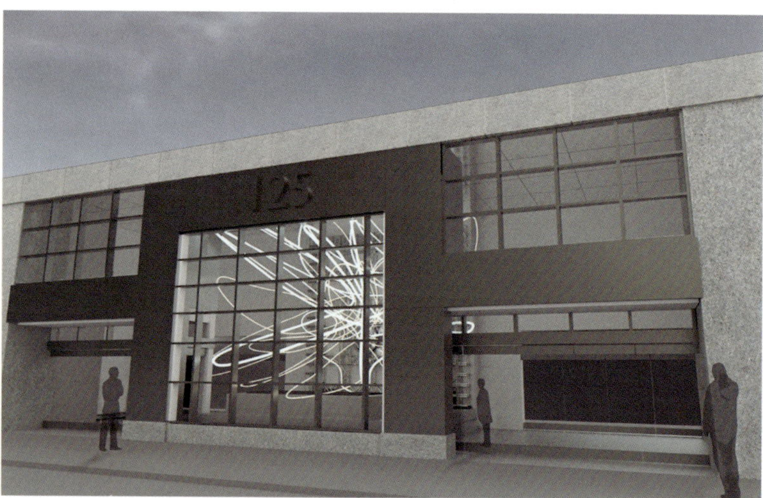

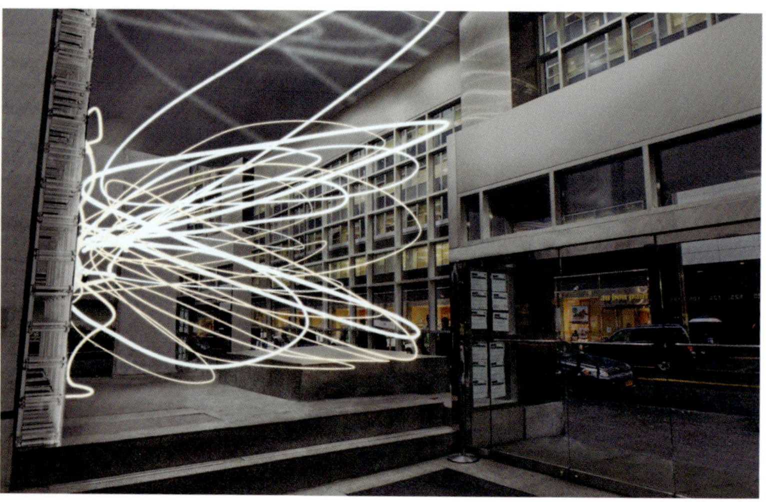

TECHNOLOGY DOES NOT DETERMINE BUT COMPLEMENTS THE AESTHETICS OF THE WORK

Set-up

Volume of the installation for illumination was a total of 230 m³ (7.6 x 5.5 x 5.5 m).
Component parts included mixed media, LEDs, diffusive reflective material, custom lighting sequence, electrical hardware, metal framework.
Used light included 25-m long, 24 V LED (85 W).
Lighting control console incorporated a DMX signal device to send data to the DMX controller, taking the data and outputting it to the appropriate LED channels and the lighting fixtures, the 24 V power supply for which was routed through the DMX controller.
Metal framework (36 m²) was assembled alongside the installation to ensure the secure positioning of the suspended sculpture in the atrium, utilising 2-cm diameter stainless steel pipes of different lengths (1 m and 6 m).
Construction time in total was 14 days; arrangement of the LED tubes were carried out on-site in a very organic process by the artist.

LIGHTING SYSTEM DIAGRAM

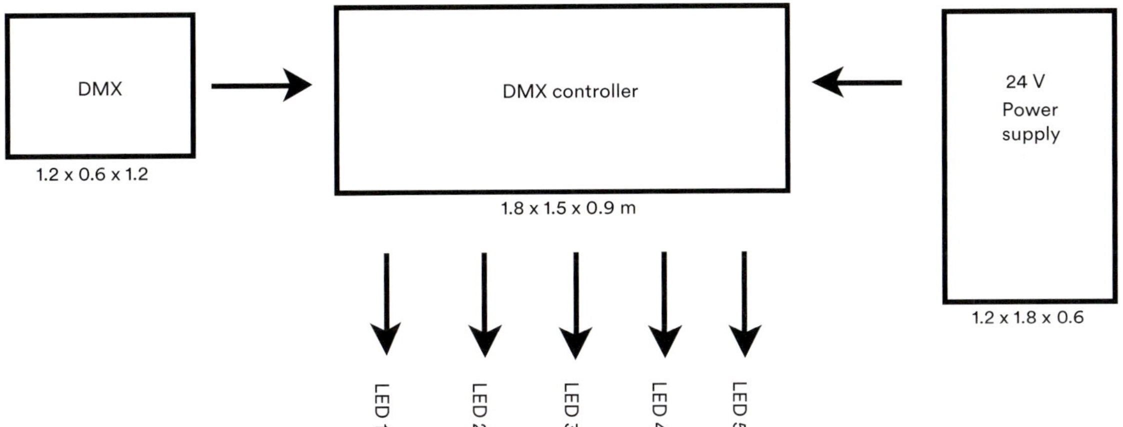

Specifics

The initial challenge came about from the positioning of the installation. Drilling into the marble walls of the grand atrium was prohibited, so it was necessary to use existing holes and screw eyes to hang the artwork.

Another difficulty was that there is no power outlets in the front atrium. Power extension cables were required to be run from the back atrium to the piece.

The vibrancy of the piece comes from the illumination creating a 'wow' factor to passers-by and users of the building. Power consumption was a consideration, as with all such projects that need to run 24/7. The solution was to incorporate a timer for the lighting, to automatically rest the piece through the night.

FRAMEWORK STRUCTURE ASSEMBLY

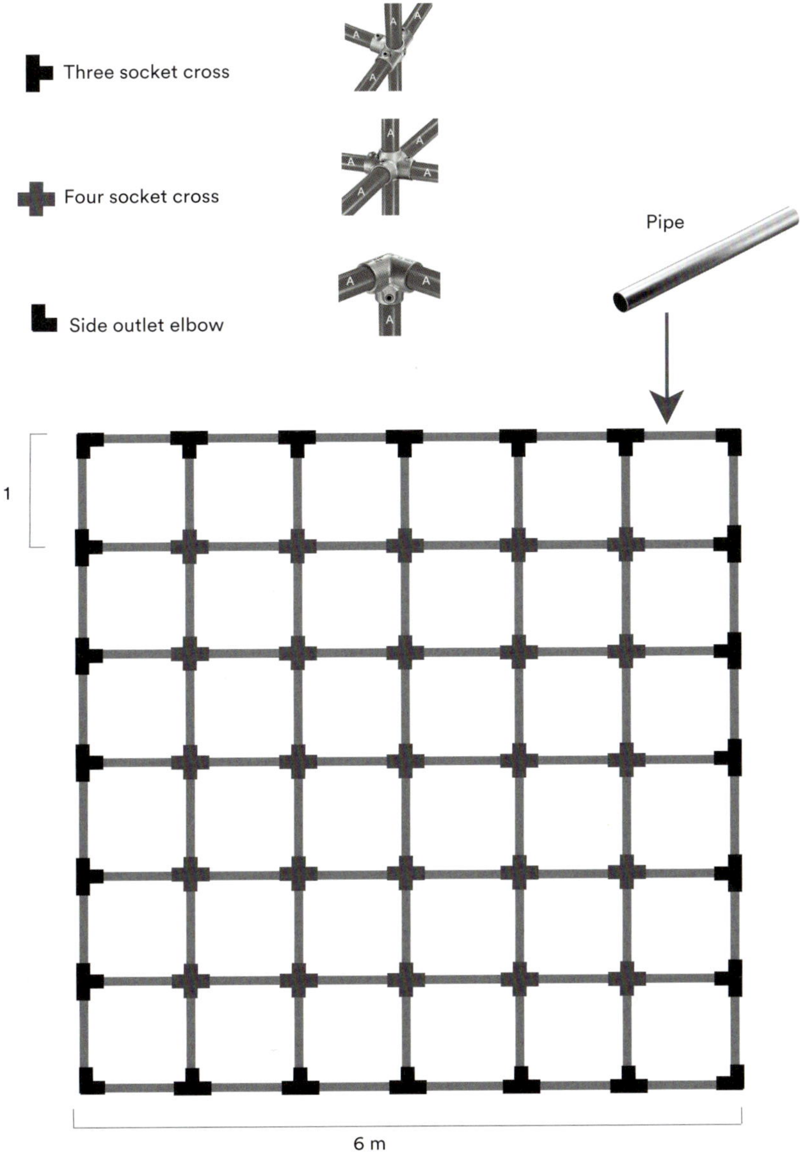

The illuminated loops are positioned by the artist on-site in an instinctive and organic process.

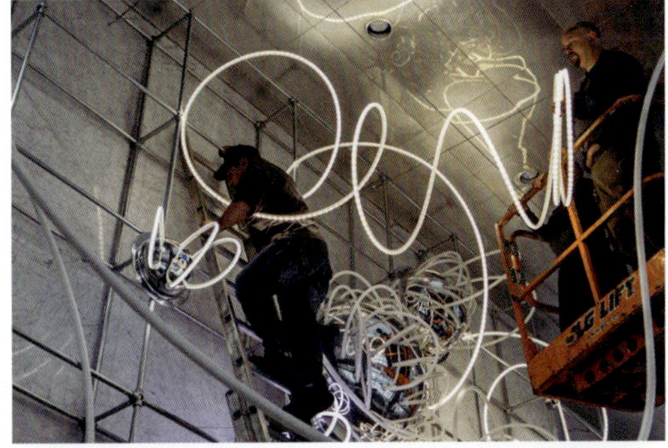

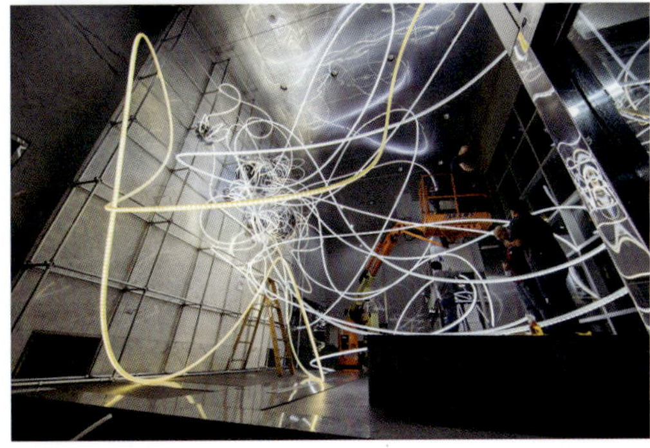

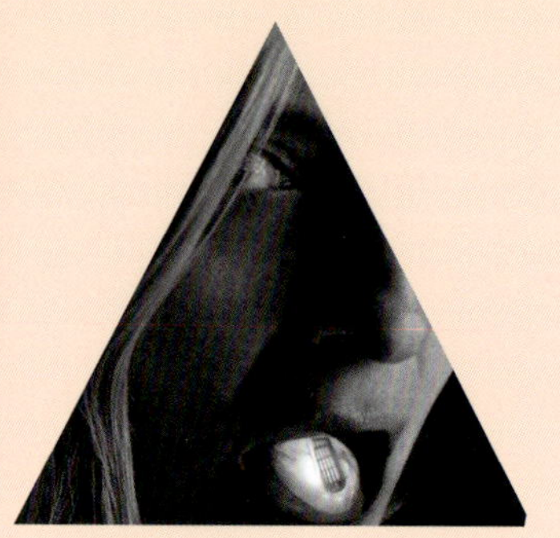

Grimanesa Amorós Studio

Grimanesa Amorós was born in Lima, Peru and lives and works in New York City. She is a multidisciplinary artist with diverse interests in the fields of social history, scientific research and critical theory, which have greatly influenced her work. Amorós researches the locations, histories and communities of the installation sites. The working process includes an intuitive relationship to technology, which is a distinctive feature of Amorós' practice. Her work incorporates elements from sculpture, video, lighting and cutting-edge technology to create site-specific installations to engage architecture and create community.

GRIMANESAAMOROS.COM

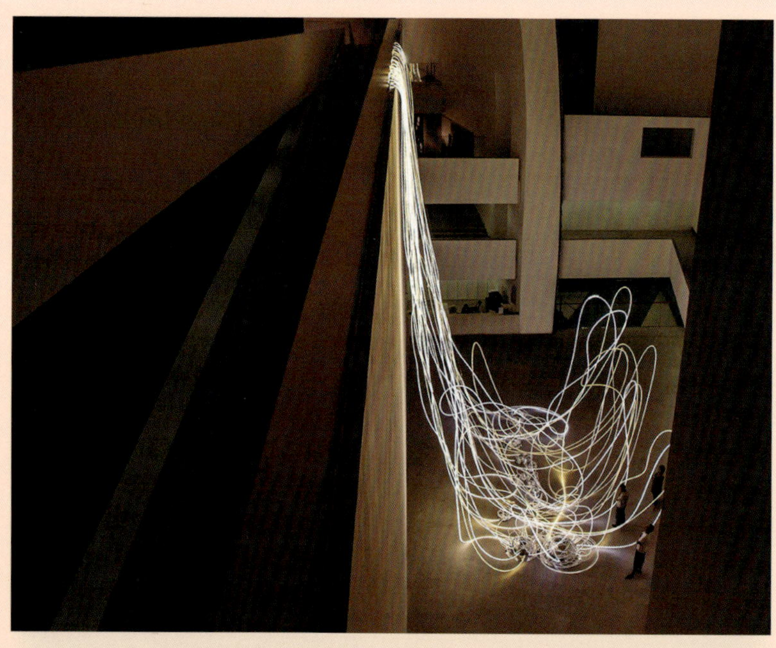

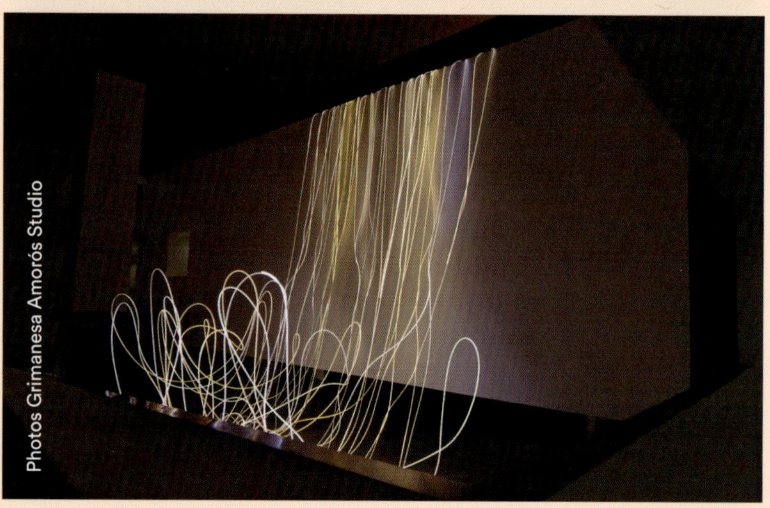

Photos Grimanesa Amorós Studio

The Mirror Connection

When **2013**
Where **Beijing, China**
Client **Museum of China Central Academy of Fine Arts**

Inspired by the luminosity, expansion and purity of the lines in the Museum of China Central Academy of Fine Arts in Beijing, designed by architect Arata Isozaki, Grimanesa Amorós took the opportunity to bring together different facets of the creative process, including an inclination towards asymmetrical balance and her search for movement and infinite lightness. 'My interests come together in The Mirror Connection, a piece where we make connections with ourselves, the space and the artwork', comments Amorós.

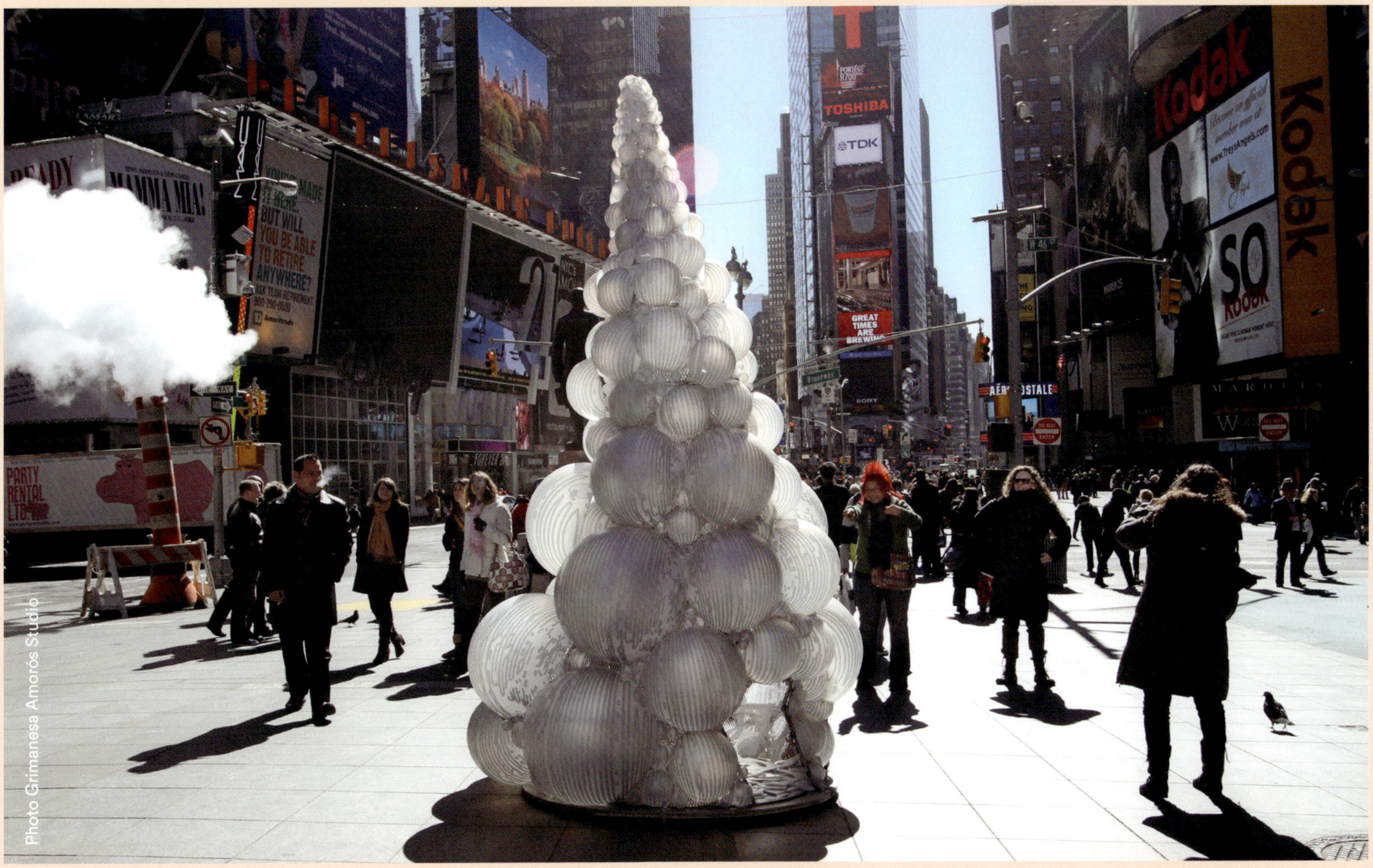

Photo Grimanesa Amorós Studio

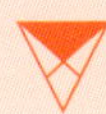

Fortuna

When **2013**
Where **Madrid, Spain**
Client **Ministry of Education and Culture of Spain**

In the art centre La Fragua Tabacalera, this light sculpture reflects the history of the location, which once was Tobacco factory in Madrid. The installation's name comes from the filters that were used for the La Tabacalera cigars. The site-specific installation comprised various lengths of tubing that extended through the space, differentiated according to thickness, with a tone and ambience created with white LEDs. The sculpture's lights shifted and changed as the viewer 'travelled' through the piece examining its architectural complexity.

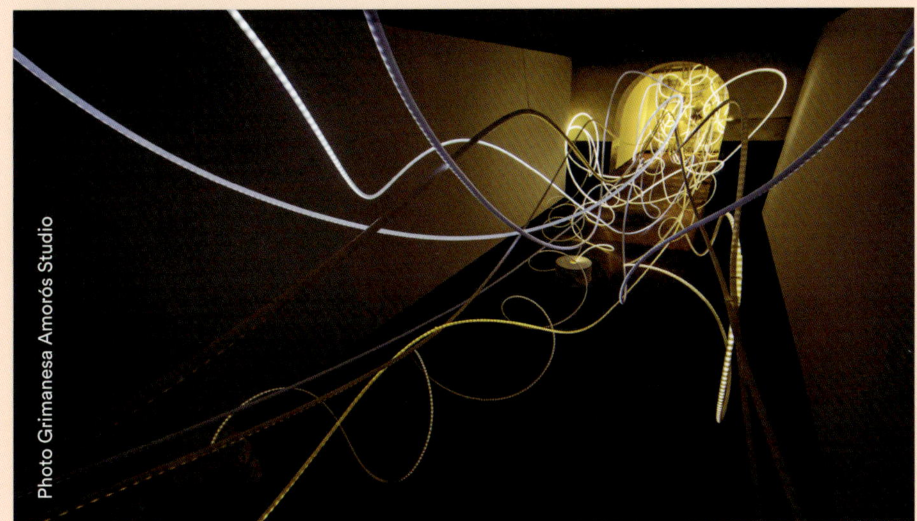

Photo Grimanesa Amorós Studio

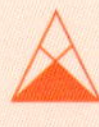

Uros House in Times Square

When **2011**
Where **New York, United States**
Client **Times Square Alliance**

Since growing up on the coast of Peru, Grimanesa Amorós has always loved the beauty of the ocean and the froth of the waves. Off the Peruvian coast are the Uros Islands – floating islets made of totora reeds that are grown in Lake Titicaca – where everything from houses to boats are made of reeds. Evolving from these two ideas, the Time Square installation embodies natural elegance and tradition. The sculpture is like a house that has bubbled up from the earth, created using diffusive reflective material and aluminium. It glows at night with LEDs programmed using a custom lighting sequence.

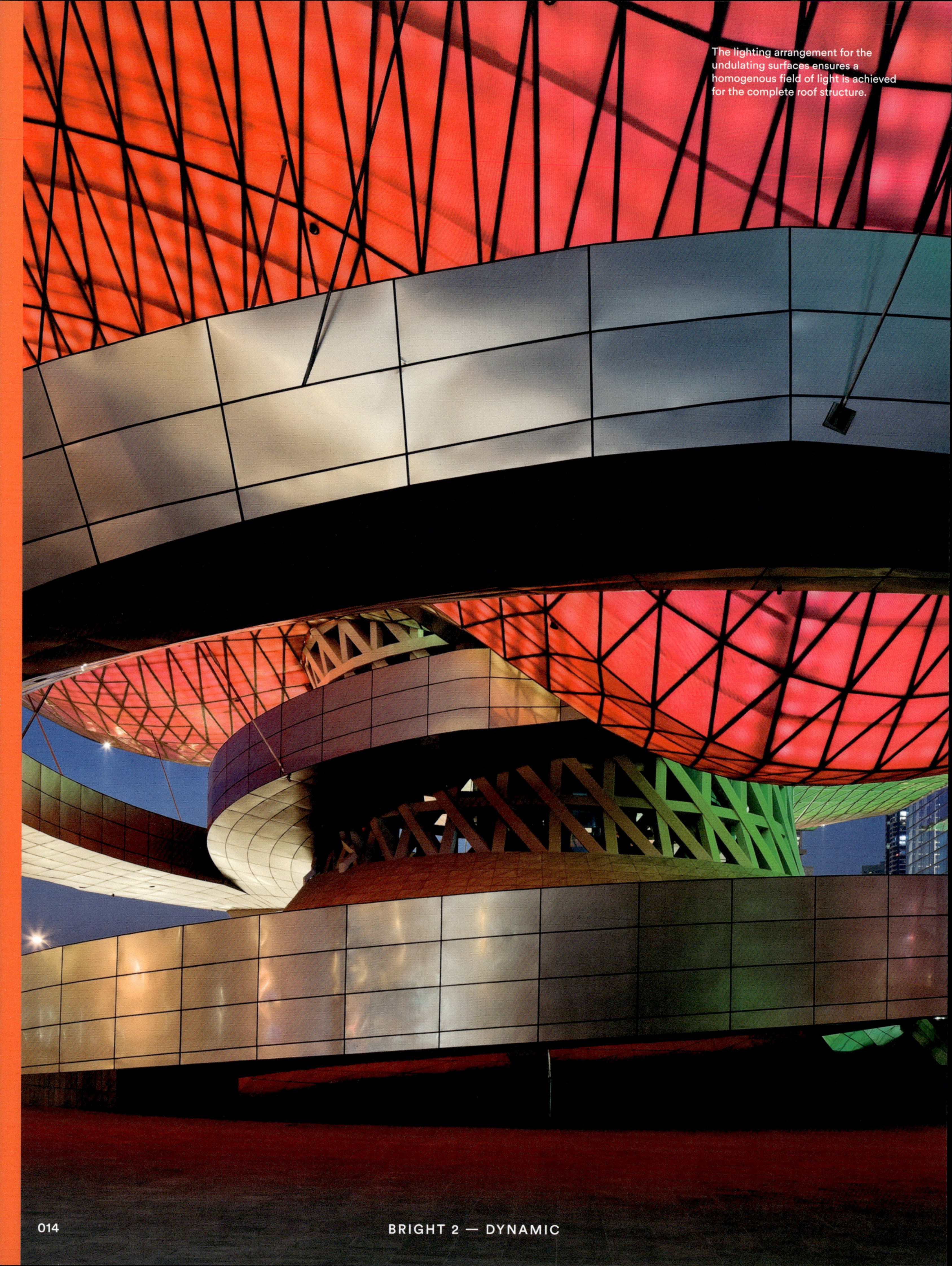

The lighting arrangement for the undulating surfaces ensures a homogenous field of light is achieved for the complete roof structure.

Busan Cinema Center

When the Busan Cinema Center became the home of Asia's most significant film festival in 2012, the vibrant lighting concept of **Har Hollands Lichtarchitect** attracted attention with its colourful and dynamic illumination of the undulating undersides of the record-breaking roof structures.

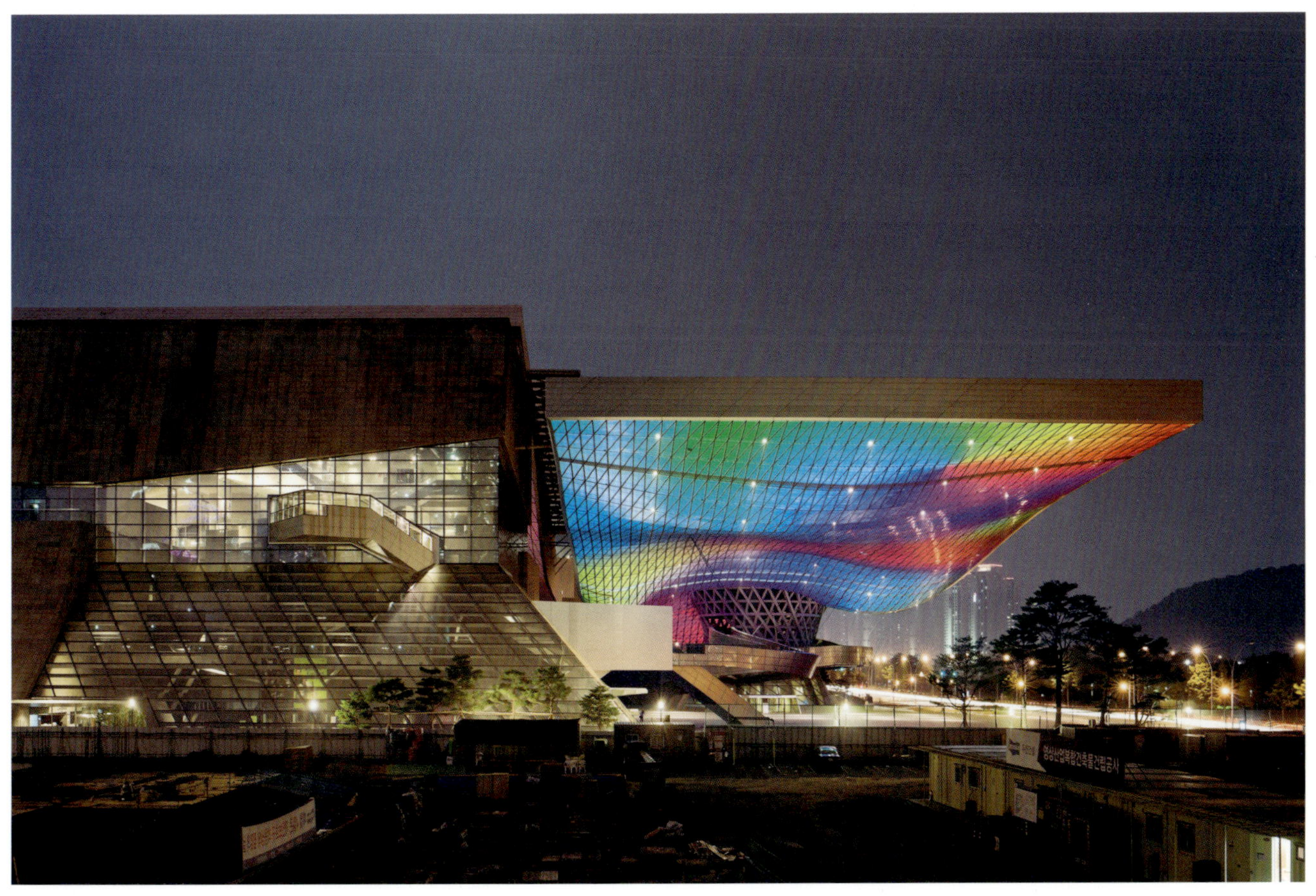

Photos Duccio Malagamba

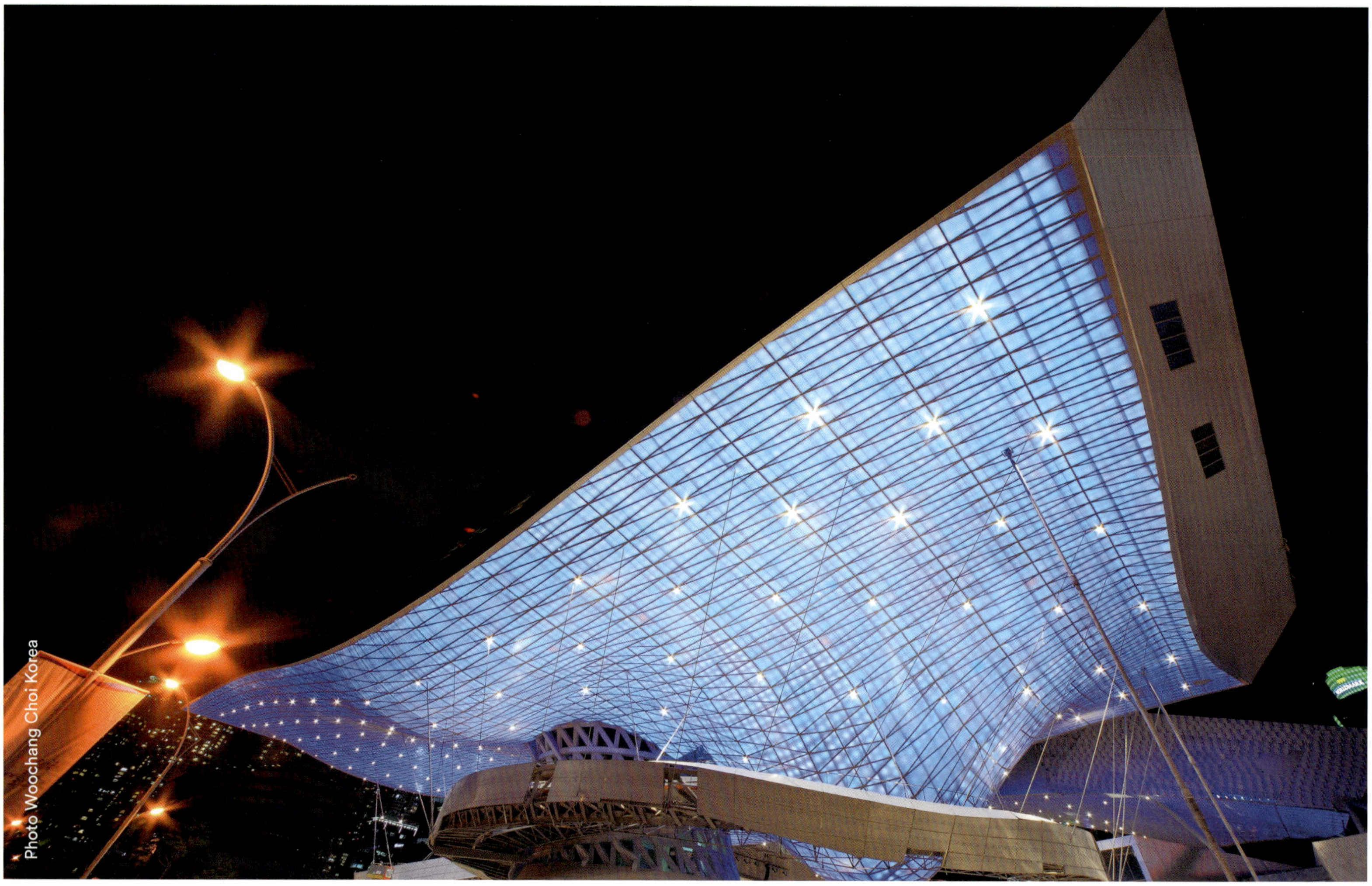

Photo Woochang Choi Korea

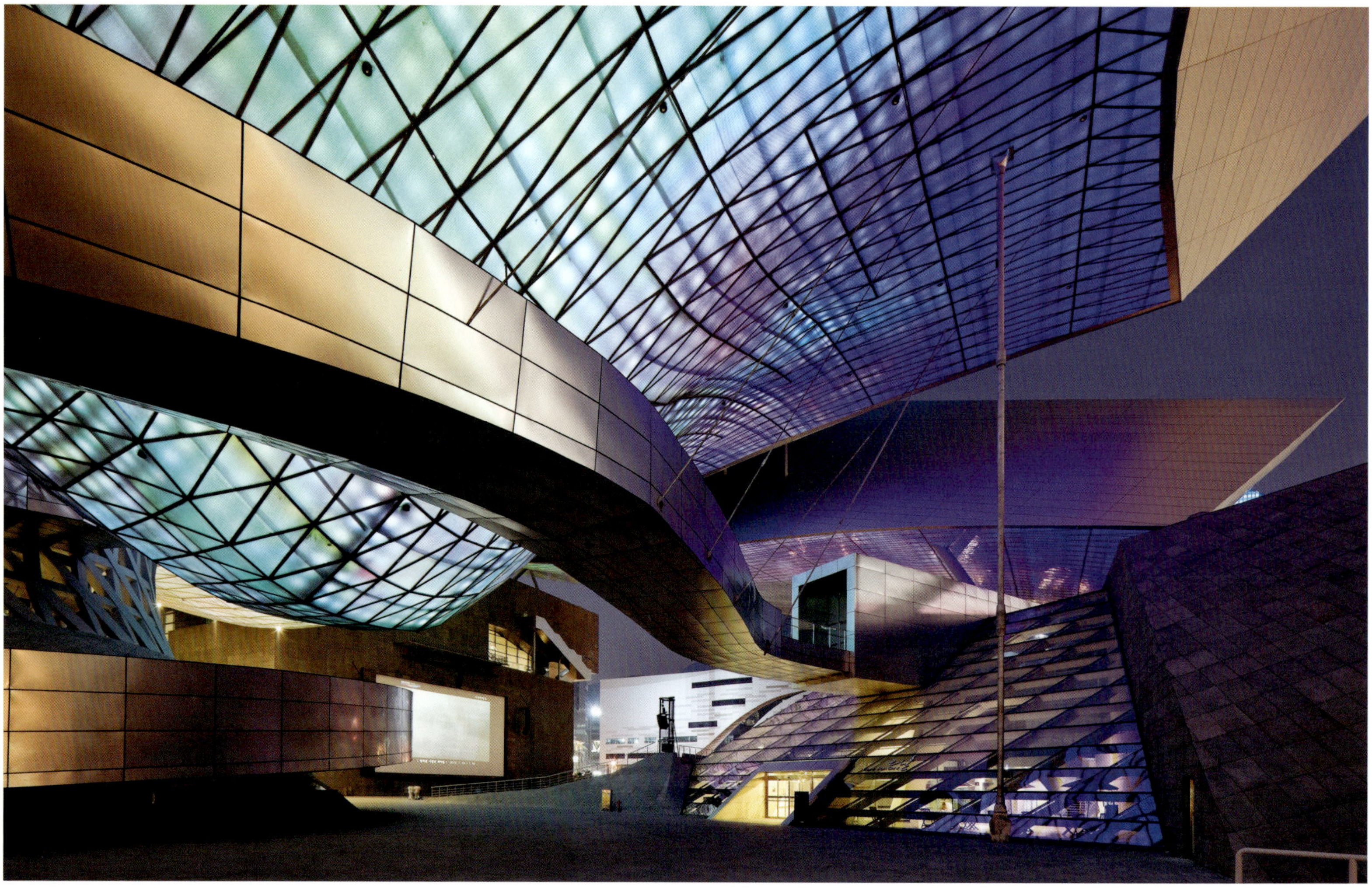

With its 3.6 million inhabitants, Busan is the second largest city in South Korea and the location of one of the most significant film festivals in Asia. Held annually, the Busan International Film Festival (BIFF) found a new home when a building dedicated solely to the art of film was commissioned in 2005. This centre had to become a prestigious, even glamorous building that could act as a new landmark for the city. The winning proposal of an international competition was conceived by the Austrian architectural office Coop Himmelb(l)au and the Busan Cinema Center (BCC) was completed in 2012.

The vast 58,000 m^2 site contains three buildings each connected by sculpturally-shaped bridges: the Cinema Mountain, a multifunctional theatre and two cinemas; the BIFF Hill, a convention hall with offices and a visual media centre; and the smallest building, the Double Cone that houses a cafe, bar and restaurant. The upper level of the Double Cone can be reached by a spiralling ramp leading to a panoramic view over a nearby park and the Nakdong River. A focal point of the complex is formed by two sculptural 'floating' roofs, known as The Clouds. One of these shelters a large outdoor cinema with 4000 seats. The undulating undersides of both gigantic roofs radiate with patterns of moving light; the essence of film. Clad with translucent material and embedded with regularly spaced floodlights for general lighting of the areas below.

The lighting concept needed to meet a number of functional requirements: to generate a feeling of safety and security; to facilitate orientation; to act as a powerful means for communication; to strengthen the unique qualities of places and moments; to create a certain atmosphere; and to act as a tool during the hours of darkness to conjure the domain of fantasy. Every effort was made to translate and integrate the functional, aesthetical and emotional aspects of the demands into a fascinating night-time experience. Steerable LED-fixtures illuminate the translucent cladding from behind and regularly-spaced floodlights are embedded in the ceiling, providing general lighting for the areas below. A total of 42,000 Luxeon LEDs were employed as the star attraction in order to create the coloured patterns of dynamic illumination. The three-dimensional articulated ceilings act like a virtual sky, which gives the cinema its symbolic and representative iconographic feature. —

Project
Busan Cinema Center
Designer
Har Hollands Lichtarchitect
Location
Busan, South Korea
Client
Municipality of Busan
Collaborator(s)/consultant
Coop Himmelb(l)au, B+G Ingenieure, Humanlitech
Manufacturer
Future Lighting Solutions
Date
Early 2012

The lighting arrangement above the undulating surfaces ensures a homogenous field of light is achieved for the complete roof structure.

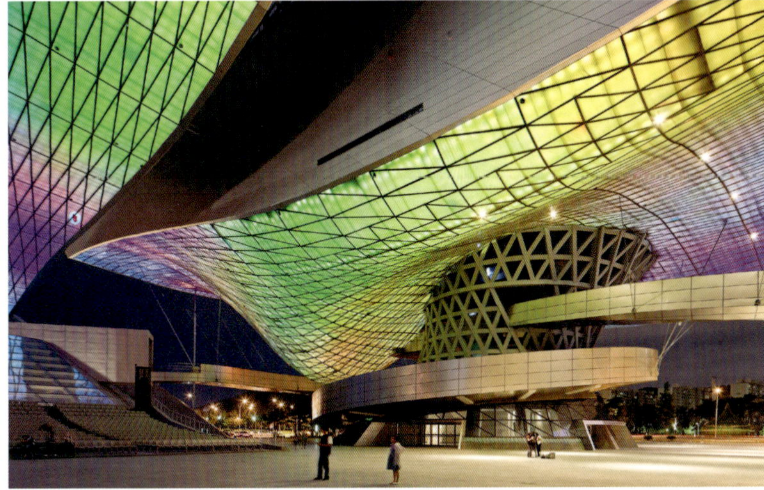

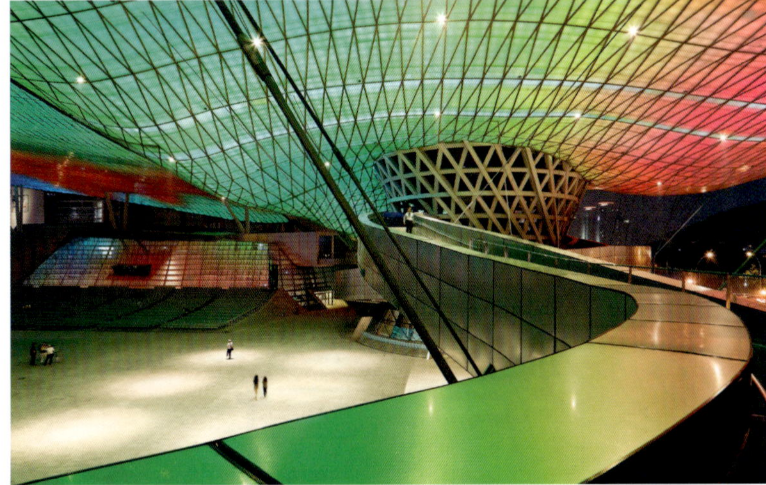

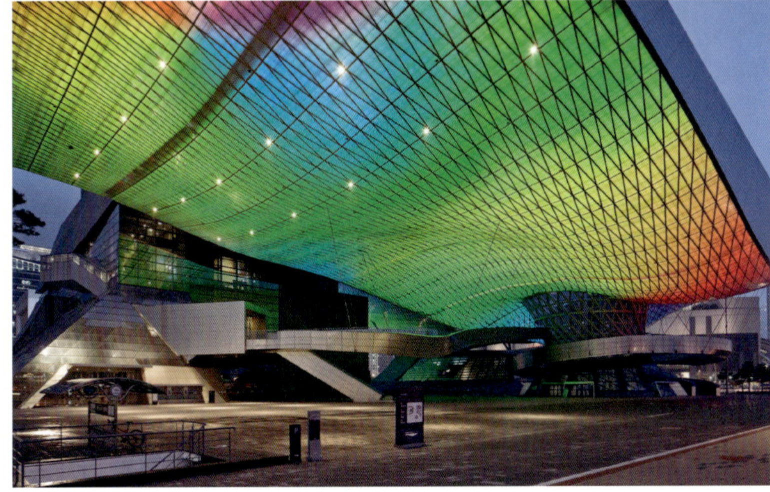

THE UNDULATING ROOFS RADIATE WITH PATTERNS OF MOVING LIGHT; THE ESSENCE OF FILM

The Contemporary Art Center in Cordoba, Spain was conceived and designed by the studio of Nieto Sobejano Arquitectos. The team at realities:united was then commissioned to design and develop the building's riverside facade (in close cooperation with the architects) to bring it to life with a unique, dynamic light and digital display.

The starting point for C4 was an analysis of the inner layout of the building. The concept was developed to utilise the creative motif in the structural design – polygonal volumes of varying dimensions – and repeat them in the unique outer skin. The tessellating inner motif was transformed into the characteristic outer topography with a surface using fibreglass-reinforced cement. Irregular-shaped indentations ('bowls') of varying density and size – geometrically derived from the building's floor plan – were strategically positioned and individually-lit to become 'pixels' of a large display system.

To transform the facade into a dynamic light and media display, without fundamentally changing its solid appearance as envisioned by the architects, turned out to be the biggest challenge in the project. The facade is accordingly designed to deliver a tactile character in the daytime whilst, at night, it morphs into a unique and dynamic communication wall that reacts very specifically to the architecture and its surroundings.

Each of the hexagonal bowls on the exterior wall serves as a reflector for an integrated, artificial light source. Utilising specially-designed software and visuals that feed into the system, the intensity of each lamp can be controlled individually, forming a huge irregular low-resolution grey scale display. The thorough immersion of the 'pixel-bowls' – like negative impressions – in the volume of the facade turns the architectural scheme itself into a digital information carrier. The tectonically-modulated surface topography is characterised by a playful composition of light and shadow and intriguing, rippling reflections in the river water alongside. —

Designer
realities:united
Location
Cordoba, Spain
Client
Nieto Sobejano Arquitectos and the Andalusian Government, Spain
Collaborator(s)/consultant
Nieto Sobejano Arquitectos/ Christian Riekoff
Manufacturer
Iluminación Lledó
Date
January 2015

AT NIGHT, IT MORPHS INTO A UNIQUE AND DYNAMIC COMMUNICATION WALL THAT REACTS SPECIFICALLY TO THE ARCHITECTURE

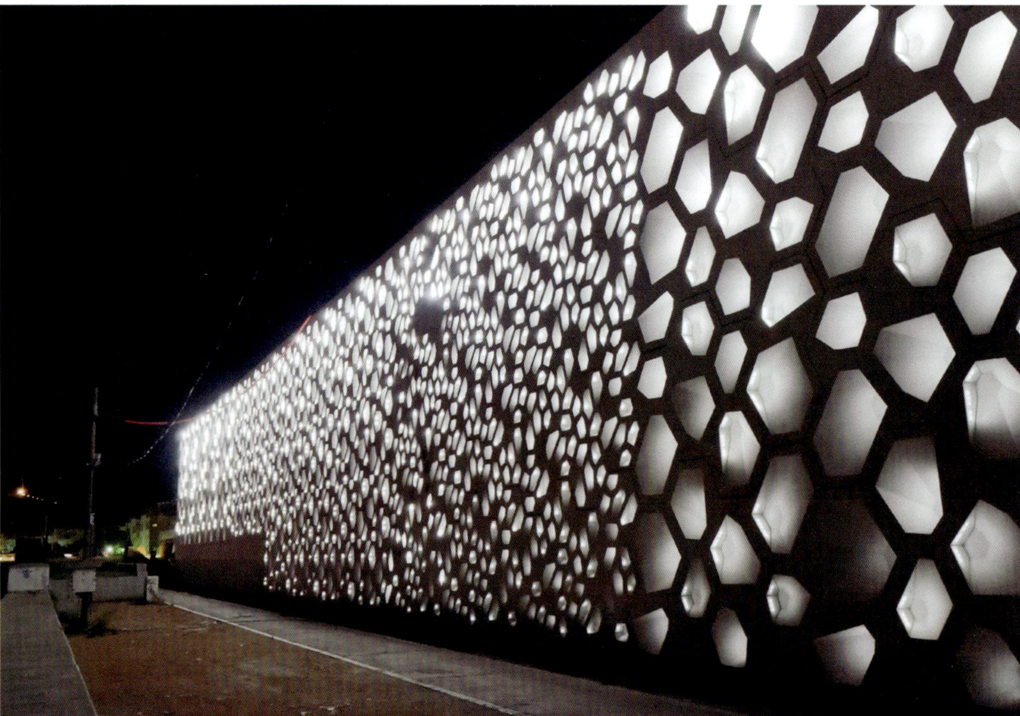

Geometric forms characterised the interior and exterior spaces.

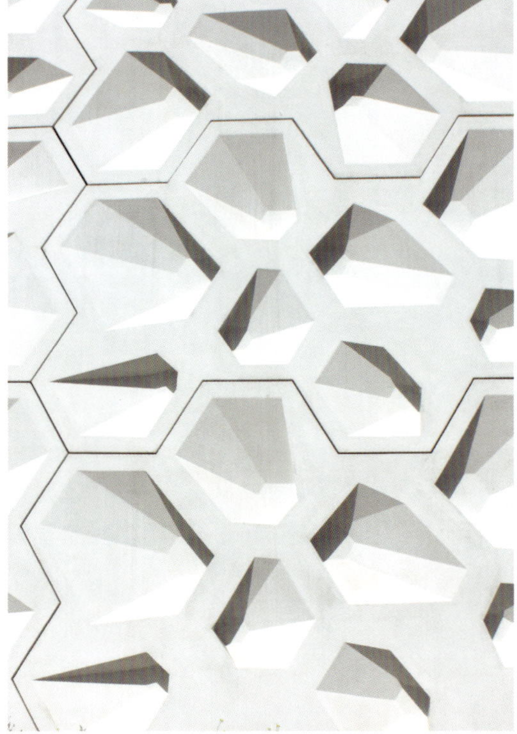

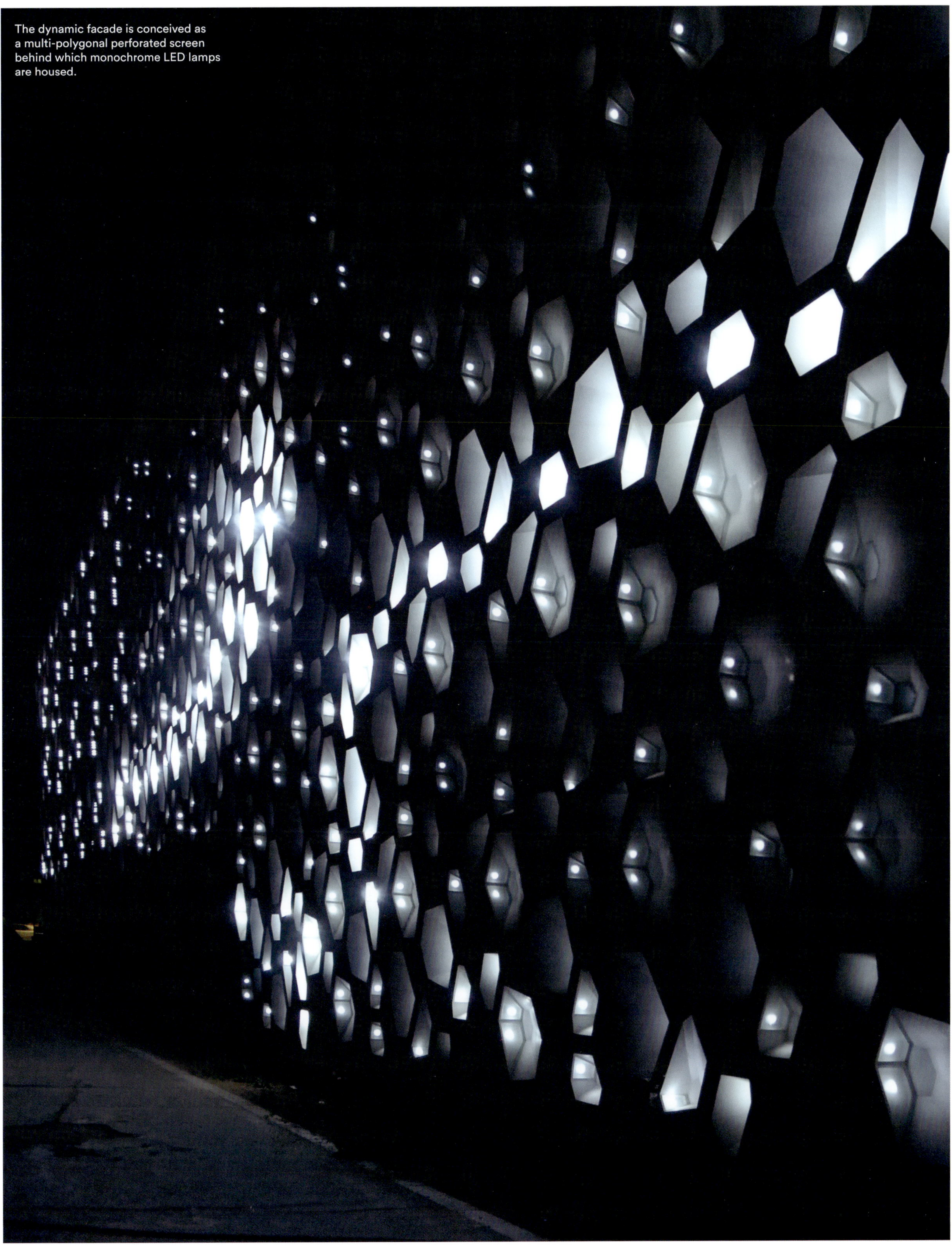

The dynamic facade is conceived as a multi-polygonal perforated screen behind which monochrome LED lamps are housed.

ELEVATIONS

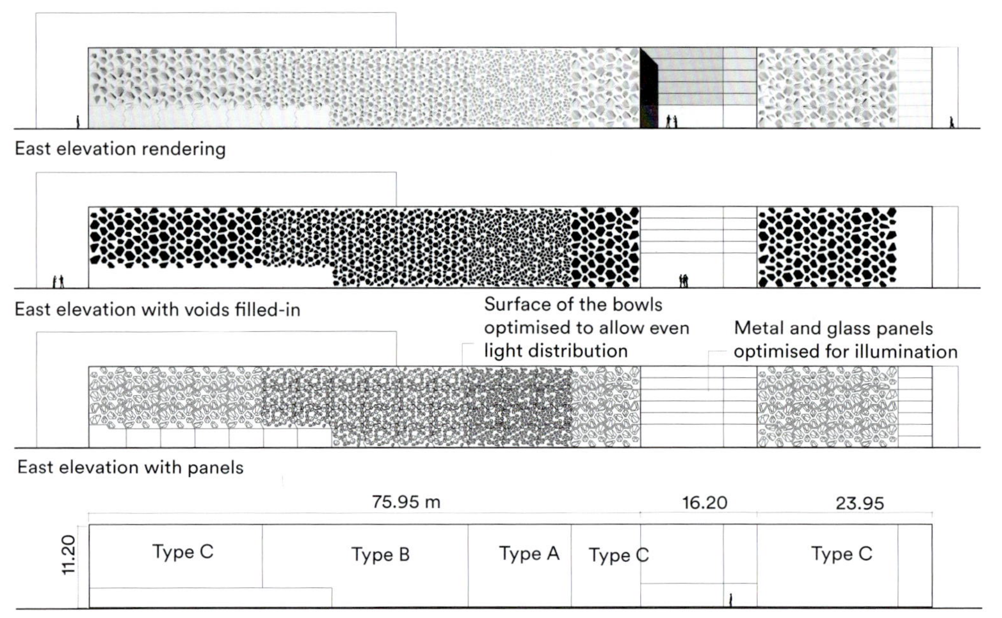

East elevation rendering

East elevation with voids filled-in

East elevation with panels

Partition scheme

PANEL TYPES

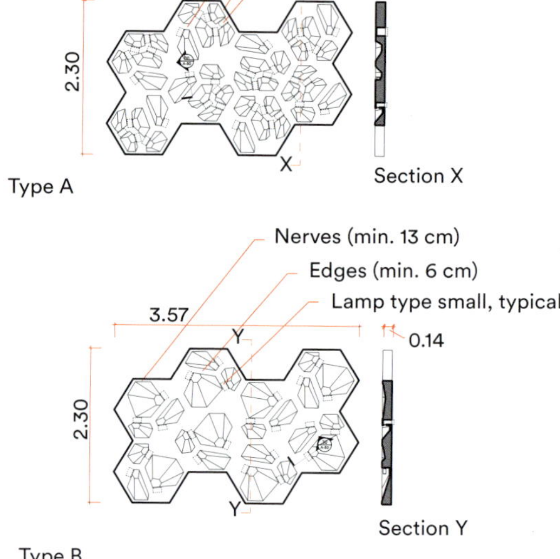

Type A

Type B

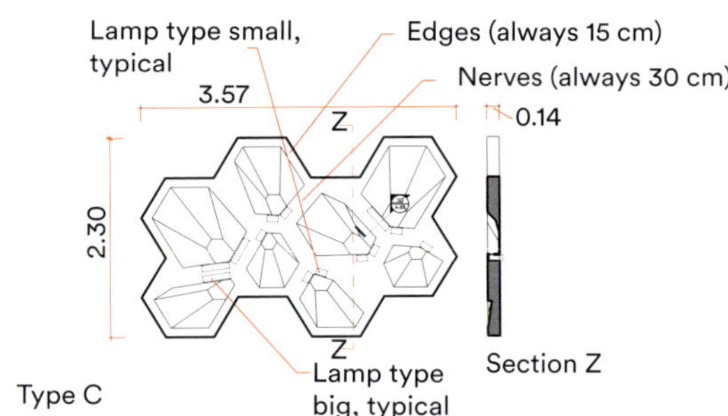

Type C

BOWL TYPES

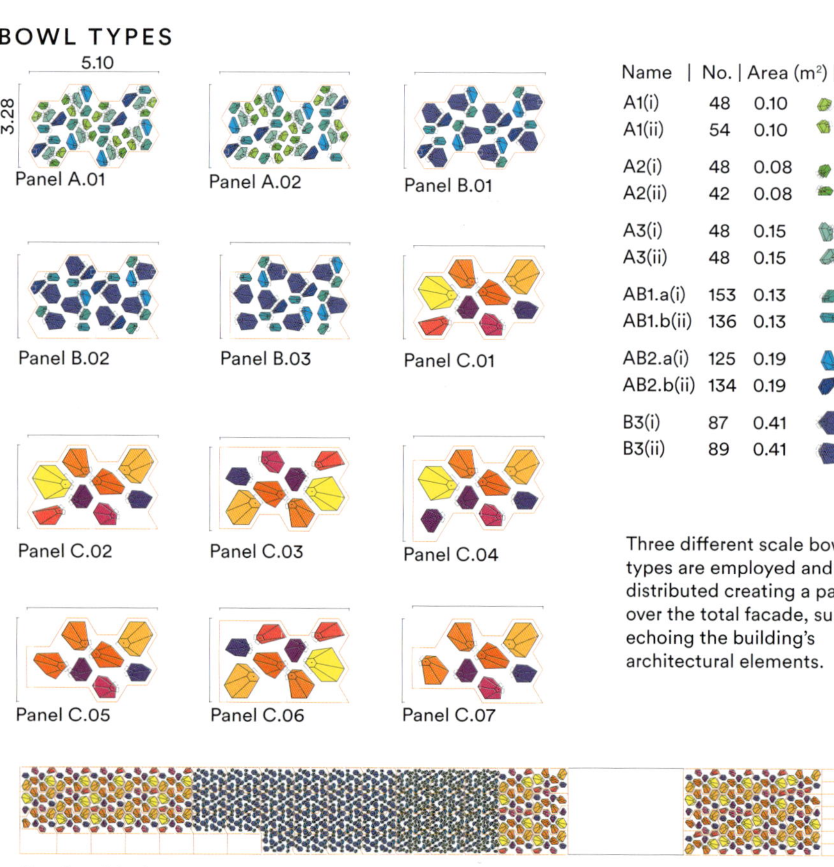

Panel A.01 · Panel A.02 · Panel B.01 · Panel B.02 · Panel B.03 · Panel C.01 · Panel C.02 · Panel C.03 · Panel C.04 · Panel C.05 · Panel C.06 · Panel C.07

Panel positioning

Name	No.	Area (m²)	Plan
A1(i)	48	0.10	
A1(ii)	54	0.10	
A2(i)	48	0.08	
A2(ii)	42	0.08	
A3(i)	48	0.15	
A3(ii)	48	0.15	
AB1.a(i)	153	0.13	
AB1.b(ii)	136	0.13	
AB2.a(i)	125	0.19	
AB2.b(ii)	134	0.19	
B3(i)	87	0.41	
B3(ii)	89	0.41	
C1	39	0.50	
C2	43	0.52	
C3	36	0.59	
C4	36	0.68	
C5	36	0.85	
C6	36	0.87	
C7	36	1.24	
C8	36	1.39	

Three different scale bowl types are employed and distributed creating a pattern over the total facade, subtly echoing the building's architectural elements.

Set-up

Facade consists of 1319 hexagonal, recessed and pre-fabricated 'bowls' of different scales.
Pixel bowls are made up of panels of fibreglass-reinforced cement.
Panels are classified into three different bowl-scale-types (A, B and C).
Light source includes monochrome LEDs housed within the perforated screen that can be individually-controlled.
Illumination comes in the form of pixel lamps of differing quantities (8, 24 and 40) and power levels (10 to 36 W), depending on the size of the panels.
Framework of each panel includes bowls that appear to be unique in shape and size, with a distribution that appears to be irregular; in fact, only the distribution density stays consistent.
Power supply with an inbuilt Ethernet controller allowed individual indentations to be addressed via processing software.
C4 software was specially-designed (by Christian Riekoff) to route and convert any kind of digital image into the special resolution and pixel arrangement of the facade.

CONCEPT DEVELOPMENT

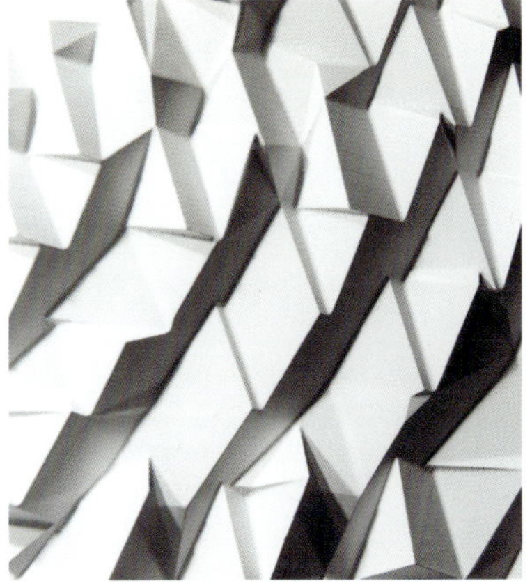

Model

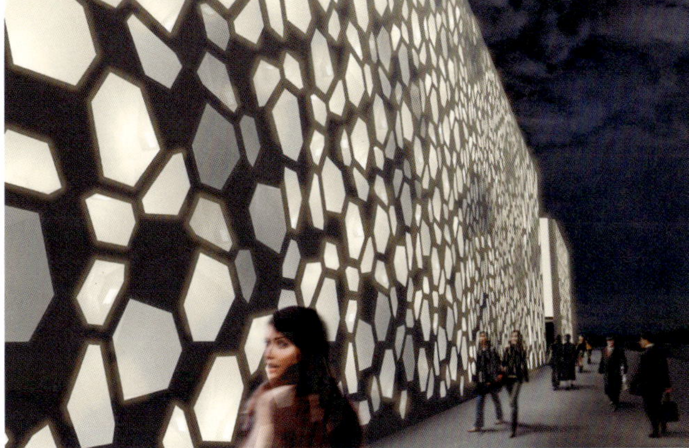

Rendering

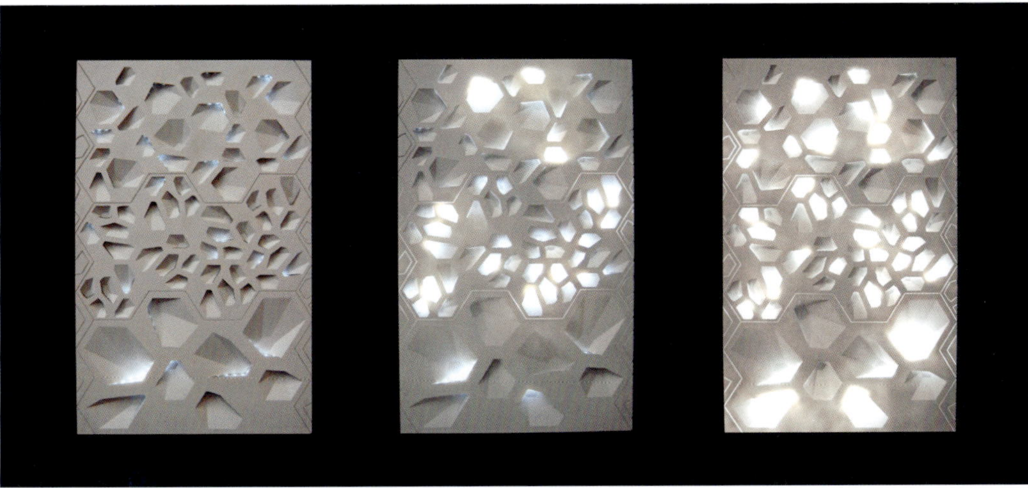

Prototype

Specifics

/\/\/\ Appropriate surface investigations for the bowls included reflective texture to be developed and tested by the use of mock-ups.

/\/\/\ Metal and glass panels were optimised for illumination (screen-printed, etc.); surface treatment and light specifications determined the appropriate reflection.

/\/\/\ Analogously to the eye's retina, the composition and layout of the bowls allows the definition of areas of varying density or 'sensitivity' on the facade. This analogy offers a certain artistic freedom: the resolution of the displayed images can stay low, fitting the blown-up scale of the screen, creating a mode of display in which the motifs are hinted at, rather than unambiguously presented.

/\/\/\ The interest in the aspect of 'visual acuity' stems from earlier projects and extensive research on the process of visual perception. For visualisations with very low resolution, the precognition of the brain determines whether an image or animation can be recognised. A motif that has been displayed at a higher resolution can be shifted to much lower resolution and still preserve its readability.

/\/\/\ The specialised software operates in real time and so this software is an indispensable simulation and preview tool for the creation process or for the selection of artistic content itself. Due to its special resolution and arrangement, it can be very difficult to imagine how certain content works on the facade.

SOFTWARE INTERFACES

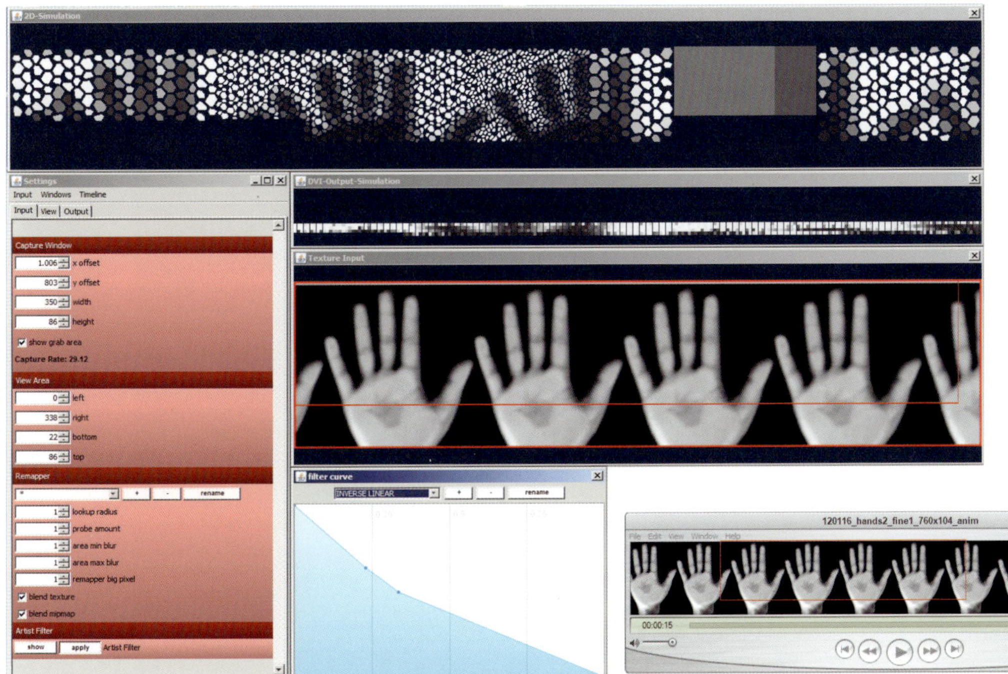

TECTONICALLY-MODULATED TOPOGRAPHY IS CHARACTERISED BY A PLAYFUL COMPOSITION OF LIGHT AND SHADOW

Photo Steffen Jänicke

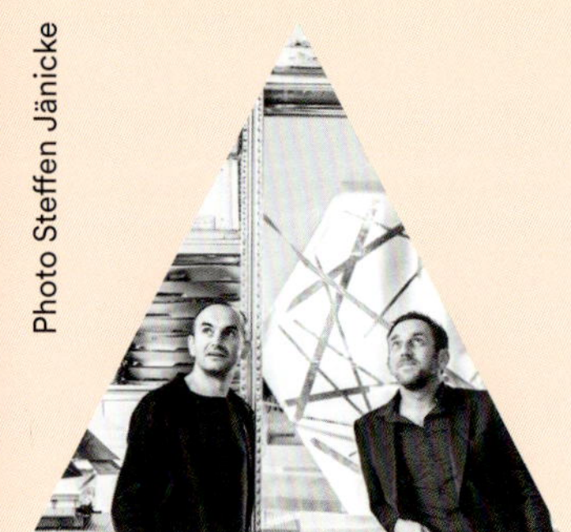

realities:united

In 2000, the brothers Jan and Tim Edler founded realities:united, a Berlin-based studio for art and architecture. Working together with some of the most prominent figures of contemporary architecture, the studio has built a unique reputation both for the initialisation as well as the realisation of spectacular art- and media-extensions to buildings and in the urban fabric across the globe. The multidisciplinary team varies in size for various projects, typically between 5 and 10 people from the fields of art, architecture and engineering.

REALITIES-UNITED.DE

Sender

When **2013**
Where **Bergkamen, Germany**
Client **Kultur Ruhr/Urbane Künste Ruhr**

Sender is a temporary, public art installation commissioned by Urbane Künste Ruhr. An industrial robot performs gestures of waving and signalling with four different 'tools' that it can pick up and lay down separately. During the day, it uses a signal flag and, at night, a light saber. In addition, the device can pick up a dumb-bell to work out and a roof when it rains.

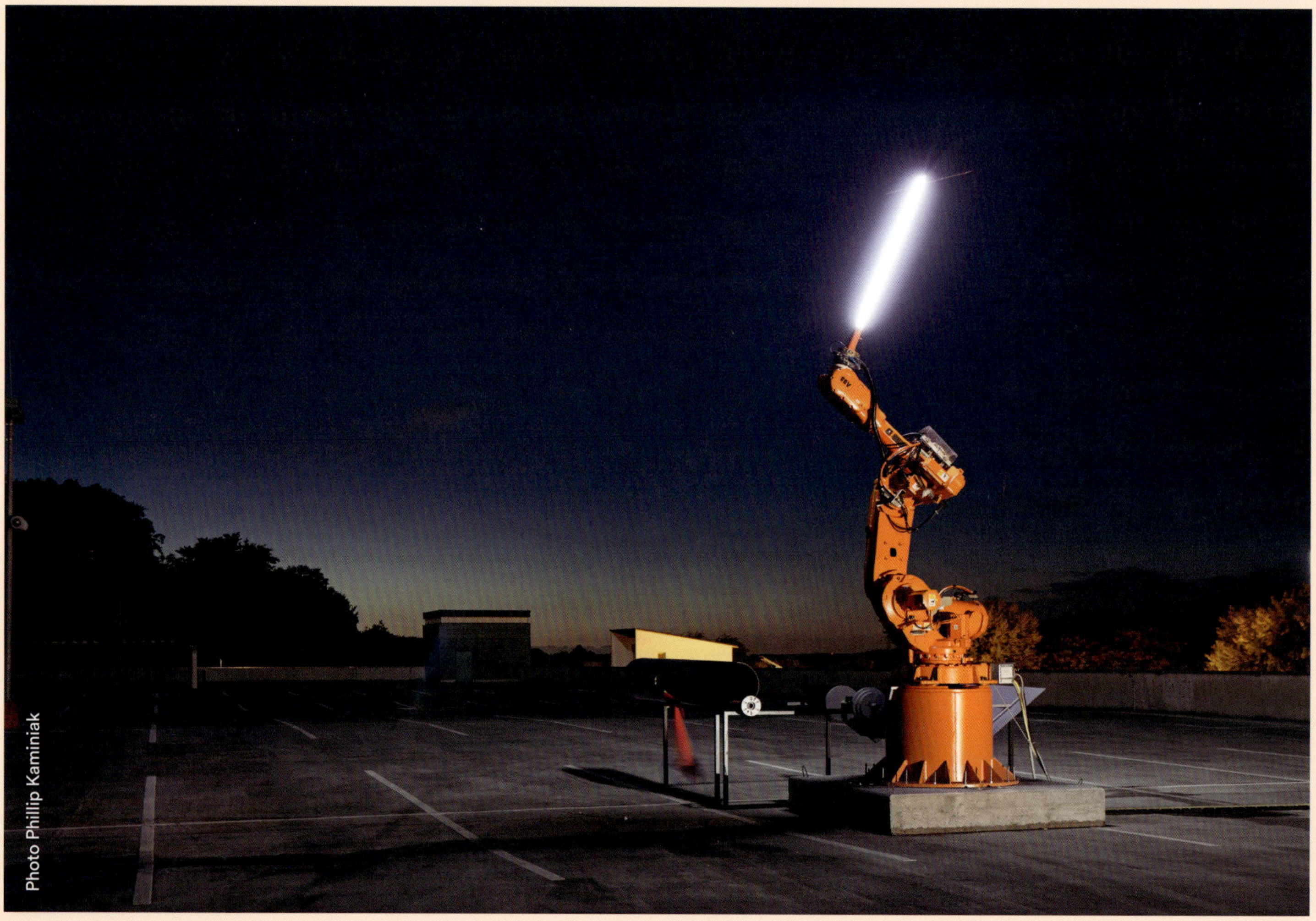

Photo Phillip Kaminiak

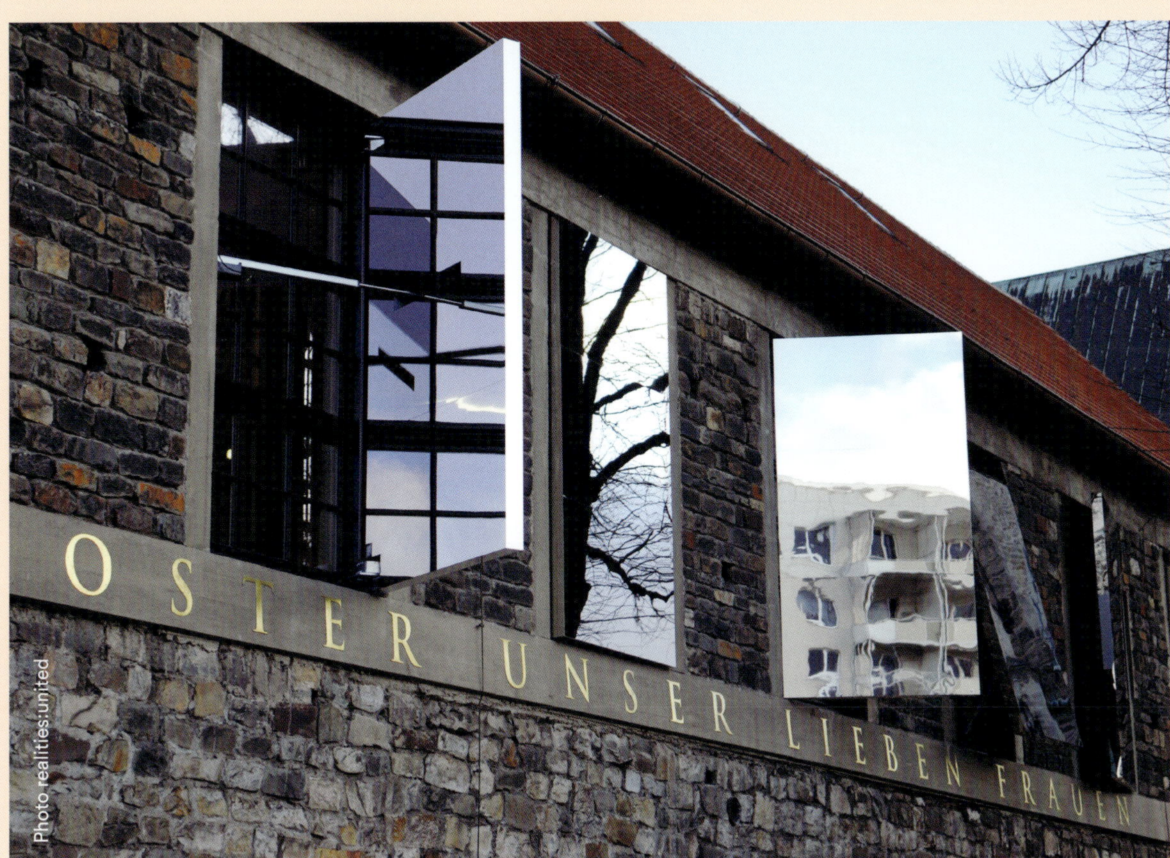

Photo realities:united

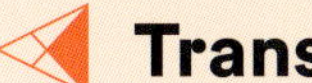

Transreflex

When **2012**
Where **Magdeburg, Germany**
Client **Kunstmuseum Kloster Unser Lieben Frauen Magdeburg**

Transreflex is a permanent, building site art installation consisting of 17 large-format mirror panels in front of the art museum's windows. The motor-driven panels move without uniformity, because the hinges are attached to different sides of the windows. With the mirrors aligned at different angles and in different directions, the installation reflects light with varying fragments of the surroundings. The resulting collage of mirror images mixes two opposite realities: the former cloister construction itself and the post-war modernist buildings in front of it.

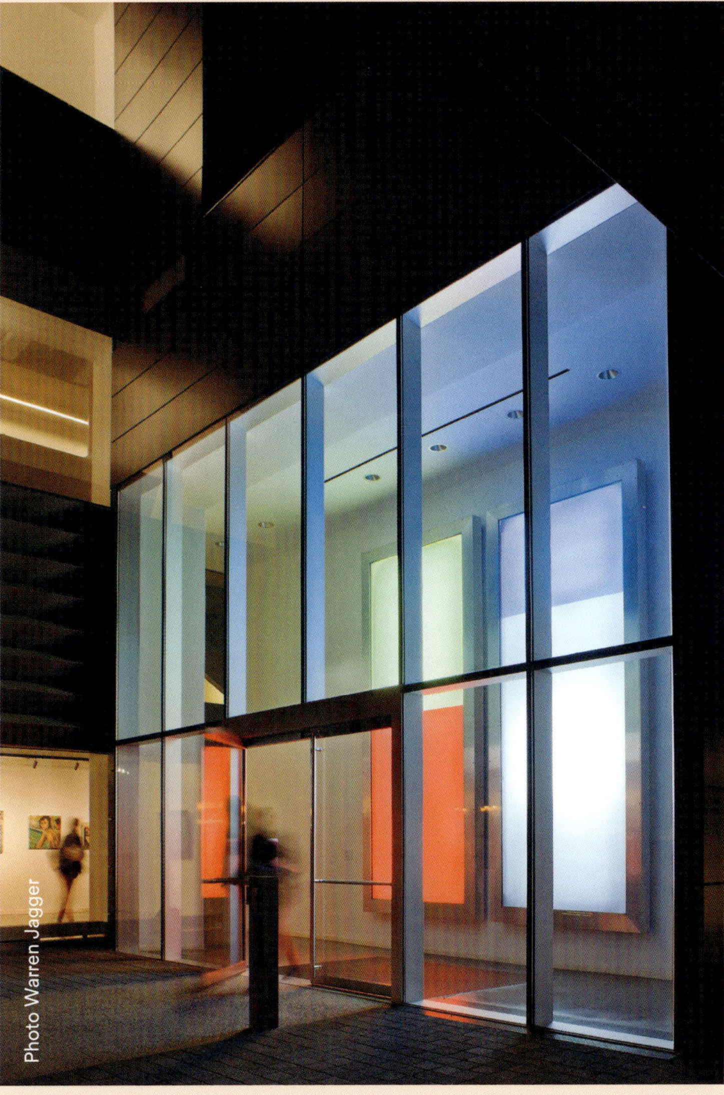
Photo Warren Jagger

2×5 (Brothers)

When **2012**
Where **Marty Granoff Center for the Creative Arts, Providence, United States**
Client **Brown University**

This is a time-based permanent, kinetic site art installation for a building by Diller Scofidio & Renfro. It makes use of an established technical advertising format: the installation consists of two mechanical 'poster scrollers' in a special format. Each of the two identical machines displays a textile print with five monochromatic colours illuminated from behind. The installation's behaviour is coupled to the progress of the academic year.

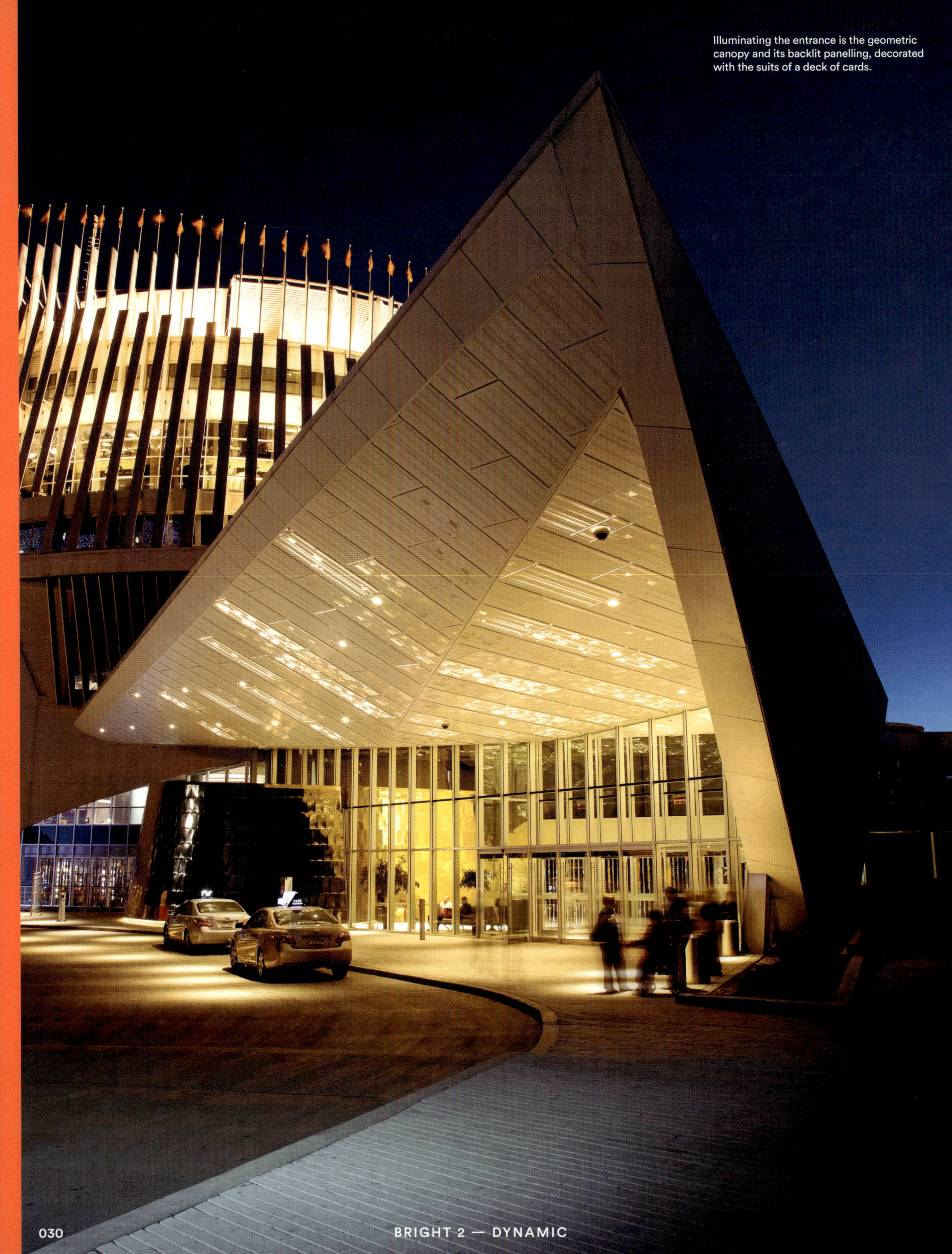

Illuminating the entrance is the geometric canopy and its backlit panelling, decorated with the suits of a deck of cards.

Casino de Montréal

An elaborate building that was first built for Expo 67 has recently been given an overhaul and now lives on as Canada's largest gambling establishment. Casino de Montréal shines like a beacon thanks to the lighting concept by Quebec-based studio **Ombrages**, which enhances the new architecture whilst surrounding it with elements in contemporary contours and colours.

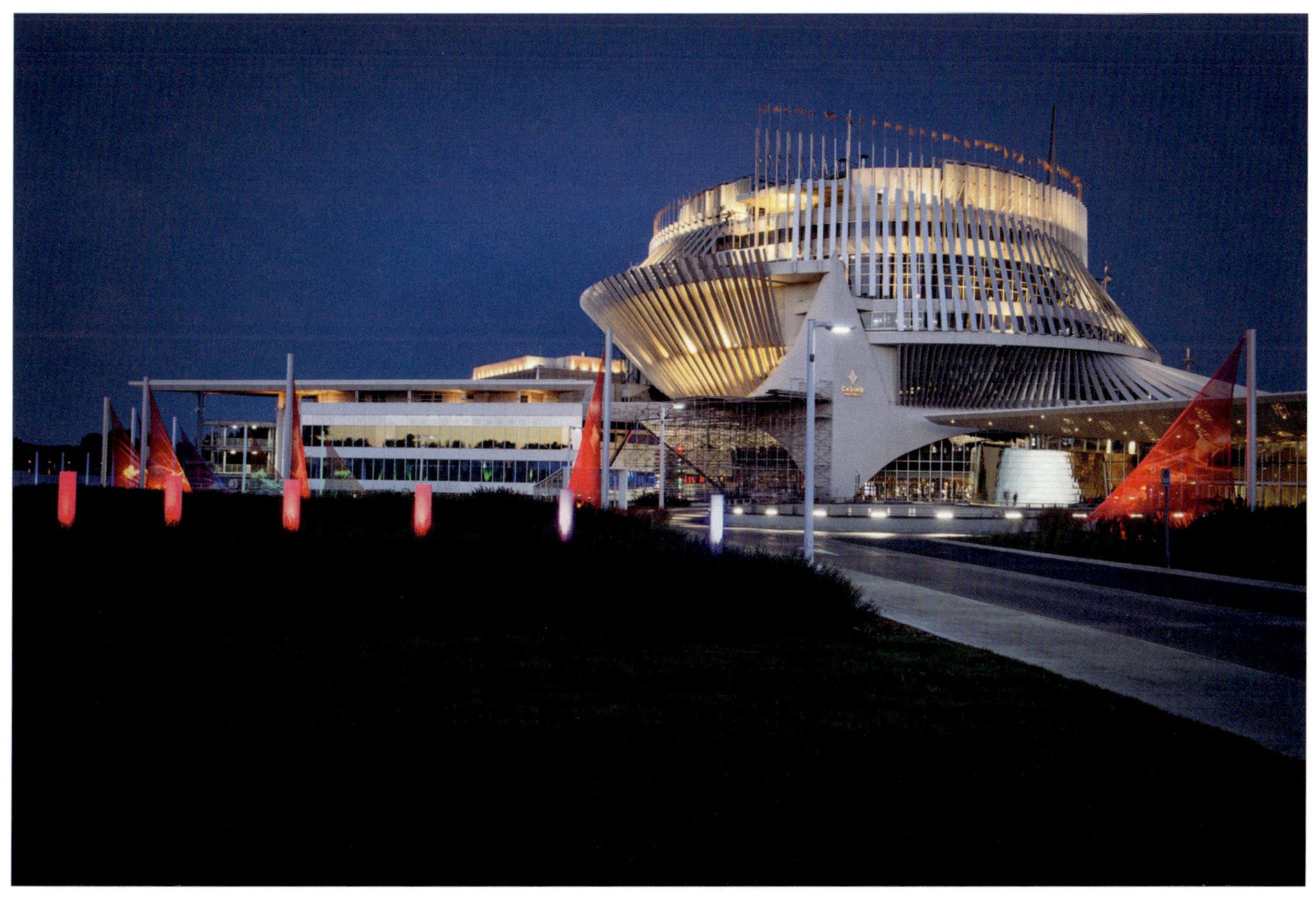

Photos LumenPulse, Christian Perreault

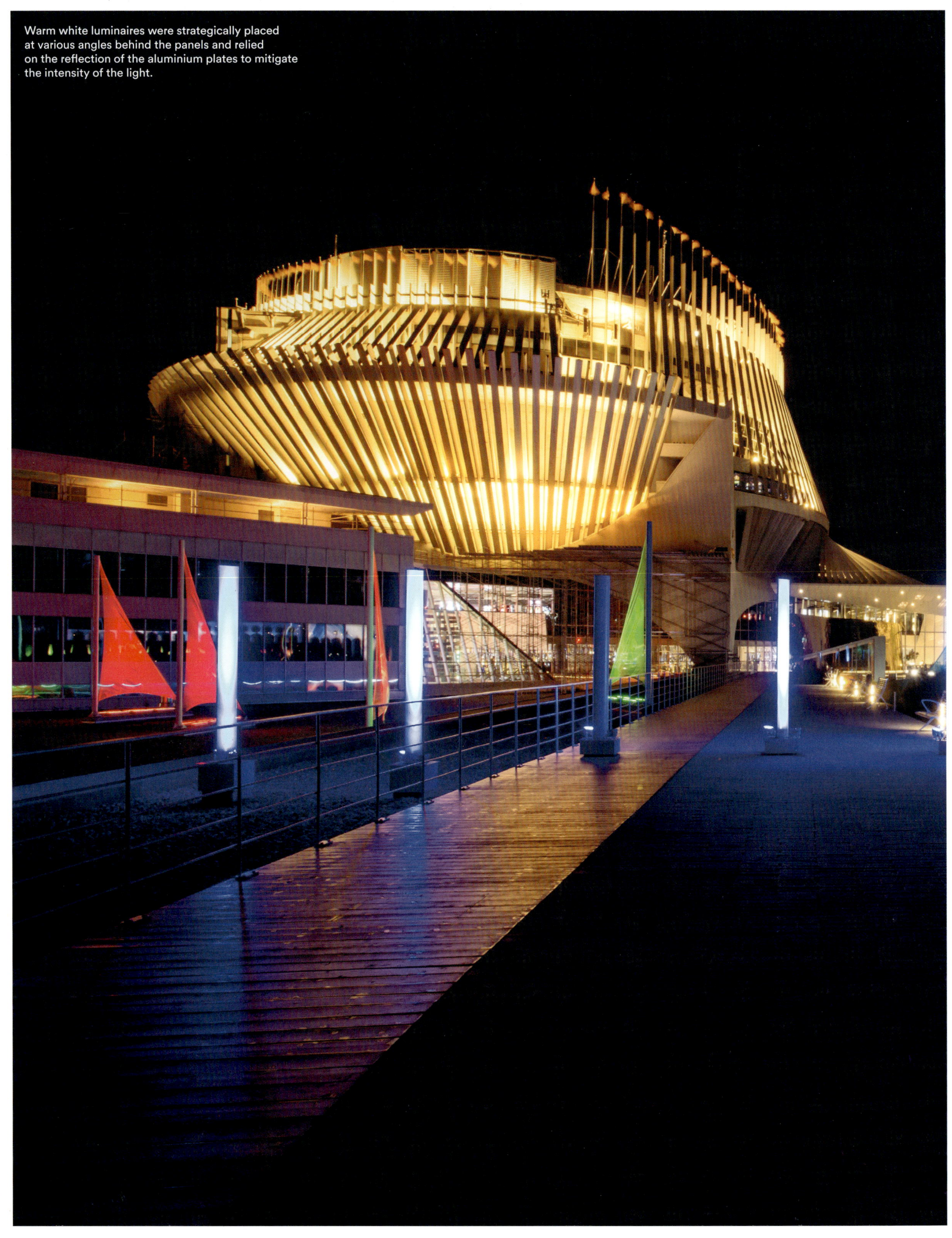
Warm white luminaires were strategically placed at various angles behind the panels and relied on the reflection of the aluminium plates to mitigate the intensity of the light.

Designer
Ombrages
Location
Montreal, Canada
Client
Loto-Québec
Collaborator(s)/
consultant
Casiloc
Manufacturers
Lumenpulse,
Delta Light, Cree,
Technilum, Pharos
Date
October 2013

Housed in the conjoined pavilions of France and Québec, dating back to the 1967 International and Universal Exposition – better known as Expo 67 – the Casino de Montréal has evolved into Canada's largest gambling establishment. Located on an artificial island in the St Lawrence River, the property recently underwent a complete overhaul, with the construction of a new lobby, entrance and landscape design. Ombrages was engaged to develop an integrated lighting scheme for the complex to modernise and embellish all its elements.

Creating curb appeal, the illumination concept of the porte-cochère canopy needed to do more than just provide light for the entrance. It was also a necessary ingredient in the master plan to make the building sparkle like a diamond shining in the distance. The lighting concept for the casino and its surroundings aimed to be dynamic, dramatic and impressive. The basic principle was, therefore, to make it seem like light was coming through transparent walls, with no apparent light fittings or fixtures. All these were concealed behind gold micro-perforated aluminium plates in the wall and canopy, which fluctuate in intensity and appear to make the entire structure shimmer in light. This creates a soft ambient glow and achieves a perfect impression. Discreet directional floodlights were integrated into the canopy to add drama and a greater degree of illumination to the building's approach. The utilised luminaires allow the possibility of enlivening the atmosphere, going from classic white shades to vibrant hues at the flick of a switch, according to pre-programmed scenarios.

Landscaping was another main focus in this project, inspired by features from the existing structure. These included lighted poles and sails, plus a series of illuminated blades. Located in a concentric circle around the building to preserve the initial geometry of the space, masts and blades of different heights create a dynamic effect. The scenography on these blades is based on stunning motions with colours and gentle sparkling effects, still with the underlying idea of a shining diamond in the St Lawrence River. Needless to say, it extends the chic aura of the casino to its surrounding environment. —

The landscaping blade elements are eye-catching in their intensity and the colours of the light, producing a dramatic effect at night.

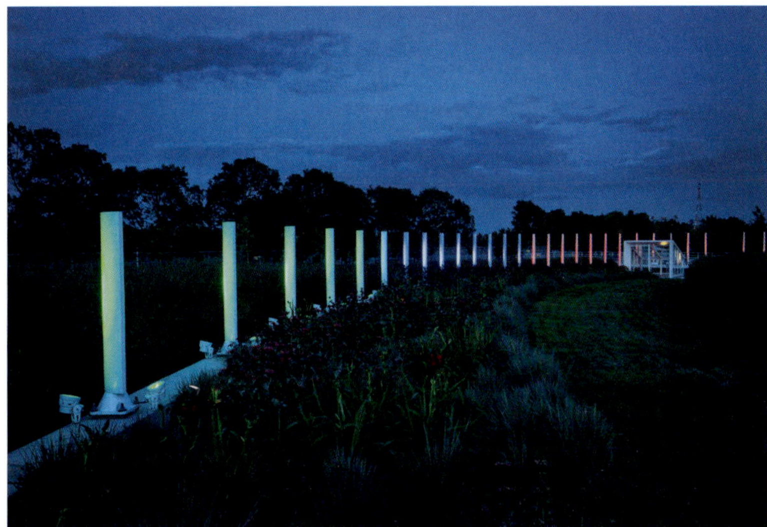

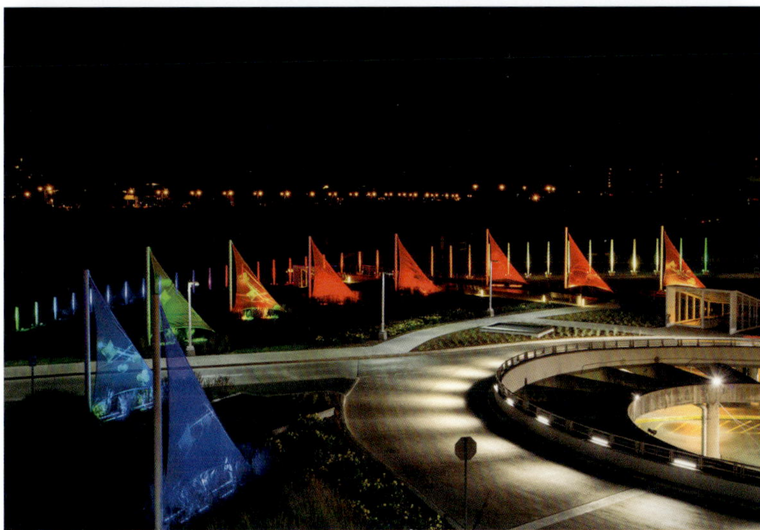

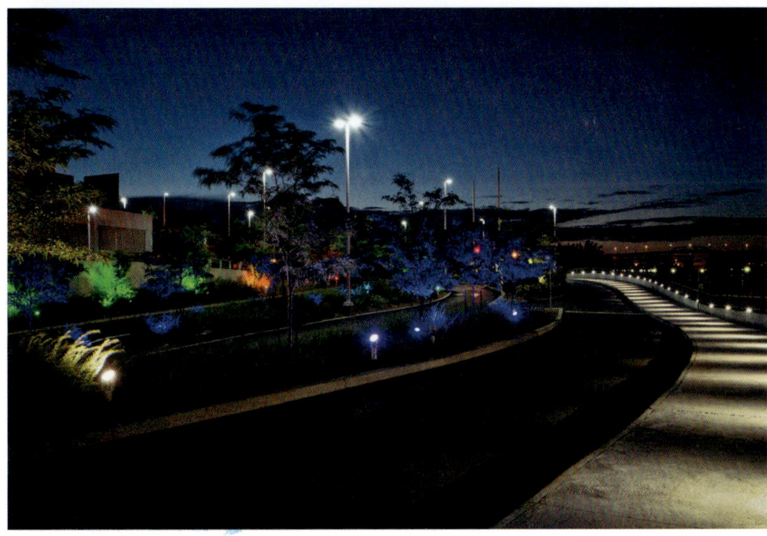

THE LIGHTING CONCEPT SOUGHT TO MAKE THE BUILDING SPARKLE LIKE A DIAMOND

PLAN

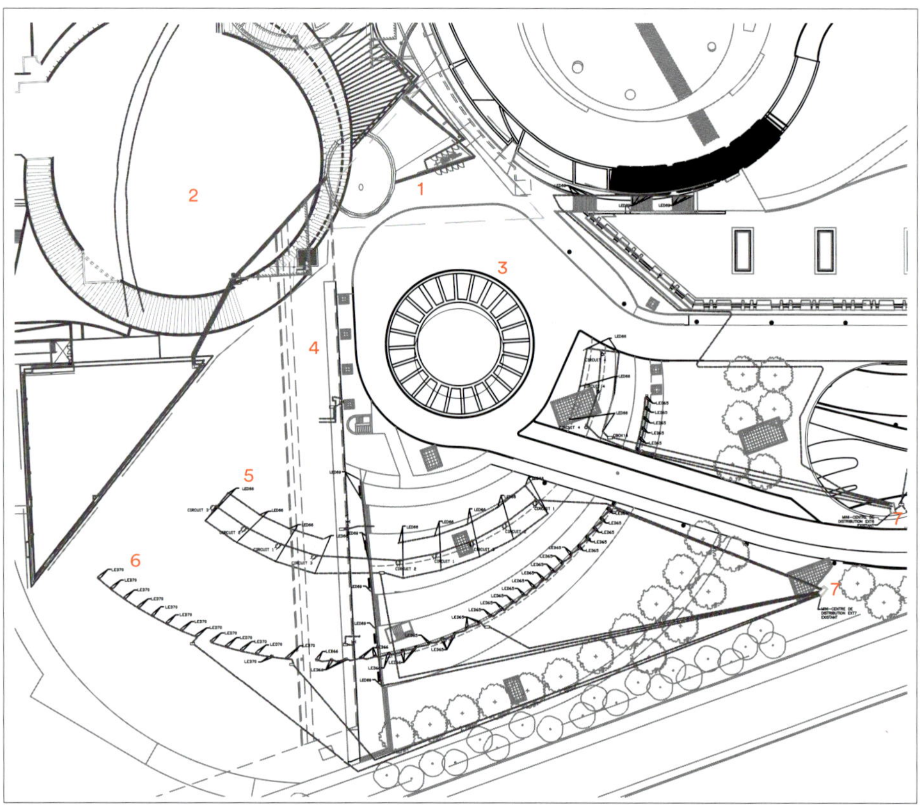

1 Entrance canopy
2 Casino
3 Central roundabout
4 Pedestrian walkway
5 Illuminated sails
6 Light blades
7 Control box

SHIMMERING IN LIGHT, A SOFT AMBIENT GLOW ACHIEVES A PERFECT IMPRESSION

Set-up

Total surface for illumination is 41,000 m. Within the landscaping, a total of 73 trees are lit-up by 146 RGB projectors.
Luminaires used are all LEDs of various types, with spotlights, floodlights, RGB linear fixtures and projectors, all having various beam angles.
Entrance canopy has a geometry made of 14 polygons, requiring 146 dynamic, warm white linear fixtures to create the shinning effect and 41 projectors to add luminance and drama to the entrance.
Illuminated poles and their associated sails are 7.3 and 6.0-m-high, respectively. The ten masts and sails in total had either four or six linear RGB fixtures on each. Additionally, the 47 illuminated blades were custom-made in aluminium with a matte white finish, lit with a total of 94 RGB projectors.
Total wattage (kW) per sector includes: canopy (6.1), pathway and decorative lighting (15.2) and roadway (6.3).
Landscaping included 88 light bollards located in the gardens and pathways, the waterfall with its 74 RGB fixtures, and the 1-km roadway lighting which required 39 light poles.
Global lighting control is based on DMX512, with a total of 20 universes. This allows numerous scenography possibilities and provides lighting colours and intensities for programmed scenarios.
Construction time from preliminary concept to delivery was 30 months.

CANOPY AXONOMETRICS

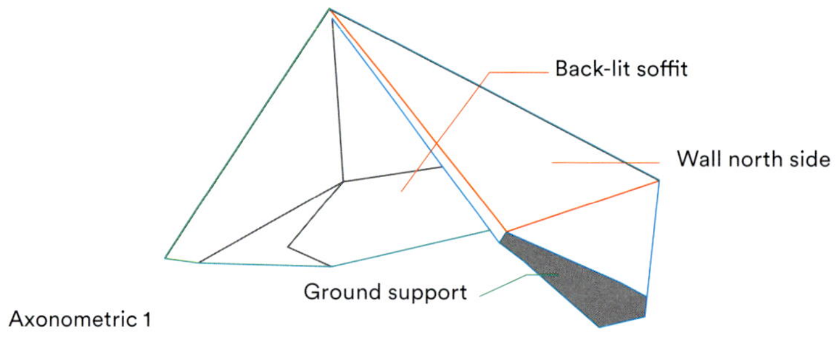

Axonometric 1

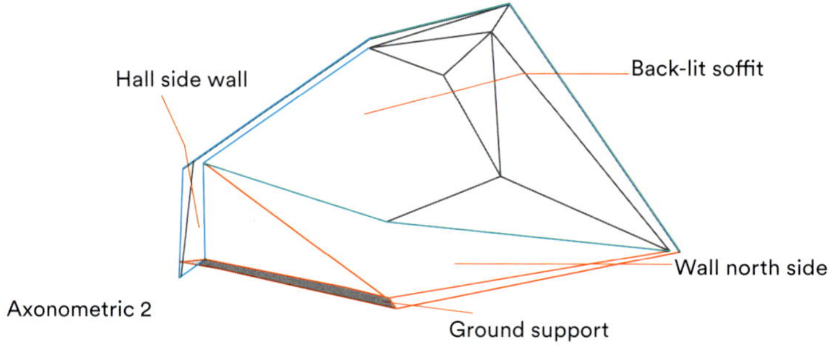

Axonometric 2

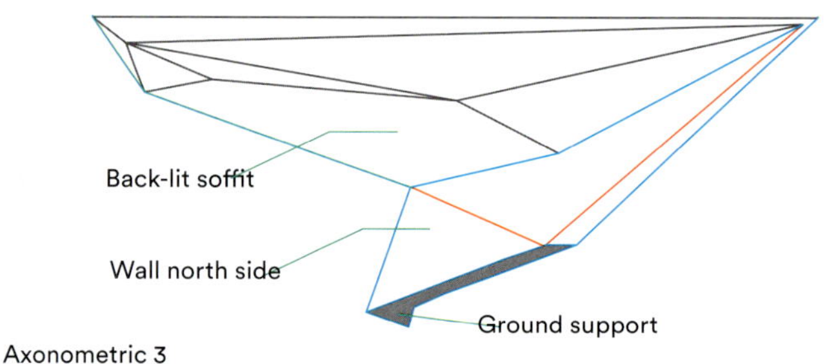

Axonometric 3

CANOPY DETAIL

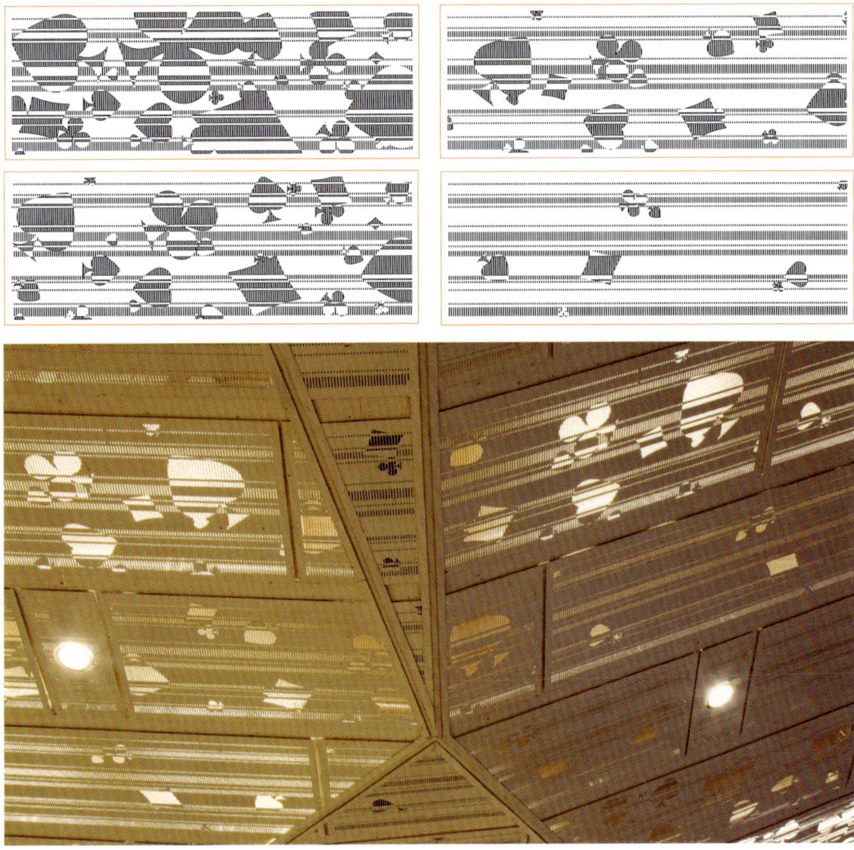

Meticulous detailing incorporated into the panels of the illuminated canopy linked in with the theme of the casino.

CANOPY ILLUMINATION STUDIES

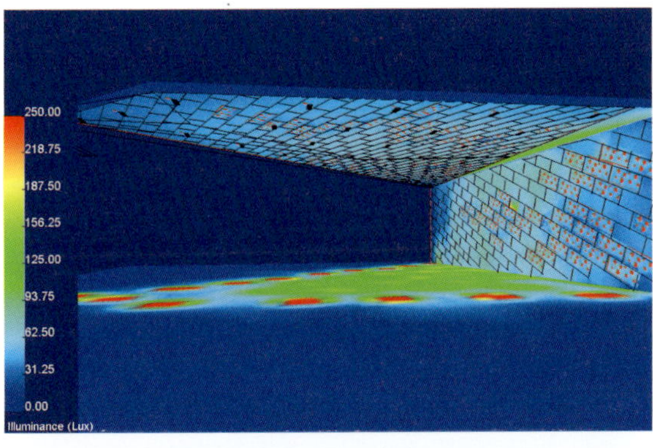

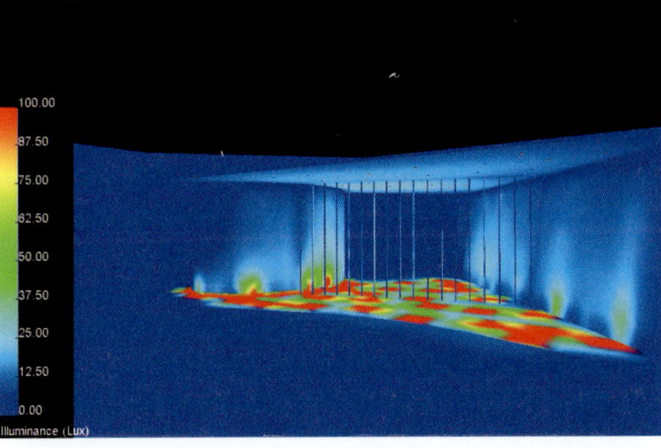

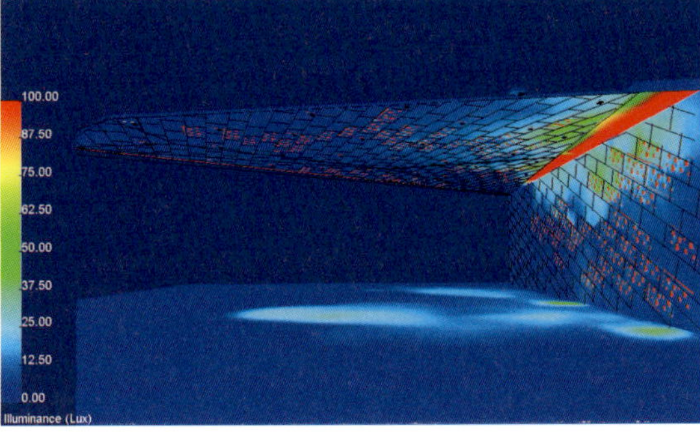

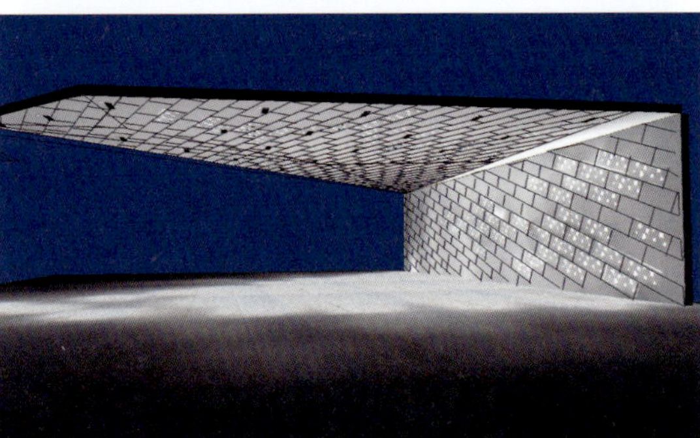

Specifics

Integrating the lighting system into the canopy proved an initial challenge, as the architects had already designed the panelled cladding without considering how the light fittings would be installed. As a result, the team had to find a way to house fixtures within the shallow box-shaped panels to enable the scheme to work.

Each aspect of the scheme was meticulously designed to enhance the harmony of the architecture without being obvious and obtrusive. The backlights in the entrance for example didn't provide enough light for security purposes. Therefore, Ombrages had to subtly integrate some directional floods into the canopy to add drama and a greater degree of illumination to the approach.

In order to create the shining effect of the canopy, the team had to work around the existing structure. The boxes used for the exterior and interior cladding were closed and the light couldn't travel from one panel to another. Also, the budget could not allow the use of one fixture per panel. The boxes were hence redesigned and the aim angles of each luminaire were strategically set to optimise reflection and a mitigated intensity of the light. This created a softer ambient glow. The mounting brackets were customised to fit perfectly into the canopy.

Originally designed in the 1960s around the idea of wind, the facade incorporated hundreds of blades moving and rotating with the power of the wind. The blades do not move anymore because of the challenging maintenance, but they still give the casino its recognisable appearance. To be reminded of this wind concept, two linking elements were incorporated into the landscape design: the lighted poles and sails, and illuminated blades.

Custom mountings and details were designed to recess all the fixtures, in order to achieve luminance, uniformity and to create stylish visual guidance under the canopy with LED projectors.

Low horizontal lighting was used in the central roundabout and the sidewalks, bringing the focus specifically and only on the casino. Achieving luminance and uniformity targets without glaring drivers and pedestrians with this design was another challenge.

Custom lighting strip was designed to create the Casino's logo on the entrance. The pitch and the position of each LED point was carefully chosen because of the shape of the sign.

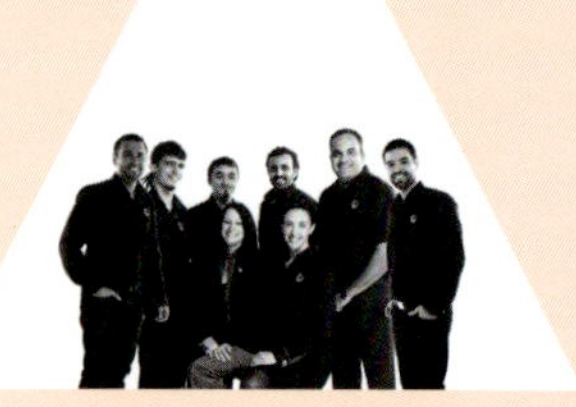

Ombrages

Ombrages creates outstanding lighting scenarios for a range of projects through the play of light and shade. The studio's concepts focus on highlighting building structures and architectural visuals. The Quebec-based company, with offices also in Montreal and Vancouver, consists of lighting designers, interior designers, scenographers, urban planners, architectural and heritage building specialists, as well as electrical engineers. Utilising state-of-the-art software, tools and lighting products, the firm works collaboratively with clients with a focus on developing comprehensive, custom solutions.

OMBRAGES.COM

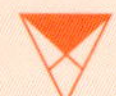

Ciel!

When **2014**
Where **Quebec, Canada**
Client **Resto Plaisir**

Ciel! is a revolving restaurant that sits at the top of the Hôtel Le Concorde in Quebec. Its rooftop views offers customers a beautiful panorama of the surrounding city. The new owners of the restaurant wanted to add modernity and appeal, both on the interior but also creating eye-catching aspects on its exterior. Ombrages' innovative lighting scheme utilised recessed linear indirect lighting which slowly rotates with the floor of the restaurant. The concept of the exterior illumination frames the windows, combining colourful accents of the building's geometry. This all adds up to a magnificent wow-factor, making it a shining beacon on the Quebec skyline.

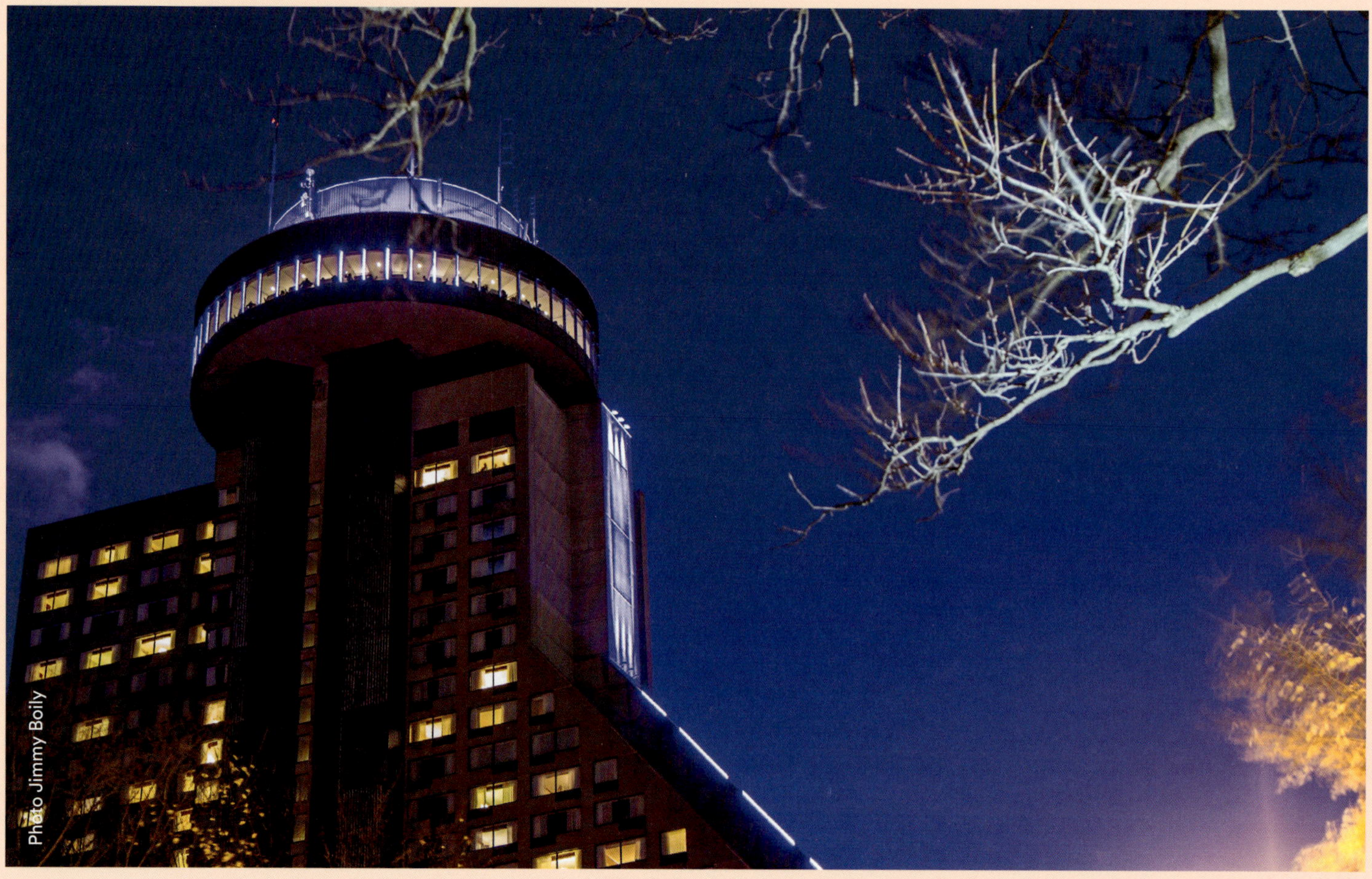
Photo Jimmy Boily

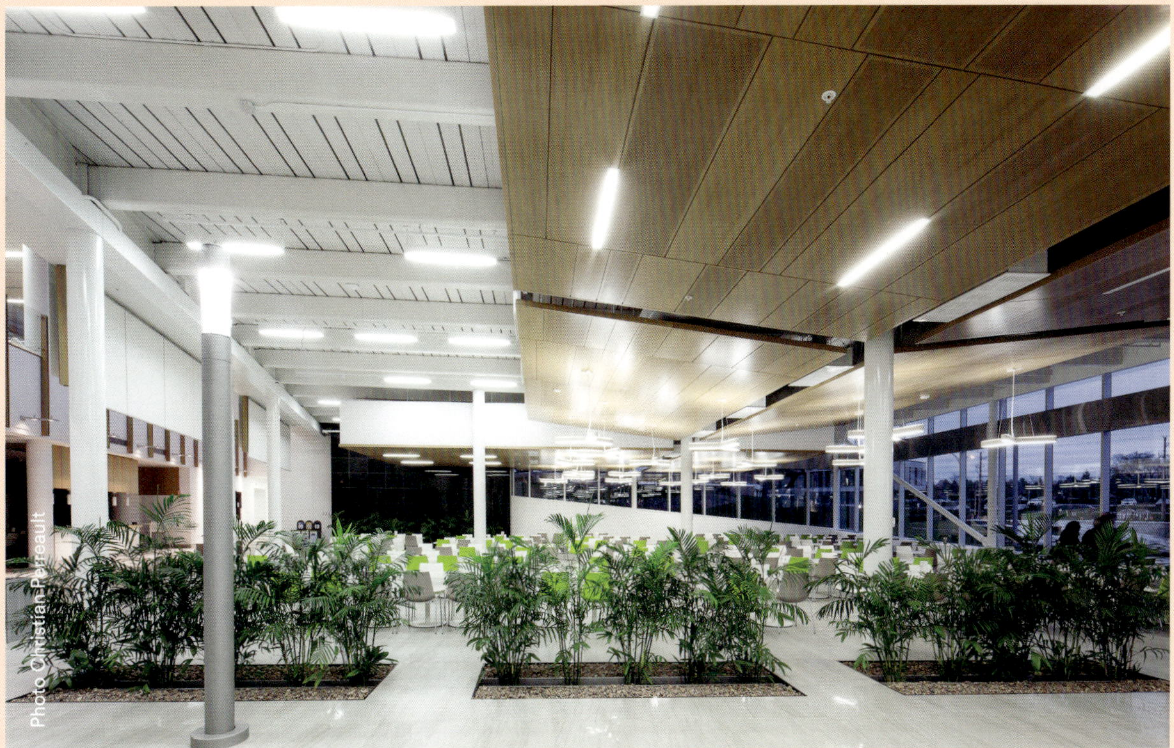
Photo Christian Perreault

La Cité Desjardins

When **2013**
Where **Quebec, Canada**
Client **Desjardins Insurance**

Ombrages was commissioned to provide landscape illumination of a restructured business district. A main focus was a wall of vegetation that serves as signage at night, with architectural lighting showcasing the wall while providing the plants with adequate light levels. Both direct and indirect recessed fluorescent luminaires were utilised for optimal visual comfort. LED accent fixtures were also integrated into the lighting design. In the plaza, lighting features blended in with the decor, and the decorative and architectural fixtures.

Monumental Staircase

When **2013**
Where **Quebec, Canada**
Client **City of Trois-Rivières**

When the City of Trois-Rivières launched its largest ever urban development project, Ombrages was contracted to redevelop the south gateway to the city's historical district. The project entailed creating two public areas to showcase historical elements. Ombrages focused on illuminating designated walkways and historic architectural aspects of the site. A main feature was the garden staircase – the steps of which are engraved with descriptions of the city's history – illuminated using a contactless LED system powered by electromagnetic induction. The main challenge was to create a strong visual impact through the illumination of one of the monuments and integrate LED modules into the structure while ensuring control over light pollution.

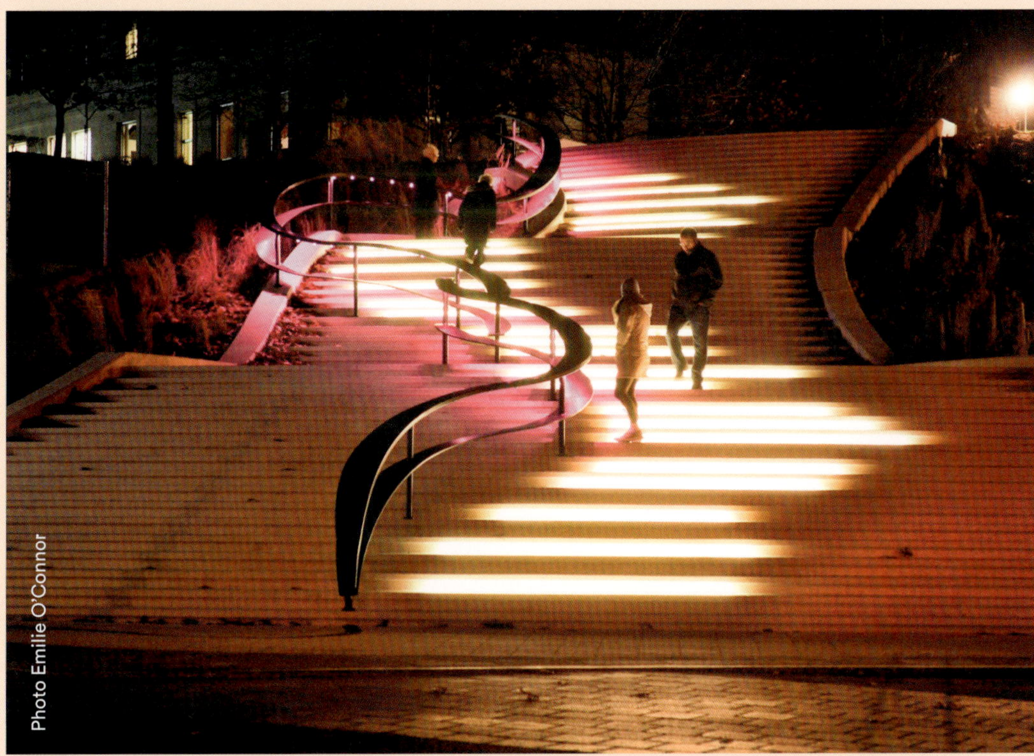
Photo Emilie O'Connor

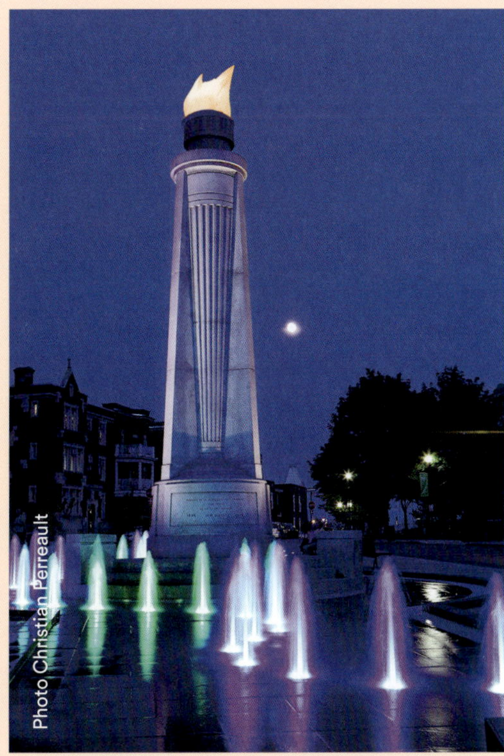
Photo Christian Perreault

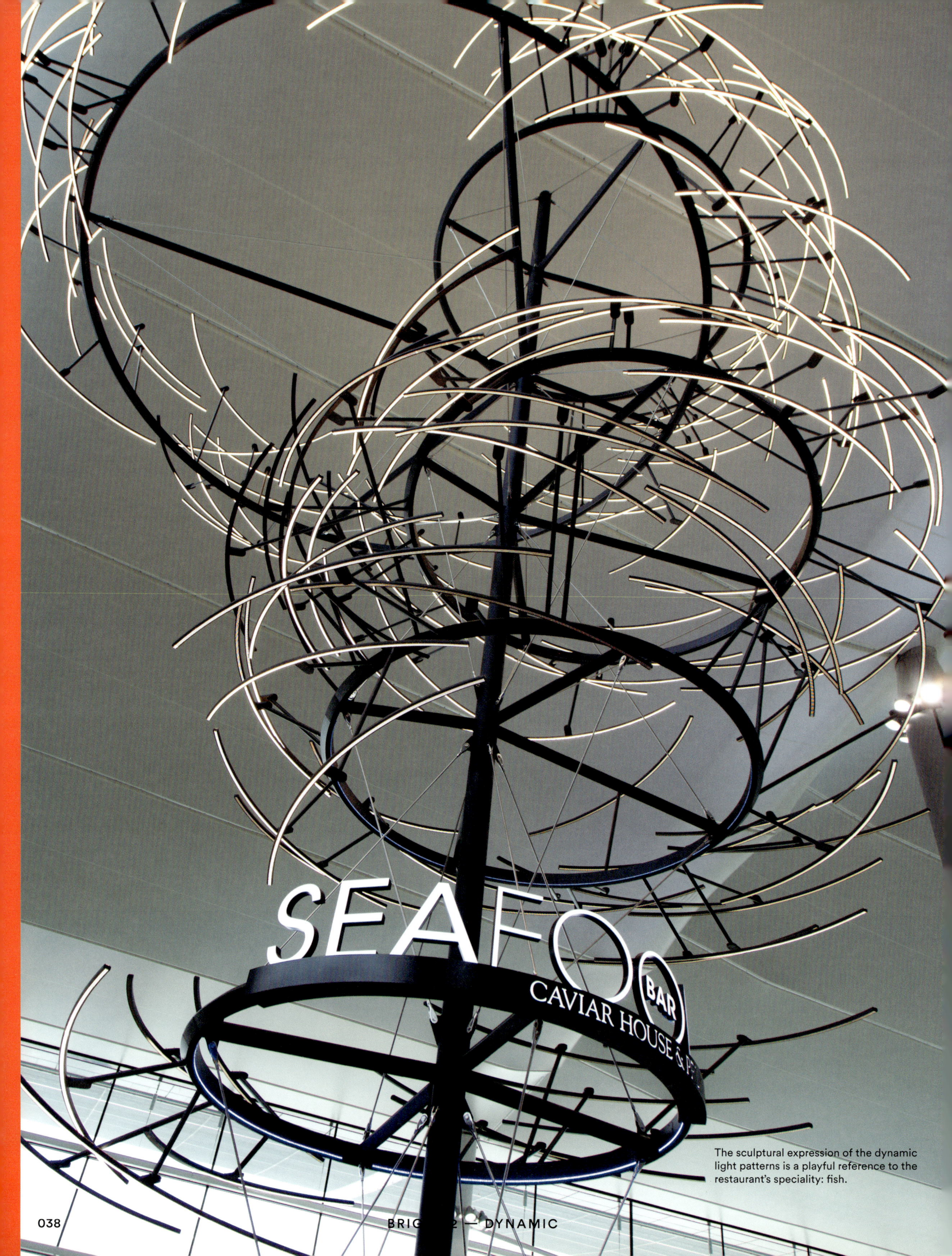

The sculptural expression of the dynamic light patterns is a playful reference to the restaurant's speciality: fish.

Emergence

A school of LED fish arrived in Heathrow Airport in June 2014, spiralling 13 m up to the ceiling above the heads of diners. This installation by **Cinimod Studio** looked to raise the profile of the seafood restaurant Caviar House & Prunier by combining a visually powerful structure with complex, tailor-made hardware, electronics and software.

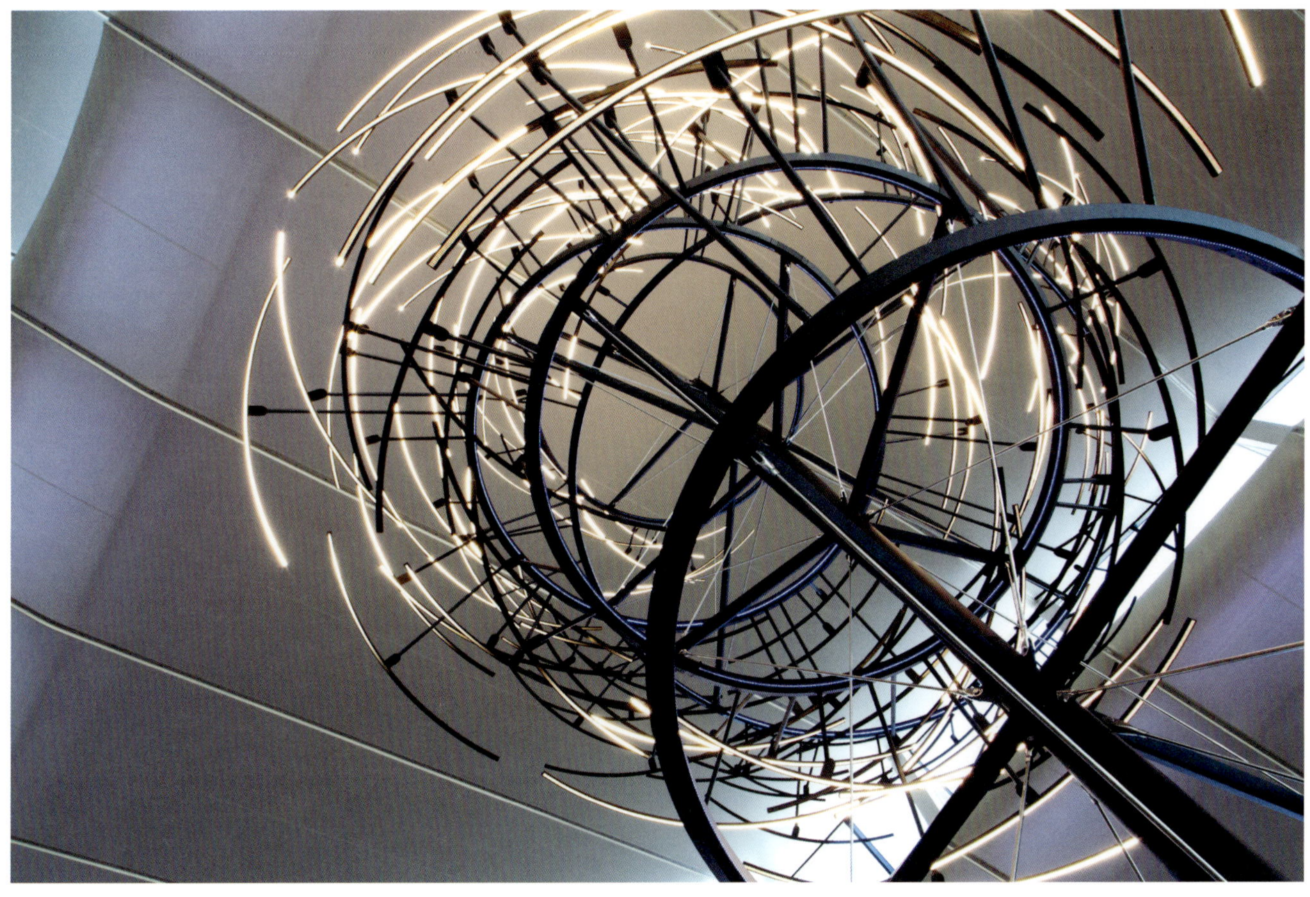

Photos Dominic Harris

Emergence, created by Cinimod Studio for seafood restaurant Caviar House & Prunier, is a lighting sculpture which re-imagines the movement of a school of fish. Individually-controllable LED arcs – made of engineered carbon-fibre composites – are animated by a fluid, synchronised and realistic array of light patterns that mimic those created as fish move underwater in perfect harmony.

This fragmented shimmer of scattered light causes the installation to radiate a mesmerising glow. The numerous arcs hold a total of 350,000 white LEDs that extend to a height of 13 m and a width of 8 m. Yet, the software and control system for the electronics developed to power Emergence is as unique as the structure itself. Its natural source of inspiration – fish – meant that it was necessary to create cutting-edge, interactive digital lighting in order to lend realism to the real-time 3D simulations. It is precisely this coming together of art, design, architecture and engineering in the creation of an icon which combines powerful aesthetics with bespoke structures and hardware, and complex electronics and software, that motors the studio's work.

The light sculpture sits above the bar, in the heart of the International Departure Lounge at Heathrow Airport, as a weightless and captivating landmark, a memorable icon that exposes the restaurant to the millions of travellers flocking through the terminal daily. The studio strived to deliver a complex and ambitious installation that is firmly routed within the commercial sensitivity of the client, significantly adding to image of the brand. —

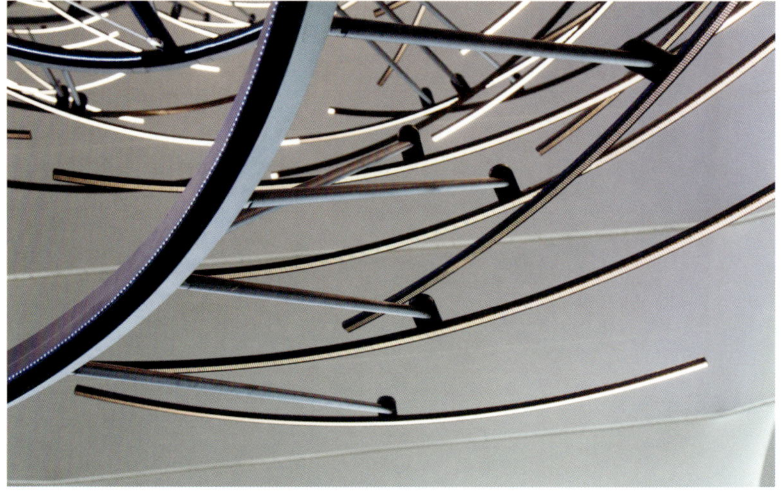

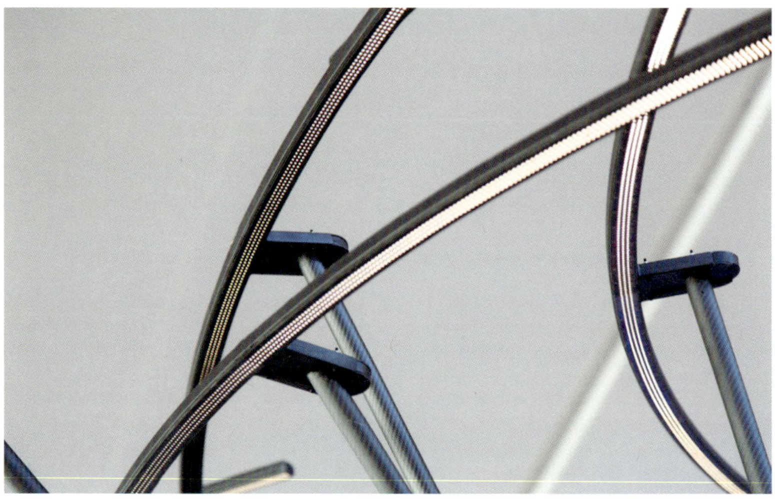

The installation rises up above the bar, creating an illuminated focal point in the airport terminal.

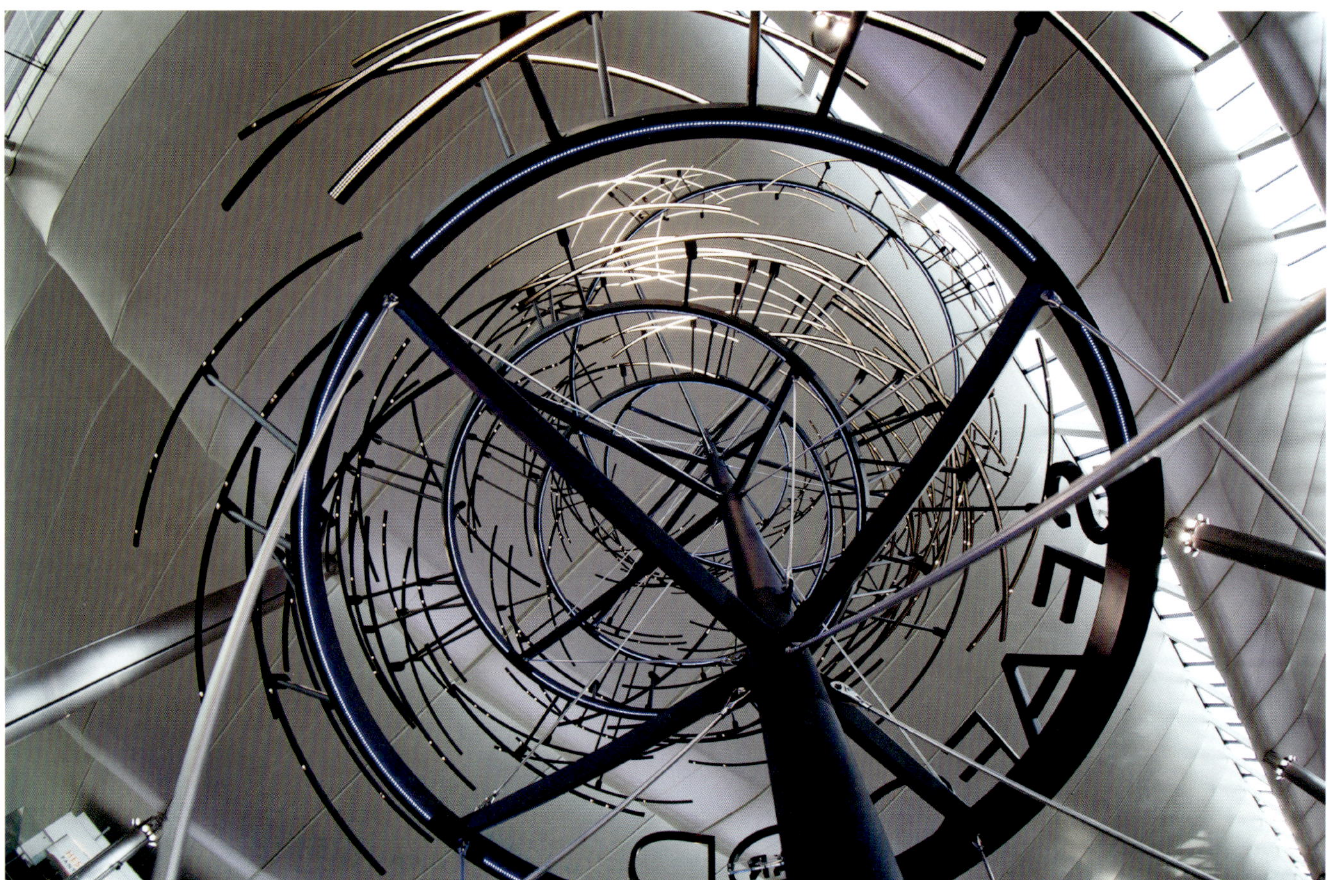

Designer
Cinimod Studio
Project
Emergence
Location
London, United Kingdom
Client
Caviar House & Prunier
Collaborator(s)/consultant
Tall Engineers, White Wing Logic
Manufacturers
Cinimod Studio, Polar Manufacturers
Date
June 2014

Set-up

Total volume of the installation is just over 865 m³. It spirals upwards 13 m with LED arcs affixed to five rings of differing diameters, layered at 2-m intervals.
Primary structure was engineered from carbon-fibre composites.
Programming language runs a complex, tailor-made, real-time 3D simulation software; the simulation of virtual schools of fish was calculated directly on the Nvidia GPU, taking cues from a variety of environmental conditions.
Used light includes 350,000 white LEDs laid out in individually controllable arcs held together through mechanical fixings.
Brightness control was achieved through an abstraction layer of manipulated video frames, stored on the LED arcs' controllers and manipulated from the control server RS-422.
Construction time was 5 months.

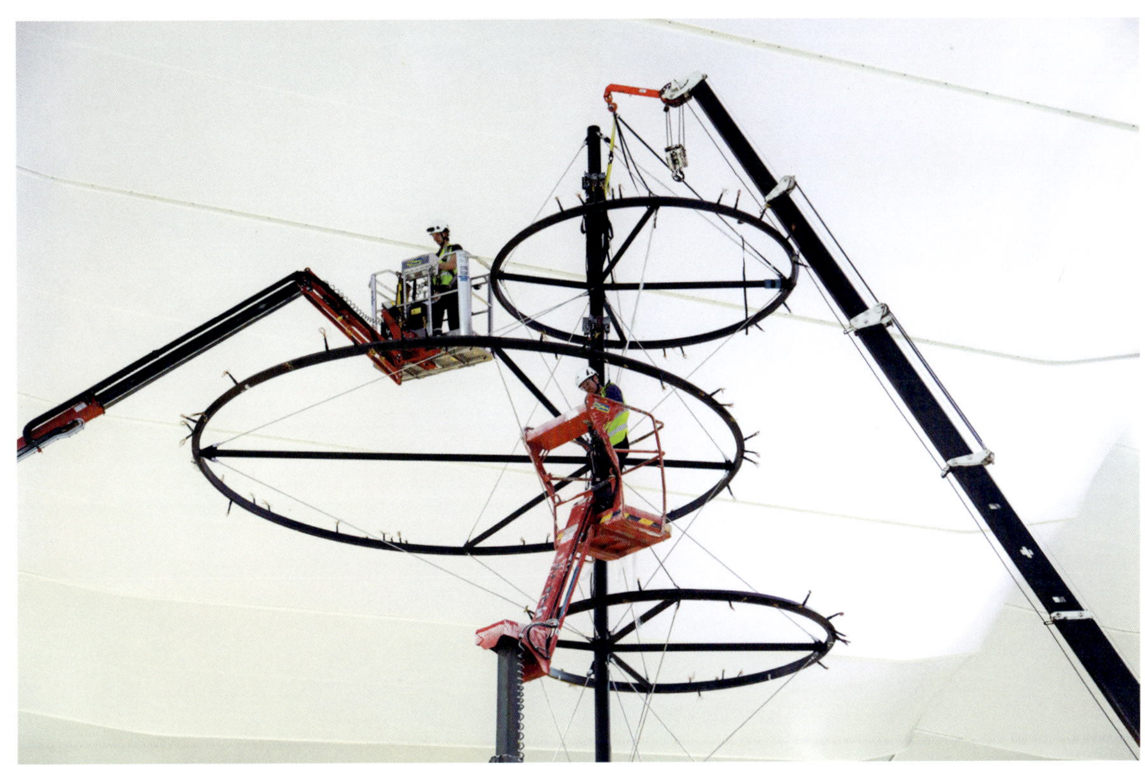

THE LIGHT SCULPTURE IS A CAPTIVATING LANDMARK AND MEMORABLE ICON

SCULPTURE ELEVATION

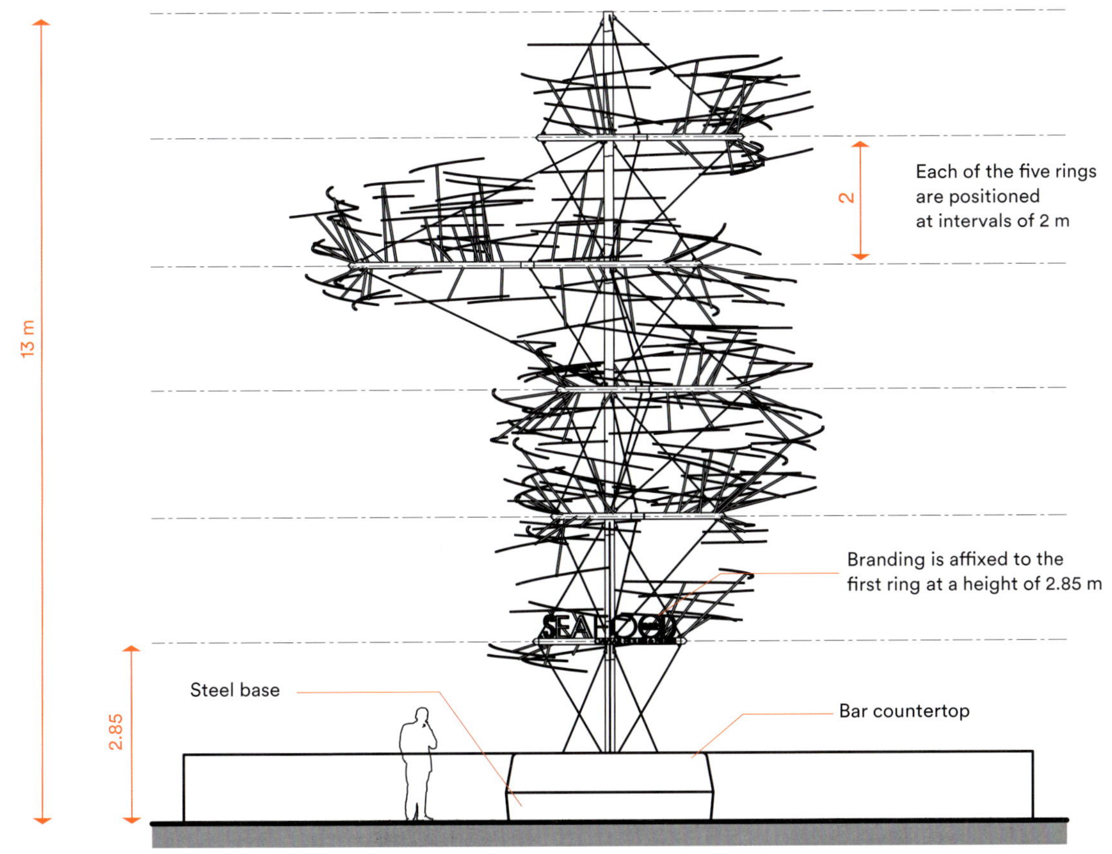

Specifics

A key challenge of this project was to create an iconic feature that would make Caviar House & Prunier stand out amongst the hospitality outfits within the airport lounge. Cinimod ensured a memorable identity for the brand with a unique and highly experiential physical form.

Airports are notoriously harsh environments within which to operate sensitive electronic equipment, due to the sheer volume of data circulating on a daily basis. From the onset of the project, it was clear that a very low bandwidth and ultra‑stable control system would be required. The team devised a bespoke system in order to control the structure's 350,000 LEDs.

Individual brightness could be monitored through the use of thousands of frames of pixel sequences within the AVI file, with five simultaneous video layers being played at a given time with arbitrary frame-rates and transformations. This meant that an incredibly fluid 60 fps can be maintained in real-time to each pixel on a minuscule overall bandwidth.

It was necessary to use a GPU instead of the conventional CPU to run highly-complex distance calculations for each LED, a process necessary in order to make the sculpture a true volumetric display.

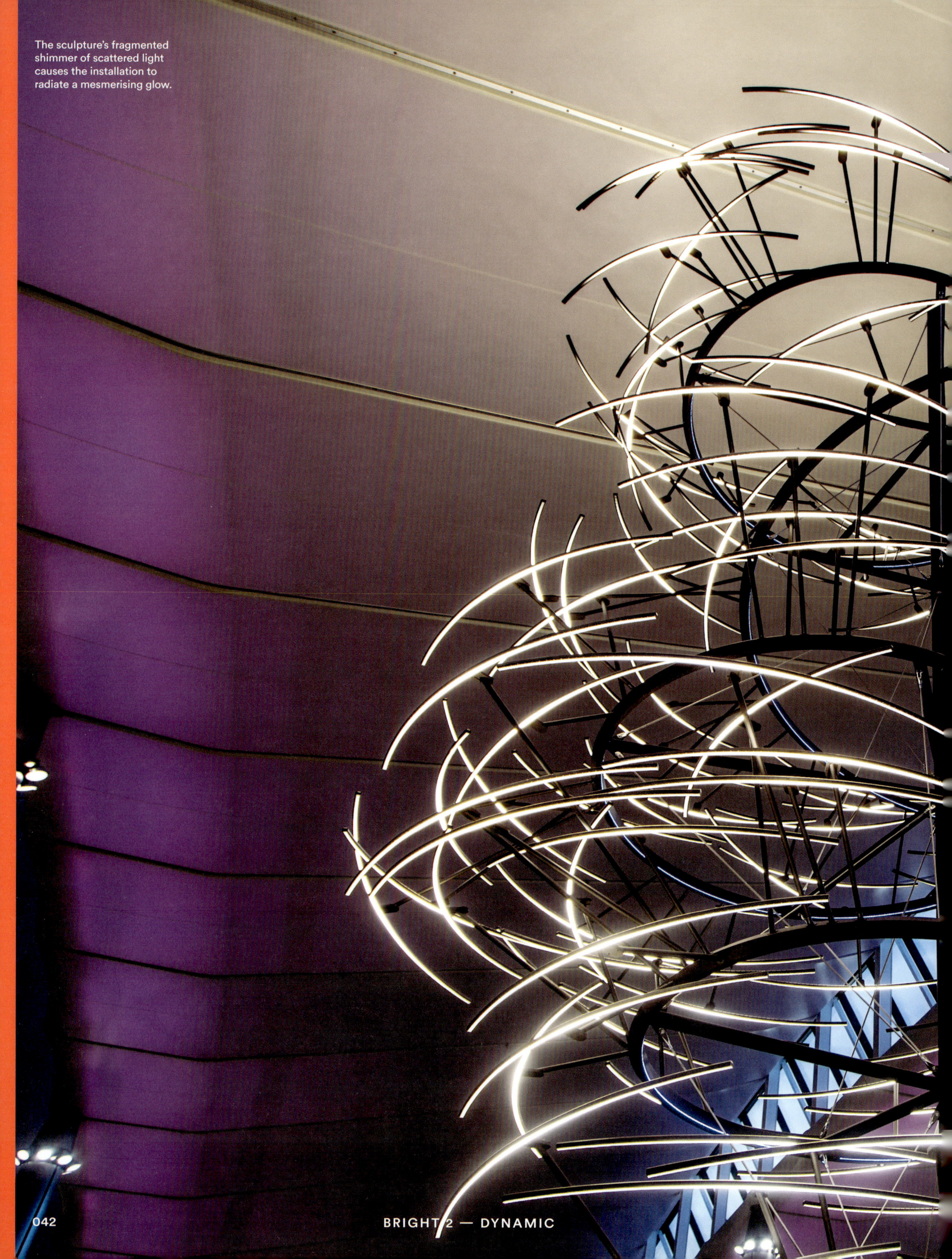

The sculpture's fragmented shimmer of scattered light causes the installation to radiate a mesmerising glow.

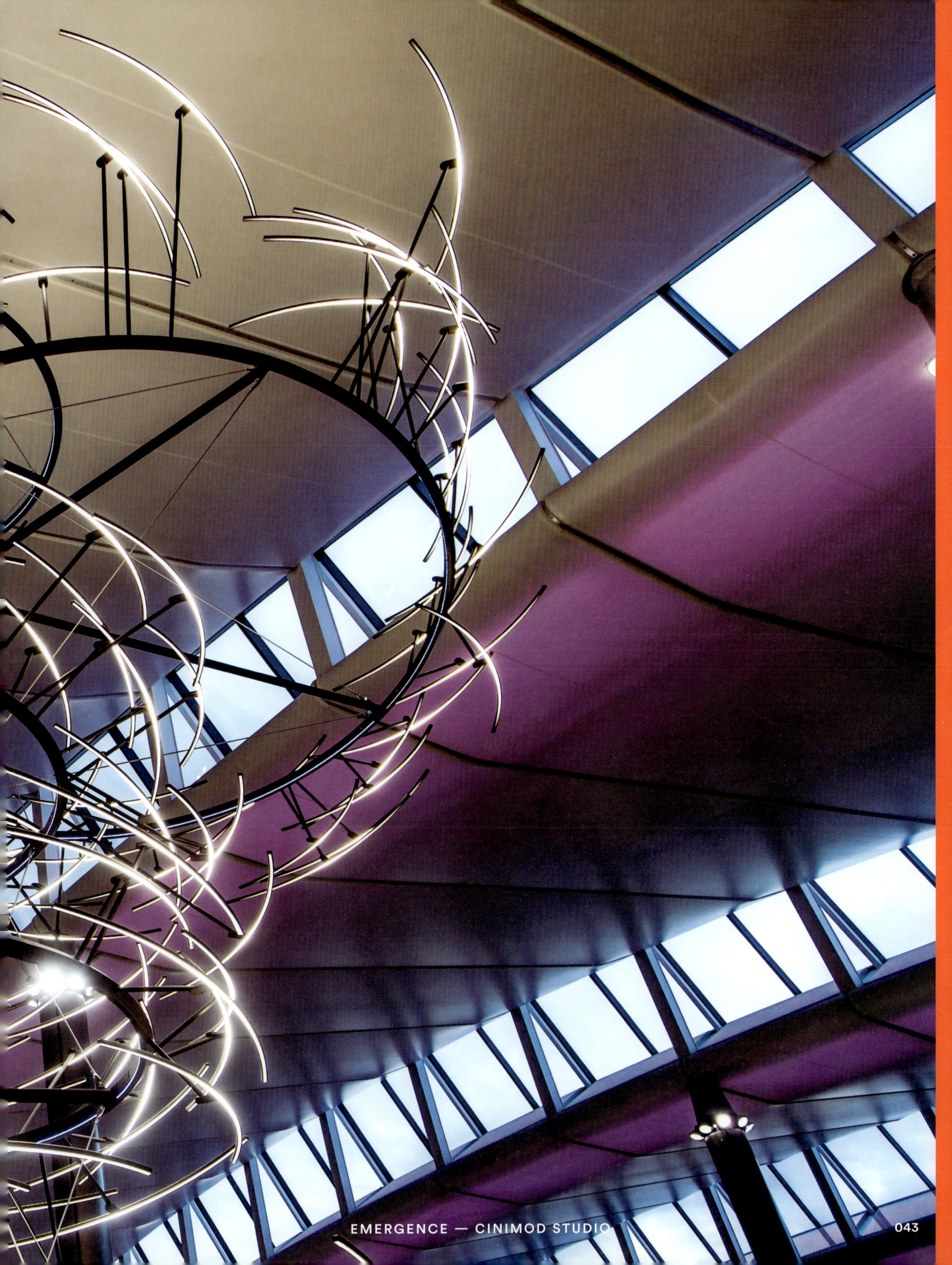

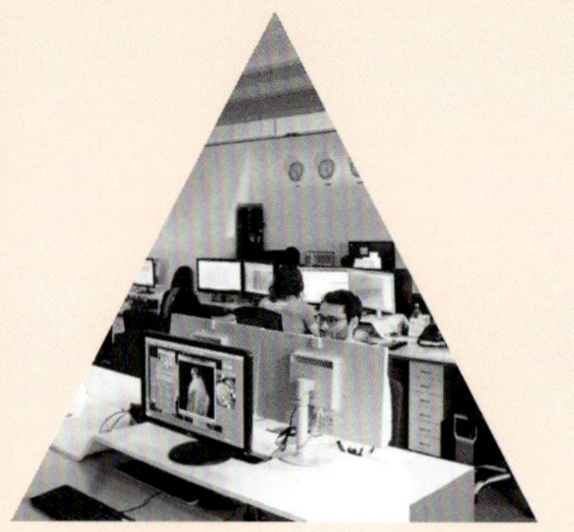

Cinimod Studio

A cross-disciplinary practice, Cinimod Studio specialises in the fusion of architecture and lighting design. The London-based firm's dedication to research ensures the desire to use the latest technologies and manufacturing techniques, thus creating work that is not only an aesthetic statement but also a technologically-advanced endeavour. Since it was founded in 2006 by artist and architect Dominic Harris, the studio has collected a number of important clients, from EDF Energy and Microsoft Xbox to the Victoria and Albert Museum.

CINIMODSTUDIO.COM

Finial Response

When **2011**
Where **London, United Kingdom**
Client **Blenheim Properties**

This permanent, interactive, public artwork for Blenheim Properties was created in collaboration with Wilkinson Eyre architects. The piece, which deconstructs the ornamental London railings into a simple and unconventional but recognisable form, has been in Soho for 3 years, blurring the boundaries between architecture and art.

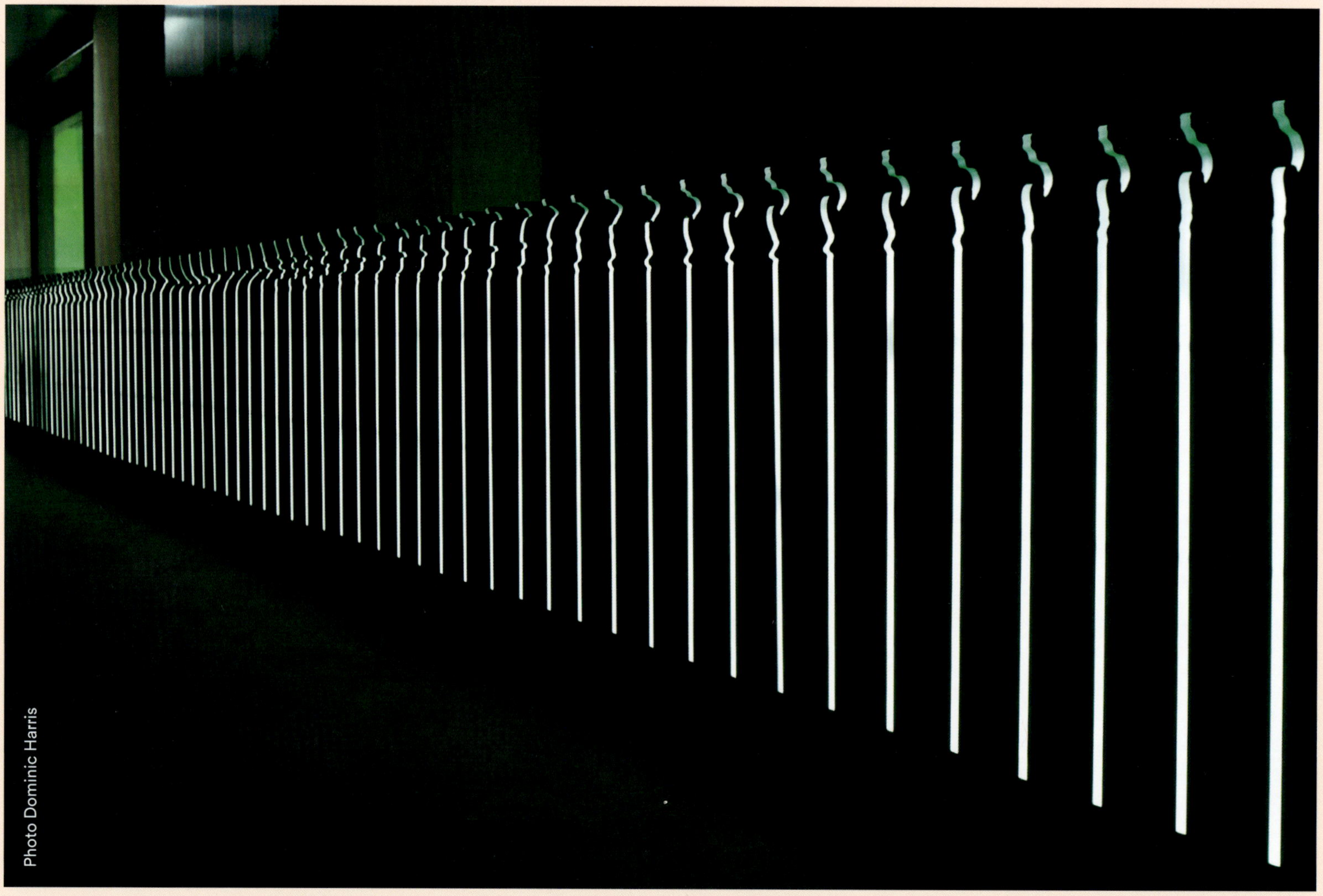

Photo Dominic Harris

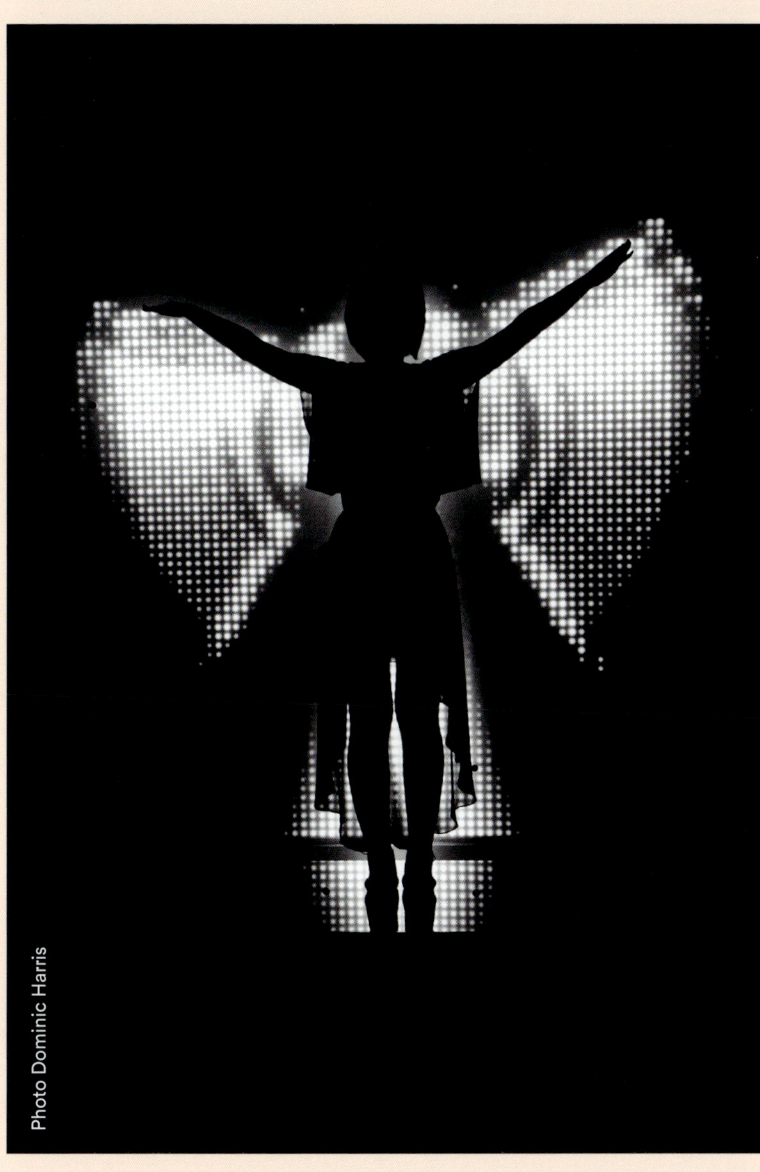
Photo Dominic Harris

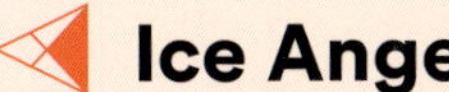

Ice Angel

When **2012**
Where **London, United Kingdom**
Client **Private Commission**

This interactive performance portrait introduces a digital twist to the making of snow angels. Virtual snow substitutes fresh snow, allowing participants to catch a glimpse of themselves as angelic beings. The artwork's memory is able to pair each participant to their specific angel identity.

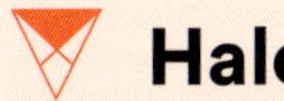

Halo

When **2012**
Where **London, United Kingdom**
Client **Microsoft Xbox**

Weighing just under 3000 kg and measuring 15 m in diameter, Halo was one of the largest and brightest man-made structures to ever fly over a capital city. The UFO in the shape of a glyph was illuminated by 113,096 orange LEDs and took off over London in celebration of the launch of the Xbox game's fourth chapter.

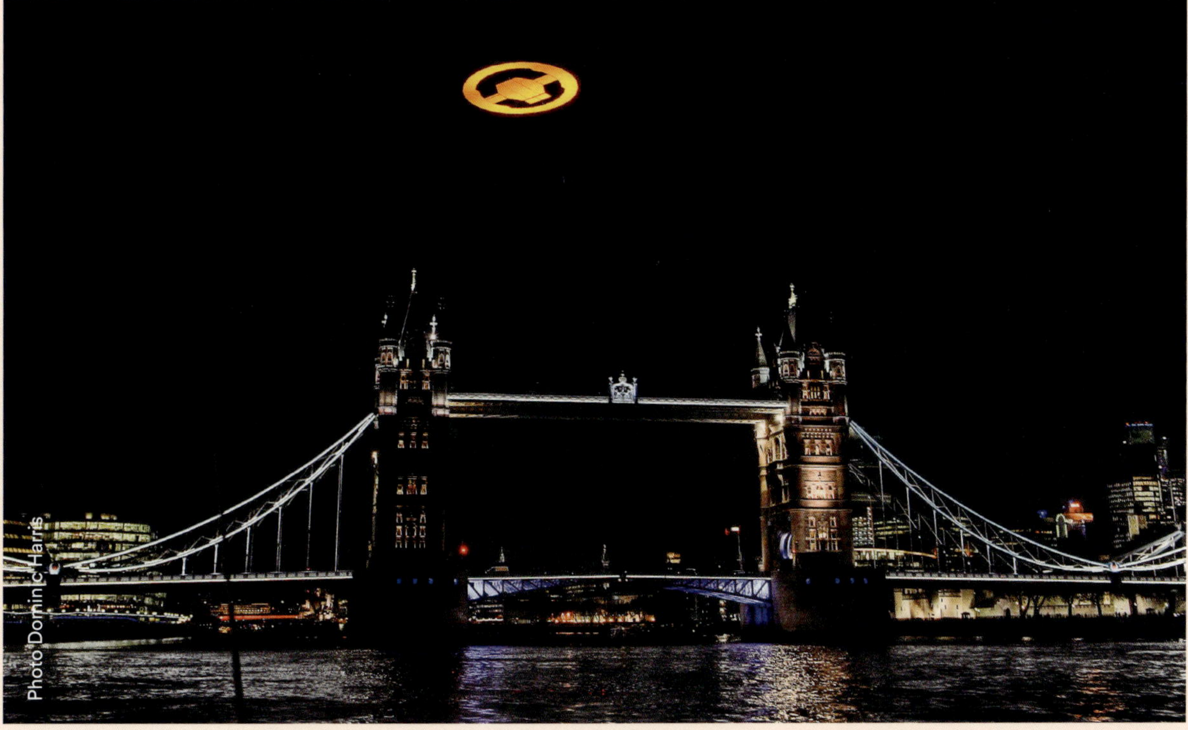
Photo Dominic Harris

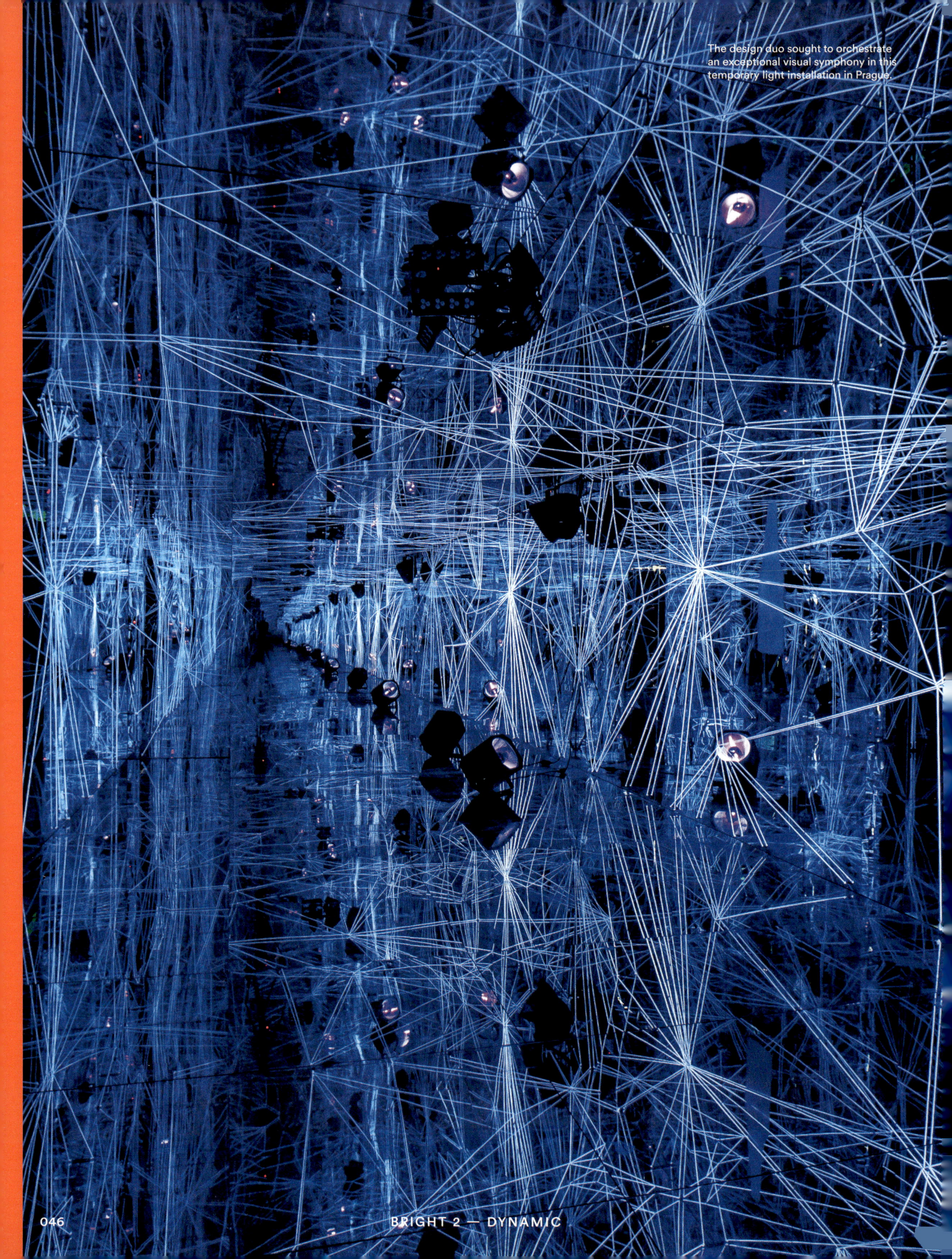

The design duo sought to orchestrate an exceptional visual symphony in this temporary light installation in Prague.

Heineken Visual Experience

For Heineken's presentation at Designblok 2013, the studio of **Jan Plecháč and Henry Wielgus** created an electrifying light installation. A perplexing mix of visual and audio bombardment of the senses, the Czech duo utilised UV light rays, LEDs, white lines and mirrors to create a disconcerting yet enticing experience for visitors, where the boundaries between real space and reflection were erased.

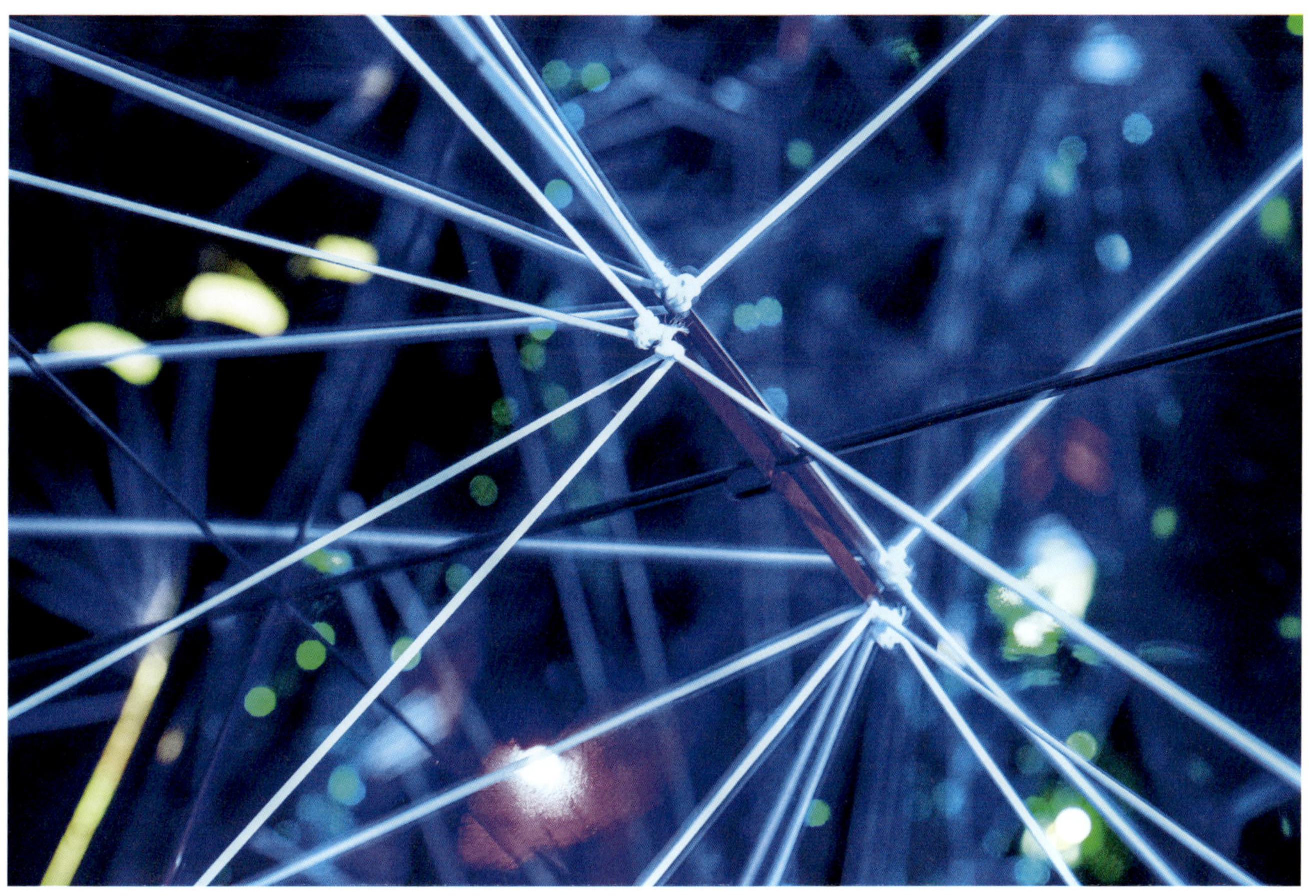

Photos Kiva

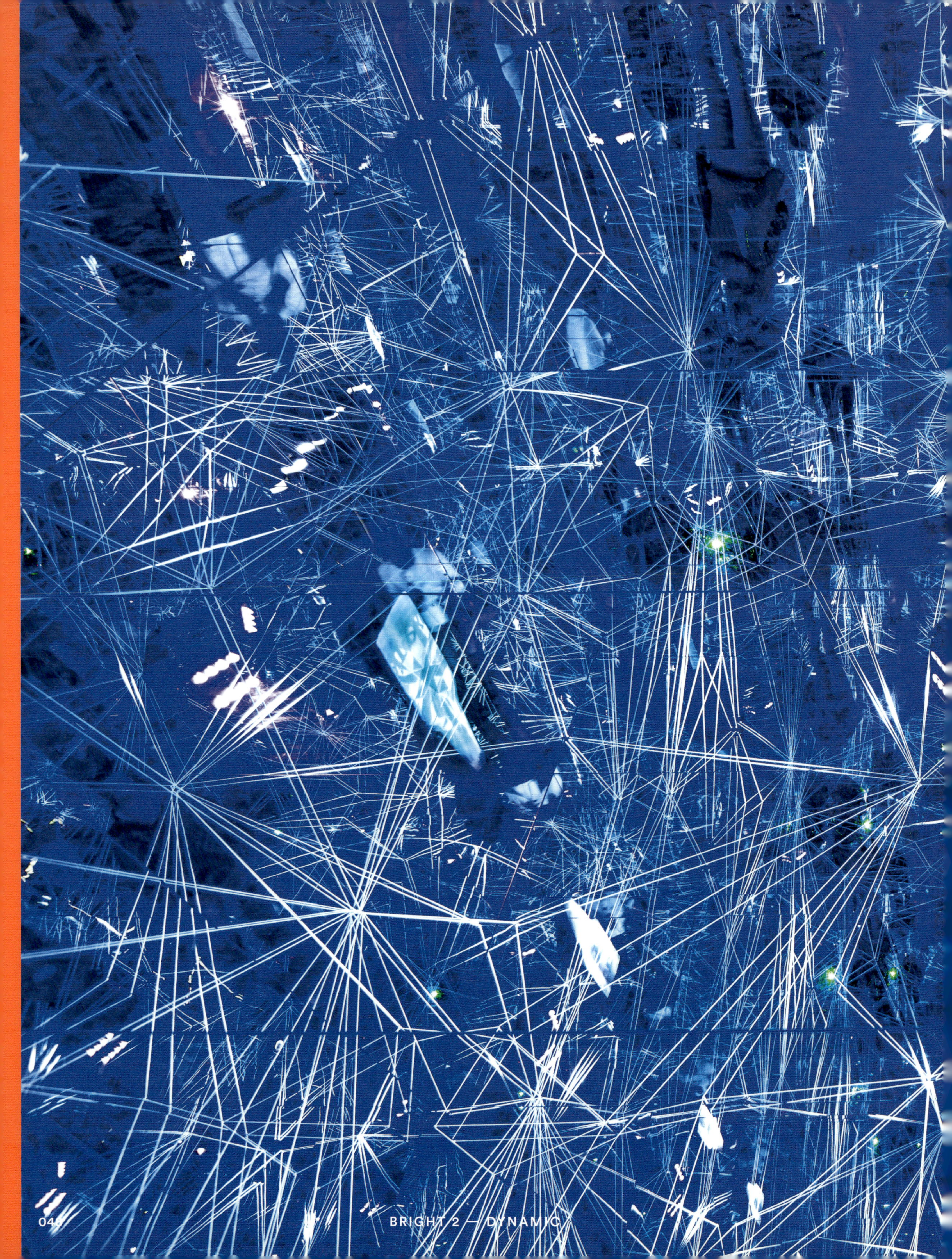

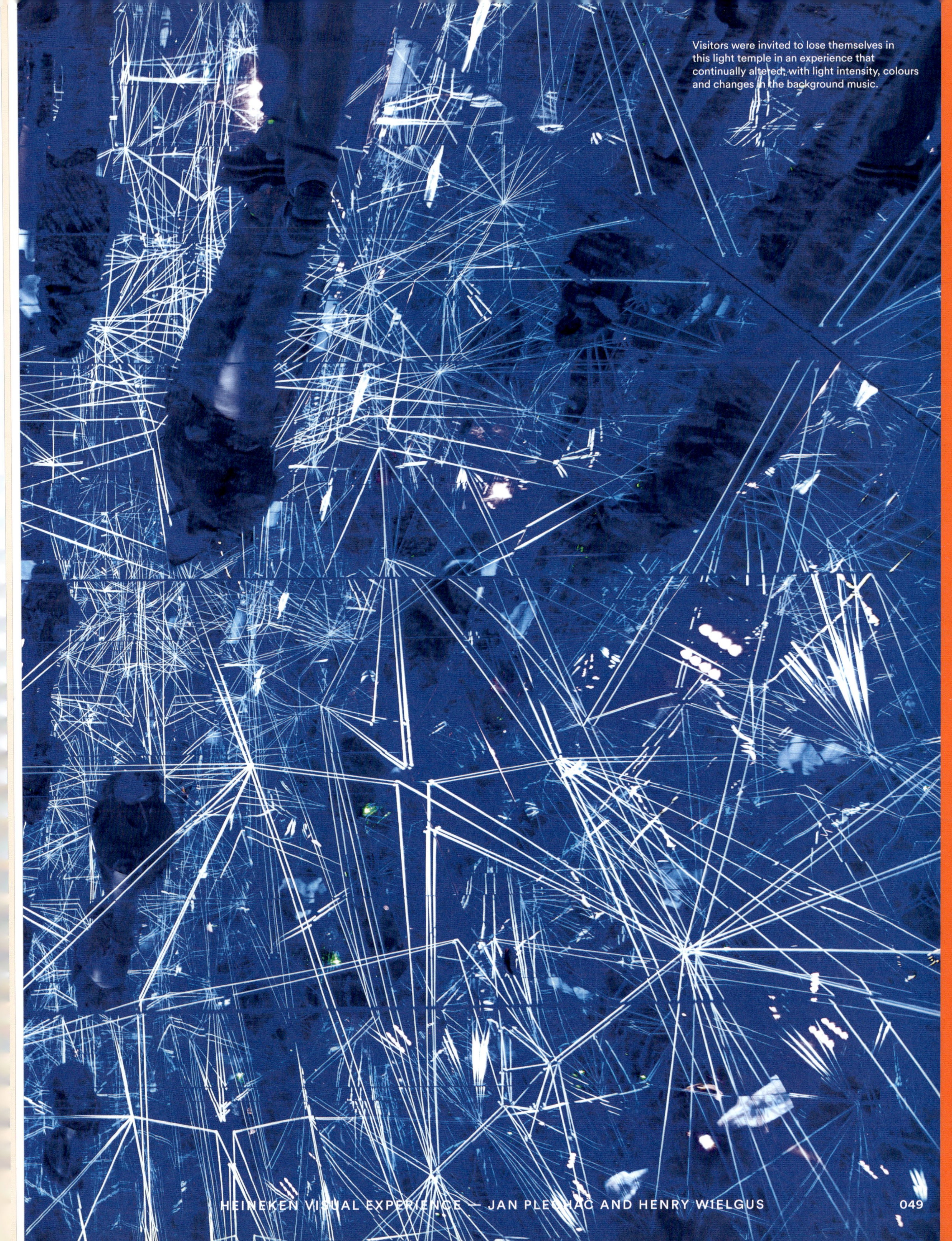

Visitors were invited to lose themselves in this light temple in an experience that continually altered, with light intensity, colours and changes in the background music.

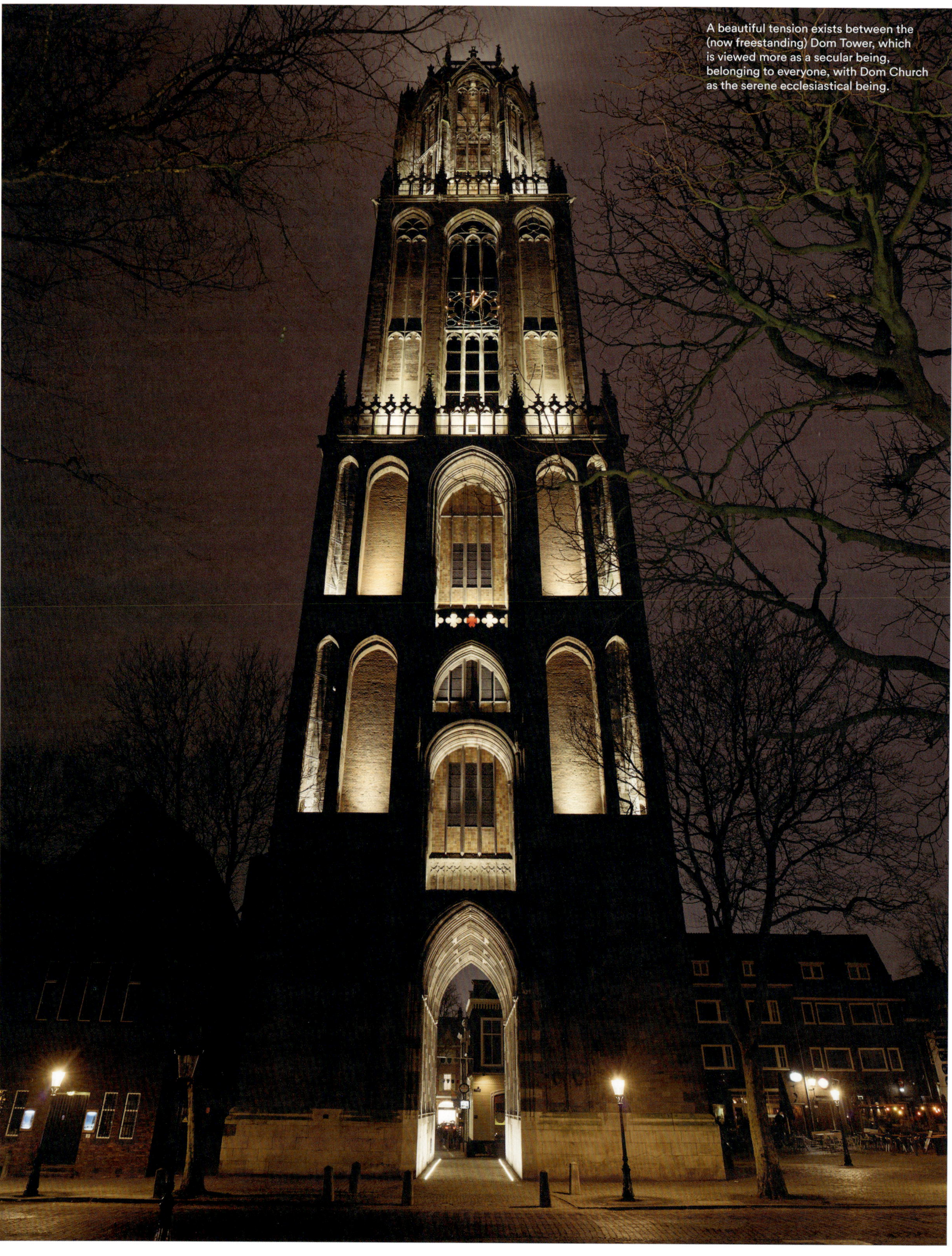
A beautiful tension exists between the (now freestanding) Dom Tower, which is viewed more as a secular being, belonging to everyone, with Dom Church as the serene ecclesiastical being.

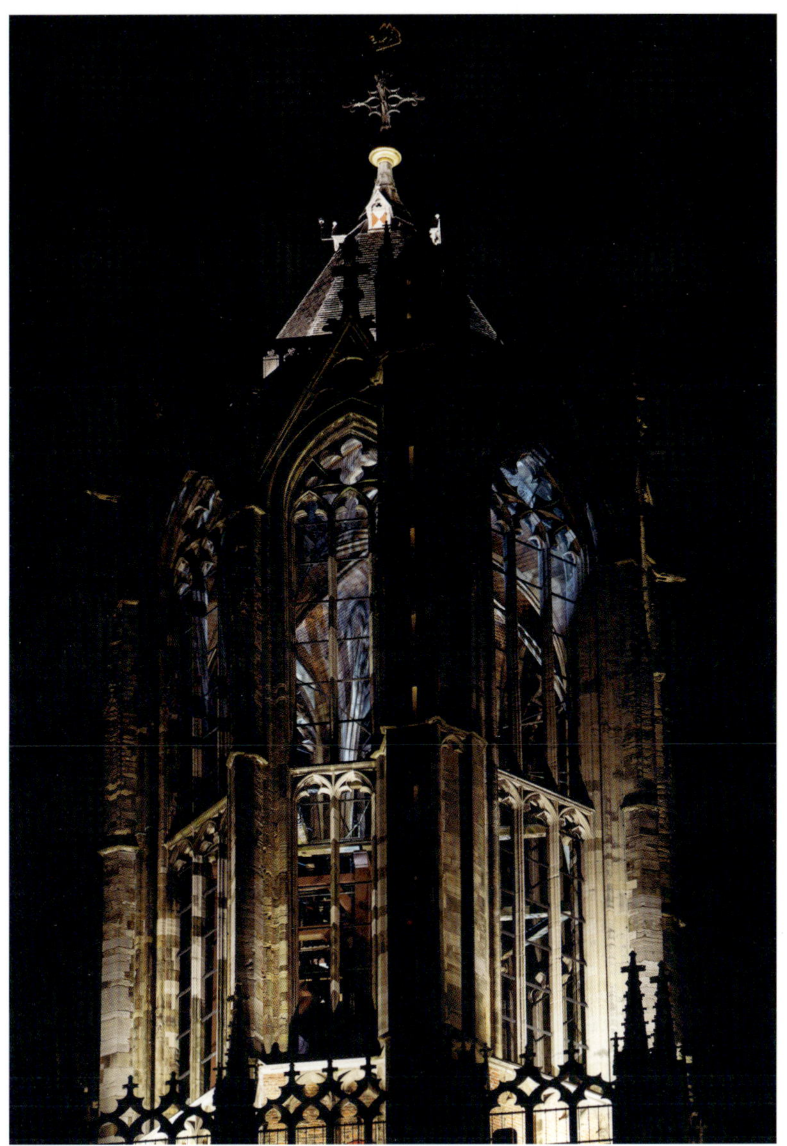

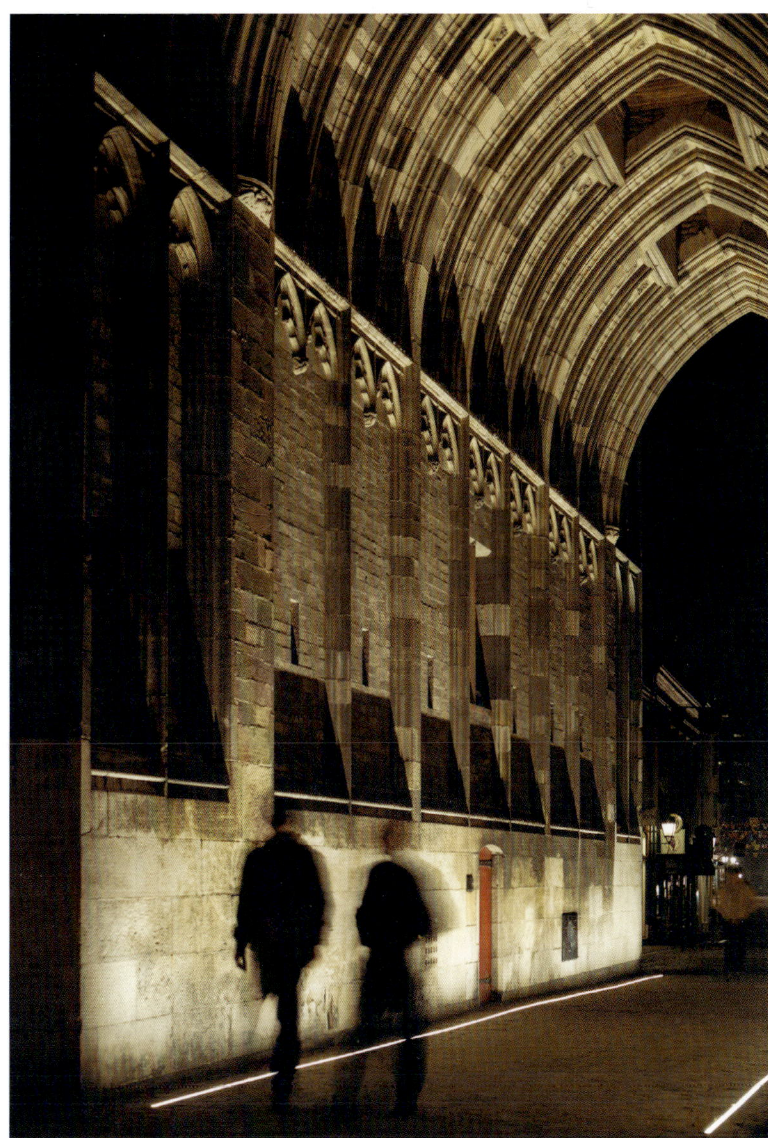

A LIGHT CONCEPT MAKING MEANINGFUL CONNECTIONS ACROSS LAYERS OF HISTORY

A unique light installation was unveiled in central Utrecht by Her Majesty Queen Beatrix in the spring of 2013 to officially open the celebrations marking the tri-centenary of the signing of the Treaty of Utrecht, a pivotal moment in history signalling the beginnings of lasting peace for Europe. The installation uses light as a narrative tool to connect the iconic Dom Tower with the Dom Church and Dom Square. The launch event that opened proceedings needed to be wider than simply lighting the buildings themselves. The piece needed to reference the celebration that would launch it, and make meaningful connections across the layers of history and Utrecht's place in the modern world.

Speirs + Major won the commission with an idea that viewed the city, district and buildings as a living organism. The site forms the body with the buildings making up its structure. Being at the heart of the city which, in turn, is in the centre of the Netherlands, Dom Church, Square and Tower can also be viewed as the hub of a huge central nervous system; a true centre of communication – with the tower itself as the 'brain', akin to a wise figure that holds on to all the memories and stories. Each of the three elements in themselves can be viewed as important members of society, living entities that have been both observers and participants in the development and history of Utrecht.

The basis of the concept was that it was the light that enabled the structures to come to life – to breathe, connect with each other, and communicate with the people in the city about the past and the present. Beginning with Dom Church, addressed with a passive, respectful design and translating the philosophy that 'light comes from within', the concept ends with Dom Tower as the show-piece: the dynamic element, the communicator. The lighting is dramatic, setting off the Gothic architecture, and is visible from many parts of the city. With a restricted palette of almost entirely LED based components and a well-finessed control program, it has a face value of simplicity belying the complexity in its execution. The end result is entrancing. The sensitive use of light brings these historic elements of Utrecht to life, celebrating their forms and recording their significance in the psycho-geography of the city. —

Designer
Speirs + Major
Location
Utrecht, the Netherlands
Client
City of Utrecht
Collaborator(s)/ consultant
Marijke Jansen, Kees Van De Lagemaat, Arthur Klink, Heijmans
Manufacturers
Proliad, Meyer, ACDC, Sill, Hungaroflash, Diversitronics, Martin Professional, Pharos
Date
April 2013

DOM TOWER ELEVATION

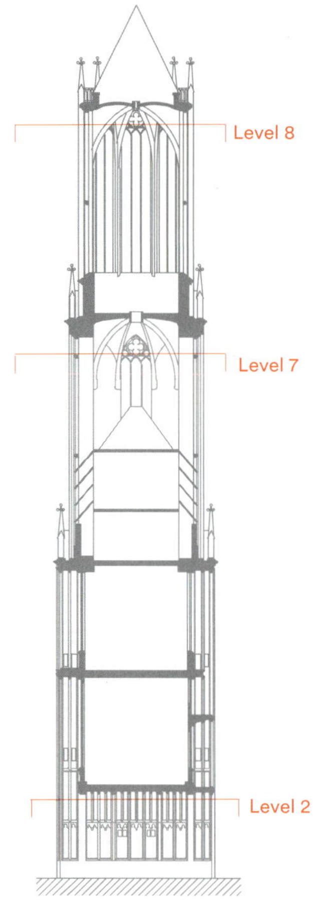

Section A

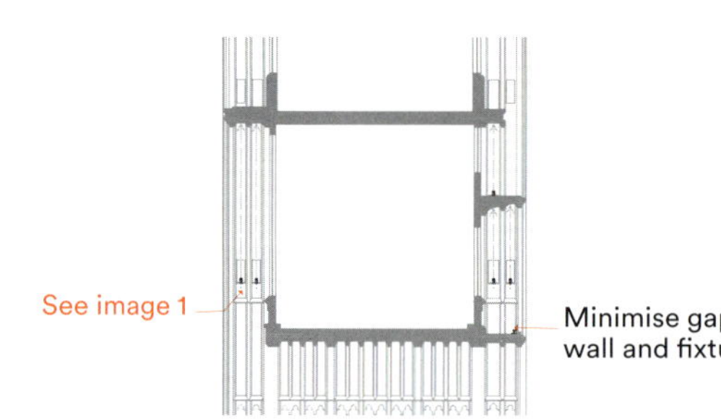

Level 2

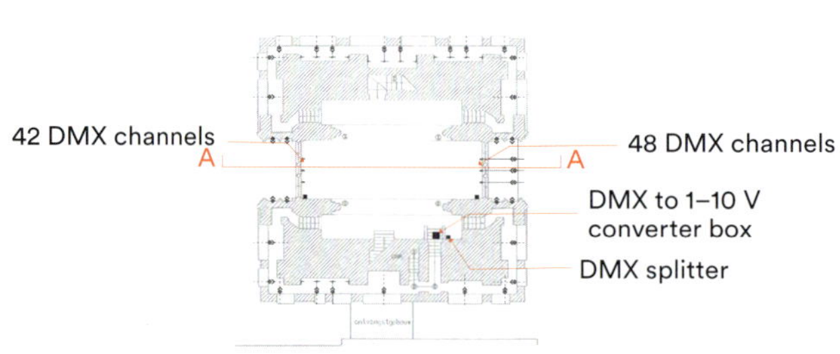

Section B

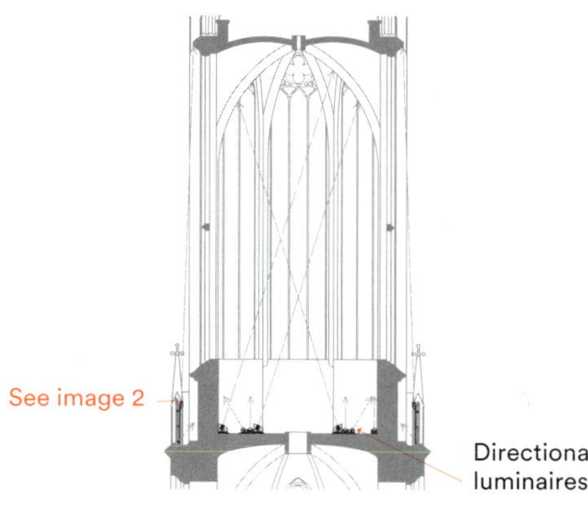

Section C

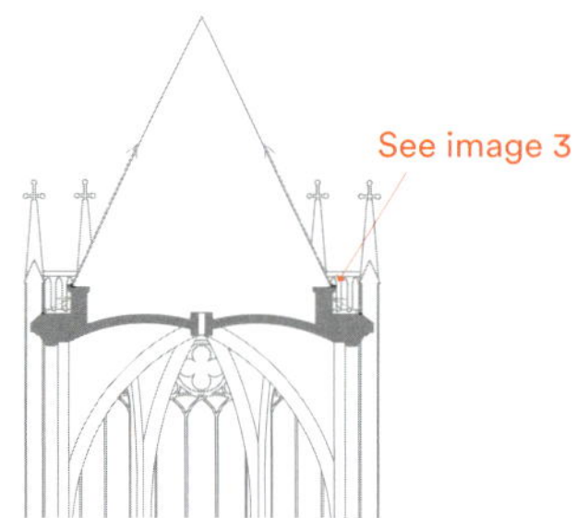

Level 7

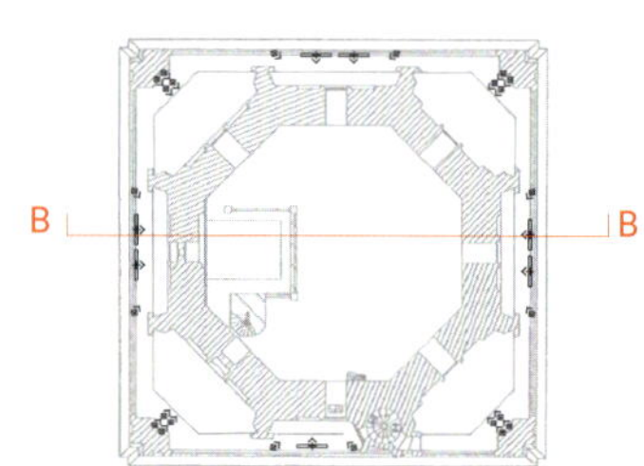

Level 8

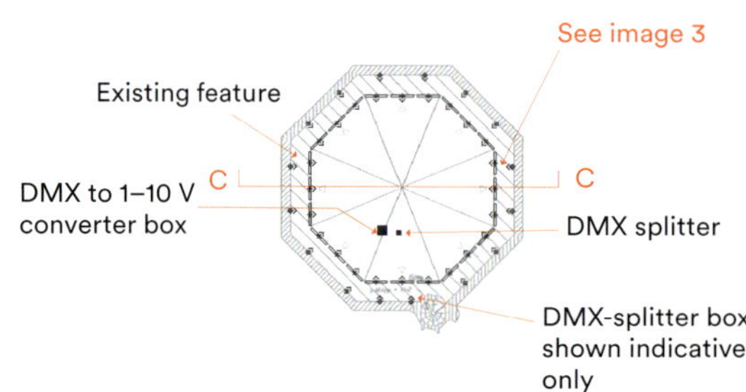

Set-up

LED based components make up almost the entire illumination, with a specified control program and a scheme that is energy efficient and maintenance friendly.

Lighting specifications

- Proliad: Xicato Spotlight 1300 LM, Spotlight 3000XL
- Meyer: Custom Superlight Compact
- ACDC: Iglu, Integrex, Fino
- Sill: 021 Series Projector, Custom inground uplight
- Hungaroflash: DMX Cap Strobe
- Diversitronics: DMX Strobe Cannon
- Martin Professional: Exterior 400 Image Projector
- Pharos: Control system

Total power output (per m²) is 0.03 W excluding strobes; 0.1 W including all strobes (strobes only flash for a very short amount of time).

Total surface area across the three sites for illumination was 4500 m².

1

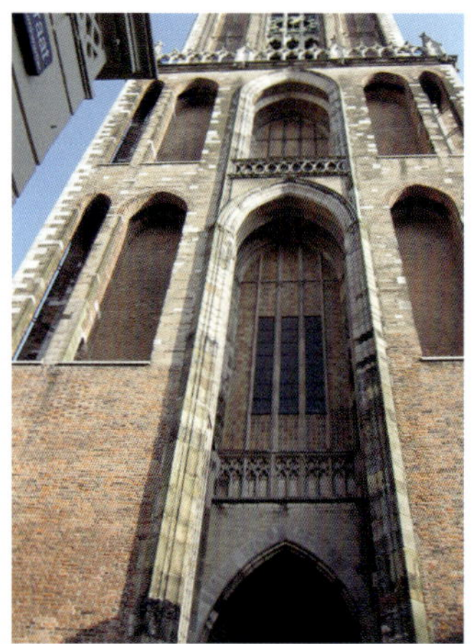

2

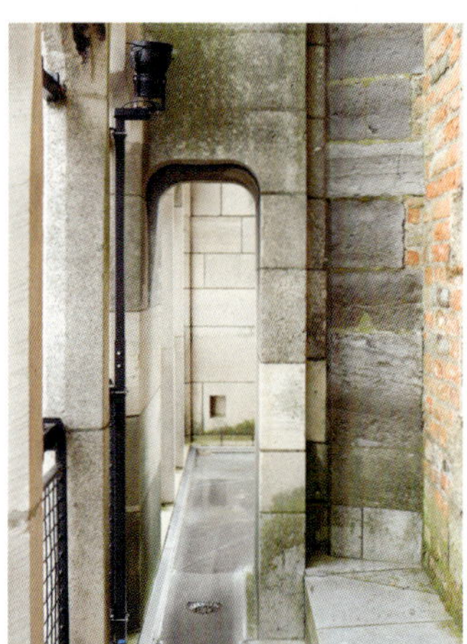

3

BALUSTRADE DETAIL

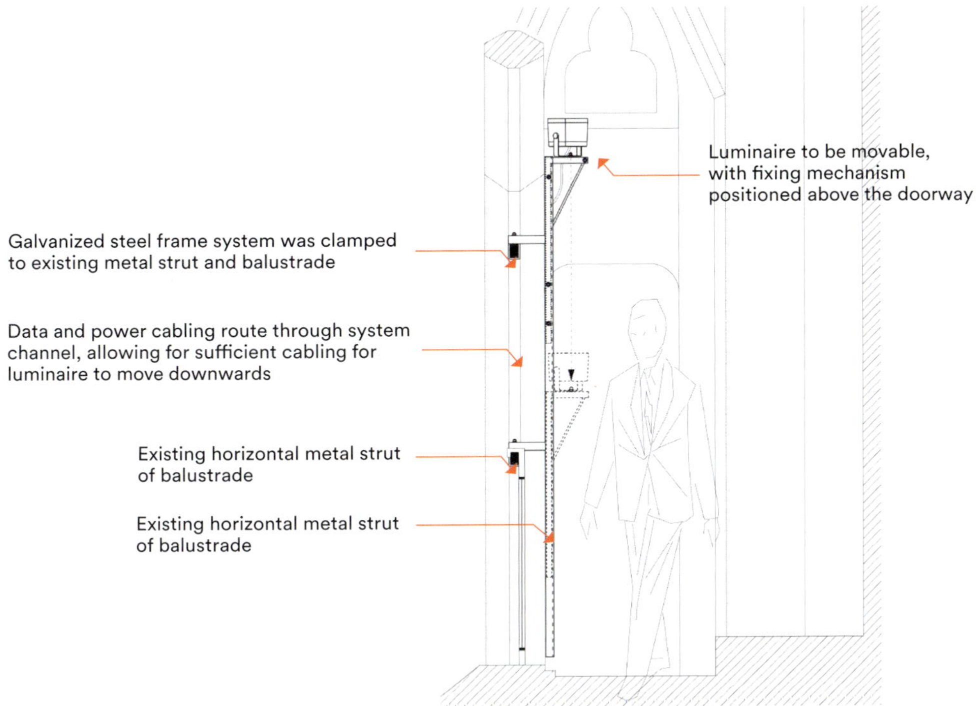

GUTTER DETAIL

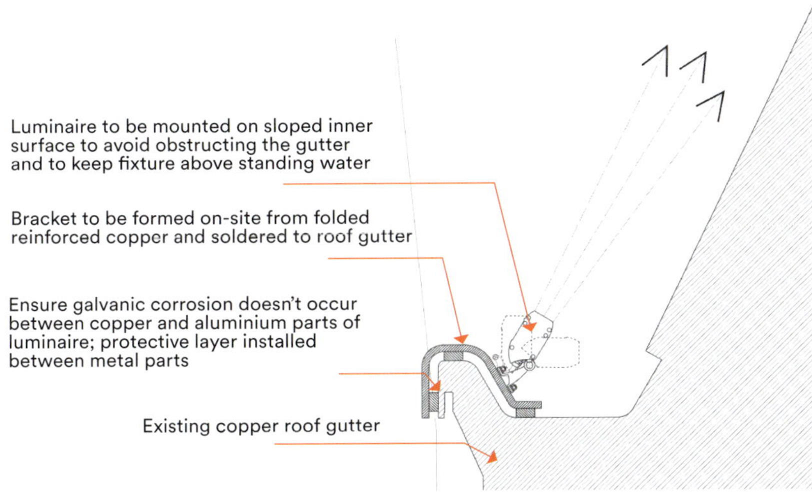

Specifics

An immediate challenge was to respect the historic treasures at the heart of the project: Dom Church and Dom Tower were built in the fourteenth century. Originally they were connected to the west wing small bridge, but in 1674 a tornado caused the unfinished west wing to collapse, and it was never reconstructed. This area became known as Dom Square. The team looked at the nature of each of these historic buildings and considered how light could make meaningful connections across the site and the wider city.

The constraints of no drilling into the stonework and no sources allowed to be in view from below, meant that mounting devices had to be custom made for each location. These details largely comprise of compression clamping arrangements, or concrete slabs not connected to the buildings, concealed where possible, and located strategically so as not to cause safety issues or impede views, but still be accessible for maintenance.

The illumination of the church is translated to a scheme where the outside faces are kept relatively dark, and the light glows through the stained glass windows and the internal faces of the buttresses, creating a lantern-like effect. The square is where the memories of the past are recalled, with illumination picking out historical aspects to create a visual connection to the arch at the base of the tower. It is the city's iconic tower that is the final show-piece, with dramatic lighting forming a backdrop to the Gothic architecture.

At the celebratory event, in time with the tower's clock, a light sequence connects the three elements which begin slowly to 'breathe' in unity, creating a conversation between each of them. The play of light accelerates, and memories, represented as bursts of light, appear to ascend the Dom Tower. The sequence culminates just before the striking of the hour with a dramatic finale in the lantern, where the memories cluster and multiple bursts of light and pattern are unleashed to the sound of the pealing of the bells.

LIGHT TESTING & FITTING

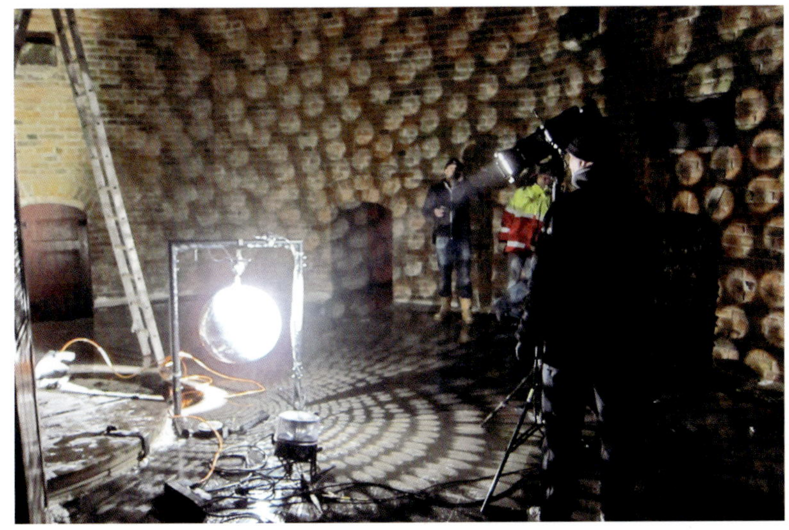

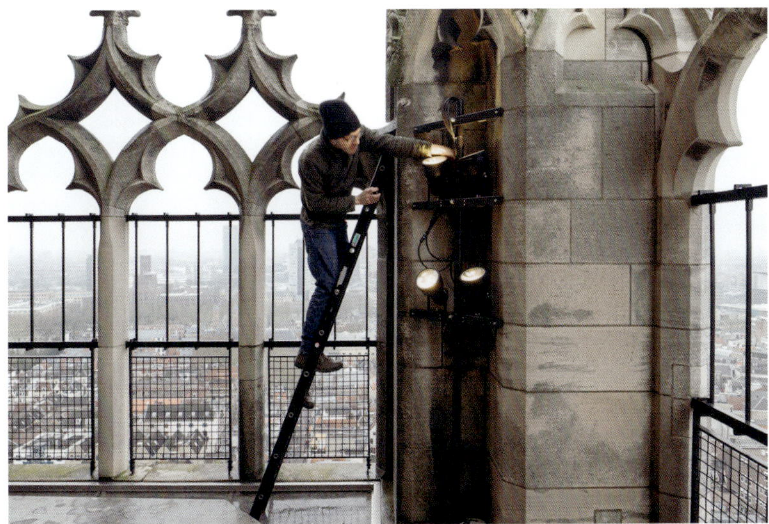

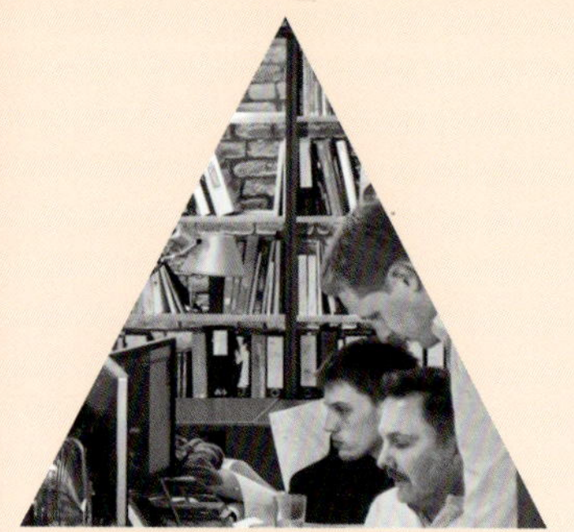

Speirs + Major

Speirs + Major is an award winning design practice that uses light to enhance the experience of the visual environment. With a wide-ranging portfolio in terms of type and scale that encompasses architecture, strategic branding and innovative product design, the firm has been credited with helping to raise awareness of the lighting design profession globally. Today, there is a team of 35 people drawn from disciplines including architecture, art, interior design, lighting, graphic design and theatre, based in two offices located in London and Edinburgh.

SPEIRSANDMAJOR.COM

Pannonhalma Monastery

When **2013**
Where **Pannonhalma, Hungary**
Client **Benedictine Abbey of Pannonhalma**

Speirs + Major have designed a simple, sensitive and flexible lighting scheme for the 13th century UNESCO World Heritage Site Benedictine Basilica of Pannonhalma Monastry as a key contributor in John Pawson's restoration team. The lighting design was considered essential both for physical reasons – to adequately illuminate the different areas – and for liturgical reasons – to concentrate attention in only those areas relevant to the celebration of a particular liturgy, and to support the rigorous focus of contemplative life.

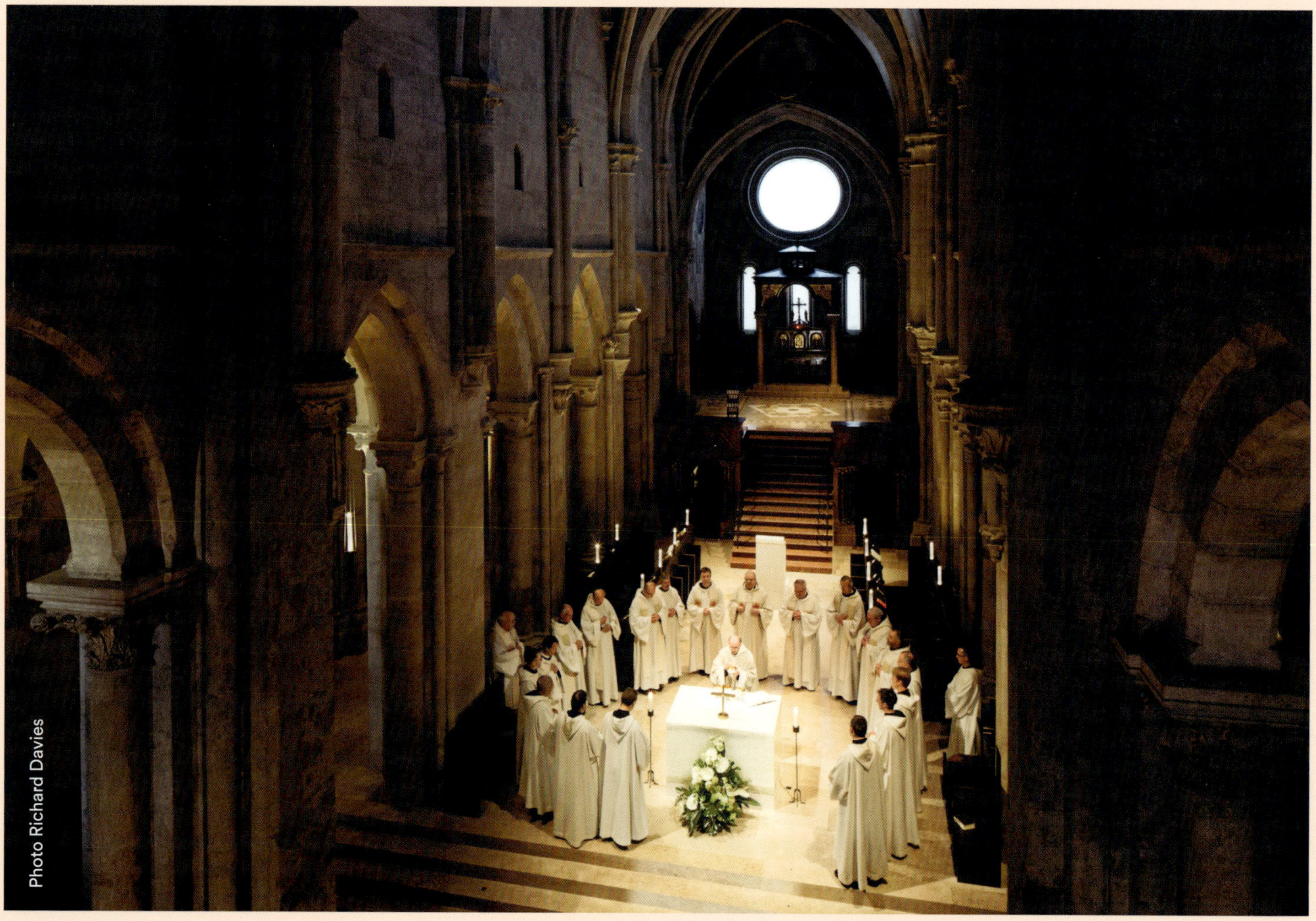

Photo Richard Davies

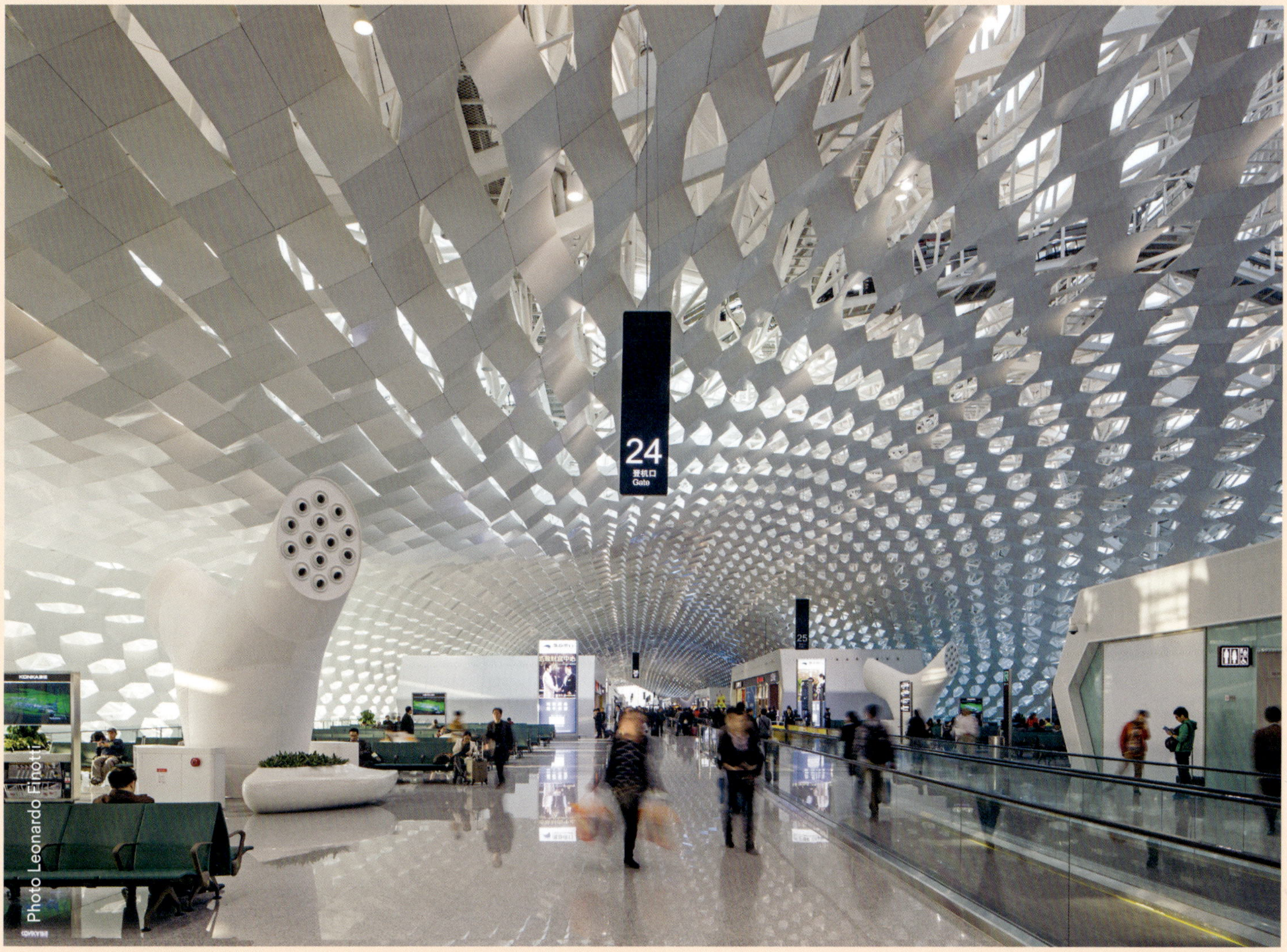

Photo Leonardo Finotti

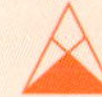 Shenzhen Bao'an Airport Terminal 3

When **2013**
Where **Shenzhen, China**
Client **Airport Authority**

Shenzhen's new terminal, built to cope with the increasing demand at its airport, was designed by competition winners Studio Fuksas. The focus for the lighting design lay in creating the right balance between the pragmatic – light that aids the passenger journey to and from road to air – and the spectacular – revealing the form and enhancing the image of the iconic design.

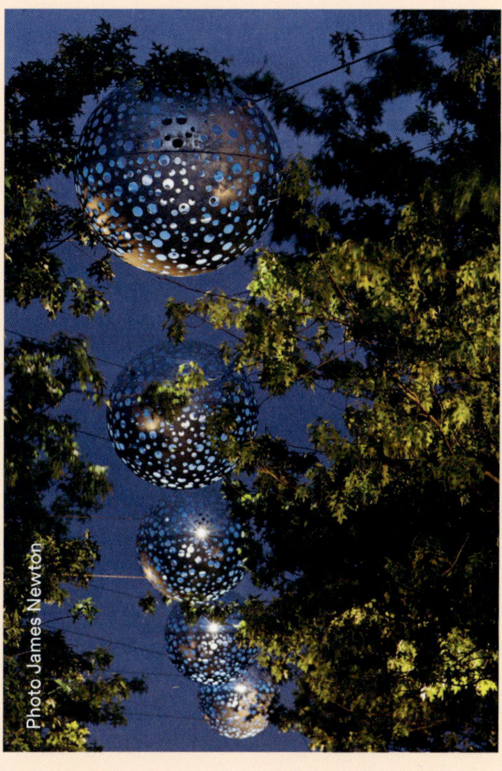

Photo James Newton

 Queen Elizabeth Olympic Park

When **2014**
Where **London, United Kingdom**
Client **LLDC**

Queen Elizabeth Olympic Park in London features an enchanting and immersive experience in light. Designed to support the joyful character of the park design, the lighting also ensures that users feel safe and secure, and can continue to use the facilities of the park as natural light fades.

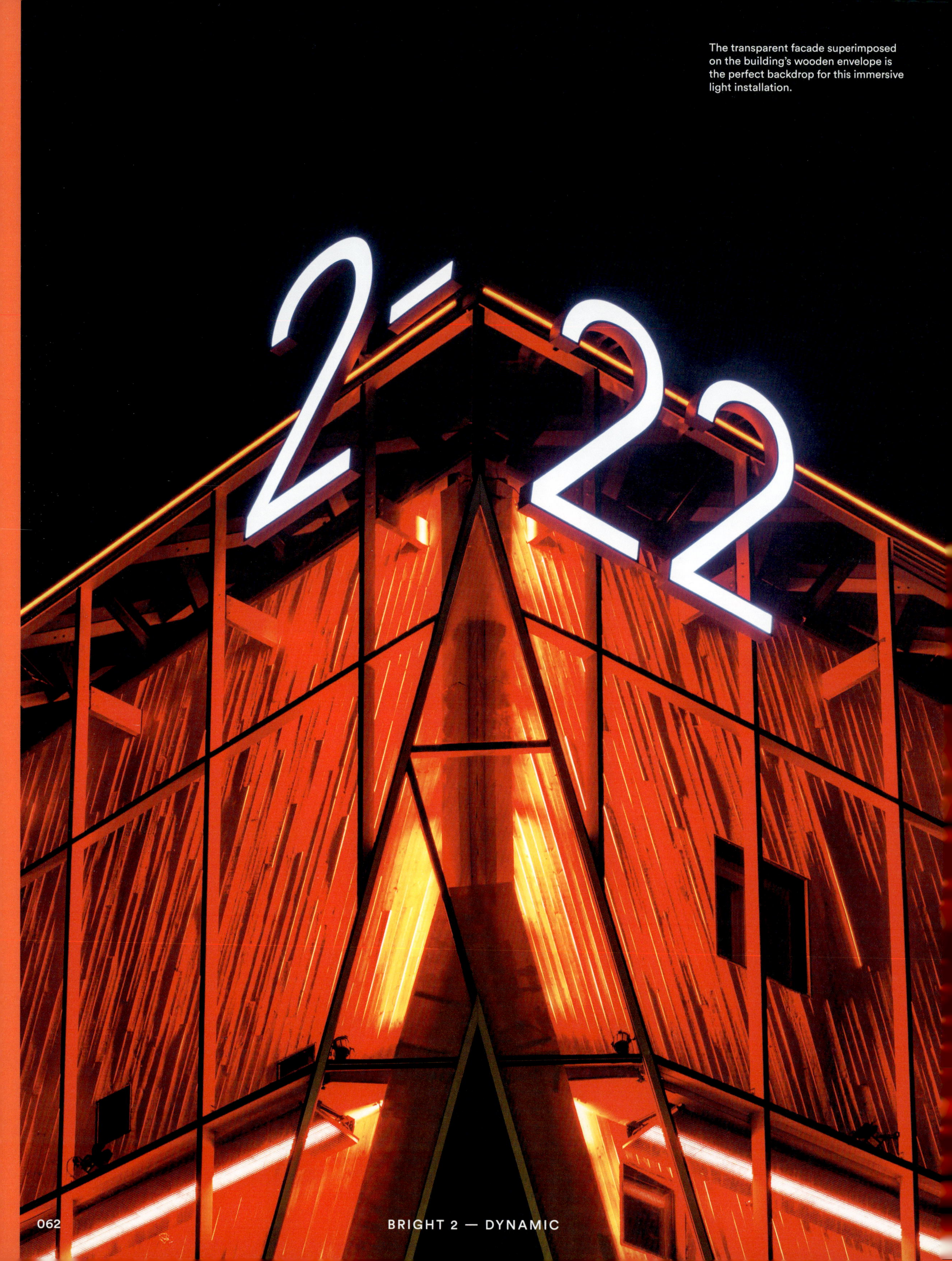

The transparent facade superimposed on the building's wooden envelope is the perfect backdrop for this immersive light installation.

La Vitrine Culturelle

To complement the striking architectural envelope of Montreal's La Vitrine Culturelle building 2-22, **Moment Factory** opted for a strong gesture by which to create a highly-narrative canvas of light.

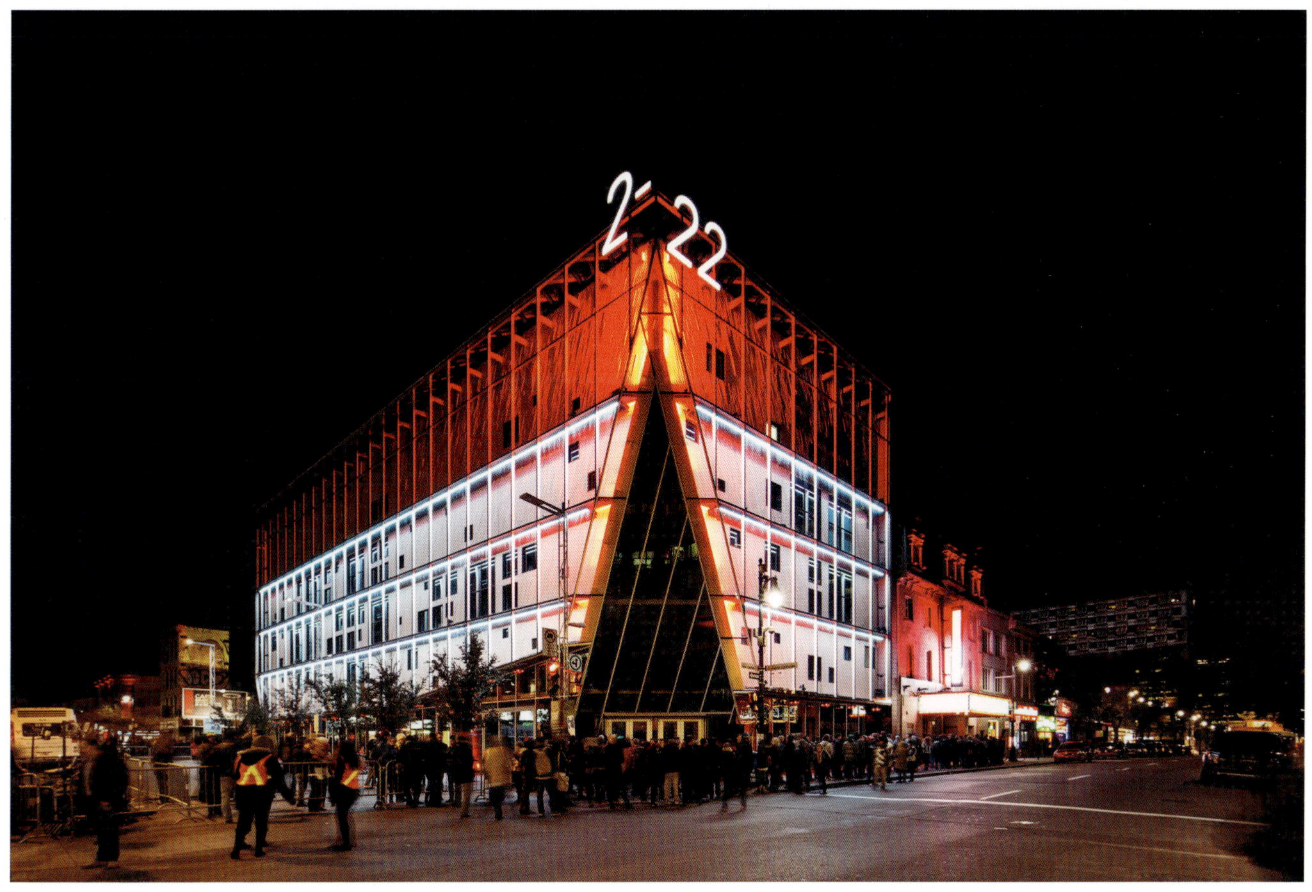

Photos Moment Factory

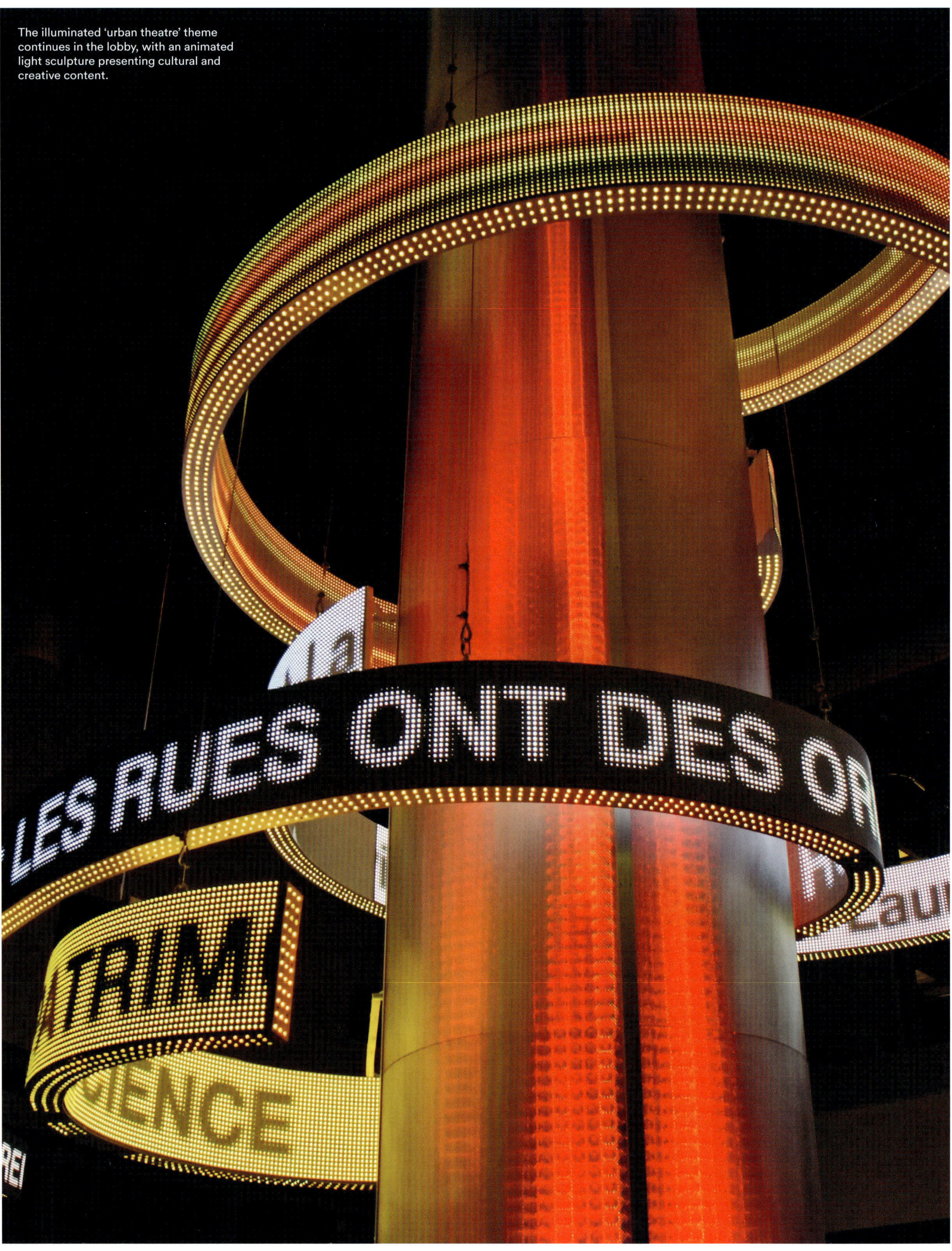

The illuminated 'urban theatre' theme continues in the lobby, with an animated light sculpture presenting cultural and creative content.

Designer
Moment Factory
Location
Montreal, Canada
Client
La Vitrine
Installation
Solotech
Manufacturers
Pledco, Enseigne Métropolitain
Date
November 2013

La Vitrine's flagship building in Montreal is a cultural landmark and information centre. It is the destination for discovering all cultural and artistic happenings in the Greater Montreal Area. Located at a busy intersection in the nerve centre of operations in the city's vibrant Quartier des Spectacles – French for 'entertainment district' – the 2-22 building was designed by Aedifica and Gilles Huot Architectes and opened in 2012.

The striking structural envelope of the building comes in the form of an innovative double wall, which sees a glass facade wrap around the inner construction. This second skin is ripe for dynamic multimedia installations to be superimposed, which is where Moment Factory come in. Commissioned to realise a beacon within the urban nightscape, the studio's luminous and highly-narrative creation sought to permanently transform the facade into a spectacular canvas of light.

The daring architecture of 2-22 inspired the creative team at Moment Factory to complement the building's strong gesture. Turning three of the exposed passageways into elongated media-enhanced spaces, the team utilised cutting-edge lighting technology in the form of unconventionally long LED panels. A 60-second capsule called 'Urban Clock' – one among many innovative features created specifically for the project – enables the vibrant LED aspects to synchronise all of the building's visual components every hour. The other atmospheric features are composed of various graphical effects that play with trompe l'oeil and optical illusion. At the pinnacle of the corner plot is a gigantic '2-22' light signature that shines brightly.

Like a real urban theatre, the building is animated 24 hours a day, 365 days a year, and has been rendered versatile enough to accommodate a diverse range of cultural programming. This new installation conforms to the original vision for the Luminous Pathway designed by Integral Ruedi Baur and Intégral Jean Beaudoin for the Quartier des Spectacles of Montreal. —

THE LUMINOUS NARRATIVE TRANSFORMS THE FACADE INTO A SPECTACULAR CANVAS OF LIGHT

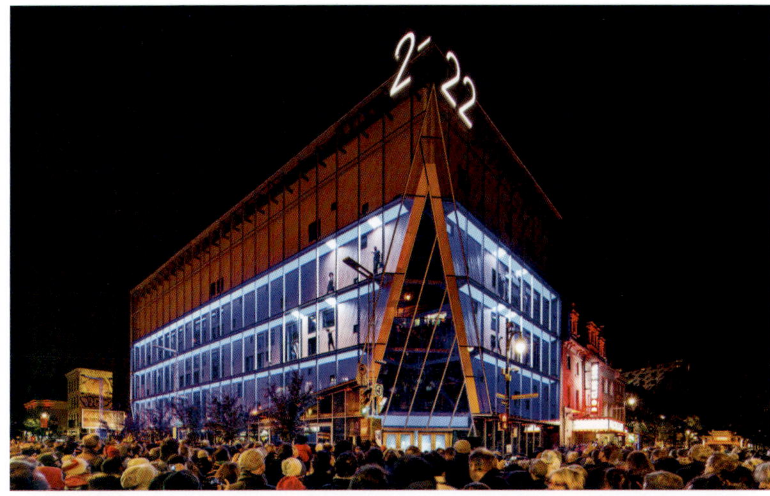

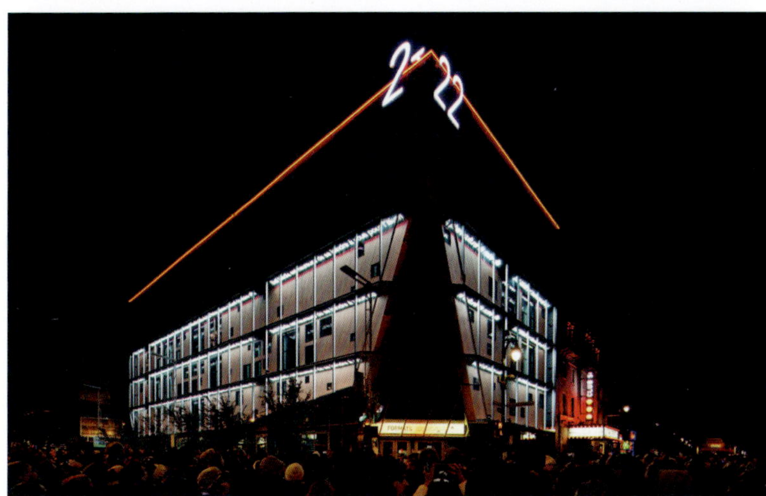

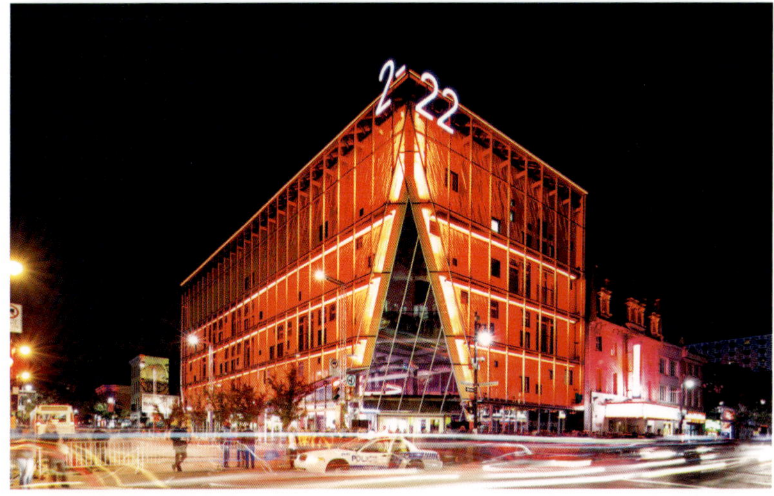

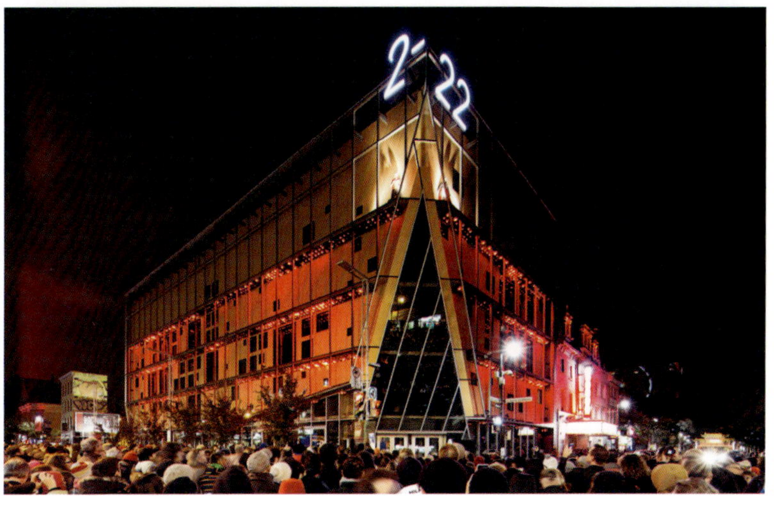

The dynamic layers of light on the corner facade, together with artistic and informative content, regard the various services provided by the local cultural body.

SIGN DIMENSIONS

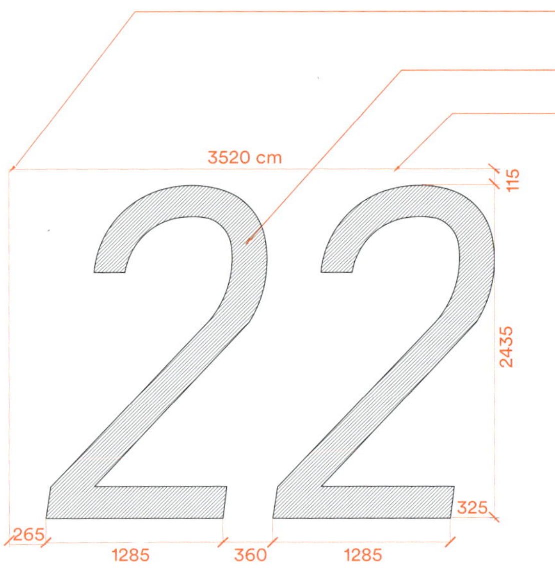

SIGNAGE

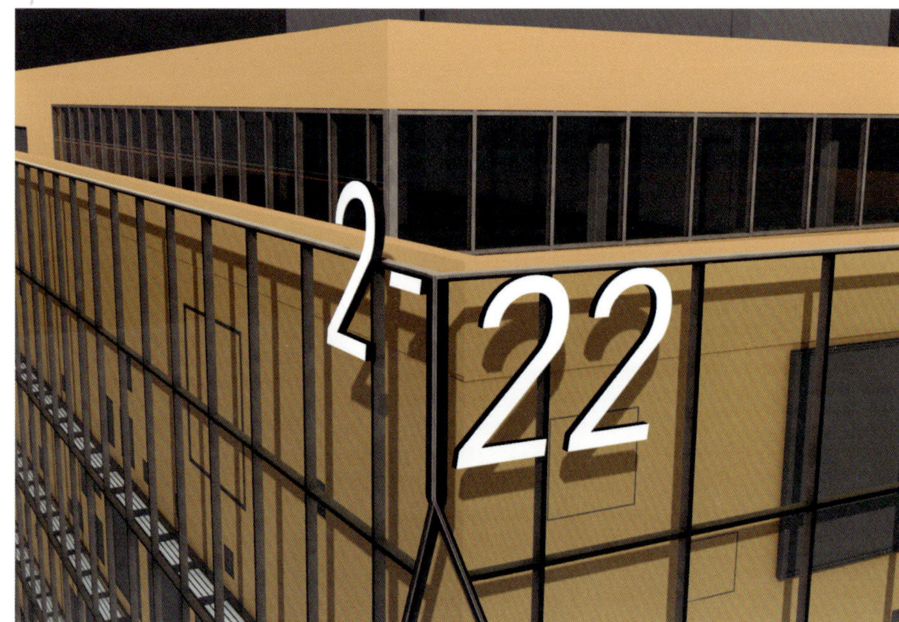

ELEVATION

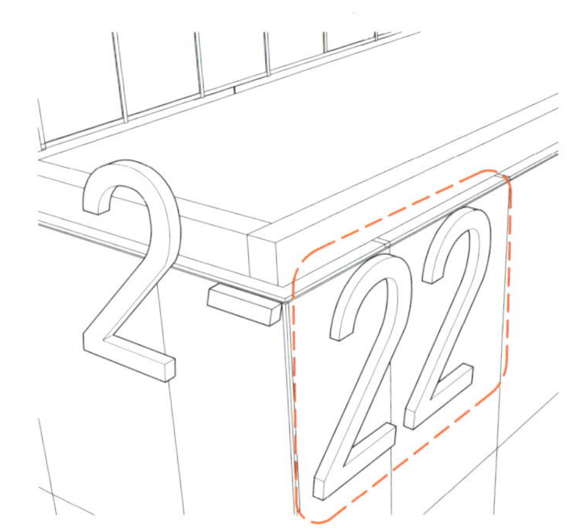

BUILDING RENDERINGS

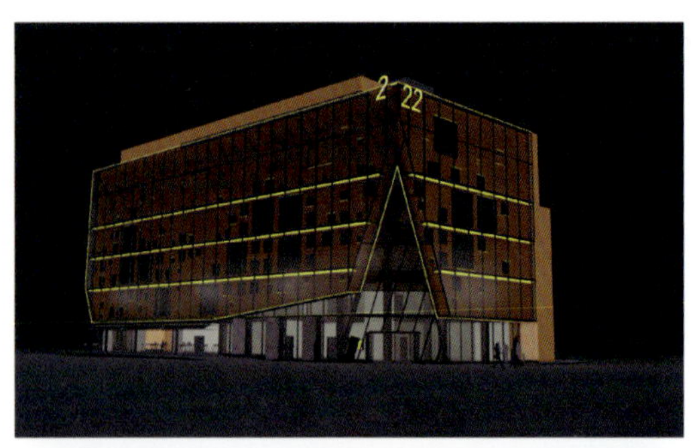

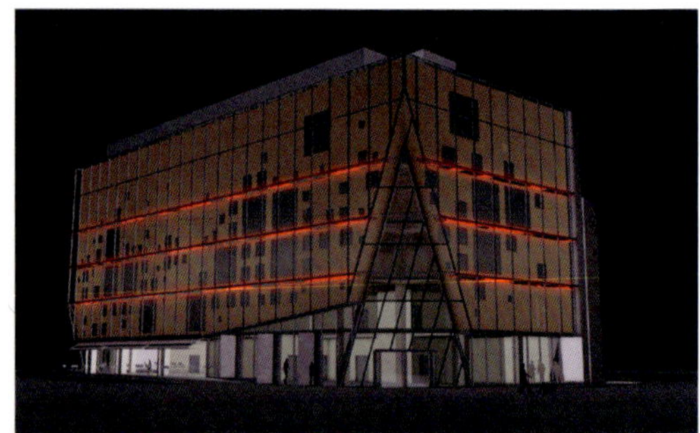

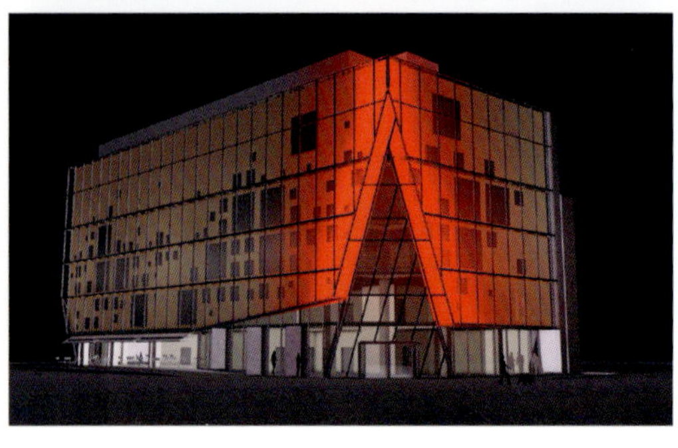

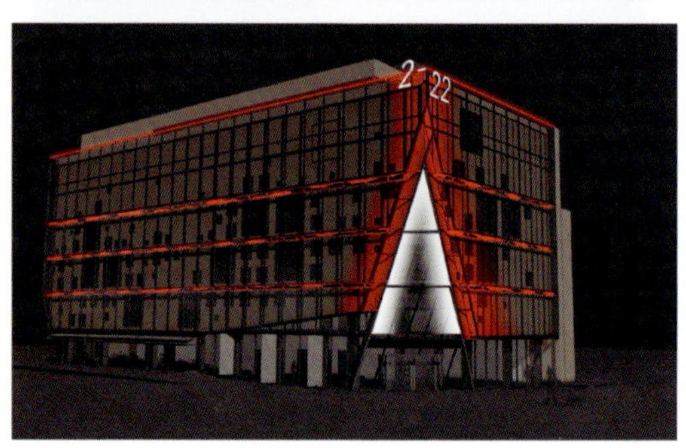

Set-up

Total area of the facade passageways for illumination was 100 m, with the linear meters of video totalling approx. 170 m.
LED tiles were custom-made with a 18-mm pixel pitch at 6000 NITS (brightness).
Grazing length was approximately 30 m, outlining and emphasising the recessed angular entrance.
Custom RGB LED fixtures had a brightness of 1350 Lux.
2-22 sign made from polycarbonate has the approximate dimensions 3.5 x 1.75 m, utilising custom RGBW LED modules.
Lighting control console utilised X-Agora, which is Moment Factory's own multimedia control and playback system that configures the video content and the lighting effects.
Construction time for the new facade installation in total was 19 months.

BUILDING SKETCHES

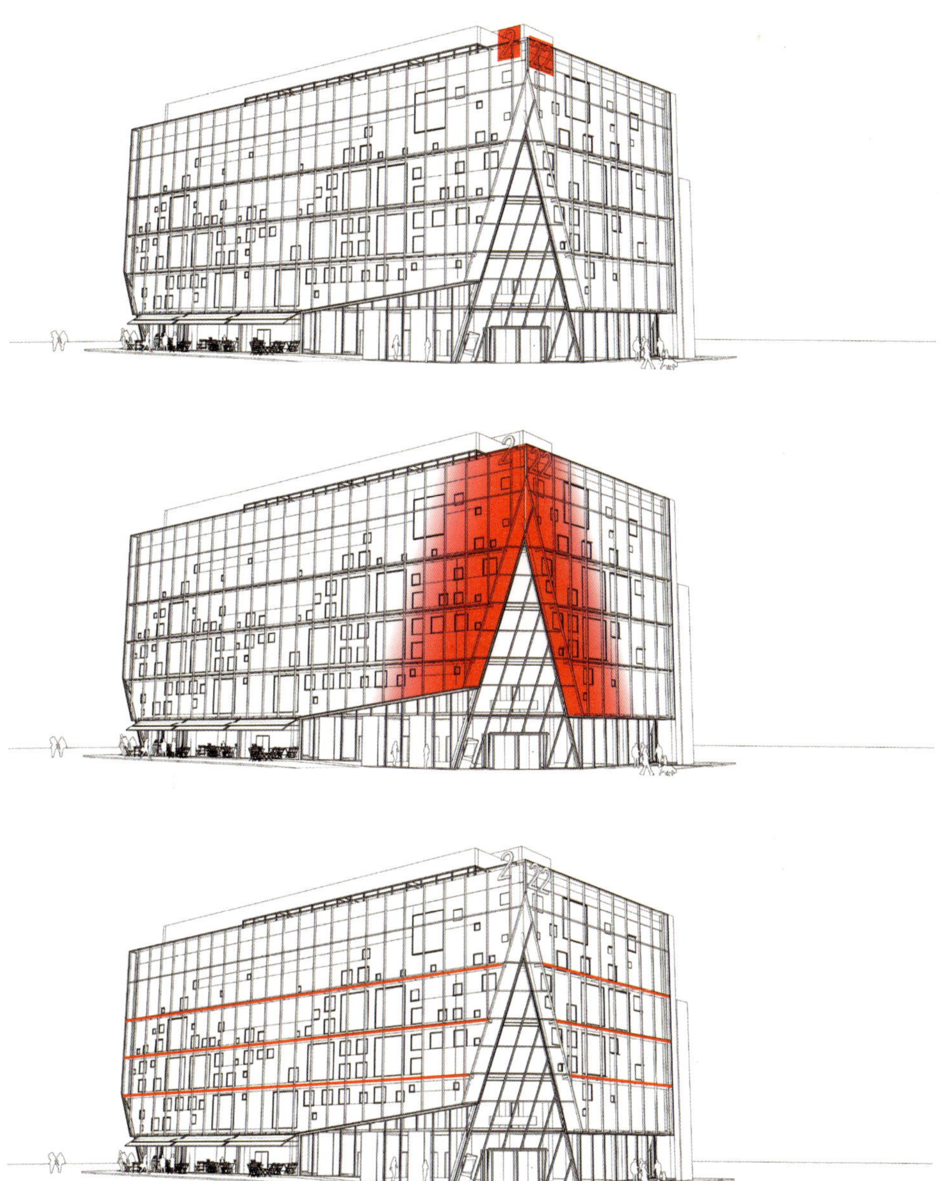

Specifics

One of the key criteria of the new lighting scheme was that each element had to be custom-manufactured to be integrated as seamlessly as possible into the building's exterior. This meant bespoke fabrication for mounting hardware (brackets, louvres, signage elements), as well as electronic specifications.

For the video format to illuminate the passageways, the team worked closely with a screen manufacturer to modify standard products to fit the building. The screen surface was designed in a way that maximises its viewing angles and brightness from the point of view of passers-by. The specified screen tile size neatly fit into existing architectural details, with cable management and mounting done in a way that caused minimal visual distraction without impinging on serviceability.

The pixel pitch chosen was based on on-site testing in the months prior to production. The specification was based on a number of factors: aesthetically balancing the light with the internal elements of the building, visible pixel resolution at ground level and cost.

The grazing fixtures were specified with narrow beam lenses to maximise the reach of light across the building facade. The light shoots horizontally from the corner elevation along each facade, designed to be controlled as a single element but for events it is possible to control at 1-m intervals. Custom louvres were required to hide the equipment from the exterior of the building and were made to match the existing finishes of the building (an unfinished pine material installed diagonally).

The contour lighting was designed to visually separate the top edge of the building elevation from the sky, therefore it was important to specify a custom mounting method so that the lighting is positioned as close to the actual top edge of the building as possible. In addition, a custom design was required for fixture housing to assure that the light was completely diffused and that fixtures could meet end-to-end without a discontinuity in lighting.

2-22 signage was based on the building's corporate identity and was specified as a custom build whereby prototypes were made to test the colour and diffusion and brightness of the light. The required parameters were given to the manufacturer, who then created a fabrication assembly and installation strategy based on the design intent.

NORTH AND EAST ELEVATIONS

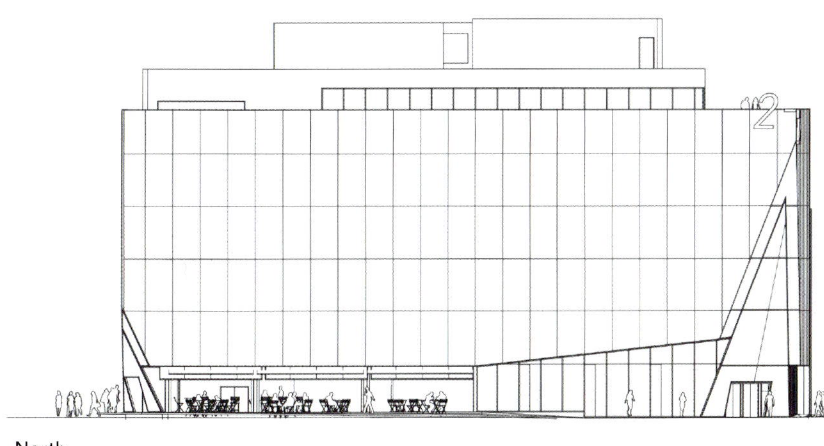

North

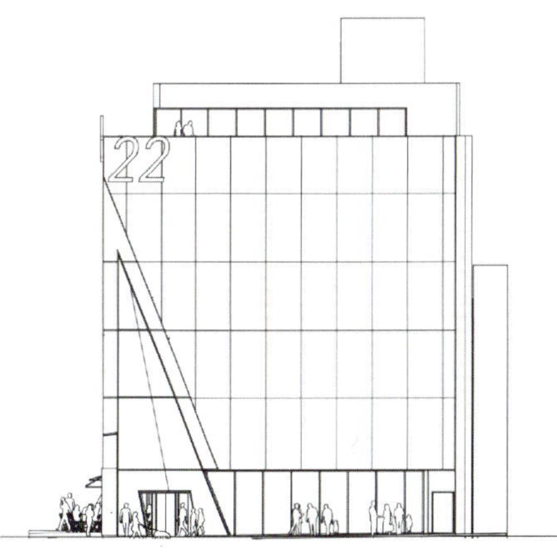

East

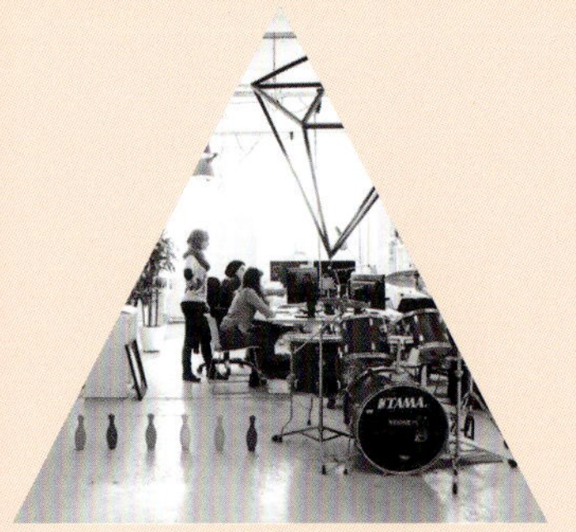

Moment Factory

Moment Factory is a multidisciplinary studio specialised in the conception and production of multimedia environments combining video, lighting, architecture, sound and special effects to create sensory experiences. Since its inception in 2001, Moment Factory has created more than 300 installations, events and shows across the world for such clients as Cirque du Soleil, Nine Inch Nails, Madonna, Disney, Microsoft and Sony, amongst others.

MOMENTFACTORY.COM

Foresta Lumina

When **2014**
Where **Coaticook, Canada**
Client **Parc de la Gorge de Coaticook**

Wanting to showcase its charms, Parc de la Gorge de Coaticook in Quebec commissioned an illuminated night walk through the forest. Visitors were invited to discover an enchanted trail winding through the woods. Meandering 2 km through the mysterious forest, they met characters inspired by the area's myths and legends, who drew them into an immersive sensory experience. The scenography, combined projections and lighting, accompanied by an ethereal soundtrack.

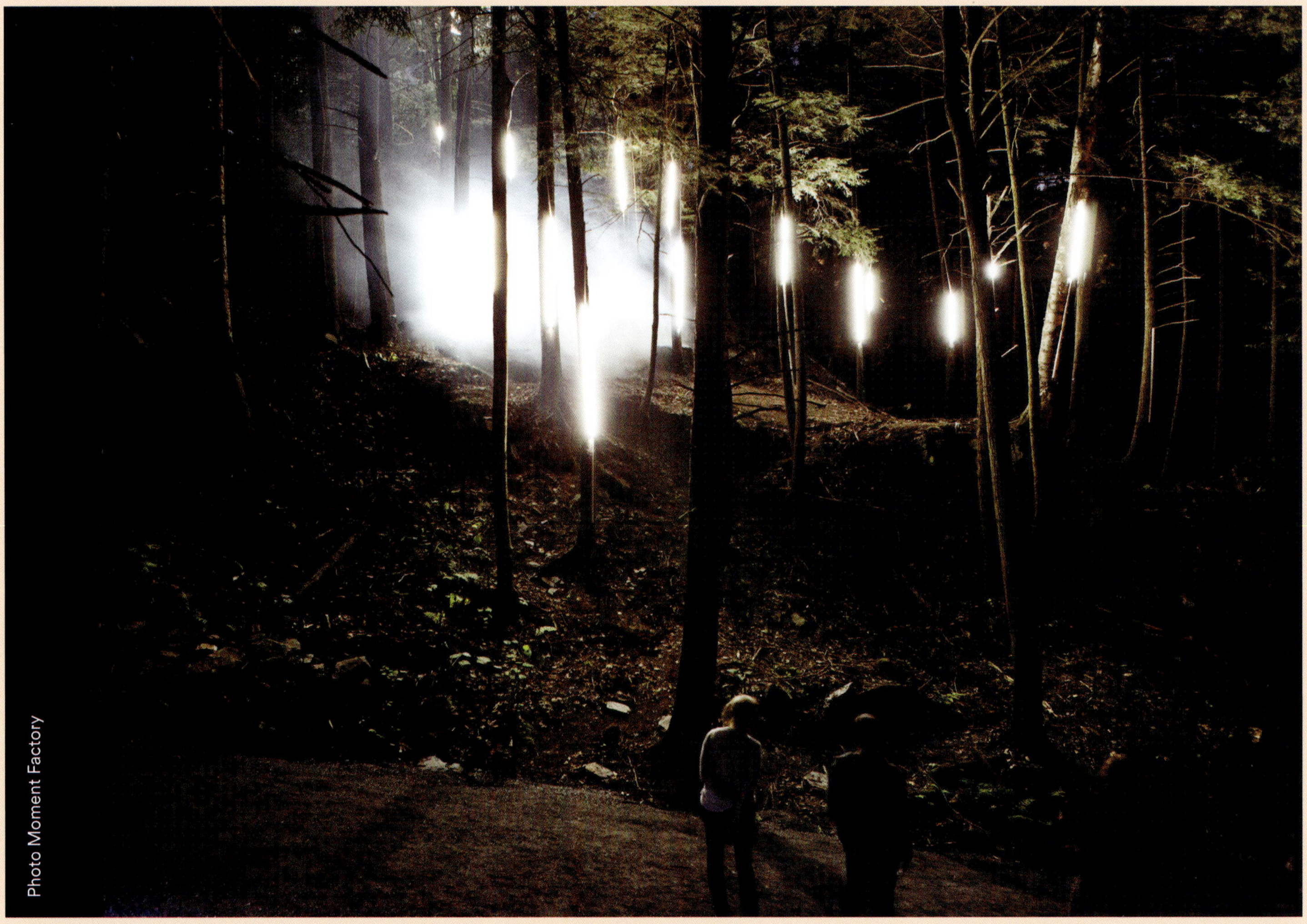

Photo Moment Factory

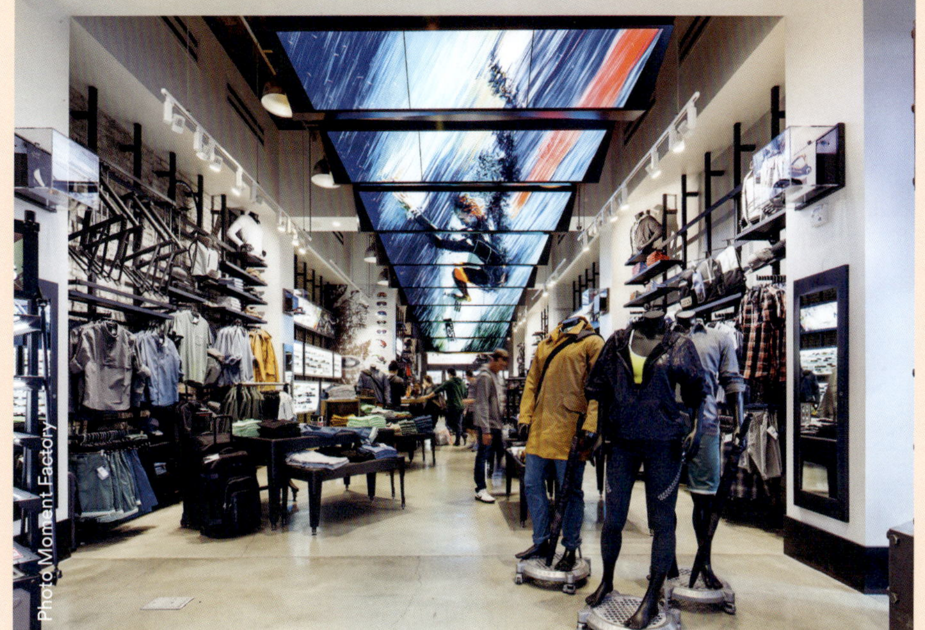
Photo Moment Factory

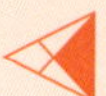

Oakley's 5th Avenue

When **2014**
Where **New York, United States**
Client **Oakley**

Oakley commissioned the studio to create a permanent installation for its flagship store in New York City. The concept had to successfully fit within the existing architectural design, while also expressing the brand's strong connection to technological innovation. The ceiling-mounted installation structure's 27 LCD screens feature footage of Oakley's performance athletes merged with a customised visual language that the studio developed for the project.

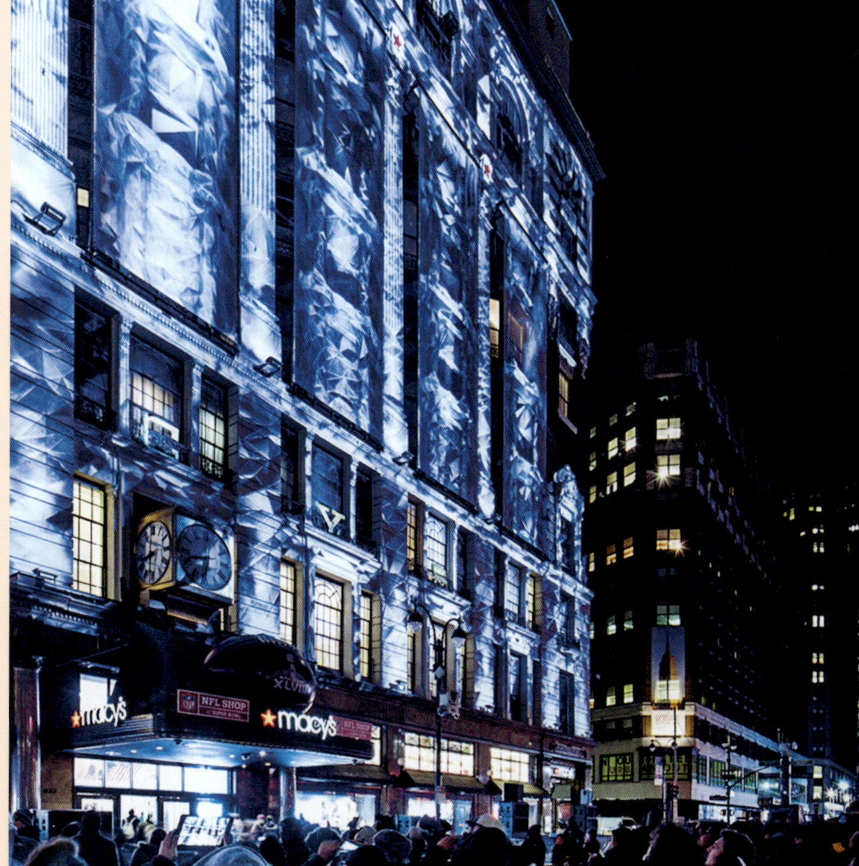

Photo Moment Factory

T Super Bowl Virtual Theater

When **2014**
Where **New York, United States**
Client **NFL and PPW**

Moment Factory was approached by the NFL and event planners PPW to create a 'Super Bowl Virtual Theater'. The spectacular multimedia show was projected onto Macy's facade in New York on the evenings leading up to Super Bowl XLVIII as a series of football-themed events spanning thirteen-blocks along Broadway. Over four nights, the 8-minute loop was projected onto the iconic storefront, treating spectators to a stunning mix of 2D and 3D animation effects together with archival NFL footage.

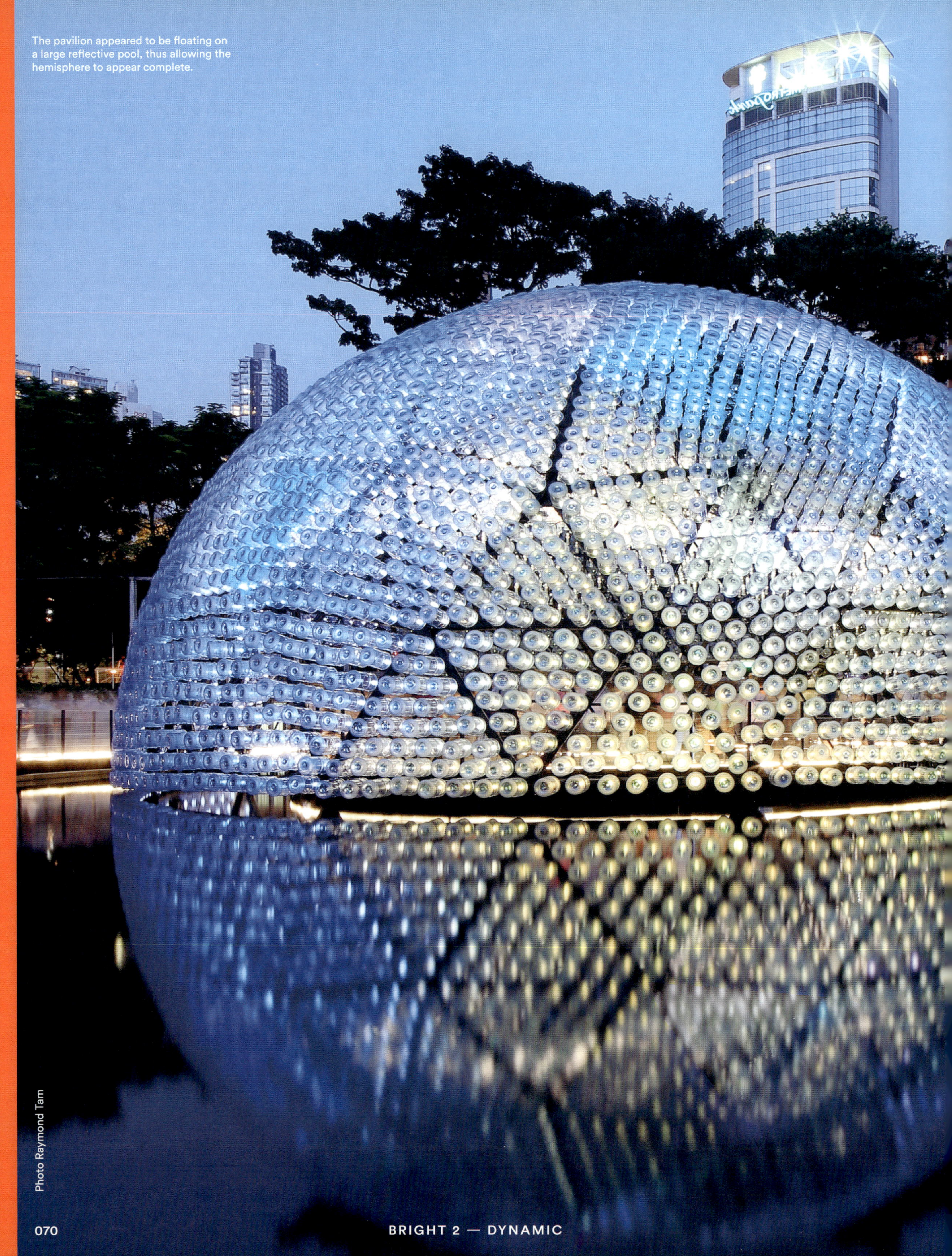

The pavilion appeared to be floating on a large reflective pool, thus allowing the hemisphere to appear complete.

Photo Raymond Tam

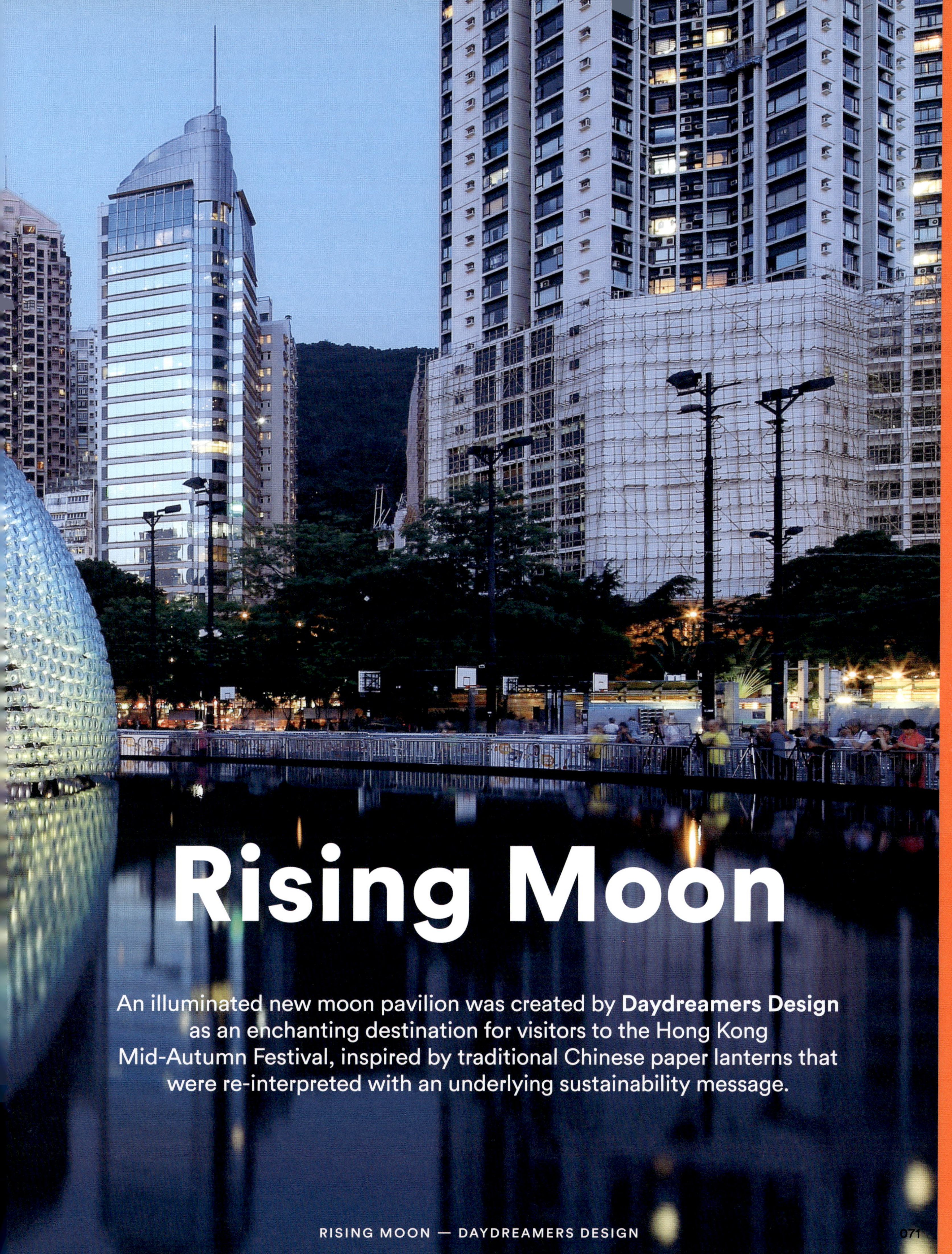

Rising Moon

An illuminated new moon pavilion was created by **Daydreamers Design** as an enchanting destination for visitors to the Hong Kong Mid-Autumn Festival, inspired by traditional Chinese paper lanterns that were re-interpreted with an underlying sustainability message.

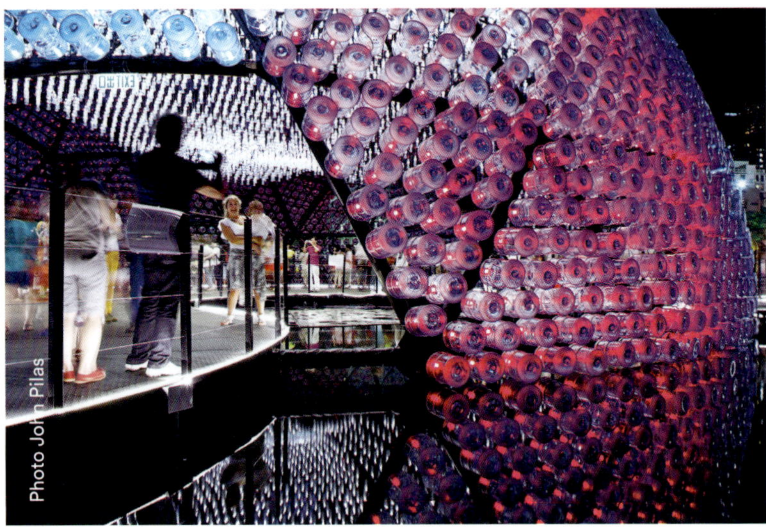
Photo John Pilas

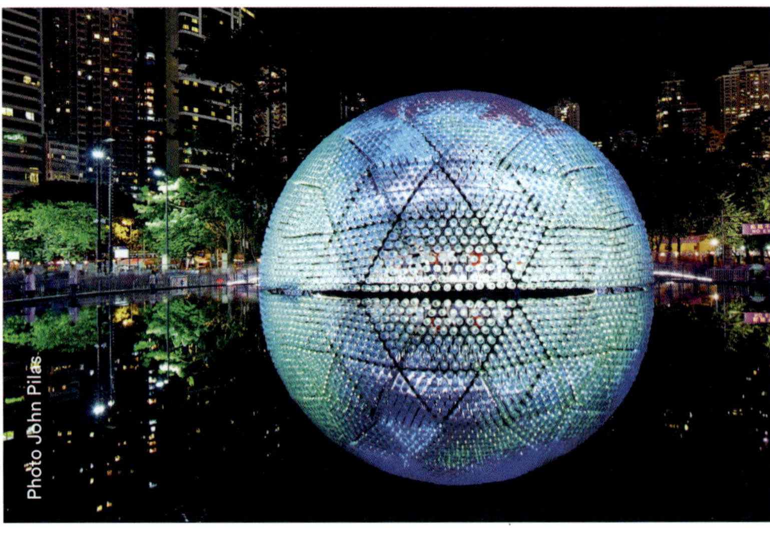
Photo John Pilas

Designer
Daydreamers Design
Location
Hong Kong
Client
Hong Kong Tourism Board
Collaborator(s)/ consultant
Outstanding Lighting Design, Takumi Works
Manufacturer
Free Form Contractors
Date
September 2013

Rising Moon was a temporary pavilion that was a major focus during the 2013 Hong Kong Mid-Autumn Festival, an event that encourages friends and family to gather under the full moon once a year as a symbol of harmony and unity. The illuminated installation was created by Daydreamers Design, the Hong-Kong based collaborative art and architecture practice, with a concept inspired by traditional Chinese paper lanterns and a desire to communicate a message of sustainability. The studio made use of thousands of recycled polycarbonate water bottles to draw attention to the year-on-year increase in their distribution and usage.

The outer skin of the dome was formed by the large bottles, with the lantern idea being re-interpreted by connecting the bottles directly to LEDs. The design team created an enchanting multisensory beacon for visitors by means of its presence and theatrical expressiveness and a programme of dynamic lighting and sound effects, both inside and out. The illuminated curved structure was positioned on top of a reflective pool, giving visitors the impression that the distant full moon had been brought down to earth.

Constructing a geodesic steel dome structure, the large water bottles were each directly connected to an LED component and affixed to triangular steel modules, each facilitating a total of 45 LED/bottle combinations. The resulting 20-m-diameter dome created a height of 10-m in the interior space, where a flowing swathe of smaller illuminated bottles formed an organic ceiling. A winding pathway directed visitors into the centre of the pavilion, leading them towards the pinnacle of their journey which – thanks to a skylight in the roof overhead – was a glimpse of the magnificent moon reflected in the internal pool. —

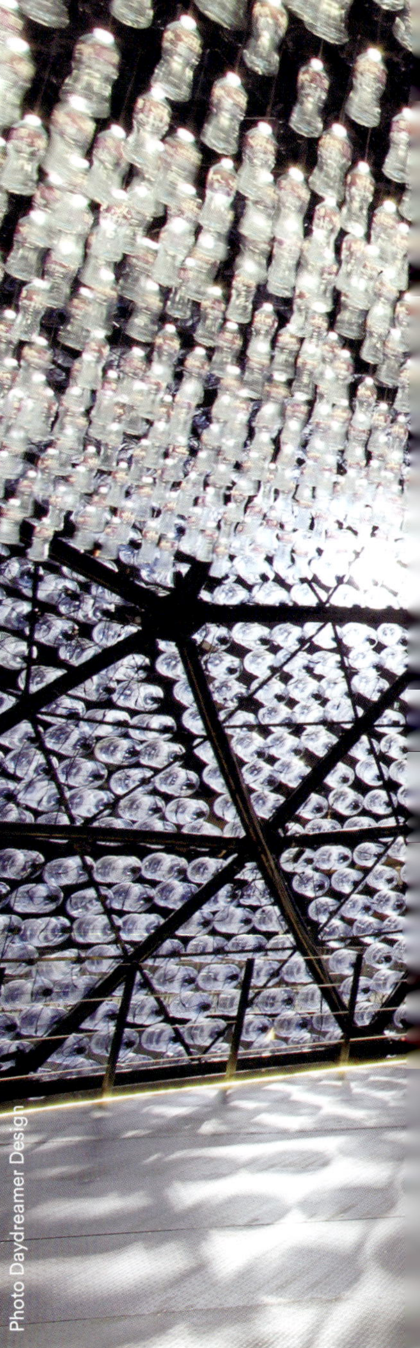
Photo Daydreamer Design

AN ENCHANTING MULTISENSORY BEACON WITH A THEATRICAL EXPRESSIVENESS AND PROGRAMME OF DYNAMIC LIGHTING AND SOUND EFFECTS

The structure is made out of recycled bottles on which are fixed a plethora of light-emitting diodes.

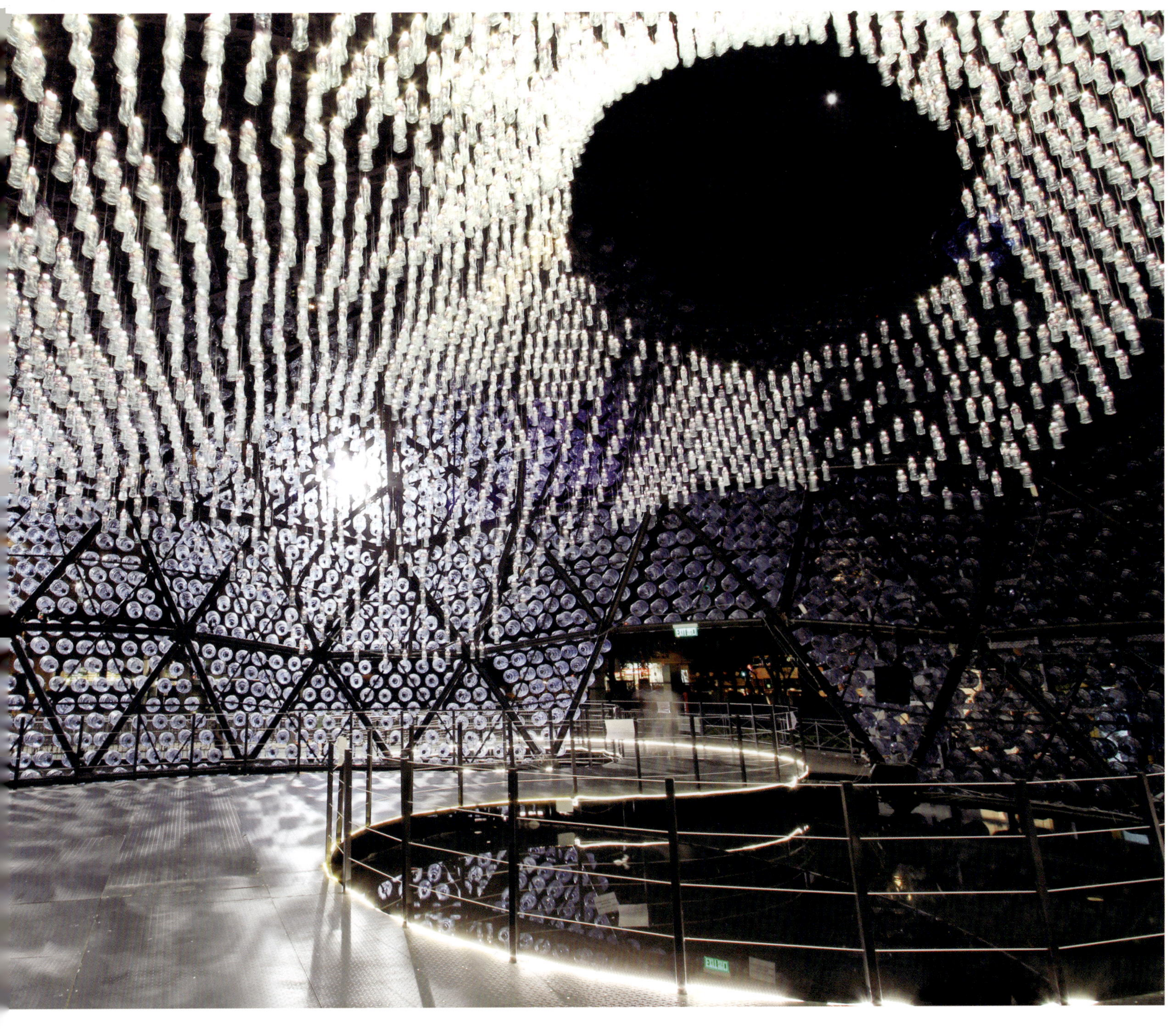

PLAN

1 Entrance
2 Pool
3 Bridge
4 Exit
5 Skylight

INSPIRED BY TRADITIONAL CHINESE PAPER LANTERNS AND A DESIRE TO COMMUNICATE A MESSAGE OF SUSTAINABILITY

EXTERNAL DESIGN FEATURES

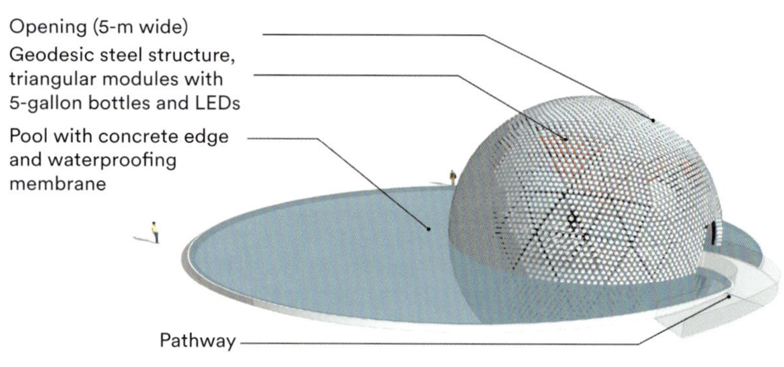

INTERNAL DESIGN FEATURES

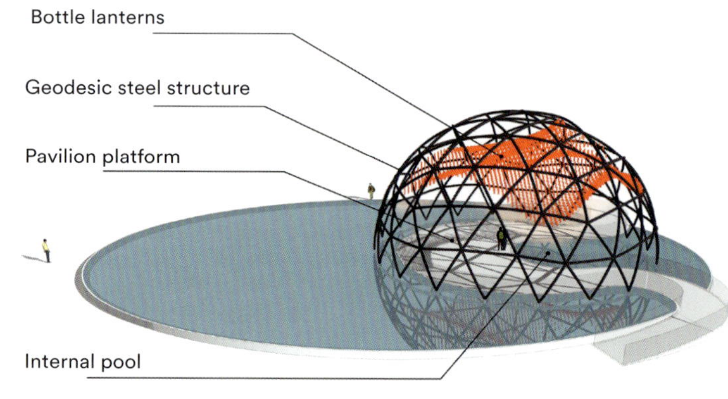

LIGHT FIXTURES

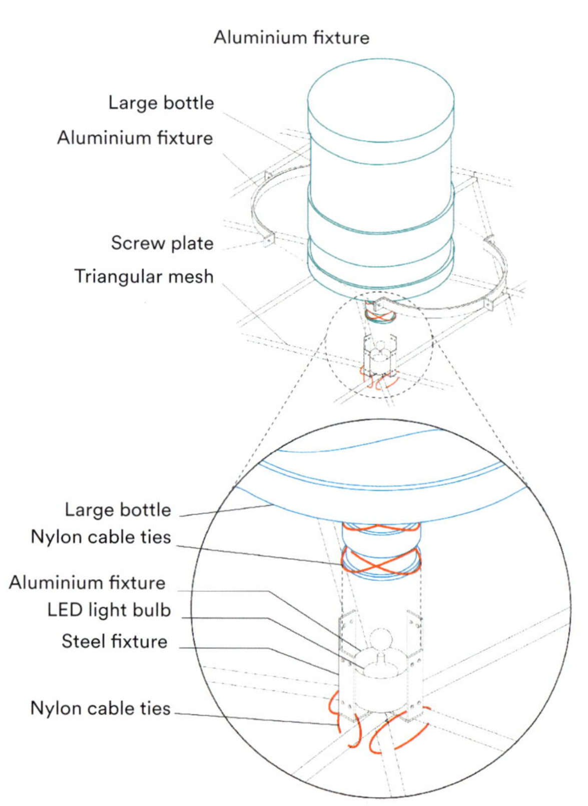

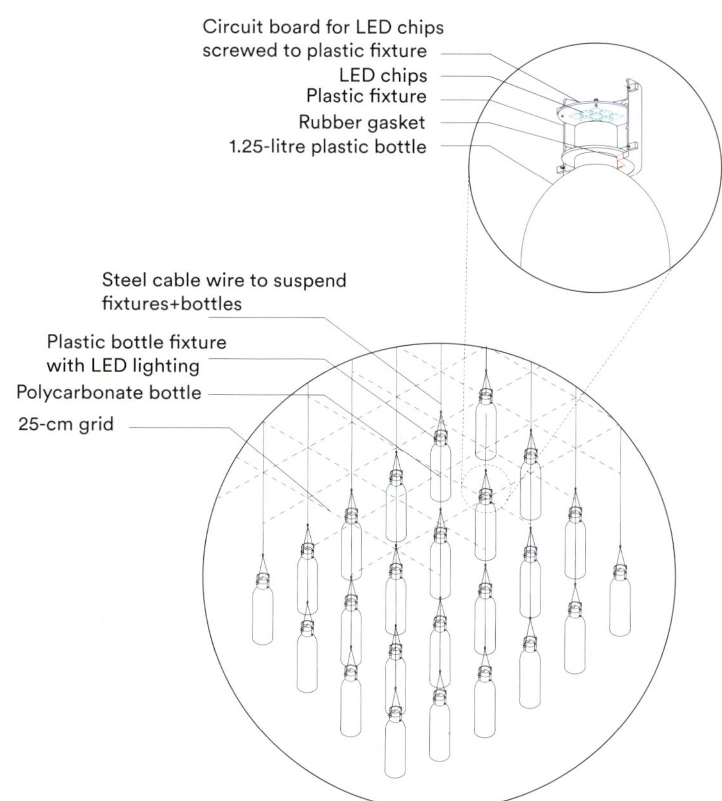

CONCEPT VISUAL

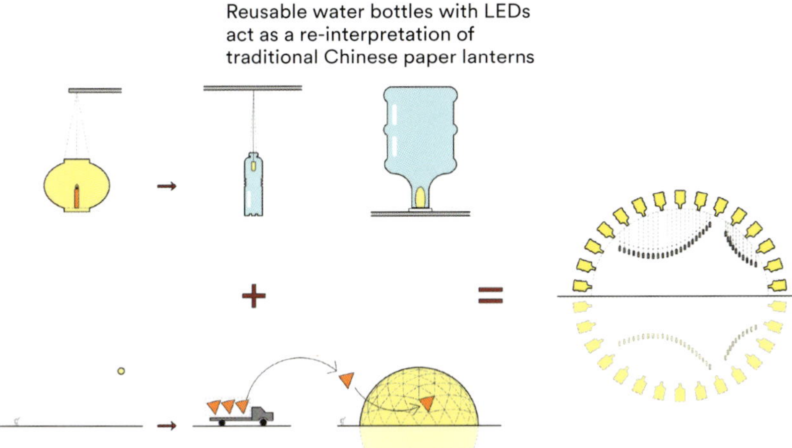

Mid-Autumn Festival is a time to enjoy the full moon in the sky

Prefabricated steel modules form a hemispheric dome, positioned atop a pool of water

Over 4800 bottles create the 'new' full moon reflected in the pool.

MOON REFLECTION DIAGRAM

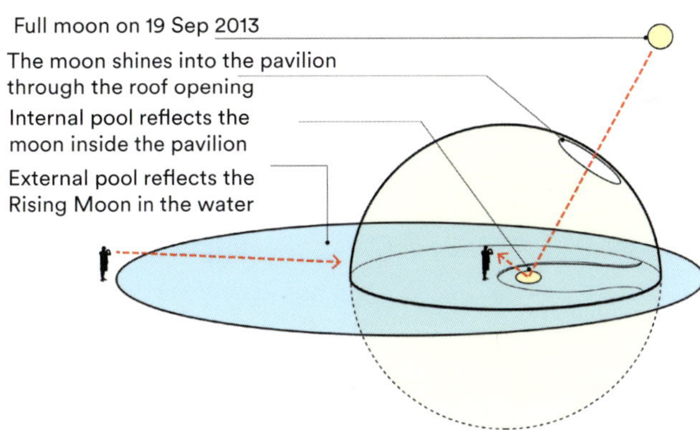

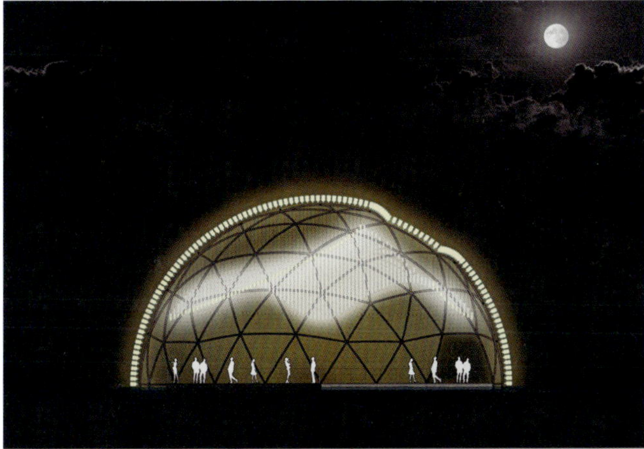

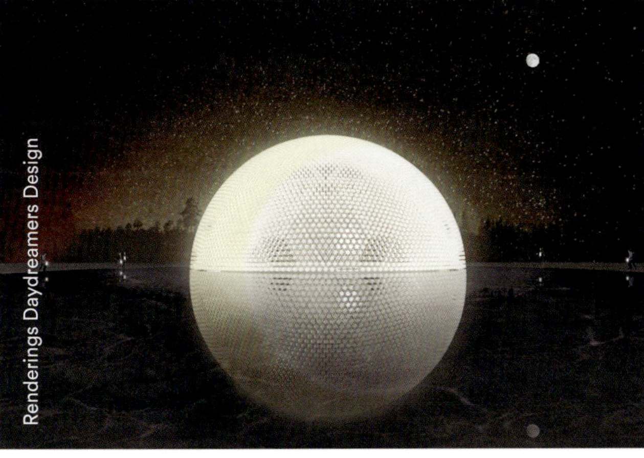

Specifics

In order to deliver the Rising Moon within a short period of time, the team had to standardise the design with triangular modules through complex geodesic calculations. It was calculated that the required 148 steel triangular modules were simplified into six different types of module.

The position of the opening at the crown of the hemisphere was also calculated carefully based on the path of the moon during the Mid-Autumn Festival by astronomy applications, in order allow moonlight to be viewable from inside.

Using 150-mm circular and hollow steel to create the frames, the constructed base utilised 'level 3' geodesic calculations. The 148 triangular steel components, each connected with cable wires forming a cable net, had total 13 different orders and each had a designated location on the geodesic structure. Three heavy cranes were operated to assemble 148 triangular components, after tests were ran by lighting technicians.

The location of every LED in the 19-litre polycarbonate water bottles was mapped through DFX programming by the lighting technician, thus lighting effects and animation could be displayed.

The specialised software operates in real time and is an indispensable tool for the creation process or for the selection of artistic content itself. In association with the lighting sequence a bespoke soundtrack by music composer Takumi Works added to the multisensory experience.

Set-up

Hemisphere structure measured 10 m in height and 20 m in diameter, sitting on top of a pool of water. With lighting effects, it joins its reflection, imitating the moon phases, forming a 'full moon' at one point.

Construction utilised 7000 reusable plastic bottles (4848 x 19-litre bottles and 2300 x 1.5-litre bottles), 150-mm circular, hollow steel frames, cable wires and energy-saving LED lights. The plastic bottles were recycled after the display.

Geodesic steel dome consisted of 148 triangular steel components, each crossed with cable wires to form a cable net. LED lights tied to plastic bottles secured to the cable net in each triangular steel frame, formed the surface of the lantern.

Organic ceiling inside the dome was made up of a 25-cm grid of steel cable wire to suspend the smaller bottles from the bamboo frame above.

Light source included 7800 LEDs connected to a circuit board, housed within a plastic fixture connected to the grid and the bottle, that can be individually-controlled.

Construction time was 12 days in total on-site, working towards completion for the full moon on 13 Sep 2013; the triangular steel modules were prefabricated to ensure minimum process time.

PHASES OF THE MOON

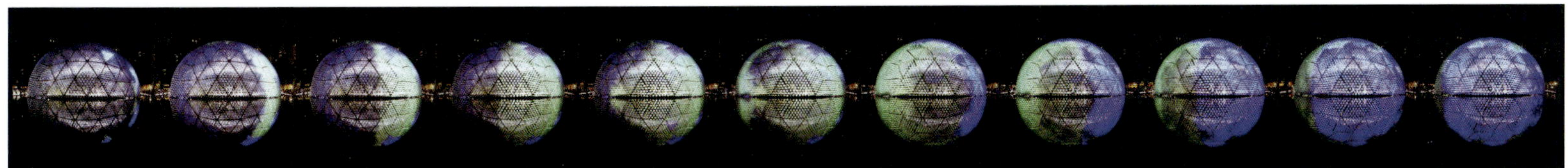

Daydreamers Design

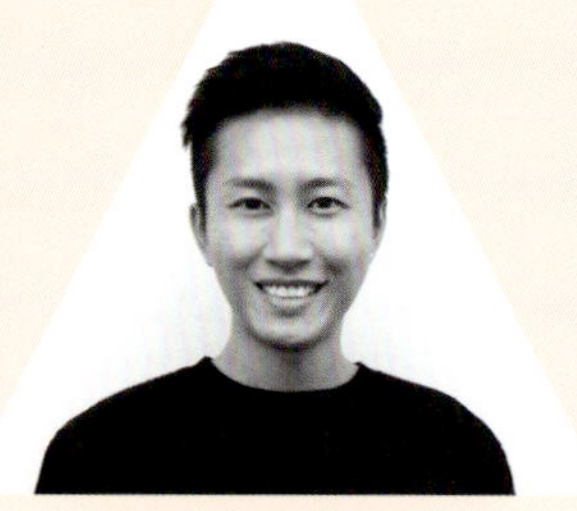

Founded in Hong Kong by the founders Stanley Siu (pictured) and Aden Chan, Daydreamers Design is an art and design collaboration constantly engaged in art, design and architectural matters, both locally and globally. The creative collective aims to synthesise individual methodologies and educe sophisticated design solutions through research, curatorship and collaboration with various artists and designers, as well as active engagements in social dialogue.

BIT.LY/DAYDREAMERSDESIGN

Exchange[2]

When **2013**
Where **Hong Kong, Hong Kong**
Client **Hong Kong Institute of Architects**

The design for this sustainable living centre was inspired by the local context of the neighbourhood of the central area in Hong Kong, where it was imagined it would be located. This was the winning entry by Stanley Siu for the Hong Kong Institute of Architects Young Architects Awards 2013.

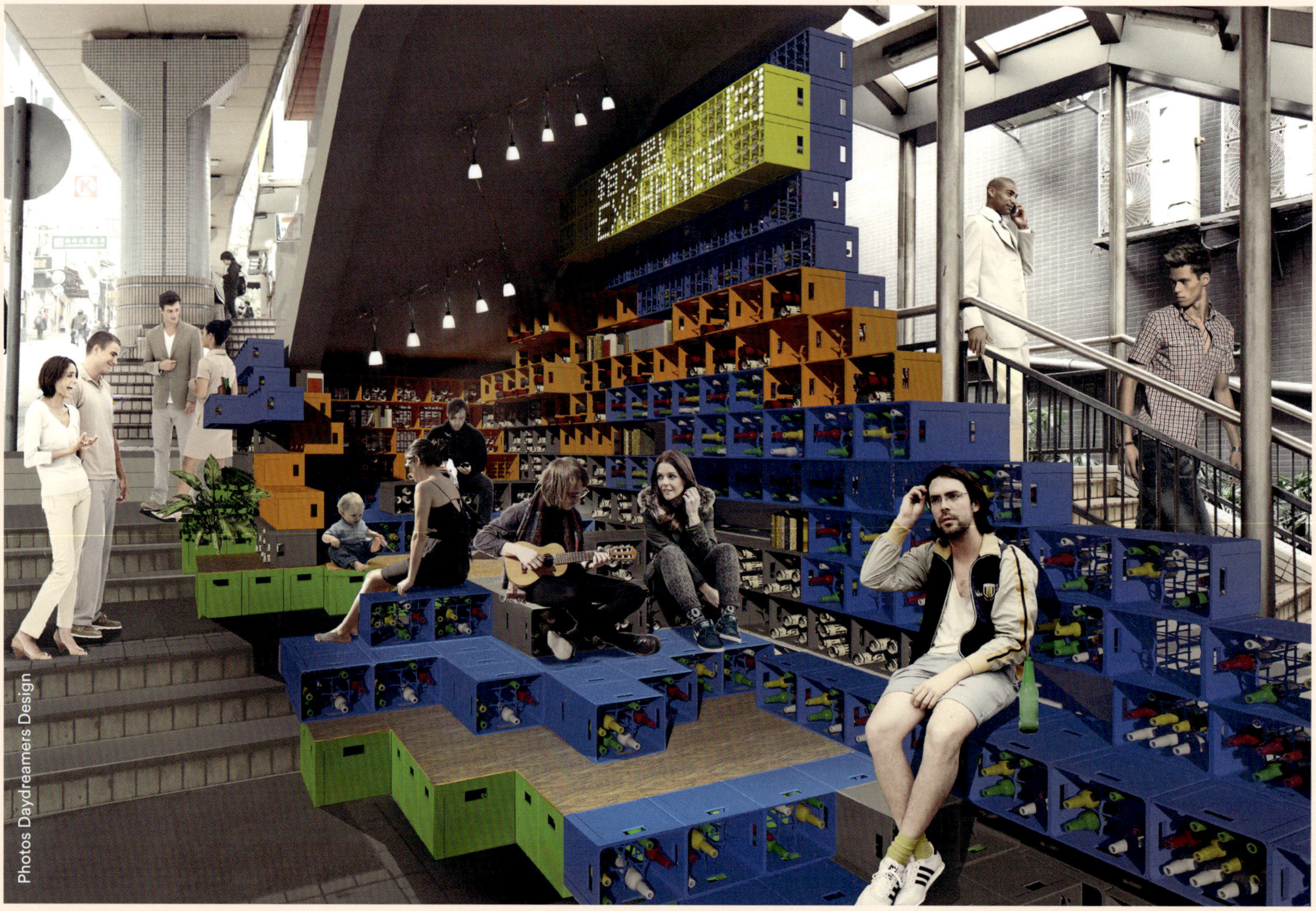

Photos Daydreamers Design

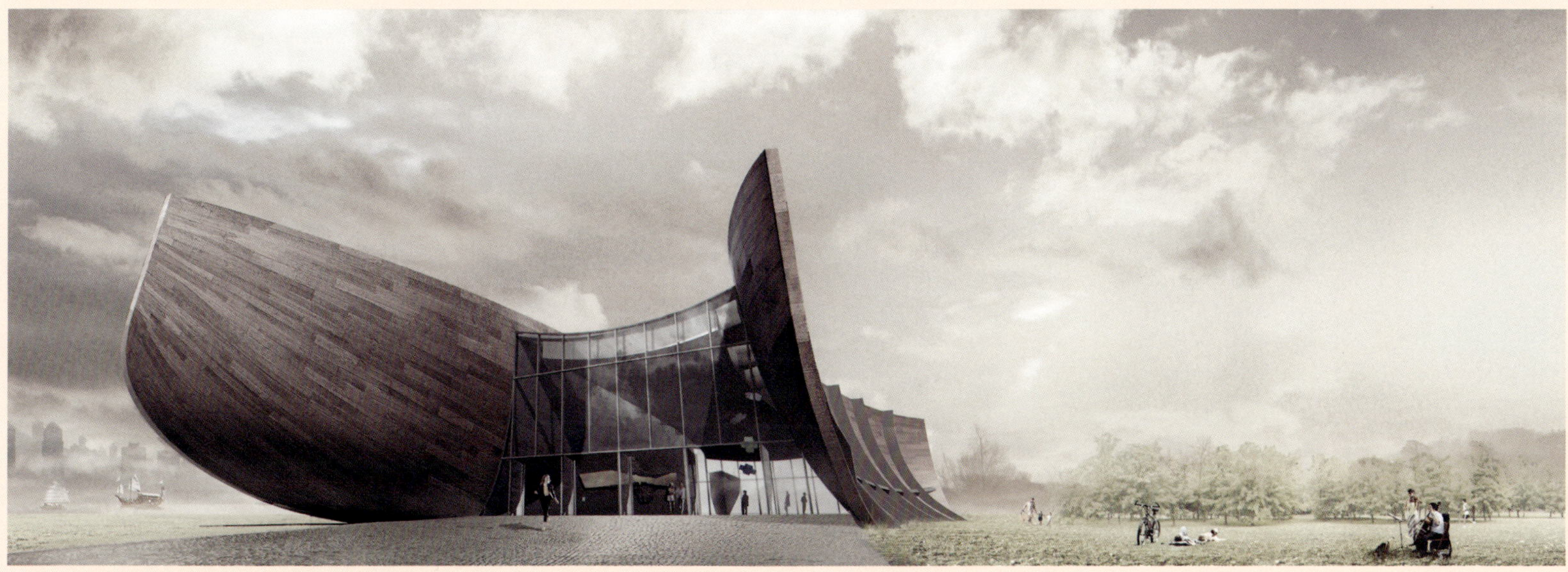

Timeslip

When **2013**
Where **Hong Kong, Hong Kong**
Client **West Kowloon Cultural District Authority**

This was an entry for an art pavilion design that was inspired by the fishing history of Hong Kong, the concept for which was created and submitted to the West Kowloon Cultural District Art Pavilion Design Competition.

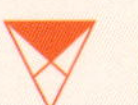

Suk[6] Muk[6]

When **2013**
Where **Hong Kong, Hong Kong**
Client **Hong Kong Environment Department**

As a sustainable landmark sculpture, this concept design was created to catch the attention of passers-by both during daylight hours and at night, thanks to dynamic illumination. The design was inspired by the impact of trees in everyday life and the design was commissioned by the new Sludge Treatment Facility in Hong Kong.

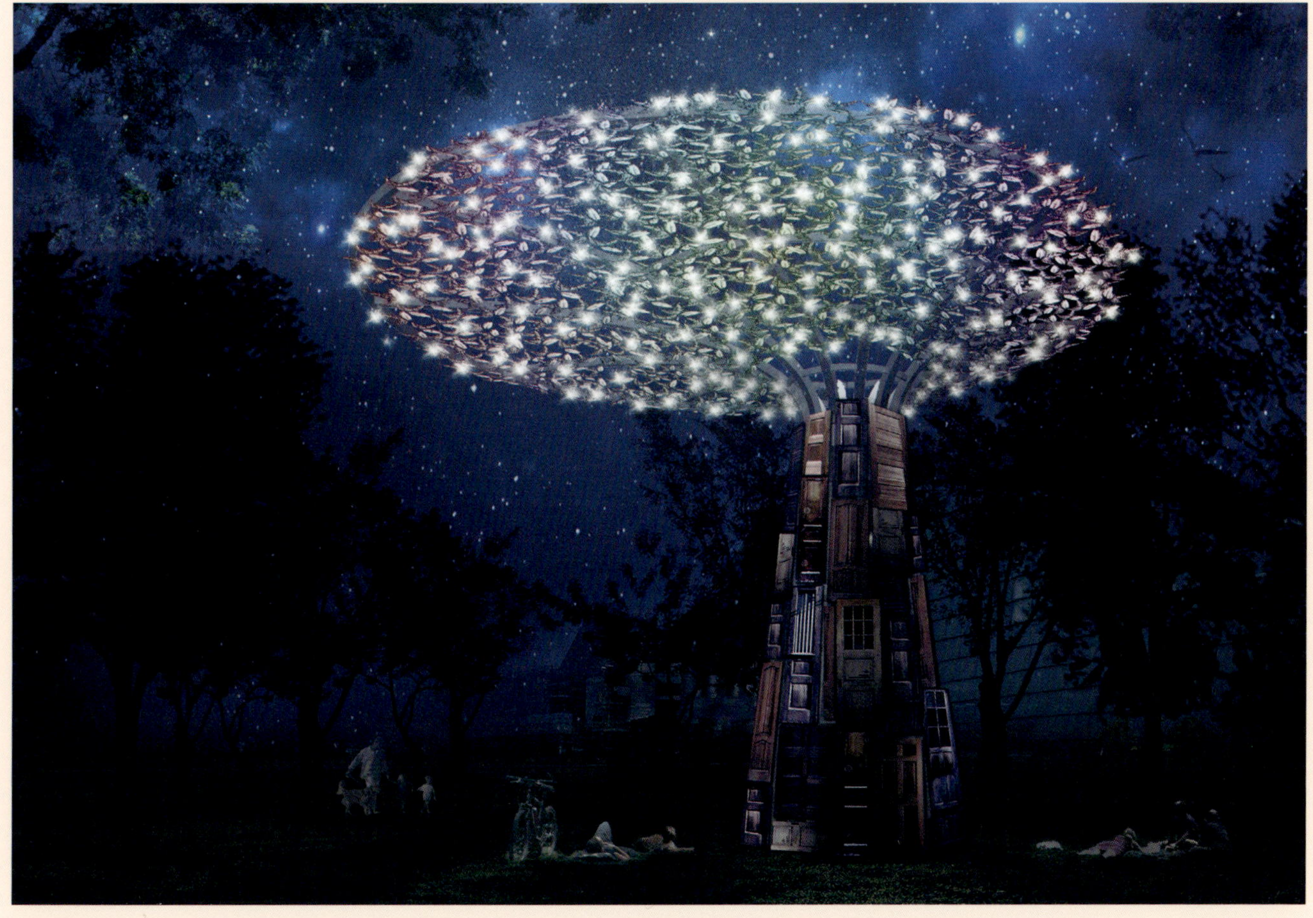

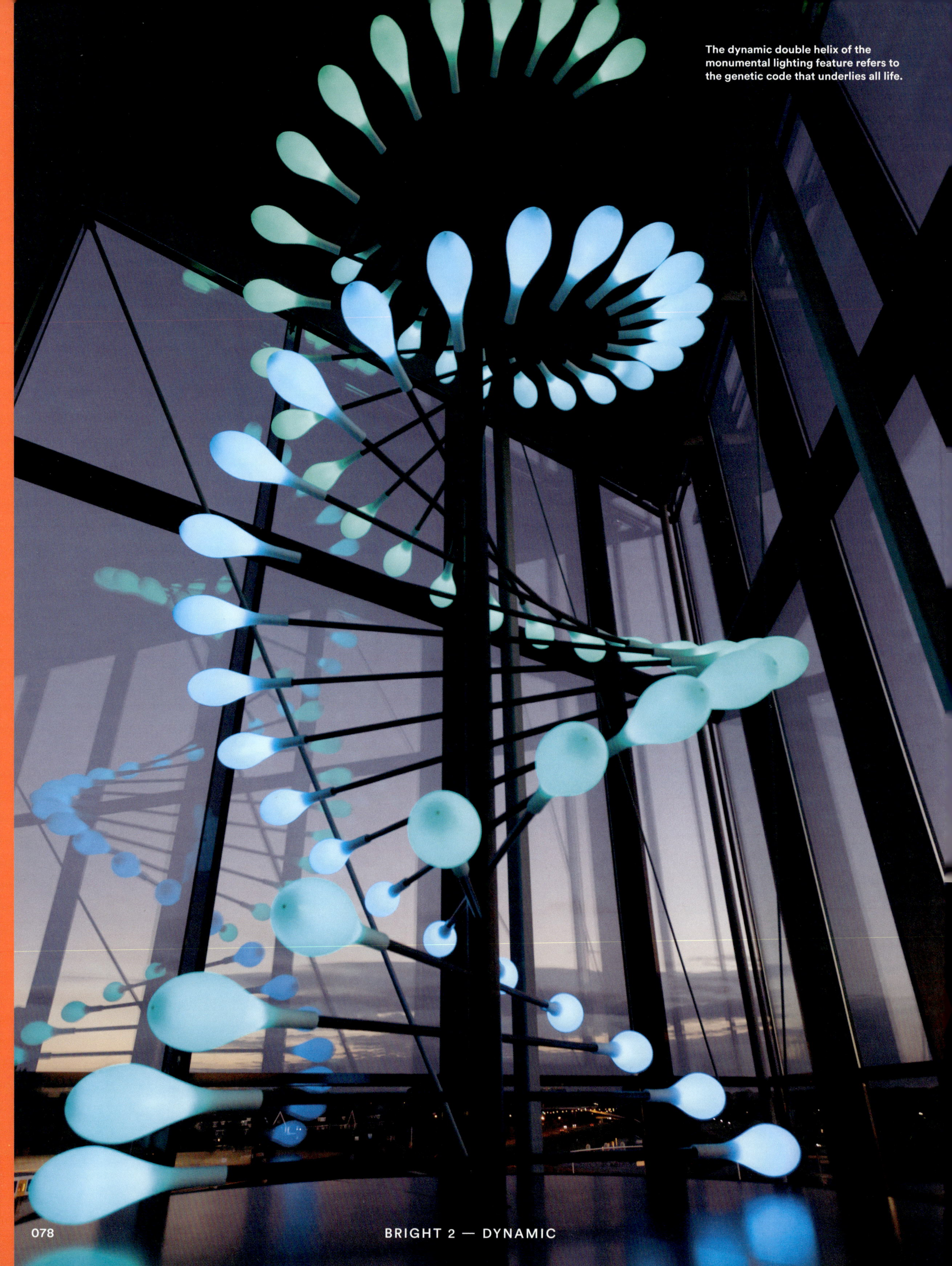

The dynamic double helix of the monumental lighting feature refers to the genetic code that underlies all life.

Seeds of Change

Engineered by **SmartLight**, Seeds of Change is a 21-m-high double-helix structure situated within the headquarters of Ricoh Netherlands. The illuminated sculpture was designed by light artist Dorette Sturm and represents the universal code of life: DNA. This dynamic installation, with its slowly-changing coloured lights, attracts attention from afar.

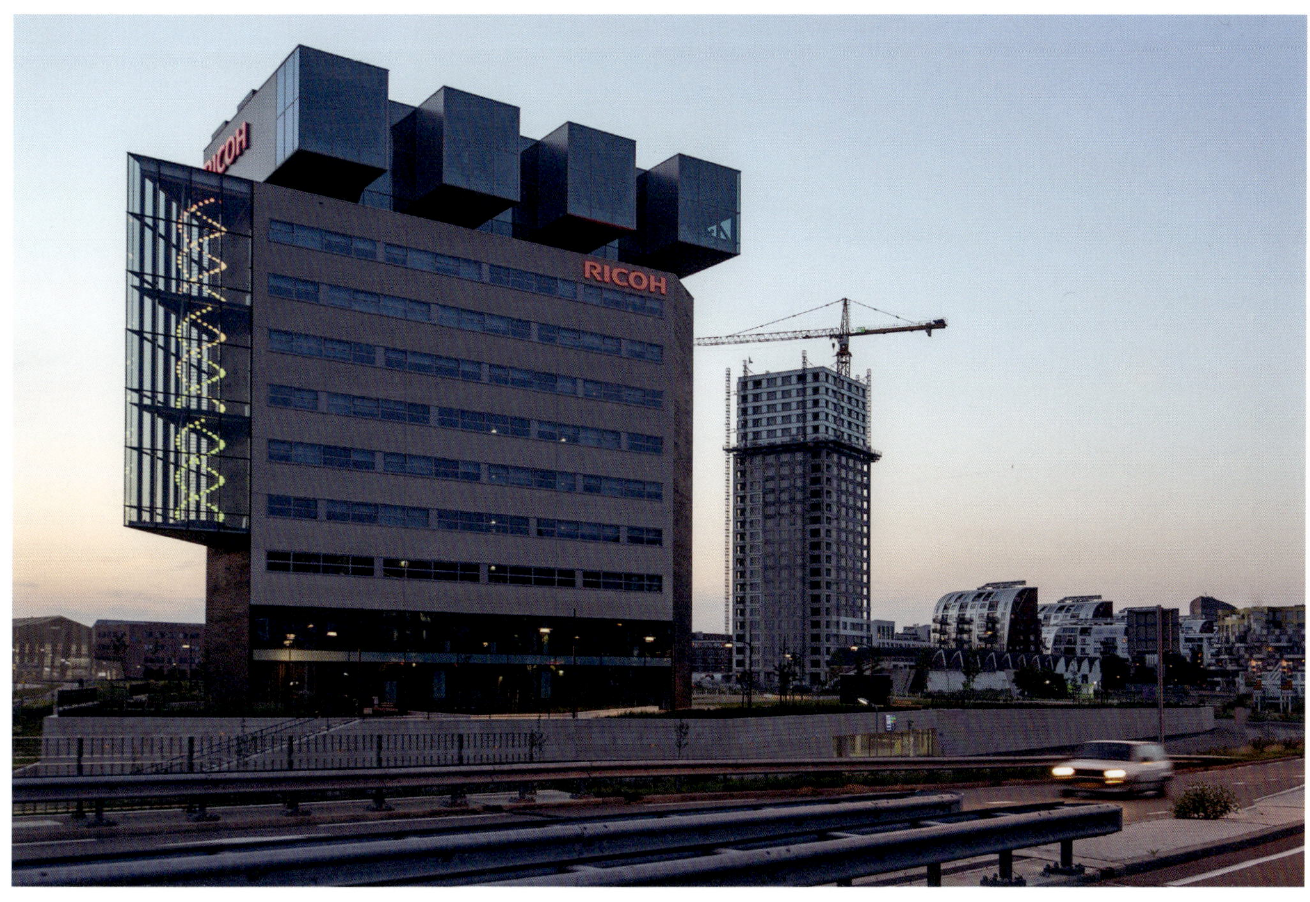

Photos Marc Bolsius

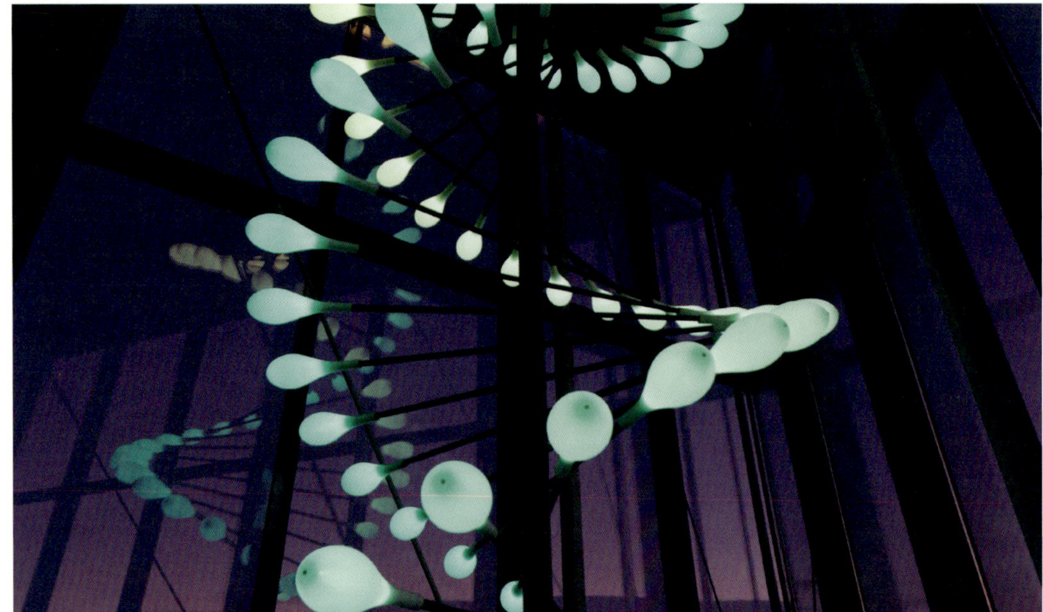

Location
's-Hertogenbosch, the Netherlnads
Client
Ricoh Netherlands
Artist/consultant
Dorette Sturm
Manufacturer
SmartLight
Date
June 2013

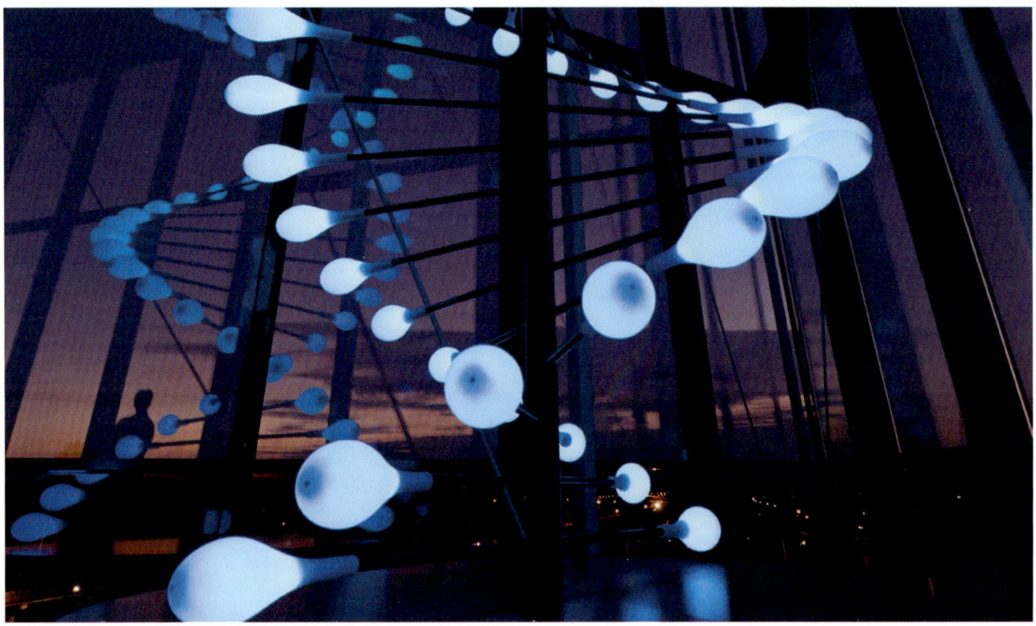

Two dynamic scenarios were created, one for the daylight hours and the other to illuminate the structure at night-time.

THE OUTLINE OF THE DOUBLE HELIX IS ACCENTED BY THE 'SEED' LIGHTS THAT SLOWLY CHANGE COLOUR

Seeds of Change is an eye-catching light installation that resembles the shape of a DNA molecule's double-helix. It is located within a glass-encased envelope, proudly making a statement at the commercial headquarters of Ricoh Netherlands. Engineered and installed by Tilburg-based SmartLight, the work attracts attention as it climbs 21 m across seven storeys, serving to bring the building together as a united whole.

Sitting in a prominent corner of the new building, the dynamic sculpture rotates and changes colour within the glass extension where it is positioned, creating a dialogue with the exterior about change and renewal, and functioning as a light beacon that draws attention to the property. DMX programming was used to control the lighting effects, which – together with the sculpture's rotational movement – helps enhance the flow and three-dimensionality of the DNA-like structure. The spiralling form of the installation is made up of a central steel column, with carefully positioned 'arms' culminating in custom-designed glass bulbs, inspired by seeds of nature. The outline of the double helix is accented by the 'seed' light sources that slowly change colour, creating synchronised lighting scenarios across the entire structure.

An important aspect of the installation, which was designed by light artist Dorette Sturm, was not only for the sculpture to have dynamic features but also a level of interactivity. It adorns the company's 'sun rooms' on the third, fifth and seventh floors, where employees can go to relax, work or simply enjoy the views. Here, motion detectors detect when visitors are in the vicinity, releasing pulses of light in response to their presence. Making the most of the variety of illumination effects that changing hues and intensity of light at different speeds allows, SmartLight programmed the dynamic scenarios developed by the light artist, which enhance the total expression of the sculpture. —

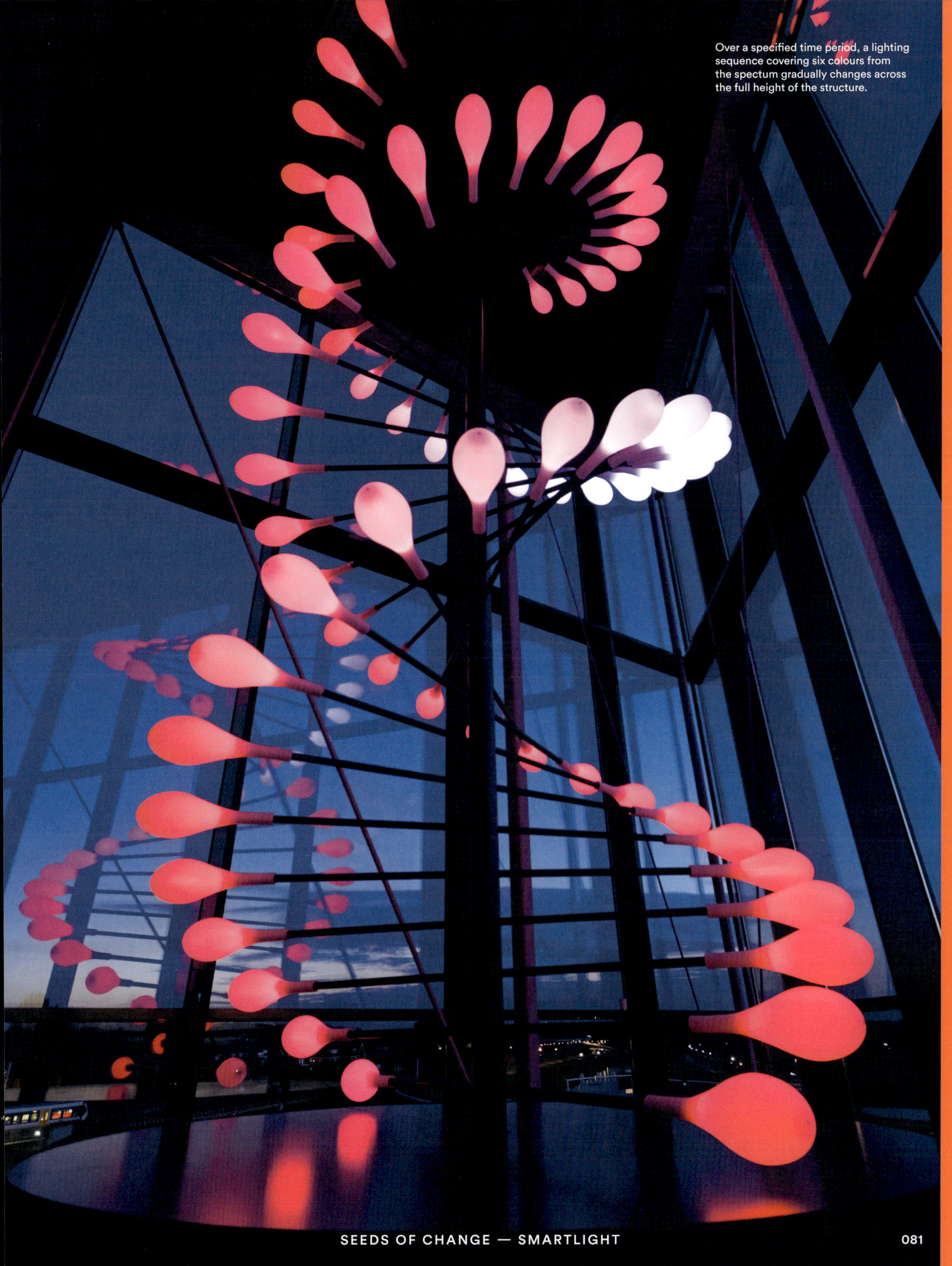

Over a specified time period, a lighting sequence covering six colours from the spectum gradually changes across the full height of the structure.

PLAN

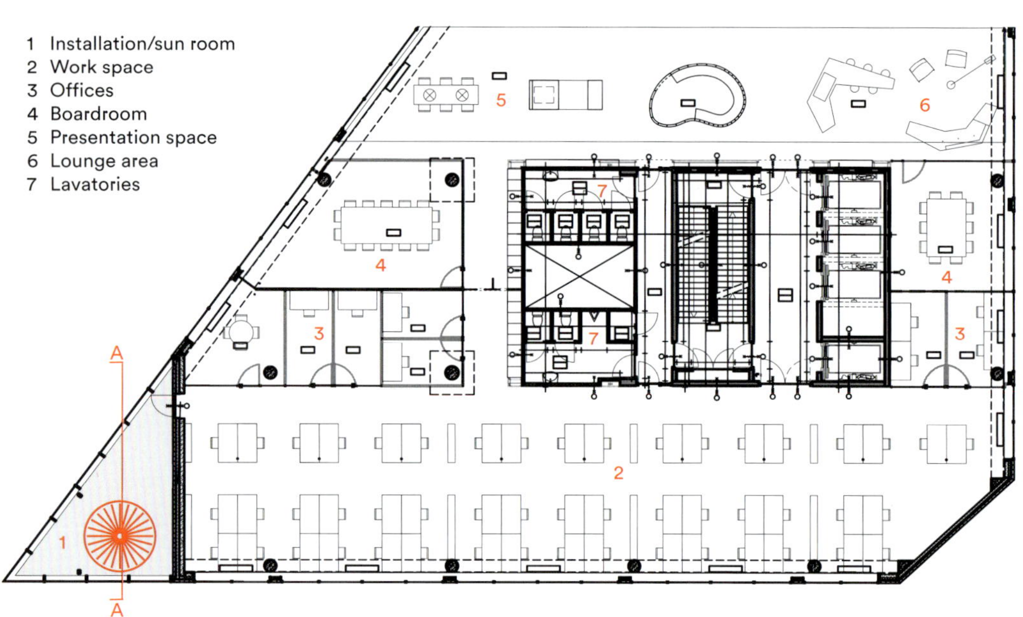

SECTION A

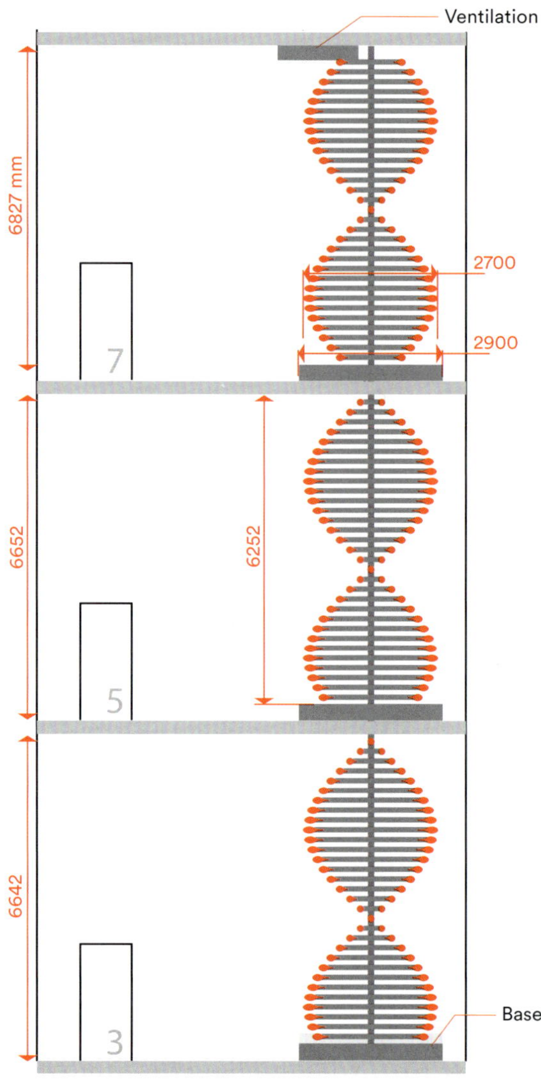

GLASS SEED DRAWING

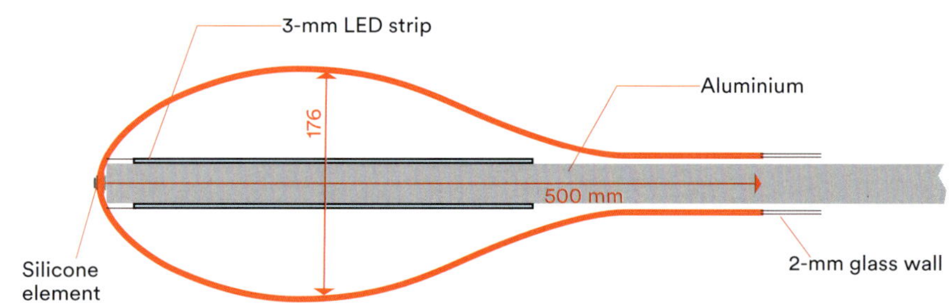

GLASS SEED CONCEPT

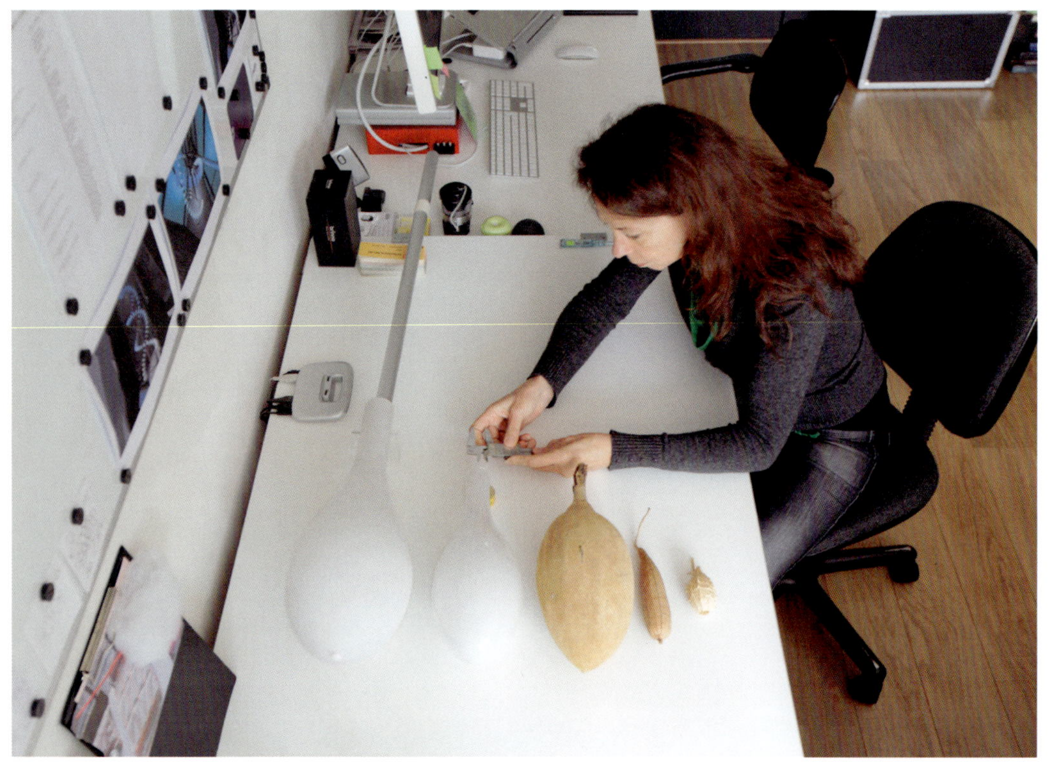

Specifics

Since the shape and size of the glass seeds were directly connected to its illumination levels and brightness, the development of this feature took a long time. In the end, the job was handed over to a country renowned for its top quality glass production, Czech Republic.

Power and data wire installation was difficult because there could be interference from data with other force fields. This was overcome through the insertion of resistors.

As all electronics must be wired with power and data, it was impossible to install sliding contacts for each connection. Smartlight came up with a solution whereby controllers and LED drivers were inserted directly on the pole so that every part rotates well.

ROTATION TESTING

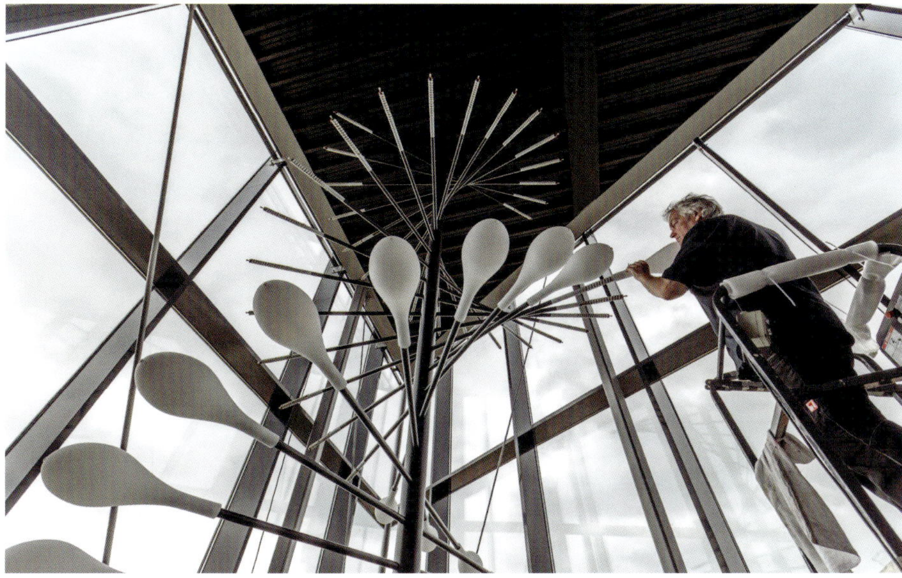

LIGHT TESTING

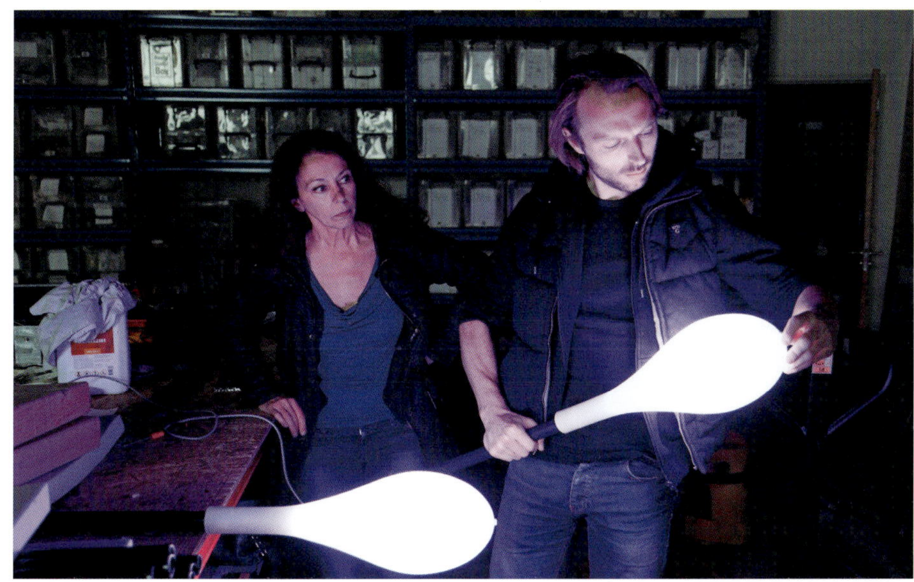

SOFTWARE TESTING

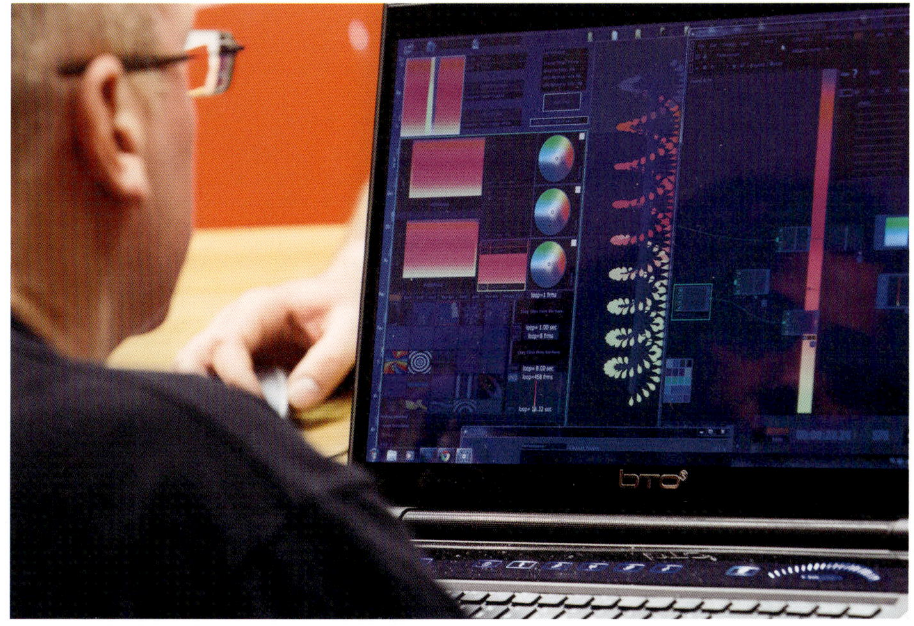

Set-up

Dimenions of the structure reached a height of 21 m and a width of 2.9 m. It was constructed from aluminium, blown glass, LEDs, sensors, propulsion, electronics and software.
Structurally, the central column (in three segments, each 7-m long) and its 'arms' are made of hollow, coated steel with 120 and 20 mm diameter, respectively.
Illumination of the sculpture is with 108 RGB SMD pixel LEDs, lighting up each 'seed' located at either end of the arms. Light armatures consists of a 13-mm LED strip, running along the outside of the alumnium tube, with a silicone element positioned at the tip. The capactiy per seed is 23 W (total 4600 W).
Glass seeds were hand-made in the Czech Republic using glassblowing and sandblasting techniques. Over 200 glass bulbs were manufactured to the optimum shape, which was inspired by seeds of nature.
Motion detectors mounted on the third, fifth and seventh floors are connected to the sculpture so that when someone enters the room it will pulse three times, welcoming its visitors.
Rotation of the structure is powered by a propulsion system is an electric motor with a worm wheel reductor, held at a steady speed. During the day the speed is slow, at 1 RPM for safety reasons, and in the evening the speed is increased to 4 RPM. The bearing keeps the sculpture rotating smoothly, with viewing platforms on the three levels made of MDF boards.
Control system utilises a wireless connection, meaning the sculpture can be programmed from a distance for a better perspective of what it is being achieved.
Software was developed and programmed to achieve specific scenarios according to outlined requirements.

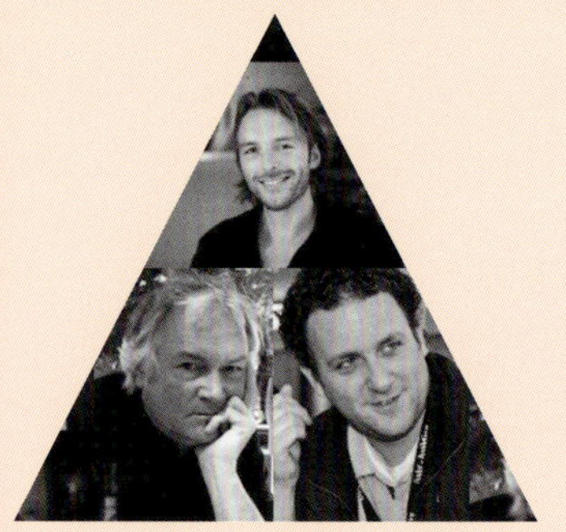

SmartLight

SmartLight specialises in creative LED lighting solutions for different market segments. Based in Tilburg, it was founded in 2009 by Bob Westerburgen, Hedwig Westerburgen and Eelco Timmermans. The Dutch company provides 'solutions' on a global scale. The team members' expertise in lighting, engineering and control systems enables them to embark on many different types of projects, from dynamic, static, exterior and interior lighting for a myriad of applications, such as artworks, architecture, urban design and retail.

LIGHTSOLUTIONS.NL

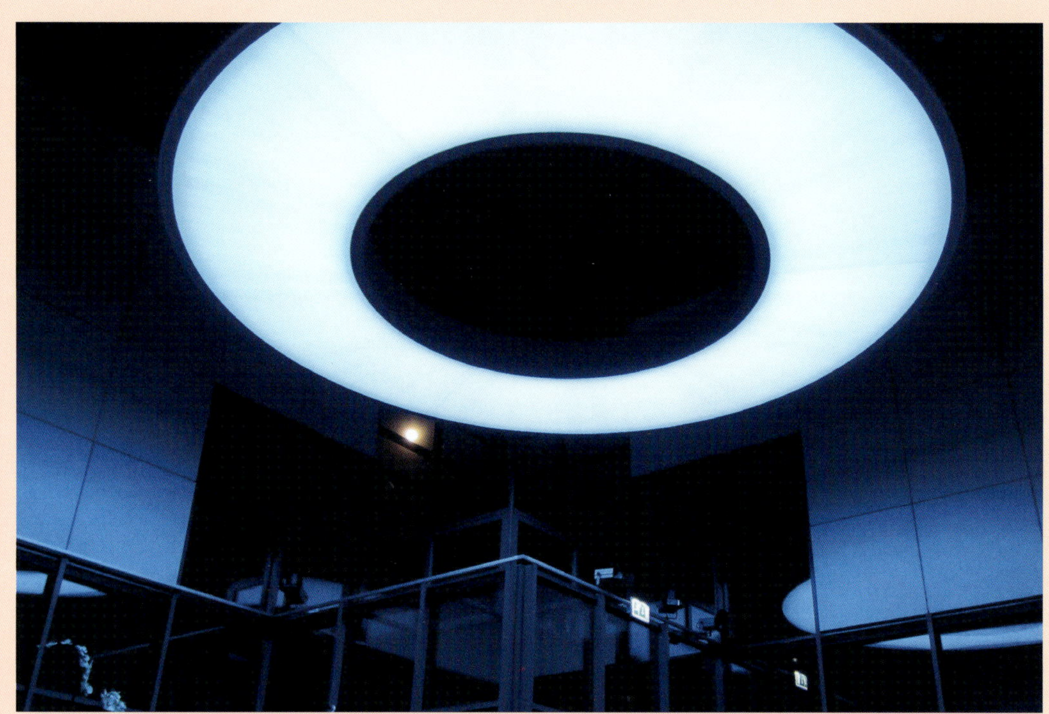

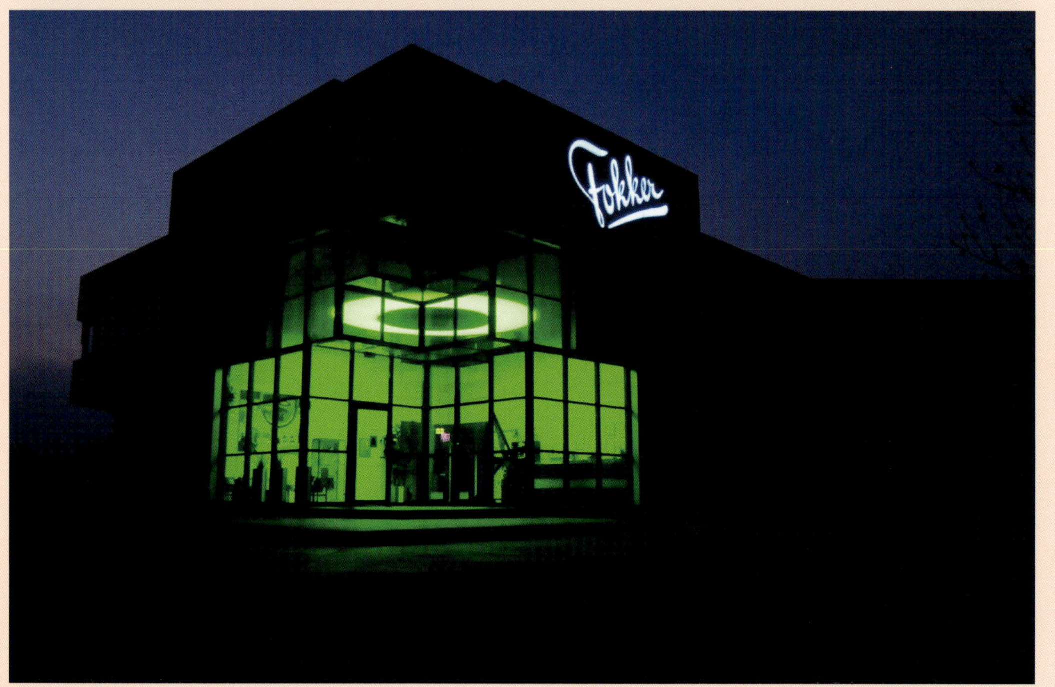

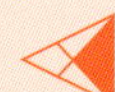

Fokker Chandelier

When **2013**
Where **Helmond, the Netherlands**
Client **Fokker Technologies**

SmartLight was responsible for installing the dynamic light sculpture, boasting a 6-m diameter, in a restaurant in Helmond, the Netherlands. Designed by Jaap van den Elzen, the dynamic and contemporary chandelier is composed of 12 linear RGB LED lighting sections, so that each part can change colours independently.

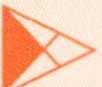

Reinders Oisterwijk

When **2011**
Where **Oisterwijk, the Netherlands**
Client **Reinders Oisterwijk**

Architect Frank Willems wanted coloured elements to be added to the envelope of Reinders Oisterwijk office building. Responding to this, SmartLight introduced RGB LED lighting behind the frosted glass panels on the facade, creating a colourful spectacle, which is replicated inside the showroom itself. Integrated into the ceiling, this feature accentuates the central space.

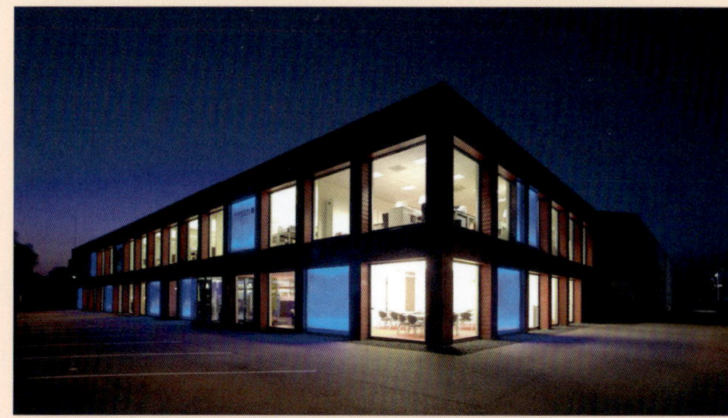

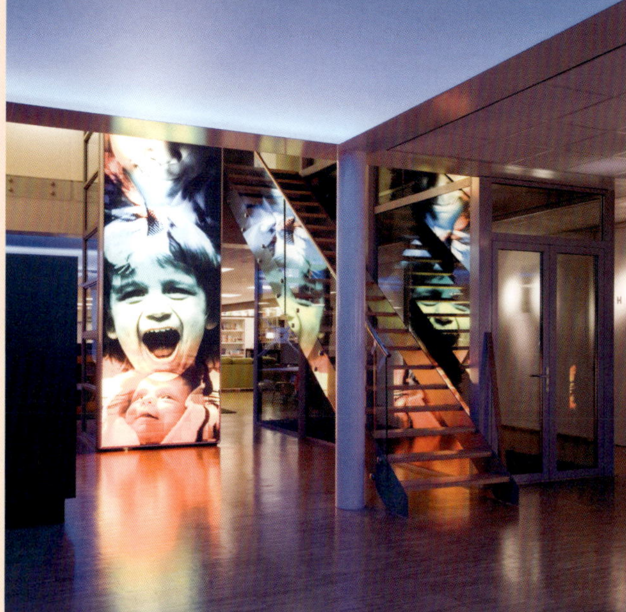

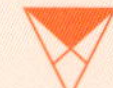

Strijp-S

When **2013**
Where **Eindhoven, the Netherlands**
Client **DNC Vastgoed**

As a part of the re-development programme for Strijp-S in Eindhoven, Har Hollands Lichtarchitect developed a lighting design that looked to re-integrate the maze of left over pipes from the long-gone Philips factories into the Leidingstraat's urbanscape. To implement this design, which called for a dynamic lighting system, SmartLight engineered, fabricated, supplied and installed a turn-key project. Overall, 780-m RGB pixel LED lines, 500 dynamic white power LED modules and linear LED lighting were used to create a lighting effect that simulates the movement of the liquids and gases in the pipes.

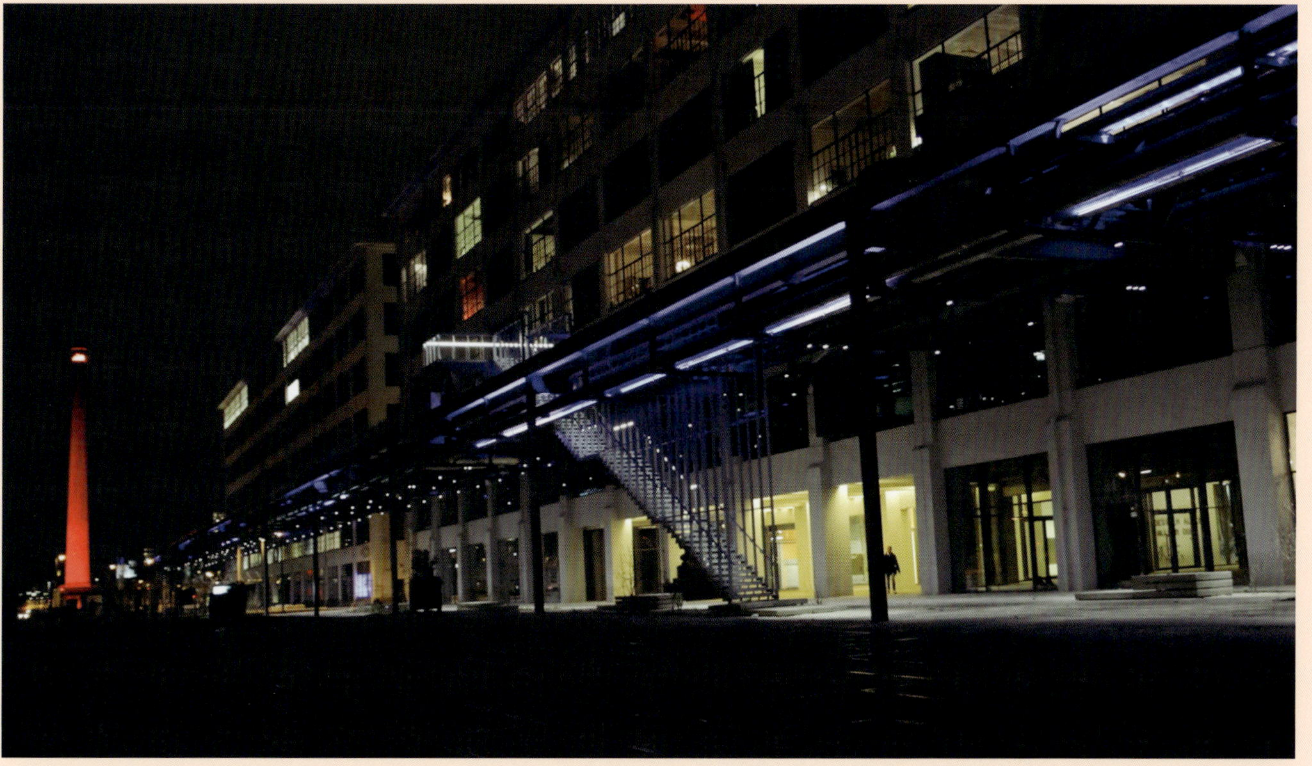

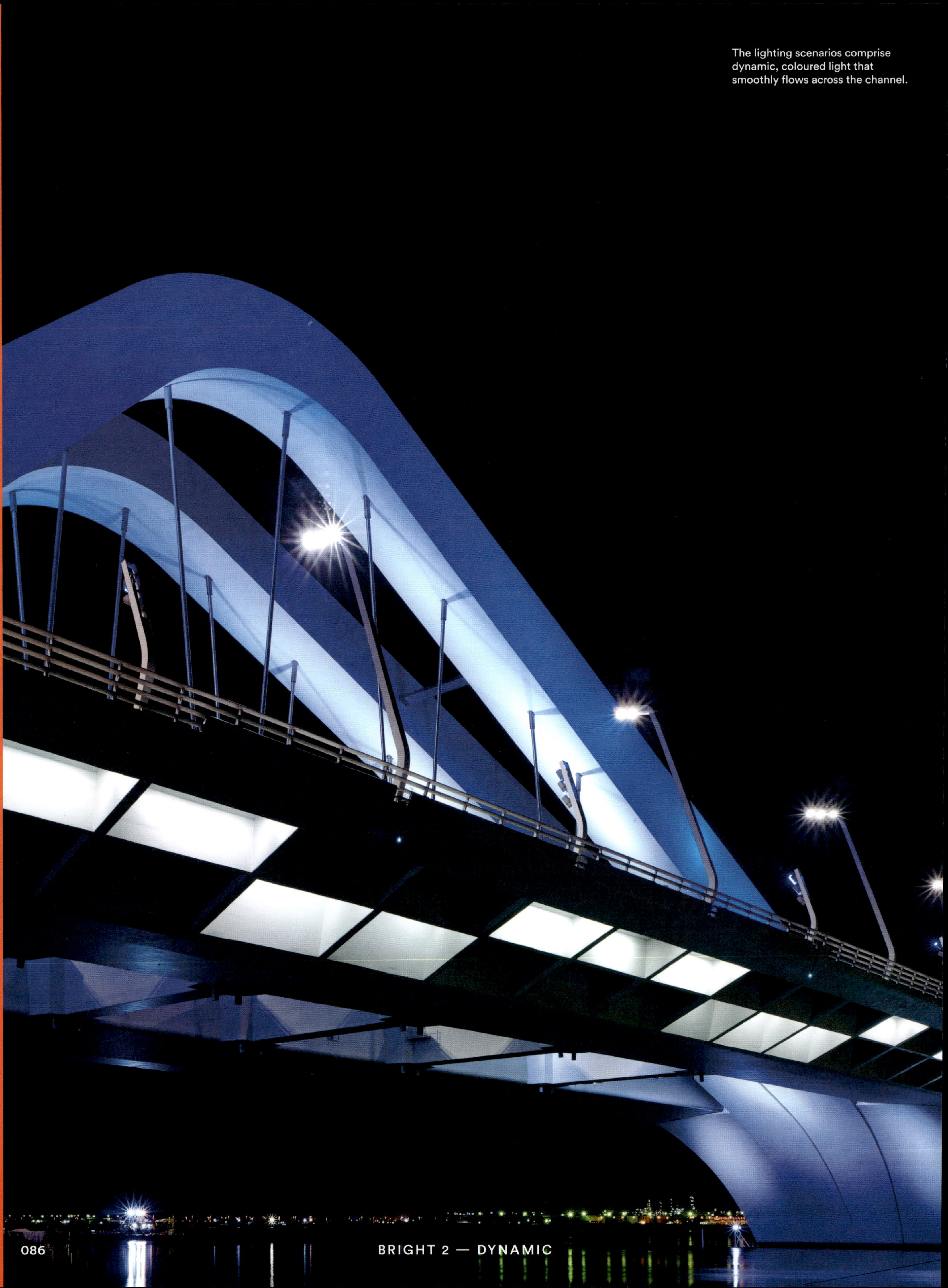

The lighting scenarios comprise dynamic, coloured light that smoothly flows across the channel.

Sheikh Zayed Bridge

The dynamic lighting scenario of **Arup** for the Sheikh Zayed Bridge perfectly illuminates the structural silhouette for the sinusoidal design by Zaha Hadid, which crosses the channel as a shining gateway into Abu Dhabi, rising 60 m above water level and celebrating this feat of engineering.

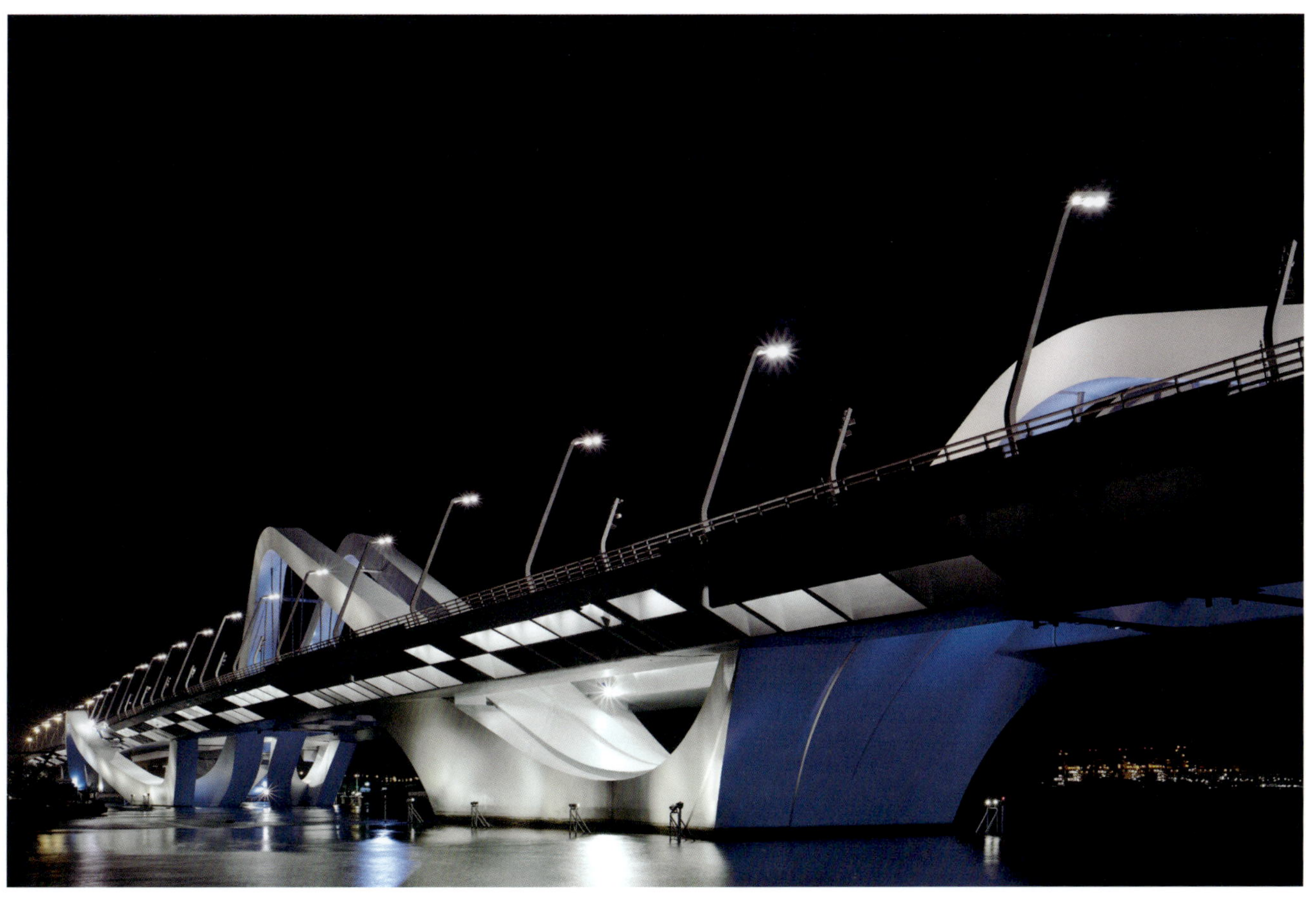

Photos Christian Richters

Designed by Zaha Hadid Architects, the 842-m Sheikh Zayed Bridge connects the mainland to the island of Abu Dhabi. Named after the country's former president, it is the gateway over the Maqta Channel that soars to heights of up to 60 m above water level and is seen as a potential catalyst to the future urban growth of Abu Dhabi.

The bridge has an arched configuration, consisting of concrete strands that gather on one shore, flowing across to the adjacent shore in a sculptural waveform. Arup is responsible for the illumination of the bridge, with the structural silhouette – the 'spine' – highlighted by dynamic coloured light, while the underside of the road deck features integrated monochrome 'cell' lighting. A lighting control system allows for animated flow movements in the spine and the cells, resulting in fluid patterns of light travelling along the bridge. The luminous gesture creates an embracing landscape of colours, up-close whilst on the bridge itself and a scenic feature from a distance.

The initial lighting concept was devised at Hollands Licht by Rogier van der Heide. The lighting design was developed further by an international design team at Arup. Collaborating closely with the Abu Dhabi Municipality, the final project phases included the programming of 13 artistic scenarios, that were translated by Martin Professional Middle East using a Maxxyz theatre console. The multiple scenarios reflect Abu Dhabi's soul. A specially-developed 'language of light' celebrates religious traditions, festivities and public events together with the people of United Arab Emirates. For instance, during a new moon the bridge lighting links with the appearance of the Grand Mosque. Once a month, both iconic pieces of architecture appear tinted in deep blue colours and create awareness of urban connectivity. —

Designer
Arup
Location
Abu Dhabi, United Arab Emirates
Client
Abu Dhabi Municipality
Collaborator(s)/consultant
Zaha Hadid Architects, Sixco, High Point Rendel, Hollands Licht
Manufacturer
Martin Professional Middle East
Date
January 2011

THE LUMINOUS GESTURE CREATES AN EMBRACING LANDSCAPE OF COLOURS, A SCENIC FEATURE WHETHER FROM UP-CLOSE OR FROM A DISTANCE

Site testing of the illumination scenarios for the bridge.

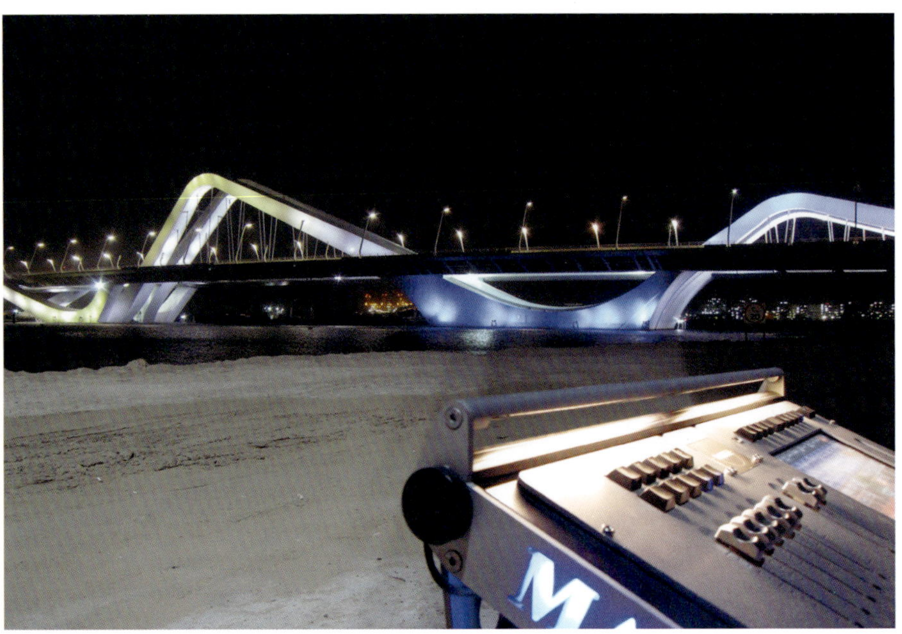

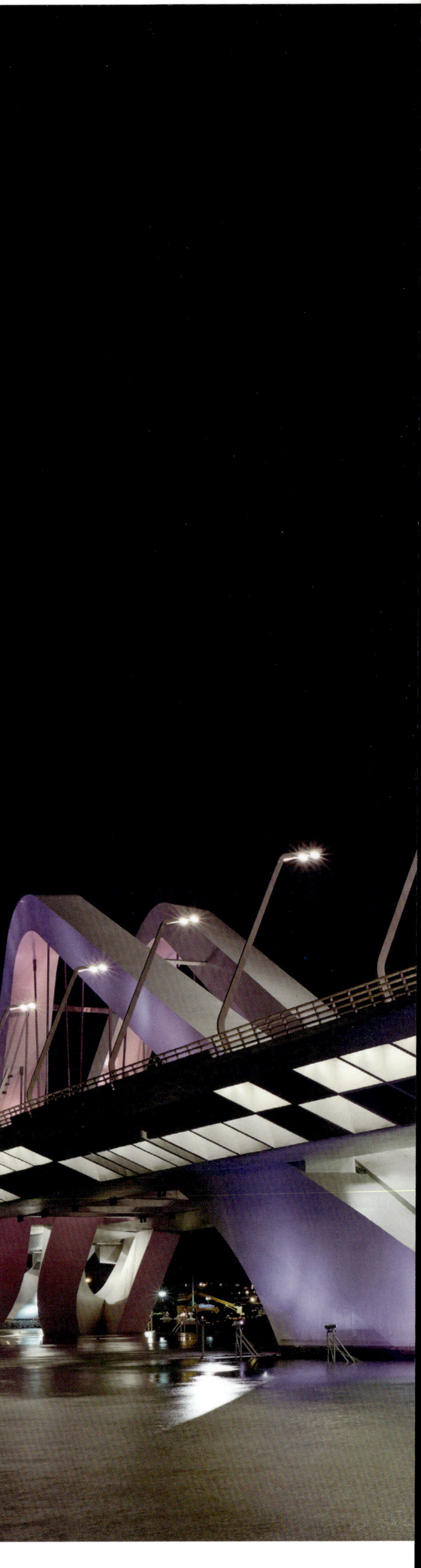

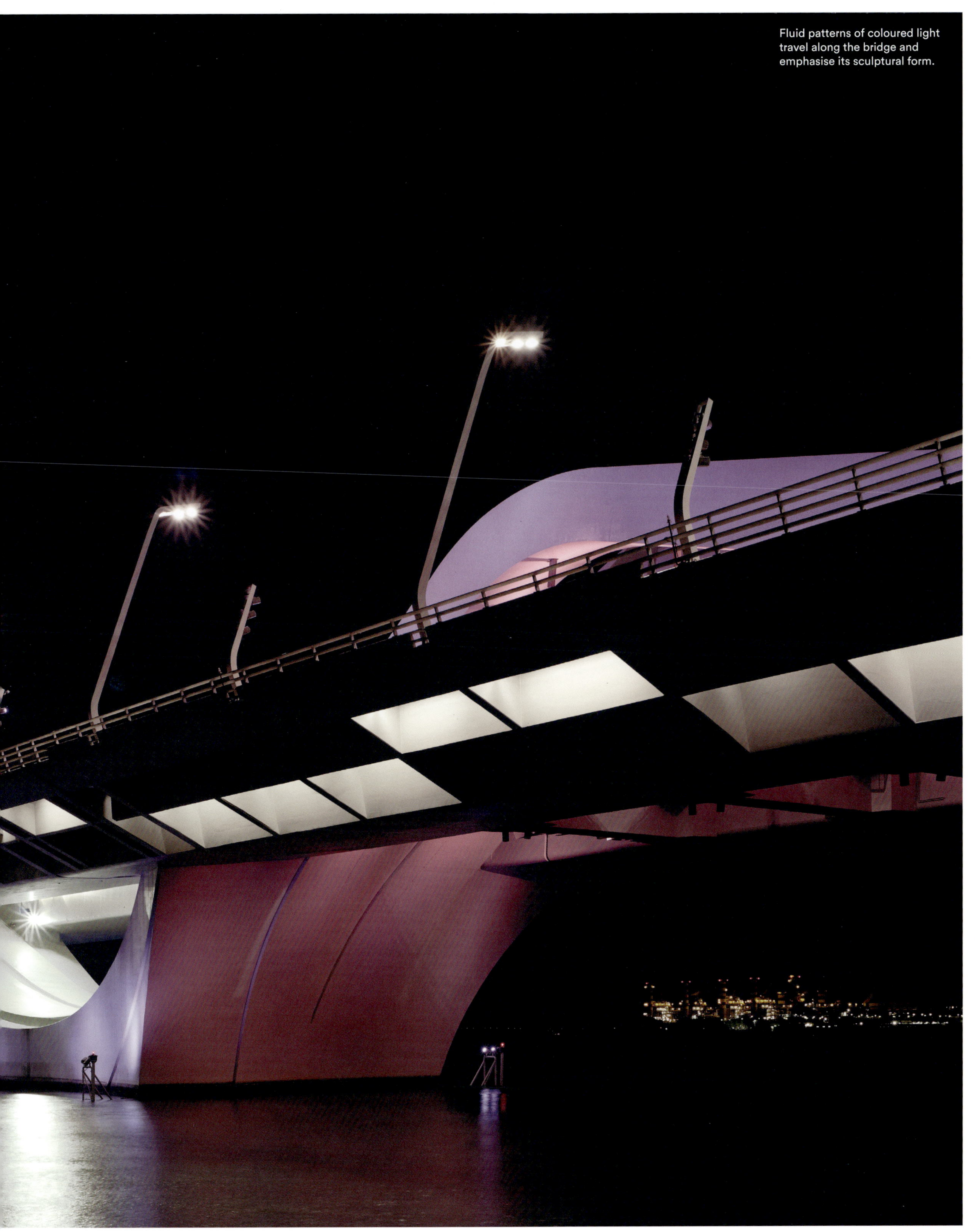
Fluid patterns of coloured light travel along the bridge and emphasise its sculptural form.

FEATURE LIGHTING MASTS

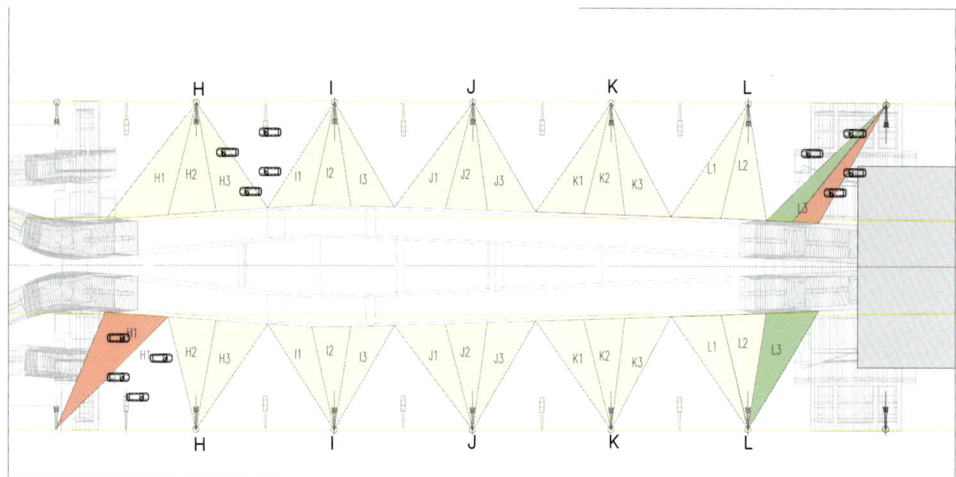

Top view

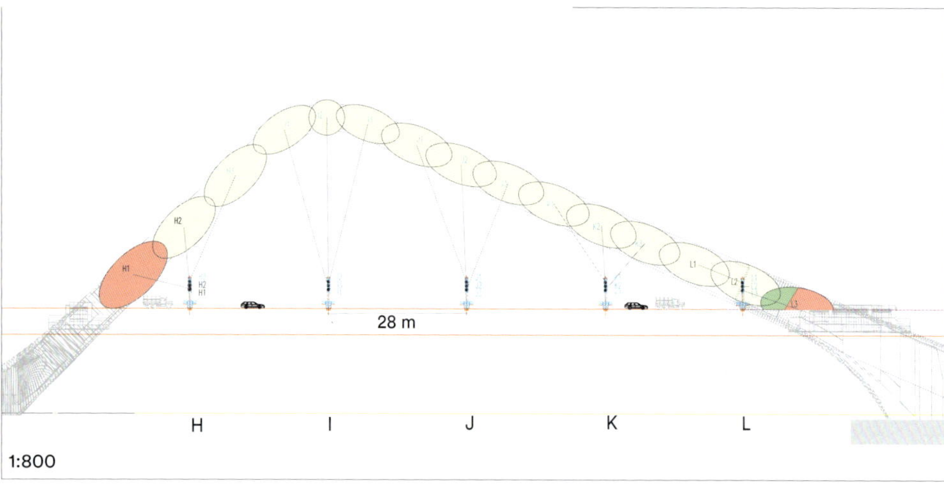

Side view

GLARE CONTROL PROPOSAL

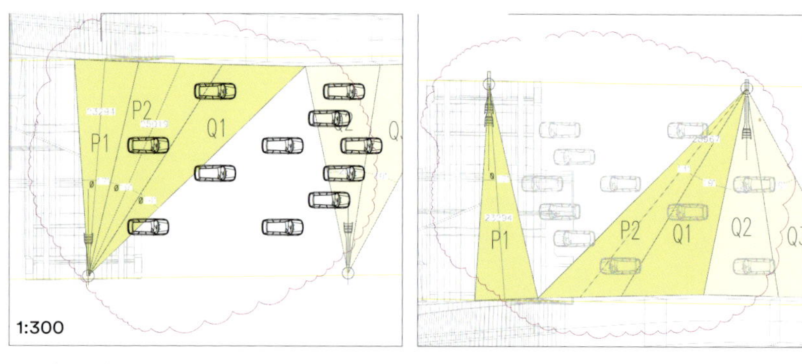

Plan (south)

Plan (north)

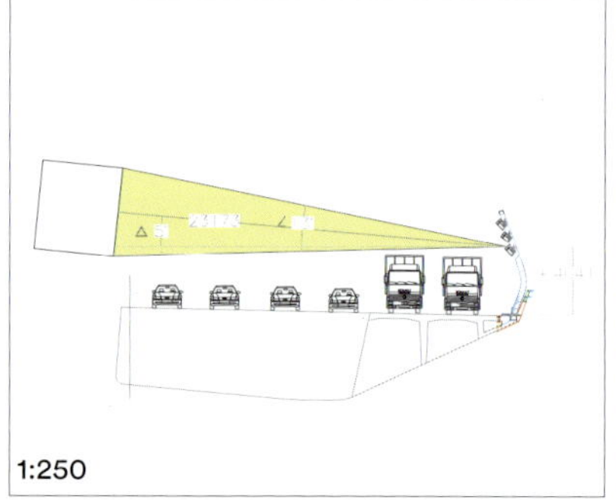

Fitting Q1

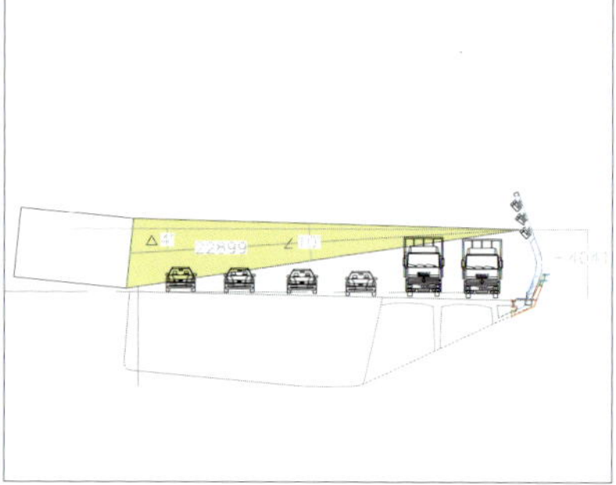

Fitting P2

Set-up

Light source needed powerful luminaires designed for projecting effects onto structures and landmark architecture. The projectors use metal halide lamp sources.
Illumination was a CMY colour-mixing system with primary colour wheels in red, green and blue, allowing for a limitless choice of colours.
Framework of the bridge was seamlessly illuminated making use of three different fixture sizes with varying power levels and lens spreads.
Lighting control was such that each luminaire can be addressed individually via DMX signals.
Software lighting scenarios were programmed using a Martin Maxxyz console.

Feature lighting products

- 204 Martin Exterior 600
- 130 Martin Exterior 200
- 956 Martin Cyclo 01

Control system products

- 1 Martin Maxxyz
- 12 Martin Ether2DMX
- 9 Martin RS-485 Optosplitter
- 10 Etherswitch.

Specifics

The fixture specifications and placement control of all luminaires was a major challenge. The design team used mock-ups to test lens-types in combination with beam spreads to achieve smooth illumination and a dynamic flow of coloured light after sunset.

In the early stages, Arup used software 3D Studio MAX and calculated the lighting effect of all luminaires, making use of photometrical data. The output was a photometrically correct animation, presenting the actual lighting effect and luminous colour flow abilities to the client.

Locations for feature lighting masts on the bridge deck, under the bridge and especially on the 'concrete islands' just above the water surface had to be carefully evaluated considering glare, access for maintenance and navy routing.

The road lighting was another element that needed particular fine-tuning; in order to not disturb the feature lighting effect, functional road lighting makes use of asymmetric wash lights.

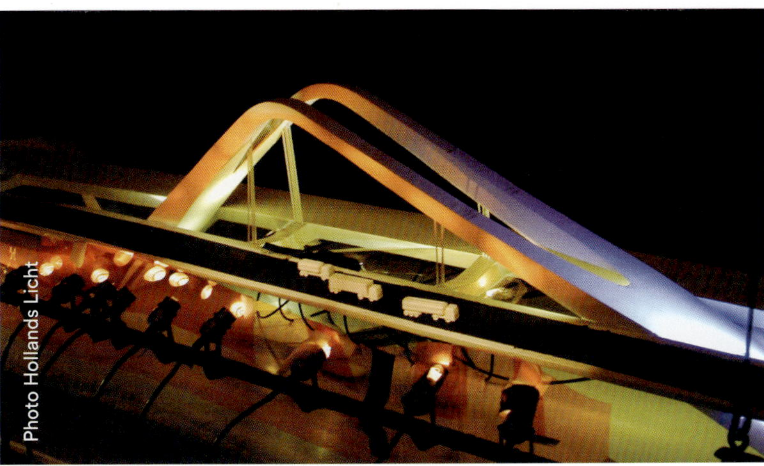

Models were used to test light projected onto the spine of the bridge.

THE MULTIPLE LIGHTING SCENARIOS CREATE AWARENESS OF URBAN CONNECTIVITY

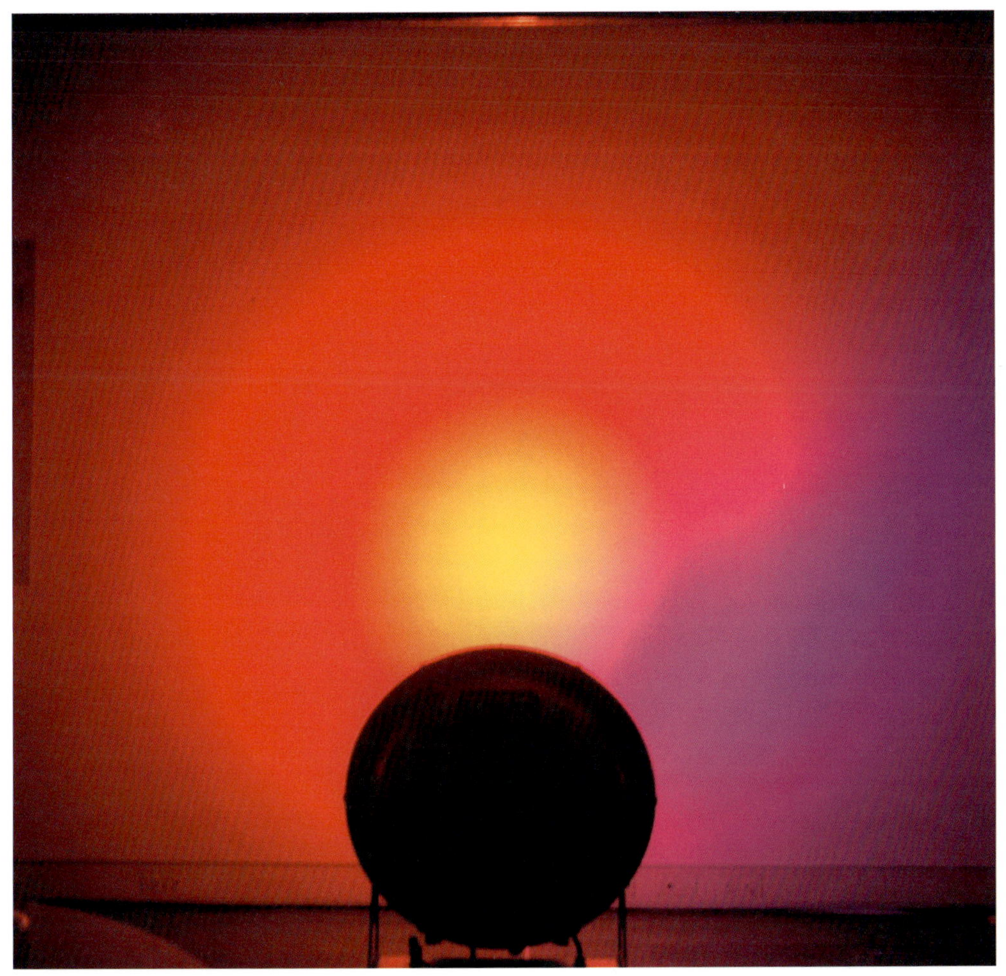

Testing of beam spreads and smooth-fade effects in Arup's Light Lab.

ANIMATION STILLS

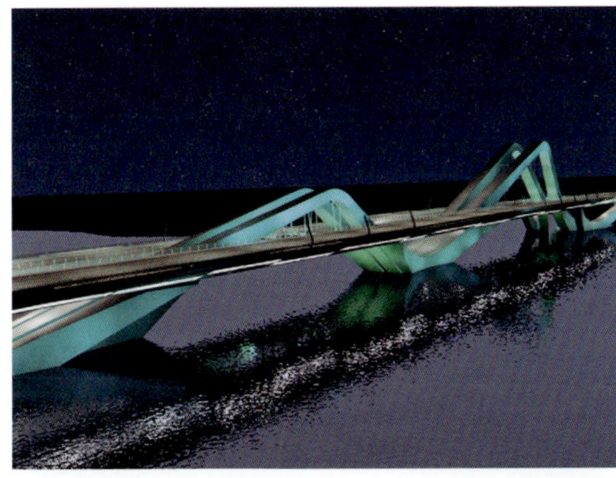

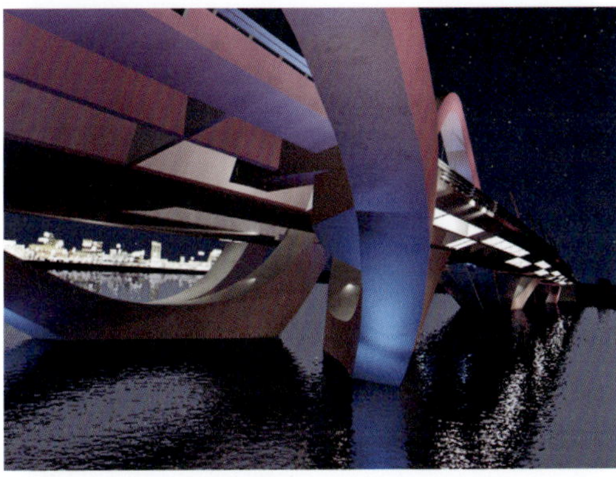

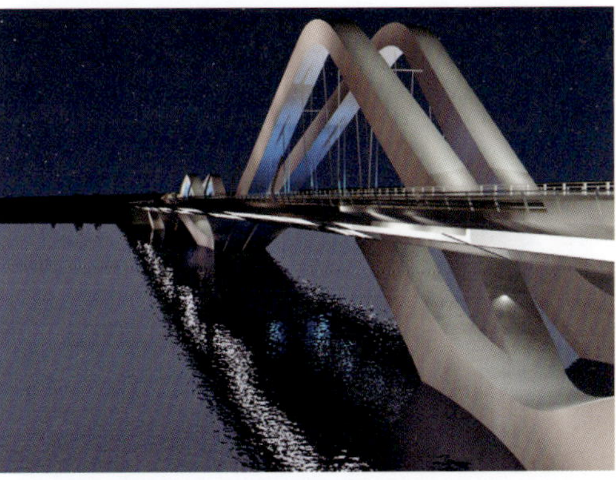

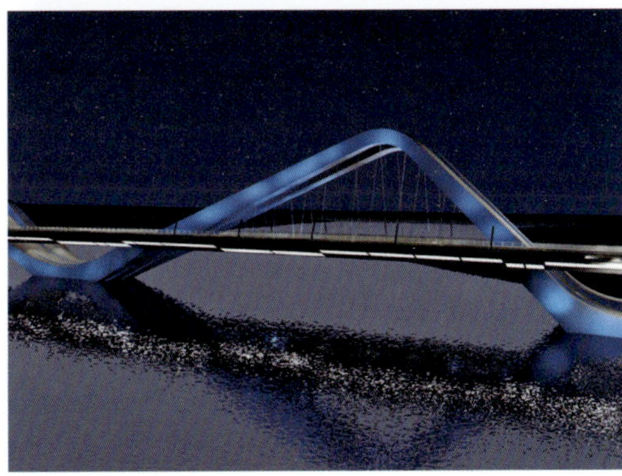

The lighting fixture's IES files were inserted into the 3D model to check the various lighting effects.

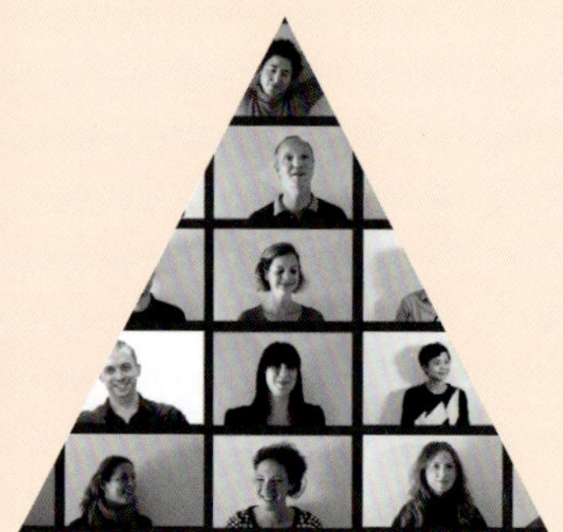

Arup

Arup is an independent firm of designers, planners, engineers, consultants and technical specialists offering a broad range of professional services. In skilled hands, lighting becomes the fourth dimension of architecture, integrating and enhancing the other design disciplines. Arup's lighting team offers a comprehensive architectural and natural lighting design service. In the independent lighting design studio, robust engineering skills and comprehensive experience are combined with award-winning creativity, flair and an innovative attitude.

ARUP.COM

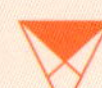

Frasers

When **2013**
Where **Sydney, Australia**
Client **Frasers Property Group**

The Frasers Broadway project is a sustainably-designed and operated development on the site formerly occupied by the Carlton United Brewery in the Chippendale neighbourhood of Sydney. Arup completed the lighting scheme for the multi-use building, including facade, entries, retail and apartment illumination. The exterior roof feature is a heliostat and reflector. Underneath the reflector, Yann Kersale's Sea Mirror is a visual lightning rod – an abstract interaction of 2880 LEDs and 320 mirrors brought to life by Arup's team of specialist lighting designers.

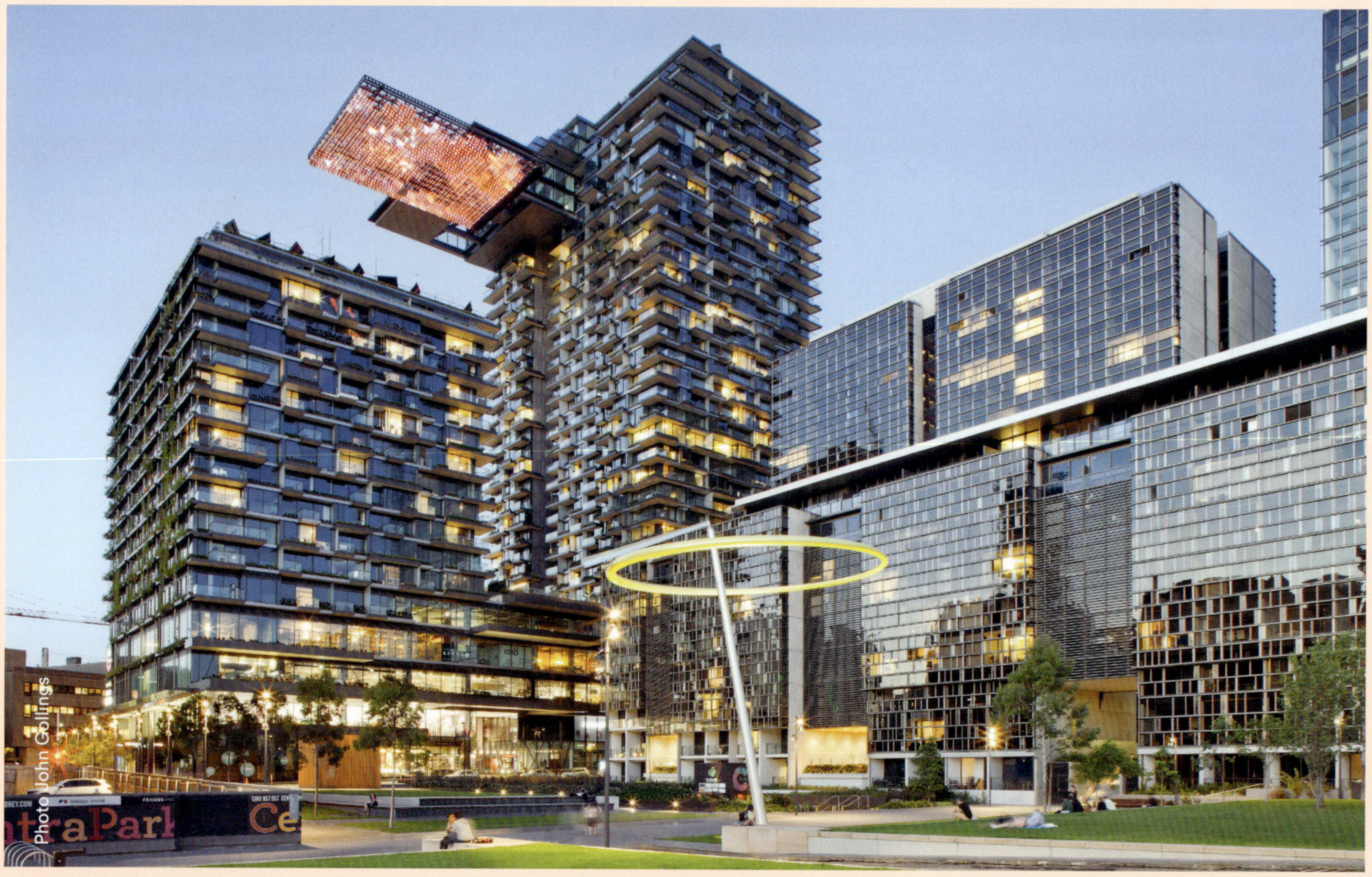

Photo John Gollings

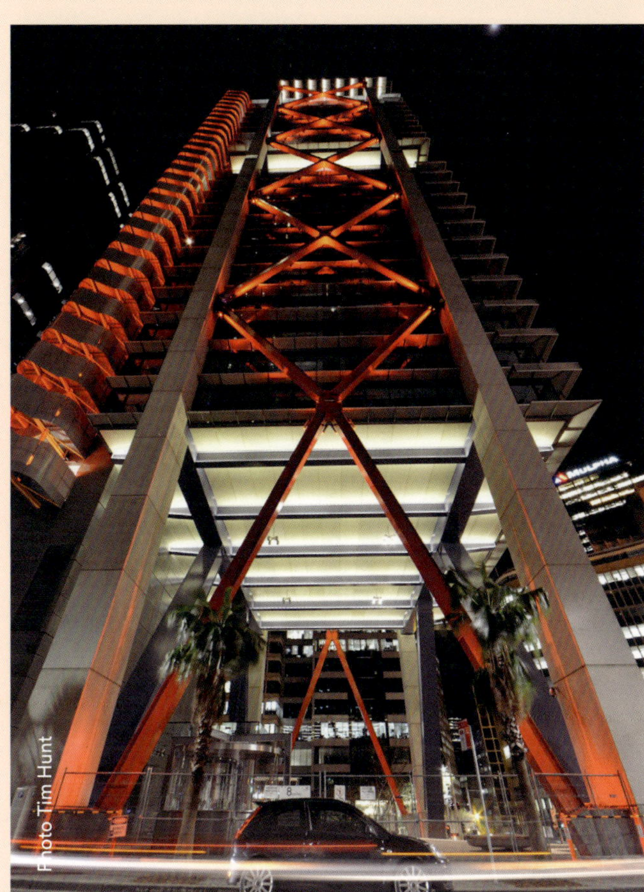
Photo Tim Hunt

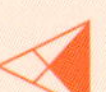

8 Chifley Square

When **2011**
Where **Sydney, Australia**
Client **Miravc Group**

The first building in Australia by Pritzker prize-winning architect Richard Rogers, 8 Chifley Square adds a distinctive aesthetic to the Sydney skyline. Through the design process, Arup lighting designers developed the brief into a concept that unified each of the structure's open spaces, creating a unified sensation that complements the philosophy of the building's design. Unique architectural features, such as the external nodes and fire stair, are accentuated with a luminous red. The design creates luminous and inviting public spaces that are connected vertically and accentuates the structure of the facade.

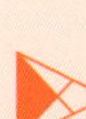

Resonanz

When **2013**
Where **Berlin, Germany**
Client **Berlin Light Festival**

The group of artists Sophie Valla, Léonard Roussel and Susheela Sankaram make up the collective known as Sound and Light Act (SLAct). The group's temporary sound and light installation transformed the courtyard of design centre Aufbau Haus in Berlin, turning it into a surreal, dreamlike and engaging public space. Overlapping layers of soft textiles, spatialised sounds and coloured illumination created a deep and stirring landscape that resonated with light and shadow amidst the flawless geometry of the complex.

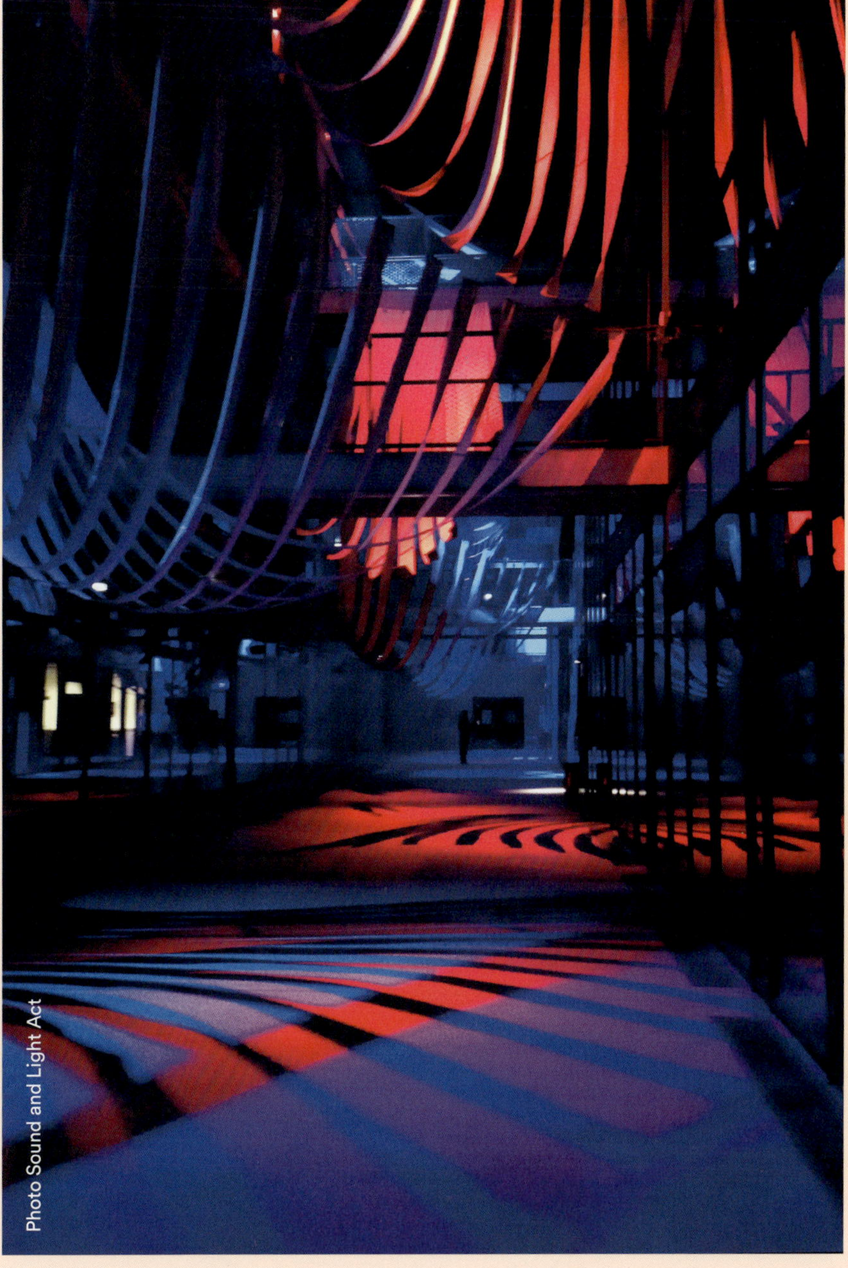
Photo Sound and Light Act

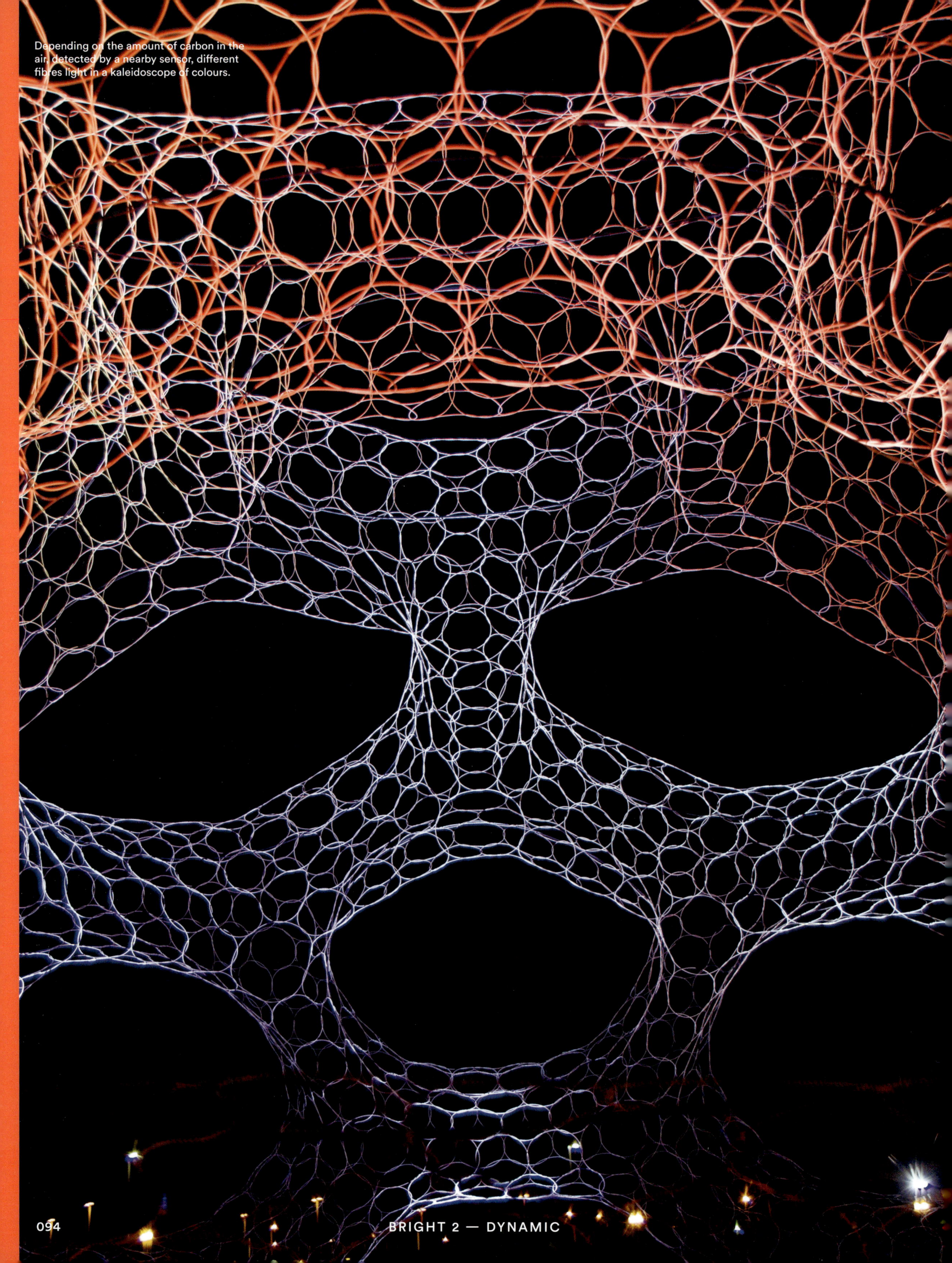

Depending on the amount of carbon in the air, detected by a nearby sensor, different fibres light in a kaleidoscope of colours.

SOL Dome

Loop.pH created a circular matrix of solar-powered LED lights to illuminate the SOL Dome in order to speculate on what the future of renewable energy could be, and how it may alter both the urban and rural landscapes.

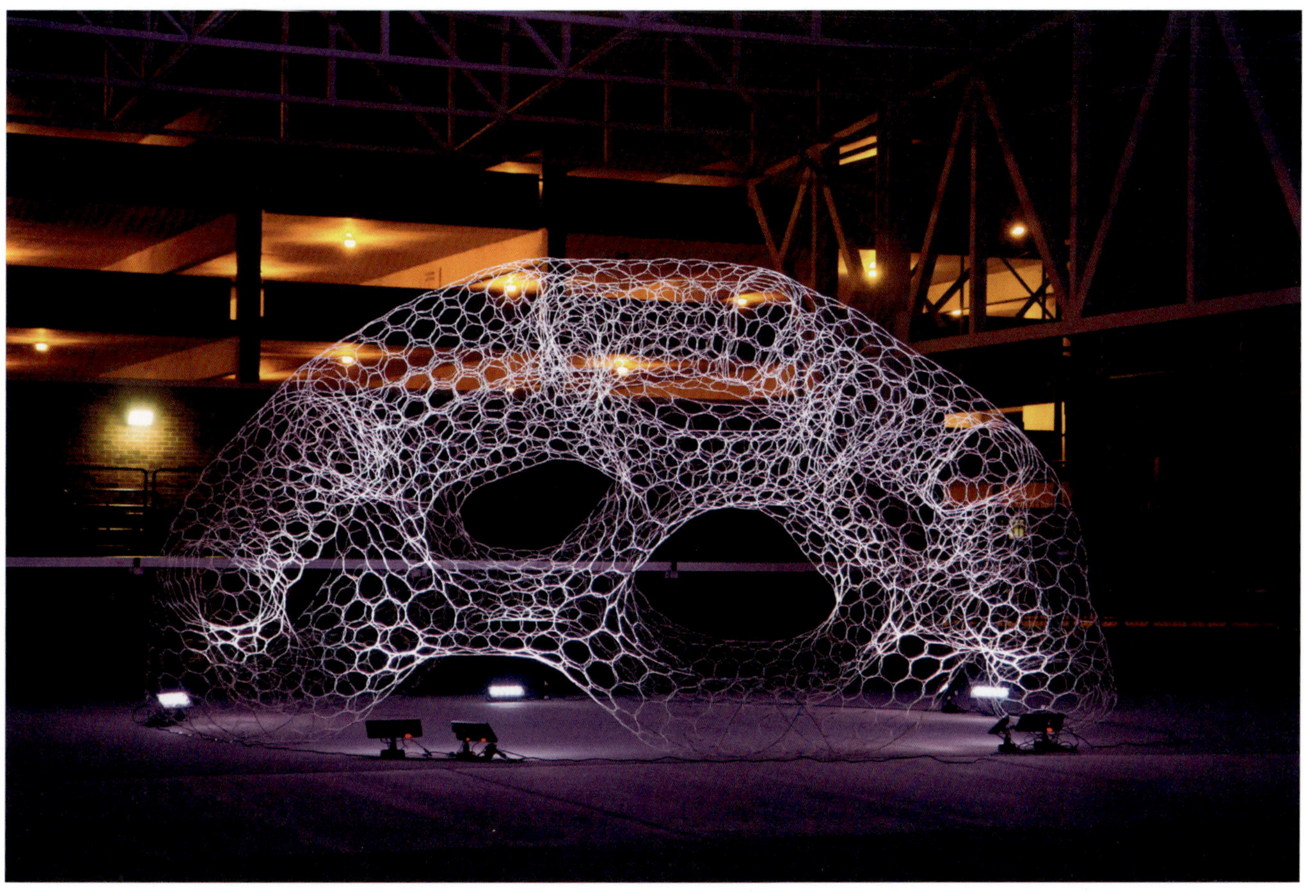

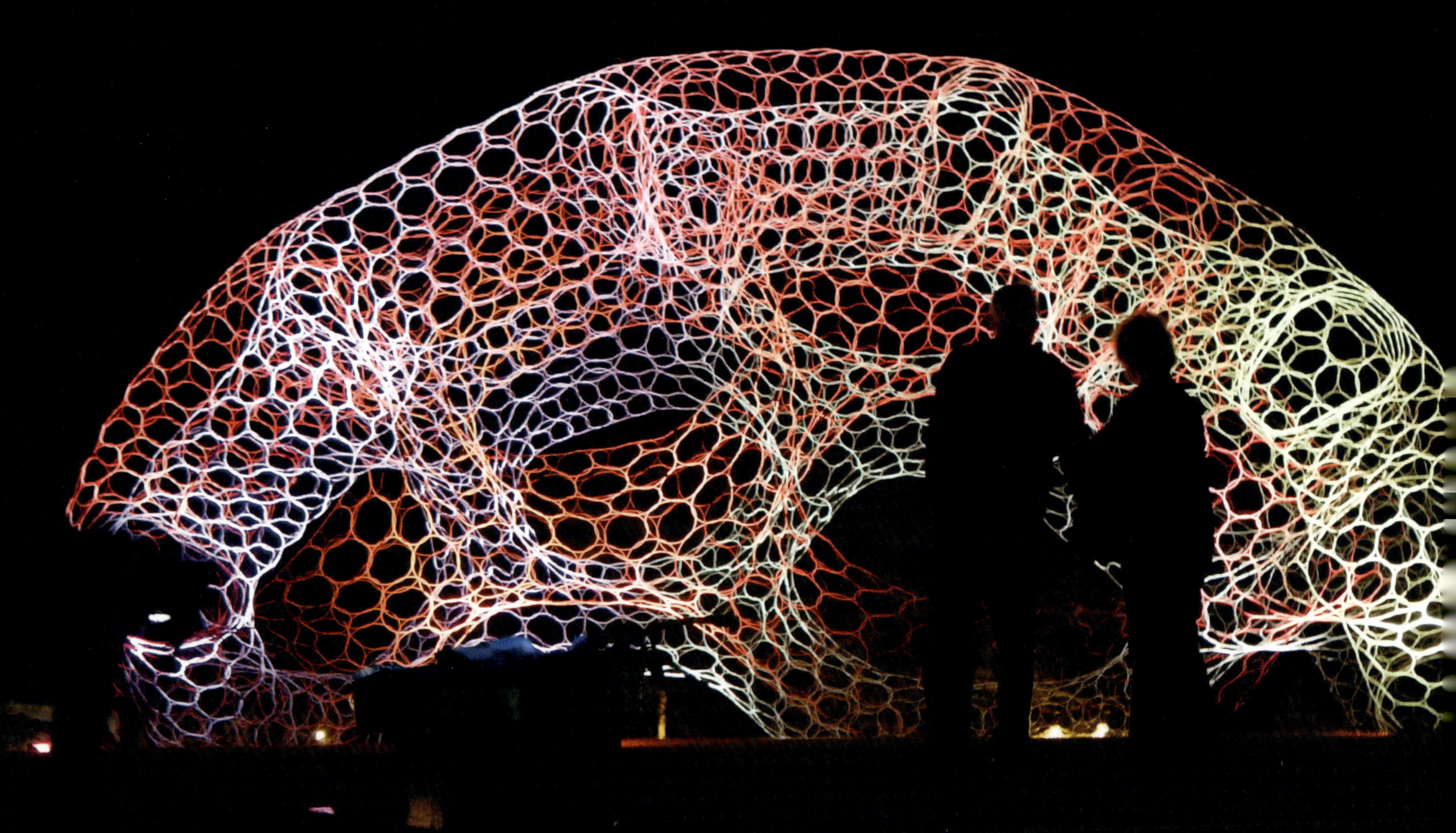

Designer
Loop.pH
Location
Michigan, United States
Client
Fall In...
Art and Sol Festival
Collaborator(s)/consultant
n/a
Manufacturer
n/a
When
October 2013

ANIMATED, RESPONSIVE LIGHTING SYSTEM MADE-UP OF A CIRCULAR MATRIX OF LEDS

The festival 'Fall In...Art and Sol' is a celebration of art, culture and science throughout Michigan's Great Lakes Bay Region in the United States. In October 2013, it featured the world's first major solar art exhibition. Creative studio Loop.PH featured a sculpture made of a circular matrix of solar-powered LED lights to illuminate the night sky.

The 4-m-high SOL Dome is a lightweight dome structure, 8 m in diameter and weighing only 40 kg. It was fabricated on-site over 4 days from thousands of individually-woven circles of composite fibre. The structure was animated and part of a responsive lighting system, lit by a circular matrix of solar powered LED floodlights. The rotational breathing rhythm of the light was driven by an on-site carbon dioxide sensor, part of Loop.pH studios ongoing research into creating environments that allow people to experience cycles of environmental data in public space. The underlying geometry and construction technique of the dome is based on chemical, molecular bonds between carbon atoms. When each fibre is bent into a circle it is like charging a battery, creating a taut, energetic structure.

Loop.pH speculates on what the future of renewable energy could be and how it may alter both the urban and rural landscapes. The studio creates environments that question what new behaviours, work forces and activity might emerge in an abundant renewable energy future – with an ultimate vision being an entirely new type of architecture that responds and adapts to its environment, similarly to a plant and its surrounding ecosystem. Dreaming of a living architecture that photosynthesises, moves and orientates in accordance to the sun; it is an architecture whereby the inhabitants can actively participate in its shape, form and function. —

When each fibre is bent into a circle it is like charging a battery, creating a taut energetic structure.

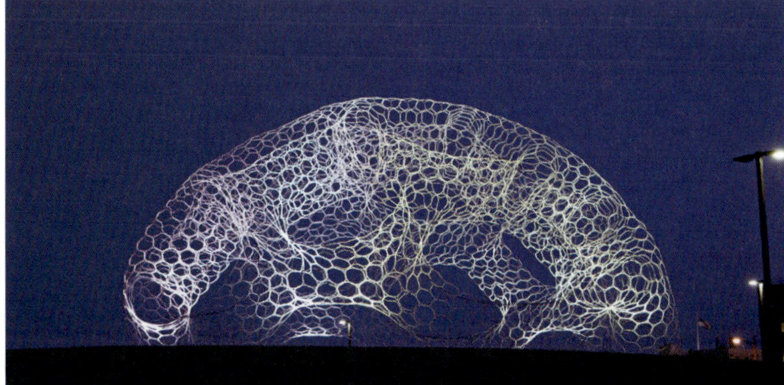

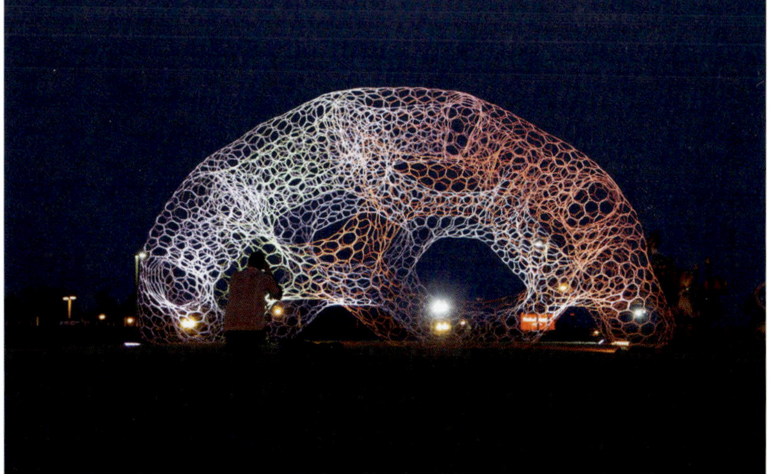

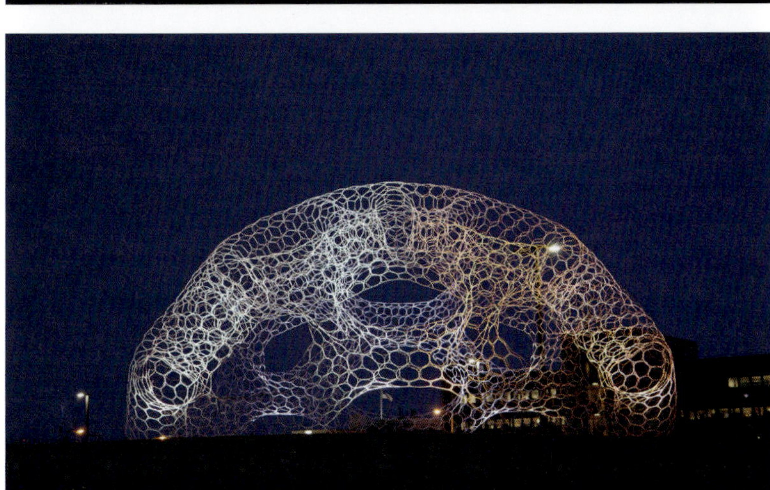

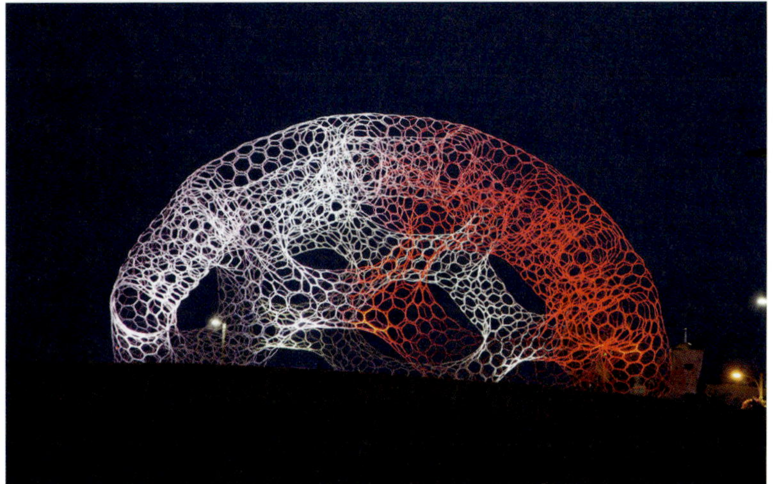

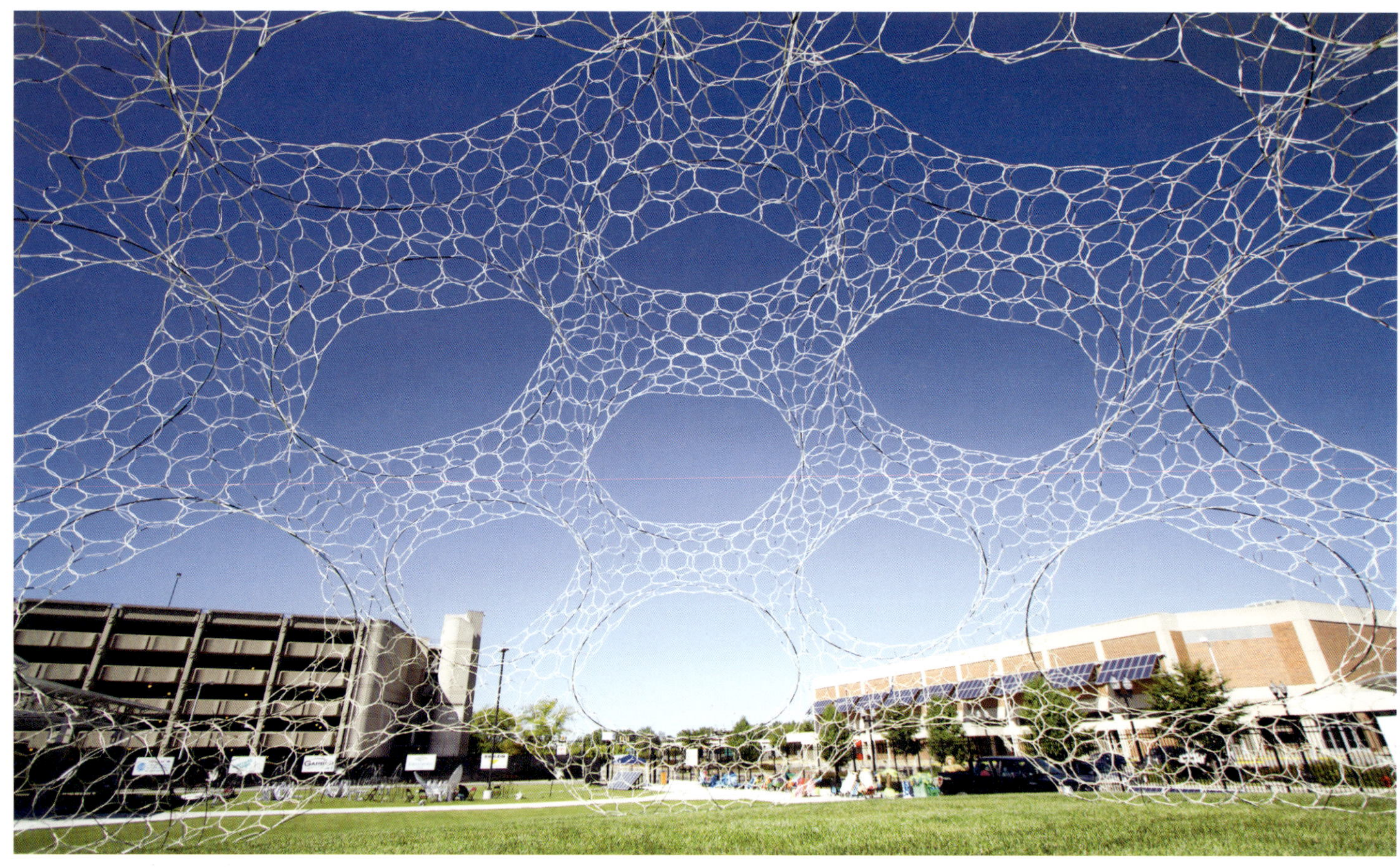

Set-up

Dimensions of this installation module were 8 m in diameter and 4 m in height. The fabricated system of composite fibres weighed-in at 100 kg.
Component parts included 'Archilace', LEDs, custom lighting sequence, electrical hardware, sensors, light-reflective fibreglass-reinforced rods with brass ferrules, and structural reinforcement carbon fibre rods.
Electronics included super-bright tricolour LED units (RGB colour change) and photovoltaic array mono-crystalline cells: Sunpower Maxeon with 22.5% efficiency).
Luminaires were positioned in a circular array of 10 RGB tricolour units that were located at the base of the structure. The maximum light output was 550 W.
LED units were used in an up-light position to animate the light-reflective fibreglass rods of the dome. Photovoltaic modules (12.5 m^2) were positioned alongside the LED units and charged with five car batteries to provide power for the whole installation.
Sensors included wireless CO_2 environmental sensors that were of scientific research grade (K-30 10,000 ppm) were mounted on nearby lampposts and transmitted live CO_2 data to the computer and DMX control. A 24-h data cycle was monitored and used to create a unique animation for each day.
Construction time was 4 days of assembly work for five people on-site.

The dome is constructed of Archilace, a unique method that is described as lace making on an architectural scale.

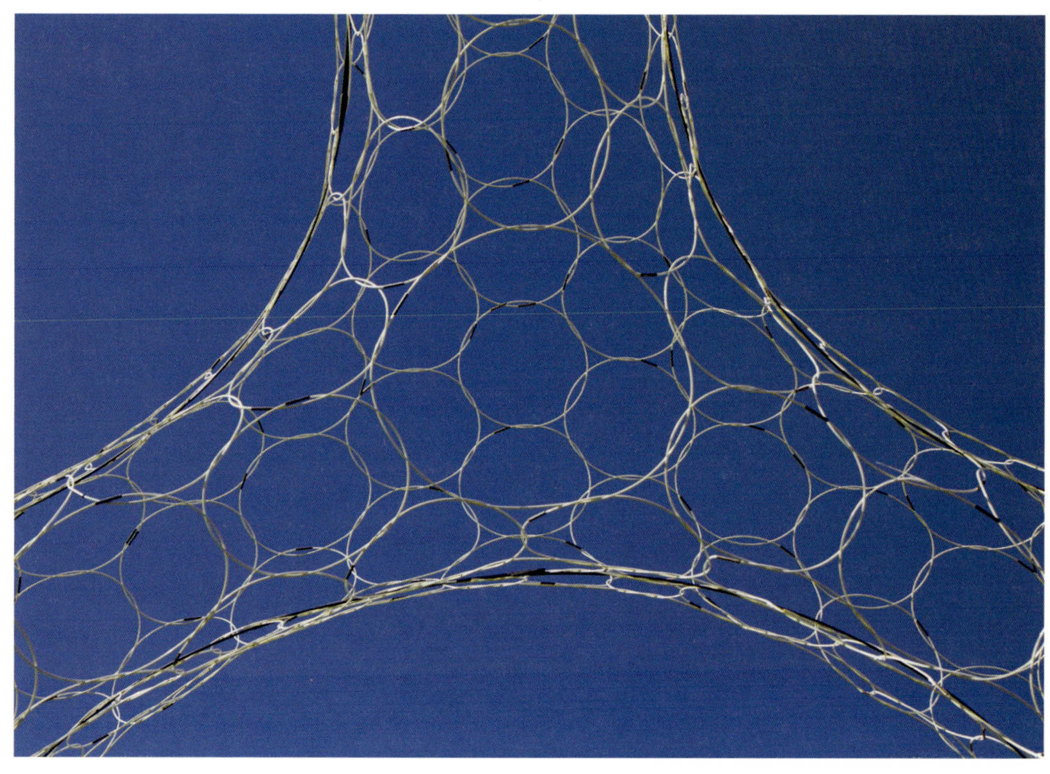

SIDE VIEW

TOP VIEW

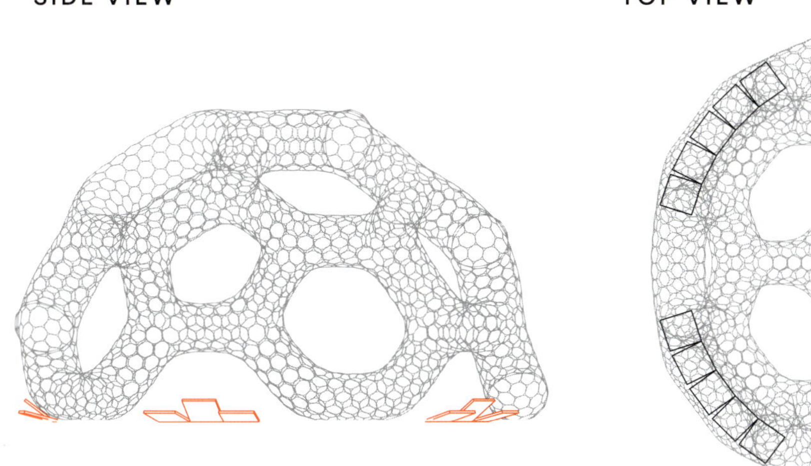

Specifics

The SOL Dome was created utilising solar-powered LED floodlights, connected to an on-site carbon dioxide sensor that caused their change in colour. Depending on the amount of carbon in the air, different fibres light up in a timed pattern, creating a kaleidoscope of colours.

The underlying geometry and construction technique is based on chemical, molecular bonds between carbon atoms. Large-scale solar energy supply will only be possible if an inexpensive storage mechanism can be found. Transferring solar energy into chemical energy (chemical bonds) is one of the most promising approaches. The dome structure is an example of this type of stored energy.

The basic fabric of the dome is called 'Archilace', a pioneering and unique method to craft space developed by Loop.pH over the past 10 years. Simply described as lace making on an architectural scale, Archilace combines a cutting-edge parametric design process with a hands-on crafting technique. The structures are fabricated on-site and woven from thousands of individually woven circles of strong composite fibre.

The dome was made as a super-lightweight structure that is modular and reconfigurable. In the development process, tests were carried out to obtain the correct form using the smallest number of modular parts that can be assembled in infinitely different ways.

Ease-of transportation and construction were also key aims for this installation. All parts fit into a small flight case. The entire structure is repairable and reusable. The composite fibres are temporarily fixed, meaning the structure can quickly and easily be taken down and rebuilt elsewhere, always being easily fabricated by hand on-site. No heavy machinery is needed to build on an architectural scale.

All electronics and lighting are powered by an array of photovoltaics. The scheme is controlled through a light sensor and is only on during dusk and the hours of darkness, using low power energy efficient LEDs.

ROTATIONAL BREATHING RHYTHM OF THE LIGHT DRIVEN BY REAL-TIME ENVIRONMENTAL DATA

DRAWING

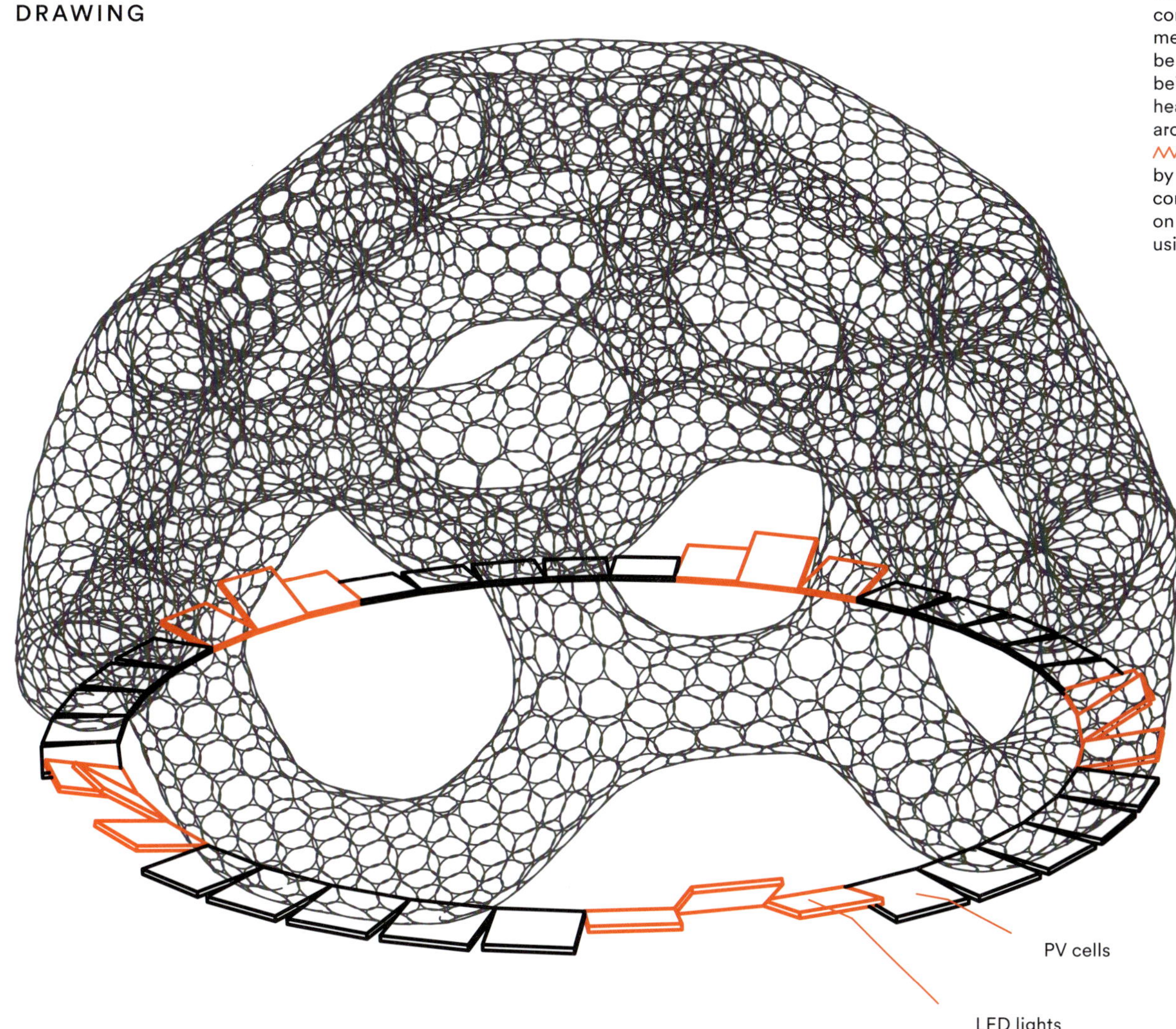

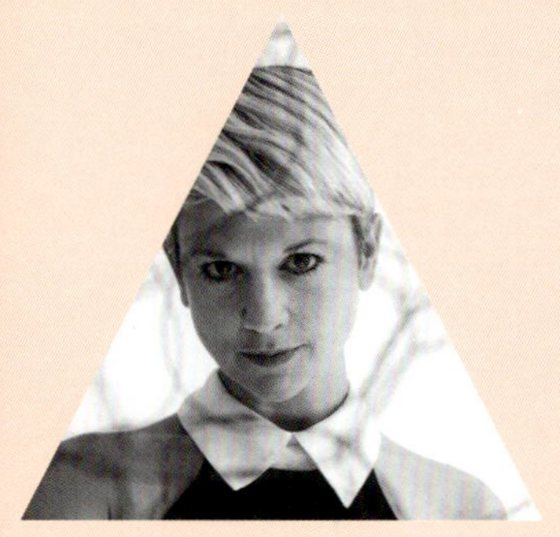

Loop.pH

Loop.pH is a London-based spatial laboratory experimenting across the fields of design, architecture and the sciences. The studio was founded in 2003 by Mathias Gmachl and Rachel Wingfield to form a new creative practice that reaches beyond specialist boundaries. The studio creates visionary experiences and environments that allow people to dream and re-imagine new visions for the future, working with clients to create brand experiences and public engagement initiatives internationally.

LOOP.PH

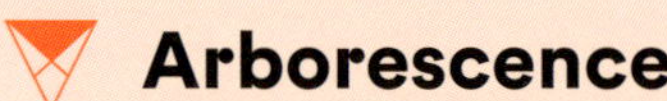

Arborescence

When **2014**
Where **Amsterdam, the Netherlands**
Client **Amsterdam Light Festival**

Arborescence is a sculpture that floated on one of Amsterdam canals during the light festival, like vegetation resembling the *Rhizophoraspecies*, luminescing and animating in response to the dynamic motion of water. The Loop.pH team refers to it as a 'bioengineered mangrove, genetically altered to bioluminesce, replacing traditional street lighting'. The project proposes a hybrid between trees and streetlights in order to create sustainable 'living lighting'. The team's goal is to bring cutting-edge scientific research out of the laboratories and into public space to inspire debate and action.

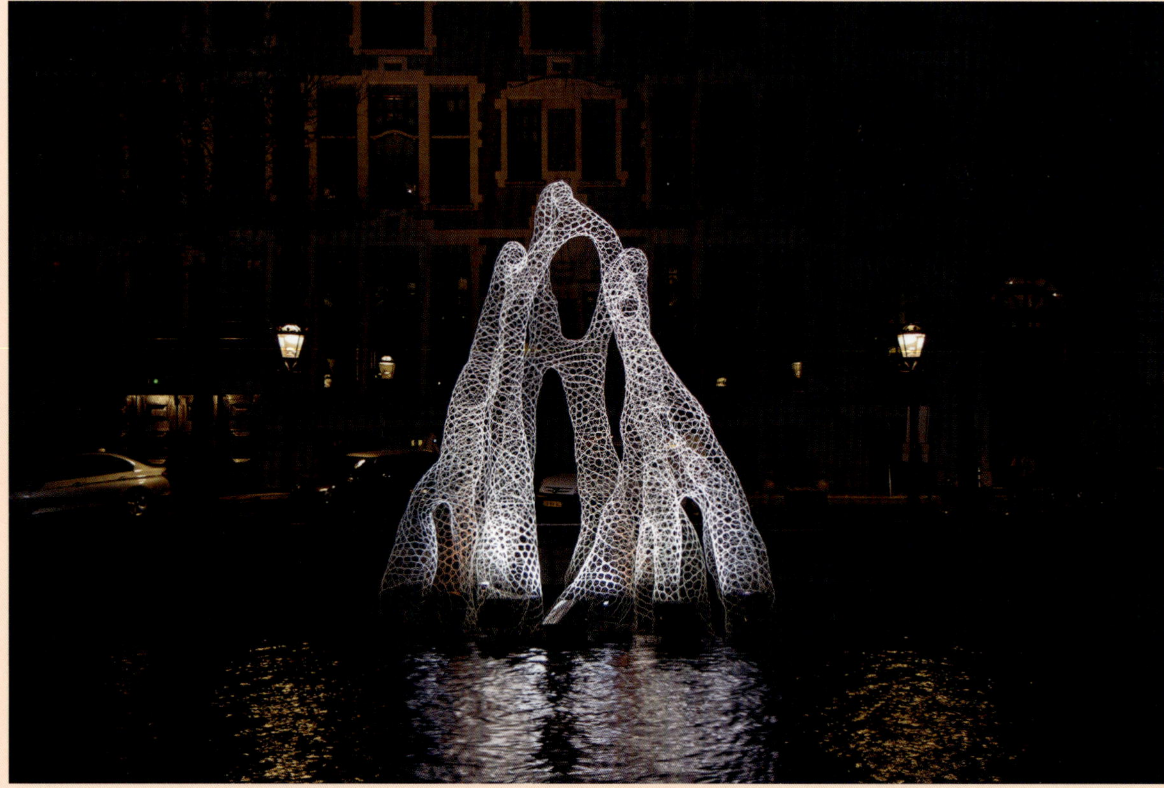

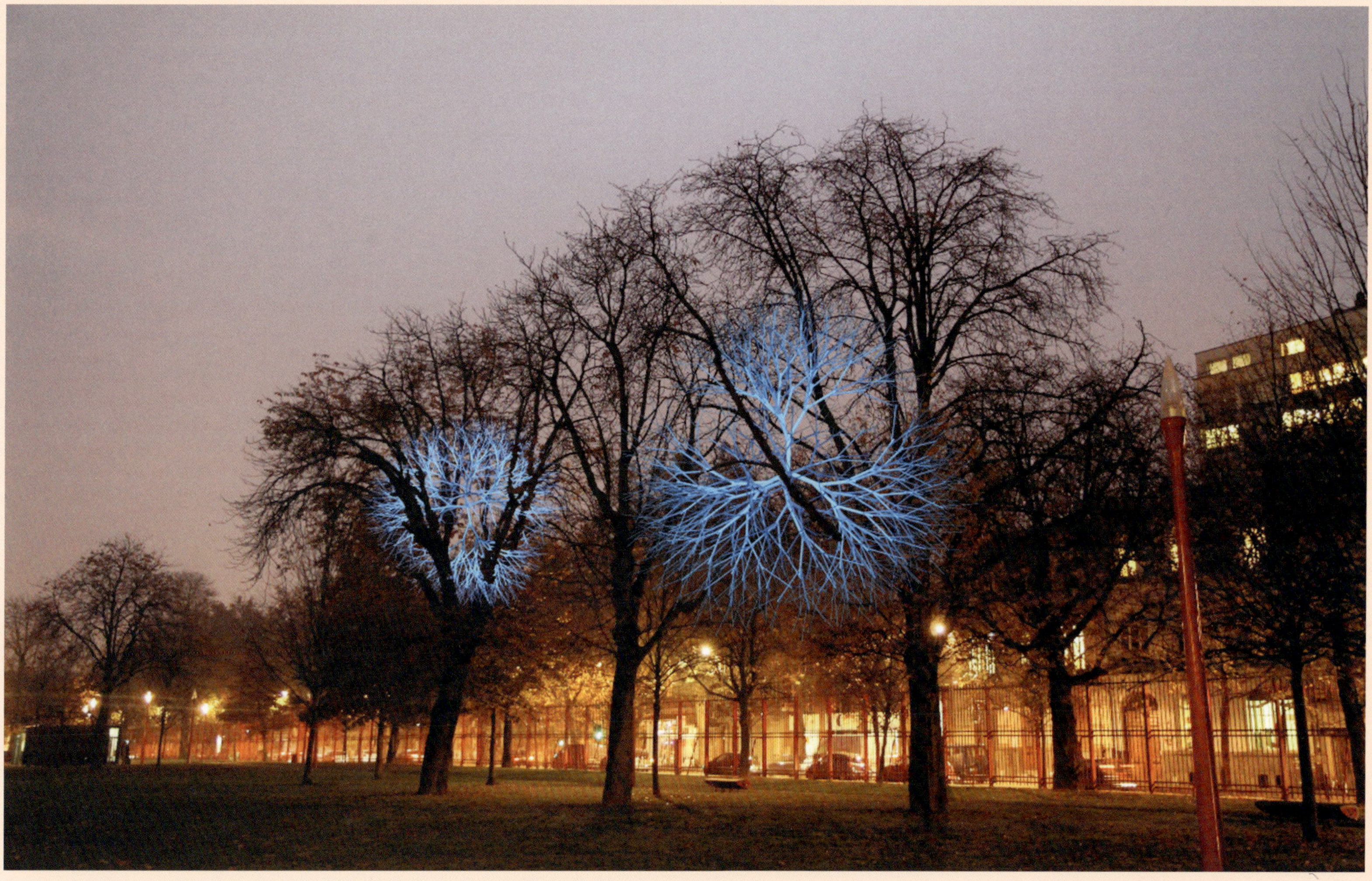

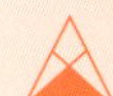

Tree Lungs

When **2012**
Where **Lille, France**
Client **EDF and Lille3000**

Tree Lungs was an installation of natural and technological materials that appeared to breathe in and out with an animated flickering of light. The breathing rhythm changed in synchronicity with the fluctuating carbon levels within the park.

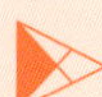

Kensington Archilace

When **2012**
Where **London, United Kingdom**
Client **Historic Royal Palaces**

Kensington Archilace is a responsive light environment, hand woven from electro-luminous fibres. Cutting edge digital tools were seamlessly combined with the ancient art of manipulating fibres into cloth. The lace includes over 4 km of electroluminescent wire and 12,000 Swarovski Crystals.

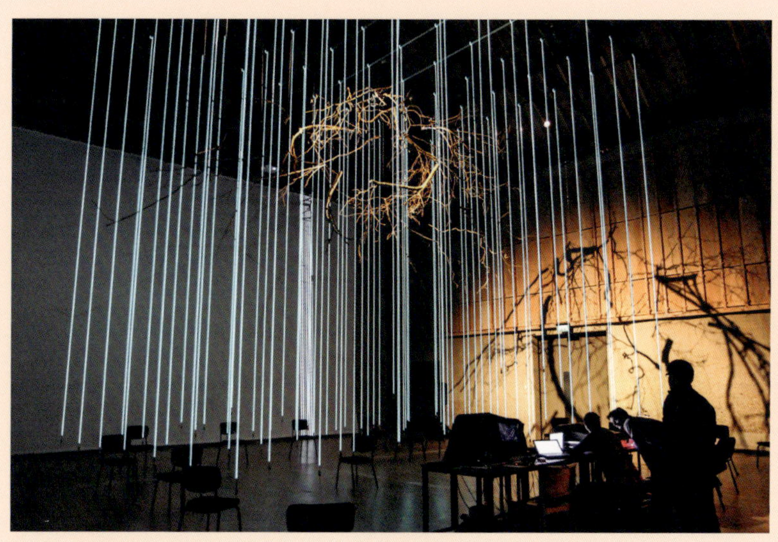

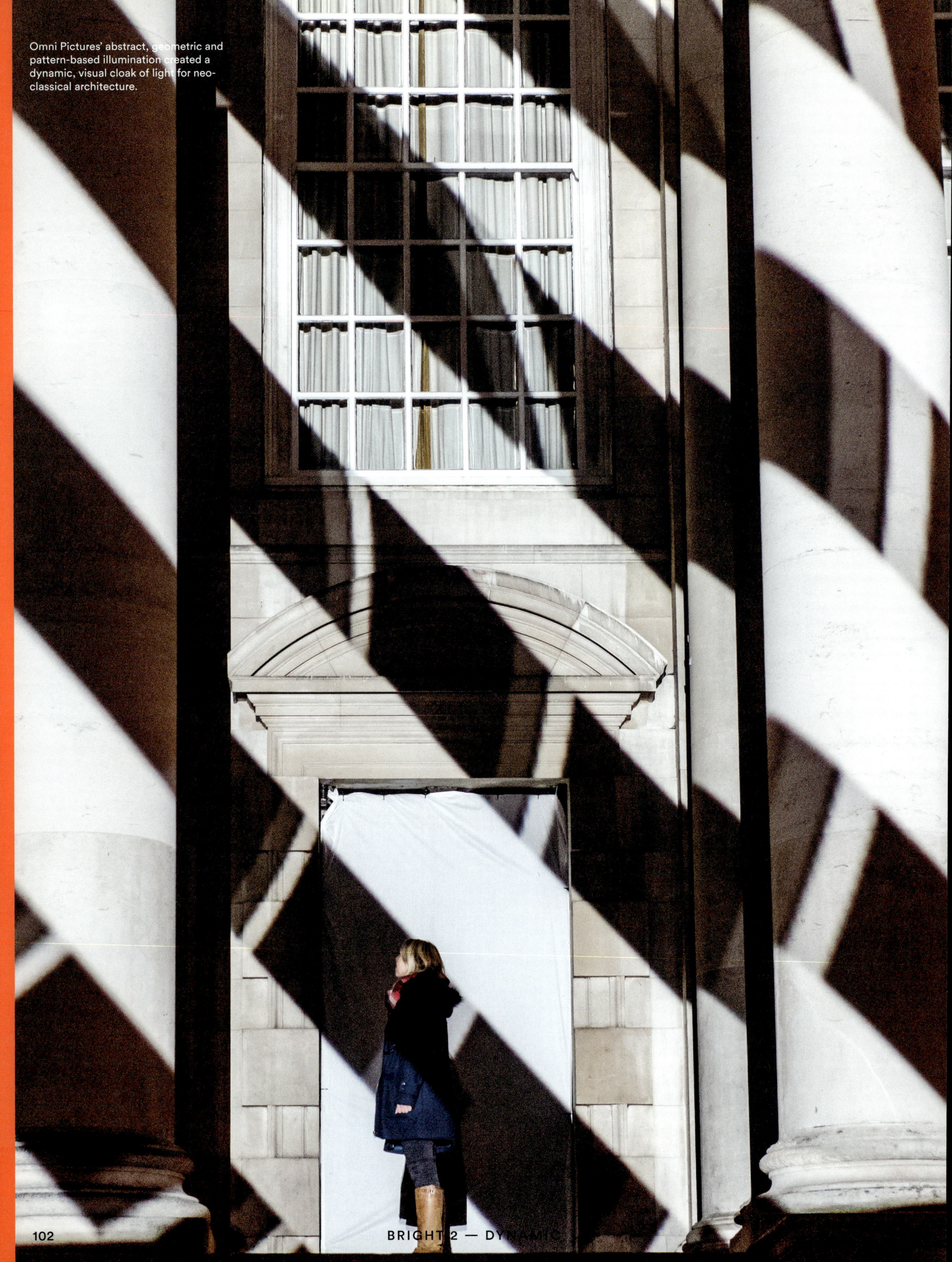

Omni Pictures' abstract, geometric and pattern-based illumination created a dynamic, visual cloak of light for neo-classical architecture.

Theatre of Illumination

An eye-catching, contemporary spectacle for Leeds' Light Night was created by **Omni Pictures**, with the team's geometric projection-mapped visuals illuminating the city's Civic Hall like never before.

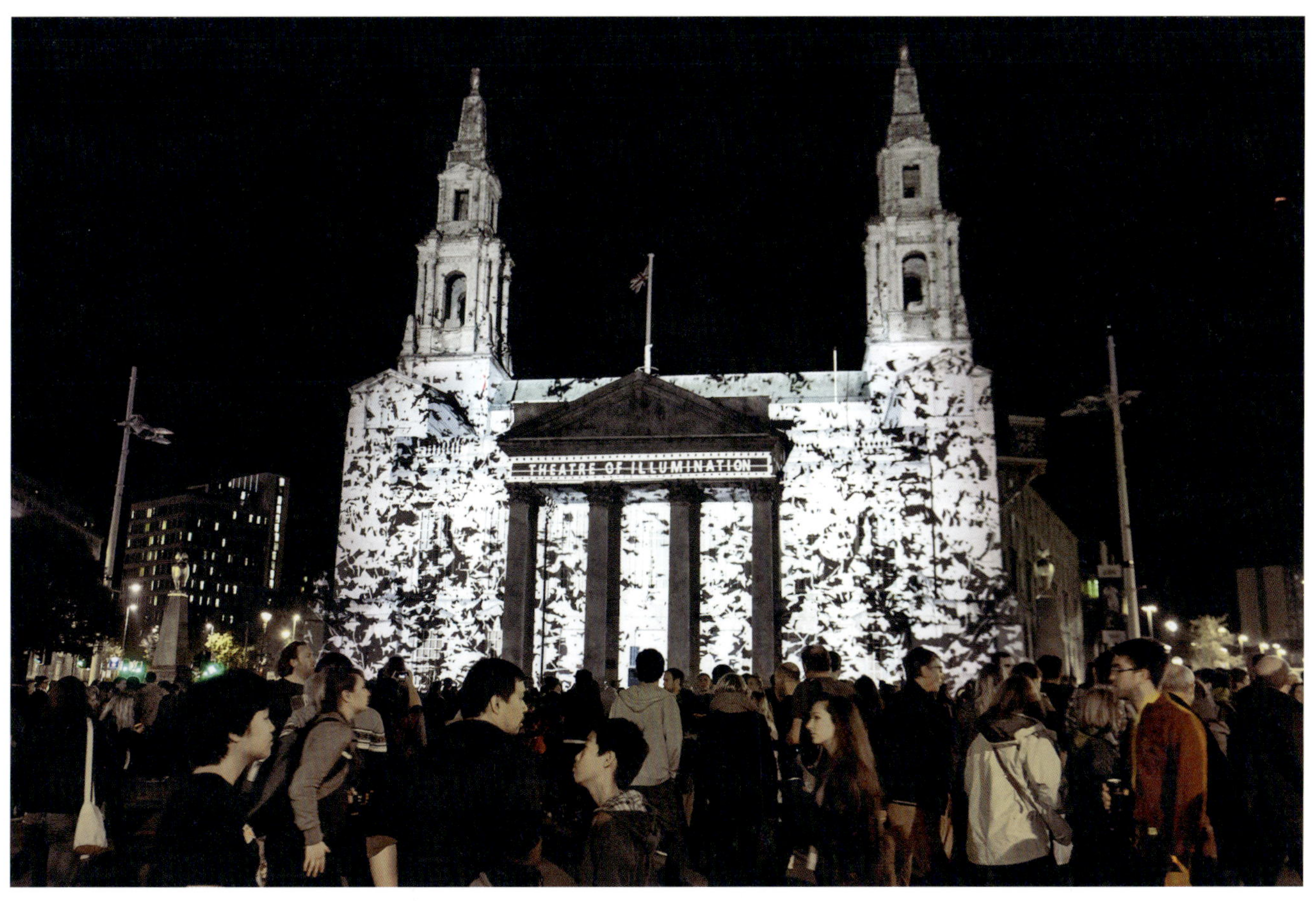

Photos Betty Lawless, Kippa Matthews, Will Simpson

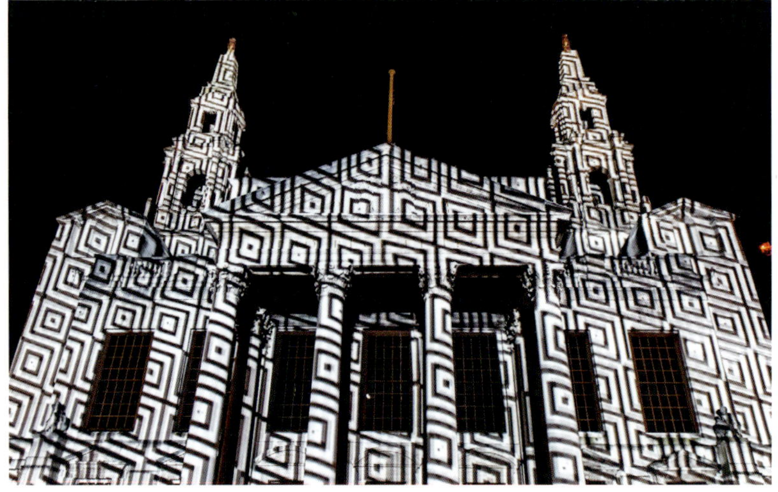

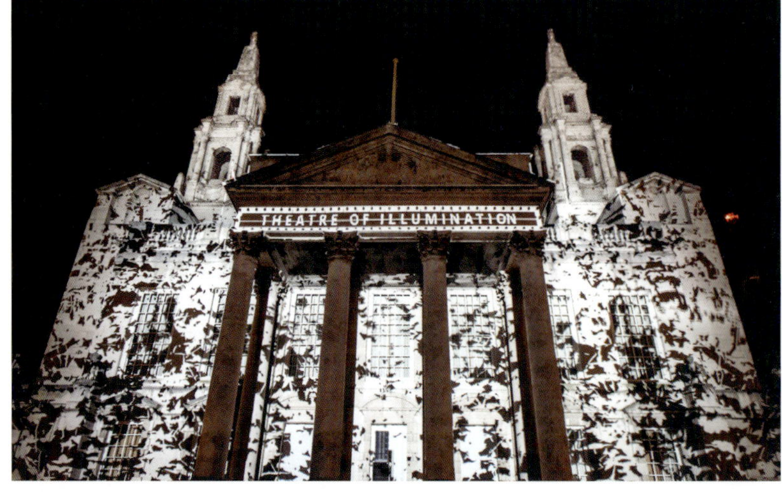

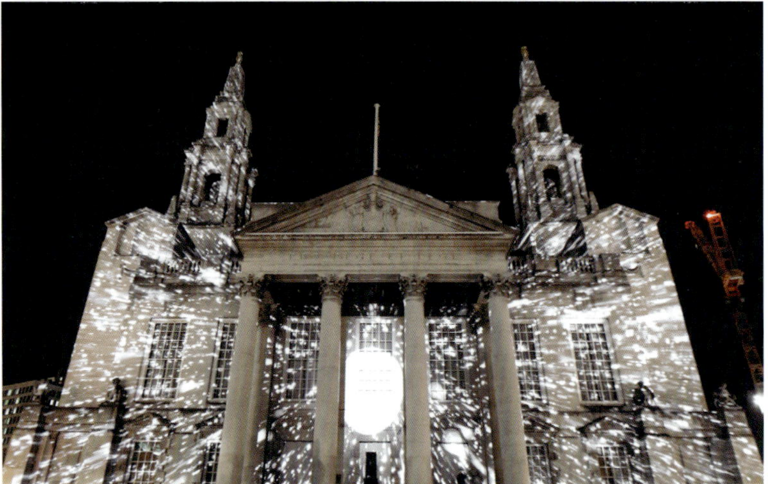

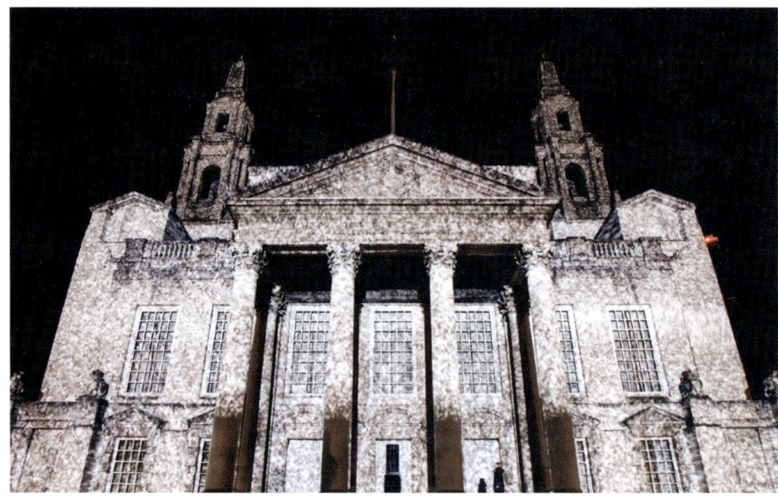

Each graphical movement required sound sync to maximise the impact and bring the piece to life.

AN EYE-CATCHING AND CONTEMPORARY SPECTACLE OF ILLUMINATING PROJECTION-MAPPED VISUALS

Designer
Omni Pictures
Location
Leeds, United Kingdom
Client
Leeds Inspired
Collaborator(s)/ consultant
QED Productions
Manufacturer
n/a
Date
October 2014

Theatre of Illumination is an ultramodern, projected white-light show that premiered on Vincent Harris' iconic neo-classical Civic Hall building in Leeds, United Kingdom for Light Night 2014. With the explosive energy of a firework display, the installation used projection-mapping to drench the architecture's facade in a monochromatic sheen. Audiences were taken on an exhilarating journey of geometric design and pseudo-optical illusions, with an accompanying musical soundtrack swirling around them. This spectacle of dynamic light and sound created an experience like no other.

Omni originally won the commission having responded to the call-out from Leeds Inspired, a regional arts funding body. It was a specially commissioned piece of work, sitting at the heart of the Light Night festival. The brief asked for the building – which was completed in 1933 – to be dressed up for the next century, mapping it in light and sound to create a futuristic new look. The team aimed to engulf the audience with visuals that created wonder in people's minds, captured their curiosity and attention, and threw them into a monochromatic world of geometric design and objects filled with life. Comments Omni director Will Simpson, 'From very early on, we decided to avoid the conventions of past.'

Theatre of Illumination utilised white light. Other projects for the studio when working on the stage are often constrained by the projection power available, visuals need to be far higher in contrast than they would normally be. For Theatre of Illumination, Omni decided to use black-and-white design to maximise the contrast and impact. The effect on the building was going to be one of revealing and concealing different architectural features. The white would reveal elements of the building whilst the black would hide it, meaning you could lose site of the structure itself when focused entirely on the mesmerising shapes and forms played out in front of the audience. —

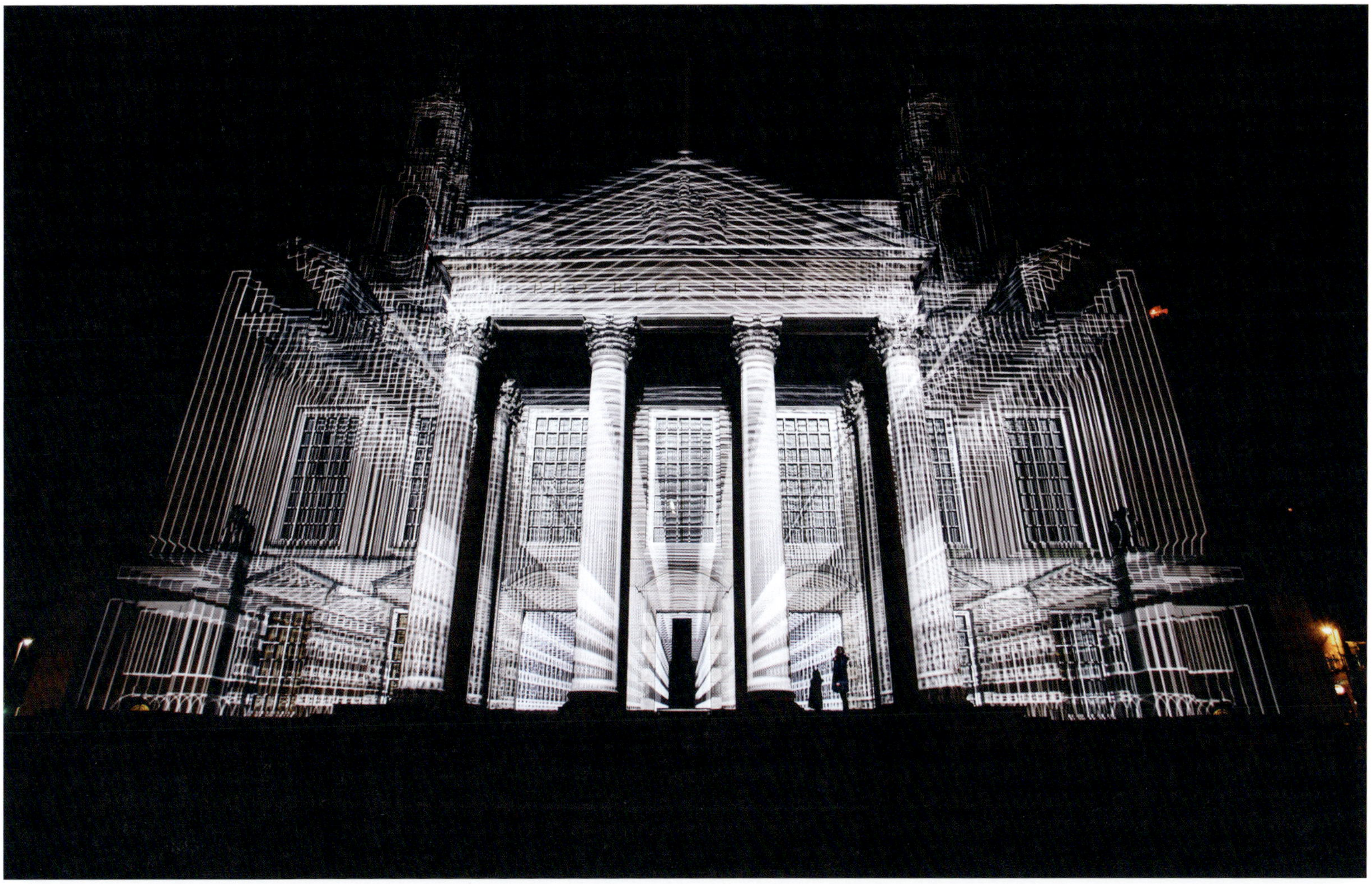

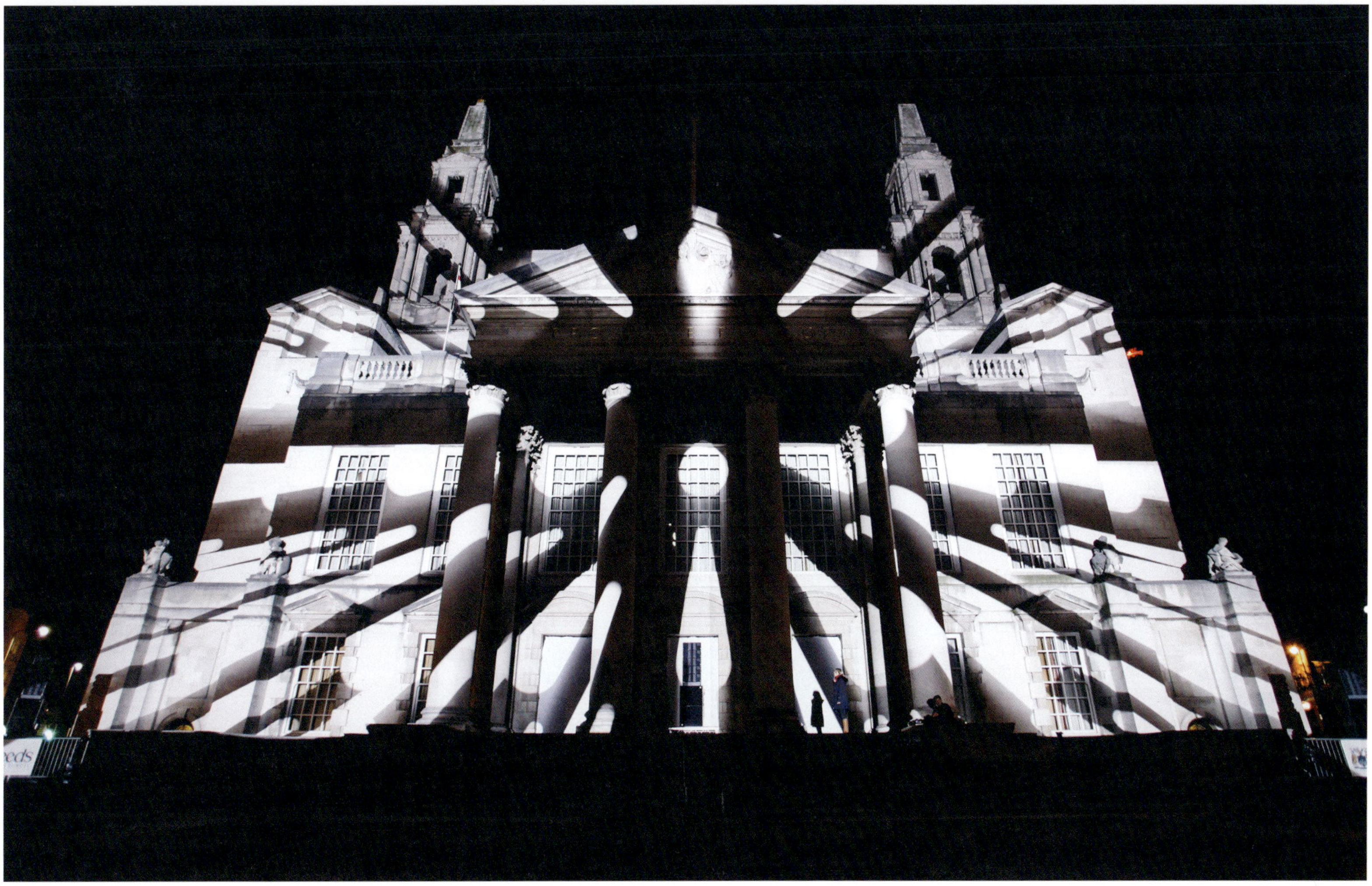

BUILDING ELEVATION

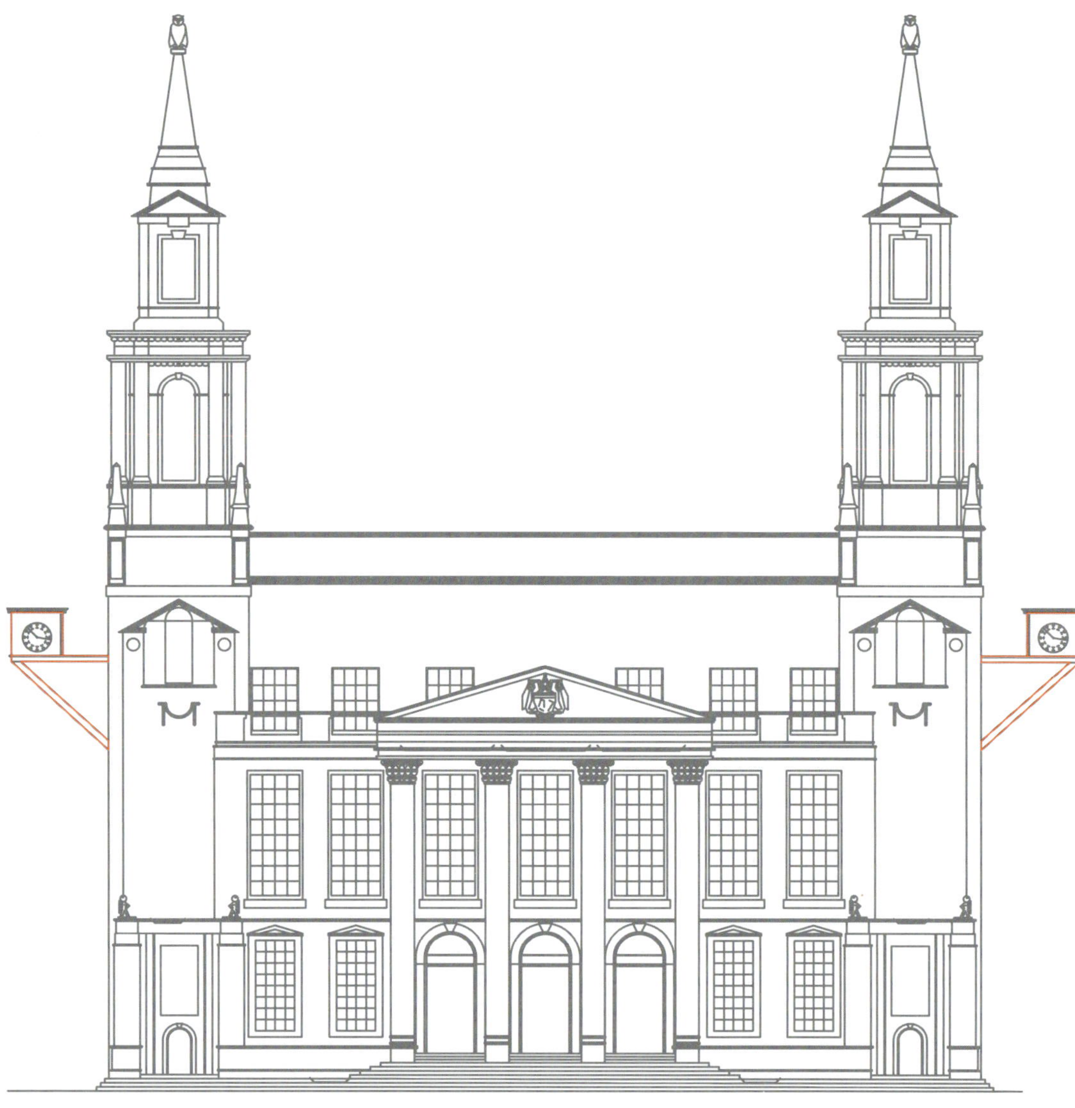

Specifics

As the bespoke video content had animations mainly comprised of monochromatic geometric shapes, it was a challenge to maximise the contrast of the black-and-white projections on the architecture in order to enhance the effect and the impact of the artwork. Negotiating the balance of projection and stage light in this instance meant the content needed to be far higher in contrast than it would normally be. In many ways the decision to use black and white was a result of the limitations of the medium.

In order for this project to be successful, the template for the content had to be as accurate as possible, to avoid line-up problems. Ideally, a 3D scan and model of the building would allow the creation of a flat UV template for the content, but budget constraints would not allow it. So instead, a head-on photo map of the building was used which suited the workflow.

The success of this project hinged on making the most of the unique black-and-white content. This meant boosting the brightness as much as possible. A 6-output/projector array was used consisting of a 2 x 2 landscape blend on the main section of the building and a projector on each of the towers in portrait. This was then doubled-up for the projectors on the main section. In total, 10 x 20,000 lumen projectors were required in order to provide enough firepower to produce the stunning results on the entire building.

A D3 media server was used as it provides the best functionality and tools for projection mapping, such as multiple layers of keystone and warping, masking and blending per output, true cross server sync and frame lock alongside instantaneous master/slave backup. In addition, the D3 systems provide playback of up to two simultaneous layers of 8k content without dropping a frame. Playback of the 4k native content allowed for an ultra-smooth show.

The bespoke soundtrack for the visuals was created in-house using a selection of city recordings and other material. Everything that happened visually was going to need very tight sound design.

The soundtrack was presented in 5.1-surround-sound with a D&B audio system which engulfed the audience in soundscapes to match the visuals and journey.

MAPPING THE NEO-CLASSICAL ARCHITECTURE IN LIGHT AND SOUND TO CREATE A FUTURISTIC NEW LOOK

CONSTRUCTION VISUALS

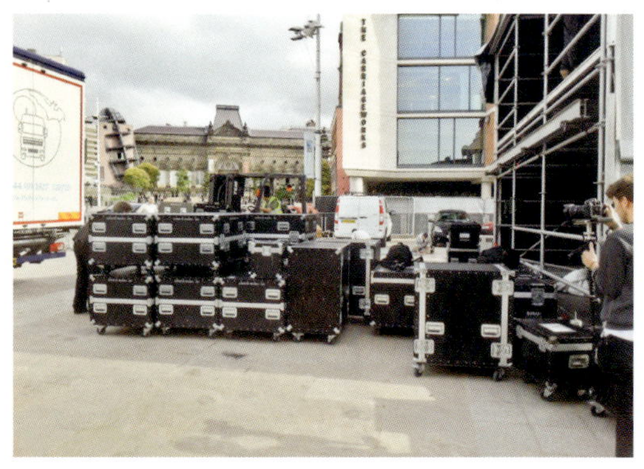

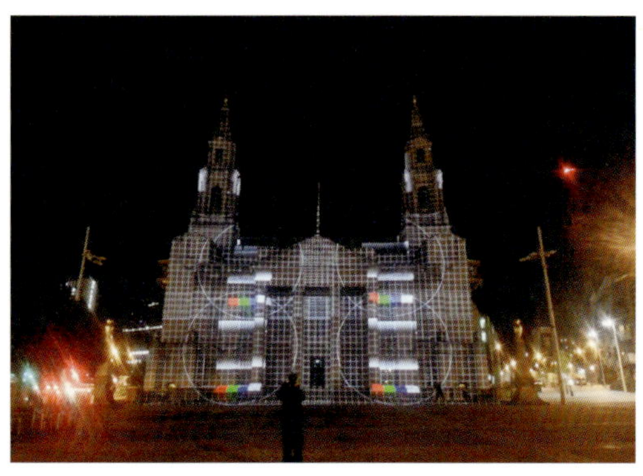

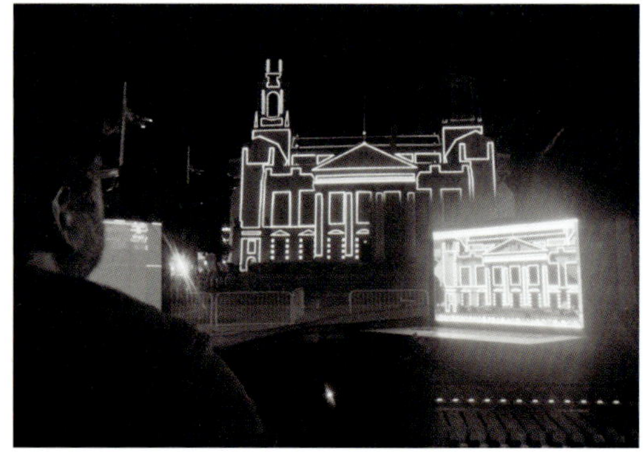

FRAME RENDERS

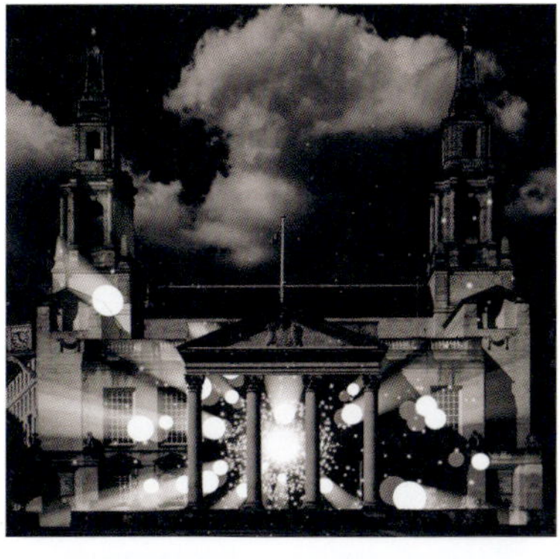
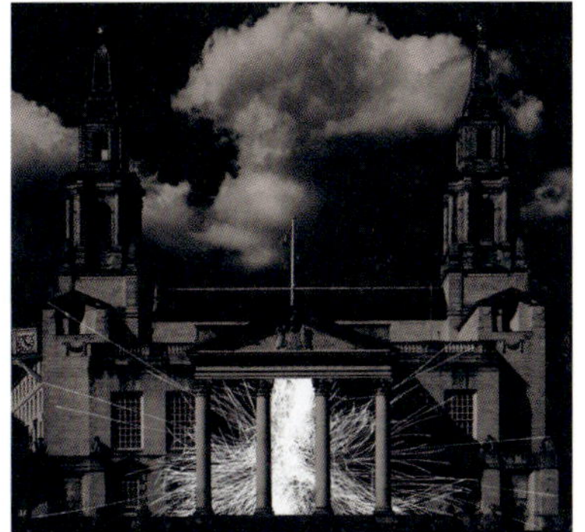
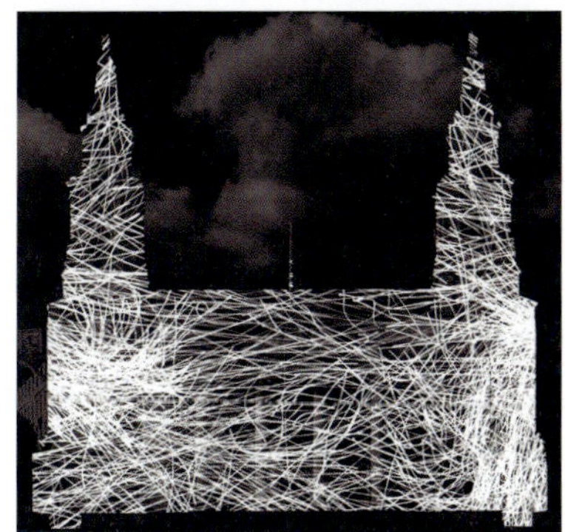
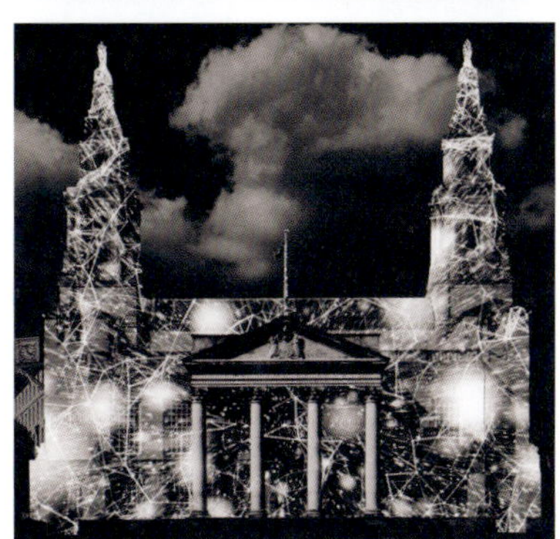
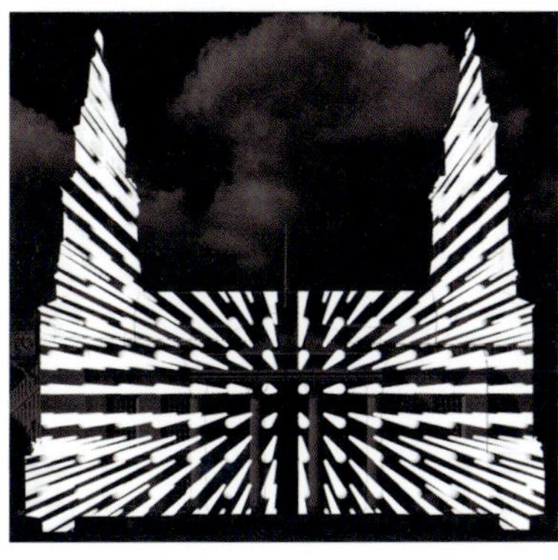
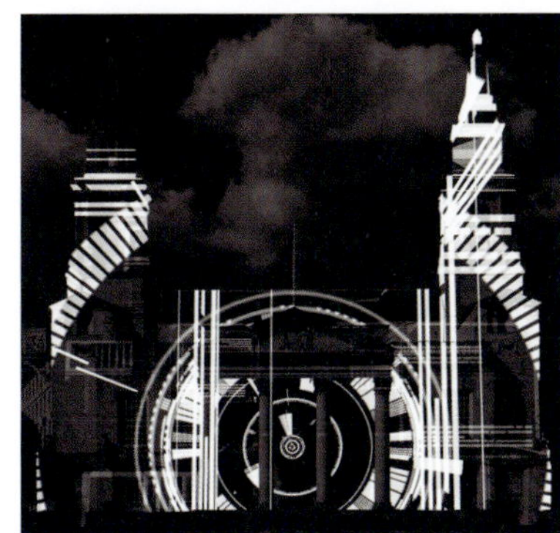
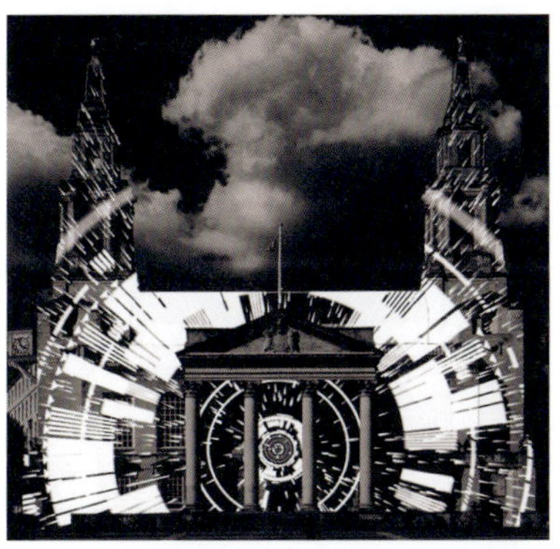
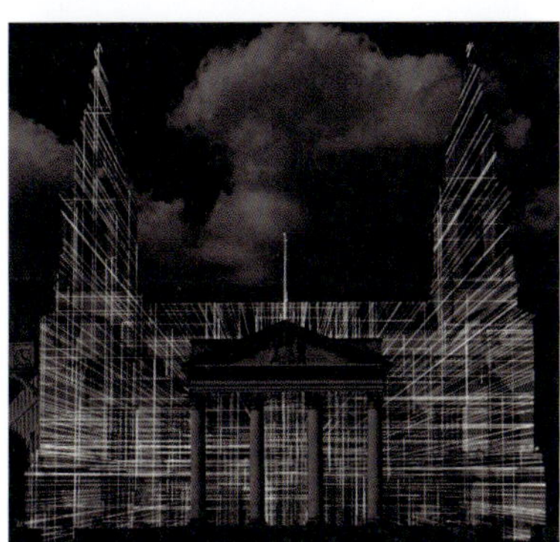
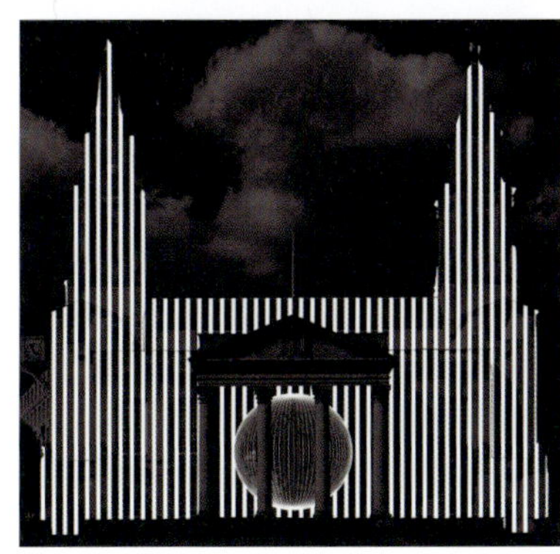
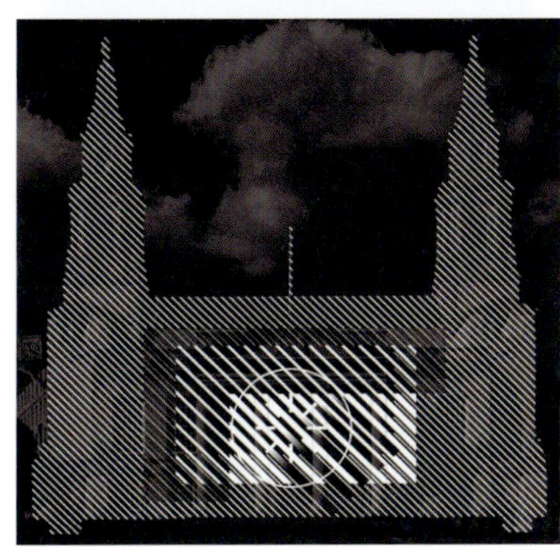

HDR VISUAL

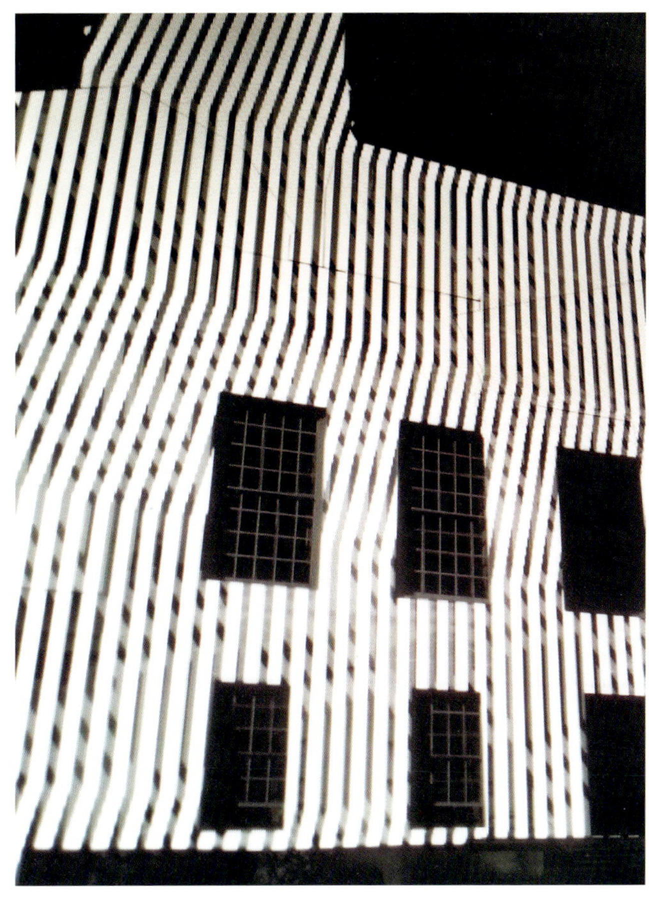

Set-up

Used light included 10 x 20,000 lumen projectors (200,000 lumens in total). Total area for illumination was 410 m^2.
Lighting was controlled by the D3 media servers.
Resolution was key so all animations were produce at 4k.
Media server type was a D3 system that, although playback was at 4k, this would allow playback of up two simultaneous layers of a total of 8k content without dropping a frame.
Project duration took 3 months for the creation of content, construction on-site was 1 day, and the event took place over 2 nights.

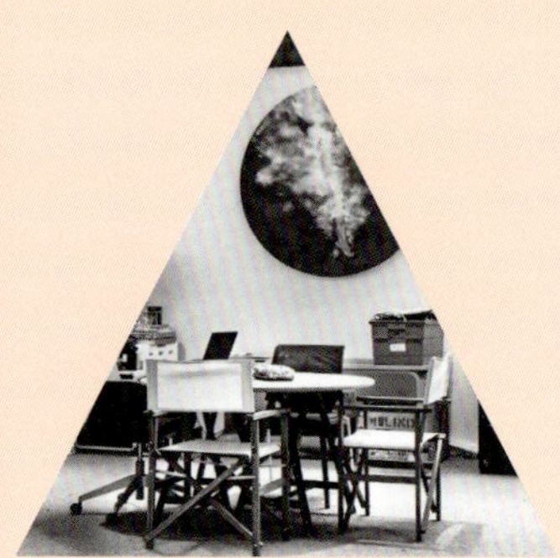

Omni Pictures

Omni is a Leeds-based creative studio that combines a wide range of disciplines, including motion graphics, film and video production, live performance and projection design. Founded in 2012 by Will Simpson, the studio uses a collaborative approach to projects with a passionate, inspired and evolving team of filmmakers, designers, artists, musicians and technologists. Creating work that is both complex in its craft and artistry, yet clear in communication, the critically-acclaimed studio collaborates with companies such as Arts Council England, BBC, Channel 4, EON, Laurent-Perrier, MTV, Orange, Saatchi & Saatchi, and many more.

OMNIPICTURES.COM

Seizure Film

When **2013**
Where **Yorkshire Sculpture Park, United Kingdom**
Client **Arts Council Collection, Southbank Centre, London**
Credit **Roger Hiorns, Seizure, 2008**
Commissioned by **Artangel and the Jerwood Charitable Foundation**

Omni was responsible for creating the accompanying film to Roger Hiorns' work as it was relocated from London to Yorkshire Sculpture Park in 2013. The immersive work was initially created in 2008 using 75,000 litres of liquid copper sulphate, which was pumped into a former council flat in London in 2008 to create a strangely beautiful and somewhat menacing crystalline growth within the abandoned dwelling. The work, weighing over 31 tonnes, was extracted from the property in February 2011 and subsequently transported to Yorkshire Sculpture Park where it is currently on display.

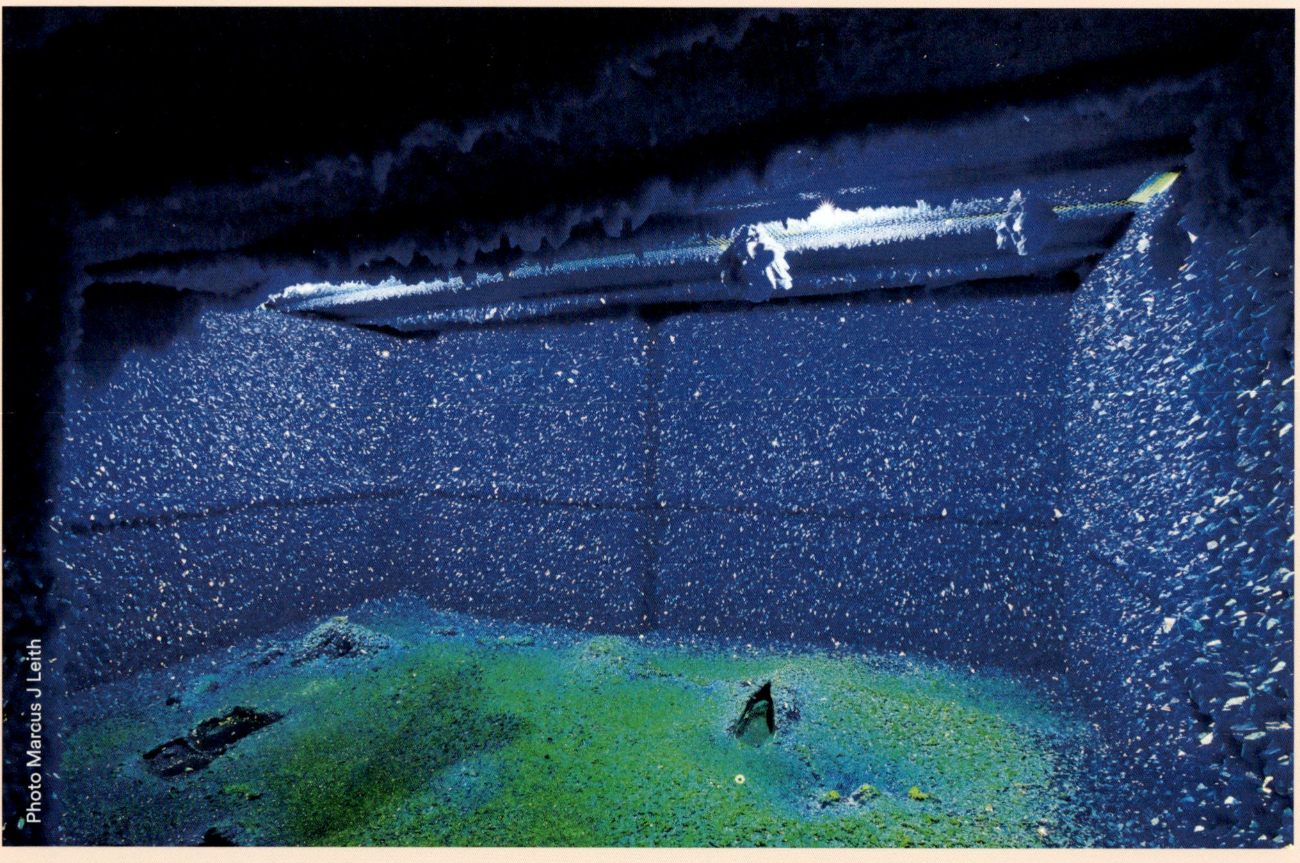

Photo Marcus J Leith

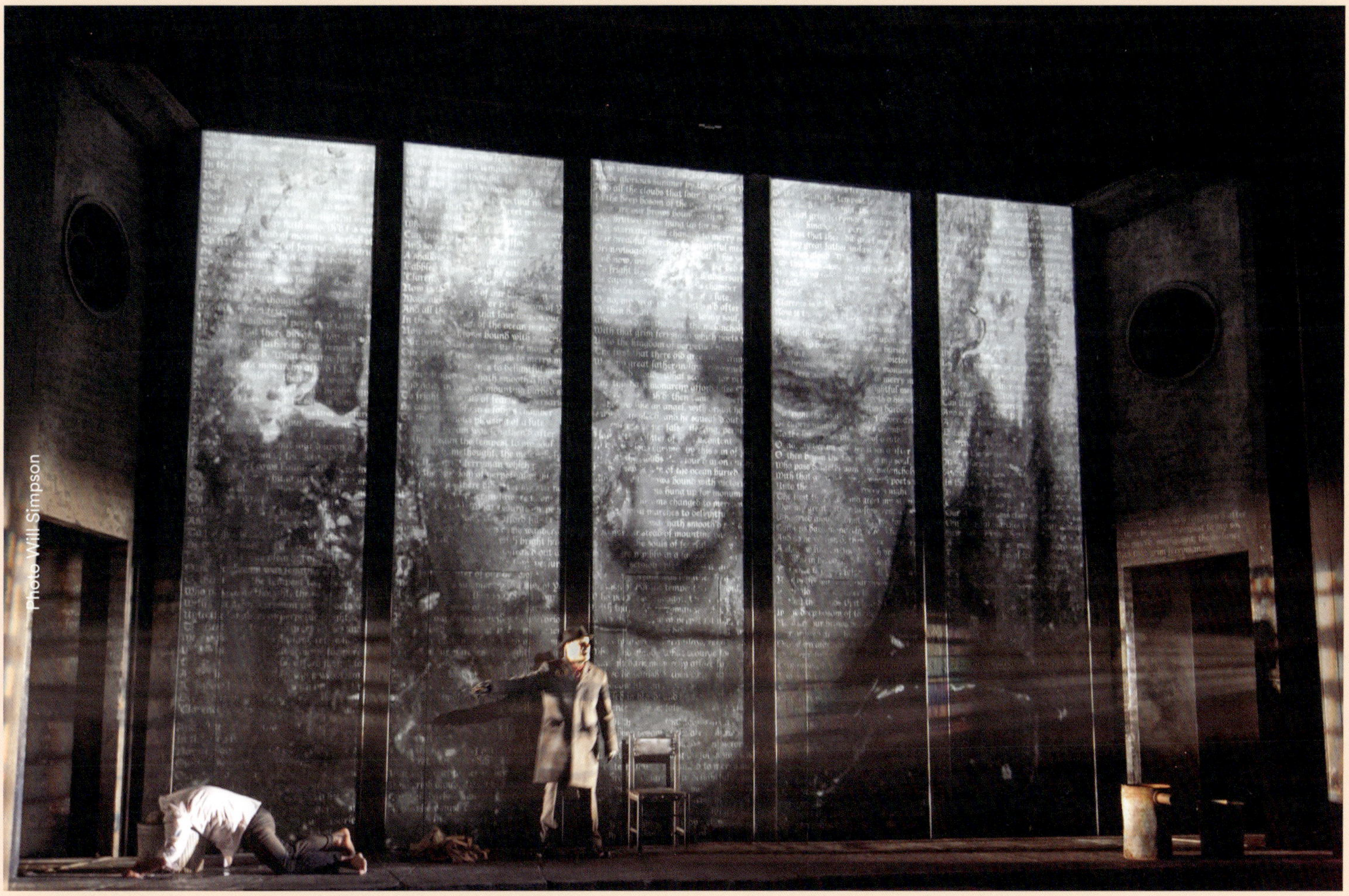

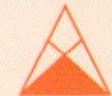

Photo Will Simpson

Richard III Projection Design

When **2013**
Where **Nottingham, United Kingdom**
Client **Nottingham Playhouse**

The first major staging since the skeleton of Richard III was discovered, Shakespeare's Richard plots, calculates and murders his way to the throne in one of the most chilling portrayals of political tyranny ever seen on stage. Omni created the mapped visuals to the set. From live cameras, to battle fields to storms and moody vistas of destruction.

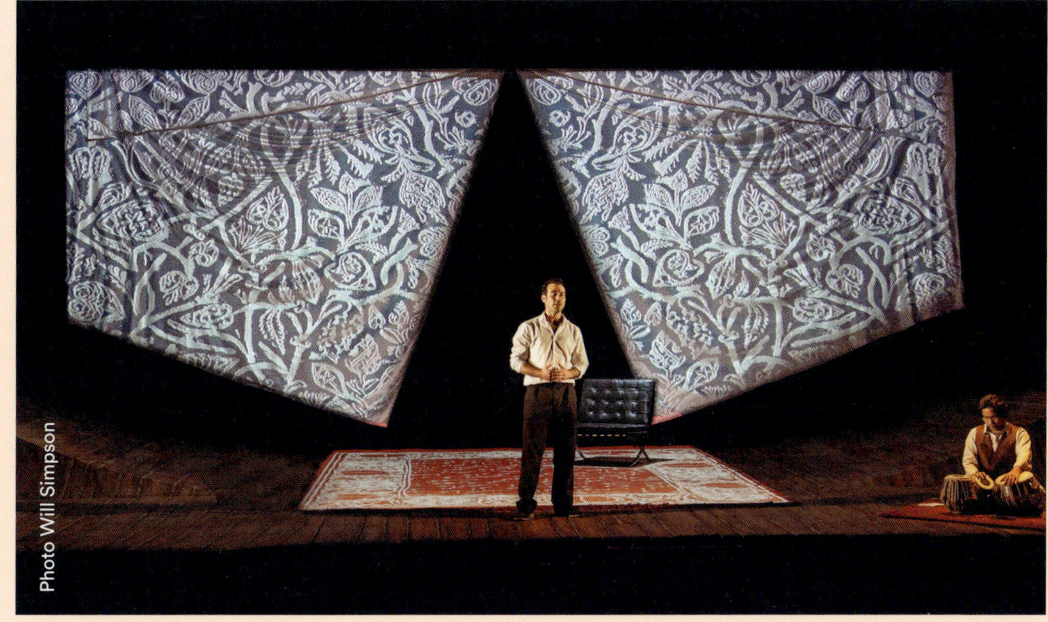

Photo Will Simpson

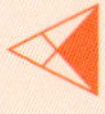

Kite Runner Projection Design

When **2013**
Where **Nottingham, United Kingdom**
Client **Nottingham Playhouse**

Omni created the projection design for the critically acclaimed European stage premiere of *The Kite Runner* at Nottingham Playhouse. It has since toured nationally. The Omni team used downward projection to provide rug patterns that change over time and place and the fluidity of the changing projection on the drapes meant the play could move at its own pace to any location without waiting for scene changes.

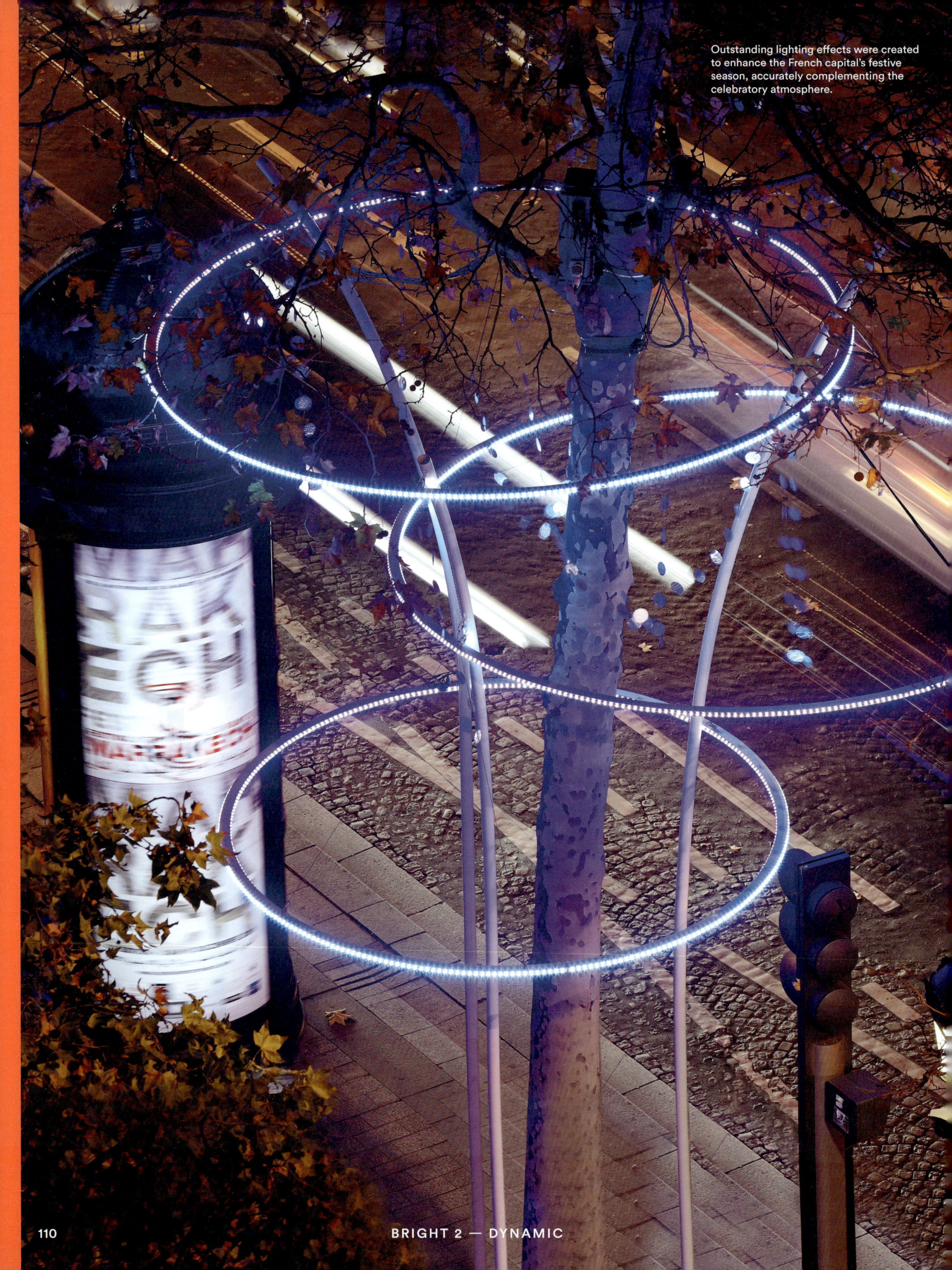

Outstanding lighting effects were created to enhance the French capital's festive season, accurately complementing the celebratory atmosphere.

Tree Rings

Created by Koert Vermeulen and Marcos Vinals Bassols of **ACT Lighting Design**, Tree Rings has illuminated the Champs-Elysées in Paris for a flurry of festive seasons, giving the most prestigious avenue in the French capital city that extra sparkle with a light installation that symbolises unity between man and nature.

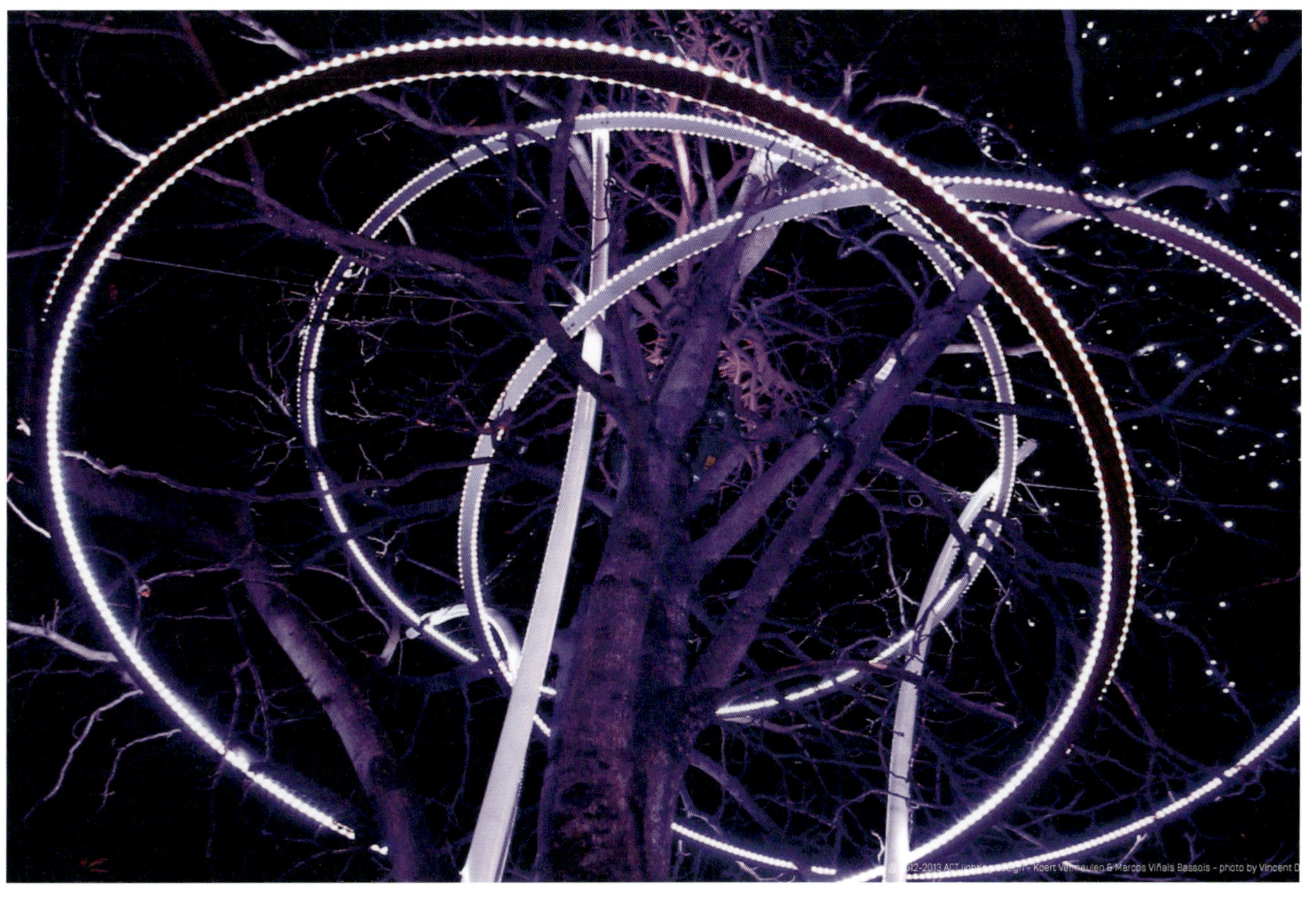

Photos Didier Boy de la Tour, Vincent Dumesnil, Eric Mercier

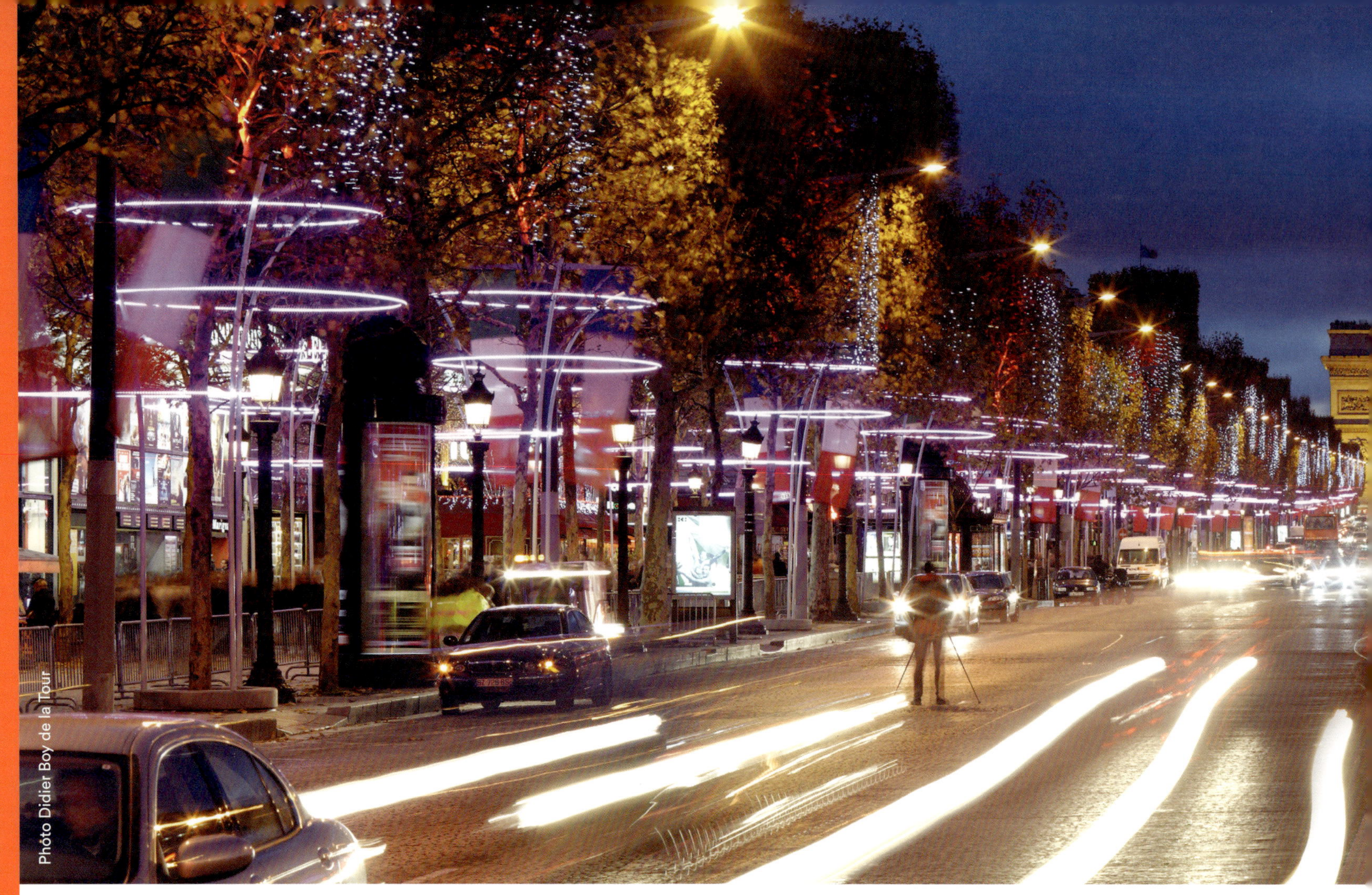

Combining solar power, eco-friendly LED strips and the savvy use of mirrored discs, Tree Rings was the first neutral-energy lighting installation at the Champs-Elysées.

THE SCHEME ENHANCED THE ESPLANADE WITH AN ILLUMINATED CHORUS OF CANDESCENT VIBRATIONS

Illuminating the Champs-Elysées during the Christmas season in recent years has been the task of ACT Lighting Design. Tree Rings is the light art installation that was conceived by ACTLD's principal lighting designer Koert Vermeulen, collaborating with artistic director Marcos Vinals Bassols. The team's creative and contemporary approach was one that saw Tree Rings become the winning entry of an international competition that ultimately led ACTLD to light-up of the famous French avenue during the festive season. Every winter since 2011, the project has draped the avenue's trees with radiating circles of light, chosen as a symbol of unity between man and nature.

From a distance, the rings of light appear to hover at various heights – forming looping and dynamic illumination around the trees' bare branches. The installation consists of three rings of differing diameters disposed around each tree, held in place by three curved-metal rods. A concrete base supports this set-up, so that the structure envelops the tree but never touches it. The rings' inner and outer circles are surrounded by programmable LED strips which sees the street and the tree's natural winter form lit-up, transforming the premier avenue into a vibrant stage for a spectacular lighting display.

The construction included each ring having a programmable RGB LED strip installed around the outside, producing a radiant glow of light. On the inner circumference, a programmable high-brightness RGB LED strip illuminated the tree at all angles. In addition, mirrored discs were hung from the tree branches which enhanced the effect, reflecting light rays in all directions. The separate programming system via DMX provided perfect precision, giving an infinite number of possibilities to control the dynamic effects, including the hue, intensity and rhythm of the illumination. The resulting scheme enhanced the esplanade with an energising chorus of candescent vibrations. The dancing pulses of colour that hopped between the two sides of the street, set the avenue alight in an unforgettable fashion. —

Designer
ACT Lighting Design
Principal lighting designer
Koert Vermeulen
Artistic director
Marcos Vinals Bassols
Location
Paris, France
Client
Ville de Paris Committee
Collaborator(s)/consultant
ASP Group
Manufacturer
Bis Lighting
Date
November 2011

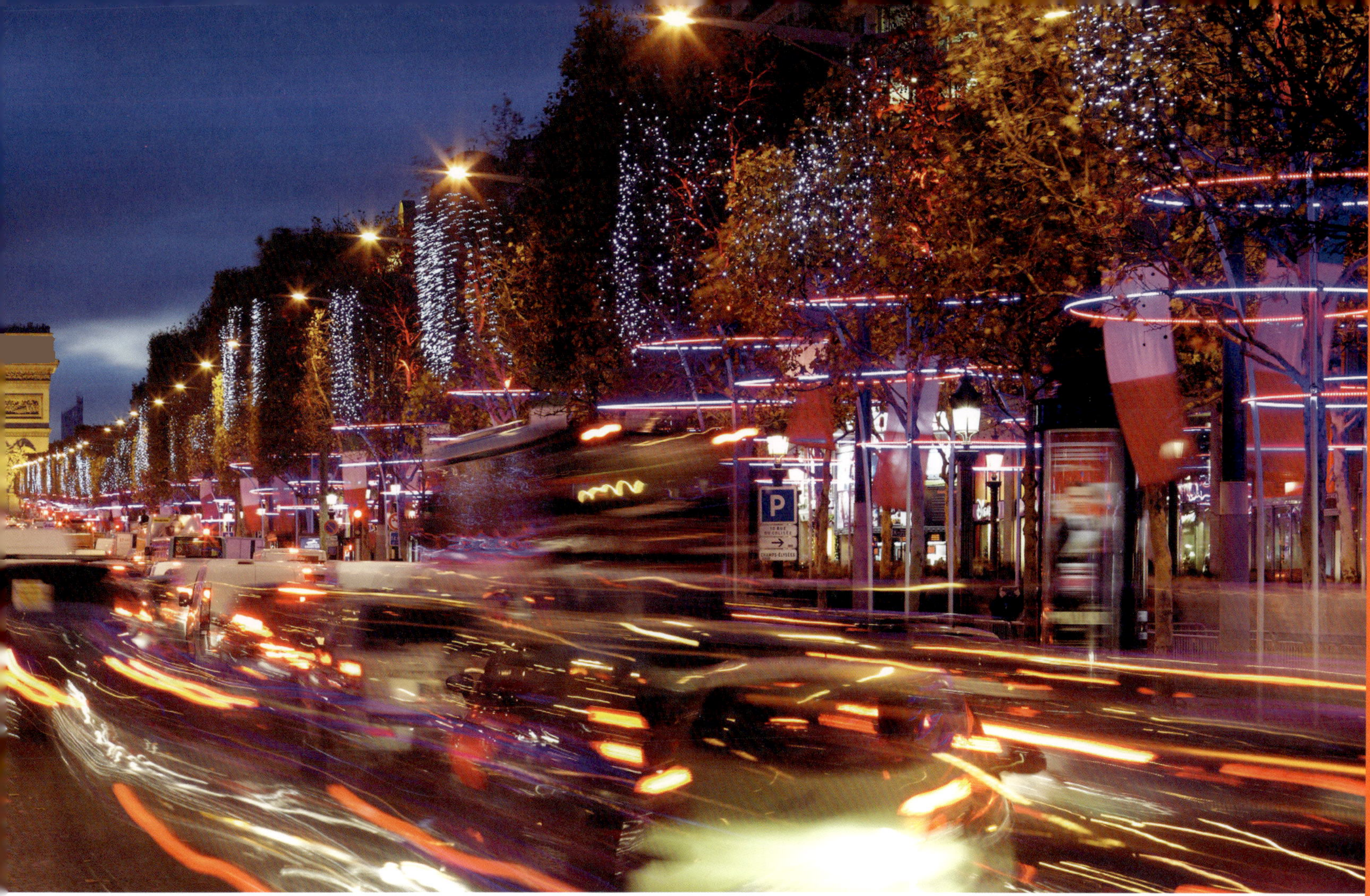

The project's main ambitions were to enhance the avenue's architecture and to highlight its prestigious role in the heart of the French capital.

RENDERINGS

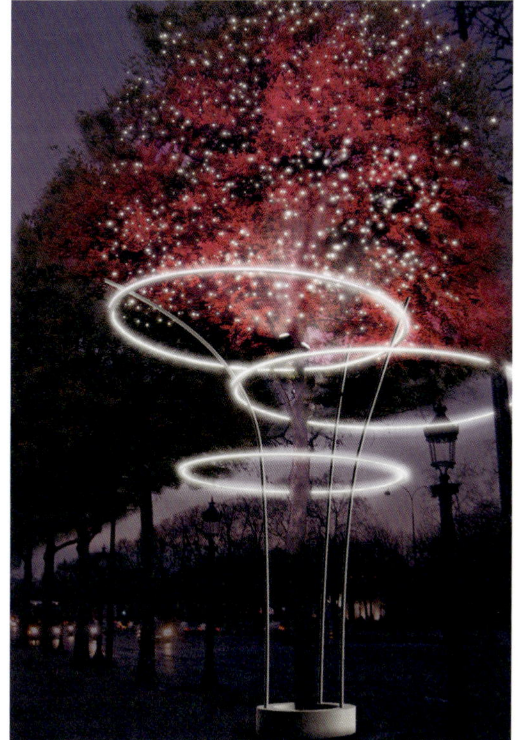

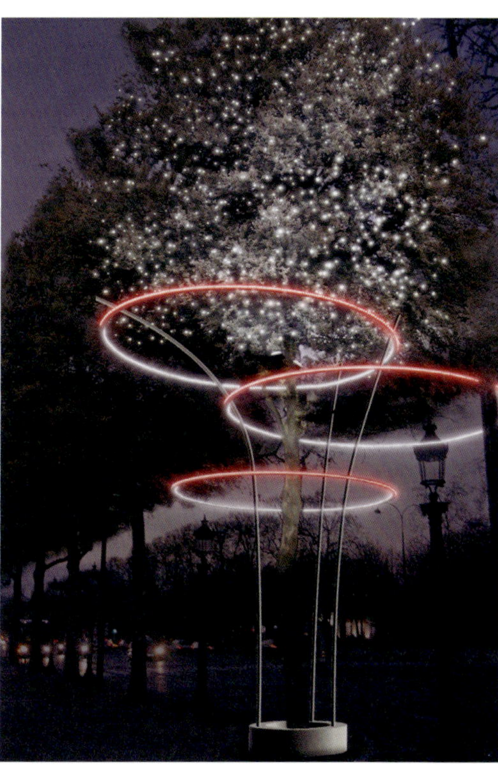

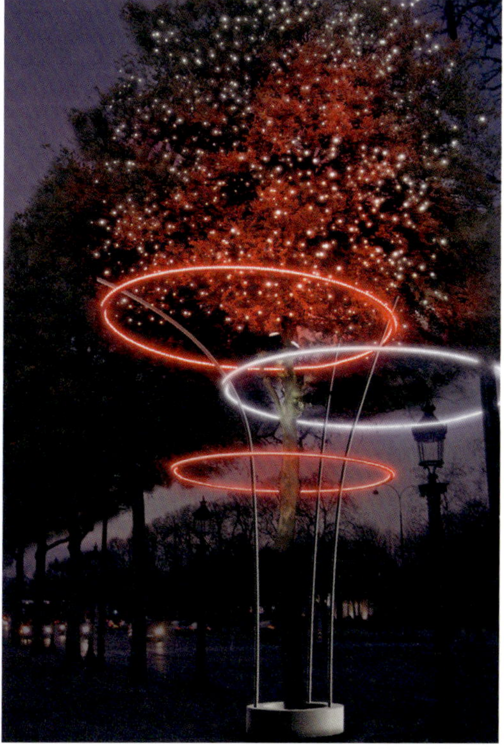

PLAN

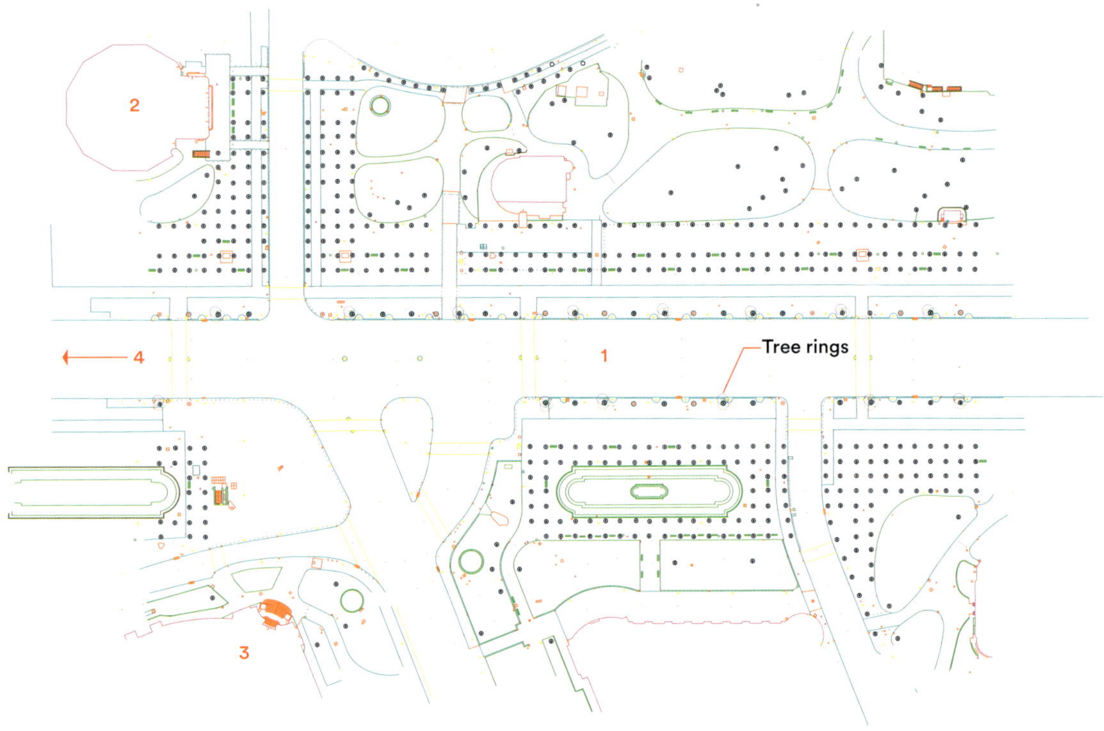

1 Avenue des Champs-Elysées
2 Theatre Marigny
3 Grand Palais
4 Direction of Arc de Triomphe

RENDERINGS (AVENUE)

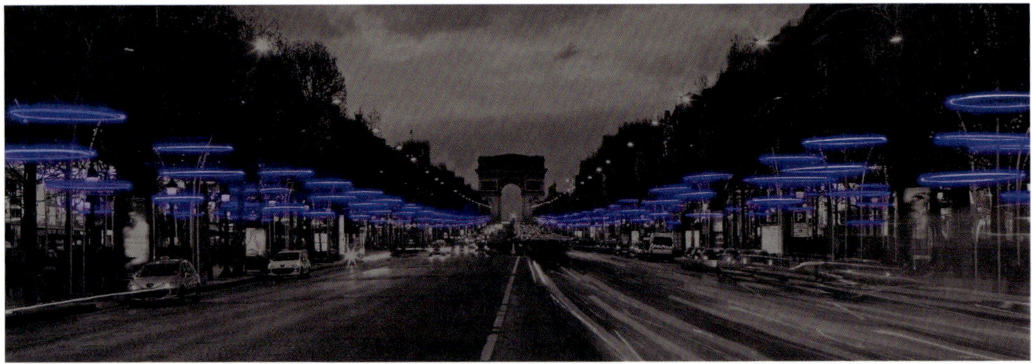

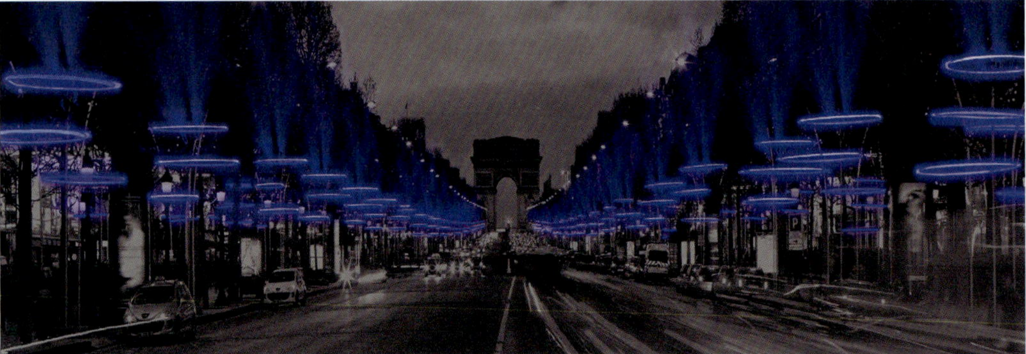

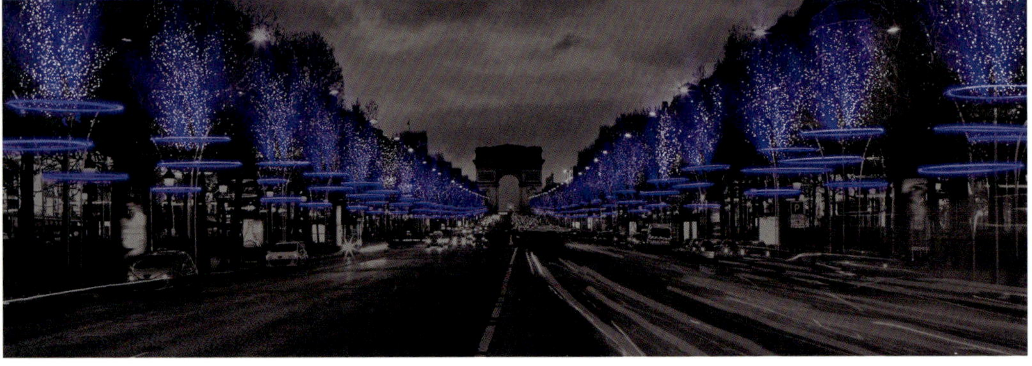

Specifics

The lighting design has been constantly developed to utilise different light effects. Changes of the light programming or physical additions to each structure enhanced the installation and provided a better perspective of the concept.

From the outset, the light art installation was based on the ACTLD eco-conscious vision. Commented Koert Vermeulen, 'We wanted the artistic vision to correspond to the respect of nature and approach of the tree as a living being, without infringing its space. We were convinced that a modern and innovative way to do that could be accomplished by the combination of specific technical solutions offered by the solar industry, with very simple geometric shape of the circle, as a symbol of gathering and union. In the end we were able to reach a concept that breaks with the past with its new design and because of its environmental-friendly features.'

Year after year, the lighting design broke with the past not only through its contemporary design, but also because of its environmentally-friendly features. The use of LED lights and solar energy from a photovoltaic farm reduced the energy consumption by nearly 90 per cent compared to the city's previous festival illuminations (comparing the 2011 data with that of 2006).

The reduced energy consumption was made possible thanks to the involvement of the French firm Soitec, which installed 30 m^2 of photovoltaic panels in the French Pyrenees, delivering 31,000 KWh of electricity directly to the Champs-Elysées lighting structures.

To control the dynamic colour effects, it was important to develop a system that could be integrated seamlessly. The programming, ignition and synchronisation mechanism was installed to ensure accurate implementation of light effects (colour, intensity and rhythm).

Special scenarios delighted visitors in the days leading up to Christmas, with very specific programming in place for New Year's Eve: a dynamic countdown was produced to illustrate the twelve strokes of midnight along the entire length of the Champs-Elysées.

DRAWINGS

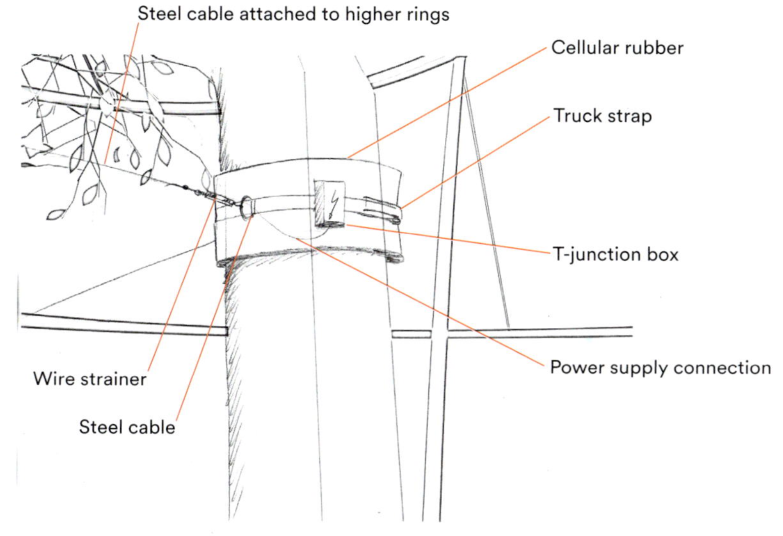

RINGS OF LIGHT FORM LOOPING AND DYNAMIC ILLUMINATION AROUND THE TREES' BARE BRANCHES

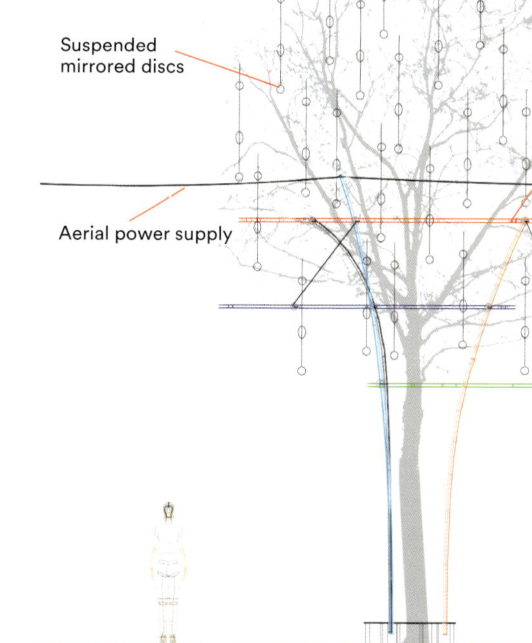

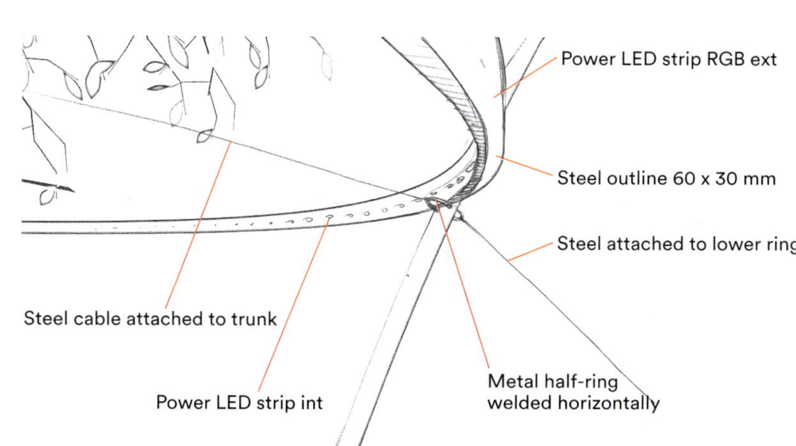

CONSTRUCTION DETAILS

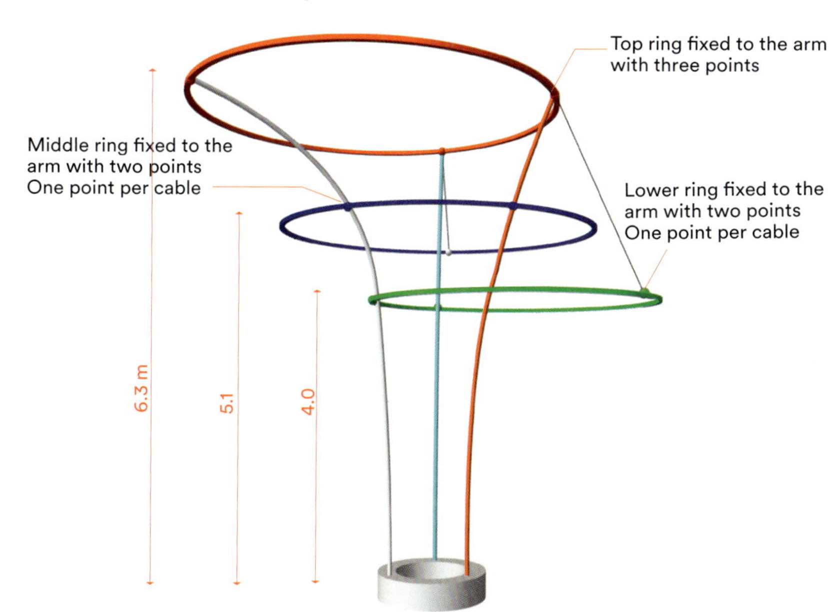

Set-up

Luminaires were programmable RGB LED strips, covering the inside and outside of the rings, illuminating the tree at a full 360 degrees. Uplights were also incorporated into the installation.
200 trees were each enveloped by a total of three LED rings, making for a total requirement of of 12,485 m of LED.
Tree rings were constructed with three RGB LED channels per circle of light, meaning that a total of 1,236,000 channels needed be controlled separately.
Used power was 457.5 KWh per tree per night (i.e. for LED rings, LED uplights and twinkling lights). In total, this was 38,800 kWh for the duration of this installation (a reduction of 90 per cent of the 2006 energy consumption).
Concrete base was constructed in two parts, with three curved-metal rods attached, to hold the rings securely in place. The highest ring was 6-m up and the diameter of the rings ranged from 2.4 to 3 m.
Programming and control of the dynamic colour effects, intensity and rhythm was carried out via DMX.
Software was developed and programmed to achieve specific scenarios according to outlined requirements.

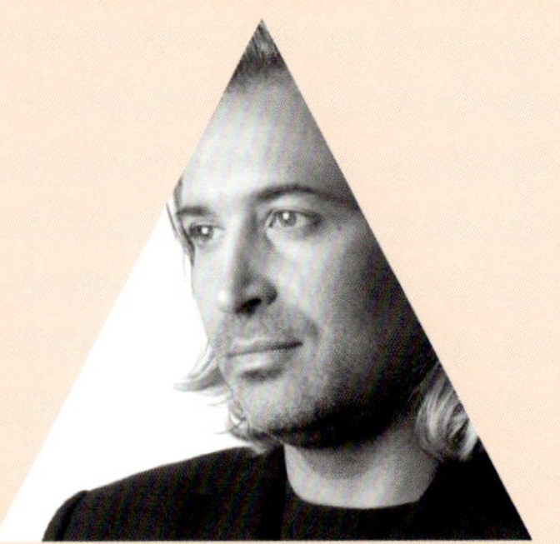

ACT Lighting Design

Founded in 1995 by Koert Vermeulen, ACT Lighting Design is a creative agency that provides lighting and visual design solutions across three areas – architectural and entertainment lighting design, and art installations. The Brussels-based firm comprises a team of 20 staff and has an international portfolio, ranging from heritage sites, urban spaces, shopping centres and city master plans to stage-set lighting and video content design for large scale events, along with contemporary light-art installations. This interchange between traditional core business and more creative endeavours presents a welcome opportunity for in-depth research and analysis that leads to innovation. The agency prides itself in devising original, integrated solutions that are simultaneously functional, flexible, sustainable, costeffective and aesthetic.

ACTLIGHTINGDESIGN.COM

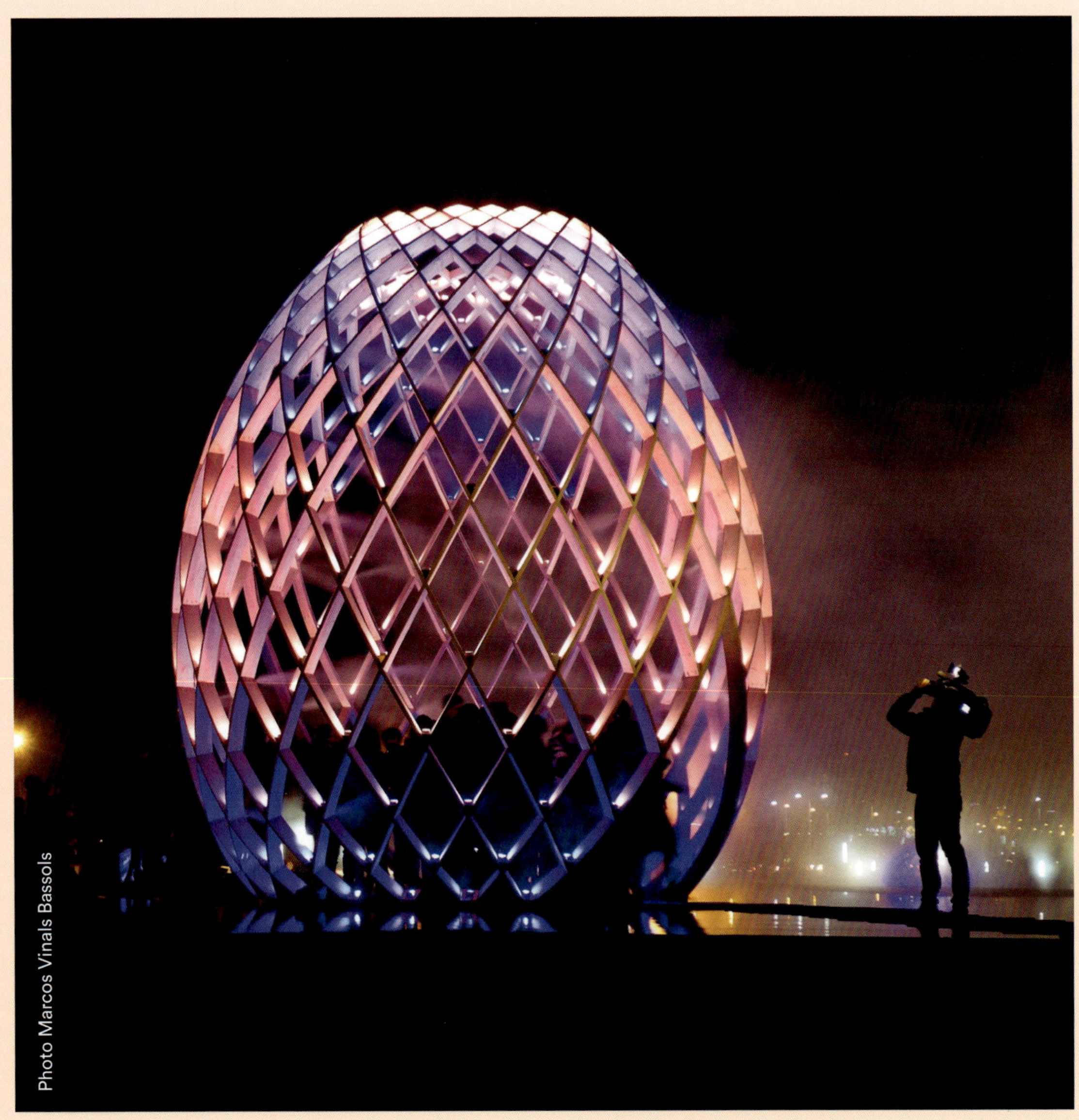

Photo Marcos Vinals Bassols

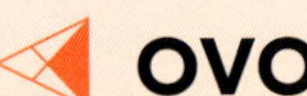

OVO

When **2010**
Where **Lyon, France**
Client **Fête des Lumières**

First presented in 2010 at La Fête des Lumières in Lyon, OVO is a multi-sensory art installation that has since been featured in a number of exhibitions close to home (Finland, France and Germany), as well as travelling across the globe to Asia (Kazakstan, Mongolia and China). Visitors are invited to 'walk on water' to reach the illuminated egg-shaped structure's interior. As if to vanish into a metaphysical mist where a combination of 24 crossed spiral pairs is based on the universe's Golden Ratio, the installation is also featured at a permanent exhibition in the Marmara Forum in Istanbul, Turkey. Collaborators on the OVO project include Marcos Vinals Bassols (scenographer) and the sculptors Mostafa Hadi and Pol Marchandise of OdeAuBois.

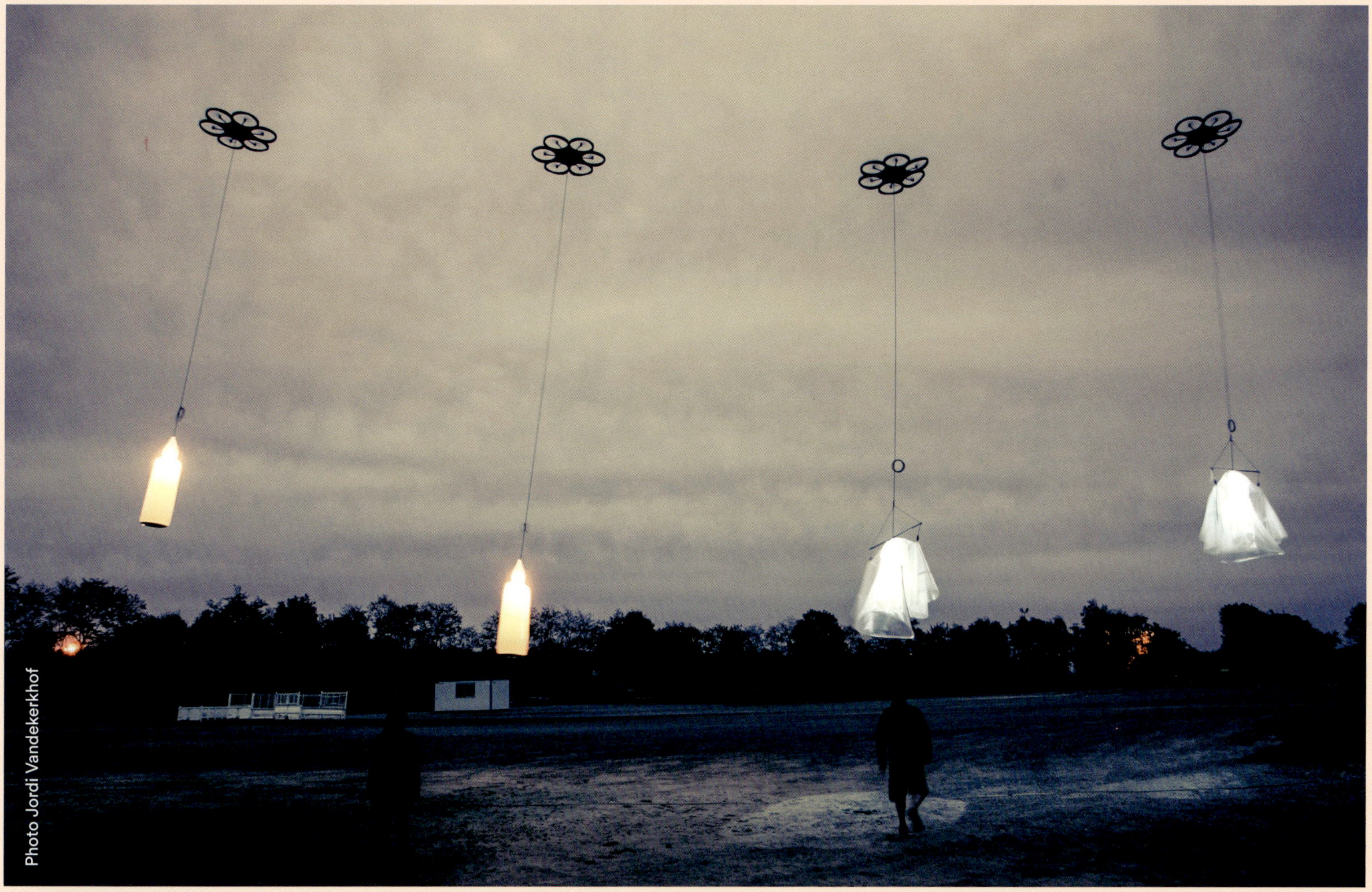

Photo Jordi Vandekerkhof

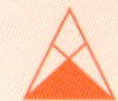

Neopter

When **2014**
Where **Les Épesses, France**
Client **Association du Puy du Fou**

Neopter is a new generation drone developed for Cinéscénie, the night-time spectacle at Puy du Fou's theme park in Les Épesses, France. After one year of development and the official approval by the General Department of French Civil Aviation, ACTLD and Puy du Fou launched the show's leading actor. Operating within an aerial fleet, Neopter's waterproof, robust technology comes to revolutionise the entertainment industry. It is able to fly autonomously at a height of 60 m. The drone performs synchronised choreographies and is capable of carrying sound, video and lighting effects, scenography and other special effects, like pyrotechnics.

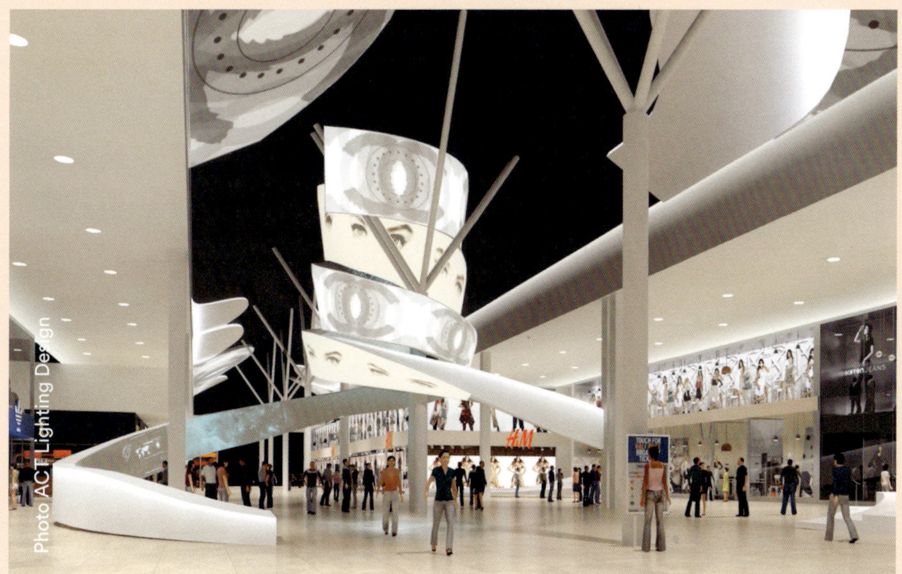
Photo ACT Lighting Design

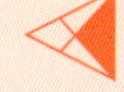

IMX

When **2015**
Where **Moscow, Russia**
Client **DG19**

In order to enhance users' spatial perception, ACTLD created an experience that allows visitors to get involved with space in unconventional ways. The project draws on a multi-sensory approach where architecture, lighting, sound, dynamics and interaction are fused together in the frame of storytelling to awaken all senses. The immersive experience of IMC contributes to the exploration of interfaces that merge the virtual with the physical as an opportunity to conceptualise scenarios for a story, a show or commercial branding.

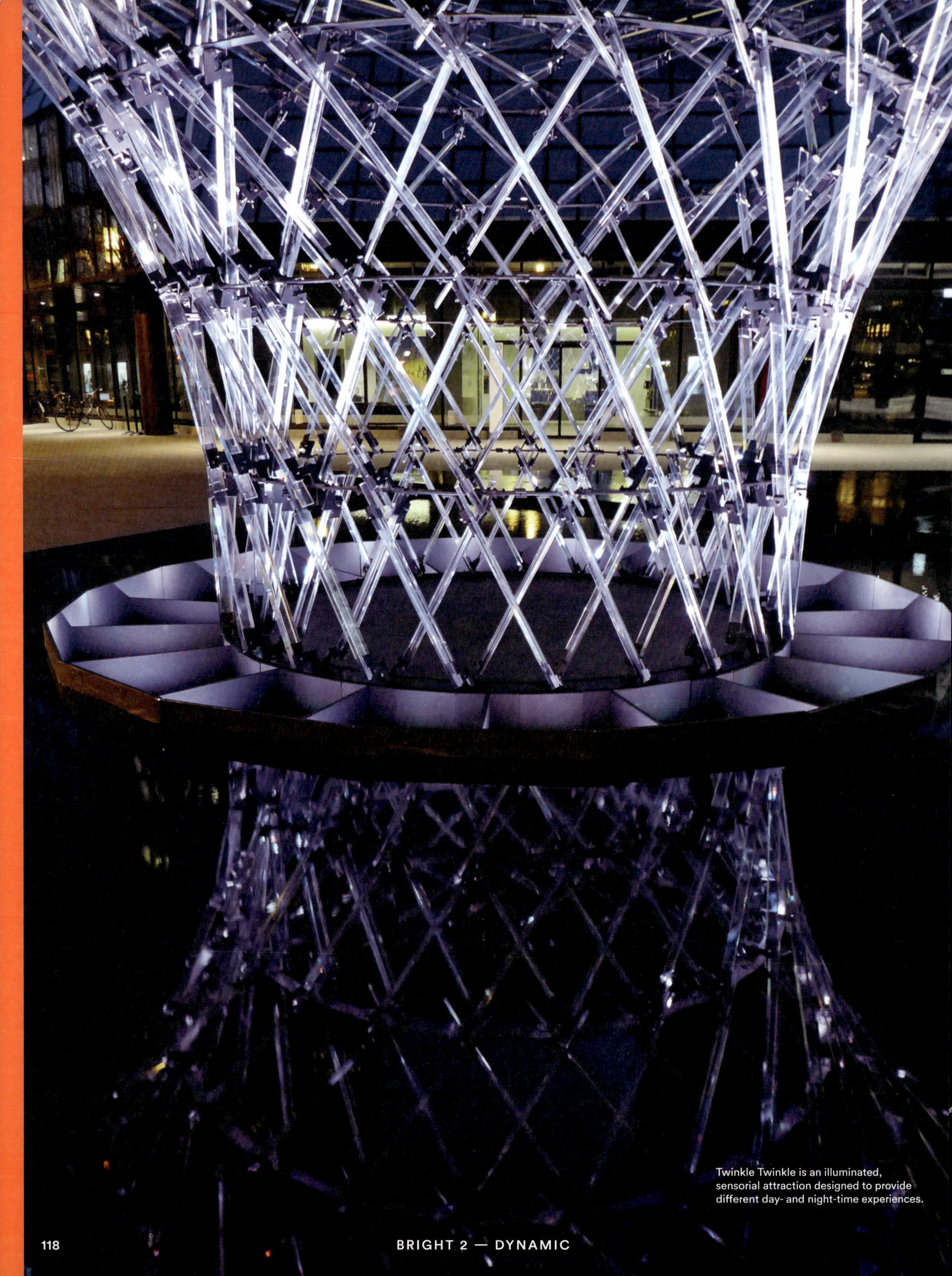

Twinkle Twinkle is an illuminated, sensorial attraction designed to provide different day- and night-time experiences.

Twinkle Twinkle

During Luminale 2014 in Frankfurt, a collaboration between **NE–AR**, **studioheyhey**, **lichtundsoehne** and **Glasbau Hahn** lent an extra sparkle to a central plaza in the city. Twinkle Twinkle – a grand, illuminated installation – used visual effects created by the interplay between light and glass in a celebration of transparency and communication.

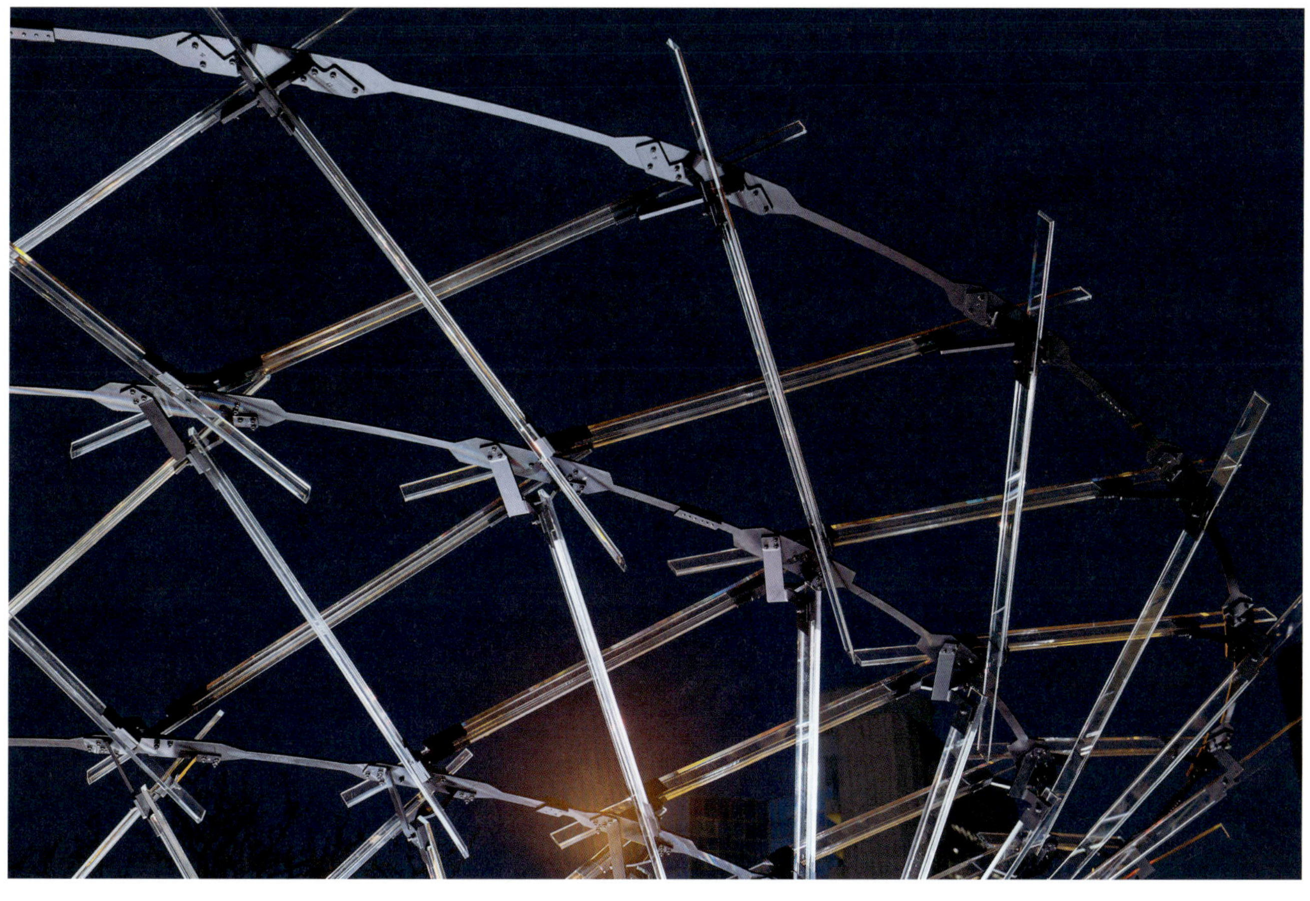

Photos Jens Ellensohn, Twinkle team

GLASS SOUND EFFECTS SYNCHRONISED WITH THE ARTIFICIAL LIGHT EFFECTS TO ENRICH THE AUDIENCE'S EXPERIENCE

Designers
NE–AR, studioheyhey, lichtundsoehne and Glasbau Hahn
Location
Frankfurt, Germany
Client
Luminale 2014
Collaborator(s)/consultant
OSD
Manufacturers
Zumtobel, Heinz-Technik, Satis & Fy, Holz Becher
Date
March 2014

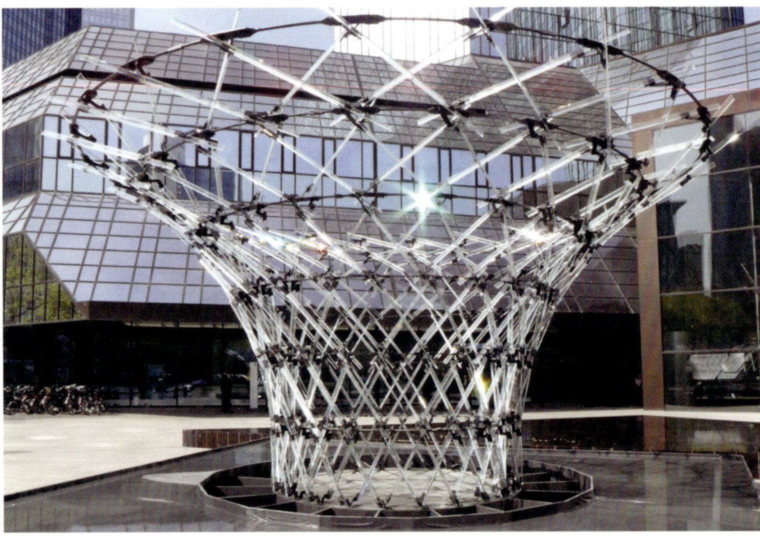

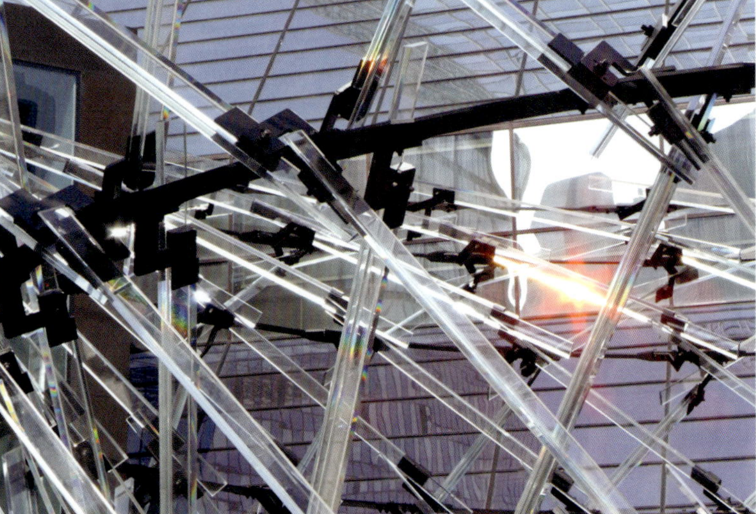

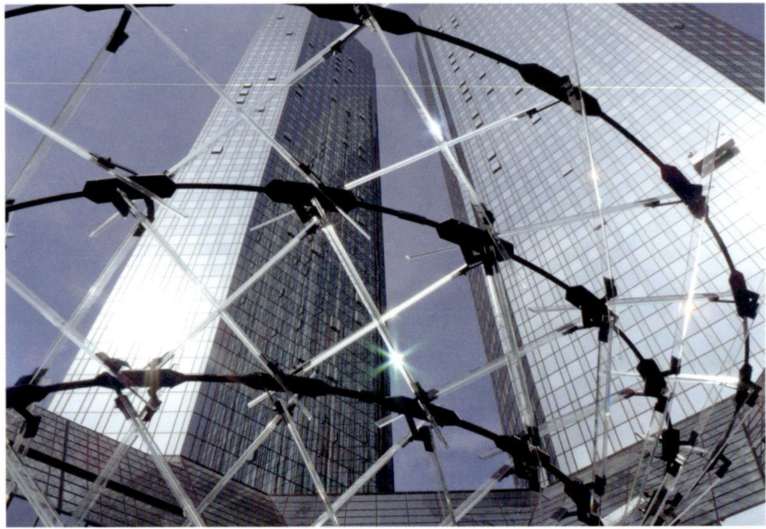

Animated by luminaires at its base, Twinkle Twinkle was an apt addition to Frankfurt's topography during Luminale 2014 – a biannual festival of light that illuminates the city. Comprising a hyperboloid form with 576 glass prisms, the installation reached 4 m up to the sky in a celebration of transparency and communication. It was developed by architects NE–AR, communication designers studioheyhey, light designer lichtundsoehne and glass developers Glasbau Hahn to transform the area in front of the Deutsche Bank towers into a dynamic stage for the visual effects created by the interplay of glass and light.

The concentric sculpture was made up of six circular sections stacked on top of each other. Within each segment, 'float-laminated glass rods' were attached to steel supports through hinged mounting bolts, forming the installation's distinctive triangulated network. Each of its glass elements functioned at once as a structural component and as an information communication medium, transporting and transforming light. The resulting sensorial experience was materialised through the interaction of natural and artificial light sources with the glass bars. Moreover, glass sound effects were synchronised with the artificial lighting enriching the audience's experience of this unexpected, enveloping multimedia installation.

During the day, the apparently fragile, transparent structure refracted and reflected its surroundings as lighting flashed in a natural interplay of colours. At night, predefined sonic and light effects animated the installation in random, short sequences – mimicking the day's visual serendipity that kept visitors absorbed.

Twinkle Twinkle's structural arrangement also played with the notion of scale and perception, further enhancing the experience by dissolving from a total shape to discrete elements as observers approached it, and providing an endless variation of effects according to the audience's attentiveness and position. —

The random succession of fast visual and sonic sequences, heightened the audience's anticipation for the next pulse of sensory stimulation.

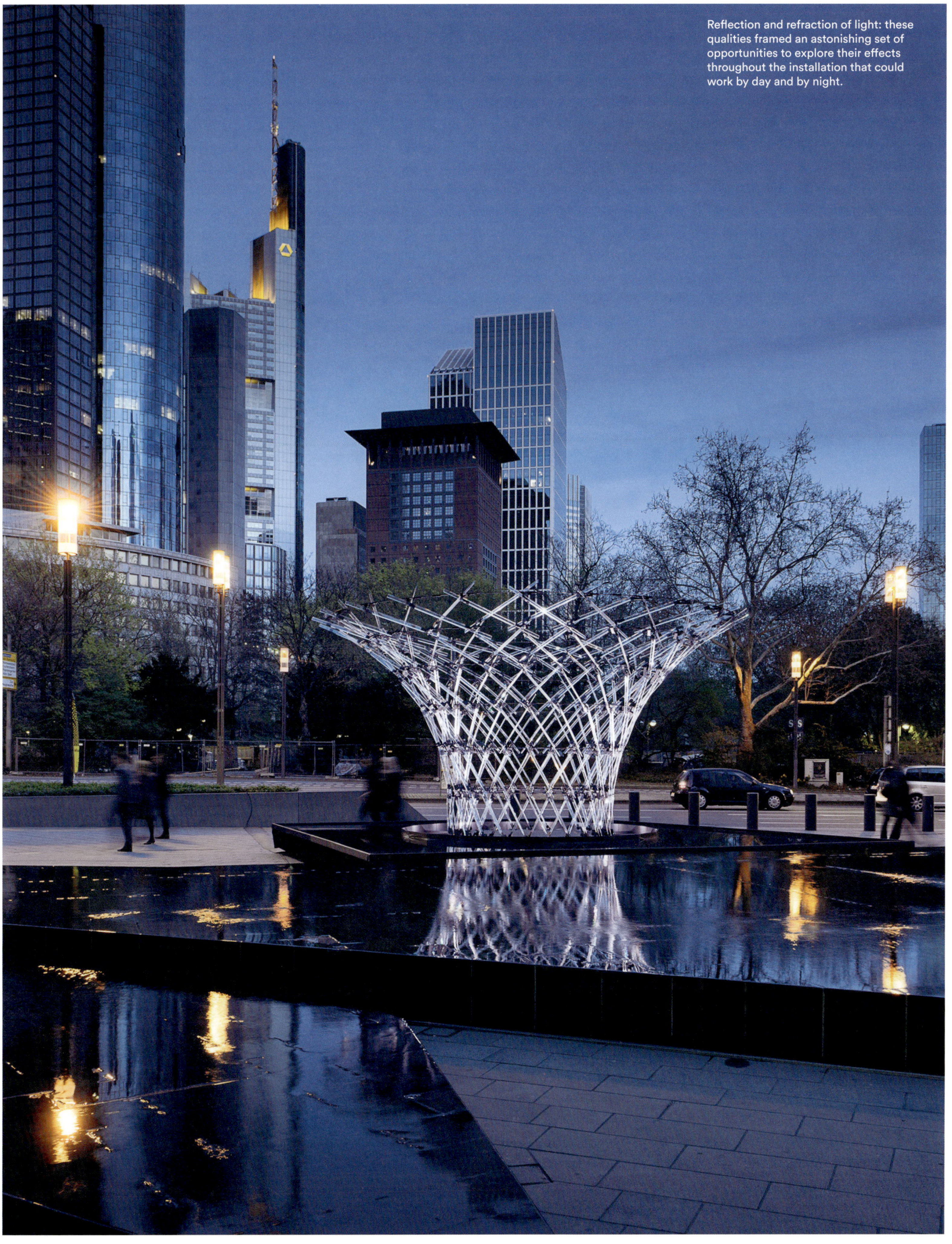

Reflection and refraction of light: these qualities framed an astonishing set of opportunities to explore their effects throughout the installation that could work by day and by night.

PLAN

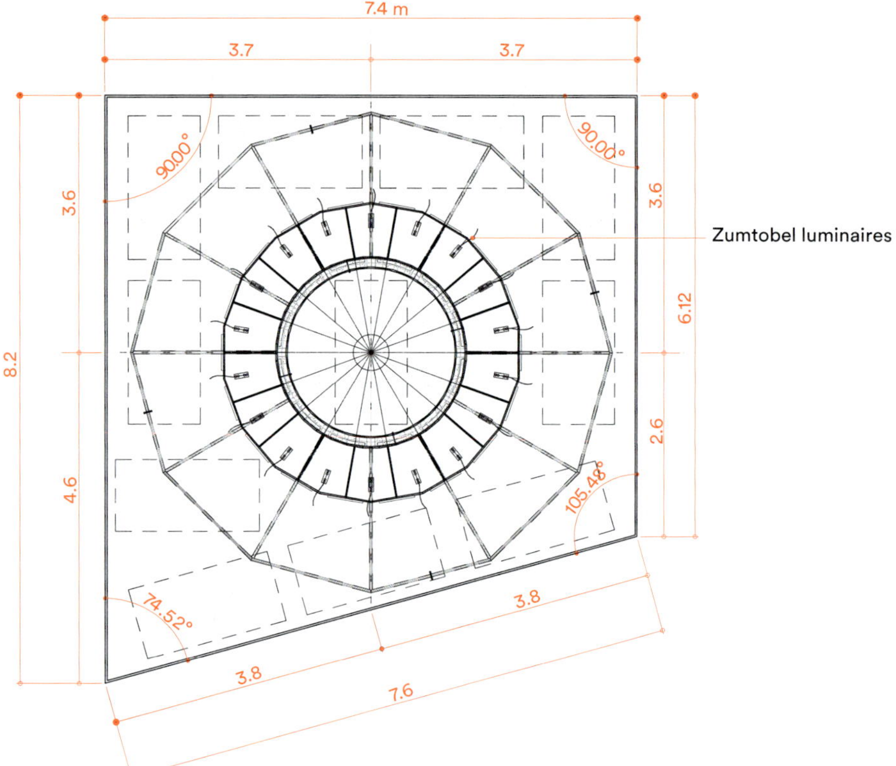

ELEVATION

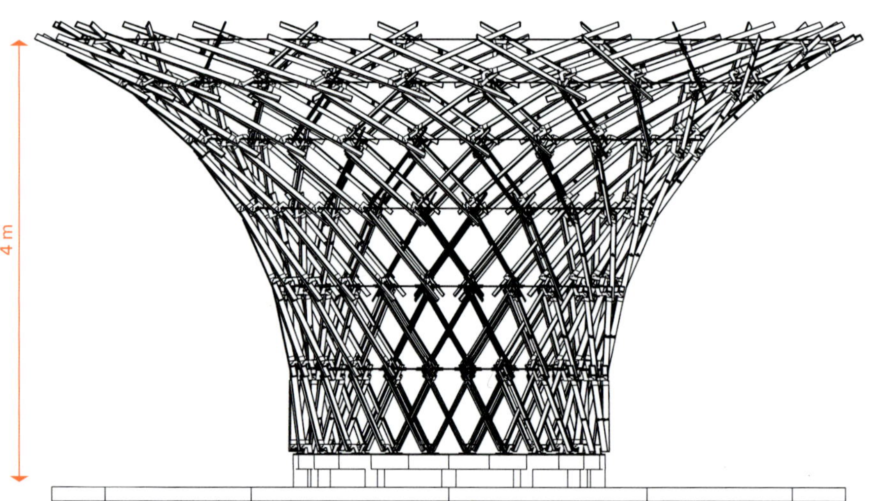

ILLUMINATION

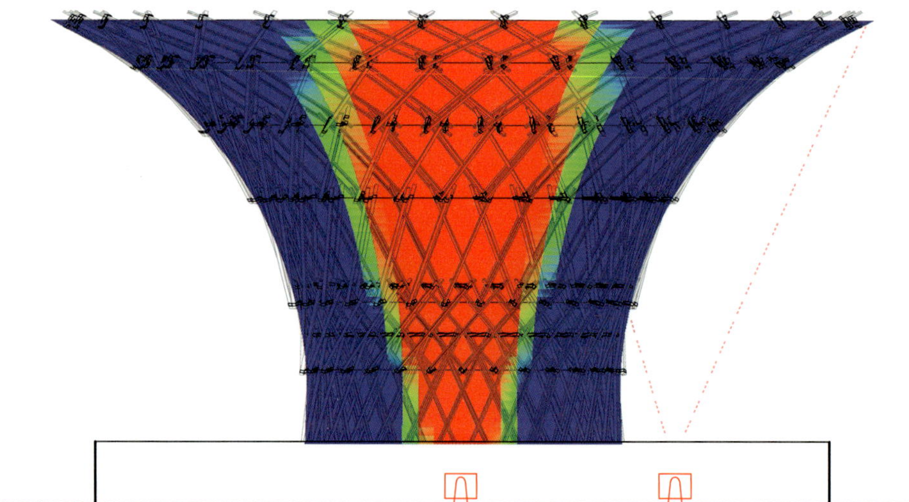

Specifics

The load-bearing components of the triangulated sculpture are the float-laminated glass rods. 'Float glass' is a sheet of glass made by floating molten glass on a bed of molten metal. Using a transparent adhesive foil that bonds two sheets of float glass together allows for structural reinforcement. For Twinkle Twinkle, 576 single rods of float glass were paired together laminating them making 288 bevelled rods used for the structure.

The material choice (flat float glass), and industrial manufacturing process derived from Glasbau Hahn expertise, was a clear choice from the outset. The structure of the installation came along through a series of brainstorm sessions with all key team players, where the focus was on the material capacities and effects that float glass has: transparency, transmission and refraction of light.

Following research, it was decided to triangulate the half of a hyperboloid in order to guarantee a coherent structural stability while at the same time offering certain dynamism to the overall shape. This dynamism was also achieved by decomposing the triangulation in two direction of glass rods: clock- and counterclock-wise. At that time, OSD (structural engineers) came on board and ran computational tests and based on geometrical and structural parameters, the optimum configuration was established.

Horizontally-extending flat sheets of the vertical sections to shortcut the ring forces and form the detail connection between the triangulated-extending truss rods of glass. The connection between the glass and steel supports act as the load-bearing detail which ensures adhesive connection with a hinged mounting bolt.

Glasbau Hahn produced a mock-up where the building procedures were tested, along with the effects of bevelled glass and the light effects with the Zumtobel artefact.

The artificial lighting design was created by lichtundsoehne, with the intention to transform the natural daylight effects and behaviour, and transform it an artistic expression for night-time, which included visual testing of different light setups and equipment, as well as precise light calculations. A special challenge, besides the task to illuminate the transparent glass material, was to specify appropriate lenses and locations for the fixtures, to fulfill the aim of well-distributed illumination for dynamic animation, without seeing the technical equipment.

The light and sound effects, created by Florian Licht, included more than 30 short light animations, not longer than 5 seconds, which were displayed in random order and intervals. These animations were rendered off-site together with a pre-visualisation and synchronised with recorded and modified glass-sounds, taken from shattering and breaking glassware, as well as sound effects from water-filled glasses and bottles.

RENDERINGS

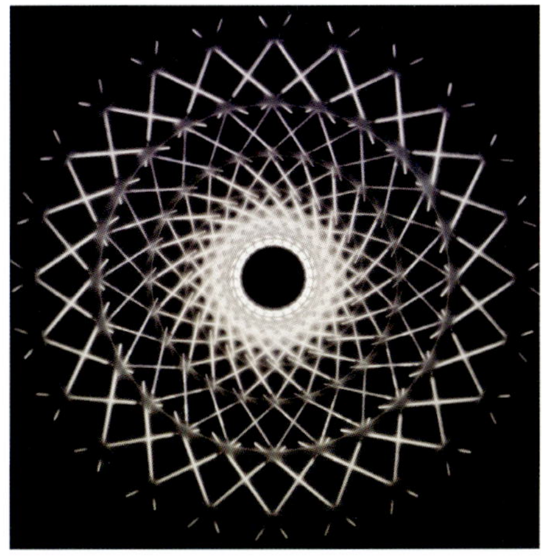

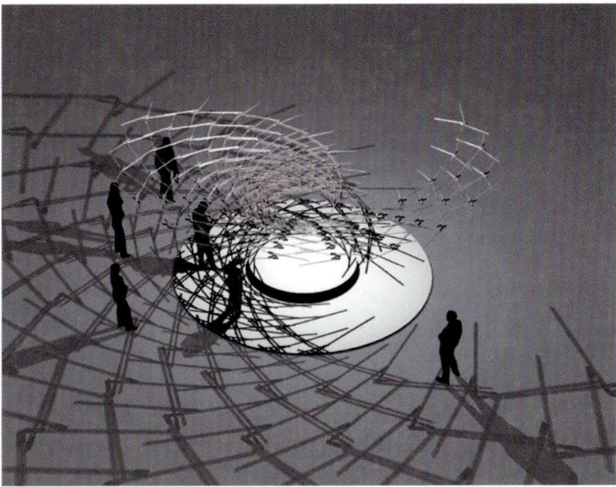

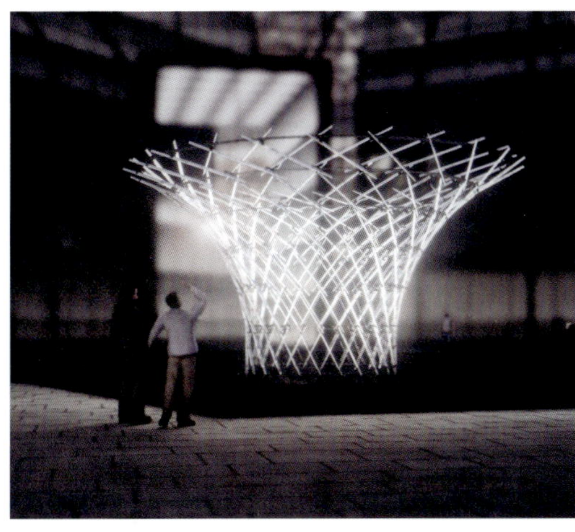

SONIC AND LIGHT EFFECTS ANIMATED THE INSTALLATION, MIMICKING VISUAL SERENDIPITY

CONSTRUCTION VISUALS

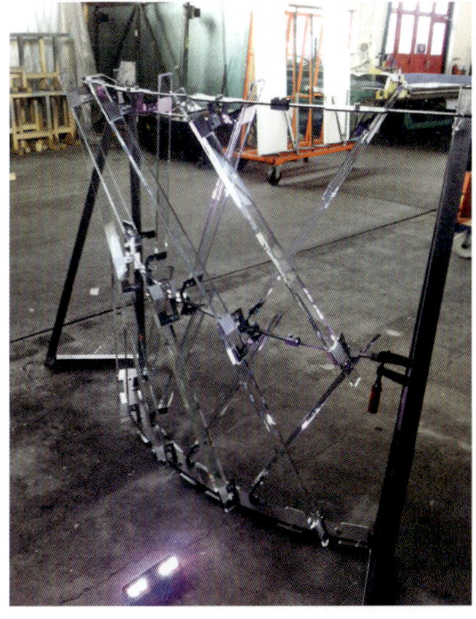

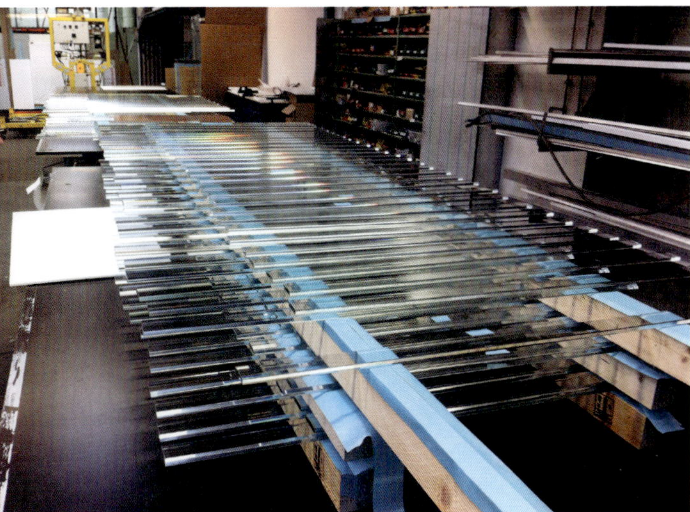

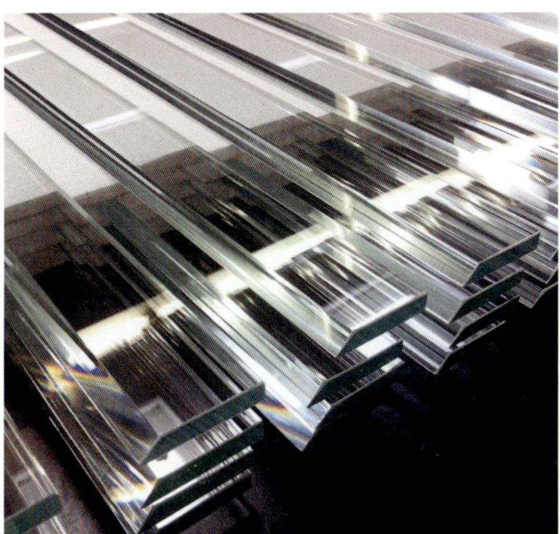

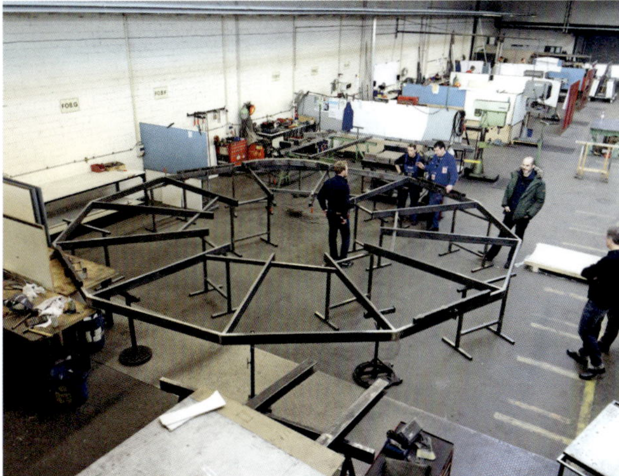

Set-up

Used light included 18 high-power RGB-LED luminaires with asymmetrical optics. These complemented and enhanced the light and colour effects, created by the interaction of sunlight with the sculpture's glass prisms, by evenly projecting white light through asymmetrical lenses.
LED animation included short and random sequences of impulses, enhanced by synchronised glass sound effects.
Light and sound system was controlled by Zumtobel and an E:cue lighting application suite, from a single computer.
Power consumption for all lighting and controlling equipment did not exceed 300 W and the maximum of the speaker system was 300 W. All equipment was either ip66 rated, or specially prepared for exterior use.
Total area of the sculpture was 60 m², reaching a height of 4 m. In total, 576 glass prisms (making a total of 288 pairs for the bevelled rods) were used in the construction.
Total weight of the installation, including its glass prisms, metal base and wooden base-cladding was nearly 4200 kg, creating a load of 76 kg/m².
Construction time was 5 days.
Light and sound system specifications

- 8 Zumtobel Elevo RGB-LED luminaires with asymmetrical optics
- One Zumtobel video control unit
- Three segment control units
- E:cue lighting application suite
- Four Speakers, 100 W.

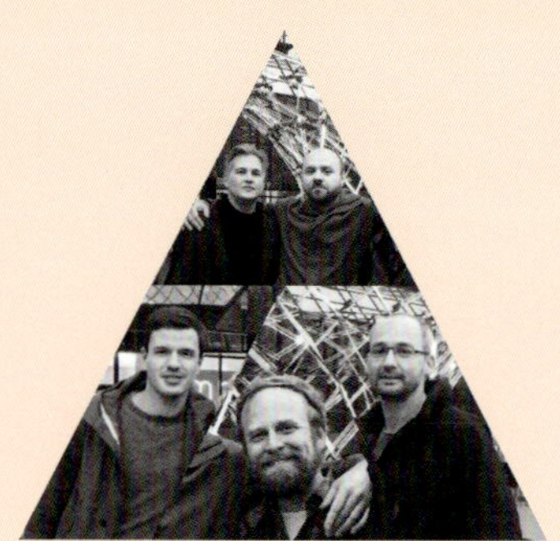

NE–AR, studioheyhey, lichtundsoehne and Glasbau Hahn

The bespoke project Twinkle Twinkle was a collaboration between three Frankfurt-based studios: NE–AR (Nixdorff Etchegorry–Architecture Research), specialising in design and architecture; studioheyhey, a communication and design agency; and Glasbau Hahn, experts in glass product design. Together, the team members brought distinctive expertise to the project. Pictured is Lars Nixdorff and Luis Etchegorry, who founded NE–AR as 'a platform for exchange within a local and global context', John Russo of studioheyhey and Tobias Hahn of Glasbau Hahn, the over 180-year-old glass experts, along with lighting designer Florian Licht of lichtundsoehne.

NE-AR.COM / STUDIOHEYHEY.COM / LICHTUNDSOEHNE.DE / GLASBAU-HAHN.DE

One Column House

Designer **NE–AR**
When **2014**
Where **Río Negro, Argentina**
Client **Undisclosed**

The core focus of One Column House is its central fireplace, derived from a study and deconstruction of the structural capacity of columns as an architectural element. Its importance comes not only from the fact that the different functional spaces and site-specific vistas are organised around it, but also because it functions as light tunnel, storage space for fire logs and as an infrastructural element, with rain water pipes being embedded within its twisted walls.

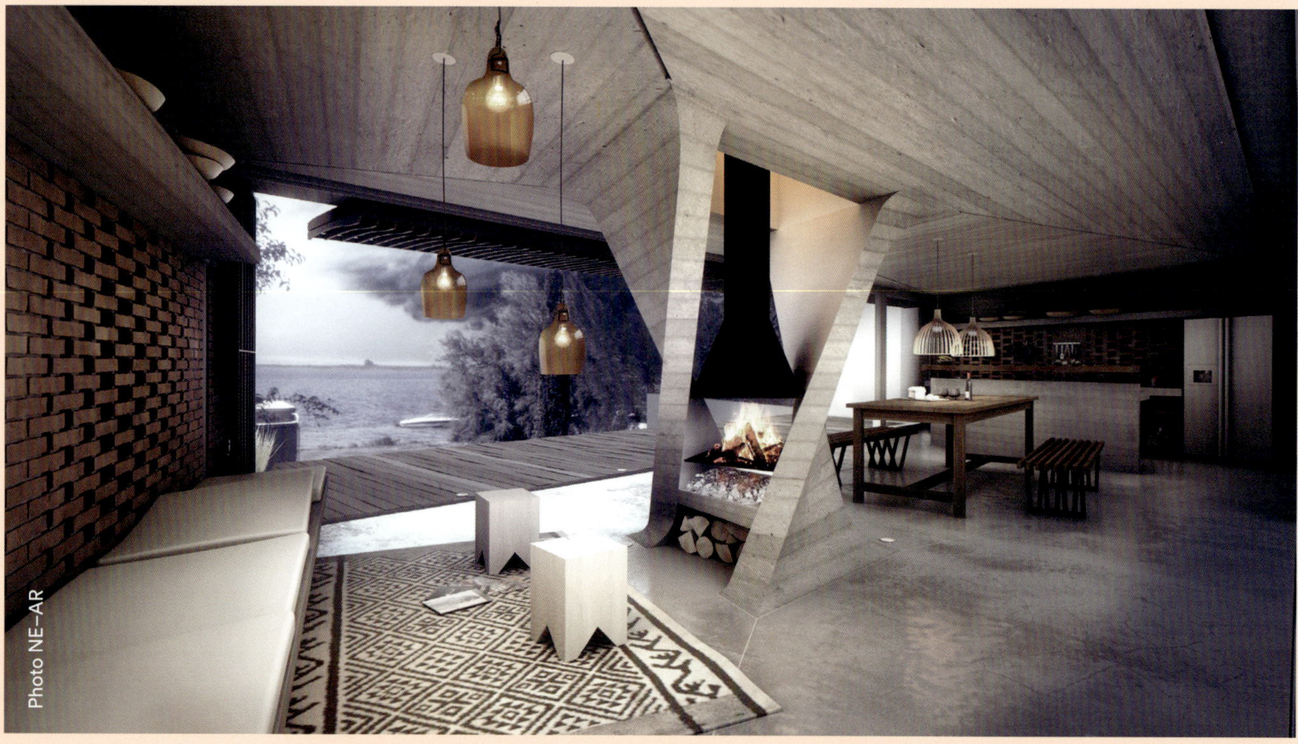

Photo NE–AR

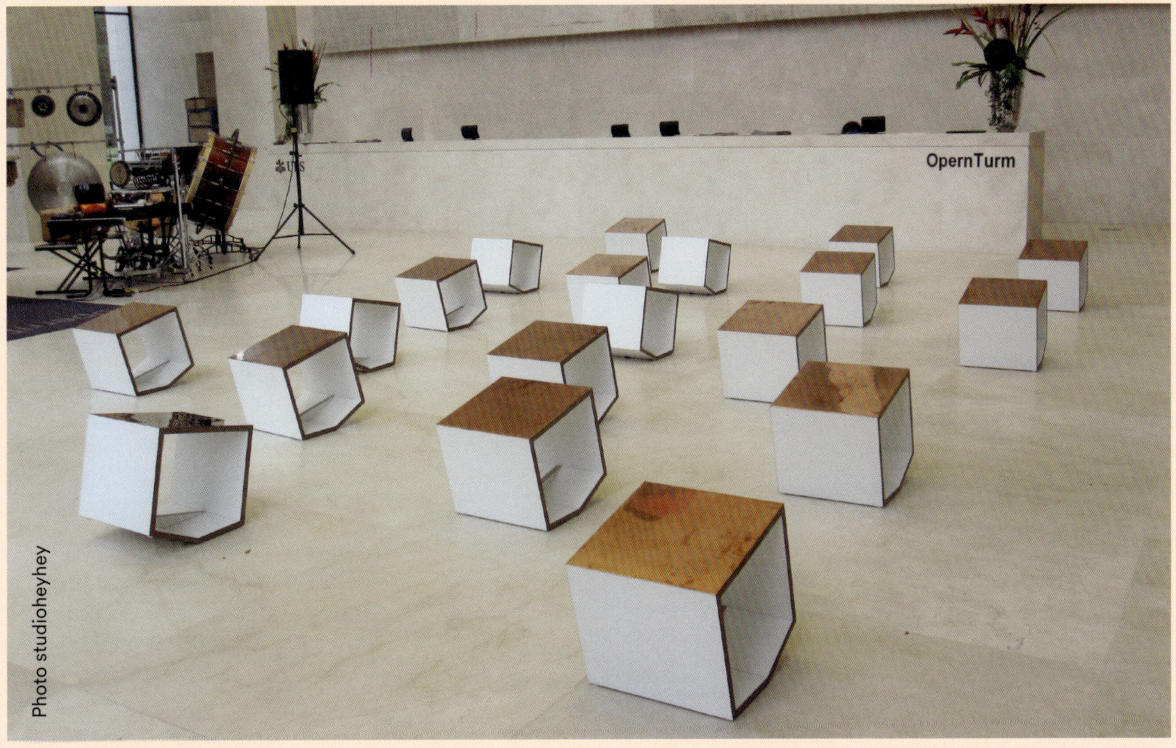

Photo studioheyhey

Balance

Designer **studioheyhey**
When **2010**
Where **Frankfurt, Germany**
Client **Junge Deutsche Philharmonie**

The motion of Balance was featured in the design of a series of stools for a duet instrumental piece. The stools consist of a seat edge which provides a firm footing while sitting. Users had to find the correct balance under their own power and balance their seating position during the period of the piece. The stools and the musical instrument were constructed with the same materials.

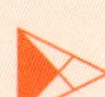

Display Cases

Designer **Glasbau Hahn**
When **2013**
Where **London, United Kingdom**
Client **V&A Museum**

The William and Judith Bollinger Jewellery Gallery in the Victoria & Albert Museum features Glasbau Hahn's curved and circular free-standing glass display cases. Each panel is equipped with anti-bandit glass and non-reflective Amiran coating. Individually shaped, the top closing panels are formed by curved, slanting glass strips.

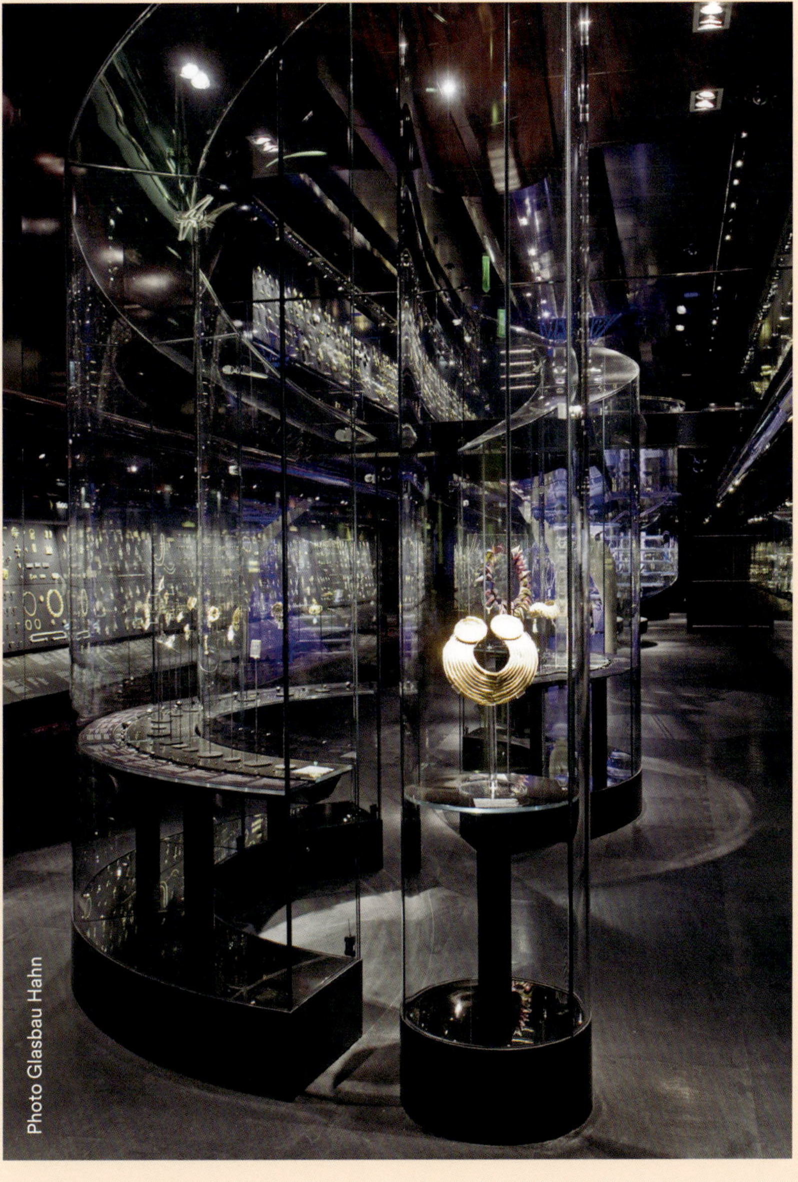

Photo Glasbau Hahn

Interactive

126

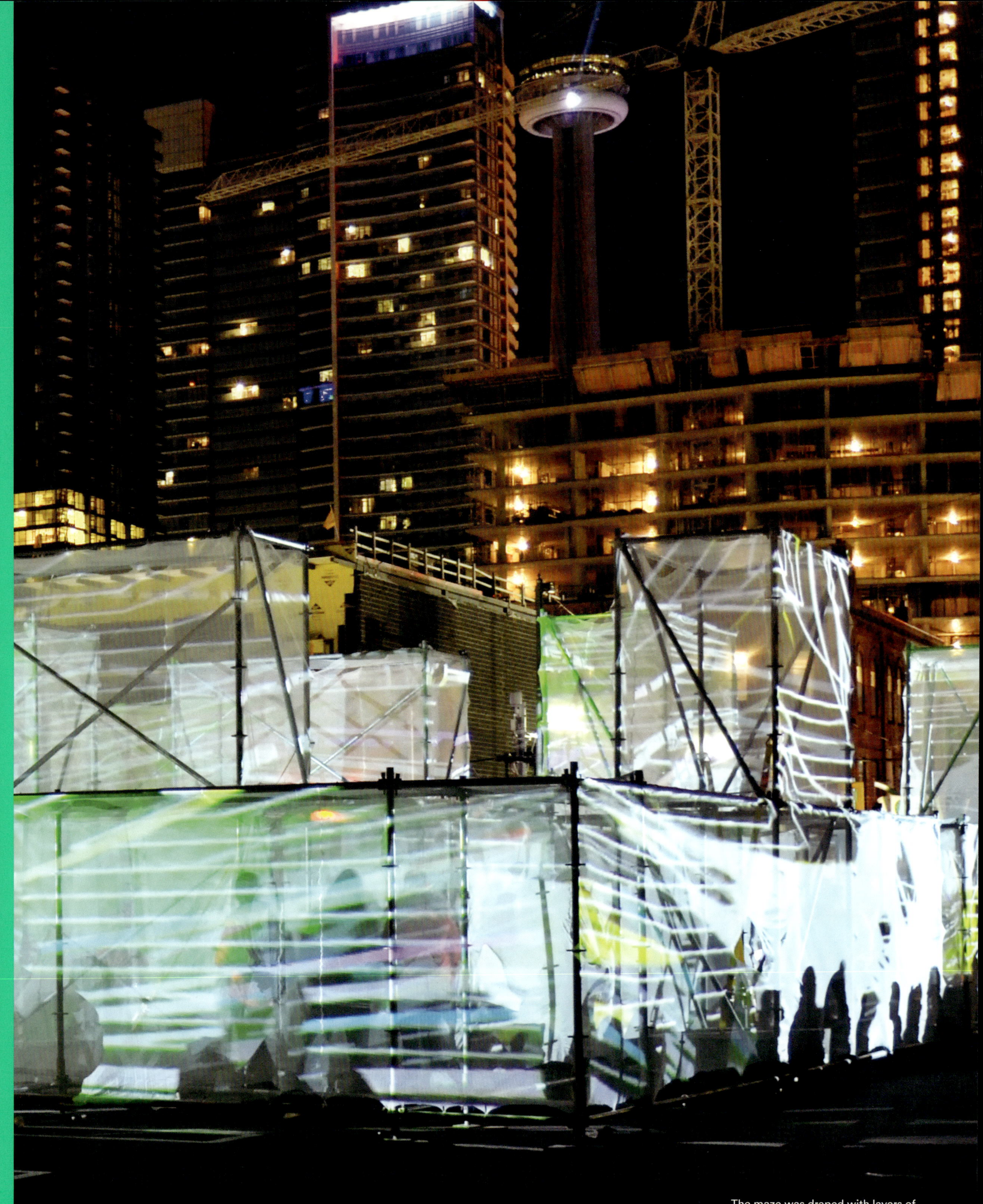

The maze was draped with layers of fabric, functioning as a translucent canvas for the spectacular light show.

Amaze

For the Nuit Blanche art festival in Toronto, the dynamically-lit architectural attraction Amaze drew attention from passers-by. Created by Reykjavik-based **Unstable** studio, the intriguing labyrinth-like installation acted as an opportunity for people to lose themselves and find new ways to think about and experience space and their city.

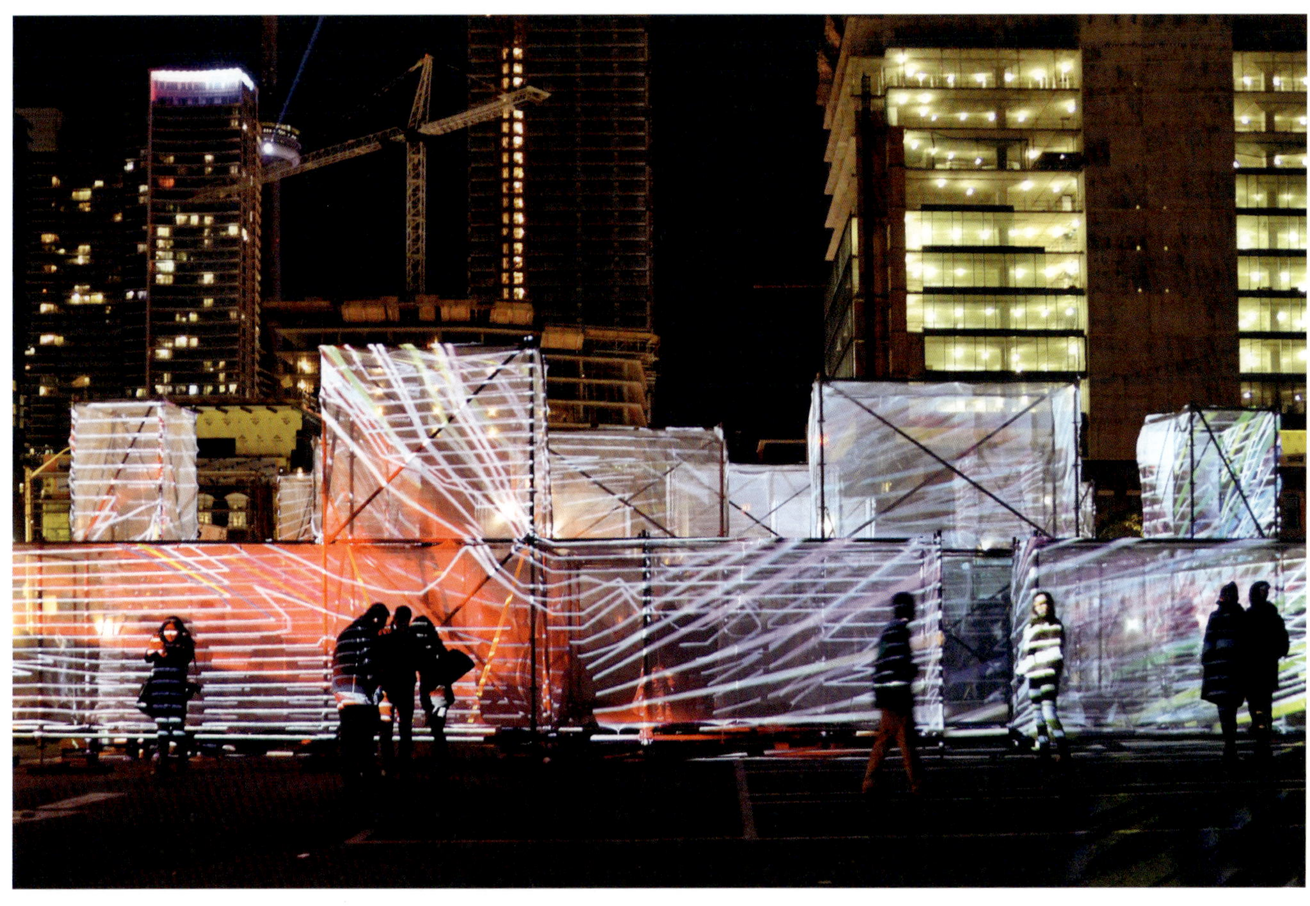

Photos Marcos Zotes

Informing the design of Amaze – an installation that draws on light and shadow to create an ephemeral, ever-changing architecture – is the idea that sometimes one needs to get lost in order to find oneself. The dynamically-lit structure was created by Icelandic studio Unstable for the ninth edition of Toronto's Scotiabank Nuit Blanche, which featured over 120 artworks.

Unstable's labyrinthinian contribution was part of an exhibition, entitled 'Between the earth and the sky, the possibility of everything'. This was a stage for innovative artworks focusing on play and interactivity to help audiences reconsider preconceptions about the urbanscape and question the present socio-political reality. Marcos Zotes, the creative mind behind Unstable, aimed to challenge visitors so they could actively engage with the installation by moving through its narrow, maze-like pathways.

Made up of a standard scaffolding system, the core structure was dressed in a translucent skin that functioned as a canvas for projection mapping. With illuminated visual components permeating the installation that mimicked the grid-like composition of the scaffolding, Amaze became a space for individual and collective exploration, disorientation and fascination. By employing light to change its structural integrity, the artwork goes beyond the use of spatial organisation to create disorientation.

Contributing to the balance between clarity and confusion of one's sense of space is the Amaze's porous skin, which allowed its audience to experience the surrounding city, as well as the illuminated shadows of other visitors. Immersed within this colourful, impermanent and transparent architecture, participants discovered space amongst the vast, disorienting sensory inputs, unravelling their thoughts to reassess the urban context and its future. —

Designer
Unstable
Location
Toronto, Canada
Client
Scotiabank Nuit Blanche
Collaborator(s)/consultant
Dominque Fonaine (curator), Umbereen Inayet (event programming), Joe Sellors (event supervisor)
Manufacturer
n/a
Date
October 2014

ILLUMINATED VISUAL COMPONENTS PERMEATING THE INSTALLATION MIMIC THE GRID-LIKE COMPOSITION

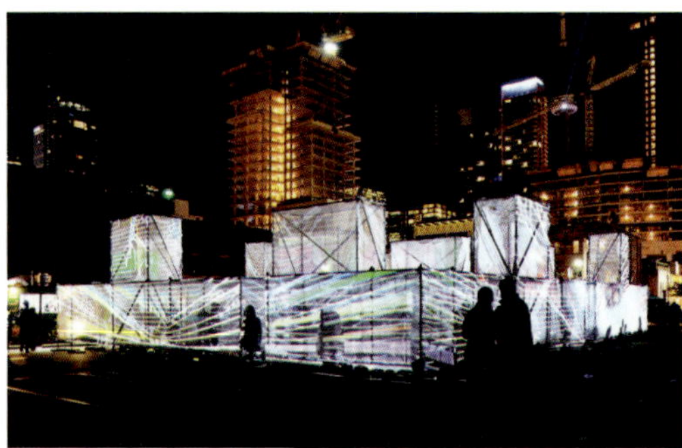

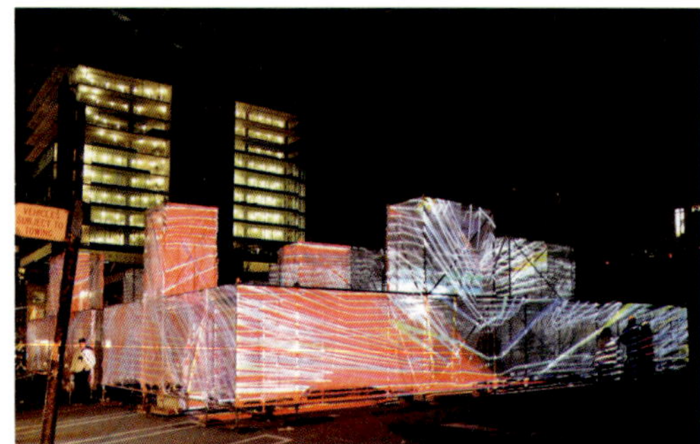

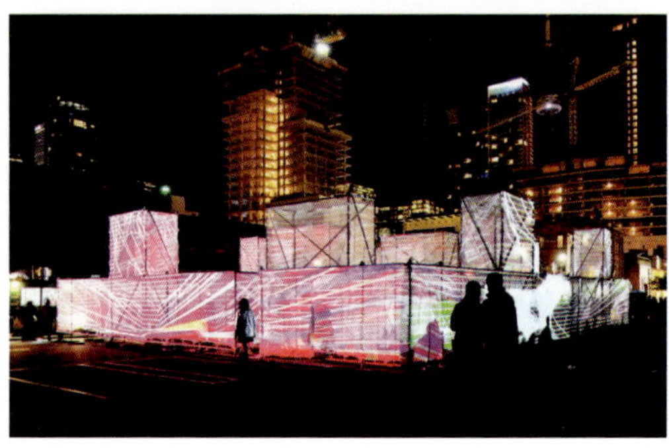

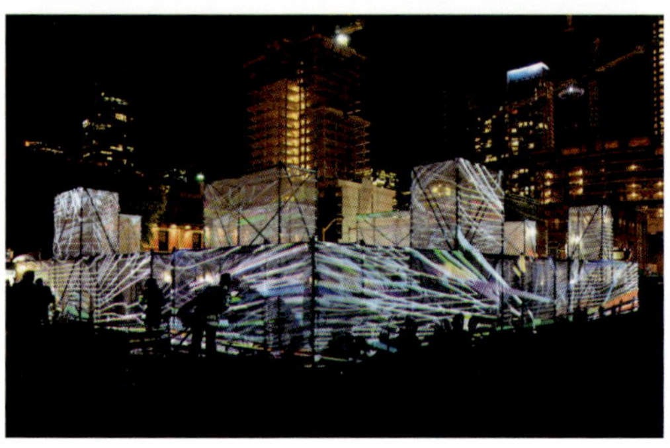

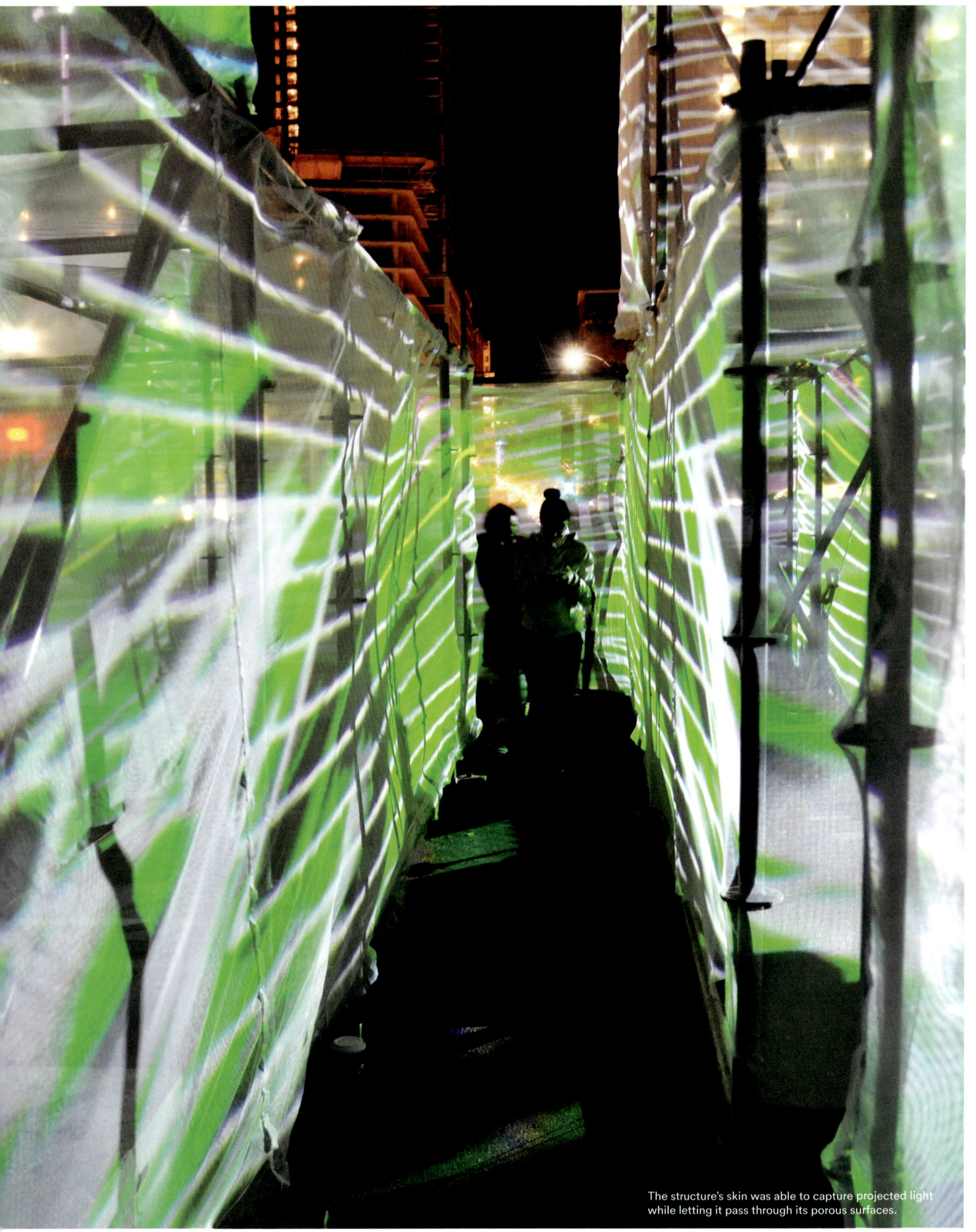
The structure's skin was able to capture projected light while letting it pass through its porous surfaces.

PLAN

17 m

18

ELEVATIONS

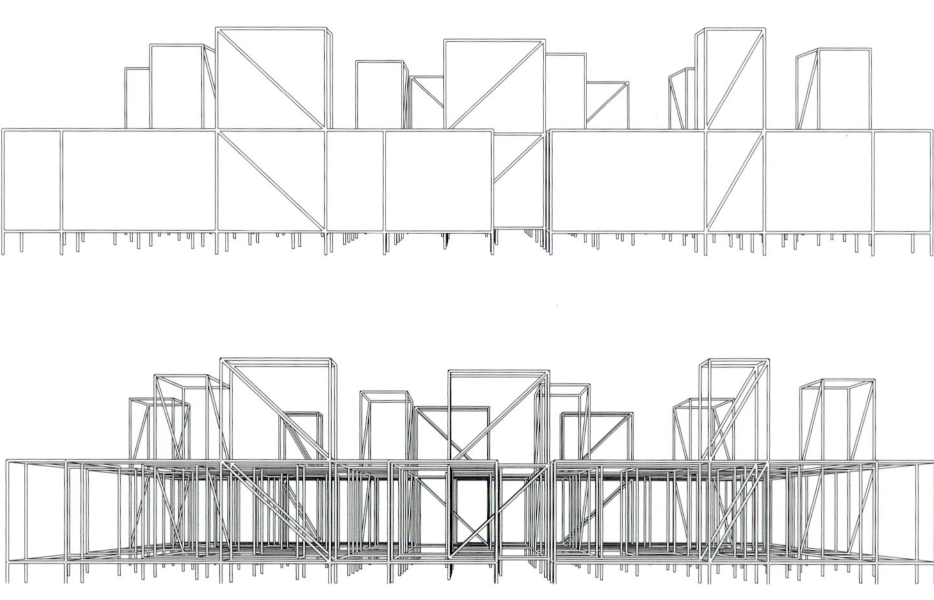

AMAZE DRAWS ON LIGHT AND SHADOW TO CREATE AN EPHEMERAL, EVER-CHANGING ARCHITECTURE

Set-up

Dimensions of the core structure were just over 300 m², with the outer walls reaching a height of 2.3 m. Twelve 4.6-m-high towers were distributed throughout the installation, mimicking the surrounding cityscape and increasing the impact of the installation.
Construction utilised a standard scaffolding system, making it versatile, affordable, safe and easy to build on-site. The envelope draped over the core structure (allowing for the light projections) was made up of multiple layers of white polyester mesh fabric. This acted a translucent skin that provided the canvas for a colourful light show.
Generative visuals were displayed using 2 x 20,000 lumen projectors positioned on distant towers at opposite sides of the maze. The visual content was adapted to the physical structure using advanced video-mapping techniques.
Power to run the projectors was supplied by a portable generator. The electronic equipment consisted of a laptop and a midi controller in order to modify the parameters of the visual content in real-time.
Build-up of the installation's scaffolding structure on-site was completed within a day.

AXONOMETRIC

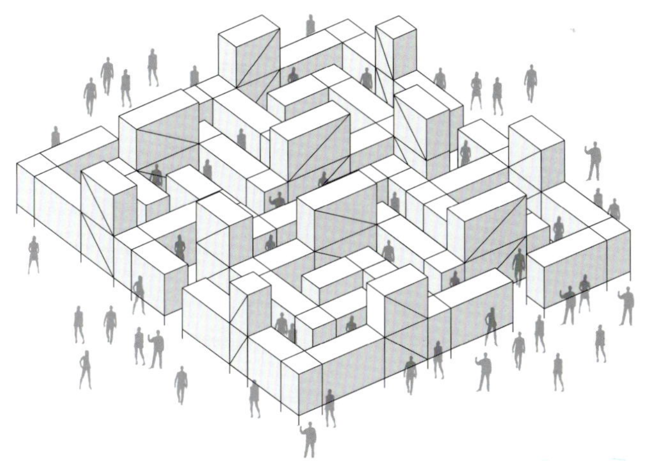

RENDERING

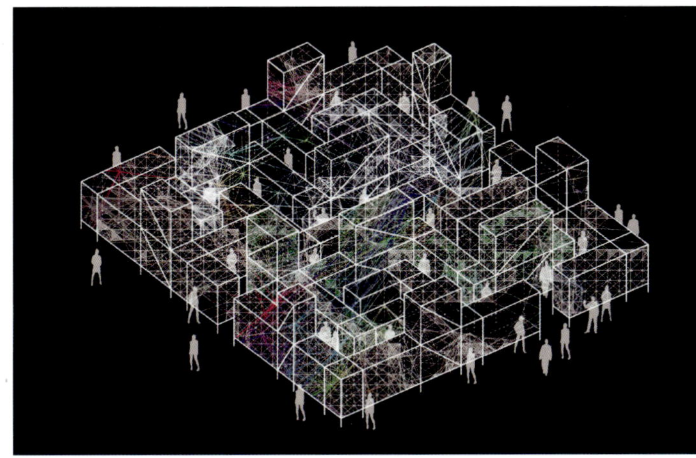

SET-UP (DAYTIME) IMAGES

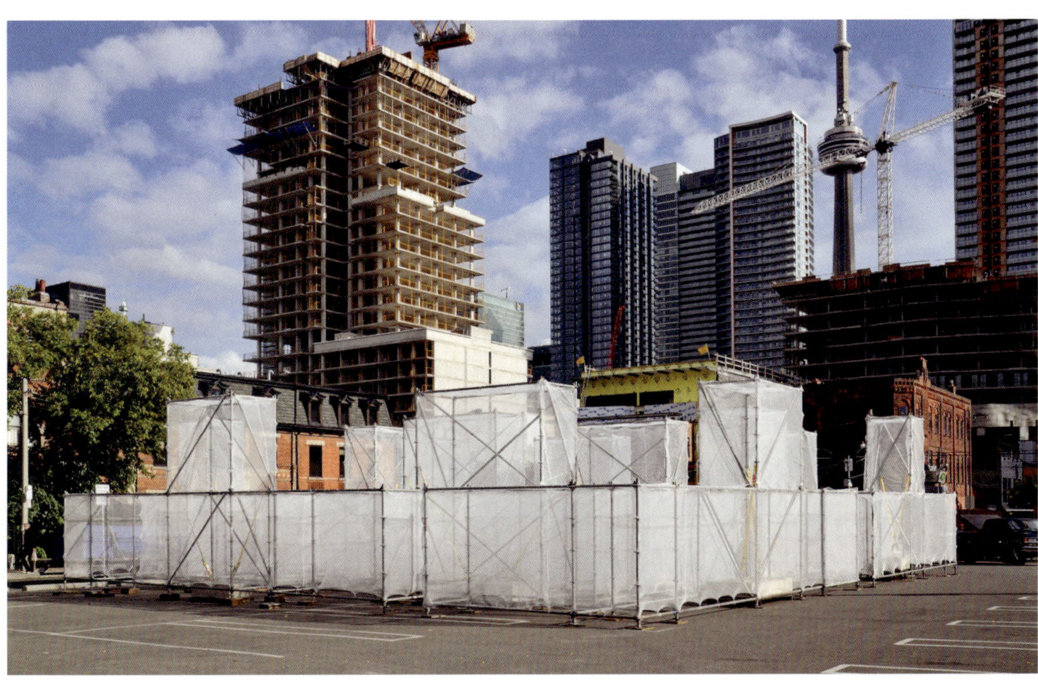

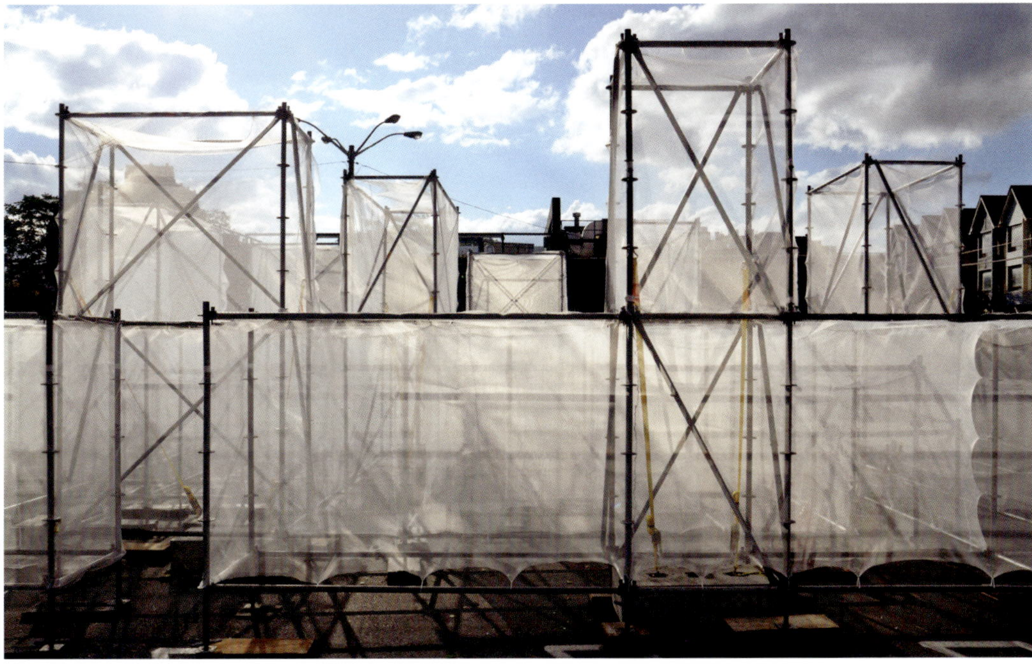

Specifics

The visual component of Amaze was derived from the concept of 'scaffoldage', a linear grid-like texture made of thin, white lines that resemble the scaffolding structure over which it is projected. This texture becomes a dynamic organism that changes colour when a series of points distributed through its body reach a certain height. As a result of this transformation, the sense of disorientation inside the maze is dramatically emphasised.

The optimum material backdrop for projections was investigated, testing the fabric for flexibility, resistance and light reflection properties. The chosen thin-mesh polyester, successfully performed as a projection screen, as well as a translucent skin that defined the narrow paths of the maze.

The installation's layout was designed using digital and physical models to test the movement of people. For an optimum flow through the maze, it was important to make the paths as narrow as possible in order to increase the performance of the video projection as it passed from one layer of fabric to the next. As a result, people walking the maze in different directions would have to come into very close with each other. The idea of bringing two strangers into such close contact, if even for an instant, was a strong component of the piece.

One important aspect of the maze was that it provided the possibility of different ways to experience the installation, depending on the chosen path for navigation. The initial design had only two entry and two exit points but in order to satisfy with emergency requirements, two additional exit points were added at the sides of the structure.

Other safety considerations included adding concrete ballast under each of the towers for structural balance; and the application of a fire resistant coating to the fabric screening to comply with fire safety regulations.

A full-scale mock-up featuring a series of distant layers of fabric within the maze was fabricated in order to test the performance of the video projection across them. Also, the shadow effects of the audience walking through the screening were tested.

The visual content was generated with the aid of parametric 3D modelling tools. An algorithm that responds to a series of pre-defined parameters drove the ever-changing qualities of the generated texture.

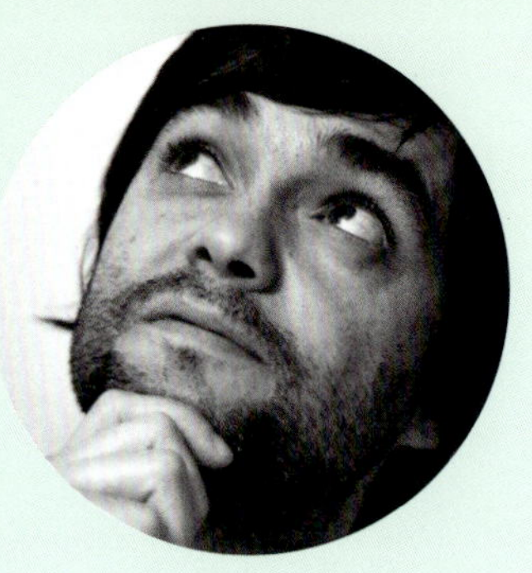

Unstable

Founded by architect and multimedia artist Marcos Zotes in 2012, Unstable is a multidisciplinary design and research laboratory that explores the boundaries between art and architecture while considering the social and political realities of the urban context. The Iceland-based studio creates award-winning interactive installations and public interventions that question one's relationship to urban spaces. The studio has exhibited and presented at a number of events and locations across the world, including Detroit, New York, Reykjavik, Roskilde, Toronto and Venice.

UNSTABLESPACE.COM

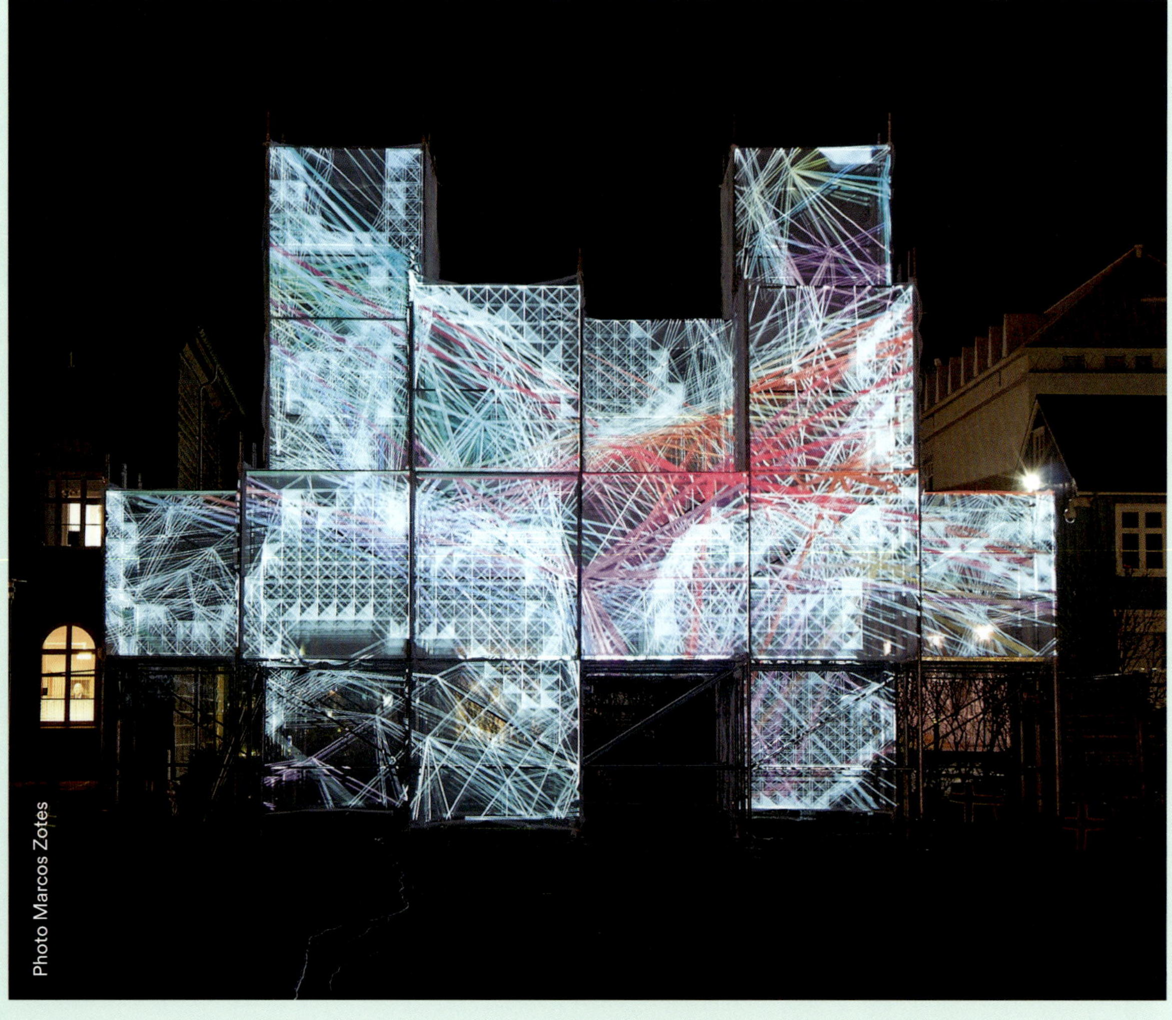

Photo Marcos Zotes

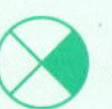

Pixel Cloud

When **2013**
Where **Reykjavik, Iceland**
Client **Reykjavik Winter Lights Festival**

Faced with the unfinished buildings and scaffolding structures that were left scattered through Reykjavik as the economic crisis hit, Marcos Zotes decided to take advantage of the incomplete cityscape. An architectural installation, Pixel Cloud turned an ordinary scaffolding structure into an immersive environment where light and sound were used to bring new life to the ordinary.

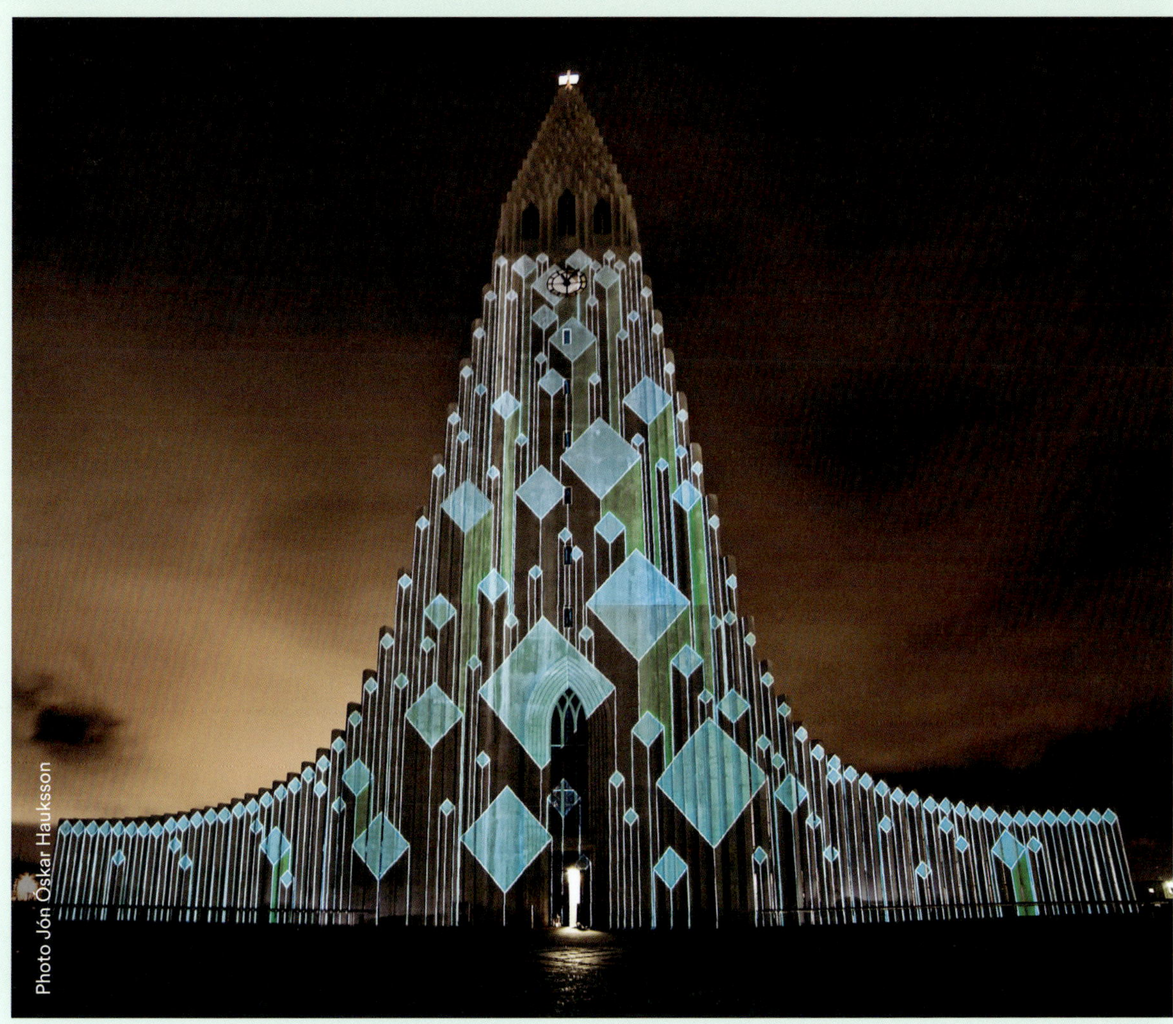

Photo Jón Óskar Hauksson

Rafmögnuð Náttúra

When **2012**
Where **Reykjavik, Iceland**
Client **Reykjavik Winter Lights Festival**

Rafmögnuð Náttúra transformed Hallgrímskirkja Church, the iconic landmark prominently perched in the centre of Reykjavik, into a community-engaging audiovisual experience. The temporary, site-specific installation drew inspiration from the country's extreme weather conditions and scenery to animate the unique architecture of the church's facade.

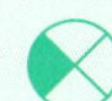

Your Text Here

When **2012**
Where **Detroit, MI, USA**
Client **Delectricity**

Looking to empower local communities to share their views and voice their opinions, Your Text Here was a site-specific, participatory media installation, that transformed people's voices into proclamations the size of buildings. Drawing on freedom of speech, this tool encouraged citizens to submit anonymous text messages which were automatically projected onto the facade of a building.

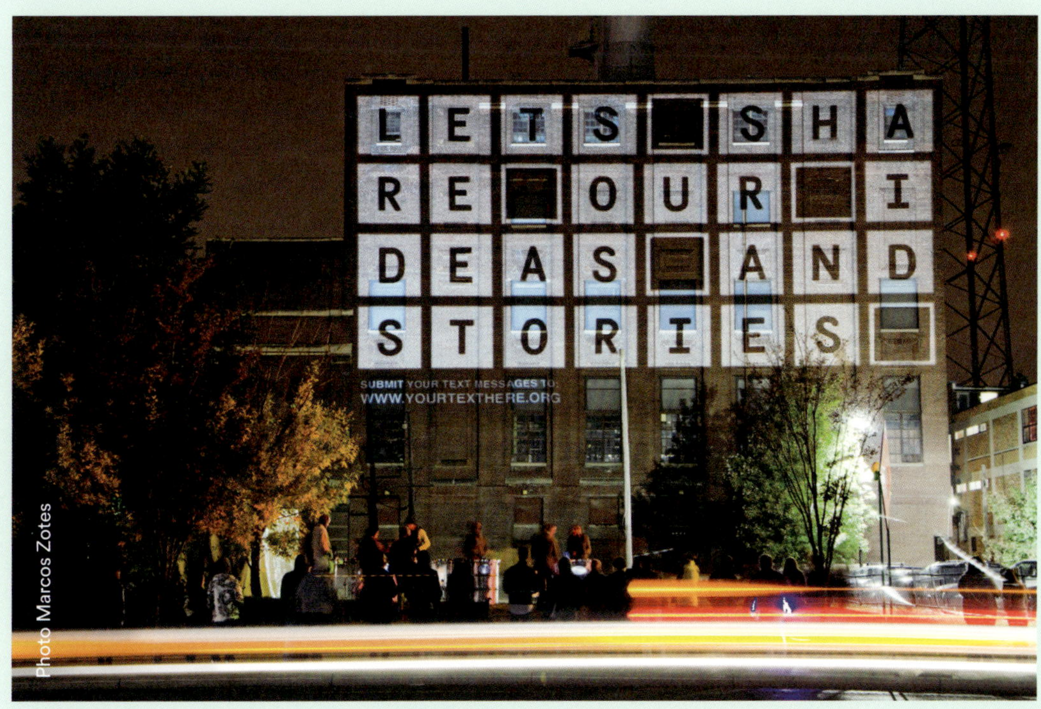

Photo Marcos Zotes

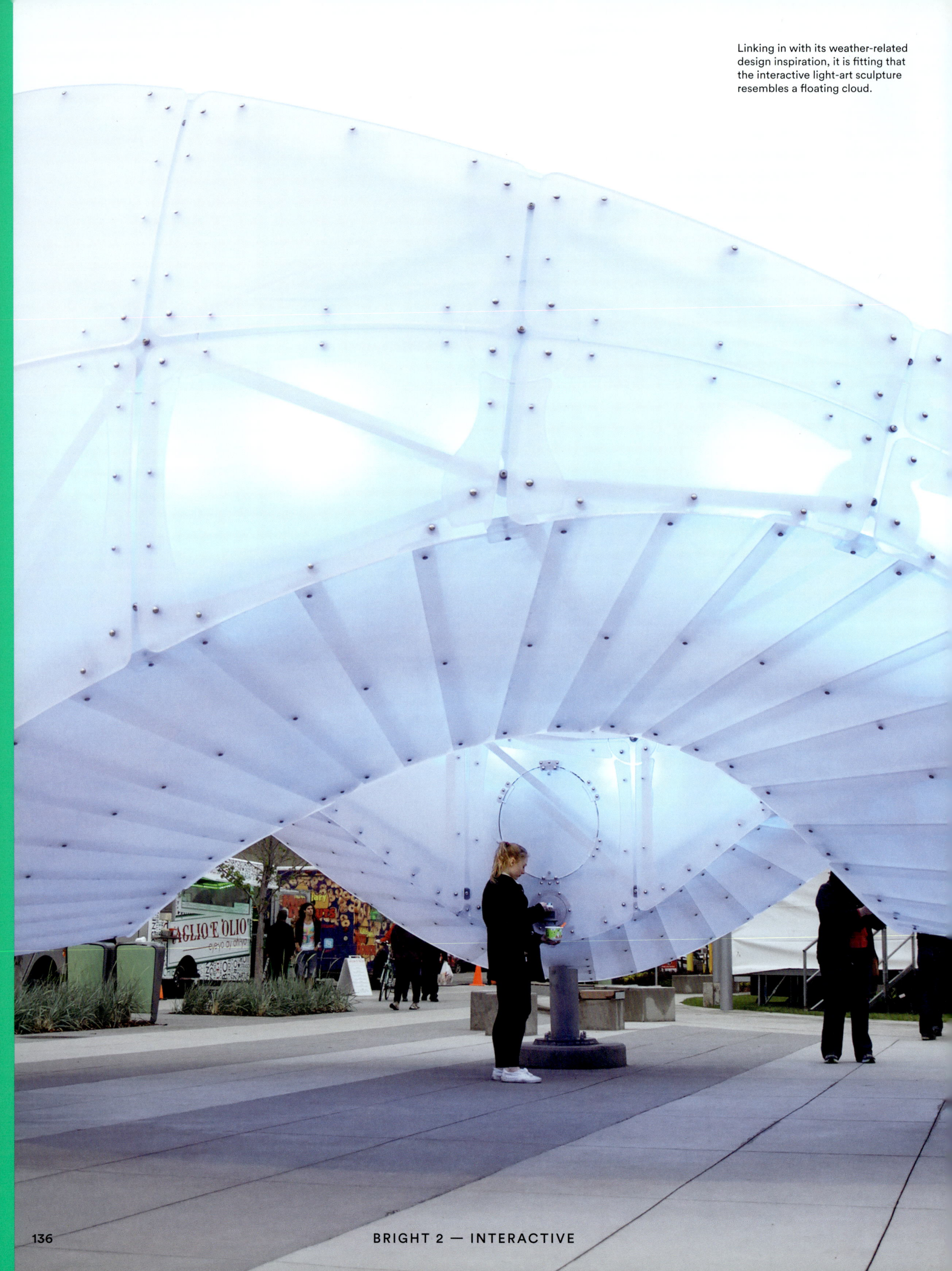

Linking in with its weather-related design inspiration, it is fitting that the interactive light-art sculpture resembles a floating cloud.

Chinook Arc

A permanent public installation, Chinook Arc by **Creative Machines**, boasts a unique visual identity that looks to reflect the local community. Located at the Barb Scott Memorial Park in Calgary, this artwork engages with its surroundings and citizens at the same time as weaving together the site's history and its contemporary context.

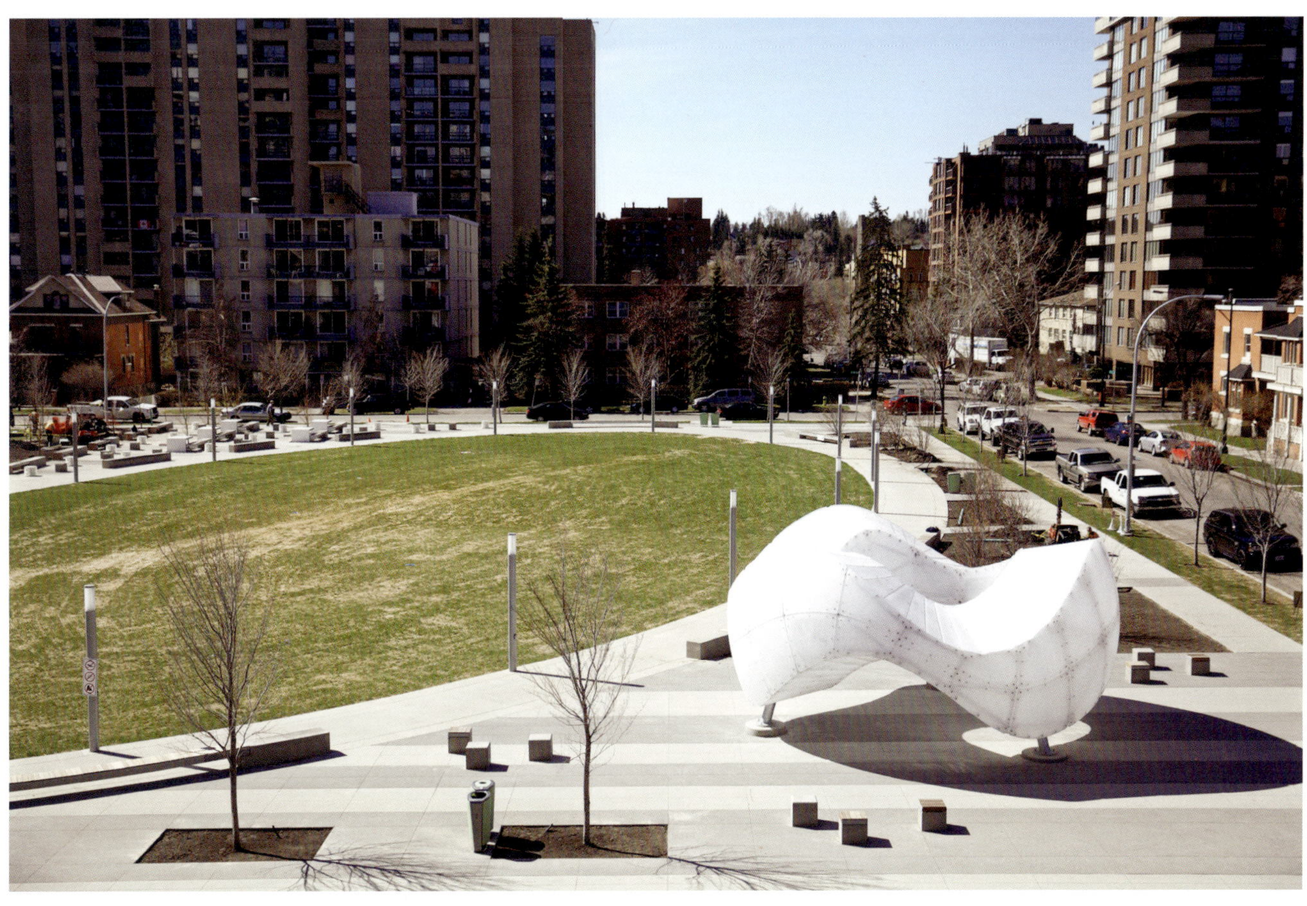

Photos Paul McGrath, Neil Zeller

Artists Joe O'Connell, founder of Creative Machines, and Blessing Hancock developed the winning design for a permanent public artwork for the Barb Scott Memorial Park in Calgary, Canada. The project sought to create a distinctive structure that would reflect the diversity of the local community. The permanent installation is an interactive, illuminated sculpture made up of a steel skeleton and a translucent acrylic skin that allows it to emit a soft, internal glow.

The structure's crisp edges and rounded curves were inspired by the historic streetcar loop that once encircled the neighbourhood and the 'Chinook arch clouds' that are a common weather phenomenon in this region. Breaking with the stern lines of the surrounding architectural landscape, Chinook Arc's soft compound shape adds to the site's visual richness. During the day, sunlight lends the sculpture an ethereal, semi-translucent appearance as it illuminates the acrylic skin. Through a play of light and shadow, the piece offers an opportunity for calm contemplation, presenting itself as a source of continuity in the midst of the city's constant growth and change. By night, the installation becomes an immersive, colourful environment. Interactive lighting invites passers-by to engage with the sculpture, and gives it an emotive, perceptual character akin to a living organism.

The artists looked to push the boundaries of public art by developing innovative structural and technological solutions. Structurally, a system of silicone risers allows the two-layer acrylic outer skin to 'float' above the steel skeleton, giving it space to contract and expand with temperature changes. Technologically, the artists created the bespoke sensor and lighting system that gives the sculpture a strong sense of personality in the process of visitors actively engaging and interacting with it. The sensor reacts to a number of visual and audio stimuli, such as waving hands or coloured clothing, or playing a music video on a smartphone. Programmed lighting sequences can also animate the sculpture at times of no public activity, meaning a constant and dynamic attraction in the neighbourhood. —

At night, the sculpture's lighting reflects its strong personality at the same time as it is open to the public's influence.

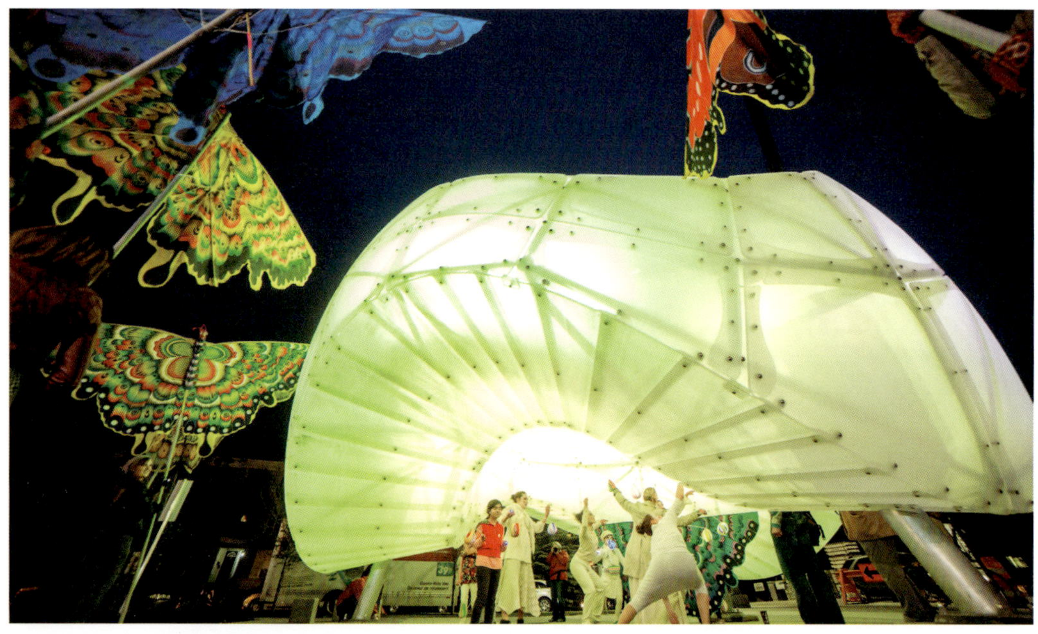

INTERACTIVE LIGHTING GIVES AN EMOTIVE, PERCEPTUAL CHARACTER AKIN TO A LIVING ORGANISM

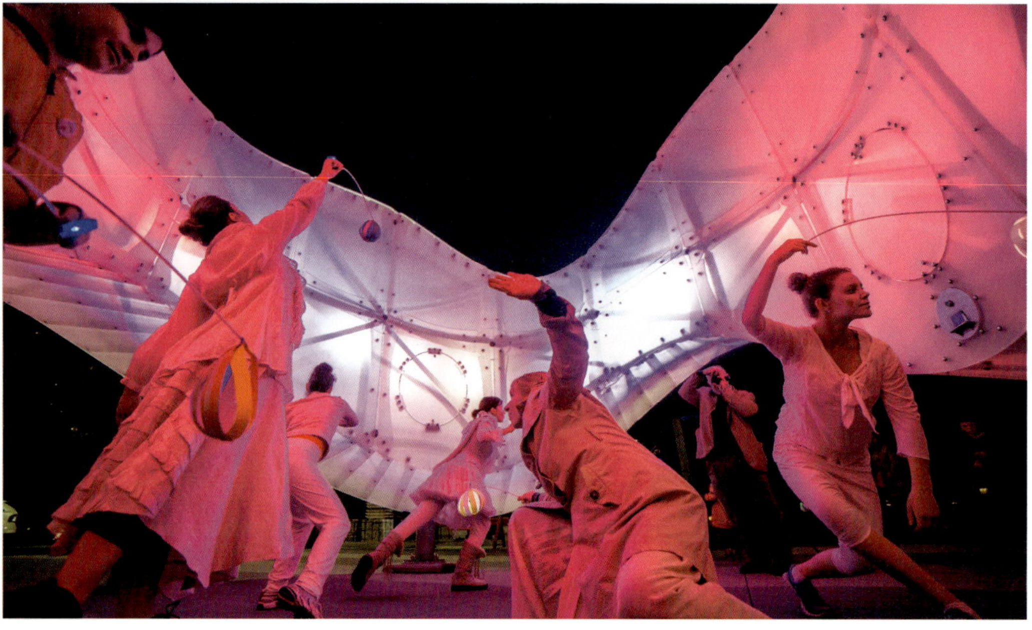

Designer
Creative Machines
Location
Calgary, Canada
Client
City of Calgary
Public art manager/consultant
Barb Doyle-Frisch, David Moses
Manufacturers
Creative Machines, Delta Structures
Date
April 2014

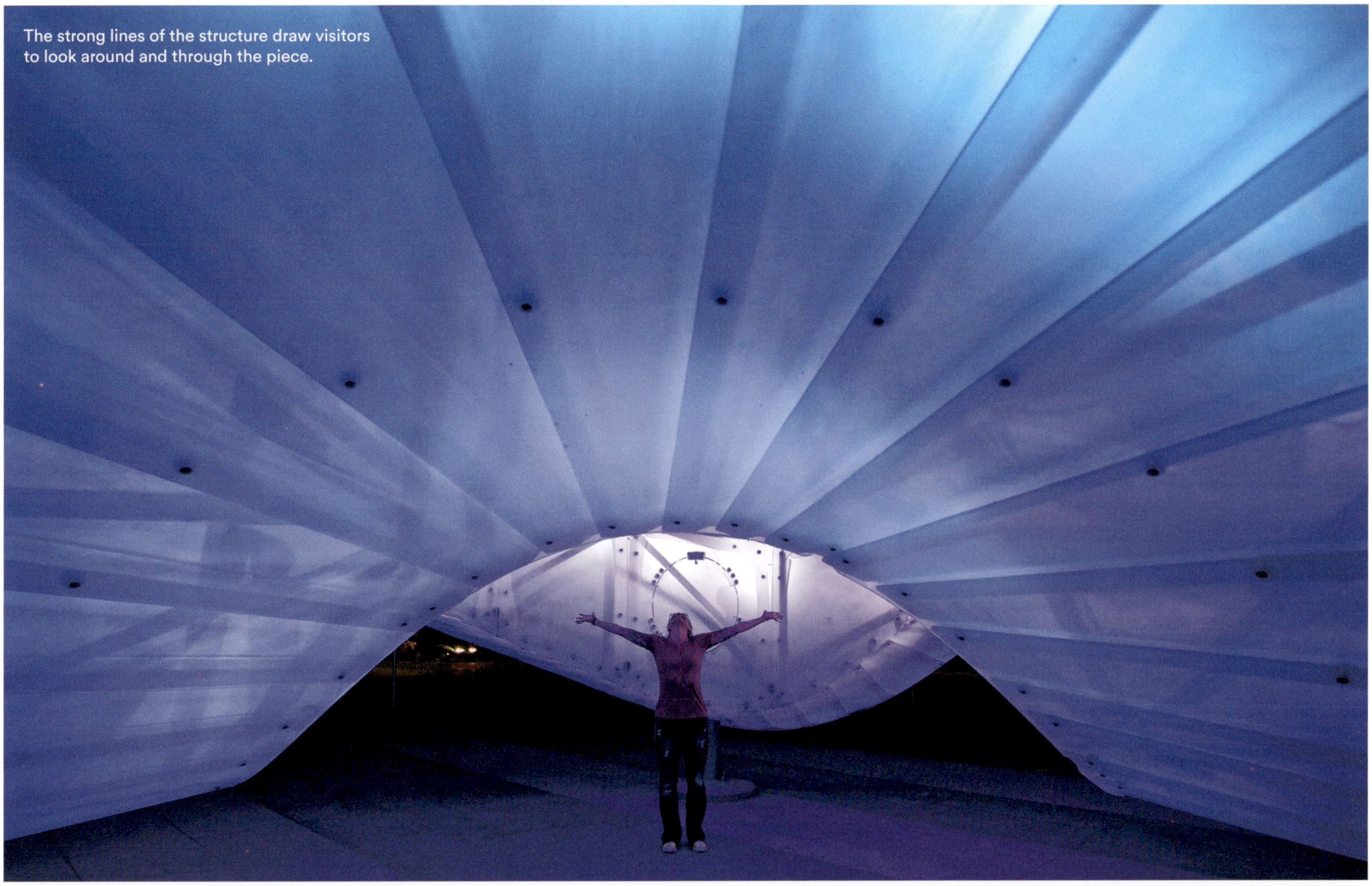

The strong lines of the structure draw visitors to look around and through the piece.

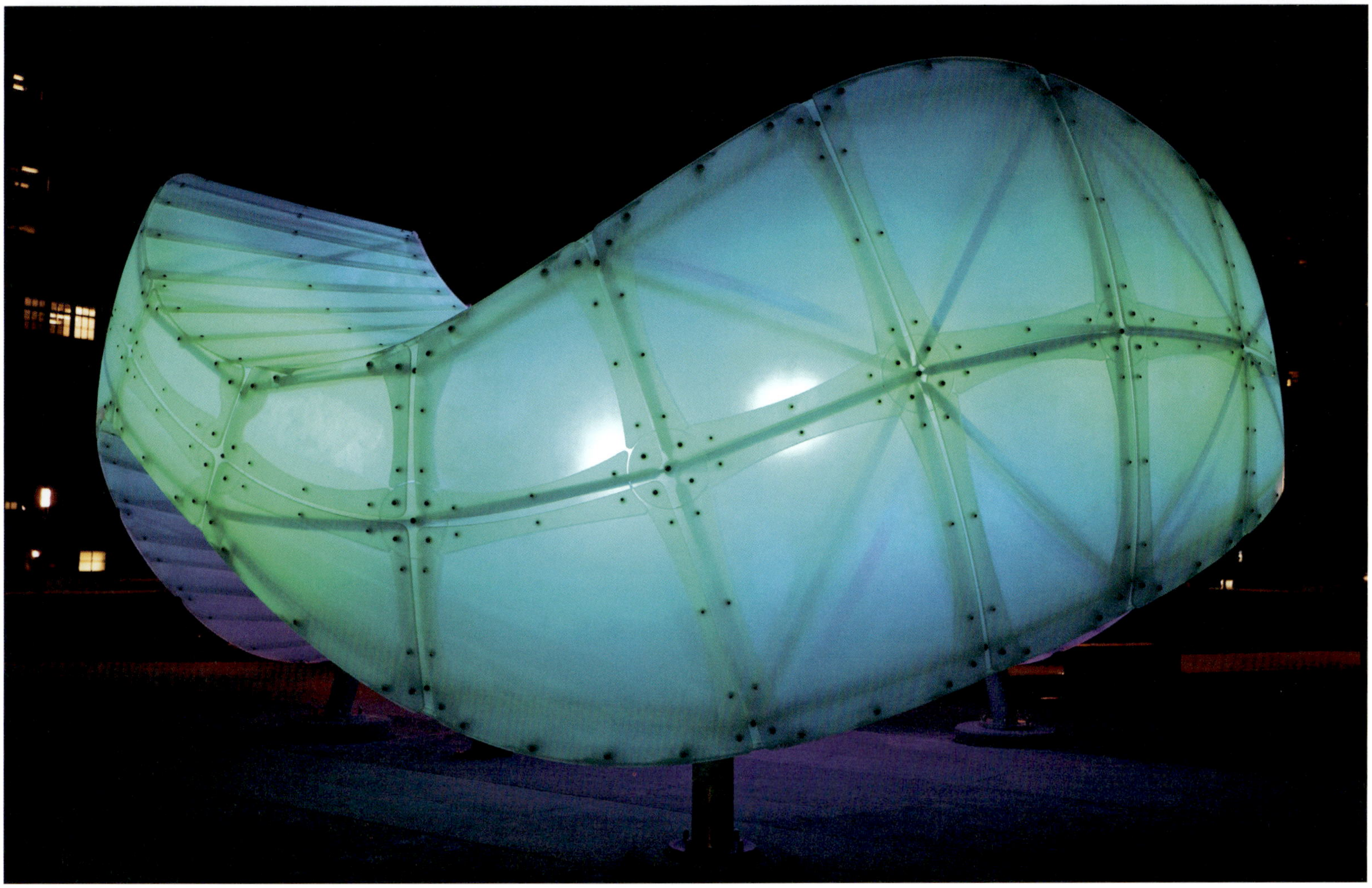

SPACE FRAME LAYOUT

CLAD AND BOLTING LAYOUT

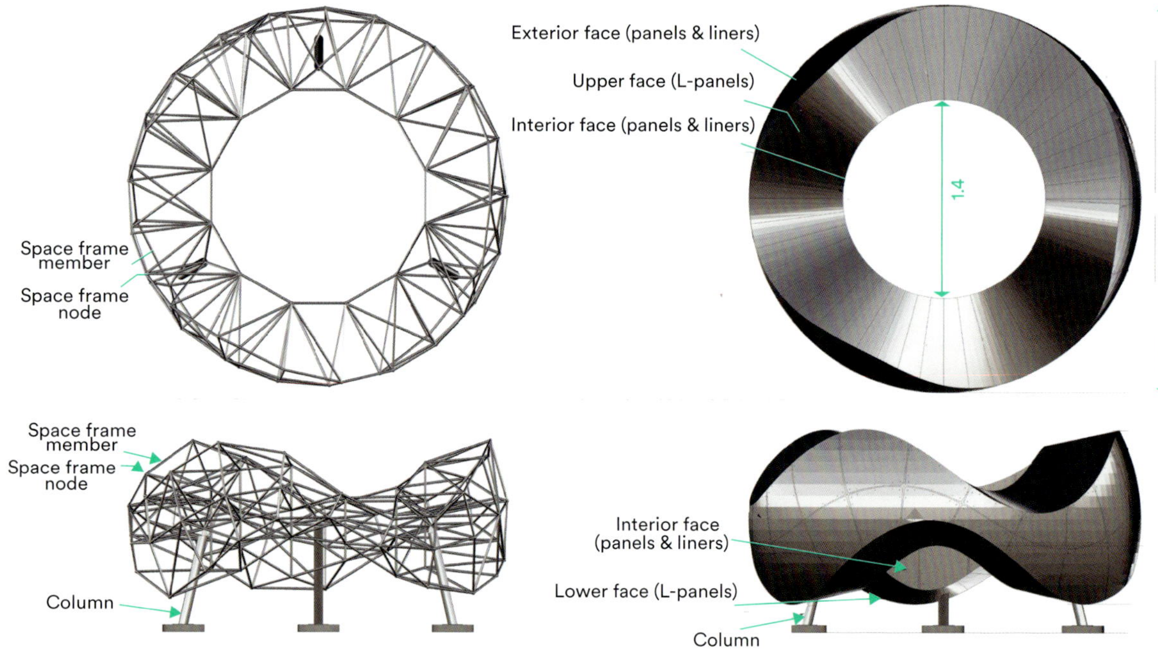

Set-up

Used light included 48 custom-designed LED point source luminaires. Each RGB LED fixture had a brightness of approximately 30 to 170 Lux, depending on the colour.
Visual and audio stimuli were detected using motion sensors and various software: a VersaLogic Tomcat Single Board Computer, an ImagingSource Industrial Camera, a custom interface board for Tomcat and a custom panel that allowed a wide range of smartphones to be placed for hands-free use.
Motion sensors were a custom-designed designed system, programmed by Creative Machines. The imaging camera assimilates visual data and the computer translates it into information that can be used by the lights in real-time.
CSA certification was given to the custom lighting and sensor system, as per the city's requirements.
Dimensions of the sculpture totalled 38.25 m^2, with a diameter of 8.5 m and reaching a height of 4.5 m.
Total area of the custom heat-formed illuminated acrylic panels was 172 m^2.
Custom-designed frame parts totalled a number of 359.
Construction time for the entire sculpture was 12 months.

ELONGATED ELEVATION

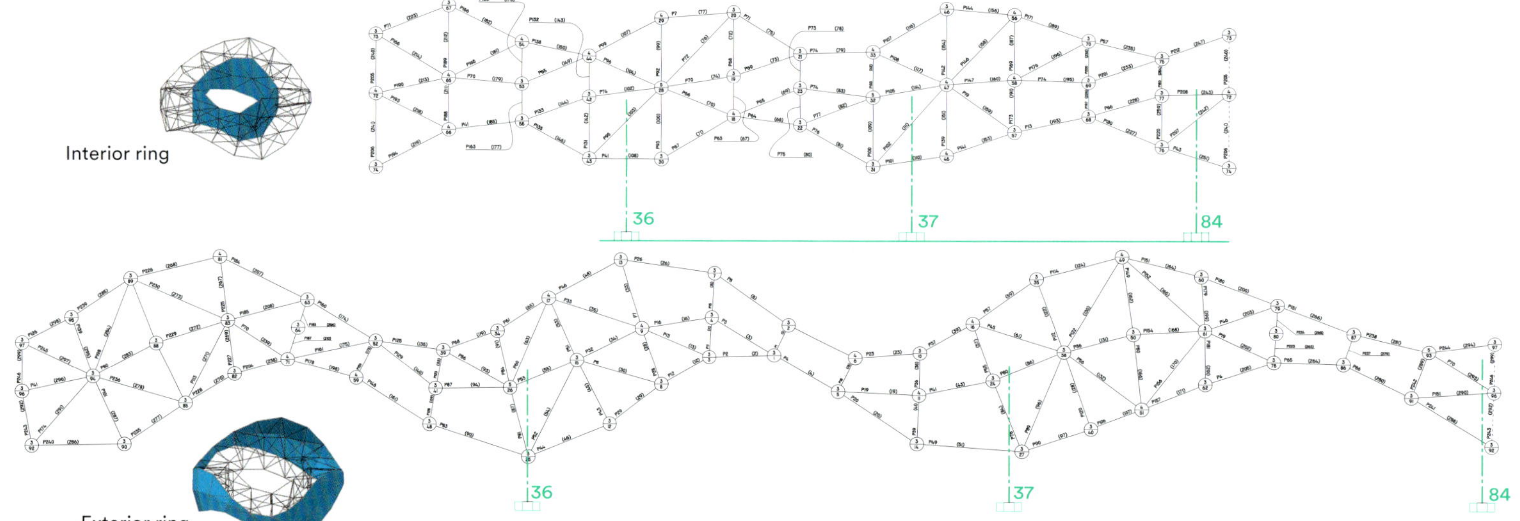

WEB DIAGRAM

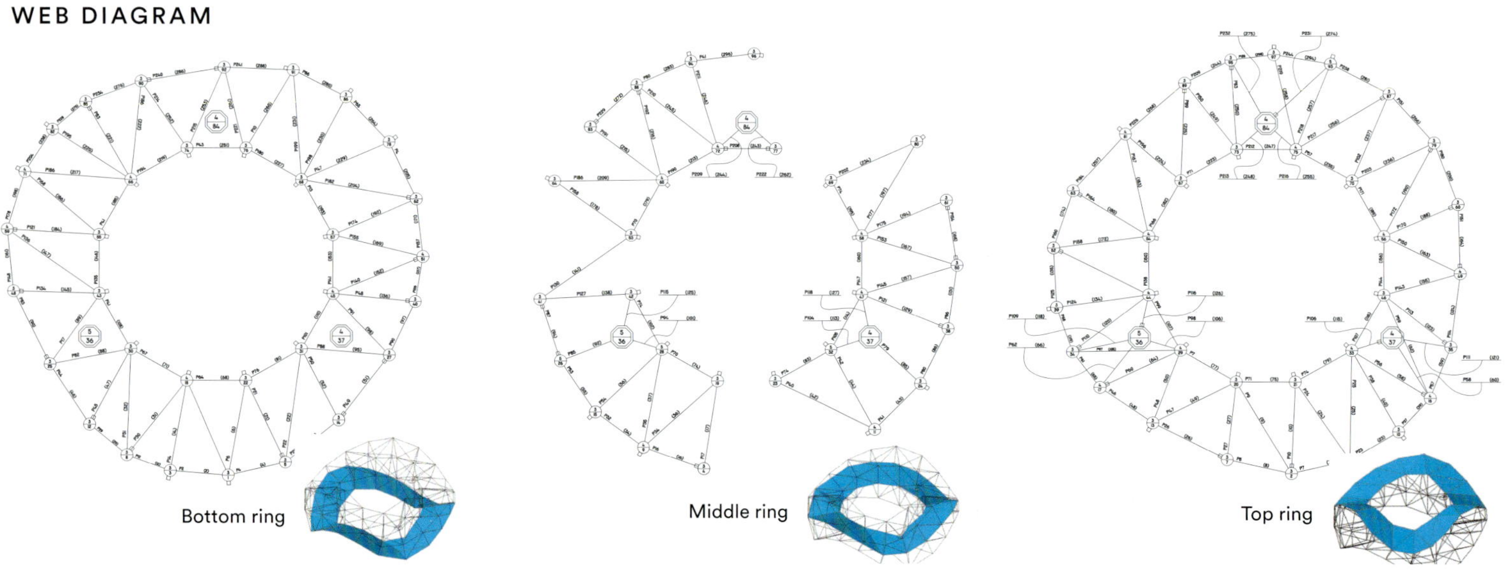

EXTERIOR FACES ELONGATED

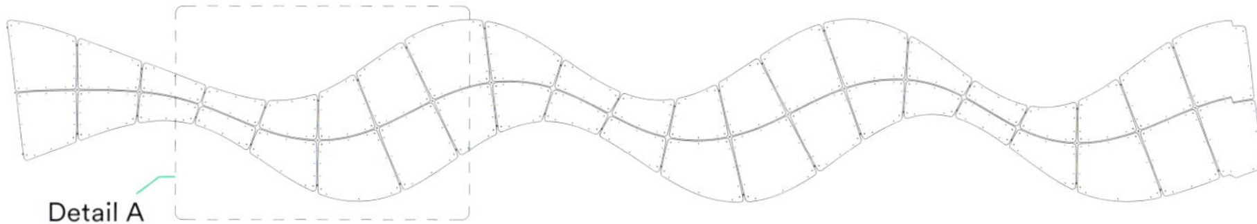

INTERIOR FACES ELONGATED

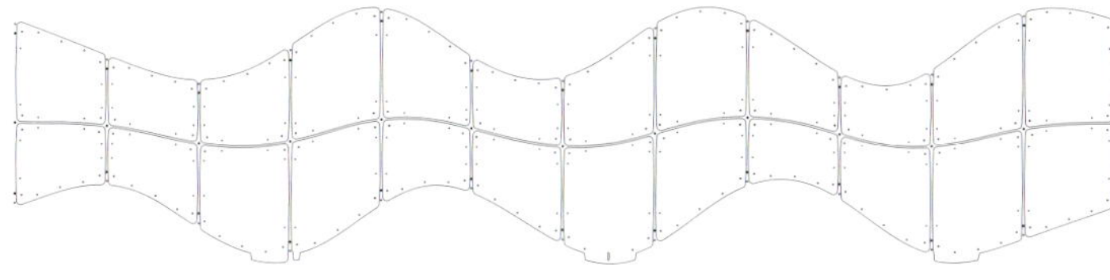

DETAIL A

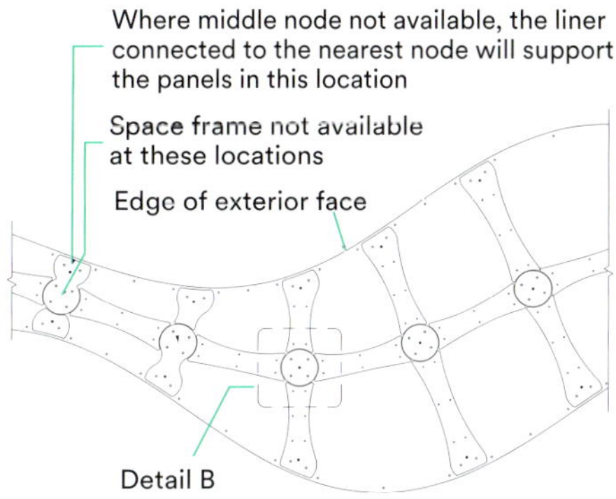

DETAIL B

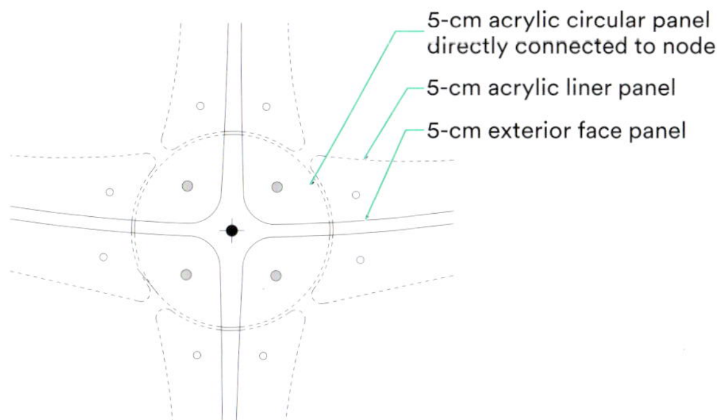

Specifics

To achieve the installation's main goal of creating a distinctive structure that would reflect the diversity of the Beltline community, the artists visited the site and spoke with local residents. Through this experience, the team heard stories about the neighbourhood and the Chinook wind phenomenon (referencing a local weather system), which all provided inspiration for the installation.

Since the sculpture had to endure Calgary's extreme weather and unsupervised public contact, the frame, brackets, cylindrical risers, acrylic panels and electronics all had to be custom-made. This presented one of the main challenges for Creative Machines.

The structure's acrylic skin and steel interior have very different thermal coefficients. This problem was resolved by making the acrylic skin in two layers that bypass each other and leads to all the acrylic components effectively 'floating' on a system of silicone risers so that there is no structural acrylic-to-metal contact and the acrylic is free to expand and contract on its steel skeleton.

Numerous physical extremes for the structure were considered, ranging from graffiti (solution: a frosted surface that can be renewed by sanding) to the wear-and-tear of people climbing on the installation (solution: rigorous engineering tests allowed for up to ten people playing on the structure at one time).

Heat-sinking, sealing and protecting the optics were three of the key challenges when dealing with the LED light fixtures for the permanent public installation. Aluminium heat-sinks, optically-clear two-part silicone capsules and clear acrylic panels were developed to solve each problem, respectively.

Since one of the design goals was that the structure's skeleton be visible through the shadows cast by point source lighting, light placement presented another challenge. To avoid distracting shadows, wire and fixtures were carefully placed and a curving all-weather cable tray was designed.

MODELS

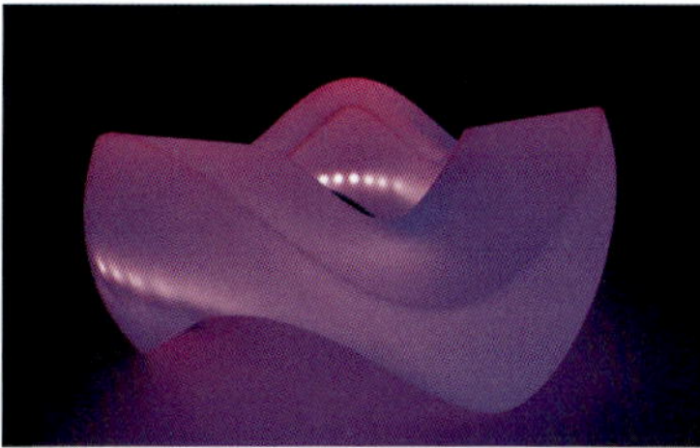

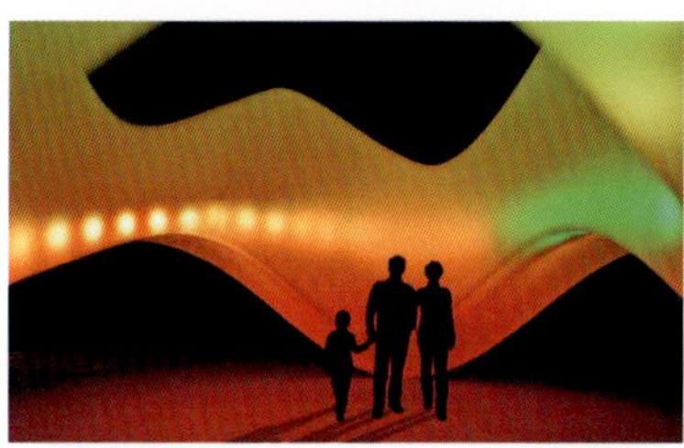

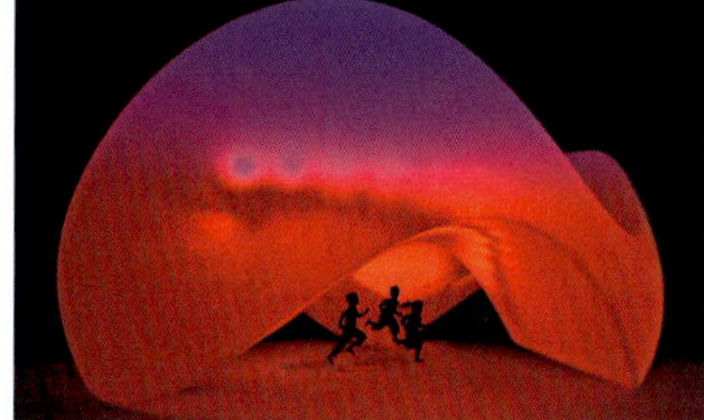

CONSTRUCTION VISUALS

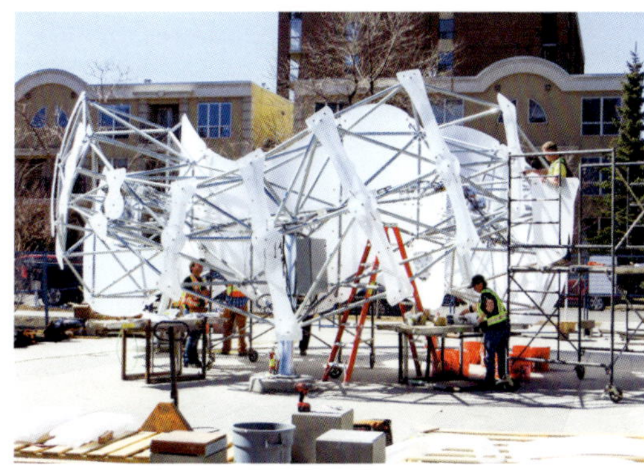

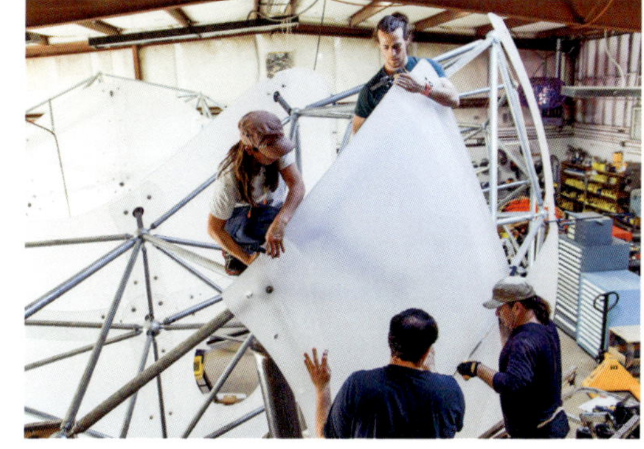

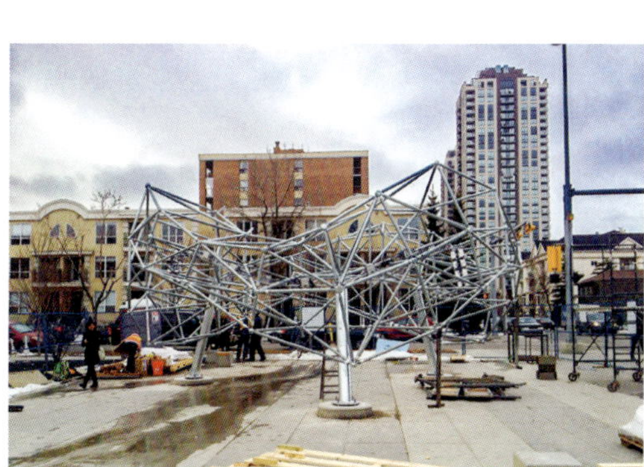

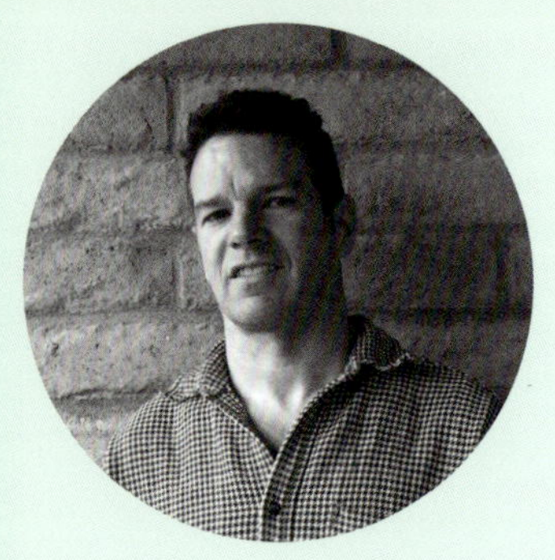

Creative Machines

Creative Machines was founded by Joe O'Connell in 1997 and now has a team of 19, including artists, designers, engineers and fabricators. As collaborators, O'Connell and the members of his design and fabrication studio work together to create art that is not just aesthetically pleasing, but goes deeper and rewards the audience's sustained engagement. Nature and technology come together in the studio's work, engaging in interactive processes that push the boundaries of existing manufacturing practices – with light as the connective yarn that is weaved throughout the work.

CREATIVEMACHINES.COM

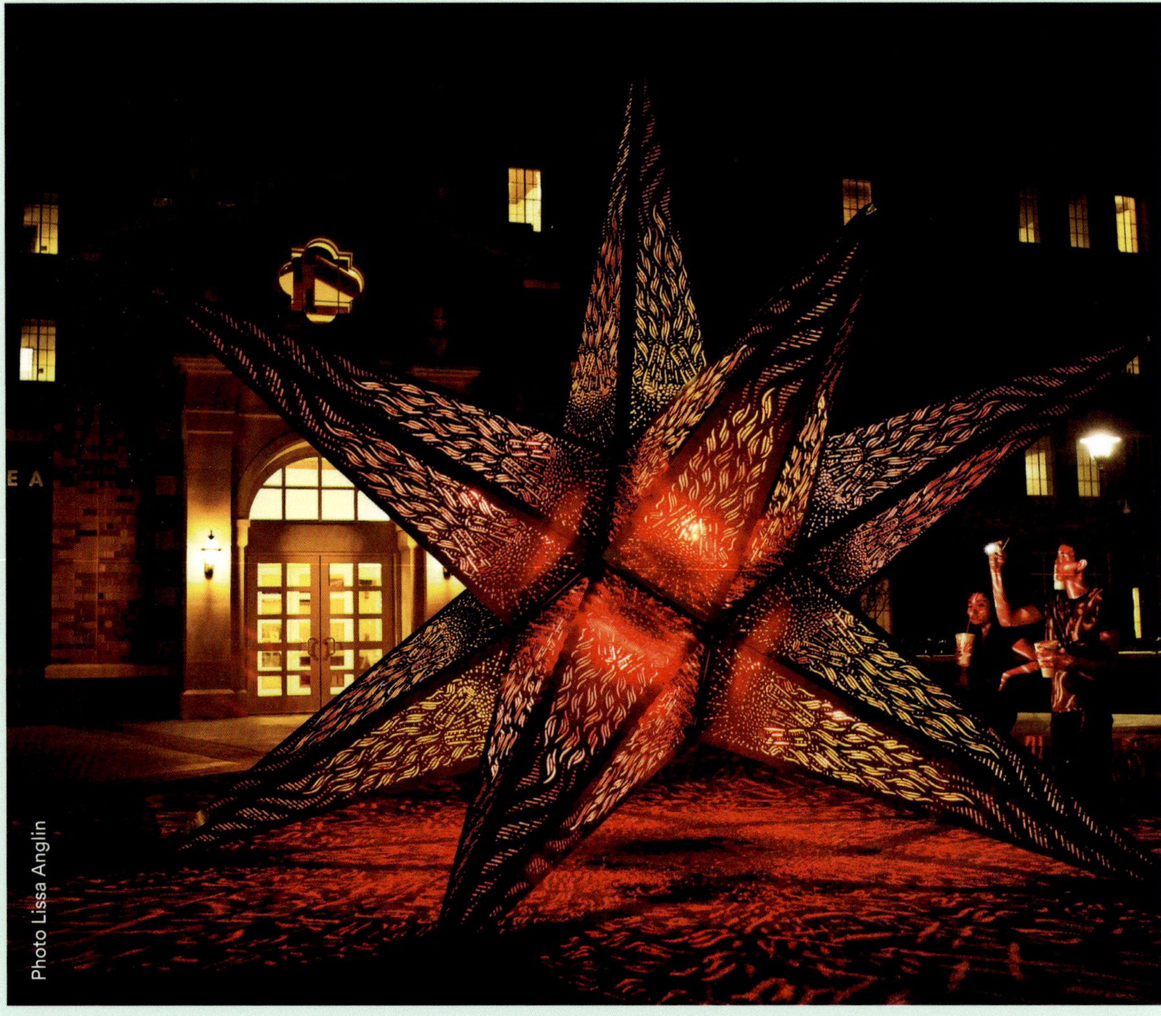

Photo Lissa Anglin

Texas Rising

When **2014**
Where **Lubbock, United States**
Client **Texas Tech University**

The five sculptures and two chandeliers that make up Texas Rising portray different stages of a star rising out of the ground. During the day, sunlight causes the stars to shine, but at night it is LEDs that make them shine, allowing the 2D patterns to cast intricate shadows on the surrounding surfaces.

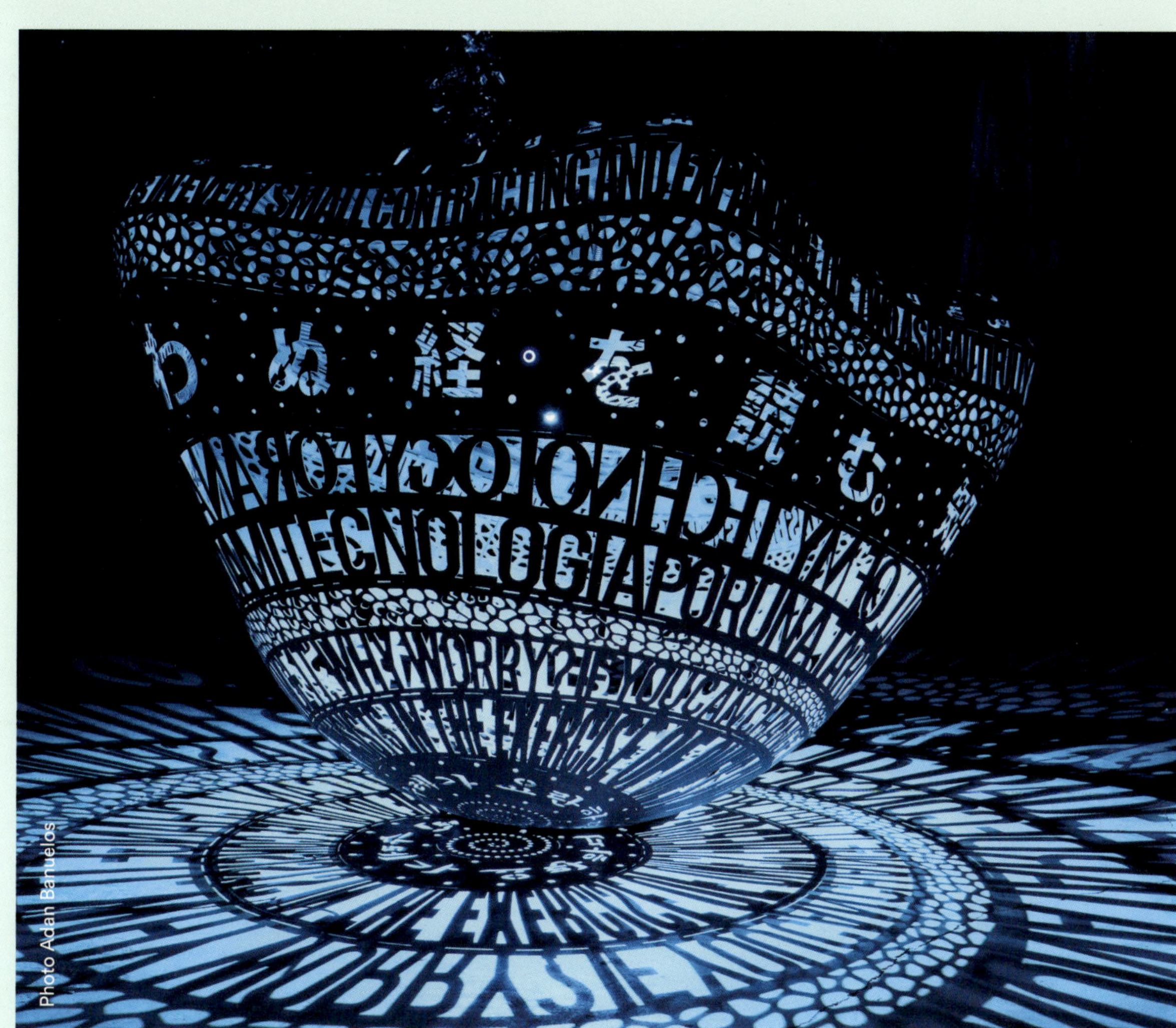

Photo Adan Banuelos

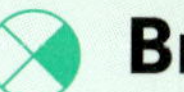

Brilliance

When **2014**
Where **City of Palo Alto**
Client **Palo Alto, United States**

Six stainless steel sculptures placed through a plaza between Palo Alto's Main Library and the city's Arts Center function as lanterns by night. Entitled Brilliance, each sculpture is made up of a series of multi-lingual phrases collected from the community. These were cut out and welded together in various lantern-like shapes.

Fish Bellies

When **2013**
Where **San Marcos, United States**
Client **Texas State University**

Large biomorphic forms, inspired in the social and biological diversity of the San Marcos river, make up Fish Bellies. The installation, translucent by day and bioluminescent by night, functions as an analogy between the river's ecological life and the varied student body of Texas State University.

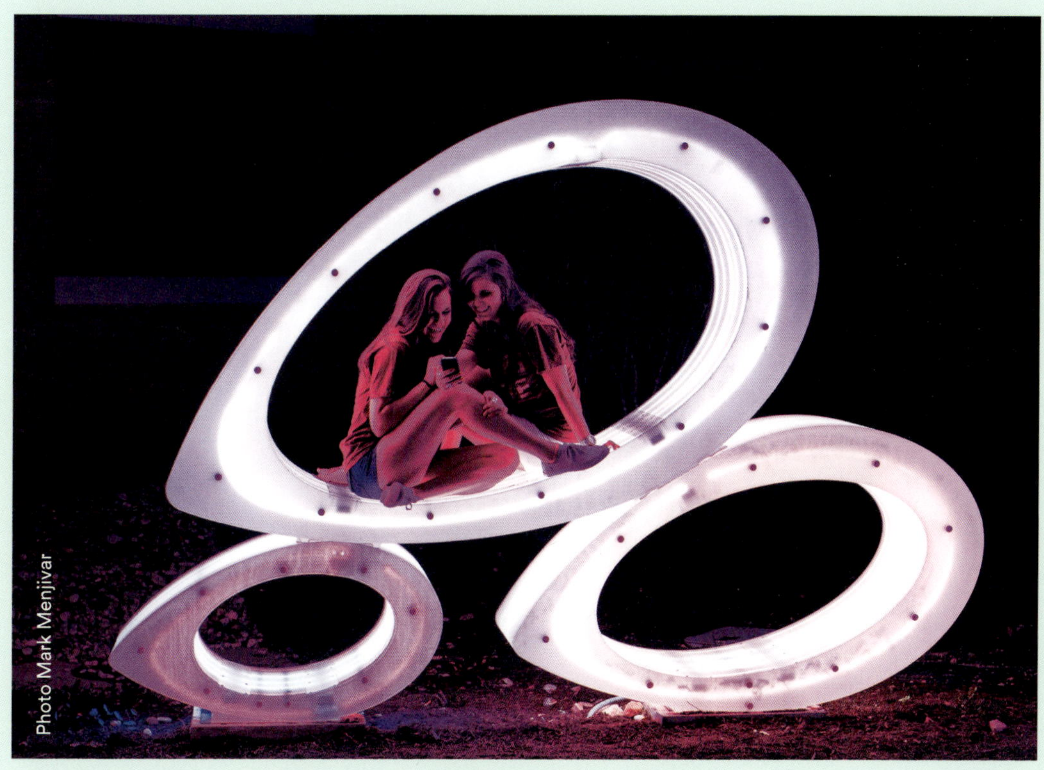

Photo Mark Menjivar

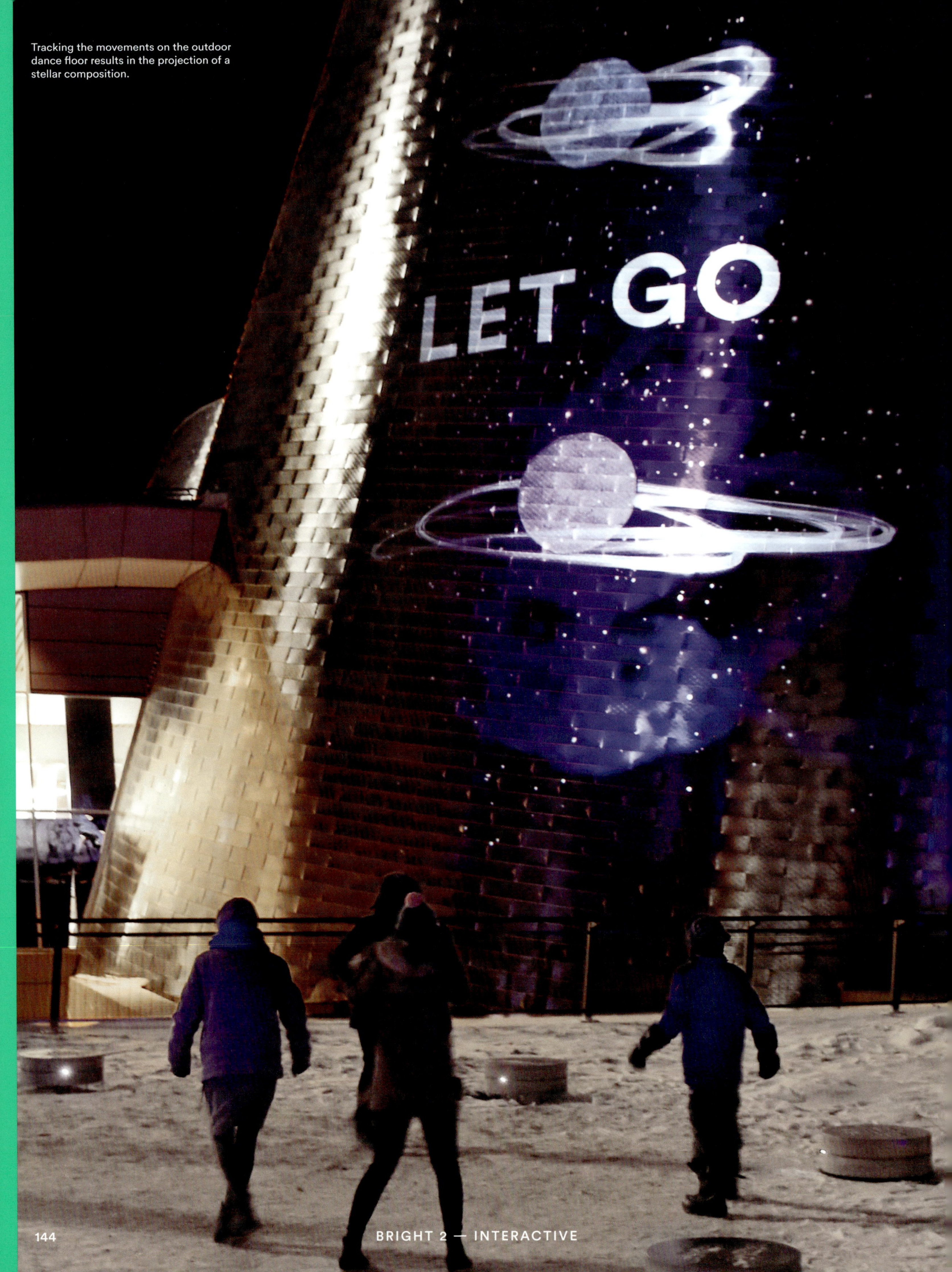

Tracking the movements on the outdoor dance floor results in the projection of a stellar composition.

Choreographies for Humans & Stars

Creating a permanent installation for Montreal's Planetarium, **Daily tous les jours** lights up the entrance area with an interactive venture based around projections on the building's facade. An open-air dance floor urges visitors to interact with each other, connecting them to the overpowering immensity of the universe by relating the movement of their bodies to celestial motion.

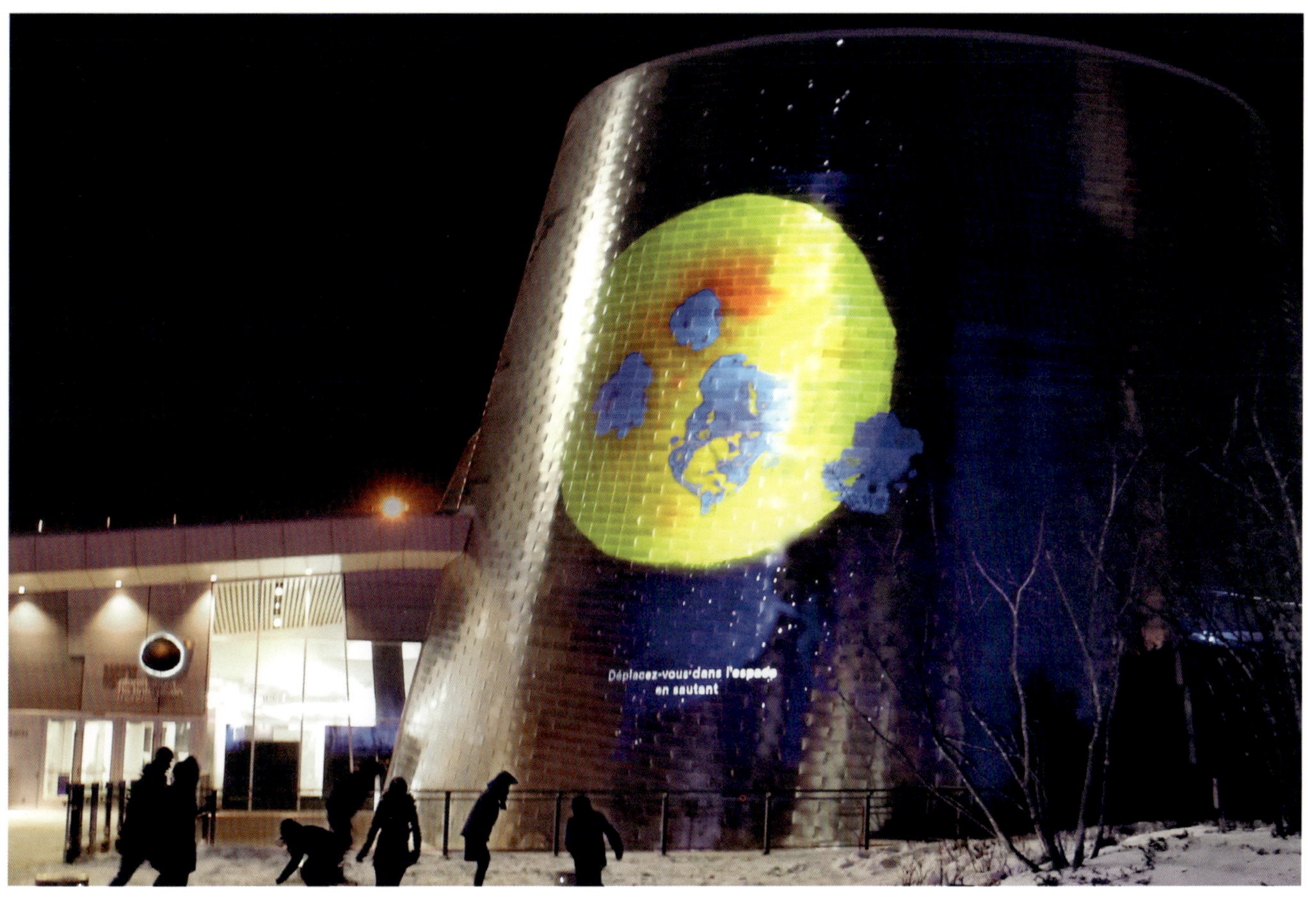

Photos Geoffrey Boulangé

Choreographies for Humans & Stars is a permanent, interactive installation that illuminates the entrance of the Planetarium in Montreal, Canada. Inspired by the immensity of space, the exploratory dance floor uses something familiar – the body – to connect people with the mysteries associated with the universe's infinity. As visitors reach the front of the building, they are invited to follow the instructions projected onto the building's facade.

The illuminated visuals turn out to be a guided choreography for the participants to follow step-by-step in the outside space that has been transformed into a dance floor, resulting in a stellar composition being projected as movements trigger different reactions that are played back on the facade. During the day, when the projection is off, the instructions can be read on the surface of the seven stones that are anchored into the ground, delineating the dance floor. Together with choreographer Dana Gingras, the design team explored different ways in which participants could use their bodies to inspire celestial dynamics. Since the responsive tracking and projection software is programmed to interpret both good movements and 'mistakes' as triggers themselves, it is able to create visual responses as much when the system tracks people as when it loses them. In this way, the dance instructions and pre-defined steps that have been devised leave space for moments of free interpretation by each participant.

Due to its permanency, it was important for the designers that the project be well received by, and integrated into, the local community. To this end, the team invited people to participate during the development phase of the project. Visuals for the projection emerged from the collaboration of visual director Patrick Péris and local youth in workshops designed to provide creative context and guidance through basic animation techniques: light painting, macro video and stop motion. —

Designer
Daily tous les jours
Location
Montreal, Canada
Client
Montreal Planetarium
Collaborator(s)/consultant
Dana Gingra, Patrick Péris, Matane Production, Société des Arts Technologiques, Matane Production
Manufacturer
M3 Béton
Date
January 2014

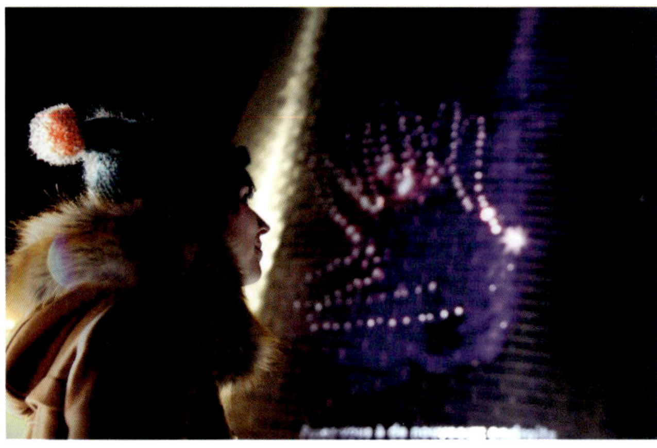

GUIDED CHOREOGRAPHY RESULTS IN AN ILLUMINATED, STELLAR COMPOSITION

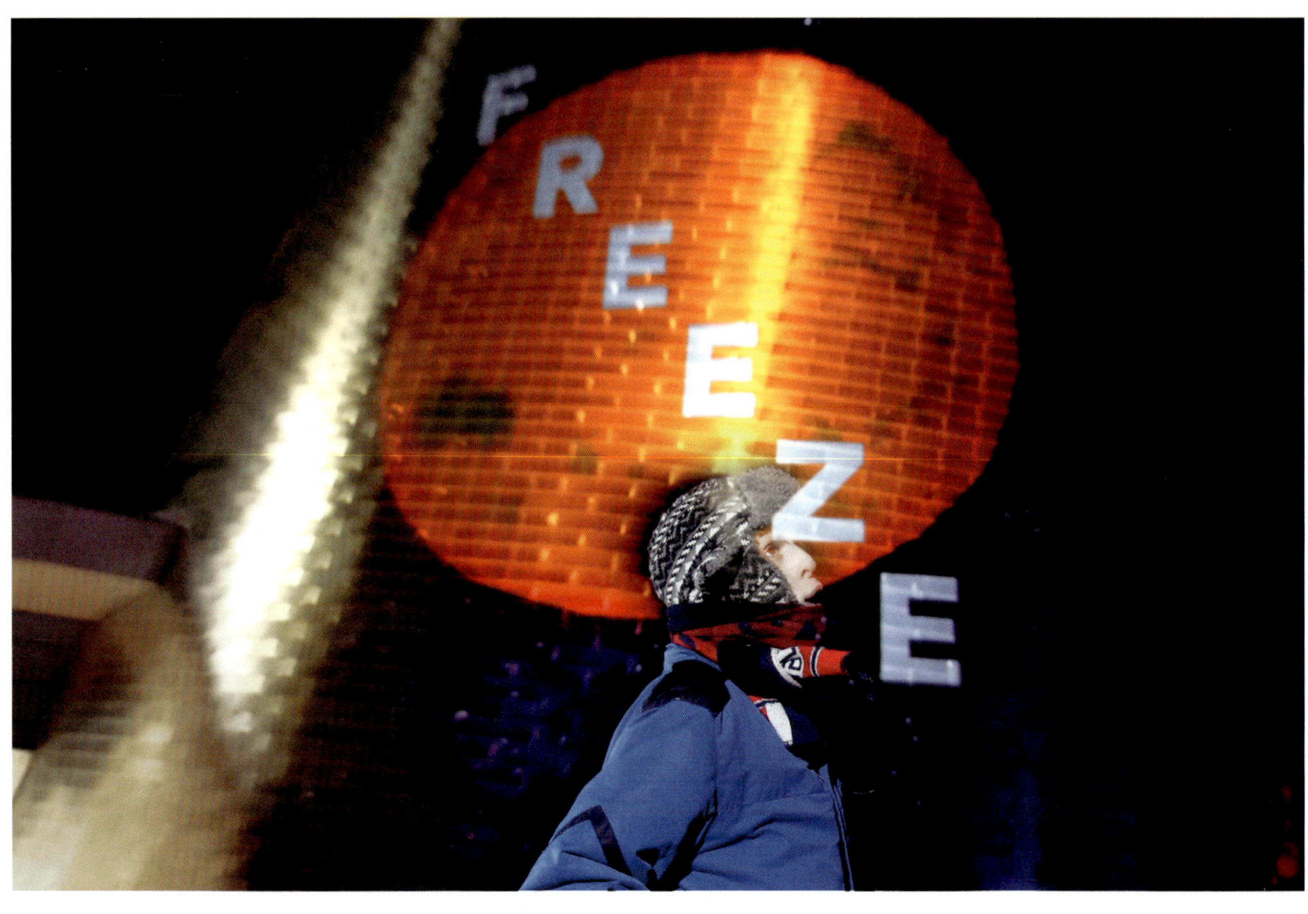

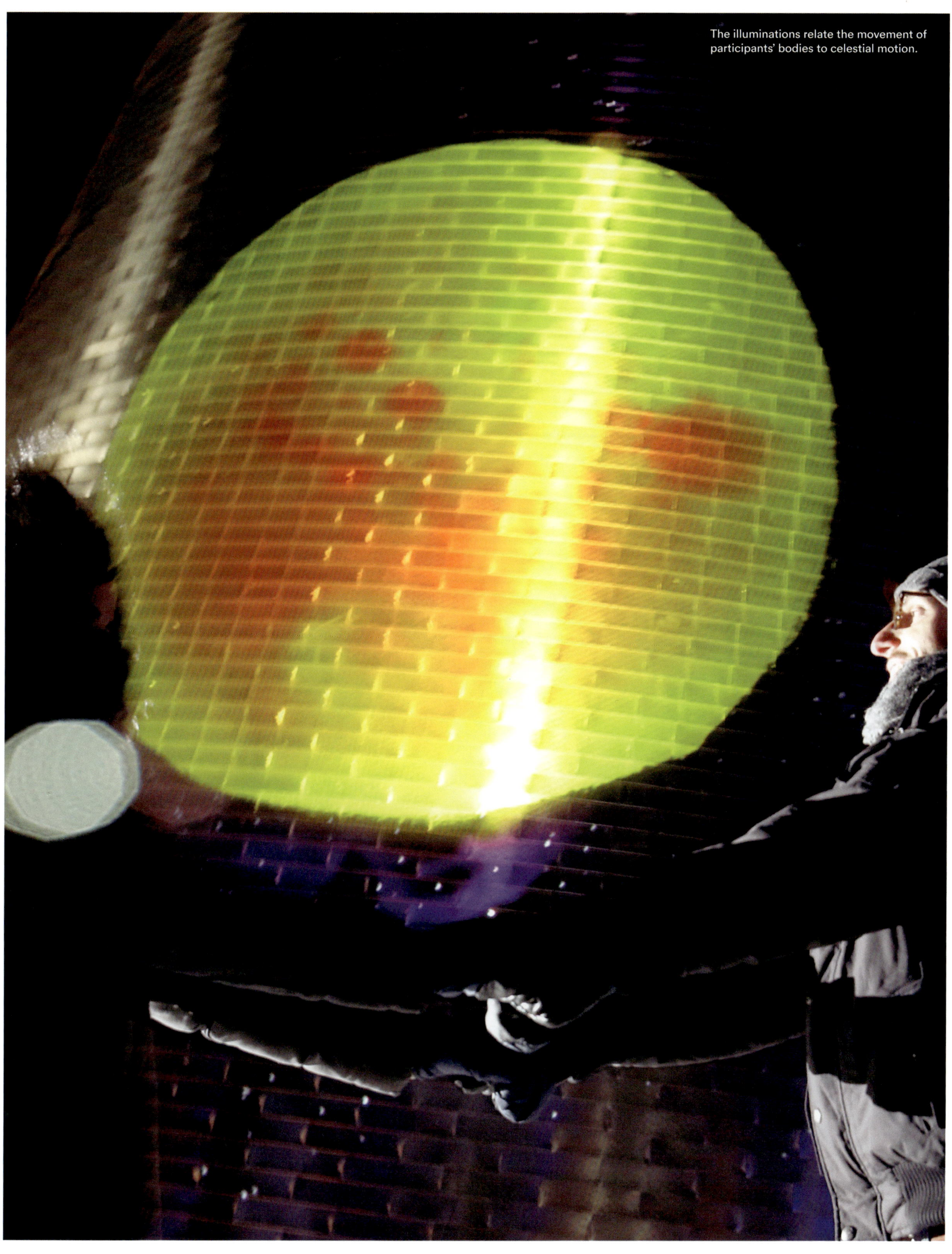

The illuminations relate the movement of participants' bodies to celestial motion.

PLAN

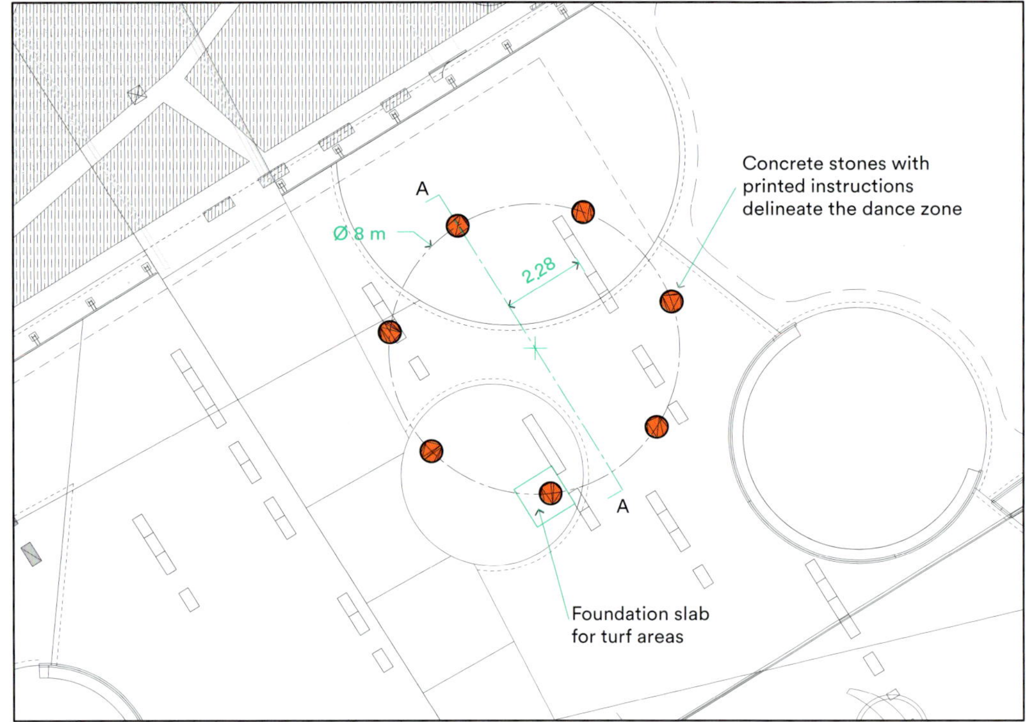

Set-up

Visualisation software used was OpenFrameworks, due to the C++ processing speed and the add-ons that were available.
Development of the visualisation software, including transformation of perspectives and implementation of video content, was done using OpenFrameworks.
Dimensions of the projection surface had a height of 20 m and a width of 8 m.
Materials used in the projection surface were aluminium tiles. Following initial tests, it was found that the shininess of the tile surface did limit the projection quality. The imagery was therefore designed specifically to work within the identified physical constraints.
High-lumen projector Barco HDX-W20 was used to cover a large area of the building's surface.
Movement-tracking camera used was a Basler Scout scA1600 with high sensitivity lens Fuji HF12.5SA-1 and a tracking solution that could convert the 2D information into 3D.
Tracking detector used was blobserver's HOG detector, which learns from a database of human shapes to detect human outlines and track their movement through space.

INSTRUCTIONS

ATTRACTION

STAND IN FRONT OF SOMEONE
MAKE EYE CONTACT
START WALKING TOGETHER
MAINTAIN THE DISTANCE
GO FOR AS LONG AS YOU CAN
WITHOUT BREAKING
EYE CONTACT.

SECTION A

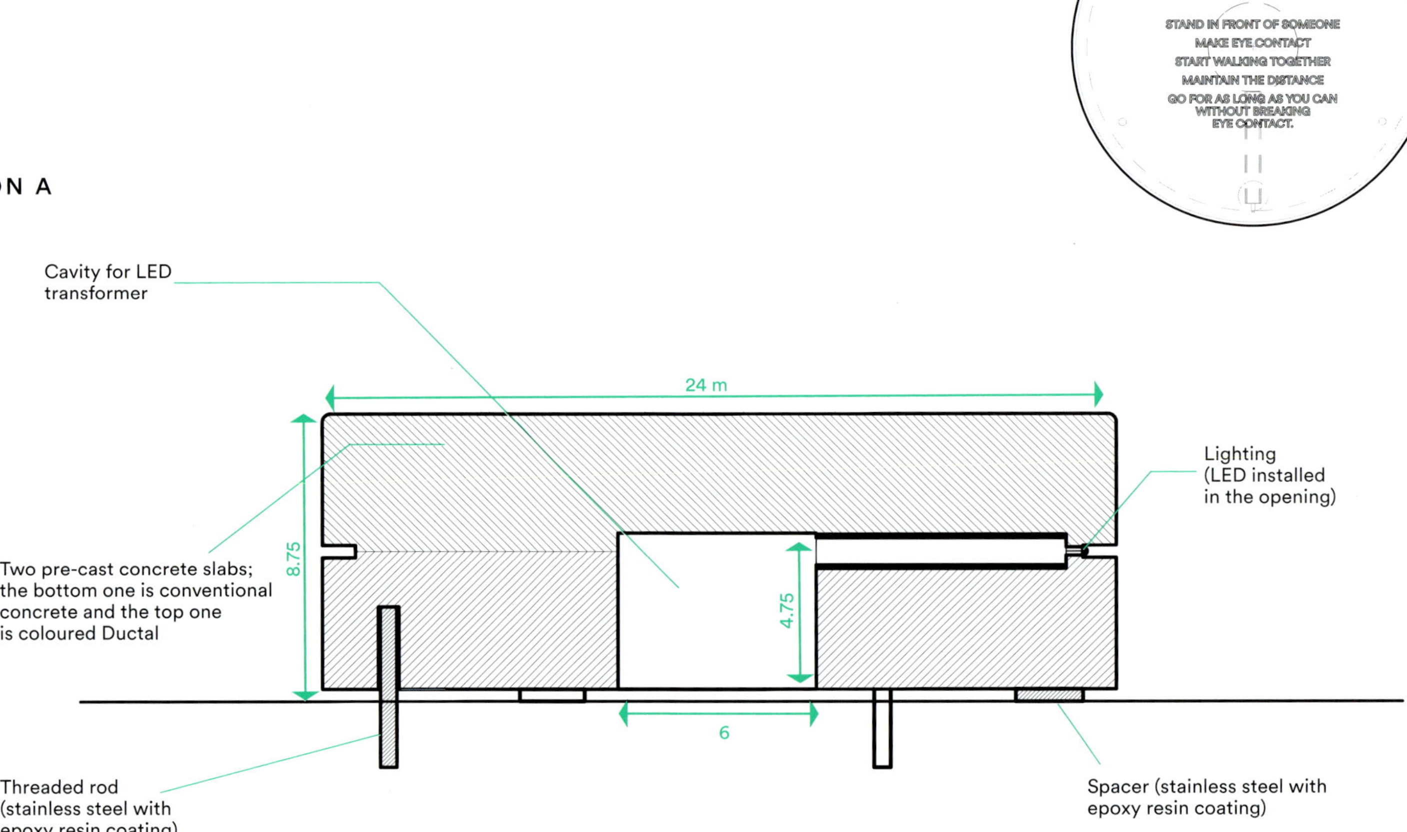

LAN

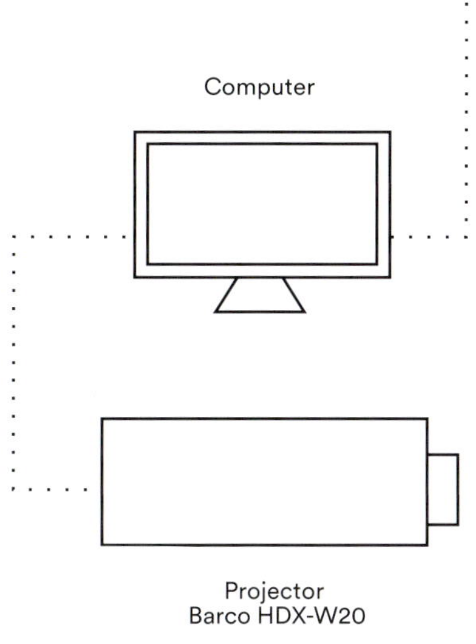

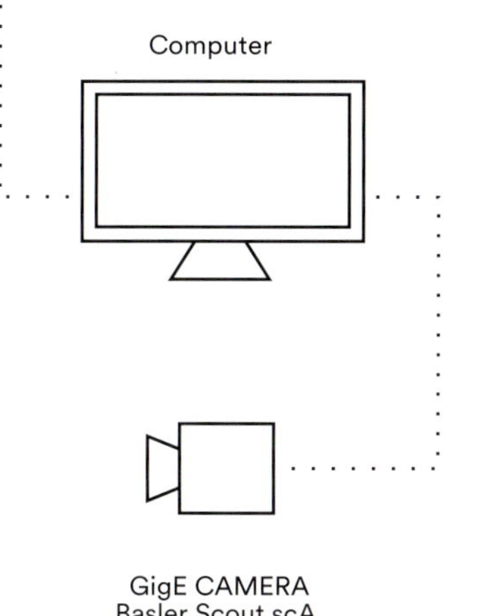

MOVEMENTS TRIGGER DIFFERENT REACTIONS PLAYED BACK ONTO THE FACADE, ILLUMINATING THE BUILDING

Specifics

The site's characteristics posed some challenges to finding an adequate movement-tracking camera for accurate responsiveness. For instance, being outdoors, the installation is subject to weather conditions that can radically change the surroundings, like snow. After thoughtful consideration of weather, ground and lighting conditions, the designers settled for a Basler Scout, a camera that responds well to the unique challenges posed by the site.

The project's biggest challenge was finding a compromise between detection precision and detection forgiveness in the tracking system. Participants were tracked in the project's dance zone by a histogram of oriented gradients, which uses a database of human shapes to detect human outlines. This system is optimal in controlled environments where ideal-case scenarios can be created, which isn't the case in public, outdoor setting of this project. For instance, garments with the same brightness and colour as the background, too many people or groups of small children, can cause the detection algorithm to fail from time to time. Making use of the system's limitations, the designers configured it to forgive mistakes and make use of unintentional noise by making these losses themselves trigger different visual reactions.

Creating relevant visuals for a planetarium proved to be a challenge when scientific imagery abounded inside the museum. Along with animator Patrick Perris, the team created an entire custom visual language that would not attempt to compete with the accuracy of the scientific imagery inside, but define a unique poetic language to showcase natural phenomena.

Working with kids to create the content while staying away from a 'children's drawings' aesthetic was also one of the challenges. The workshops were designed to direct the result while still leaving room for creative expression.

Visual director Patrick Péris at work in one of the workshops to create the animations and visuals for the project.

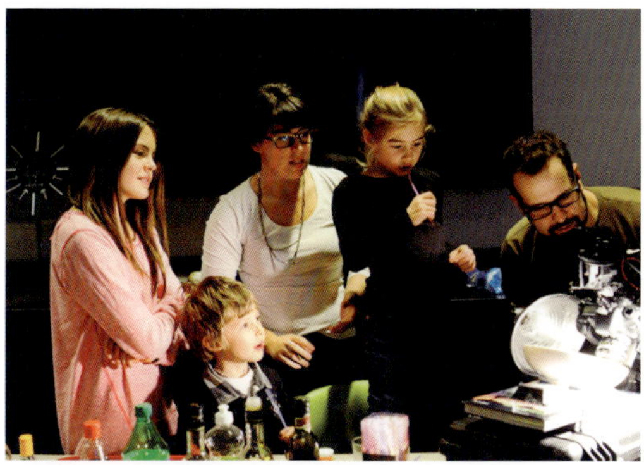

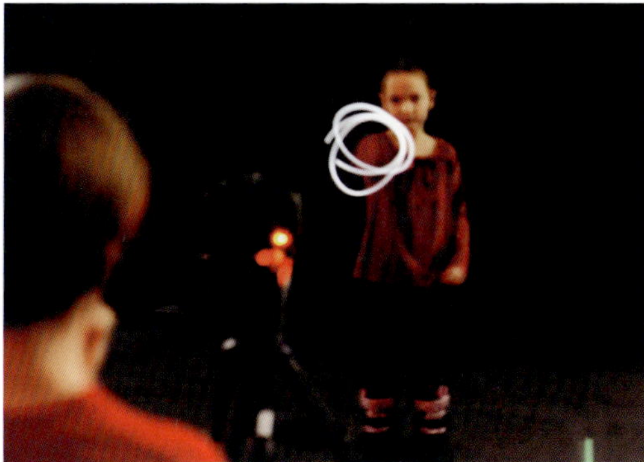

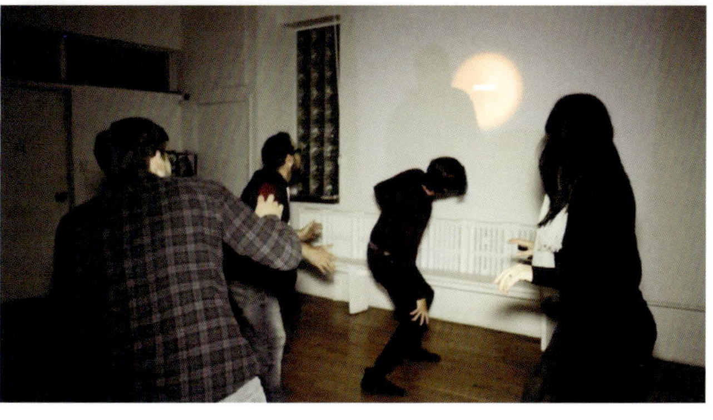

The design team worked with a choreographer to explore different ways in which participants could use their bodies to inspire celestial dynamics.

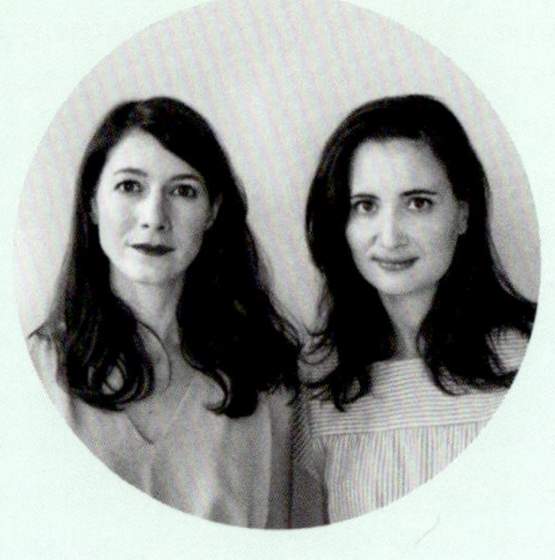
Photo Nicolas Fonseca

Daily tous les jours

Melissa Mongiat and Mouna Andraos founded Montreal-based studio Daily tous les jours in 2010. The duo's interventions look to bring an element of surprise and playfulness to everyday life and encourage communication and interaction amongst people. Having developed projects with clients worldwide, the team of ten staff, including key designers Pierre Thirion, Eva Schindling and Michael Baker, and producer Antoine Clayette, work within the fields of interaction design and narrative environments to create collective experiences.

DAILYTOUSLESJOURS.COM

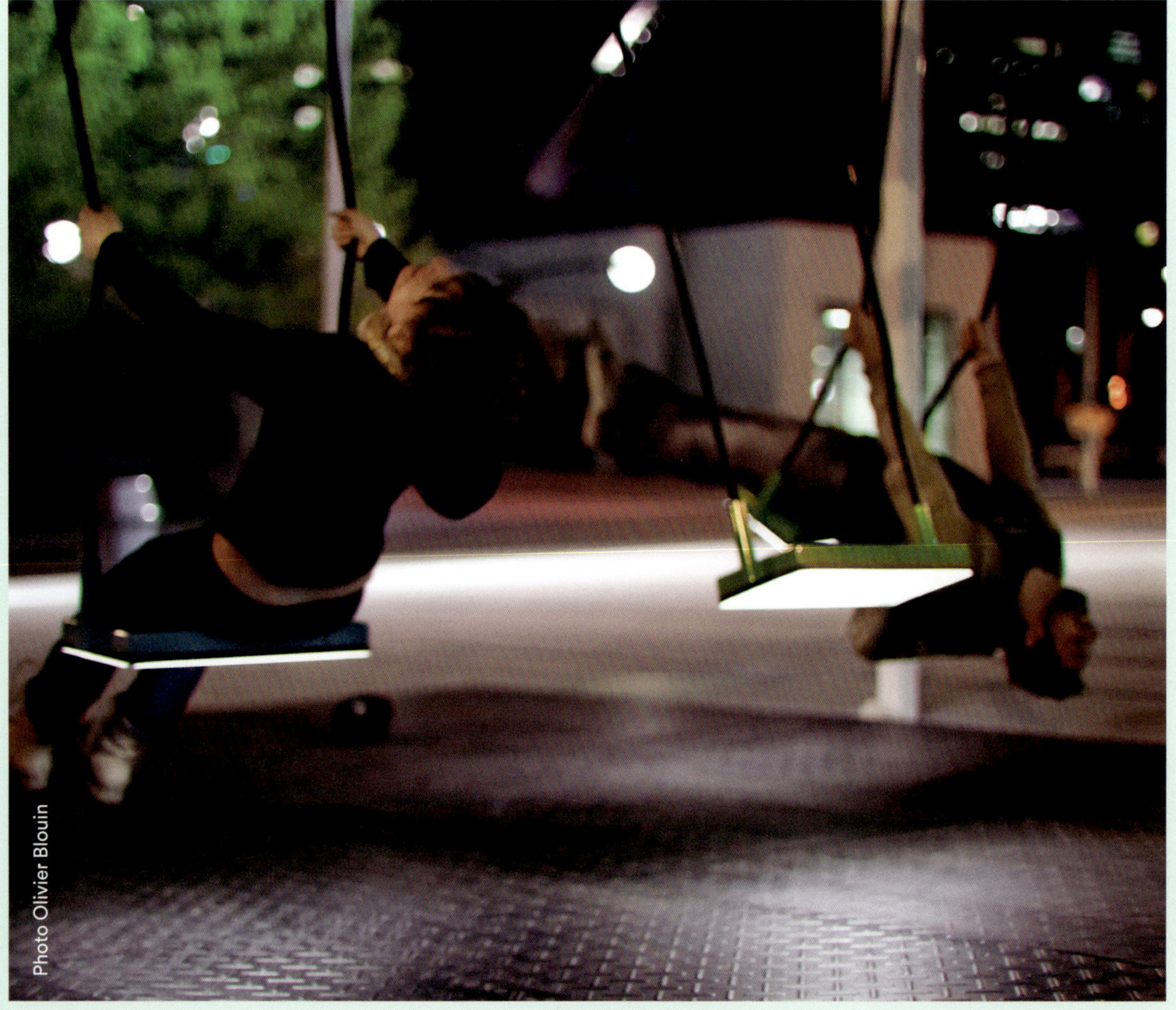
Photo Olivier Blouin

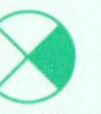

21 Balançoires

When **2011**
Where **Montreal, Canada**
Client **Quartier des Spectacles de Montréal**

An interactive installation, 21 Balançoires (which translates to '21 swings'), is an illuminated exercise in musical cooperation. When in motion, each swing is lit-up as it triggers a different note, and as more people participate, the complexity of the musical composition grows, with certain melodies emerging only through cooperation. Ever since spring since 2011, inspired by this project, a new interactive installation – 'The Swings: An Exercise in Musical Cooperation' – has toured the globe, popping up at various locations, events and festivals.

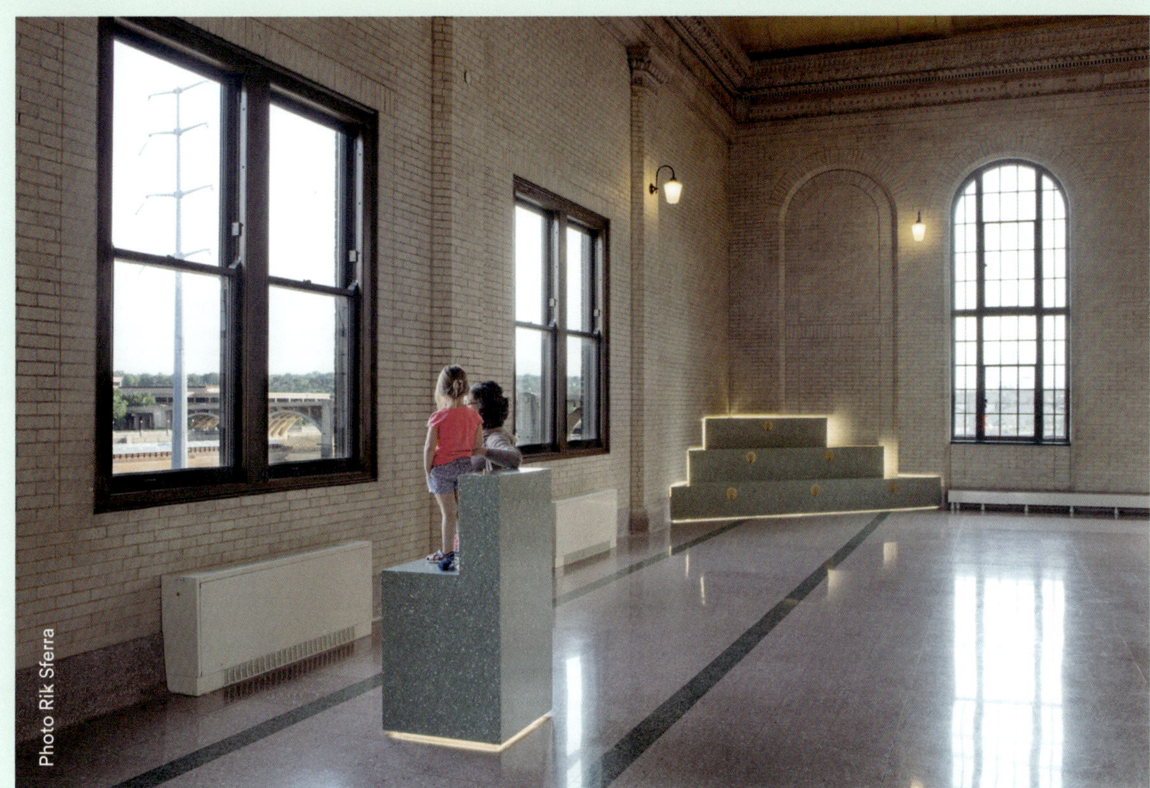
Photo Rik Sferra

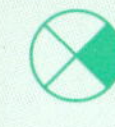

Amateur Intelligence Radio

When **2014**
Where **Saint Paul, United States**
Client **Northen Lights.mn**

St Paul's Union Depot, a historic landmark considered by many to be the city's living room, is the host of radio station Amateur Intelligence Radio (AIR). The building narrates activities within its walls as people pass by or linger on. AIR prompts people to connect with the space and each other.

McLarena

When **2014**
Where **Montreal, Canada**
Client **Quartier des Spectacles and the National Film Board of Canada**

Inspired by the 1964 film *Canon* by Norman McLaren and Grant Munro, Daily tous les jours created a collective cinematic experience that invites visitors to reproduce the film's original choreography. The installation was commissioned by the Quartier des Spectacles and the National Film Board of Canada for the McLaren Wall-to-Wall tribute, celebrating the centenary of the Scottish-born Canadian animator and film director.

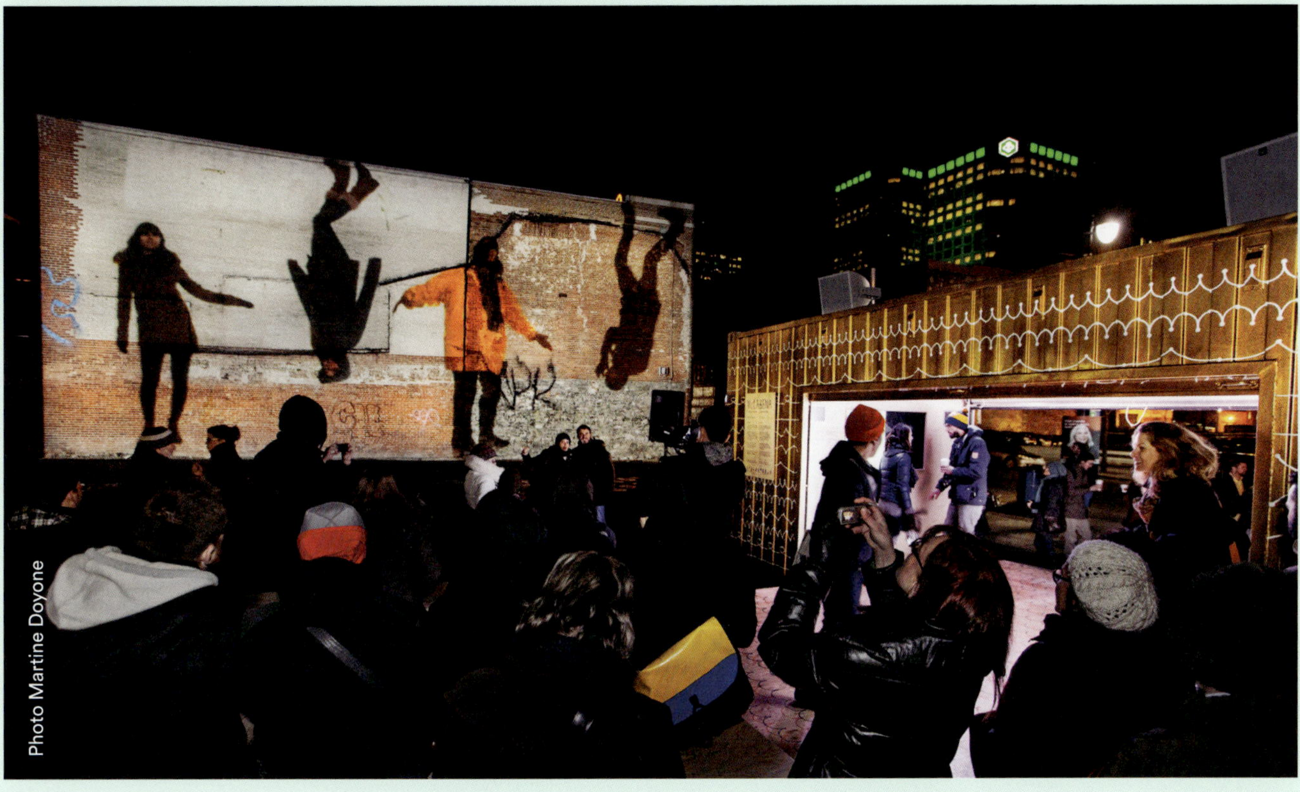
Photo Martine Doyone

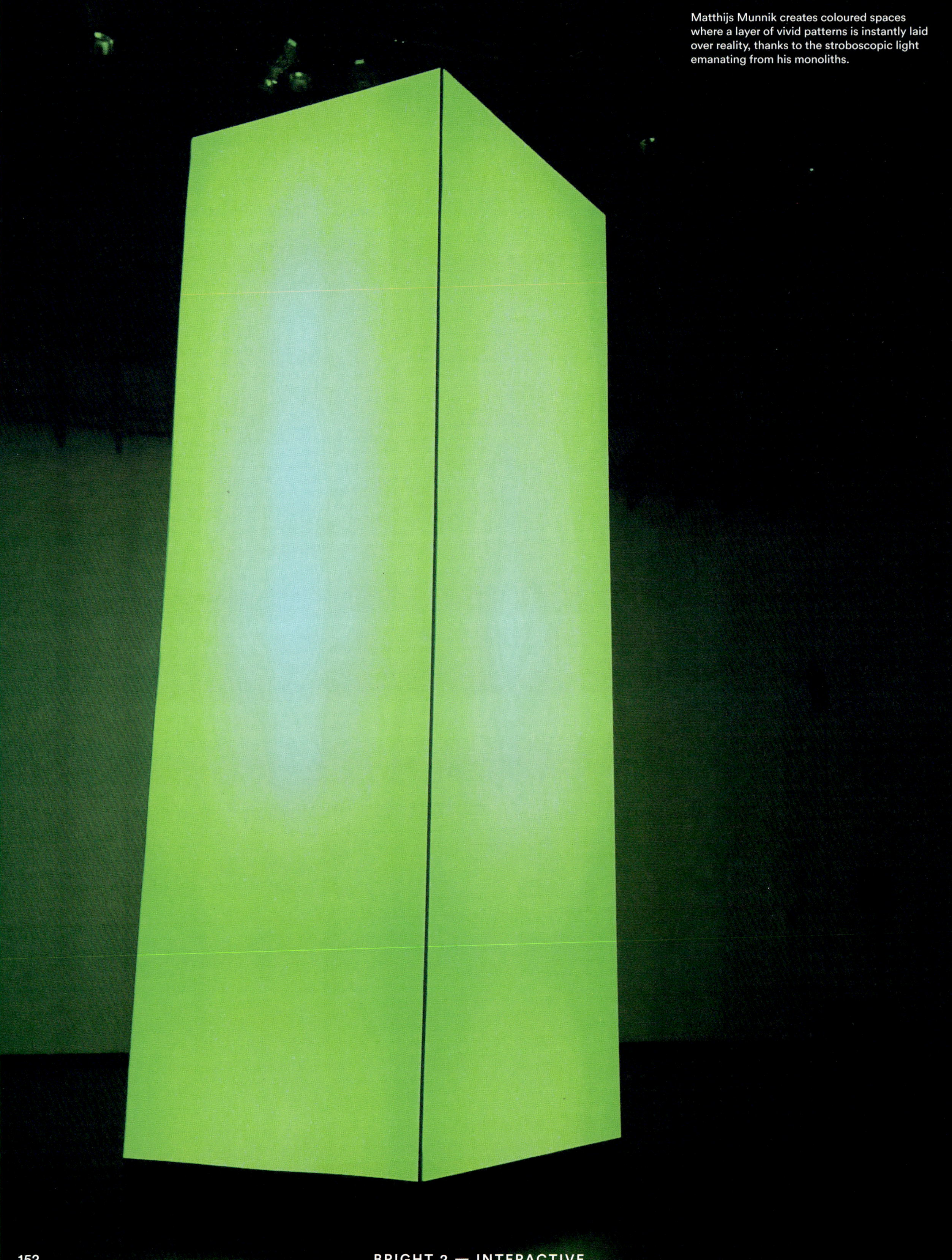

Matthijs Munnik creates coloured spaces where a layer of vivid patterns is instantly laid over reality, thanks to the stroboscopic light emanating from his monoliths.

Citadels

In the Citadels series, **Matthijs Munnik** uses monolithic sculptures that emit coloured stroboscopic light to induce hallucinations in spectators, and at the same time transform space into an observatory of the visions that manifest themselves in the eyes of the viewers.

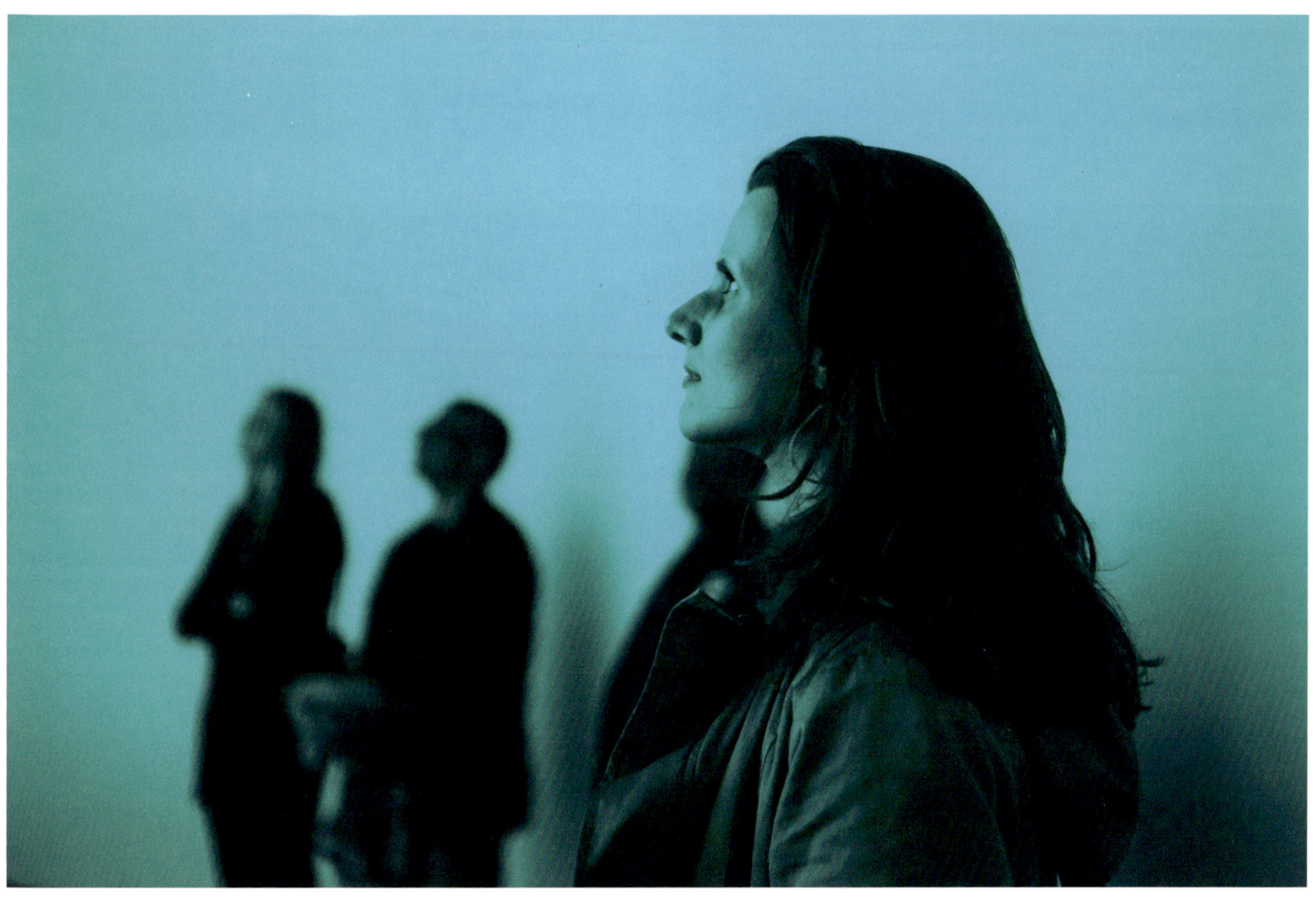

Photos Pieter Kers, Ed Jansen, Ayako Nishibori

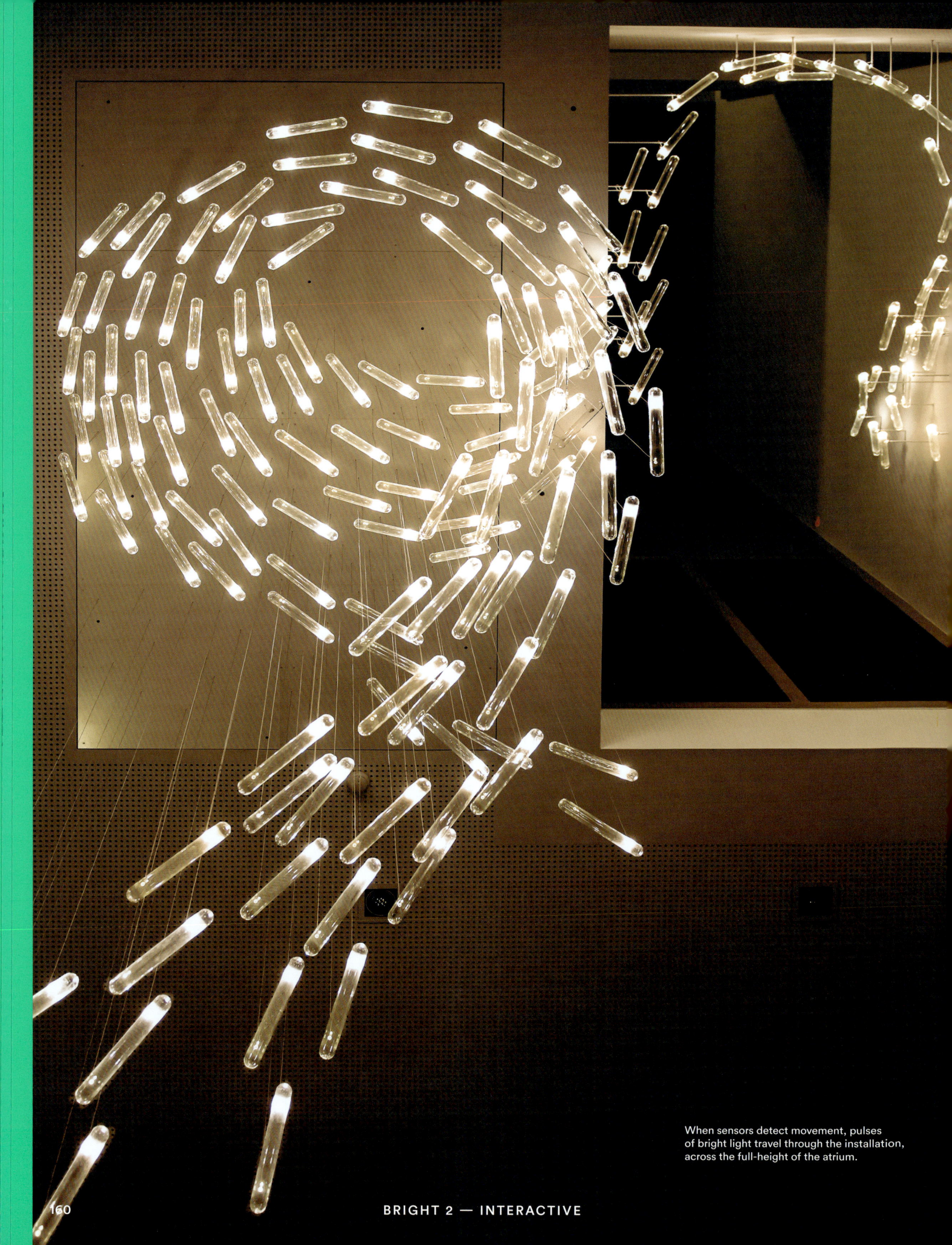

When sensors detect movement, pulses of bright light travel through the installation, across the full-height of the atrium.

Flylight

Flylight is **Studio Drift**'s site-specific light installation that directly interacts with its surroundings. Originally launched with a halogen light source, this new edition in Eindhoven has custom-made glass tubes incorporating LED lighting.

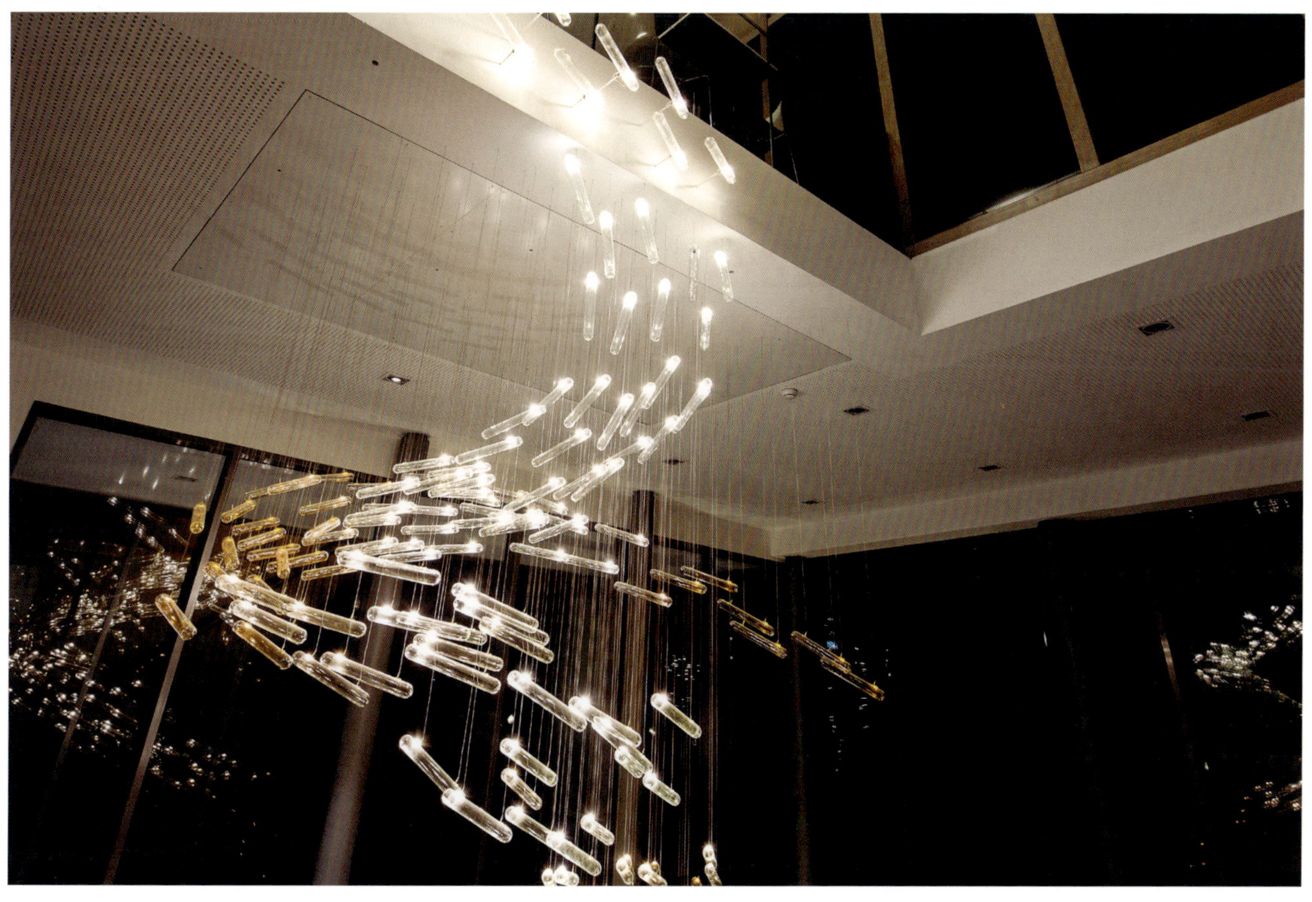

Studio Drift was commissioned to create a Flylight installation for the main hall of the Beta Building of the Hi-Tech Campus in Eindhoven. It now takes pride of place in the lobby, lighting up the entrance to the space in an illuminating and eye-catching fashion. Flylight runs over three floors connecting the movement and scientific exchange between the companies based in the building.

The Amsterdam-based atelier created its first Flylight in 2007 as a site-specific light installation that directly interacts with its surroundings. It consists of delicate glass tubes that light up in an unpredictable way, partially responsive to external stimuli. The light mimics the behaviour of a flock of birds in flight, symbolising the conflict between the safeness of the group and the freedom of the individual.

The Eindhoven installation is the first to incorporate LED that mimics the exact colour and behaviour of Studio Drift's previous halogen Flylights. This is also the first installation of this type that runs partly over a wall, which is a development to be implemented further in the studio's future public-space installations.

Flylight is a poetic and interactive light installation, monumental and fragile, illuminating and deploying patterns at the discretion of the viewer. —

Designer
Studio Drift
Location
Eindhoven, the Netherlands
Client
Brainport Development
When
October 2013

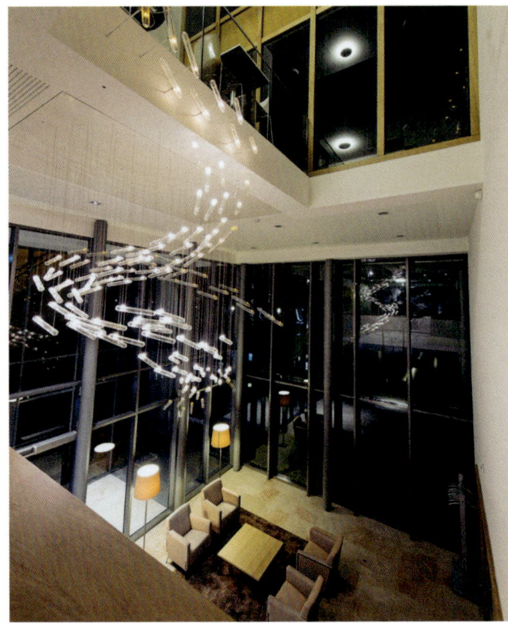

This installation welcomes each visitor to the building, creating a dramatic and dynamic display in the entrance area.

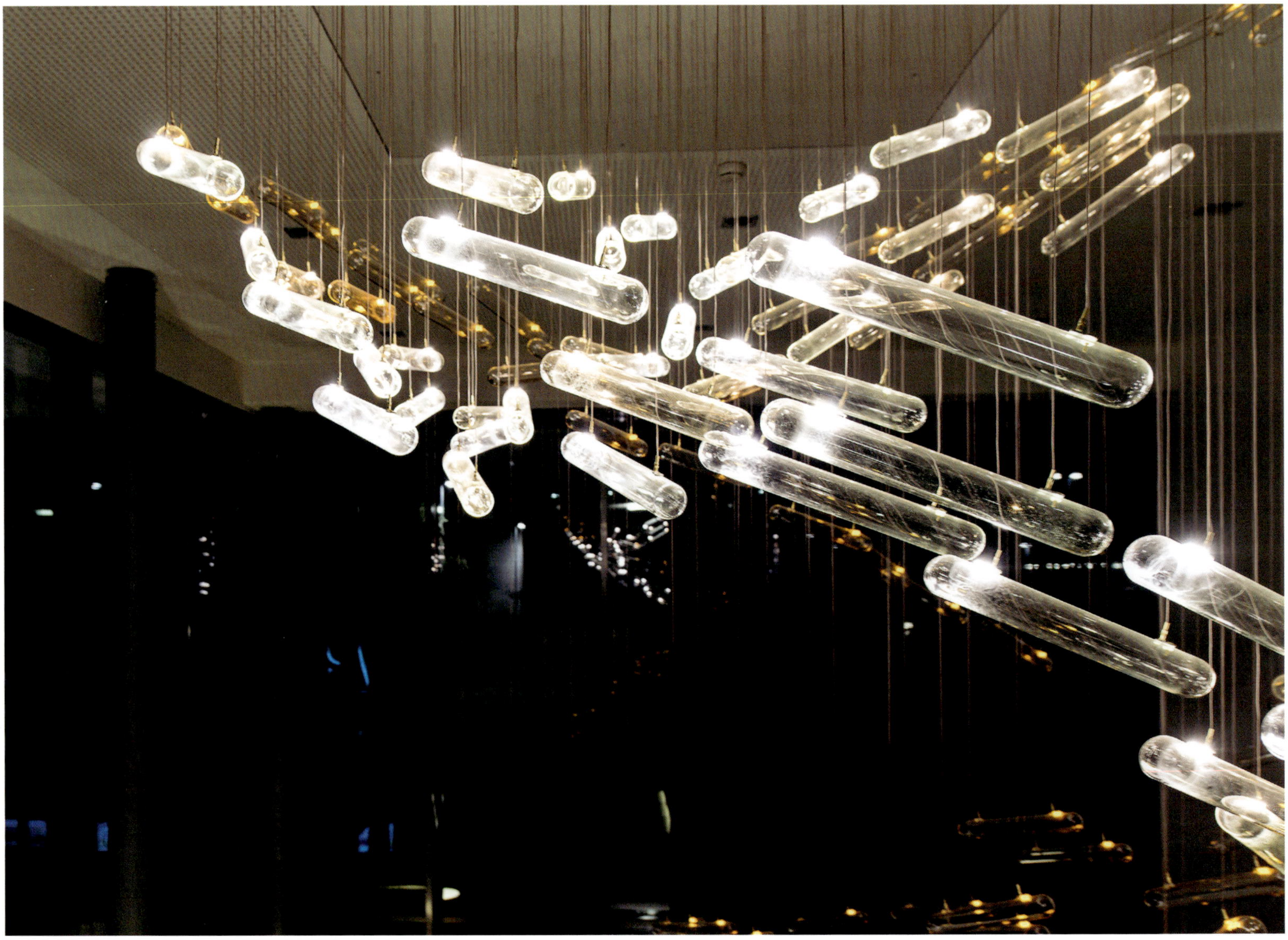

ELEVATION

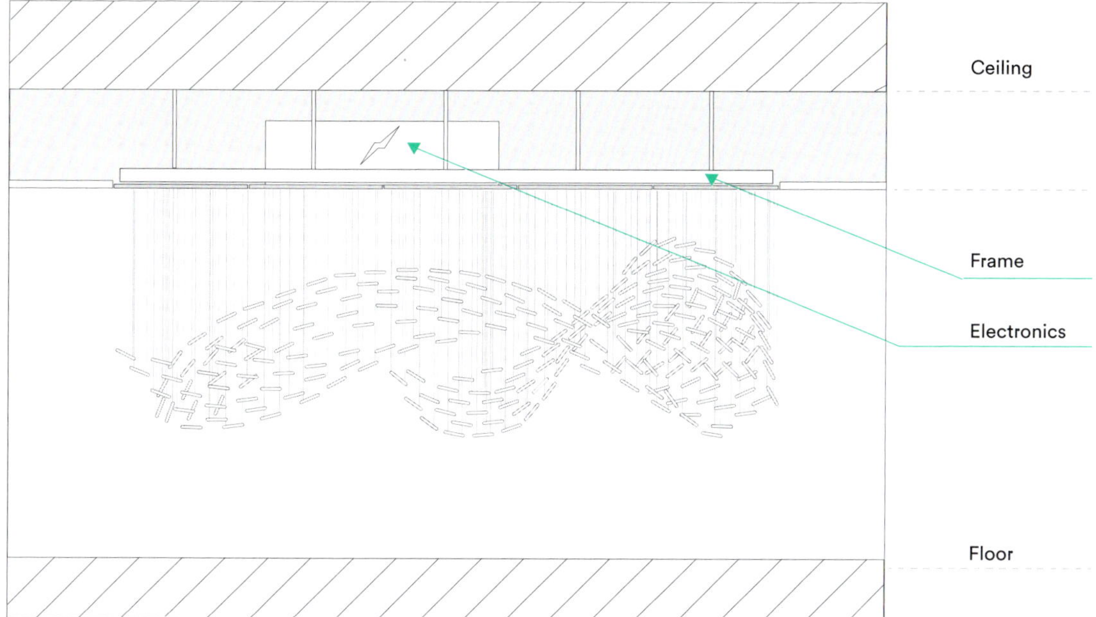

Set-up

Used light is in the form of LED lights: max. 1.7 Watt (at 100% dutycycle), with max. nominal value of 0.54 A. In the brightest moments, the LEDs have a colour temperature of 2700 K. In the most dimmed moments, the colour is a mix of warm white tones and amber.
Component elements included glass tubes, steel frames, fittings and light sources, all custom-made for Flylight: the glass is handblown borosilicate (300 x 40 x 40 mm); each glass tube holds one LED light source and consists of several steel frames (30 x 30 x 2 mm steel bars) connected together; the maximum size of one frame is 1600 x 1800 x 30 mm.
Ceiling panels are made of aluminum plates with dimensions 800 x 900 x 3 mm. These panels can be powder-coated in the same RAL-colour as the ceiling of the space. Each panel holds several glass tubes.
Electronics are positioned on top of the ceiling panels and consist of LED dimmers, transformers, controller and sensors. Sensors are placed according to the specific location and can detect movement up to 7 m distance from ceiling.
Cables are shielded, 1-mm thick (including isolation). Two electrical cables hold one glass tube with strain reliefs. The cable conducts 12 V for the LED bulb inside the glass tube.

Specifics

/\/\/\ This installation was only realised after spending time in the actual space and making physical models. An object like this cannot easily be seen in a 3D drawing as the structure gets lost. Typical architectural plans also do not provide sufficient information.
/\/\/\ The self-developed light modules needed some early revisions to ensure the LEDs did not touch the glass and cause vibrations. Through testing this purely mechanical problem in this highly-electrical installation was alleviated.
/\/\/\ A key challenge was actually integrating Flylight with the building. Each light is connected individually to the dimmers and had to pass various floors, rooms and walls to be connected in the right sequence.
/\/\/\ Setting the sensors was also challenging. At the start, all lights are in dimmed mode. There is one aspect that is more bright and this part 'flies' through the installation: whenever a movement is detected by a sensor, the bright light starts travelling to the area around this person who has entered the space. In this movement, the light makes its own choices and reacts to its neighbouring light and to the next and the next, etc. Only the DNA to move towards the person is programmed; how it exactly behaves is within the Flylight algorithm that makes its own choices.
/\/\/\ The process always started with a technical computer animation translating the behaviour of the light; then this was checked with the reality before making the final iterations, once everything was installed in the space. Light is not something that can be planned exactly in advance; the environment, natural light and surrounding materials all have influences on the final result.
/\/\/\ All software has been specially-especially developed for Flylight, so that the patterns in which the installation is illuminated are not preprogrammed but have an interactive component.

MIMICS THE BEHAVIOUR OF A FLOCK OF BIRDS IN FLIGHT

EXPLODED VIEW

Cable

Light source

Glass tube

4 cm

30

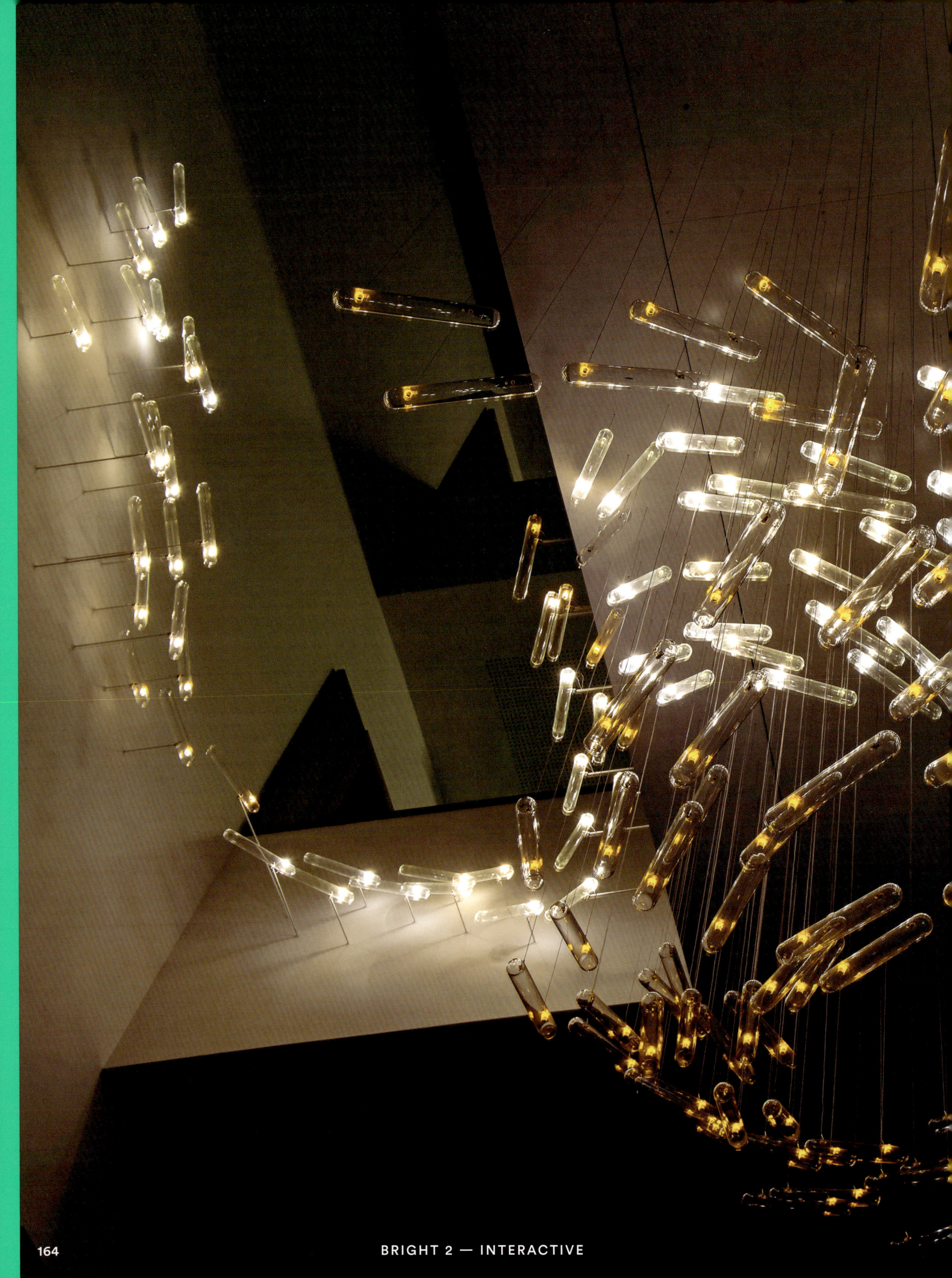

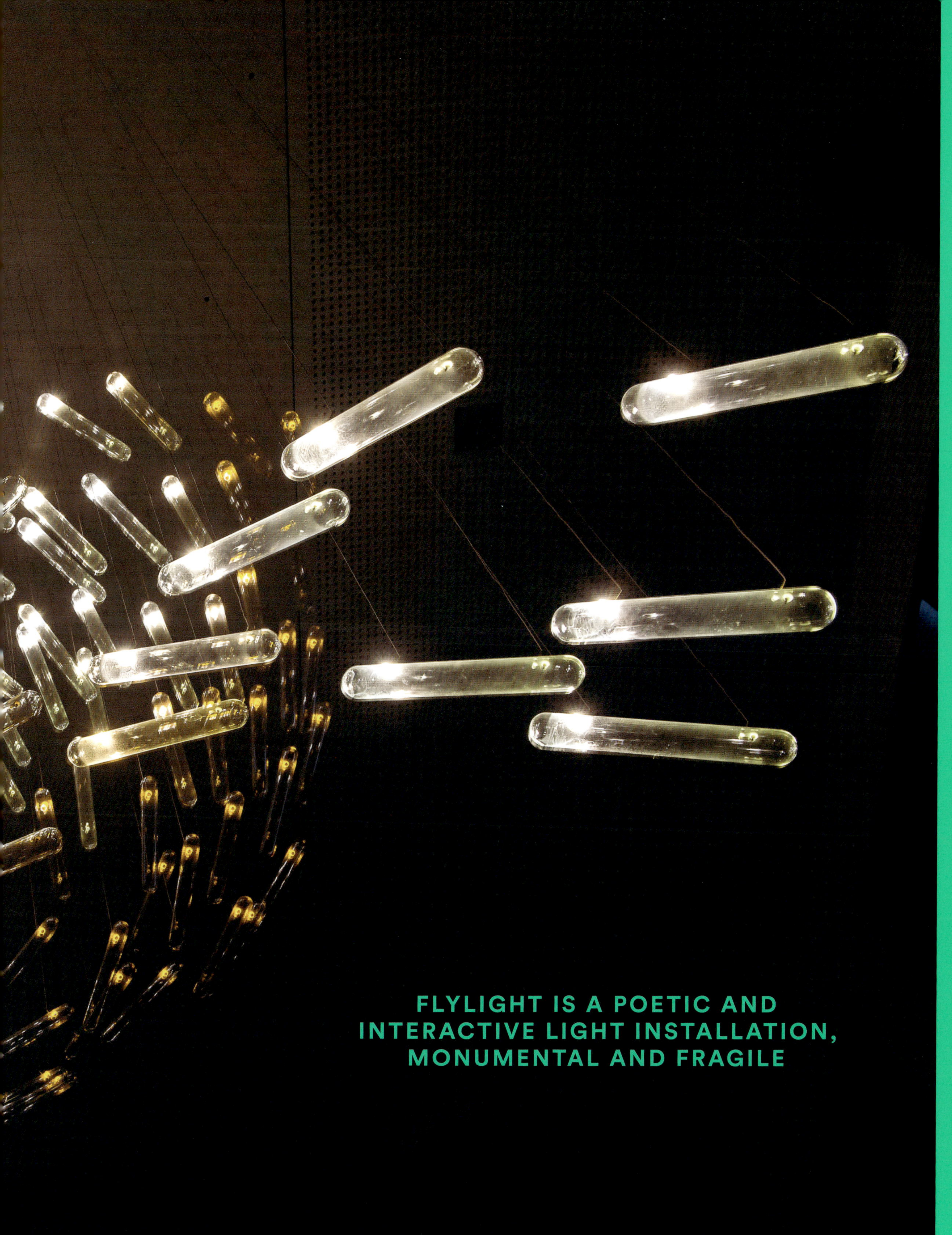

FLYLIGHT IS A POETIC AND INTERACTIVE LIGHT INSTALLATION, MONUMENTAL AND FRAGILE

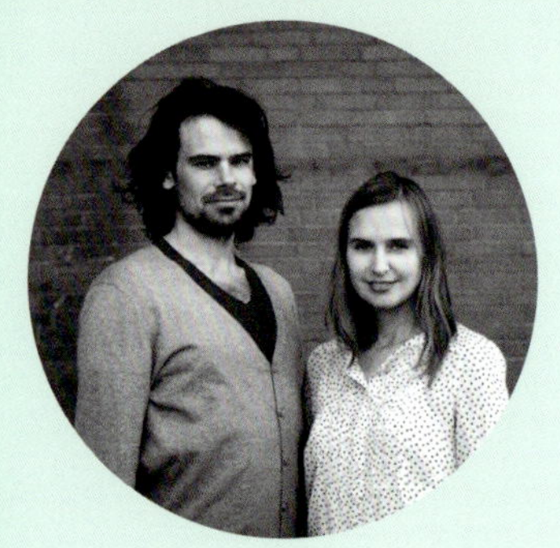

Studio Drift

Studio Drift's creations aim to establish a point of balance between information overload and human sensibility, with a goal to create a dialogue between nature and technology, creating a new synergy. The Amsterdam-based practice was founded in 2006 by Ralph Nauta and Lonneke Gordijn, both graduates of the renowned Design Academy Eindhoven. Light is a key element in their work, but is always treated as an artistic ingredient rather than a functional tool to illuminate. Studio Drift creates projects that balance on a fine line between art and design.

STUDIODRIFT.COM

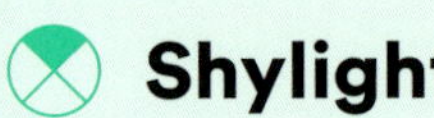

Shylight

When **2014**
Where **Amsterdam, the Netherlands**
Client **Rijksmuseum**

Certain flowers close at night, which is a natural mechanism that inspired Studio Drift to create Shylight – a light sculpture that unfolds and retreats in a fascinating choreography. Created out of many layers of silk, robotics, aluminium and LED modules, the movement of the lights can be controlled to the millimetre. The performative sculptures induce a spontaneous and emotional response from viewers, for instance at the permanent installation in the Philips Wing of the Rijksmuseum in Amsterdam.

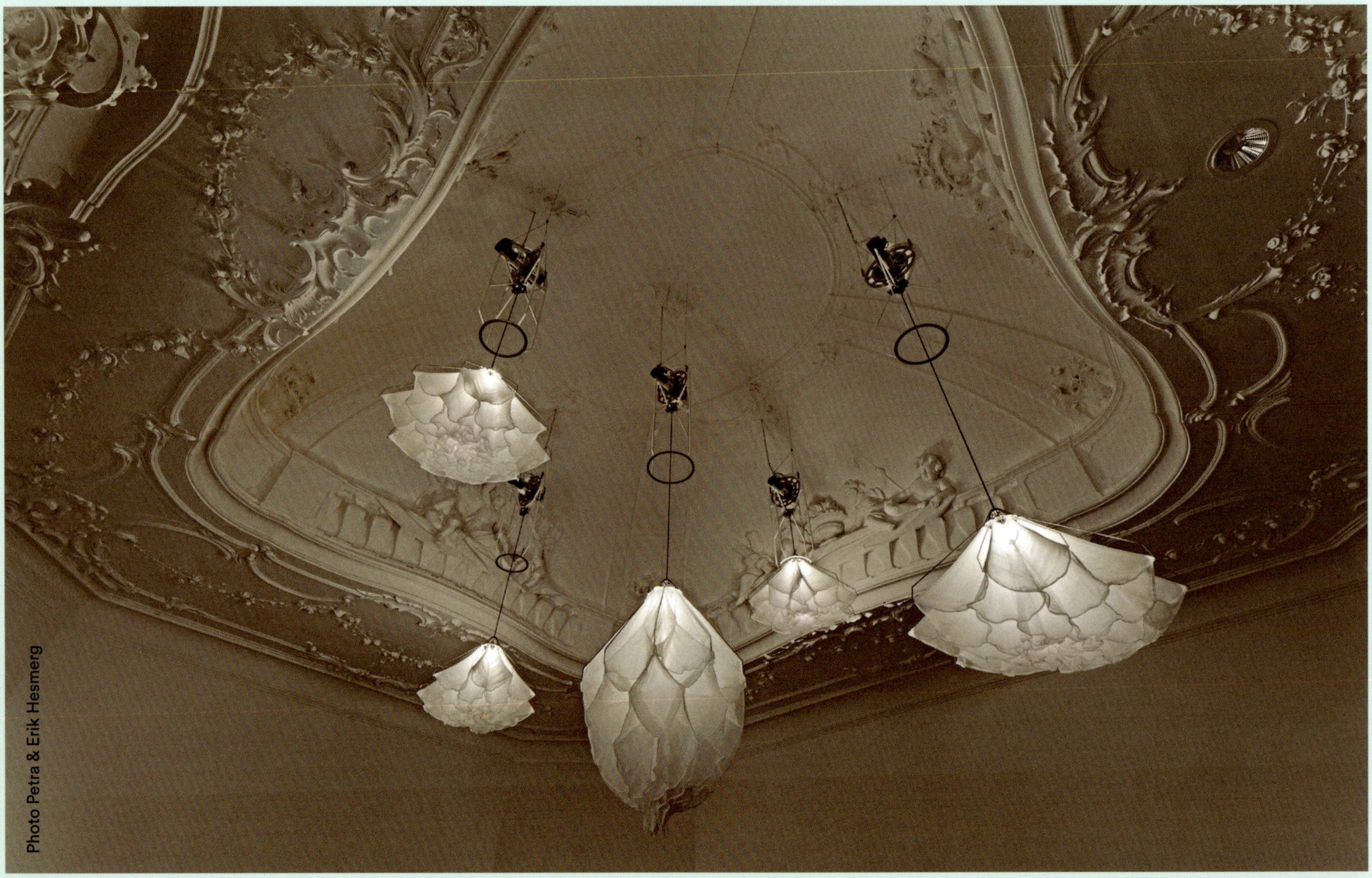

Photo Petra & Erik Hesmerg

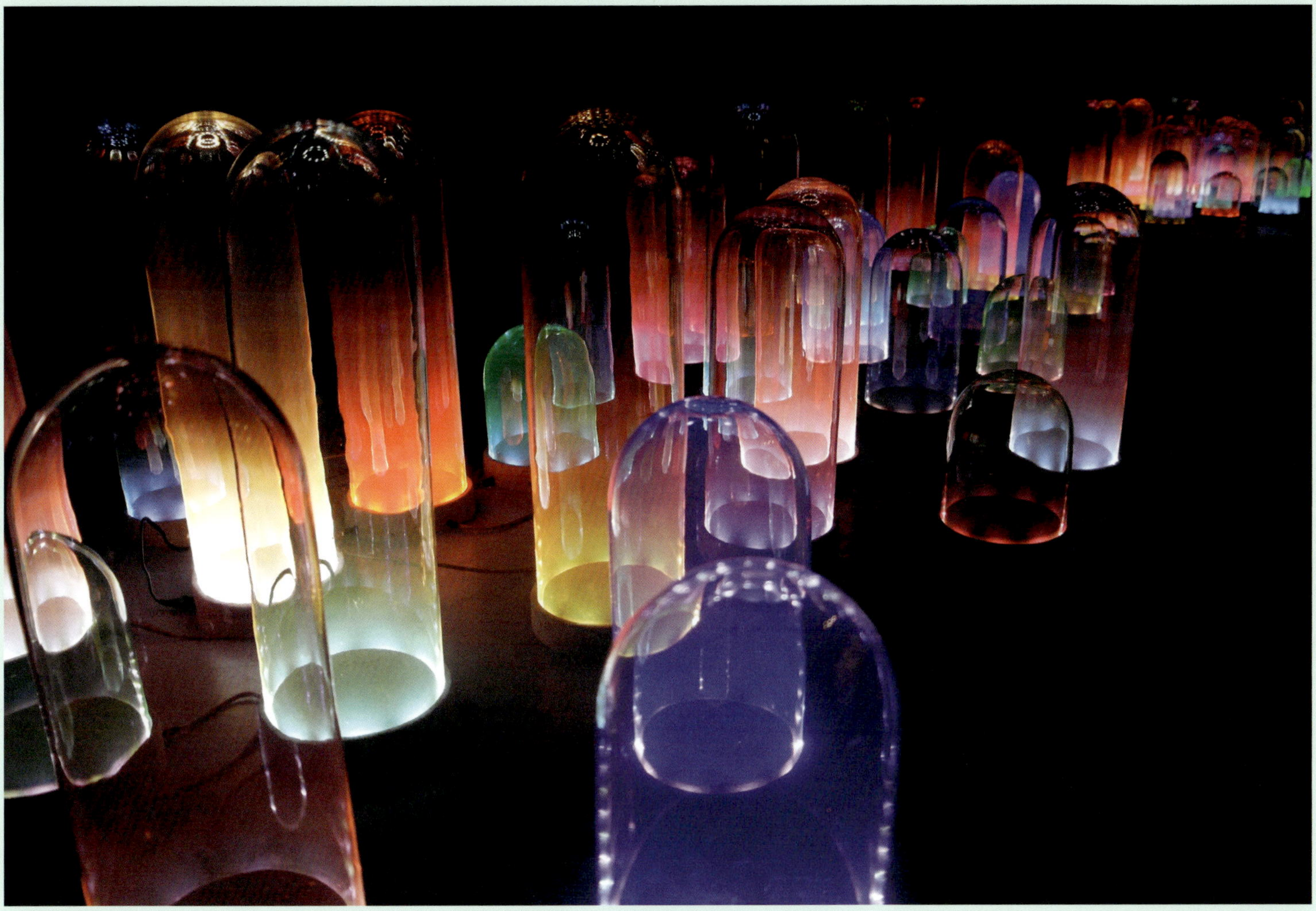

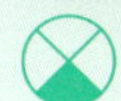

Nola

When **2013**
Where **Eindhoven, the Netherlands**
Client **DDW and Buhtiq 31**

Nola creates a landscape of light, in pastel-coloured glass bell jars. Through the combination of LED lights integrated in the base of the bell jars, a unique colour palette can be created. The light intensity and colour mixture can be adjusted at the press of a button.

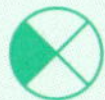

Fragile Future

When **2013**
Where **Weill am Rhein, Germany**
Client **Vitra Design Museum**

Fragile Future III is a combination of nature and technology, consisting of 3D bronze electrical circuits connected to light-emitting dandelions. It is a modular work, with one bronze module in a visible circuit with three 'dandelights'. A module can be attached to the next one (in unlimited different ways).

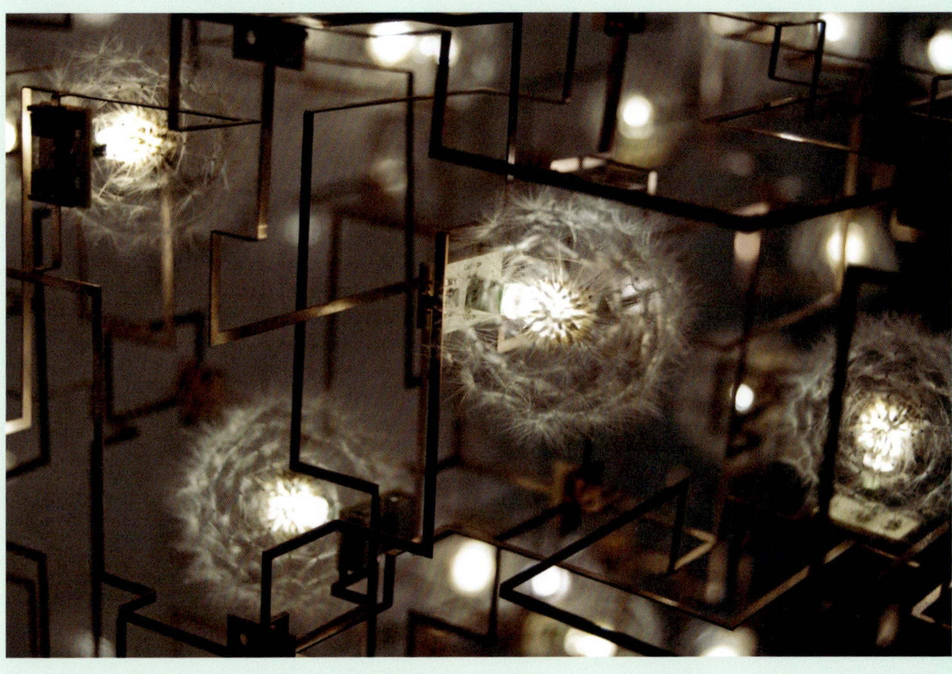

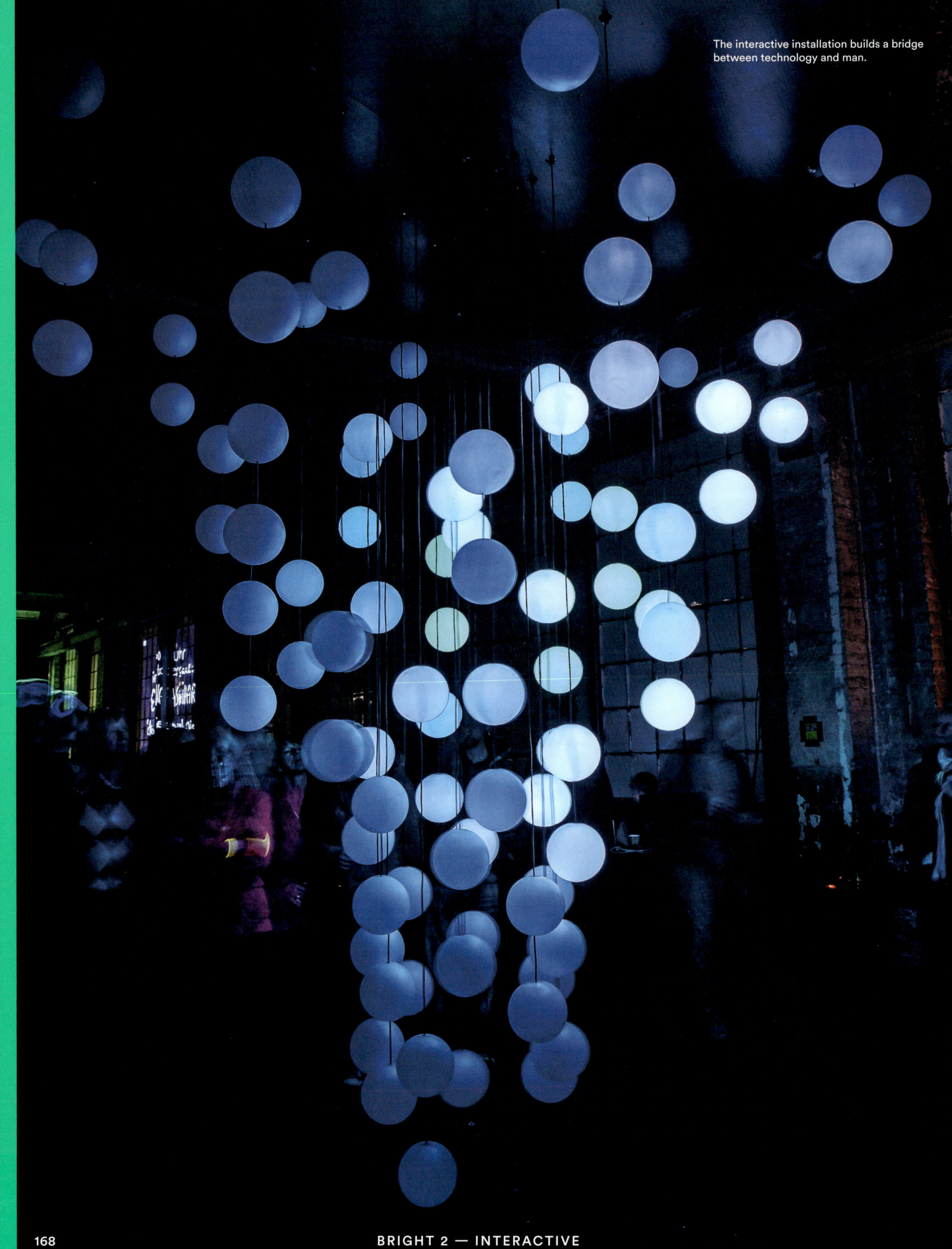

The interactive installation builds a bridge between technology and man.

Genariya

An interactive light installation by **Christine Lyschik** bridges the gap between technology and people by conveying moods and emotions through movement and colour, creating the illusion of sentience. The concept was developed around questions about what we perceive as living beings.

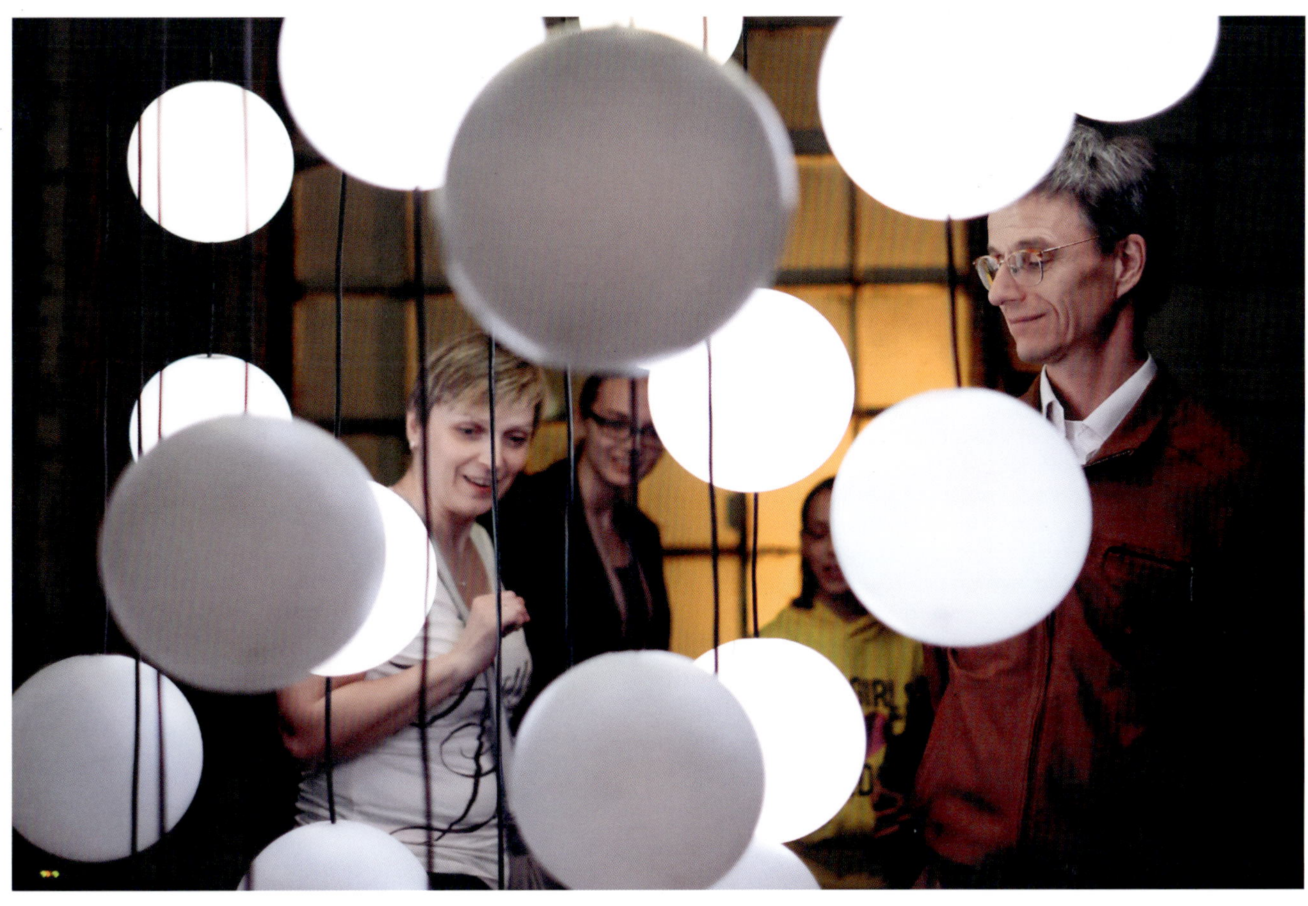

Photos Hans-Peter Fuchs, Christine Lyschik, Patrick Zeller

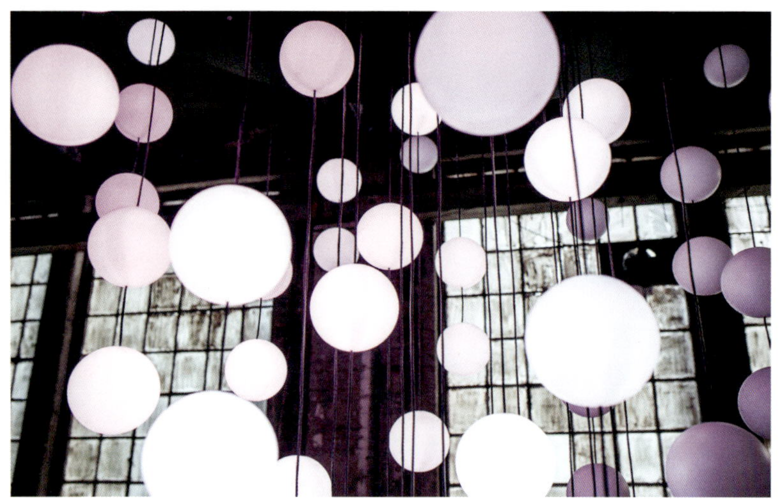

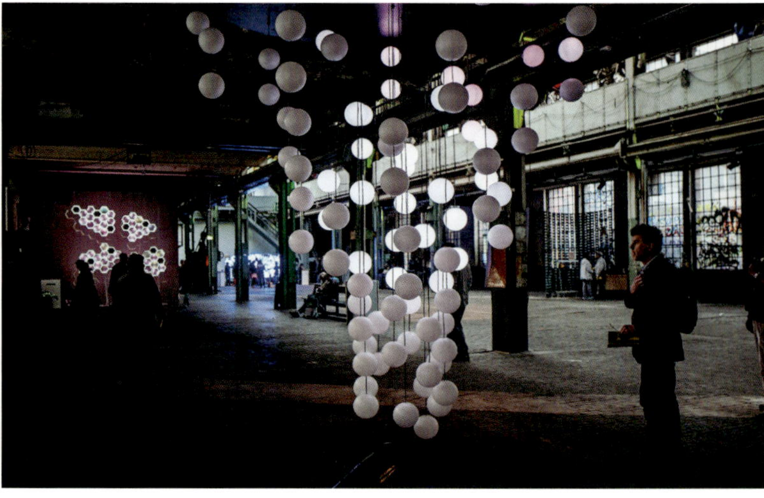

Designer
Christine Lyschik
Location
Frankfurt, Germany
Client
Hochschule Darmstadt
Consultants
Kyrill Fischer, Thorsten Greiner, Nils Weger
Manufacturers
Christine Lyschik, Jan Mayerhofer
Date
March 2014

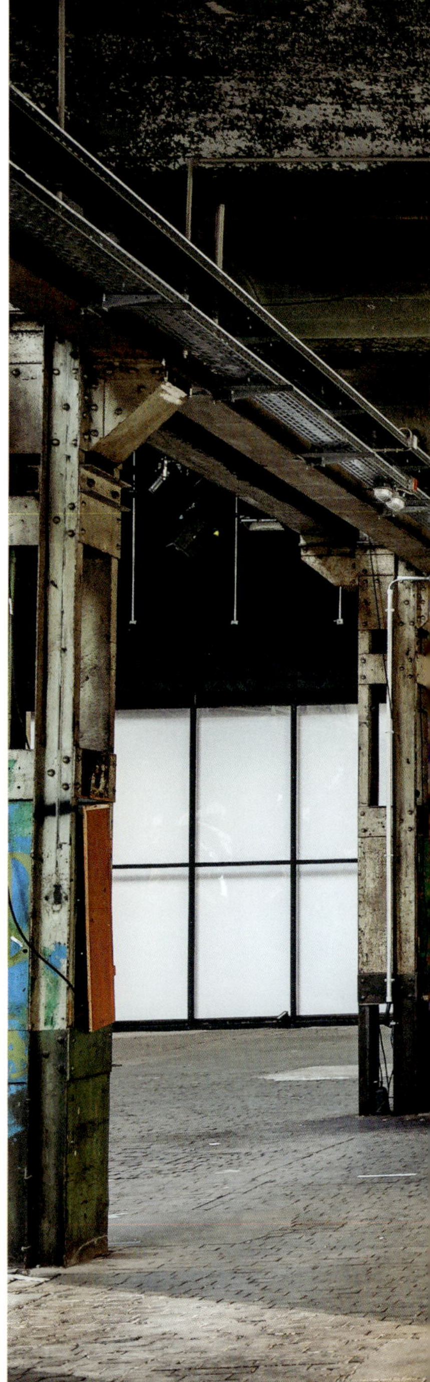

Questions around what we perceive as living beings and how a machine can convey sentience formed the conceptual roots of Genariya, an audiovisual interactive installation by Christine Lyschik for Frankfurt's biennial lighting fair, Luminale. The artwork invited the audience to abandon all preconceptions about life and embrace the possibility of a sentient 'light being'.

Studies of movement and video analysis allowed Lyschik to understand that the way a movement is executed is more important than a sequence of actions, when it comes to our perception of what 'alive' means. With this in mind, the artist transferred recorded human movement – an improvised dance by German dancer Victoria Söntgen – to the installation. Dance was chosen for its singular power of conveying unspoken feelings and emotions as a universal language, meaning that participants from around the globe were able to intuitively grasp Genariya's vitality and emotions.

A complex real-time responsive system animated the 96 LED spheres arranged in six intertwined spirals that made up this 'being of light'. For its emotions to be represented, the dynamic of the lights' movement was accompanied by melodious sounds triggered as the lights went on and off, thus creating an audiovisual harmony.

Genariya was able to interact with its audience by conveying different moods – frenetic, angry and demanding, quiet and reserved and jittery – according to the number of visitors and the way they interacted with her. Using infrared scanners (RadarTouch devices) meant that she could detect humans approaching and respond to their presence. —

THE ARTWORK INVITES THE AUDIENCE TO ABANDON ALL PRECONCEPTIONS ABOUT LIFE AND EMBRACE THE POSSIBILITY OF A SENTIENT 'LIGHT BEING'

The objective was to find a way of representing vitality and conveying emotion through an audiovisual medium.

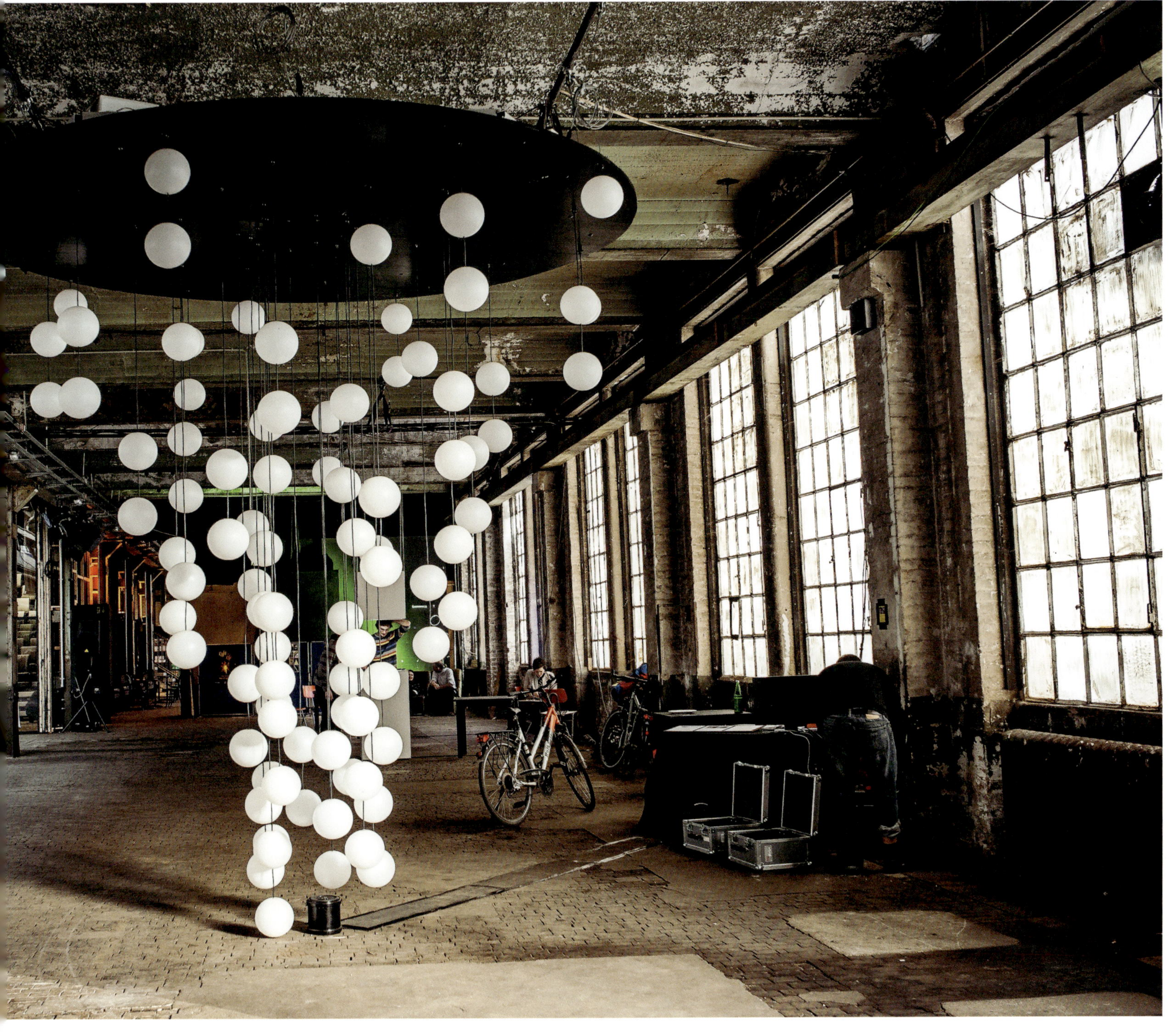

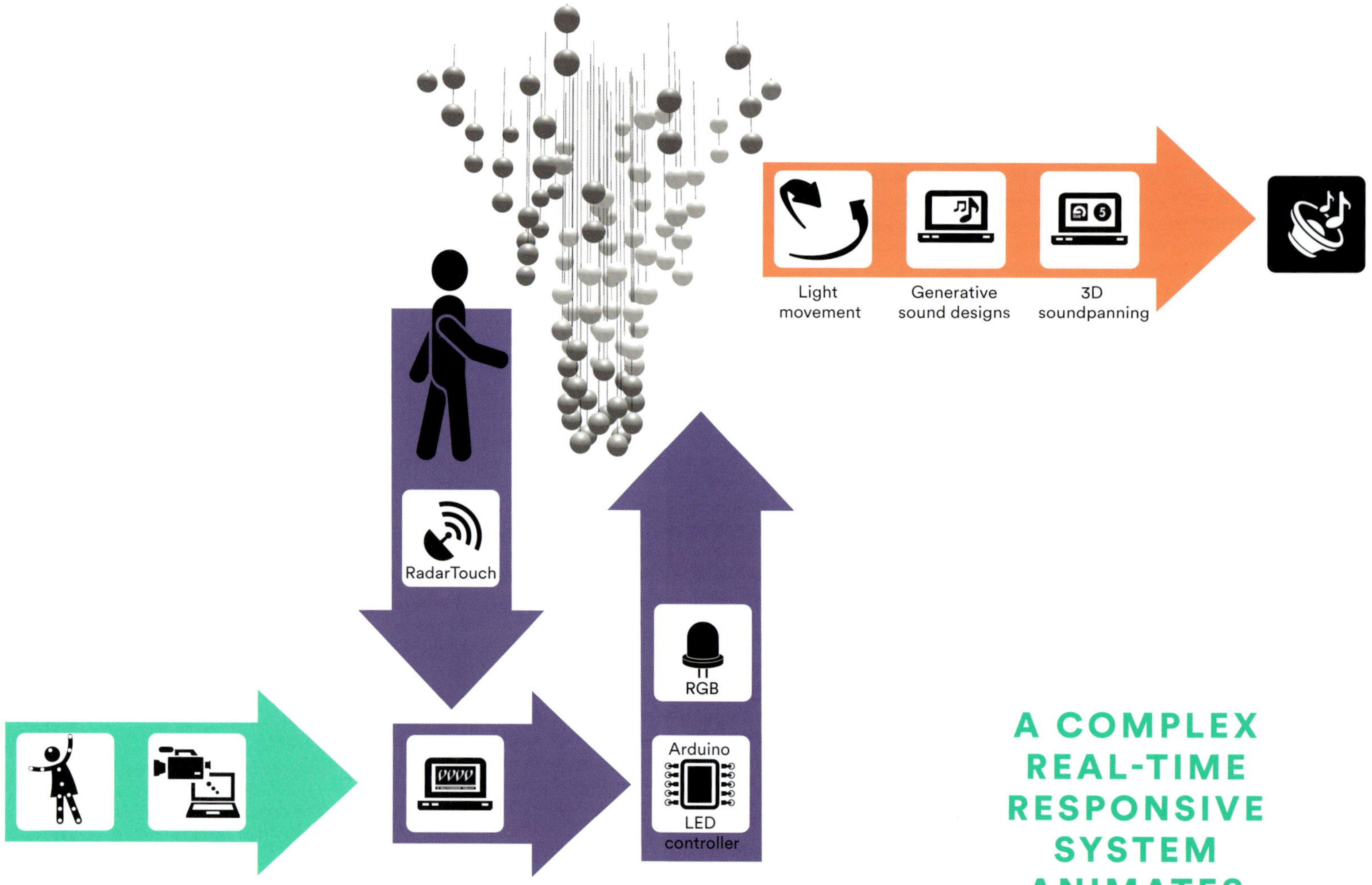

Light movement

Generative sound designs

3D soundpanning

RadarTouch

RGB

Arduino

LED controller

A COMPLEX REAL-TIME RESPONSIVE SYSTEM ANIMATES THE 96 LED SPHERES

Specifics

A wider challenge of the project was to transmit human emotions to a machine, to attempt to bring it to life.

The use of mathematical algorithms to convey the dancer's abstract movements to Genariya's body was a laborious task and with it came the responsibility of making the sculpture's body language understandable and accessible to the public. The result was a highly complex generative lighting and sound system guided by human emotions.

Another challenge was generating harmonic sounds in real time, by transforming light movement into melodies. Sound designers Jan Mayerhofer and Sebastian Hohberg created an orchestral soundtrack with a composition that had its origin in the choreography of the dancer (Victoria Söntgen), emphasised with 3D sound panning; an absolute symbiosis of light movement and sound experience.

Genariya was developed as a modular construction system. This was always a key aim for the team, so that it could be reinstalled in any exhibition hall. Further installations are planned, which will see Genariya interacting with visitors at numerous events and locations.

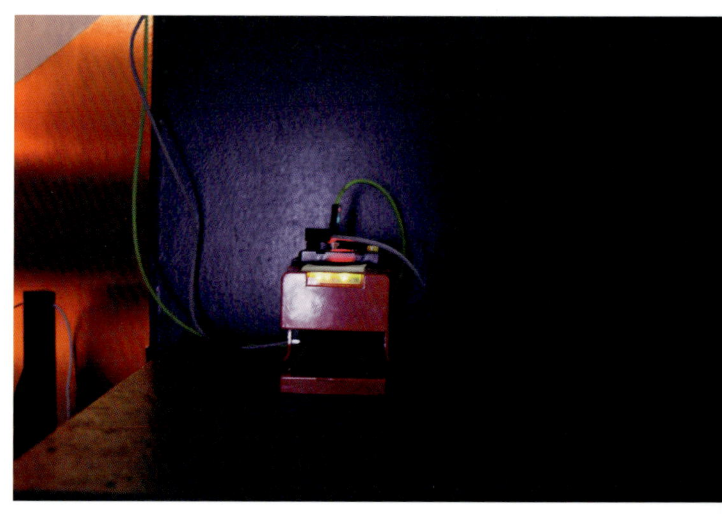

RadarTouch, the 2D infrared scanner used in the project, was developed by Realtime Department/Lang.

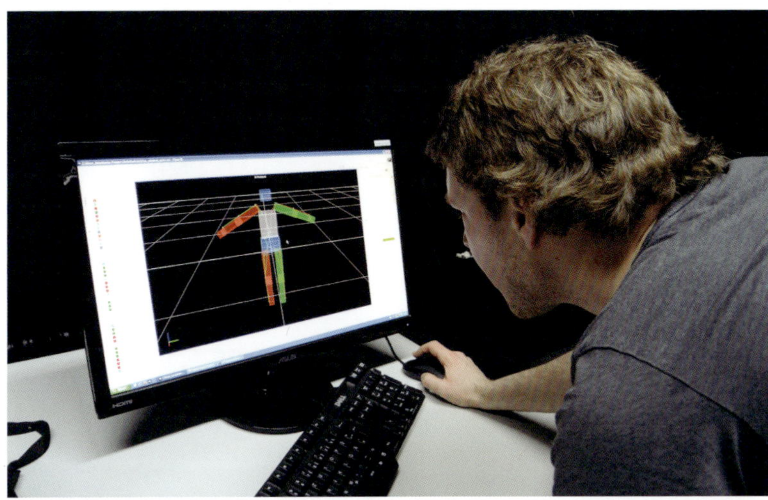

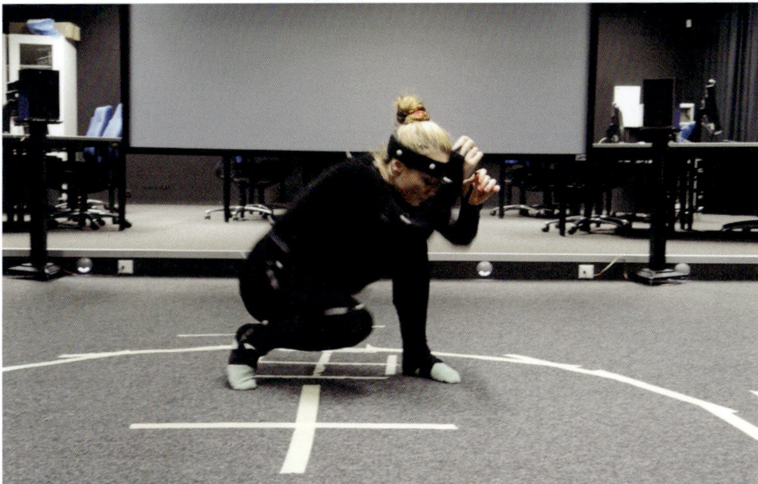

Set-up

Used light was digital RGB pixel. The white lights, with a 20-cm diameter, were hand-crafted from plastic spheres and RGB LED-printed circuit boards.
Total area for illumination was 4 x 4 m, with 350 m of cable being used.
Installation was based on VVVV. This programming language was used to design Genariya's responsiveness, including motions, sculptural form, colours and hardware management.
Choreography was improvised by dancer Victoria Söntgen, filmed in a 3D motion capture studio, saved in a database and the movements were adapted for the light sculpture.
Infrared scanners (two RadarTouch devices, sponsored by Realtime Department/Lang) served as 'eyes' by tracking the positions of the people in the room in real time. This technology is based on an invisible infrared laser, scanning the positions and movement of the audience with 25 fps.
Generative sound design was created with Max4Live (Max/Msp) in conjunction with Ableton. Individual lights were assigned different pitches and instruments. The tonality, instruments and acoustic colour varied according to the mood or type of interaction, resulting in a symbiosis of light and orchestral music. Moreover, five integrated speakers generated 3D sound panning, so that the location of the music followed the movement of the light.

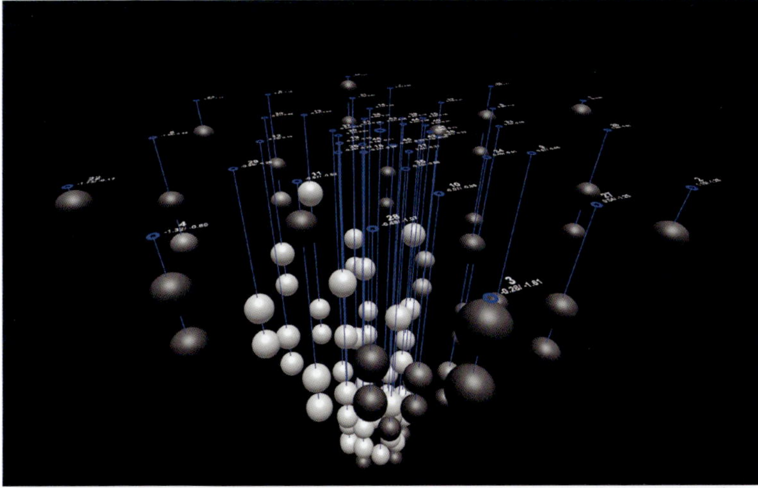

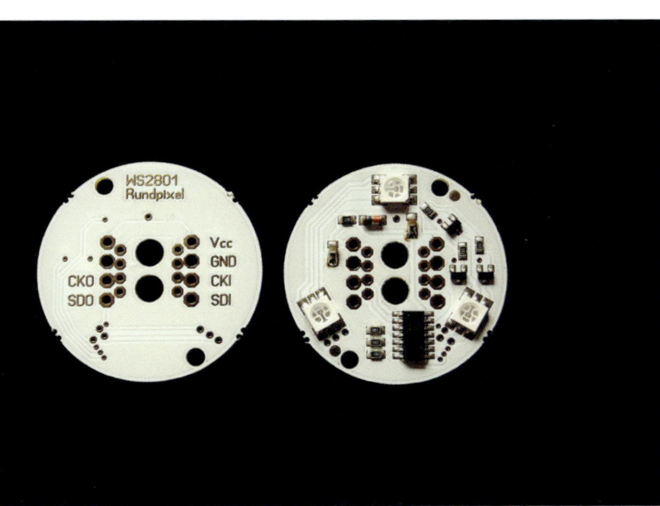

Digital RGB Pixels, machine-printed circuit boards (led-studien.de) were positioned in the plastic spheres.

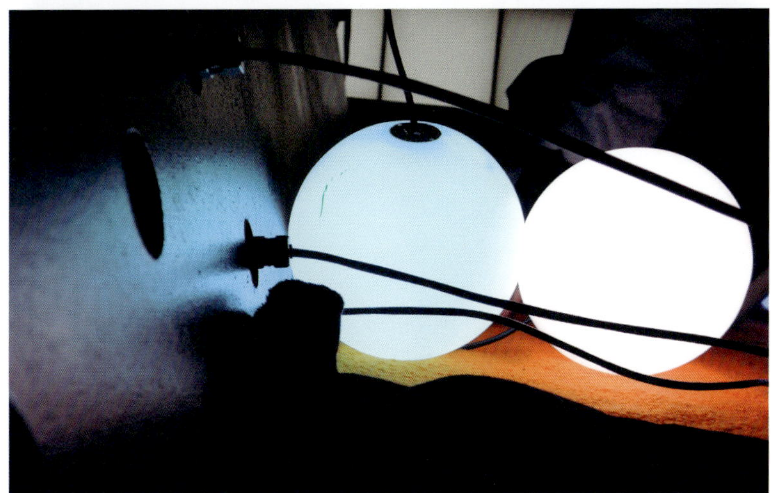

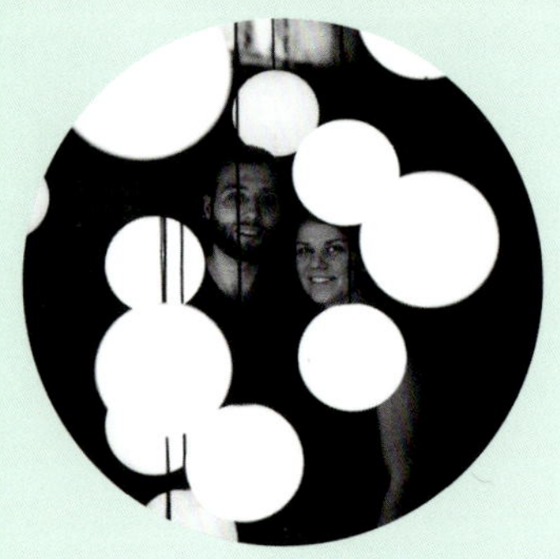

Christine Lyschik

Christine Lyschik founded her creative studio in Offenbach, Germany in 2009. She works with electrical and sound engineer Jan Mayerhofer to develop communication services for a number of spaces using 3D visuals and programming. The team works from the conceptualisation to the implementation of interactive multimedia installations, exhibitions and light installations, always with an important goal in mind: to introduce digital art into spaces, creating connections between complex technology and the human environment.

CHRISTINELYSCHIK.DE

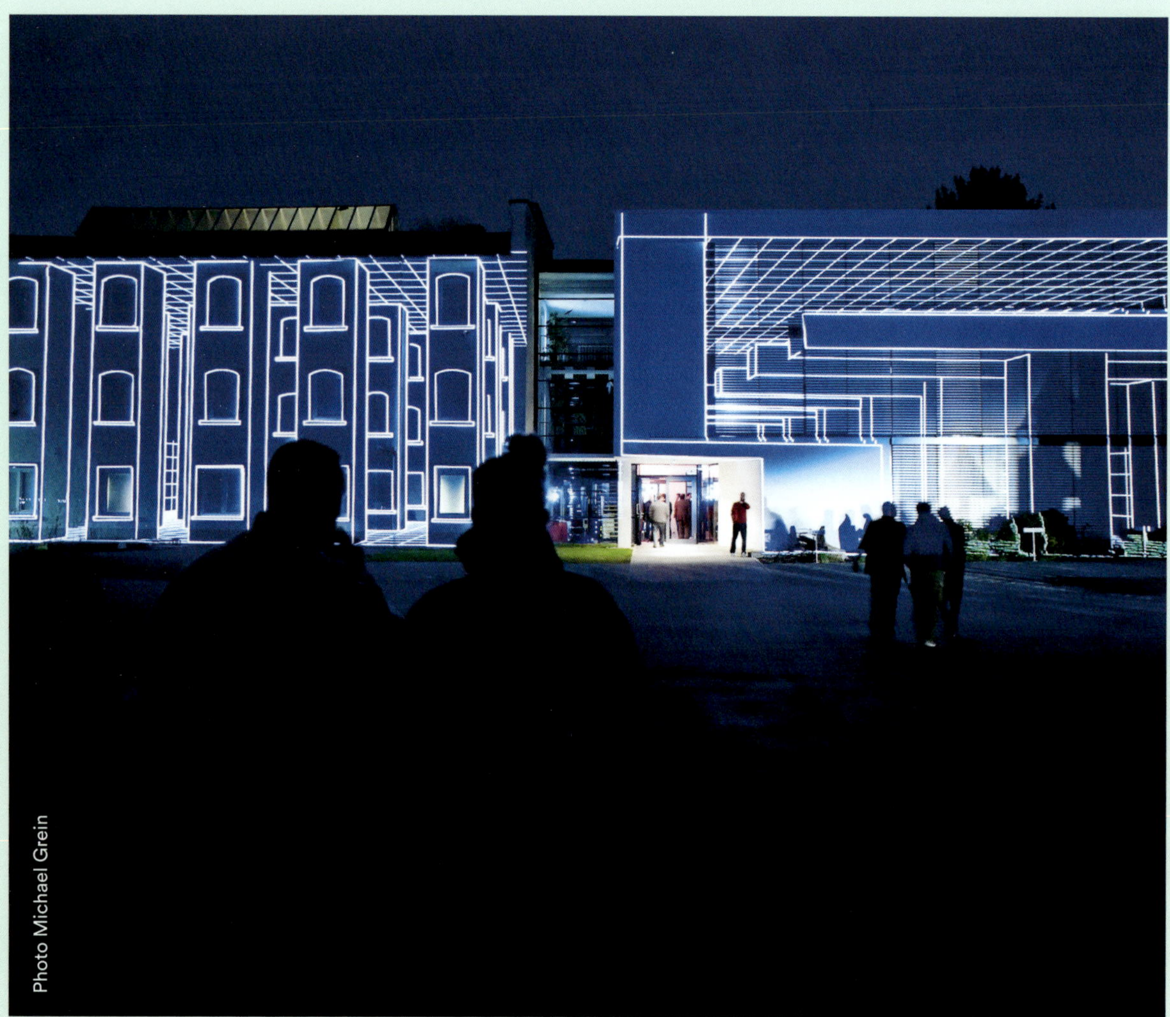

Photo Michael Grein

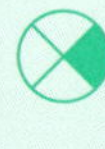

2 × 4 K Projection Mapping

When **2014**
Where **Karben, Germany**
Client **satis&fy**

Lyschik collaborated with the video department of satis&fy in the development of the first high-resolution building projection mapping at 60 fps in 4 K format, taking the creative lead and directing a team of internal and external staff for the video's content production. Two native 4 K prototypes from manufacturer Barco were used as projectors.

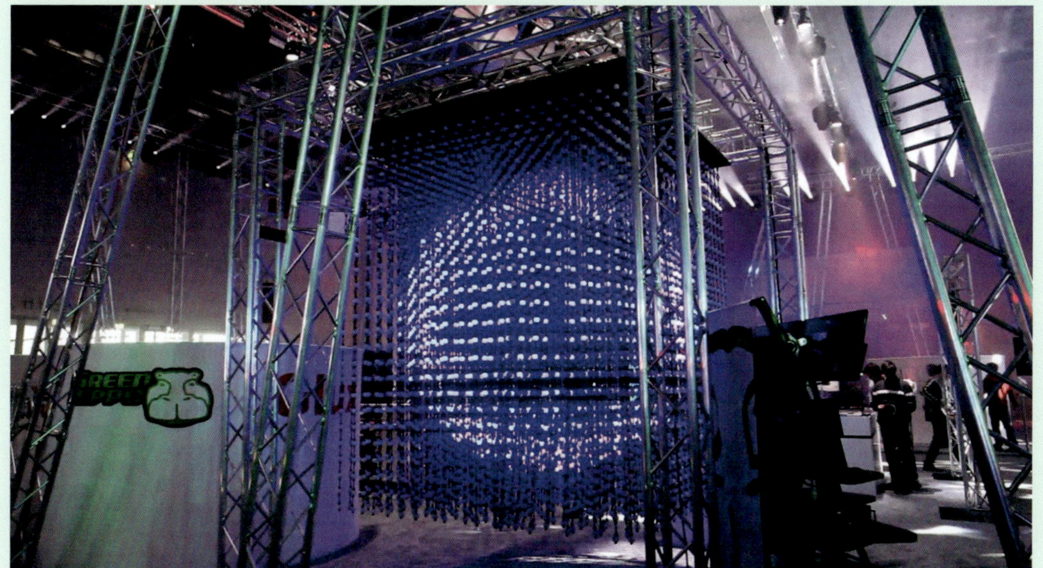

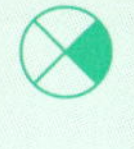

3D Visuals for the G-LEC Solaris Cube

When **2013**
Where **Frankfurt, Germany**
Client **bright! creative event solutions**

Christine Lyschik was commissioned by bright! to realise the 3D visuals for the G-LEC Solaris Cube, one of the company's interactive concepts that was installed at the Pro Light and Sound Fair. It is a matrix of 9600 LED spheres that were hung from the ceiling within an aluminium framework. The concept was developed with audioreactive, generative 3D visuals, and the challenge was to adapt a series of 3D motion graphics for a matrix of spheres.

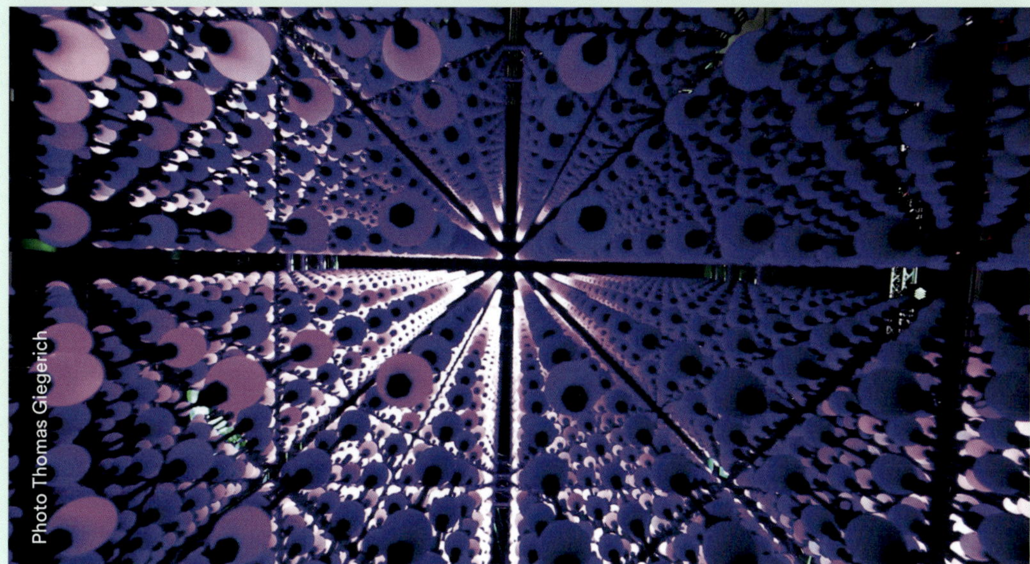

Photo Thomas Giegerich

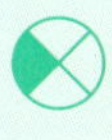

Audioreactive Visuals, Xavier Naidoo Open Air Tour 2014

When **2014**
Where **Germany, Austria, Switzerland**
Client **Cue Design**

In collaboration with Franz Schlechter (visuals) and Gunter Hecker (stage design), Lyschik created the generative motion graphics for the 130 m^2 LED screen. The cinematic LED display and accompanying dynamic light show and generative motion graphics was part of the stage design for Xavier Naidoo's Open Air Tour 2014.

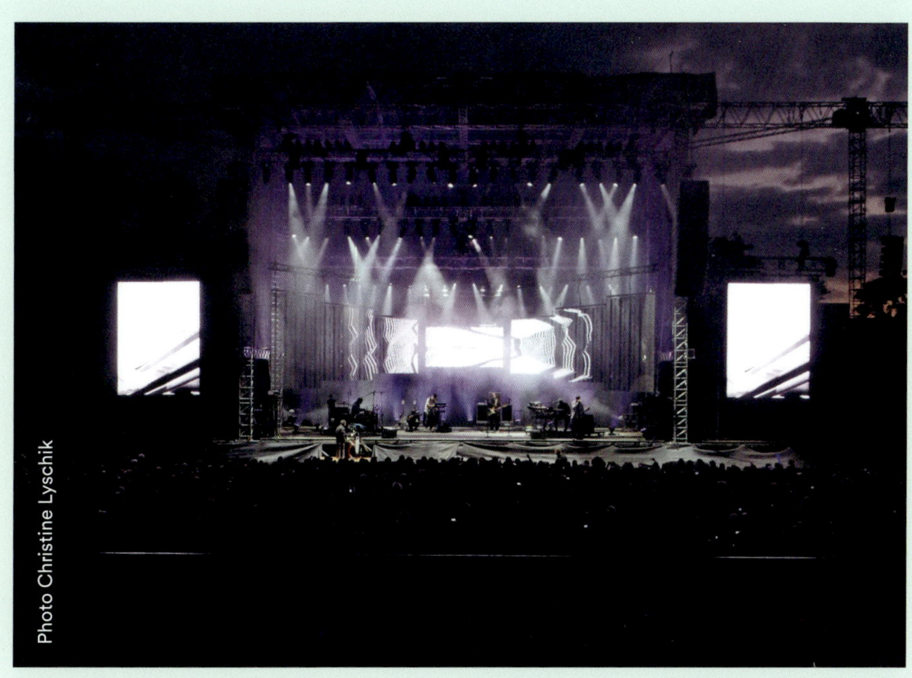

Photo Christine Lyschik

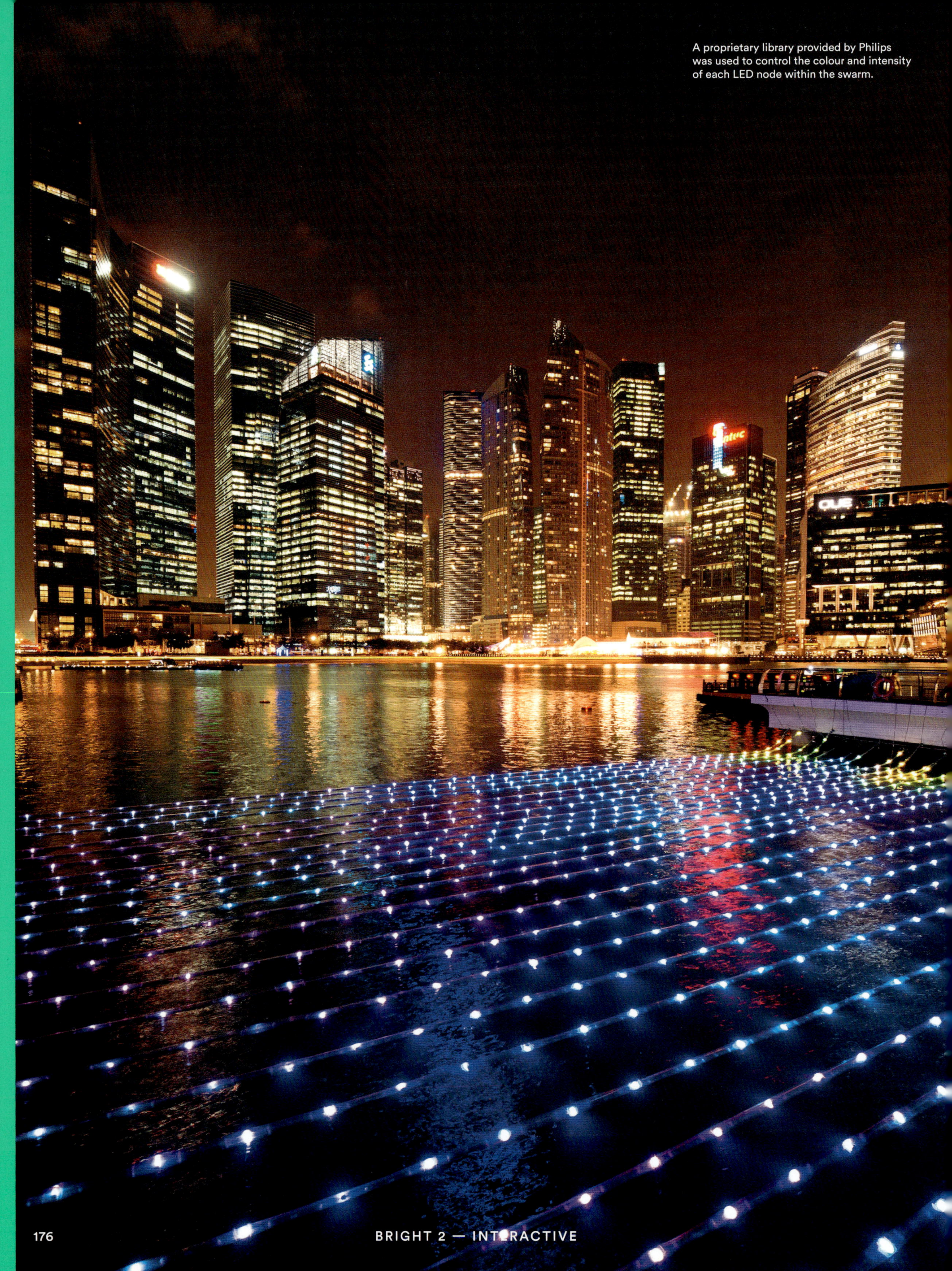

A proprietary library provided by Philips was used to control the colour and intensity of each LED node within the swarm.

iSwarm

Two research groups at **SUTD** collaborated to present iSwarm, illuminating Singapore's Marina Bay with an installation reminiscent of such natural phenomena as bioluminescent algae floating on water, that reacts to visitors with a subtle modulation of its light patterns, triggered by motion sensors positioned around the periphery.

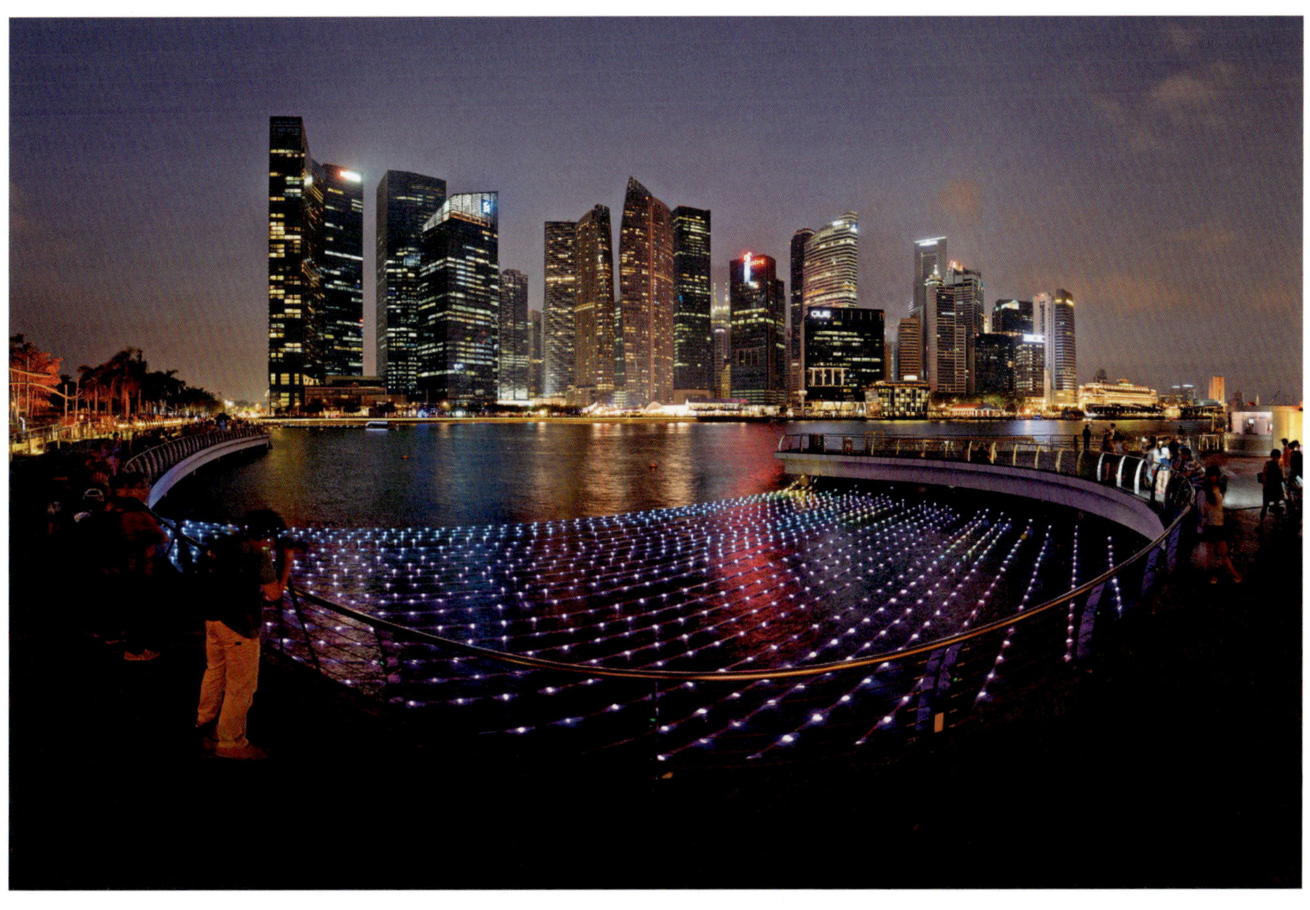

Photos Richard Koh, SUTD AAL

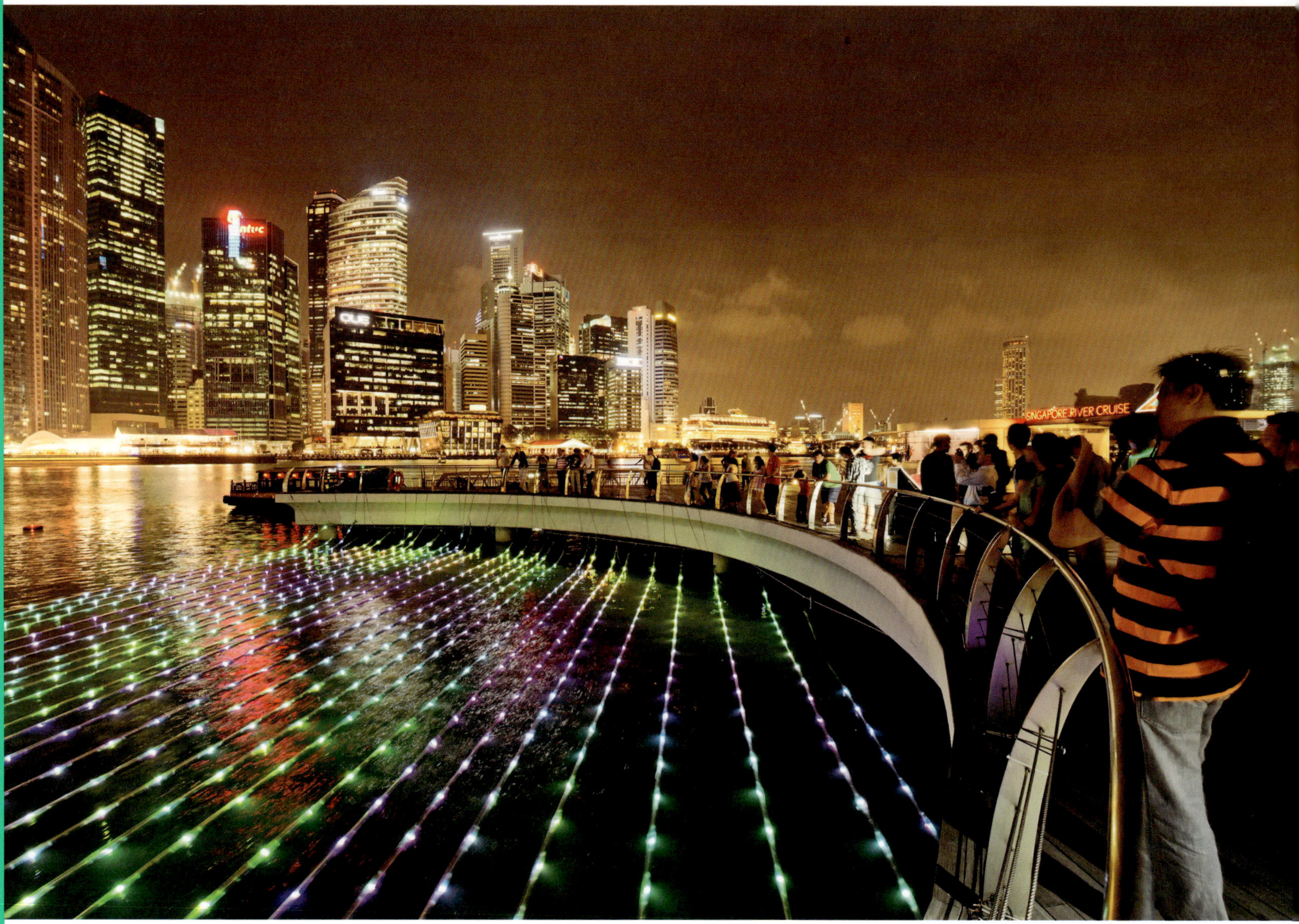

The installation had LED flexible strands laid in a linear arrangement, which differed from the original proposal (see below).

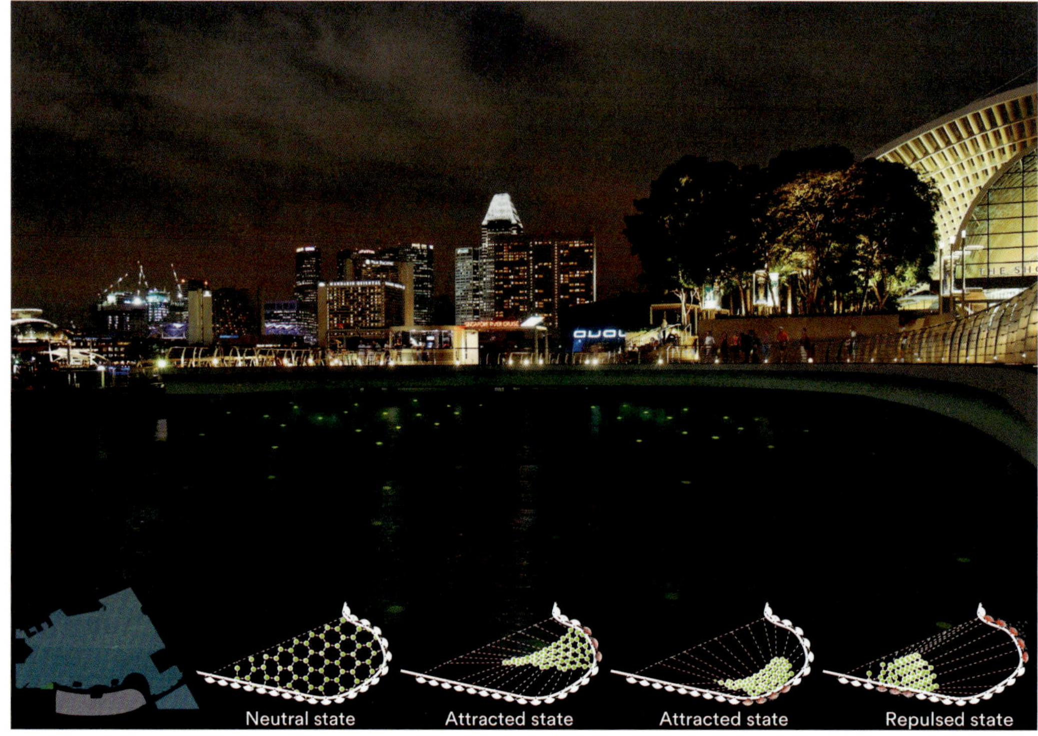

Designer
SUTD Advanced Architecture Laboratory & Augmented Senses Group
Location
Singapore
Client
Singapore Urban Redevelopment Authority
Collaborator(s)/consultant
Philips Lighting
Manufacturer
Philips Lighting
Date
March 2014

Lifeless and hardly visible by day, iSwarm is night-active, like a throng of luminous, mechanical sea creatures that playfully interact with passers-by. The Advanced Architecture Laboratory (AAL, led by Thomas Schroepfer) and the Augmented Senses Group (ASG, led by Suranga Nanayakkara) at Singapore University of Technology and Design (SUTD) worked collaboratively to develop the installation. It engages the audience through a sense of wonder and curiosity, yet is deeply rooted in cutting-edge technology. A series of LED lights, reminiscent of natural phenomena such as bioluminescent algae floating on water, reacts to the presence of approaching visitors with a subtle modulation of its light patterns, triggered by motion sensors around the installation.

Presented at an urban sustainable light design festival held in Singapore's Marina Bay in early 2014 – an event which is a testing ground for innovative projects that bridge architectural design, technology and urbanism – iSwarm was created by a team that believes designing for liveable 21st-century cities cannot merely be about ever-smarter tools and ever-greater efficiency. Rather, it needs to engage human experience as a whole, including all our senses, to encourage meaningful interactions with the environment around us.

The project comprised a network of flexible strands with full-colour LED nodes deployed just below the water surface. After exploring a variety of waterproofing methodologies, the design team settled on deploying the lights within customised extrusions of clear PVC hosing. The team worked directly with the manufacturer to design a bespoke hose extrusion that was both large enough to encase the lights and flexible enough to be easily workable on-site. The layout of the lights was initially designed such that they created an evenly dispersed hexagonal grid. This layout accommodated the most even movement of the 'swarms' across the field of lights when the sensors (deployed on land at the perimeter) detected the movement of people around the installation. —

THE LIGHTS ARE LIKE A SWARM OF LUMINOUS SEA CREATURES THAT INTERACT WITH PASSERS-BY

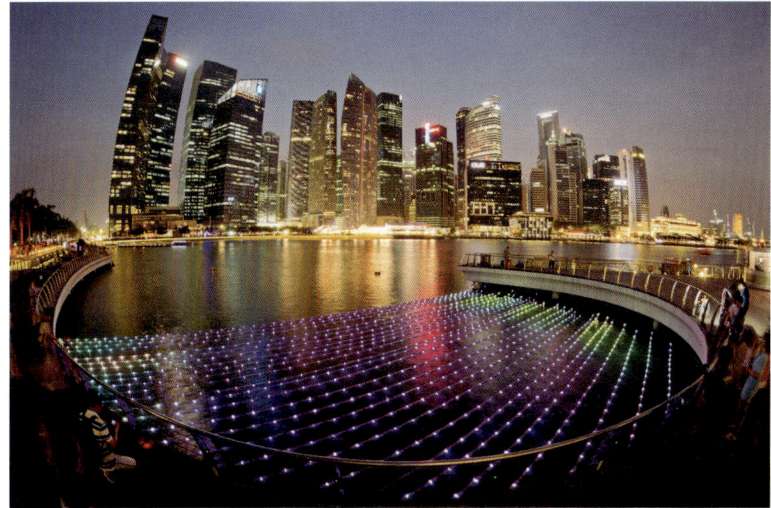

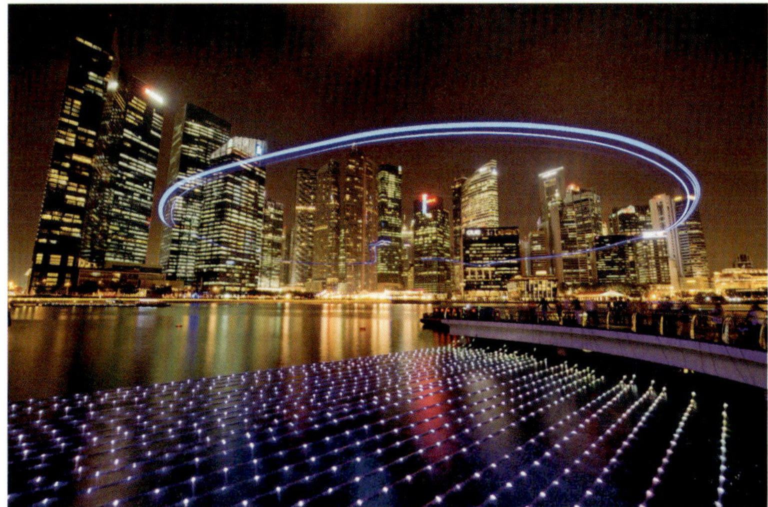

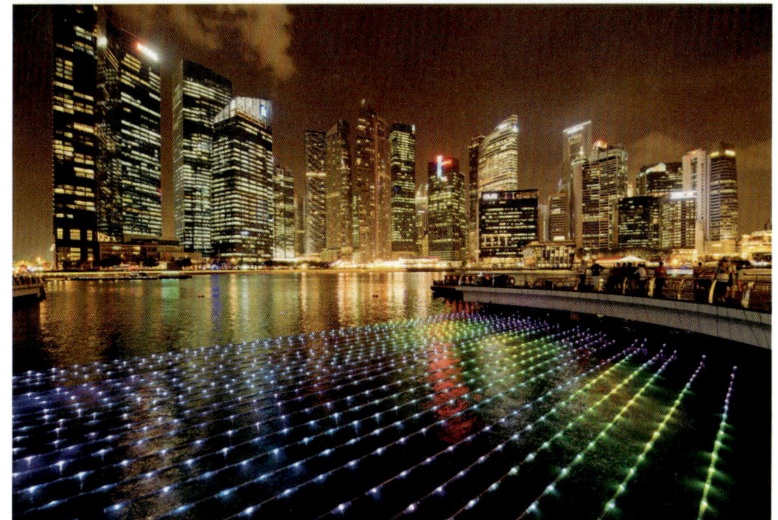

In over-crowded or noisy situations, iSwarm would retreat in search of a more peaceful location.

SITE PLAN

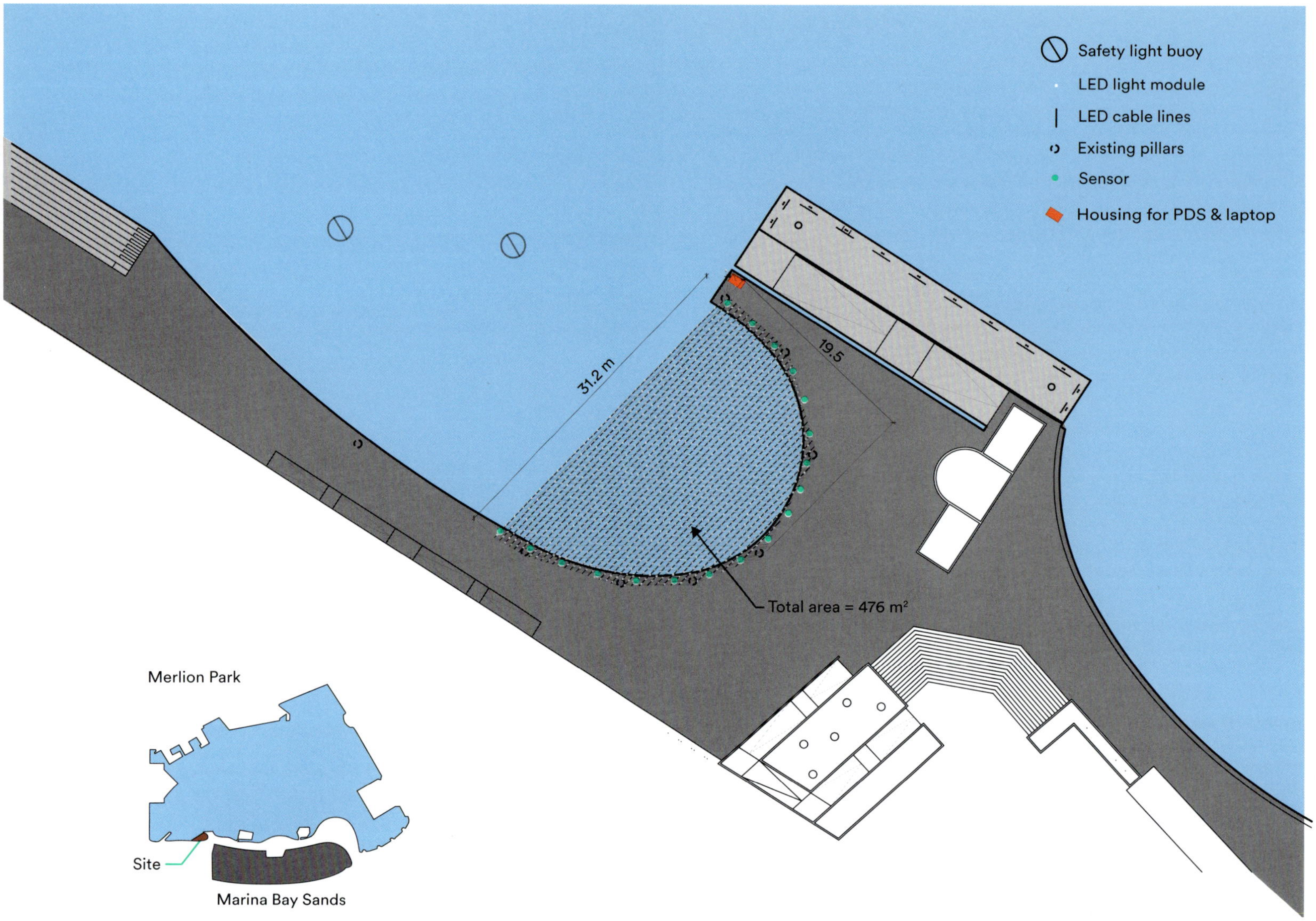

Safety light buoy
LED light module
LED cable lines
Existing pillars
Sensor
Housing for PDS & laptop

31.2 m
19.5
Total area = 476 m²

Merlion Park
Site
Marina Bay Sands

INITIAL STRAND MODULE DESIGN

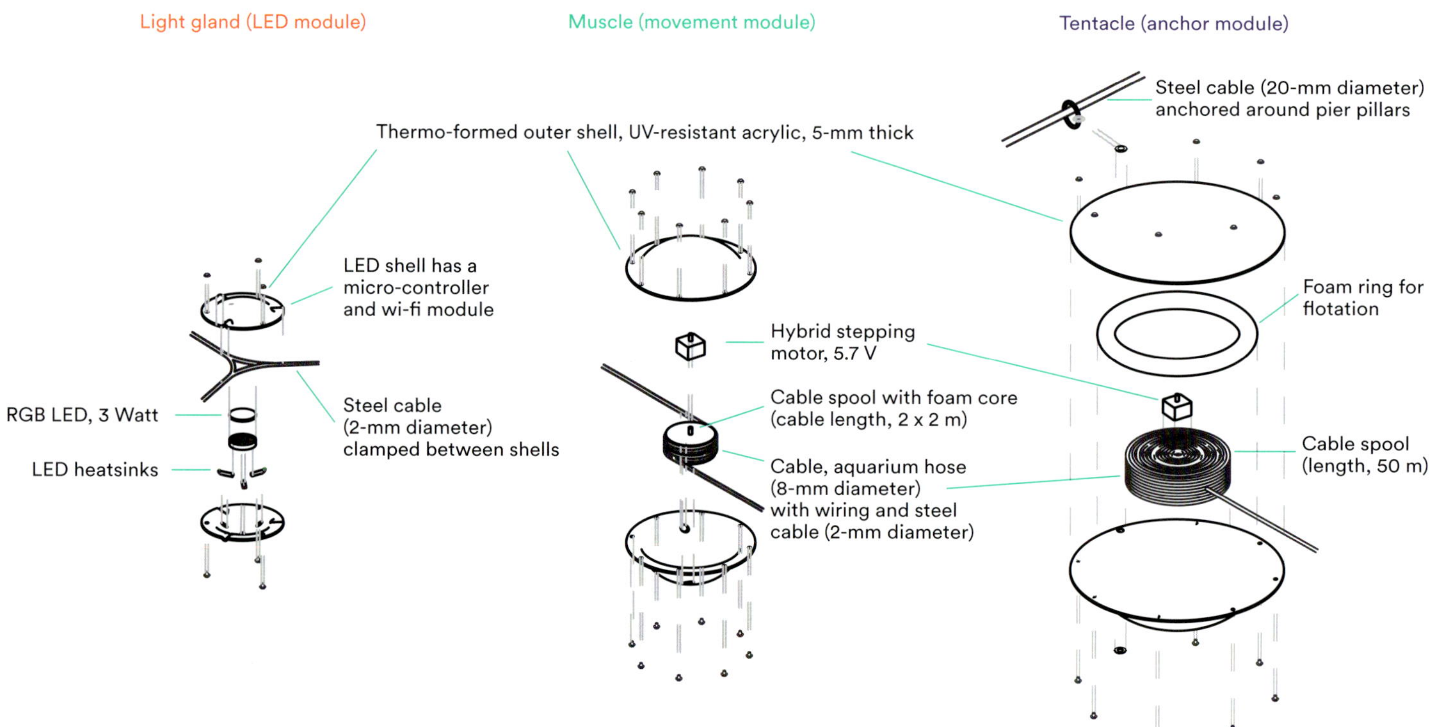

Set-up

Light source included 1500 LEDs strung through a 1-km-long PVC hose.
Illumination using iColor Flex LMX – flexible strands of 50 large full-colour nodes, individually accessible and can produce a maximum of 6.56 candela of light output while consuming 1 W of power.
Framework of flexible strands was arranged in a linear arrangement to fill an arciform layout that had a total area of 476 m².
Motion was detected by 32 sensors that were deployed on land, surrounding the perimeter of the installation. Sensor data was acquired by an Arduino Mega.
Passive infrared sensors were designed and a total of 32 units were 3D-printed at SUTD.
Power supply with an inbuilt Ethernet controller allowed individual nodes to be addressed via processing software.

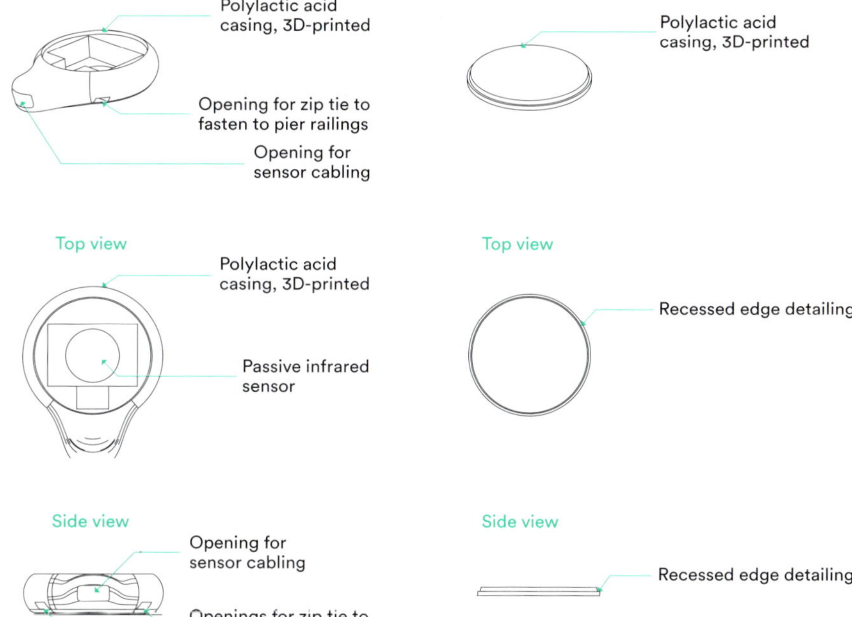

CONSTRUCTION VISUALS

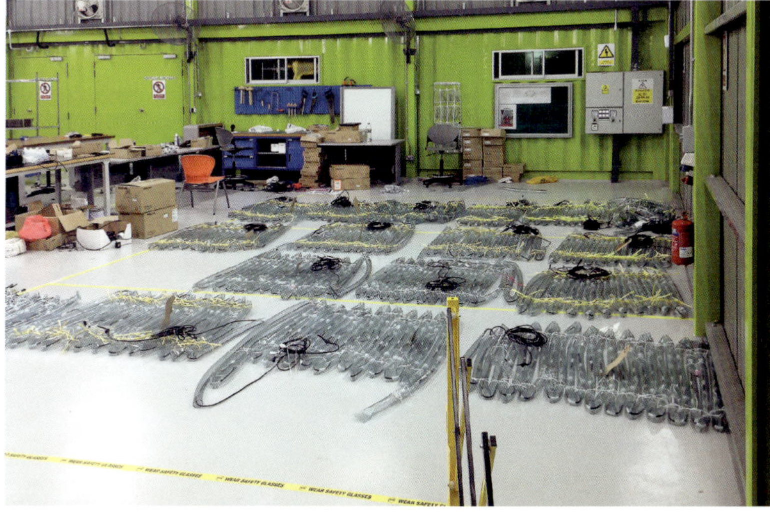

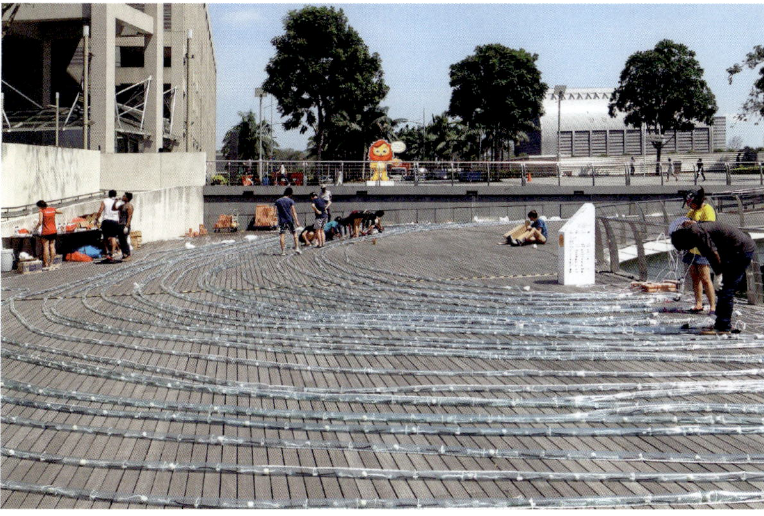

Specifics

The LEDs were deployed just below the water surface, which meant many investigations to explore a variety of waterproofing methodologies, including vacuum-formed housings, cast silicon casings and 3D-printing; the design team settled on deploying the lights within customised extrusions of clear PVC hosing used in the medical and food service industry.

The passive infrared sensors on the land detected the movement of people around the installation; the sensor data was acquired by an Arduino Mega, which communicated with the processing program. Waterproof housings for the sensors were designed and 3D-printed in-house at SUTD.

The individual light strands of iSwarm, fabricated at SUTD by researchers and students over the course of 6 weeks, in total covered an area of approx. 500 m². Changing the spacing of the lights on the strands to accommodate the final design turned out to be a major task, as the team had to rewire 10,000 connections.

Waterproofing the 500 m² covered with lights was a big challenge. For iSwarm's installation in Marina Bay, the team designed pier-to-pier rigs with hooks for attaching the light strands. However, as the actual distances between the piers differed from the as-built drawings the team had received prior to the event, many adjustments had to made on-site.

Further changes to the original design were due to the water level being higher than expected. The team had to spend about 12 hours more to install the project than originally planned.

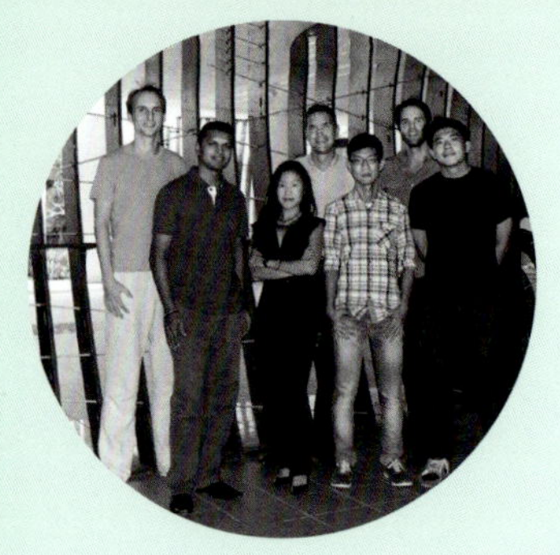

SUTD AAL & ASG

Working collaboratively at SUTD along with their respective research groups are Prof. Thomas Schroepfer, leading the Advanced Architecture Laboratory (AAL), and Prof. Suranga Nanayakkara, leading the Augmented Senses Research Group (ASG). AAL investigates the increasingly complex relationship between design and technology in architecture. Its research and design projects relate to advances in building structure and form, environmental strategies, performance and energy, computer simulation and modelling, digital fabrication, and building processes. ASG focuses on creating 'enabling' human–computer interfaces as natural extensions of our body, mind and behaviours and has been exploring and pushing state-of-the-art advances in this field.

SUTD.EDU.SG

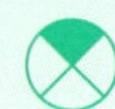

Dhoby Ghaut Green

When **2013**
Where **Singapore**
Client **Singapore Institute of Architects**

Dhoby Ghaut Green is a pavilion designed as a temporary, outdoor recreational space, which reinvents the experience of a public place in Singapore by providing a comfortable, unique and multi-sensorial experience of the city to be enjoyed by everyone. The project uses minimal material to create optimal comfort for the tropical climate.

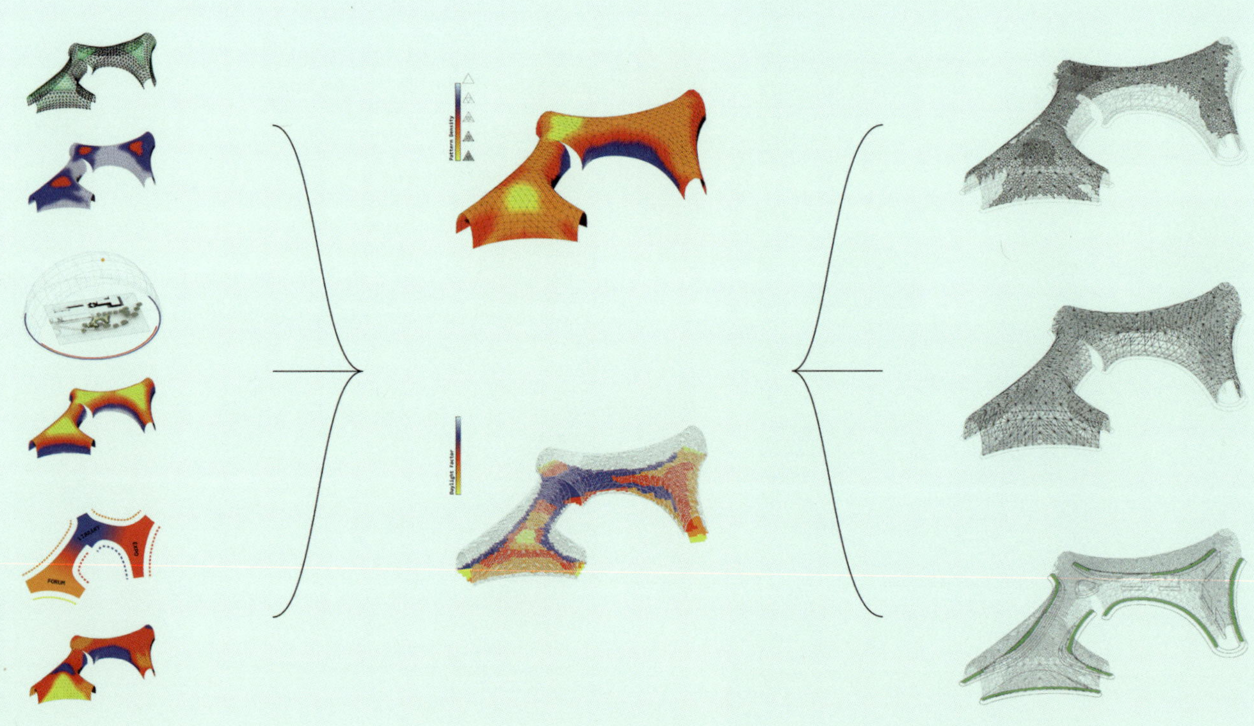

Drawing SUTD AAL

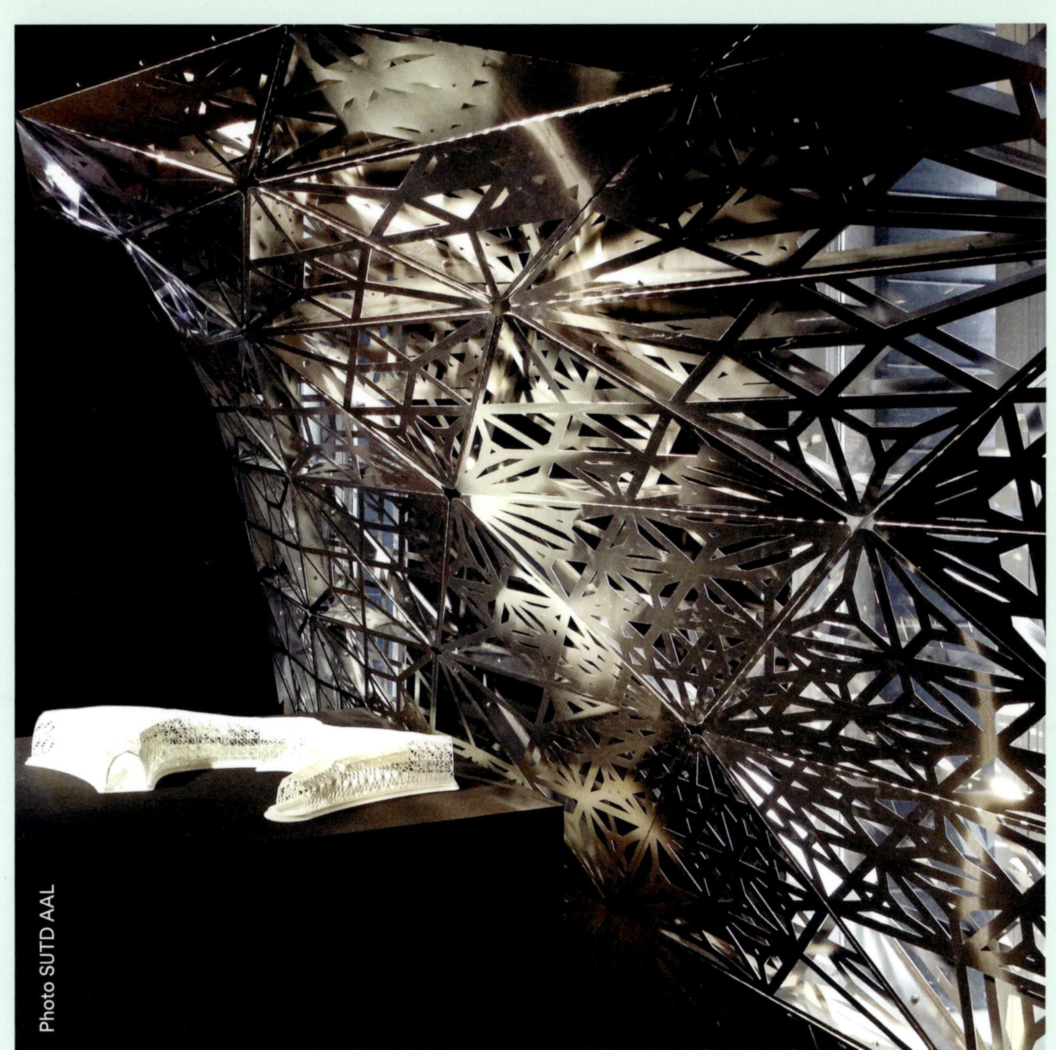

Photo SUTD AAL

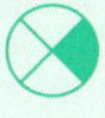

Tropical Urban Living

When **2014**
Where **Venice, Italy**
Client **Venice Architecture Biennale**

The project showcases AAL's work with state-of-the-art computational simulation tools that allow for analysing daylight factors and structural displacements to produce building skins that respond to their tropical environments, producing intricate geometric patterns of varying porosity that complement the intricate play of light and shade provided by nature.

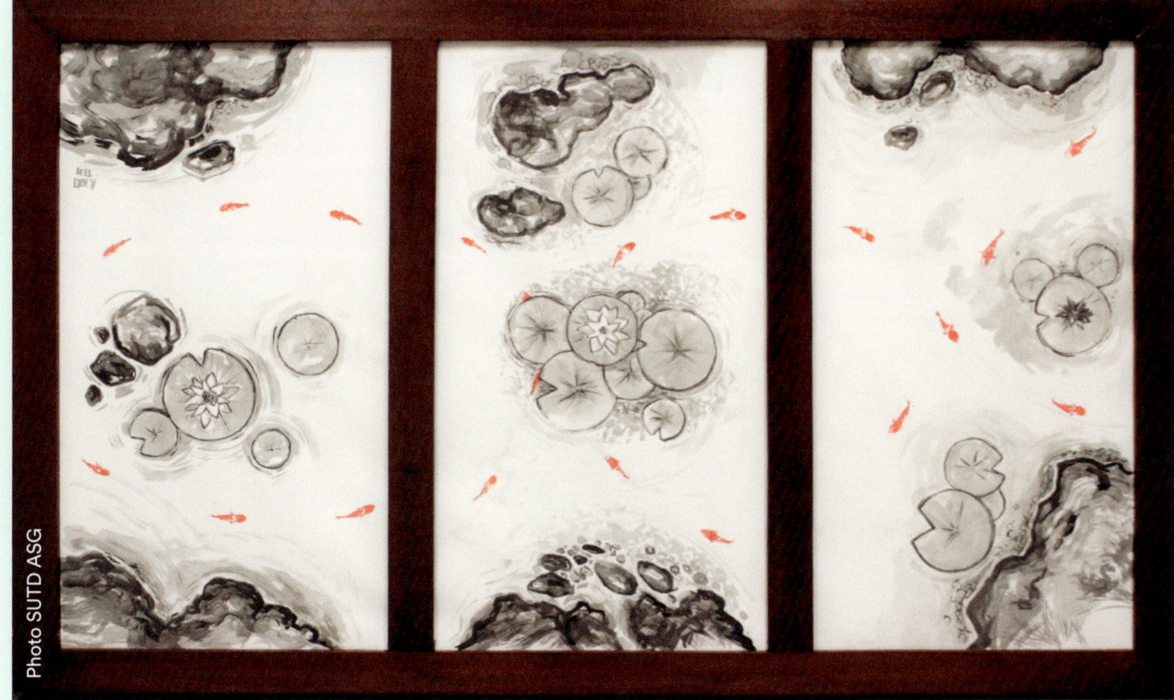

Photo SUTD ASG

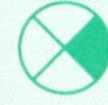

PaperPixels

When **2014**
Where **Singapore**
Client **National Design Centre**

PaperPixels uses a patented thermal mechanism to allow users to animate content on regular materials, such as paper and textiles. Kinetic Canvas is an instantiation of PaperPixels by an artist to release the static nature of artistic representations into a dynamic form. This particular piece represents one such animation of a fish swimming across the canvas (only a single fish is visible at run time).

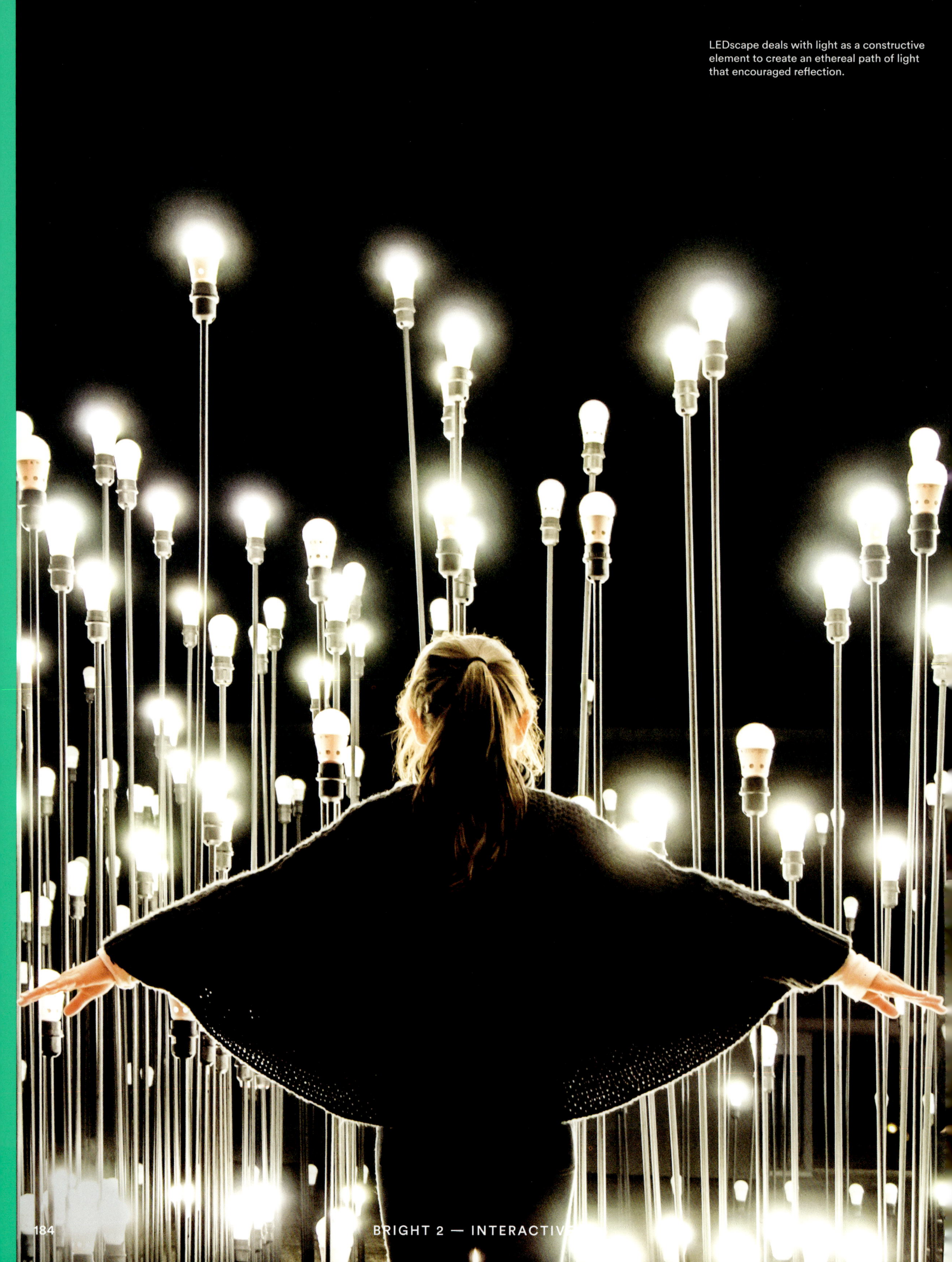

LEDscape deals with light as a constructive element to create an ethereal path of light that encouraged reflection.

LEDscape

The temporary light installation LEDscape by **Like Architects** was a response to the challenge set by the client for the Porto-based architects. Presenting the Ledare lightbulb to the public in a unique manner, the team created an enticing scene where thousands of light spots were used to create an ethereal, illuminated pathway.

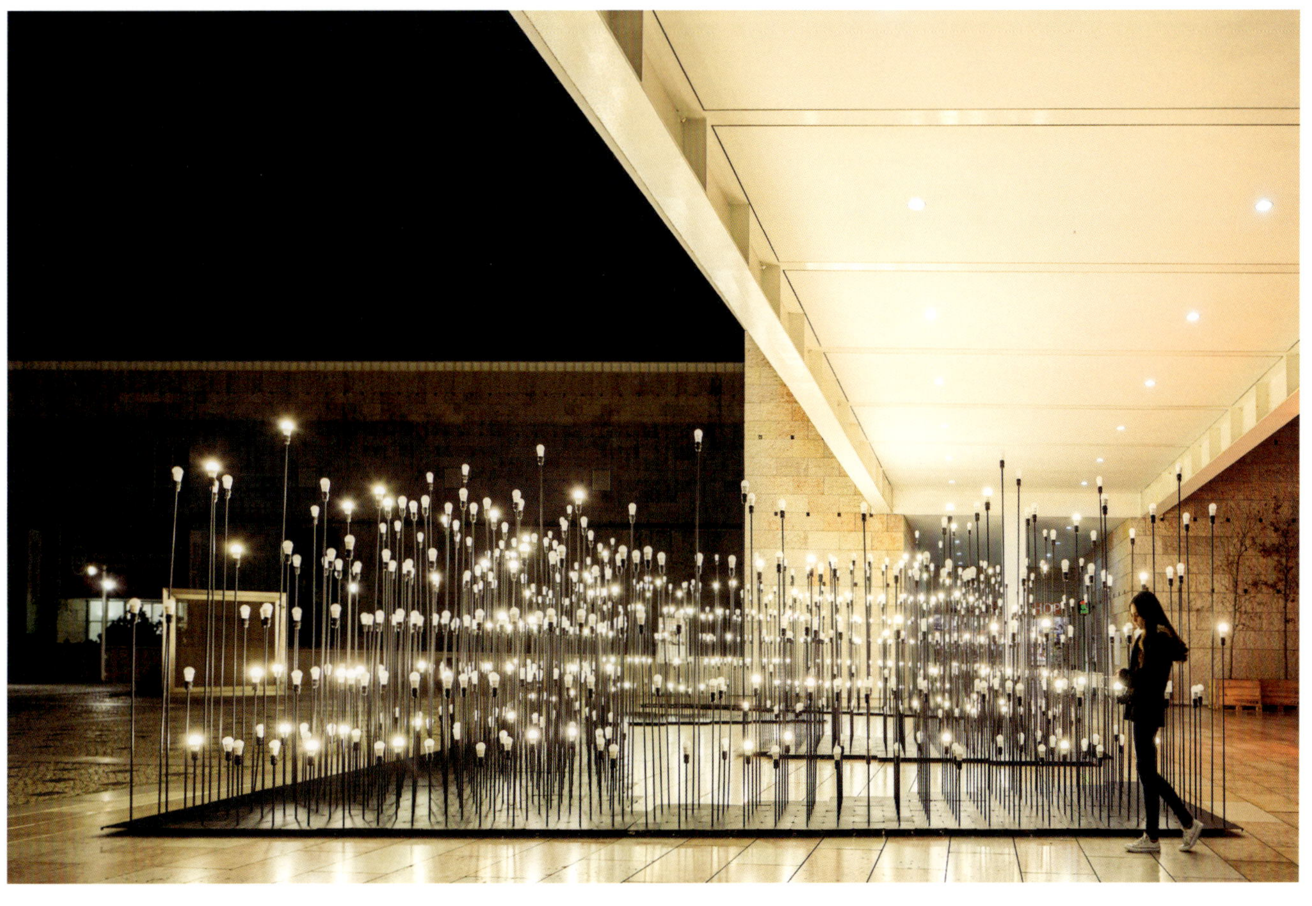

Photos Fernando Guerra, Sérgio Guerra

Designer
Like Architects
Location
Lisbon, Portugal
Client
IKEA
Consultant
Sérgio Sousa
Manufacturers
Like Architects, Artisforum
Date
November 2012

In 2008, Teresa Otto and Diogo Aguiar joined forces and won first prize of a competition amongst the students of the Faculty of Architecture of the University of Porto to design a temporary bar for the city's yearly academic festival. This award-winning project was the first installation for which the architects used IKEA products as a modular aspect within a project.

Fast-forwarding a couple of years, the duo joined forces with João Jesus to establish the studio Like Architects. In 2012, it was this creative team that was invited by the Swedish brand IKEA to introduce the (then new) Ledare LED lightbulb to the Portuguese public in a unique manner. LEDscape was the resulting temporary installation: a sea of light that was open to all to be traversed.

Made up of 1200 light spots and 1200 connected floor lamp bases of varying heights, the installation formed an enticing pathway that encouraged visitors to stop and take an illuminating journey along its length. LEDscape used light as an element that builds space, as well as crafting a landscape. Through this piece, installed in the central courtyard of Centro Cultural de Belém in Lisbon, the architects looked to comment on the importance of LED technology for a more sustainable future.

Passers-by were invited to engage with the installation: walking through it and appreciating the soft changes to the space's atmosphere as lamps pulse in different rhythms, the architects challenged its visitors to appropriate the piece for individual reflection. As the brightness of the LED bulbs changed in response to stimuli, different perceptions of space were created in an involving sequence that cemented a bond between the participants and the bulbs. The success of LEDscape was such that the duration of the light installation was extended, so that animations of the display continued to spread the sustainable message throughout the entire festive season. —

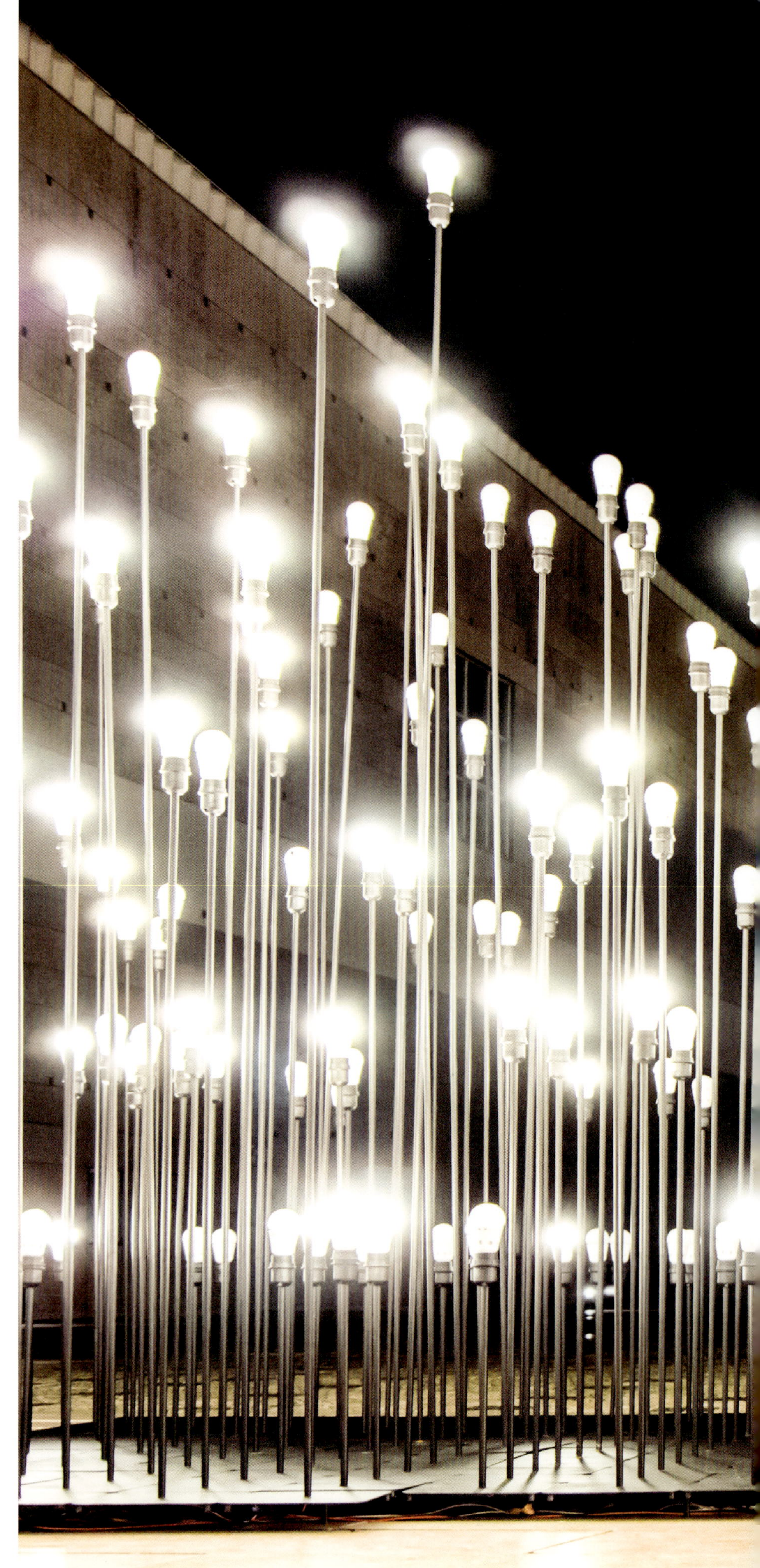

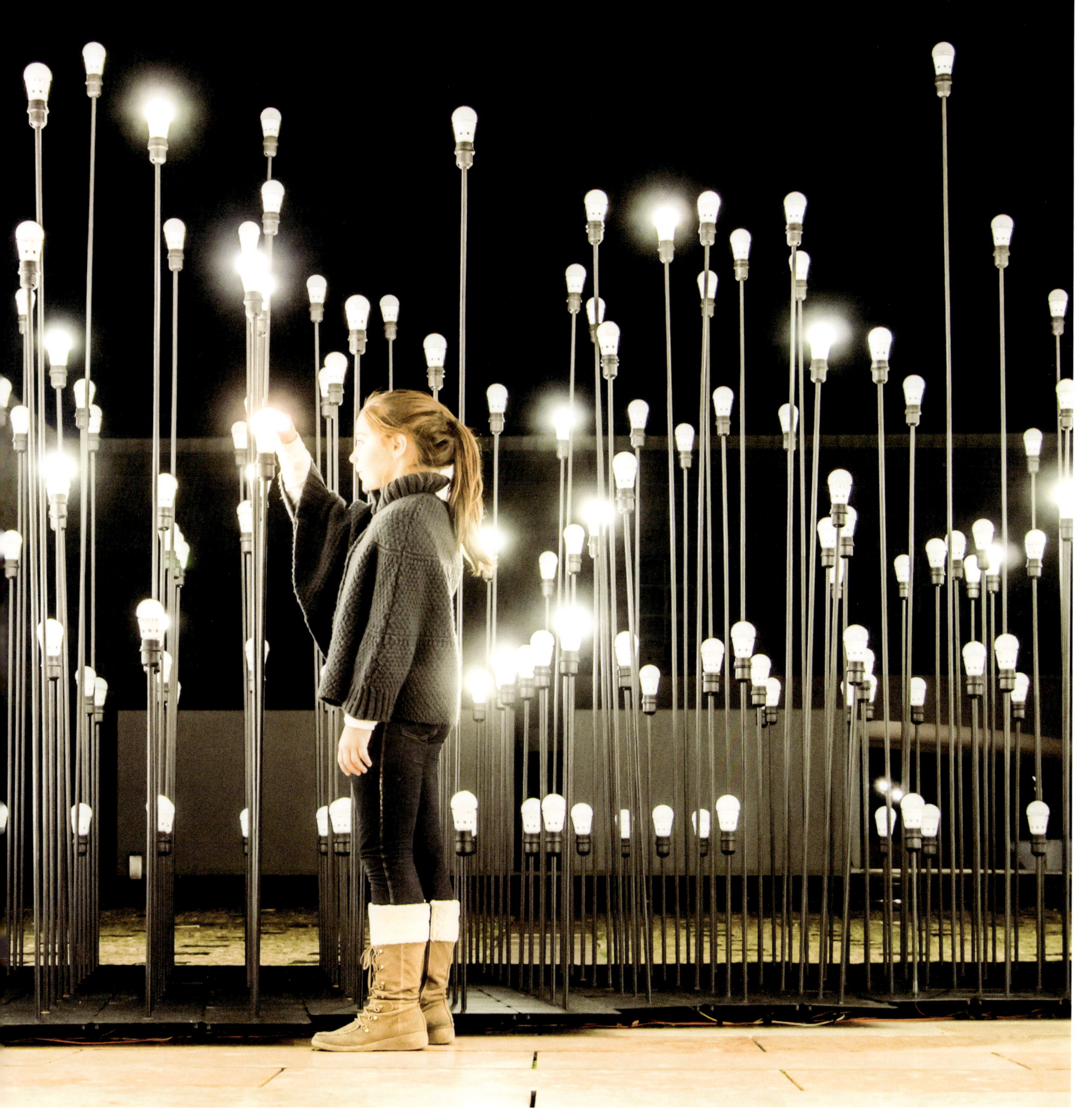

USING LIGHT AS AN ELEMENT THAT BUILDS SPACE, AS WELL AS CRAFTING A LANDSCAPE

PLAN

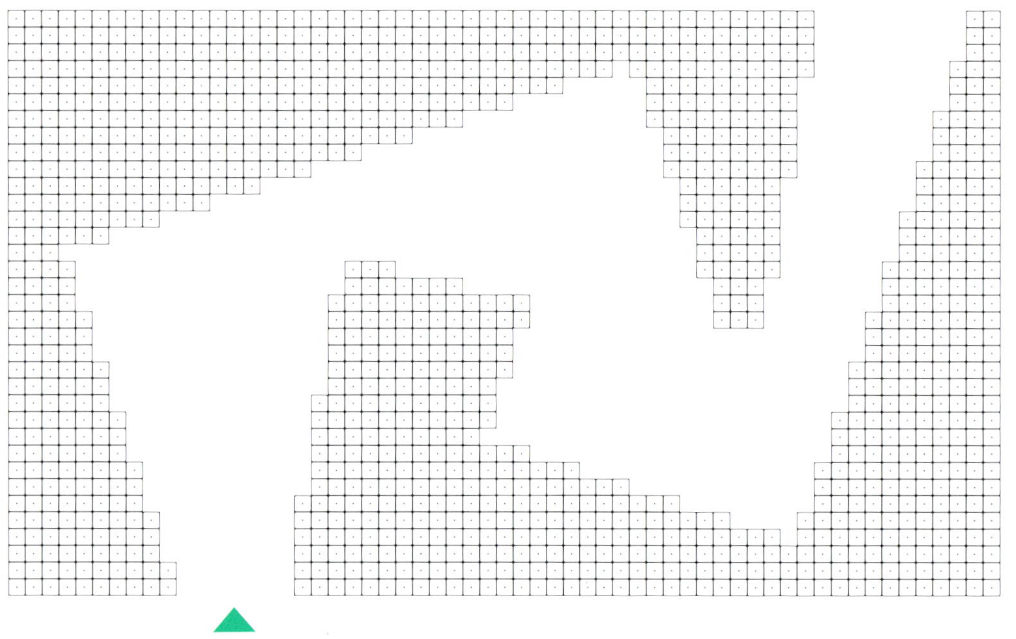

SENSOR POSITIONING

Sensors
sector A – 110 lamps
sector B – 110 lamps
sector C – 105 lamps
sector D – 105 lamps
sector E – 100 lamps
sector F – 80 lamps
sector G – 80 lamps
sector H – 50 lamps

Sequencers
GROUP 1 – 115 lamps
GROUP 2 – 115 lamps
GROUP 3 – 115 lamps
GROUP 4 – 115 lamps

Total: 1200 lamps

SIDE ELEVATION

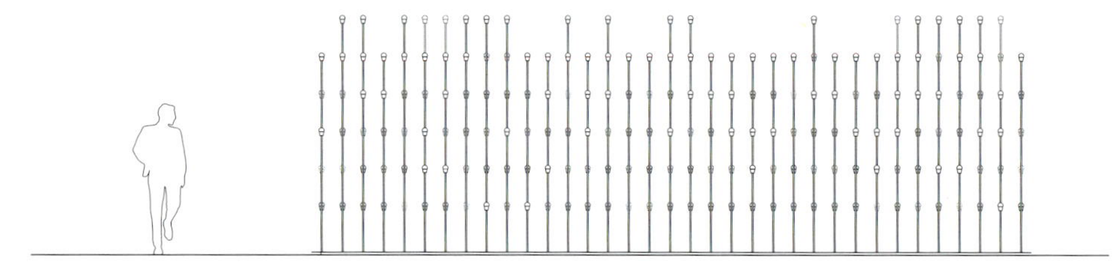

Specifics

Despite being designed for interior use, Hemma lamp bases and Ledare bulbs have good specifications regarding mechanical and water resistance. Without manufacturer warranties, however, the architects tested a unit of the materials under demanding conditions, namely different amounts of water and weight loads. Installing rubber rings on each lamp holder and replacing the base screws with longer ones to improve the structure were some of the minor upgrades made to the materials, following on from the tests. These upgrades warranted the safety and eligibility of the materials for outside use.

To create a direct relation between the user and the LED lights – and emphasise the LED technology as part of a household appliance – the architects created a pathway that was lit-up specifically by the movement of people within the installation. Eight motion sensors were strategically placed by the bases of the lamps and calibrated to react on marked areas. When someone approached, it triggered each sensor to illuminate a group of lights. For the installation to remain dynamic even when there was noone walking on the path, a 4-channel sequencer box was pre-programmed to ensure an animated light sequence illuminated the installation at all times.

In addition to the prototypes made for waterproof and mechanical resistance tests, digital models were made for the pre-programmed lighting sequences and for the investigation of sensor response to passers-by, which were then tested and calibrated on the final installation.

Since the LEDs were not dimmable at the time the installation was designed, the sensors were used only with an on/off set-up.

HEIGHTS OF LIGHTS

1 0.56 m
2 0.98 m
3 1.40 m
4 1.82 m
5 2.24 m
6 2.66 m

AXONOMETRIC

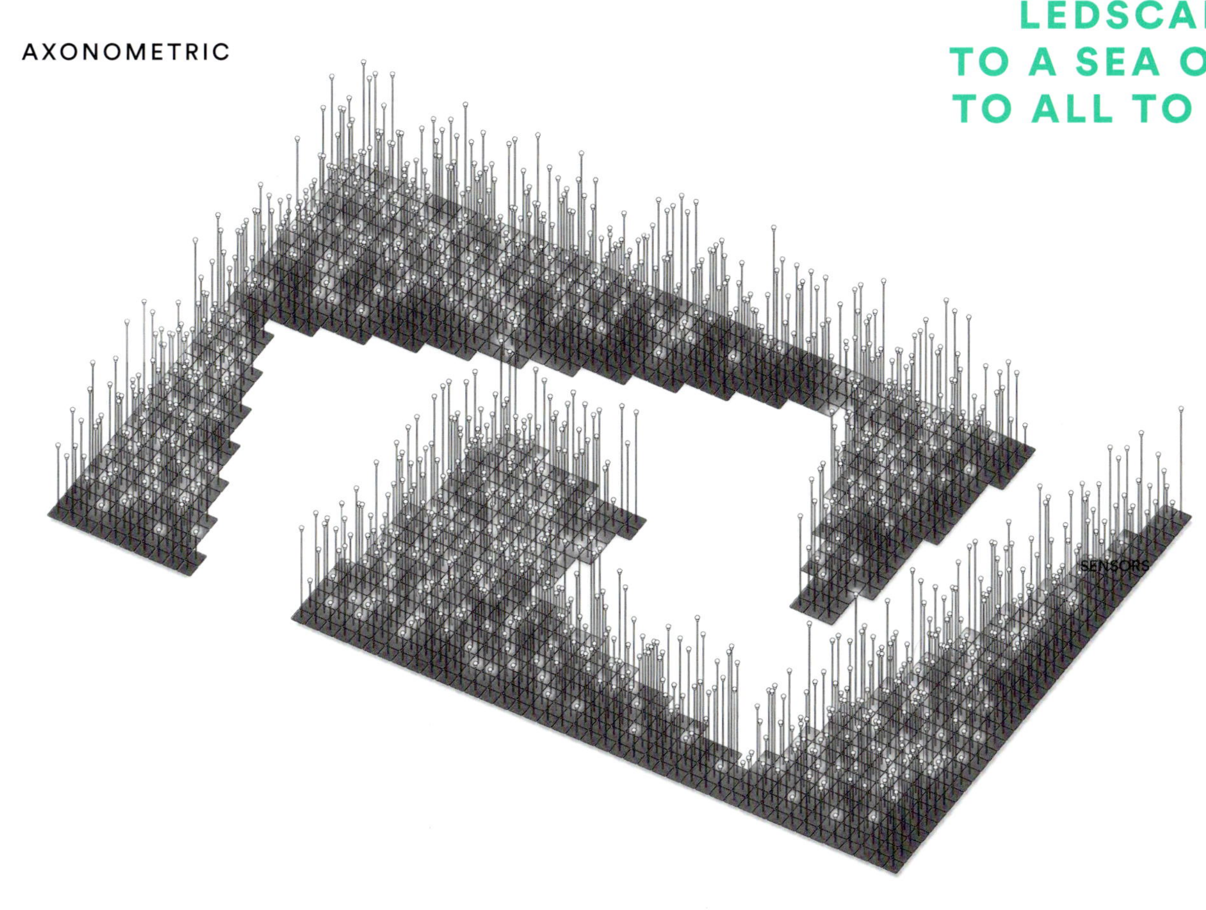

LEDSCAPE EQUATES TO A SEA OF LIGHT, OPEN TO ALL TO BE TRAVERSED

FRONT ELEVATION

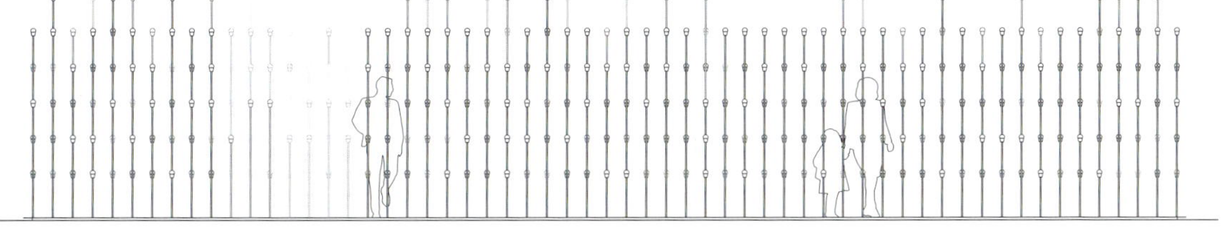

Set-up

Light source were 1200 Ledare bulbs and 1200 Hemma floor lamp bases. There were six different heights that formed the installation, ranging from 0.56 to 2.66 m.
Total area was 112 m^2.
Motion sensors were hidden under the lamp bases and activated by the movement of visitors through the pathway.
Luminaires were positioned in different groups of 80 to 110 light bulbs associated to 12 different sectors and sequences.
Lighting control system was simultaneously a predefined program run on a 4-channel sequencer box (groups 1 to 4) and 8 channels of motion activated sensors (sectors A to H).
Construction time was about 3 days with a team of four people.

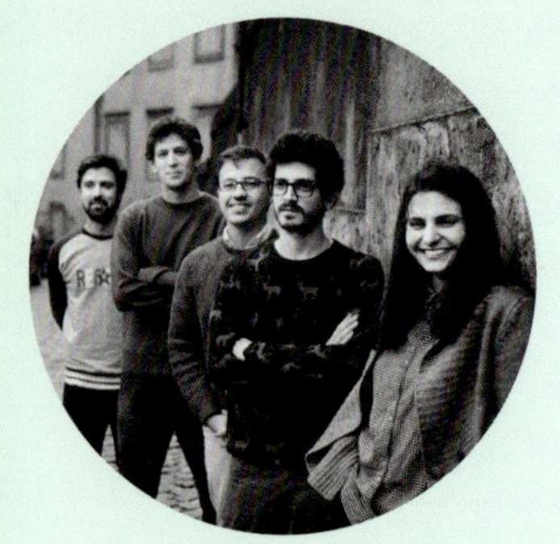

Like Architects

This young, provocative design atelier was established in 2010 by Diogo Aguiar, João Jesus and Teresa Otto. The Porto-based architects have been working together since 2008, when they were still students at the Faculty of Architecture of the University of Porto. The first collaborative project was an acclaimed temporary bar for an academic festival, which formed a good base for them to establish their studio 2 years later. There is currently a team of ten members who like to use transience, spontaneity and urban art as primary mediums to create architecture that instills critical dialogue, enhances experiences of space and is sensitive to the present socio-economic climate.

LIKEARCHITECTS.NET

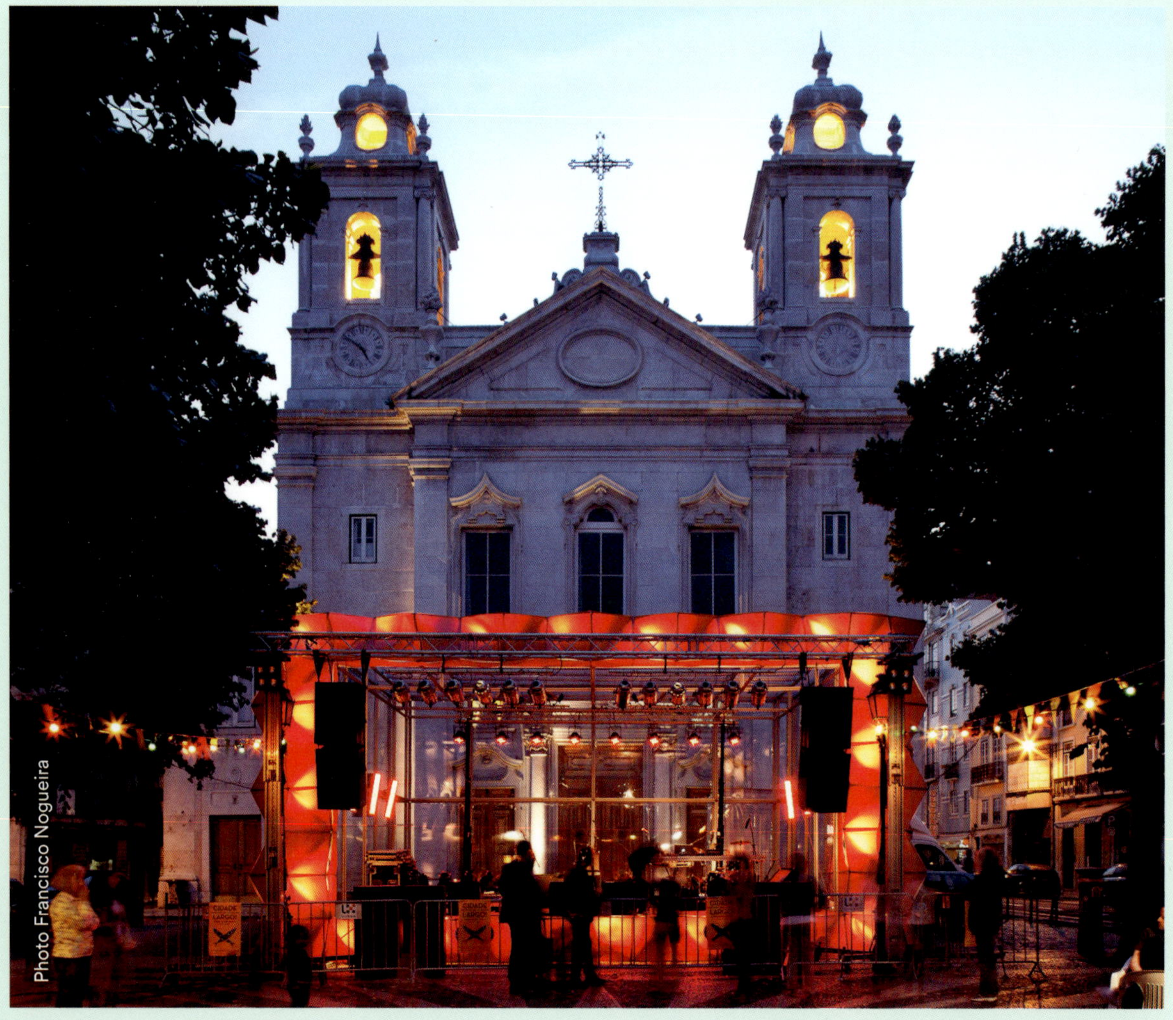

Photo Francisco Nogueira

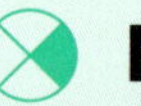

Party Animal

When **2011**
Where **Lisbon, Portugal**
Client **Ordem dos Arquitectos**

A temporary stage, Party Animal looked to put a central Lisbon square into the cultural spotlight. The front of the lightweight, low-cost stage was made up of prefabricated modular parts which at night became an illuminated geometric frame. The glowing red light illuminated many of the performances that animated the square during the summertime celebrations of the city's patron saint. Serving as 'an iconic urban catalyser', the installation revitalised the area and replaced it in the routes of city-dwellers as a destination.

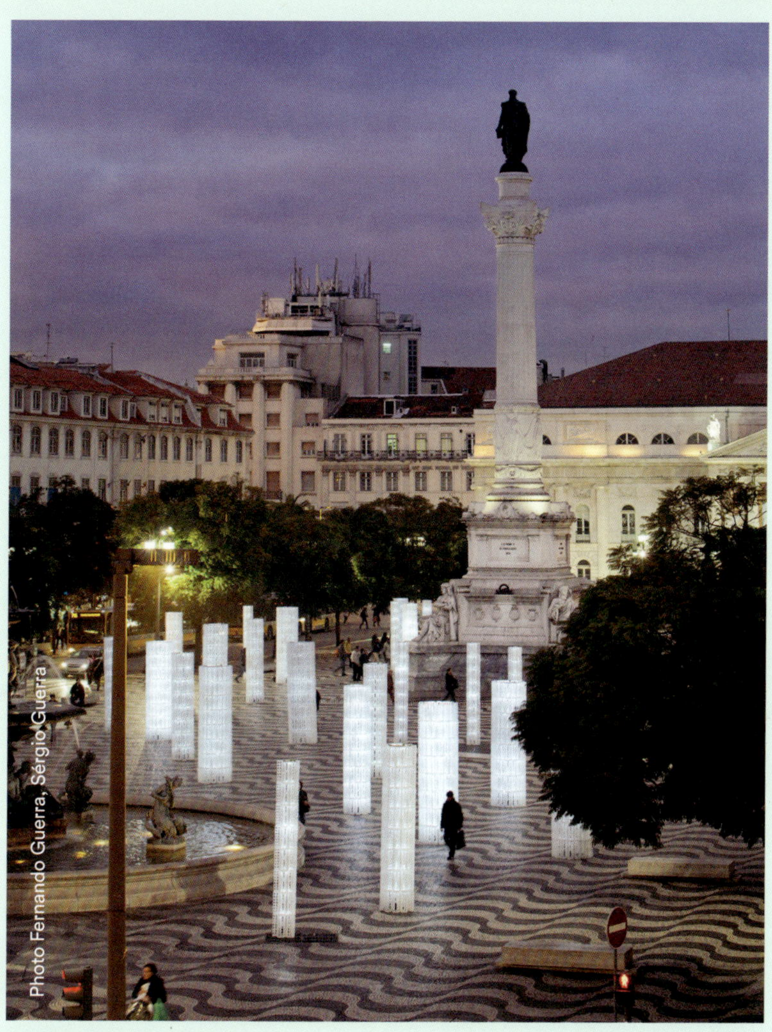
Photo Fernando Guerra, Sérgio Guerra

 Frozen Trees

When **2011**
Where **Lisbon, Portugal**
Client **Lisbon City Hall and Fashion and Design Museum**

Frozen Trees illuminated a Lisbon plaza during the festive season of 2011. Thirty cylindrical columns, made up of IKEA's Rationell Variera plastic bag dispensers, functioned as frozen, self-sufficient streetlights that affected the square's atmosphere and the path of passers-by, inviting them to engage in new spatial experiences. By using a domestic object and disassociating it from its meaning, the architects looked to call attention for the need of creativity in the current climate of crisis.

 Constell.ation

When **2013**
Where **Lisbon, Portugal**
Client **Museum of the Presidency of the Portuguese Republic**

Constell.ation adorned the gardens of Palácio de Belém, the official residence of the Portuguese chief of state, during the 2013 Christmas season as part of an initiative that opened up the gardens to the general public. Multiple red arches illuminated by a LED system – a reinterpretation of the lighting elements that usually adorn the city during festivities – were placed throughout the classical garden, creating a sense of continuity. The installation has since appeared at various international locations.

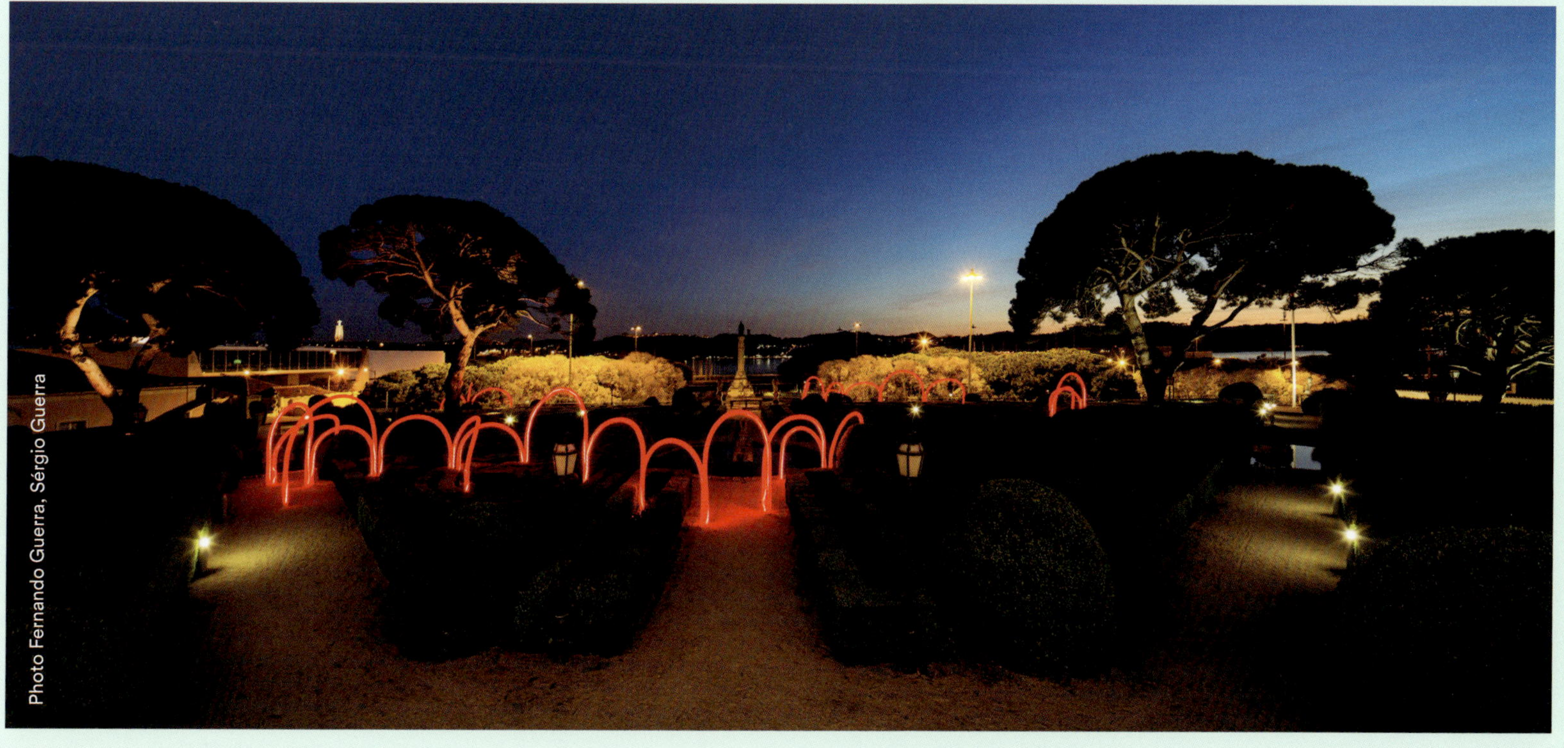
Photo Fernando Guerra, Sérgio Guerra

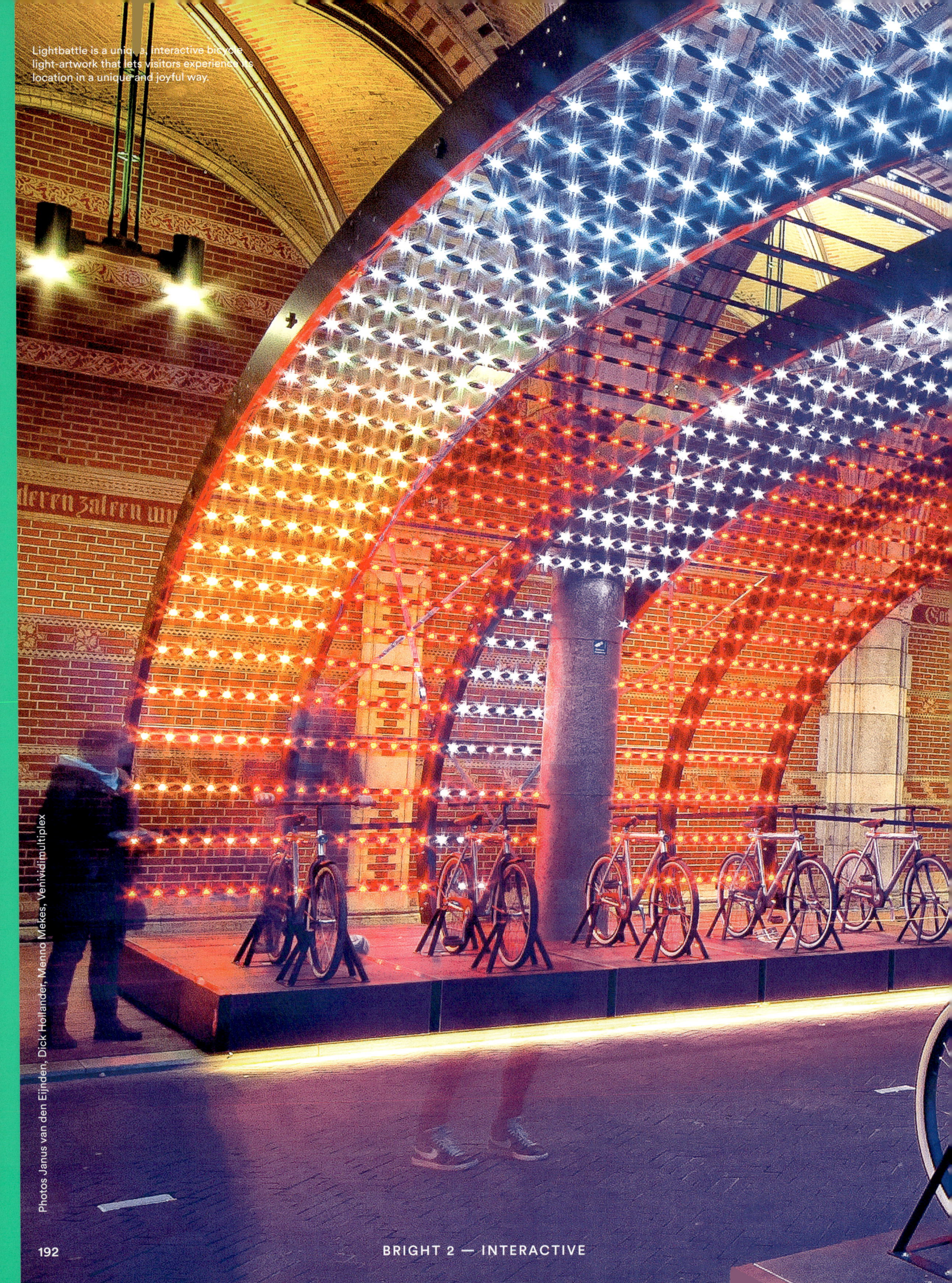

Lightbattle is a unique, interactive bicycle light-artwork that lets visitors experience its location in a unique and joyful way.

Photos Janus van den Eijnden, Dick Hollander, Menno Mekes, Venividimultiplex

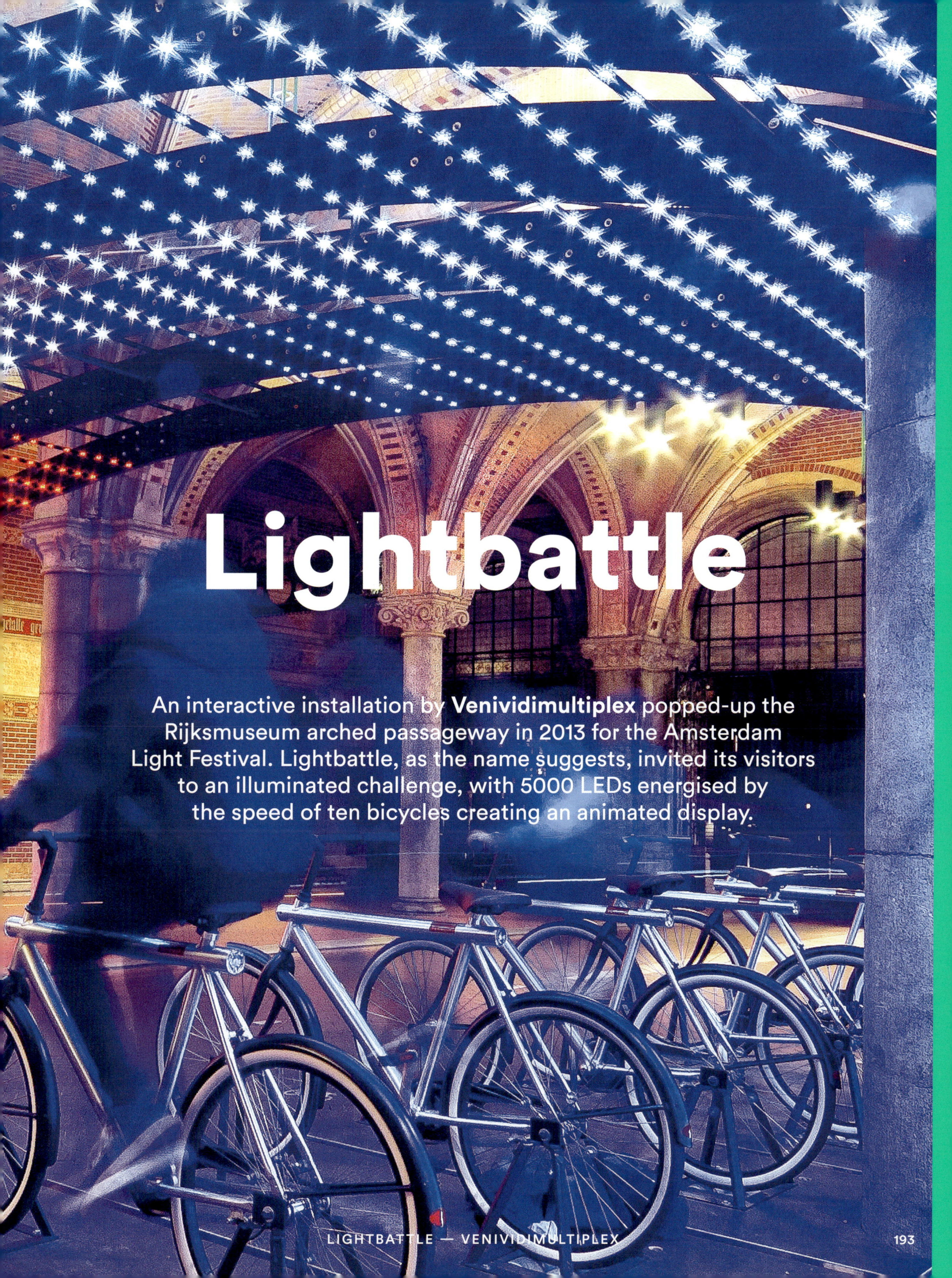

Lightbattle

An interactive installation by **Venividimultiplex** popped-up the Rijksmuseum arched passageway in 2013 for the Amsterdam Light Festival. Lightbattle, as the name suggests, invited its visitors to an illuminated challenge, with 5000 LEDs energised by the speed of ten bicycles creating an animated display.

The majestic arched passageway for cyclists that runs through the heart of the Rijksmuseum in Amsterdam was the destination for an illuminating installation during the annual winter light festival in 2013. It seems appropriate then, that this project was radiantly lit-up by pedal-power.

Dutch studio Venividimultiplex created Lightbattle which invited visitors to initiate a duel of light. The challenge was open to two teams of participants, with the energy created by the speed of their pedalling leading to LEDs being animated overhead. Consisting of an arched curvature (mimicking the shape of its surroundings), the structure spanned the width of the passage. On either side were five bicycles, with the pedal-energy created triggering each of the 5000 LED lights attached to the arch as the battle of light unfolded. Participants were directly involved in the message that the architects sought to convey, as their energy physically prompted the illumination.

Functioning like an arcade game, Lightbattle attracted visitors by creating a visual spectacle as red and blue lights illuminated the entire archway passage, challenging audiences to take part. Each participant controlled a specific arch of light, linked to one of the bikes. When one team overpowered the other, the LEDs would all turn to the team's colour and flash, creating an even more vibrant display.

Utilising Dutch designer bikes and a dazzling light display, the installation had a high-tech feel. It was built from reusable materials using a proven triple-layered wood construction method. Six wooden arches, anchored at either end to platforms for the bicycles, were lined with the aluminium LED strips. All the elements for the lighting effects, electrical cabling and control devices were connected to six pre-programmed DMX computers. The overall outcome was dazzling, yet power consumption was kept to a minimum by the use of low energy-consuming LED lights. The result was a low cost attraction that had an extremely high-impact and wow factor. —

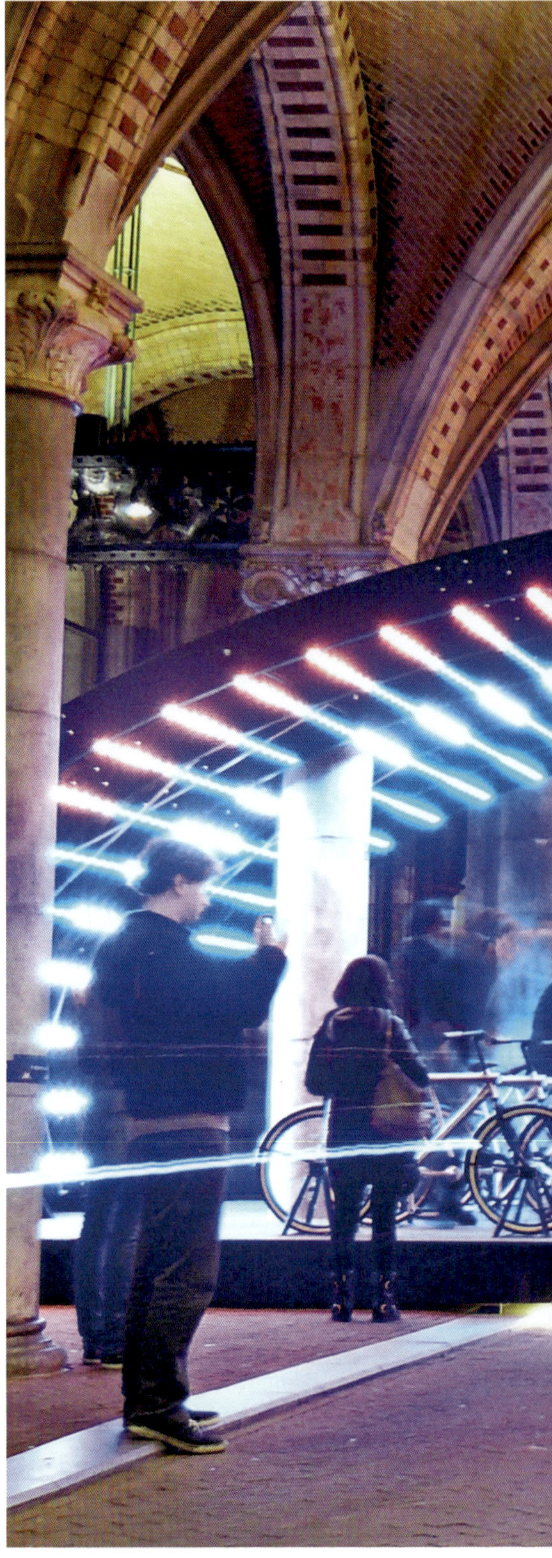

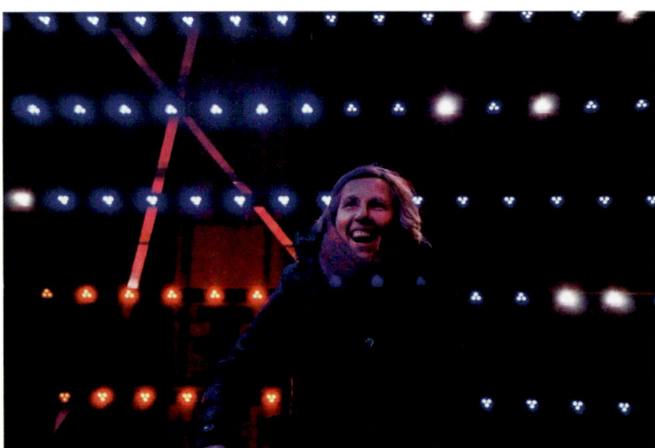

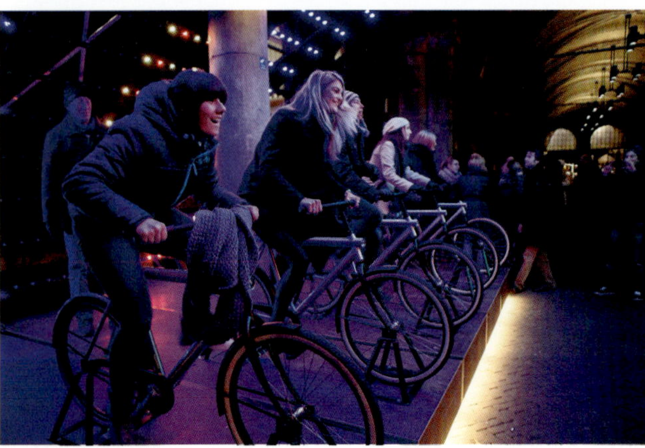

Each participant controlled an arch of light, so that the 'game' could be played individually or as a duel, but the lighting effects became more epic as more people joined in.

Designer
Venividimultiplex
Location
Amsterdam, the Netherlands
Client
Rijksmuseum and Amsterdam Light Festival
Collaborator(s)/consultant
InventDesign, DHL Express, Van Moof
Manufacturer
Van der Lee Interieurbouw
Date
December 2013

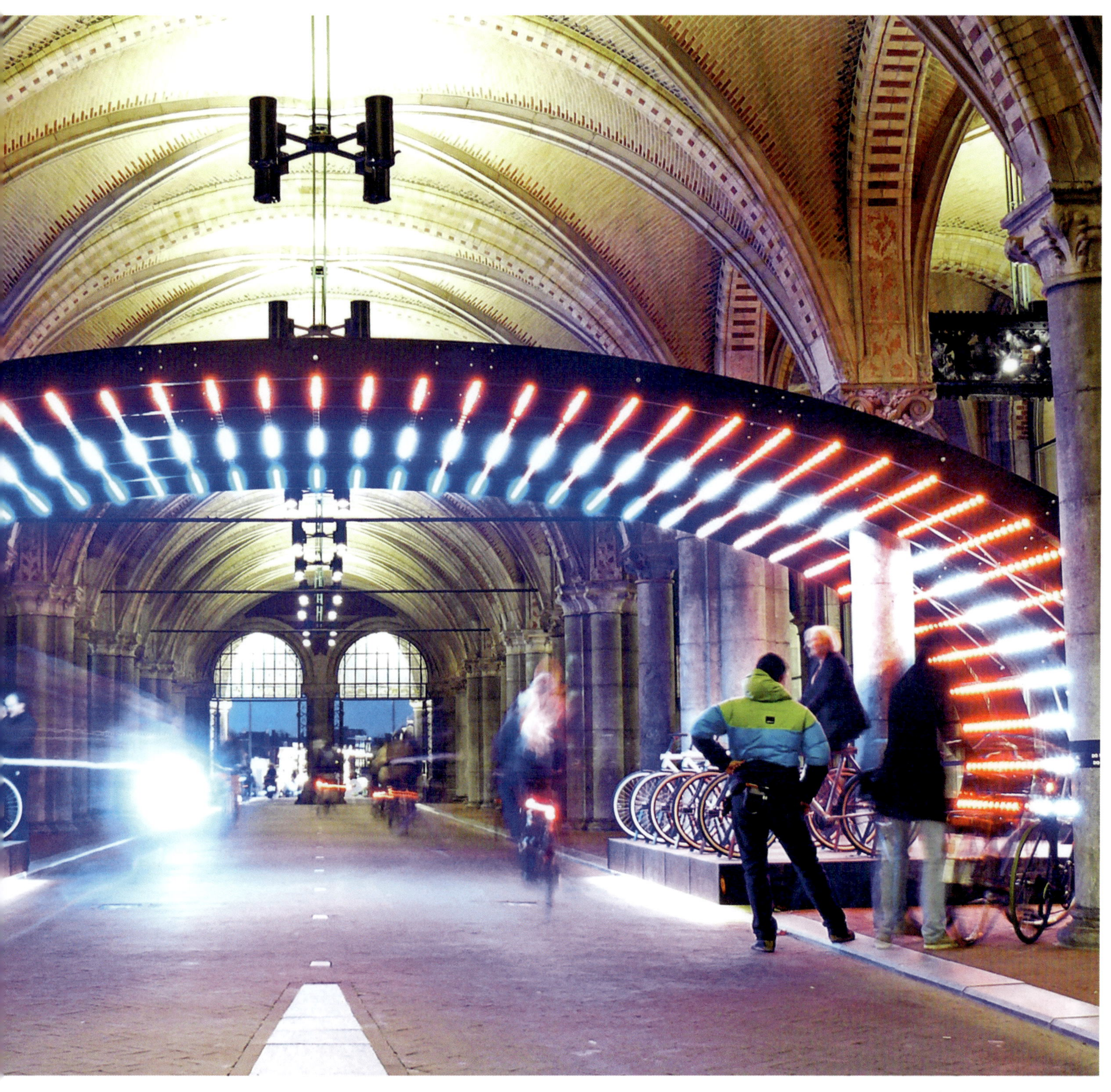

Attached to the arch were 5000 LED lights, individually triggered by state-of-the-art computer systems.

THE INSTALLATION'S DAZZLING LIGHT DISPLAY HAS A HIGH-TECH FEEL

PLAN

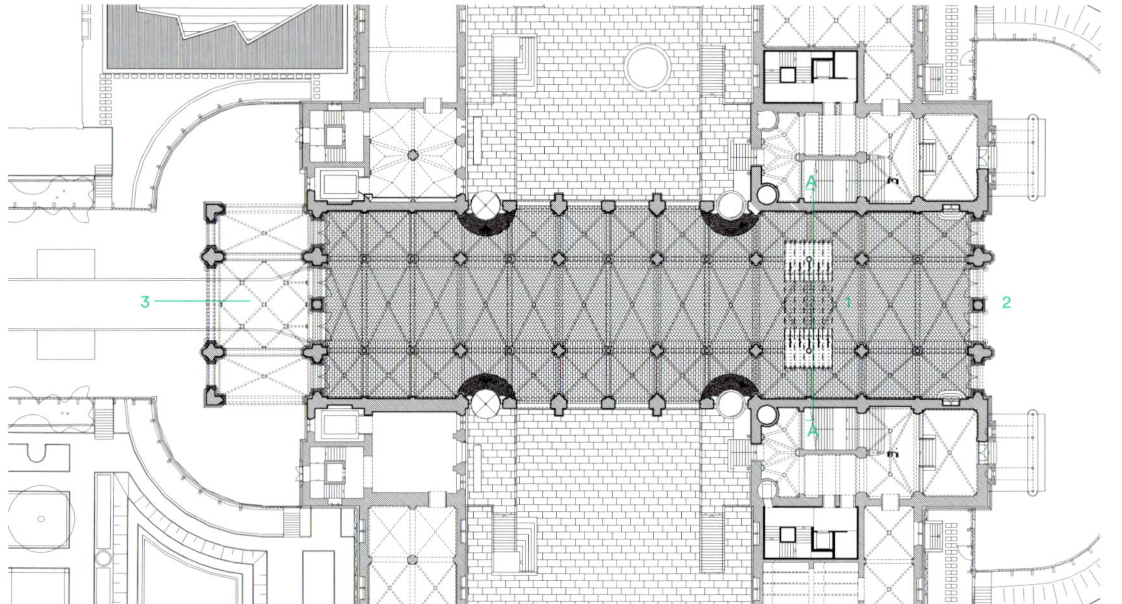

1 Installation
2 East entrance of passageway
3 West entrance of passageway

SECTION A

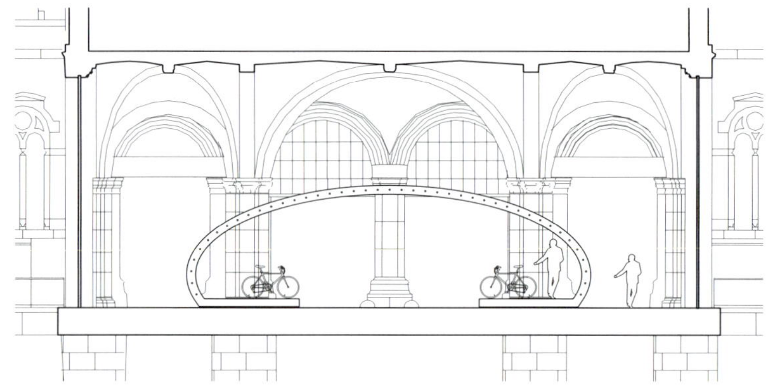

DRAWINGS OF INSTALLATION

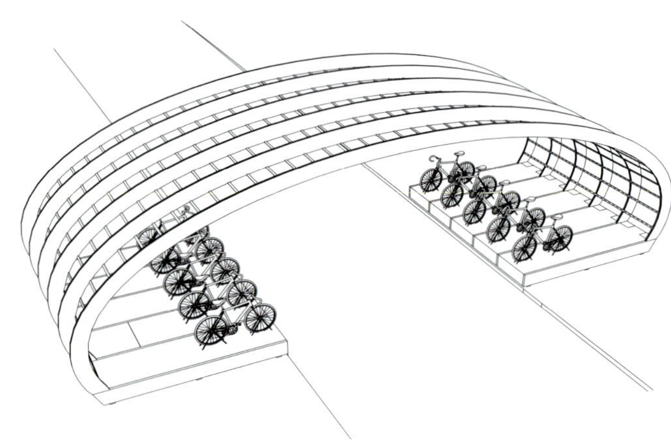

Set-up

Dimensions of the arch-like construction reached a height of 4.5 m, with a width of 6 m and a length of 13.5 m.
Prefabricated wooden segments formed the arch: six wood layered arches (made up of seven 3-m segments), three 6-m cross bars (built in six segments) and two 17.5 m² platforms.
Bicycles were supplied by Van Moof and secured to the platform via custom-made brackets, rolling friction components and speed and distance sensors.
Illumination consisted of 5000 ultrabright LED pinlights in custom-made Digidot B3 modules mounted on aluminium strips (total length 225 m).
Lighting control was achieved through six Matrix Plexus DMX computers, six Audrino controllers and six power backup and supply packs.
Cabling measured in total 1000 m (600 m for DMX and 400 m for power).

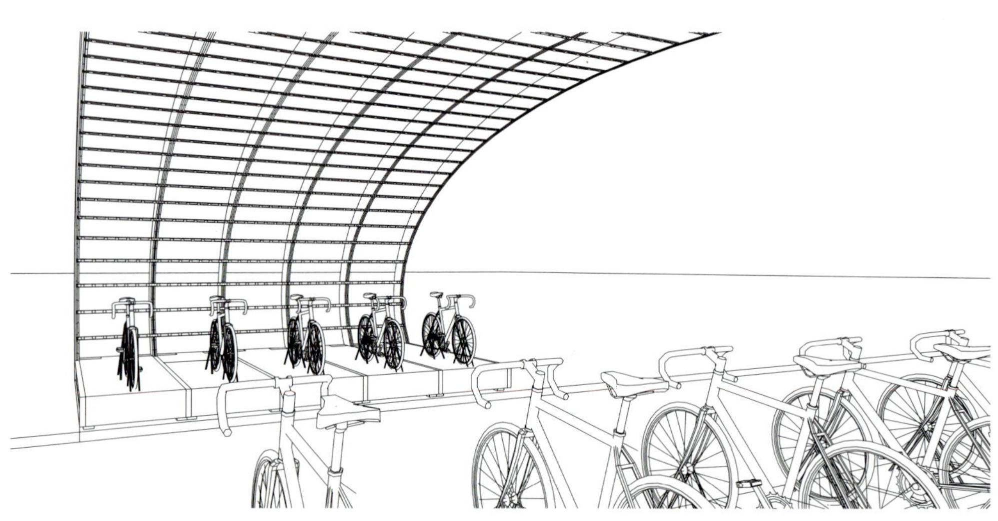

EXPLODED VIEW

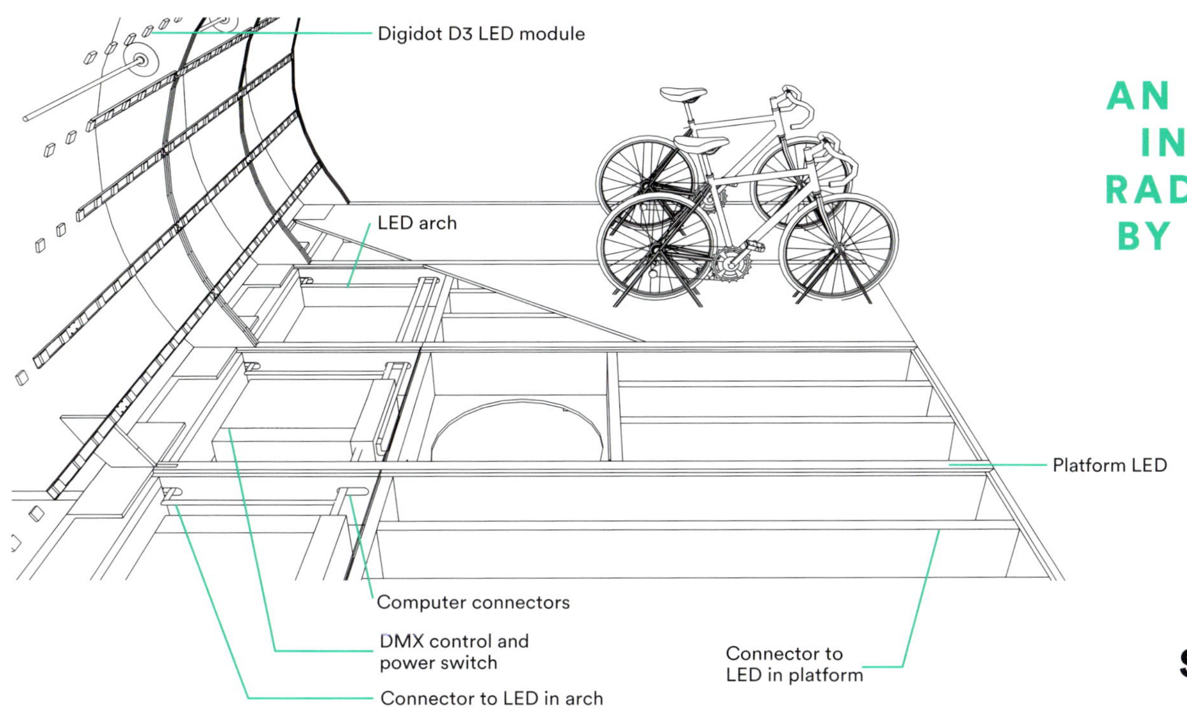

AN ILLUMINATING INSTALLATION, RADIANTLY LIT-UP BY PEDAL-POWER

BIKE FIXTURE SECTION

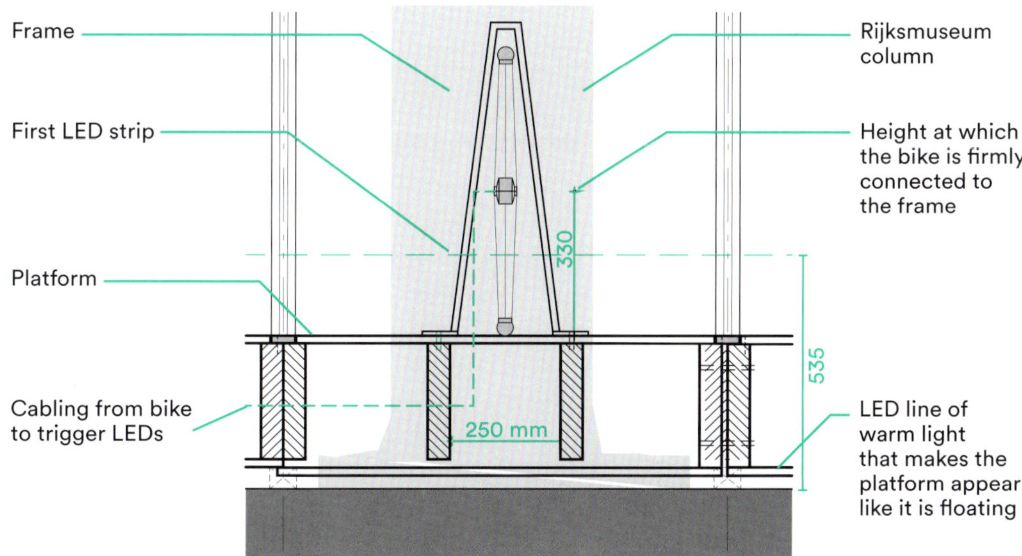

INSTALLATION PROTOTYPE

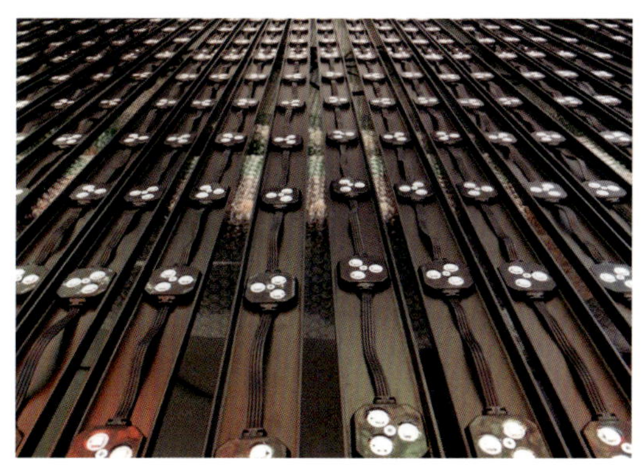

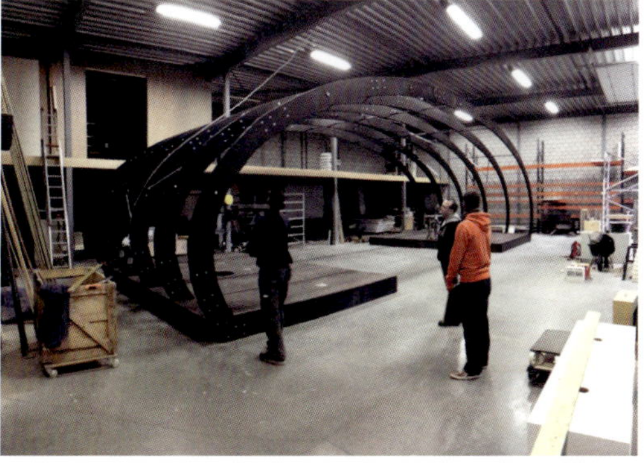

Specifics

After initial funding fell through, this installation only saw the light of day thanks to the intervention of Wim Pijbes, the director of the Rijksmuseum. Originally designed to be positioned over a canal, Lightbattle had to be rethought for the museum's majestic archway passage.

The new vision for the artwork needed to still entail sustainable and cost-effective methods. Here, the architects collaborated on the construction with Van der Lee Interieurbouw to create the stabilised, multilayered, computer-milled wooden beamed structure.

Due to the historical location, there were restrictions relating to attaching fixtures to the Rijksmuseum infrastructure. The location of the passageway columns therefore needed to be taken into account in the development phase of the structure's construction to enable implementation on-site, but be flexible in the future.

The six wooden curved beams needed to be built-up in a pre-fabricated process, in workable segments. Following investigations, three cross bars were fixed to the lightweight structure in order for it to maintain its broken-ellipse shape.

Achieving the intended lighting effect in a cost-effective manner was another challenge. The architects initially started with full-colour DMX-controlled waterproof LED strips but were forced to change tactics in line with the budget. Following in-depth research on lighting techniques, a collaboration with lighting company InventDesign resulted in a custom-made system using DigiDot B3 ultra-bright LED modules mounted in coated aluminum strips.

The original driving mechanism of the bicycles were adjusted to control the lights. The cycling mechanism is as simple as it is clever, with the added advantage of being low in maintenance.

All electrical connections and control devices were weather resistant (or encased in weather resistant containers) and designed to be out-of-sight to maximise the impact of the light effects.

Its modular construction system means the installation can relocate easily. In October 2014, it travelled to the Berlin Light Festival and the installation is always ready for for the next battle.

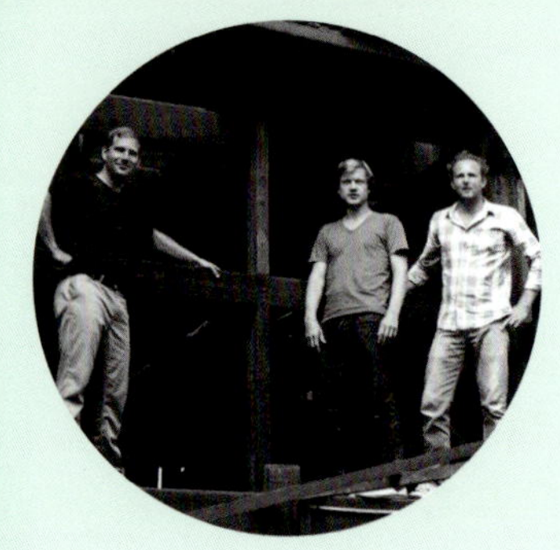

Venividimultiplex

Joost van Bergen, Dirk Schlebusch and Onne Walsmit met at the Academy of Architecture in Amsterdam and have collaborated on a number of projects ever since. Three years after graduating, they founded Venividimultiplex in 2012 to embark on a number of architectural, interior and conceptual design projects. Using their combined experience, the three architects create interactive interdisciplinary installations, always with the aspiration of producing 'positive messages in a fun and interactive way'. The team has worked with clients worldwide, including the Amsterdam, Berlin and New York light festivals.

VENIVIDIMULTIPLEX.COM

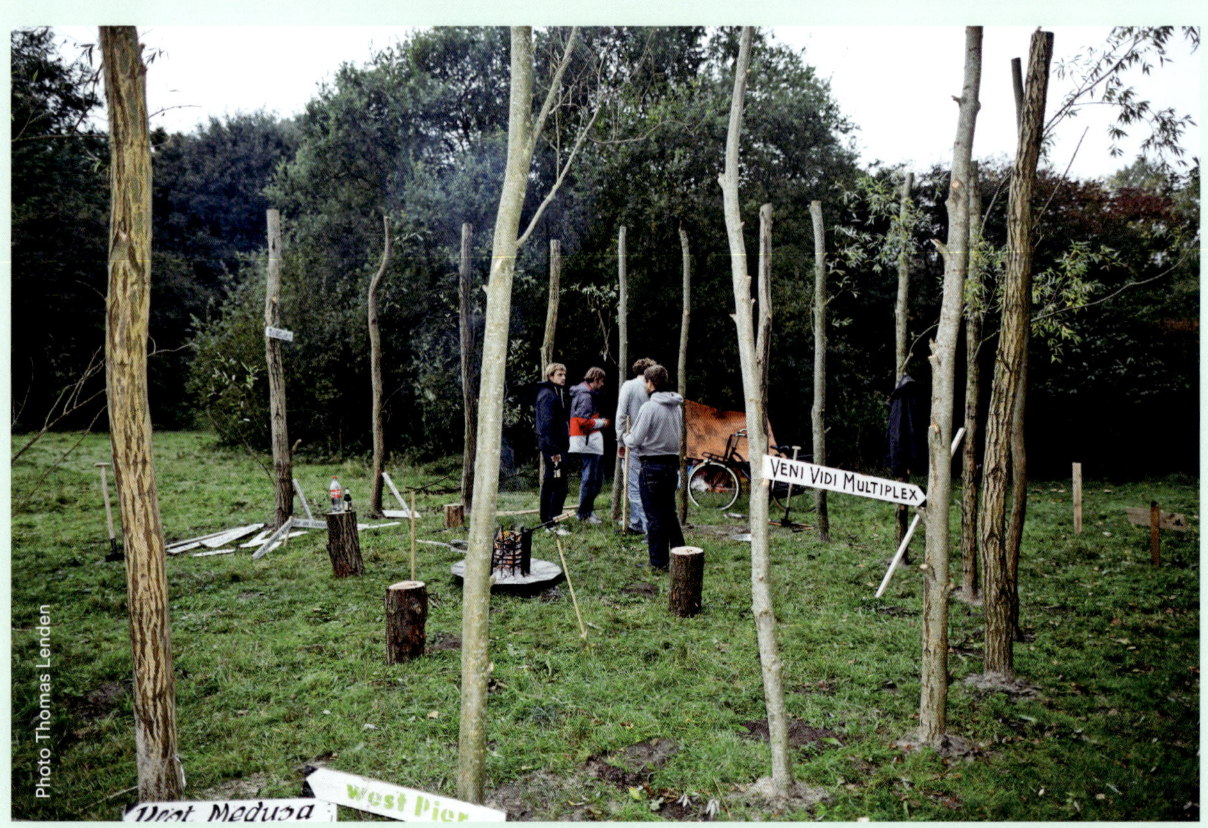

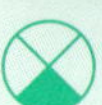

Treevillion

When **2013**
Where **Amsterdam, the Netherlands**
Client **Ecofestival Landen**

In this project, culture literally grew out of nature. A series of fast-growing willow trees were planted to define a circular perimeter in Oeverlanden Park – the central location of Ecofestival 2013. The trees were ceremoniously trimmed, in an act that supported the natural development of the park. Over time, the trees' branches will grow back again, and the cultural object will be as natural as it used to be.

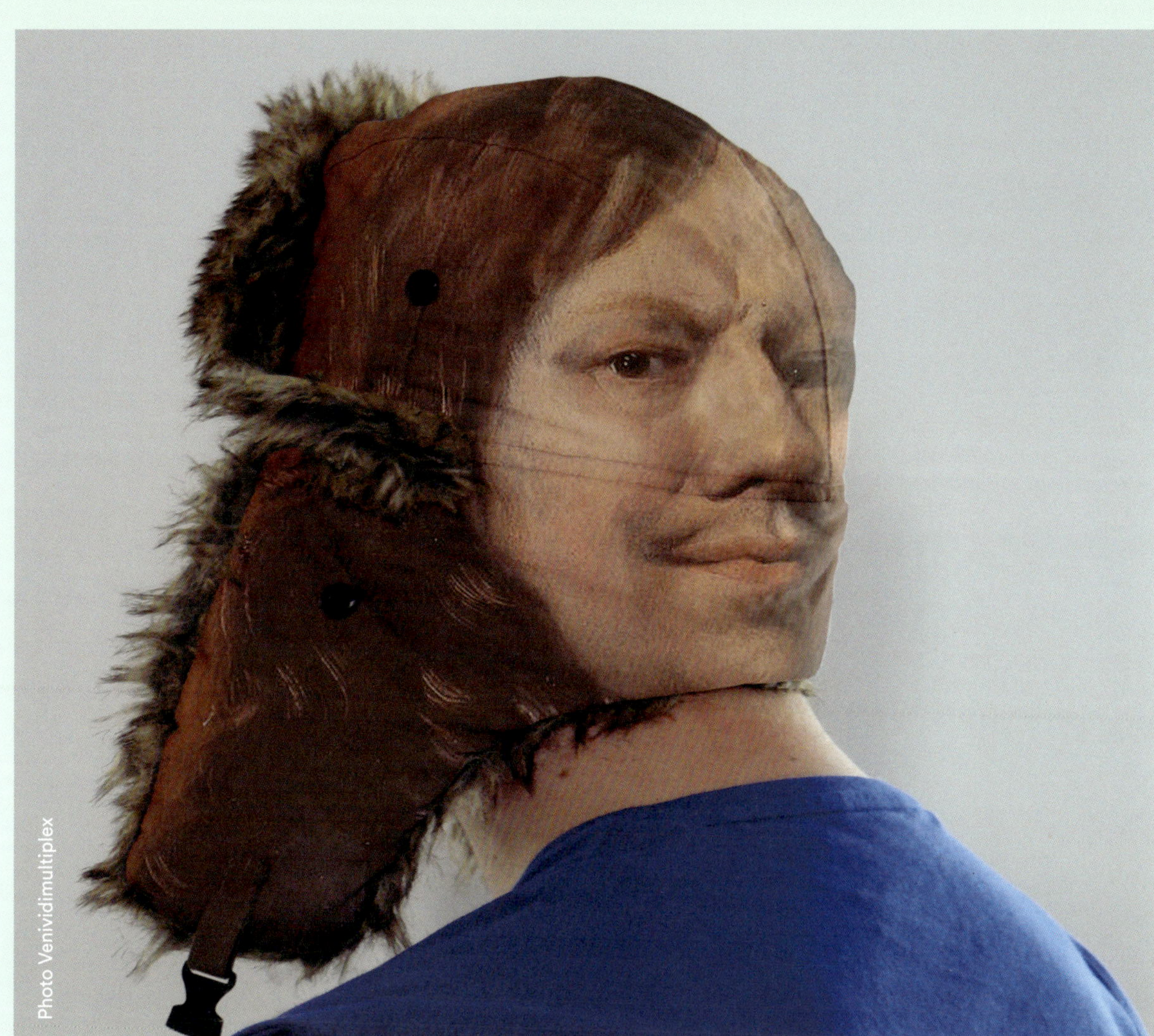

Masterhat

When **2014**
Where **Amsterdam, the Netherlands**
Client **Rijksmuseum Amsterdam**

Rijksmuseum's Masterhat is a series of products that include fun keepsakes for tourists to take away as reminders of their trip to Amsterdam. The designers wanted to create something that stepped outside the standardised styles that people might expect to find in souvenir shops. Masterhat features depictions of Dutch Masters' portraits on the back, so that when seen from behind it looks like someone from the Golden Age is walking around the modern day city.

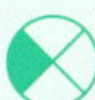

Superfire

When **2013**
Where **Amsterdam, the Netherlands**
Client **Urban Campsite**

A versatile power source, Superfire is a first-class fireplace that goes beyond one's expectations. As the fire burns smoothly, functioning as a normal campfire or a grill, a power supply can feed its energy into a number of 'add-ons' to heat up pillows, beds and even charge mobile phones.

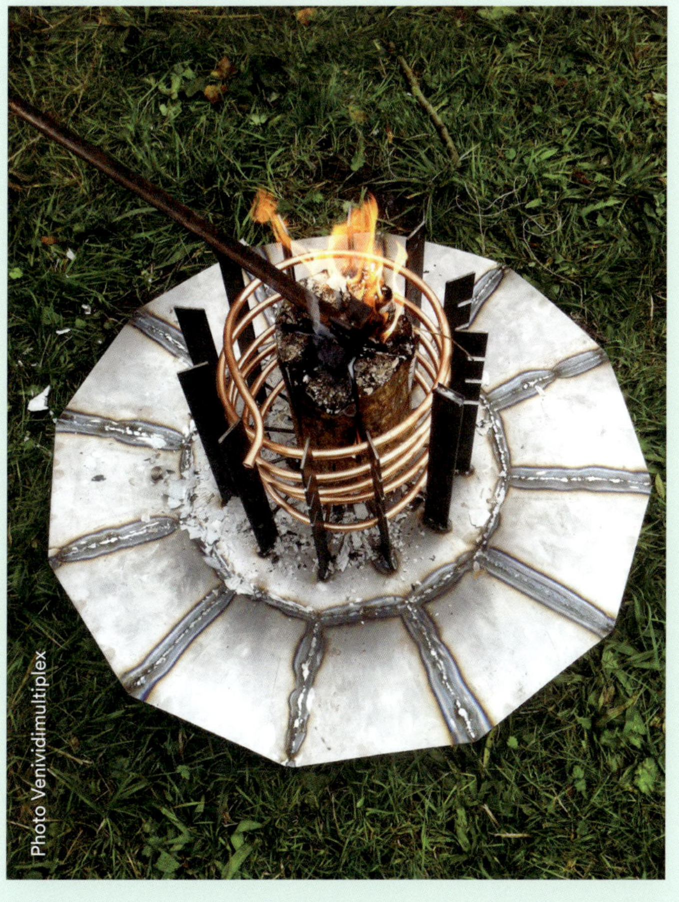

Interactivity shines along the Amsterdam waterways thanks to the Light Bridge installation from studio Tjep.

Light Bridge

As a celebration of illumination that defines urban activity, the Light Bridge installation by studio **Tjep.** aimed to dazzle passers-by with its contemporary contours outlining architectural aspects that are the heritage of the Dutch capital.

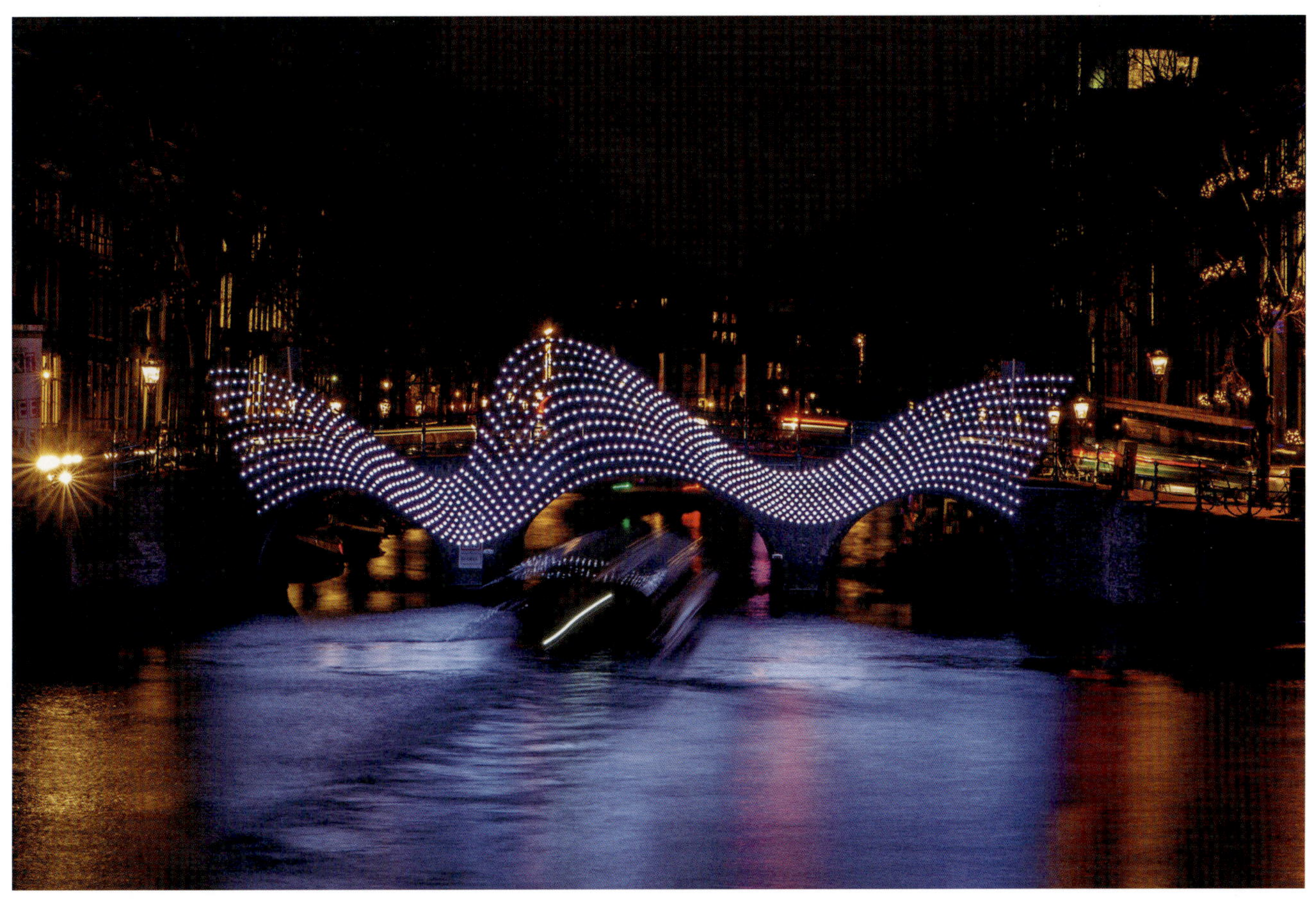

Photos Tjep.

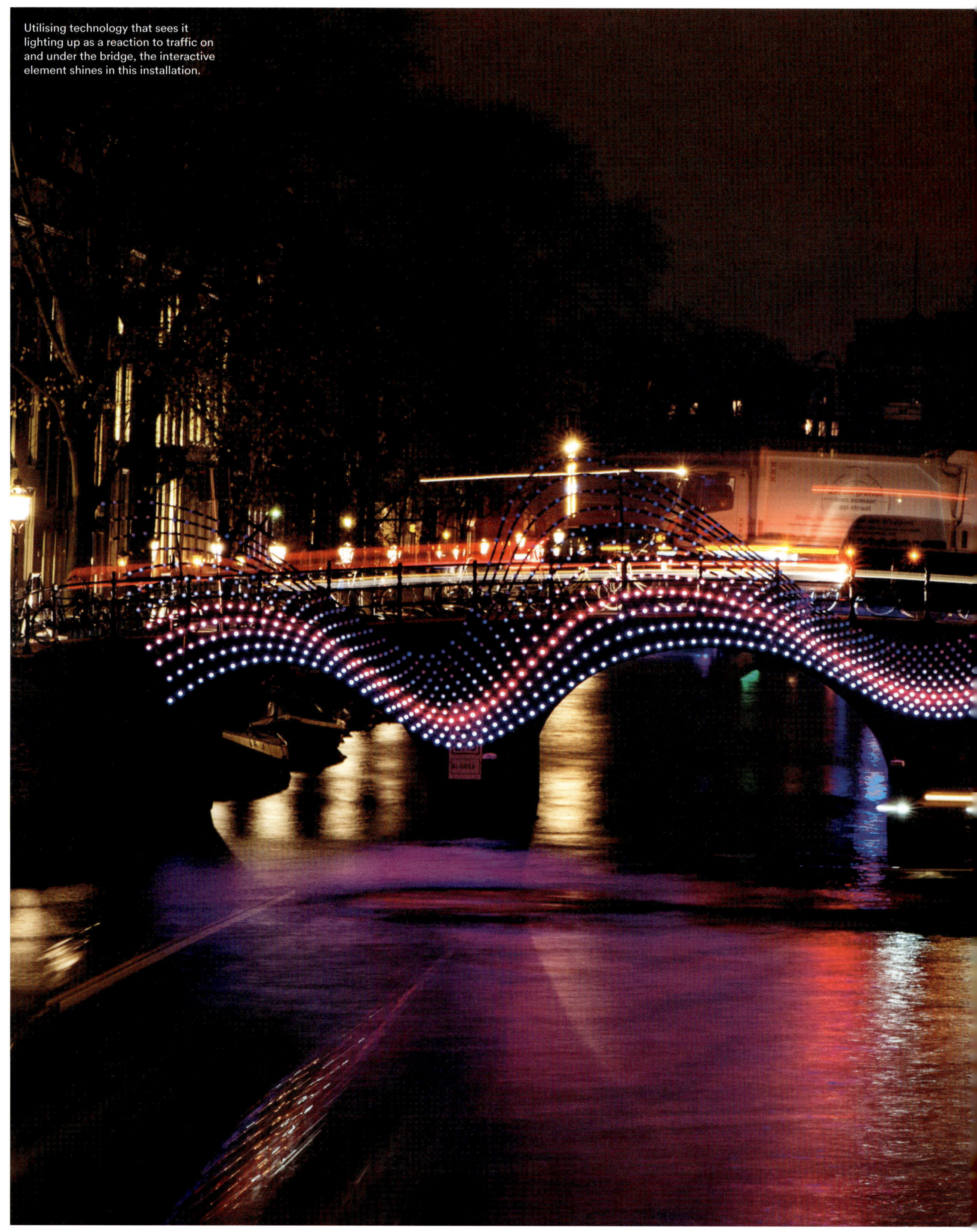

Utilising technology that sees it lighting up as a reaction to traffic on and under the bridge, the interactive element shines in this installation.

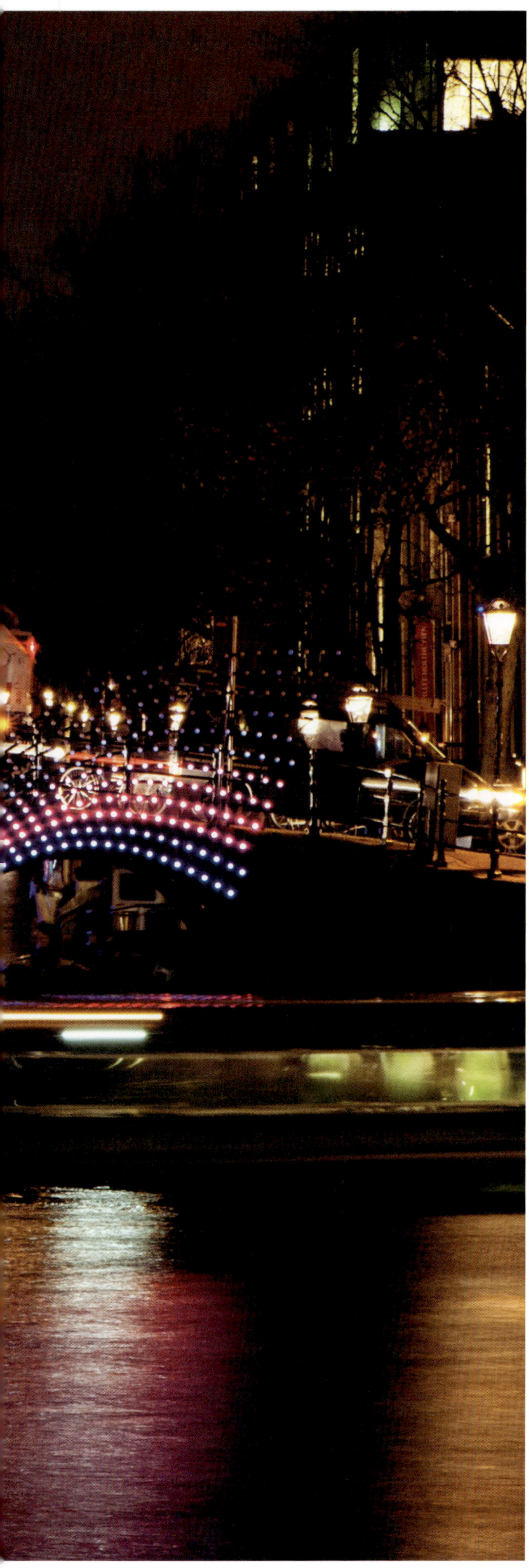

Designer
Tjep.
Location
Amsterdam, the Netherlands
Client
Amsterdam Light Festival
Collaborator(s)/consultant
Bruns, Ata Tech
Manufacturer
Bruns
Date
November 2014

THE ULTIMATE DYNAMIC ARCHITECTURAL ELEMENT SYMBOLISING CONNECTION IN THE MUNICIPALITY

The annual festival of light transforms Amsterdam into a glittering illuminated showcase over the course of the winter months. The city is draped in swathes of light and movement for more than 50 days – and nights – which is when the attractions begin to emanate light and come into their own. Brightening the night skies was one of the many eye-catching stops on the festival's Illuminade (illuminated promenade through the city): the prominent light installation by the Amstel designed by studio Tjep. The sculpture, Light Bridge, gained a position on the last bridge of the Herengracht as it meets the Amstel, opposite the Hermitage Museum.

Amsterdam has more than 1200 bridges crossing the city's hundreds of canals and one can consider these structures to be the ultimate dynamic architectural element symbolising connection in the municipality. Bridges reflect the historical growth of Amsterdam and are often the most visually impressive points on every visitor's tour. Inspired by the architectural heritage of the Dutch capital, Frank Tjepkema comments on his team's vision for the project, 'We imagined this art installation as the celebration of two essential, yet separated, characteristics that define urban activity in Amsterdam – life on the streets and life on the waterways. Light Bridge is an ode to the special dynamics that take place here and also forms a celebration of the transformation from classic architecture (the old city centre) to modern architecture (like the Eye and the Stedelijk Museum).'

The starting point for the design was the lighting that has been used over the centuries on many of the old Amsterdam bridges: a series of simple bulbs that outline the contours of the arches. Using almost as many lightbulbs as the number of bridges in the city, the work adds a contemporary colour scheme to transform this simple element into a monumental, dynamic and interactive intervention that both reacts to movement on the street, as well as the motion of boats on the water underneath. When there is no traffic, the installation is animated with the reflections of water ripples to create a subtle mirage of reflections echoing between the bridge and the water surface below. —

RENDERINGS

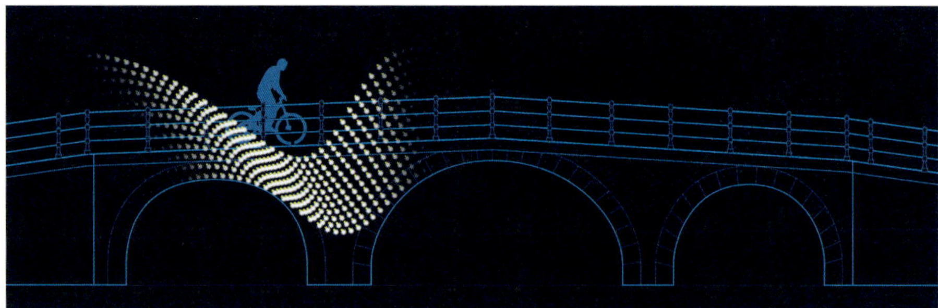

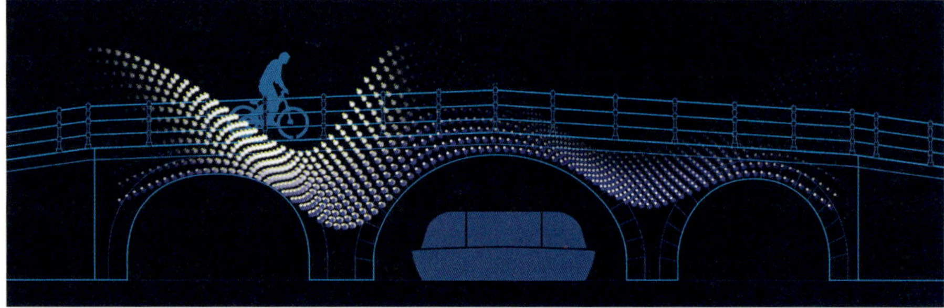

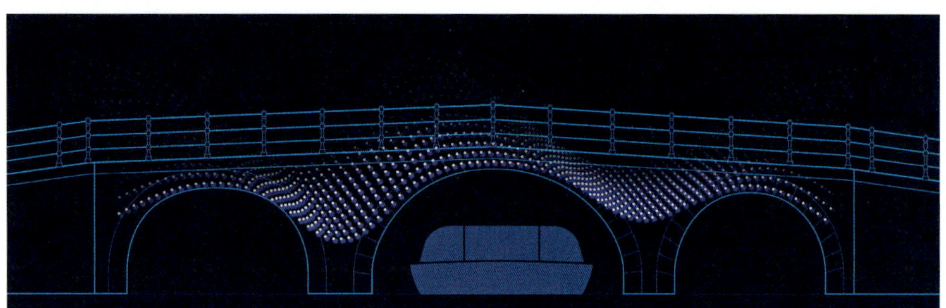

Set-up

Total volume of the installation for illumination was 315 m³ (21 x 5 x 3 m).
Component parts included mixed media, LEDs, diffusive reflective material, custom lighting sequence, electrical hardware and metal framework.
Used light included 25-m long, 24 V LED (85 W).
Lighting control console incorporated a DMX signal device to send data to the DMX controller, taking the data and outputting it to the appropriate LED channels and the lighting fixtures, the 24 V power supply for which was routed through the DMX controller.
Metal framework (60 m²) was assembled alongside the light pattern to ensure the secure positioning of the suspended sculpture to the bridge, and is composed of a combination of metal sheeting (3 mm) and metal rods (30 x 30 mm). The whole structure was powder coated to protect from rust.
Construction time of the structure took 6 weeks (manufactured by Bruns). In the first phase, Light Bridge was worked on off-site in order to test and programme the lighting under optimum conditions. The process of mounting it to the bridge on-site was completed in 2 days.

A CELEBRATION TO DEFINE URBAN ACTIVITY WITH A CONTEMPORARY, ILLUMINATED COLOUR SCHEME

MODEL

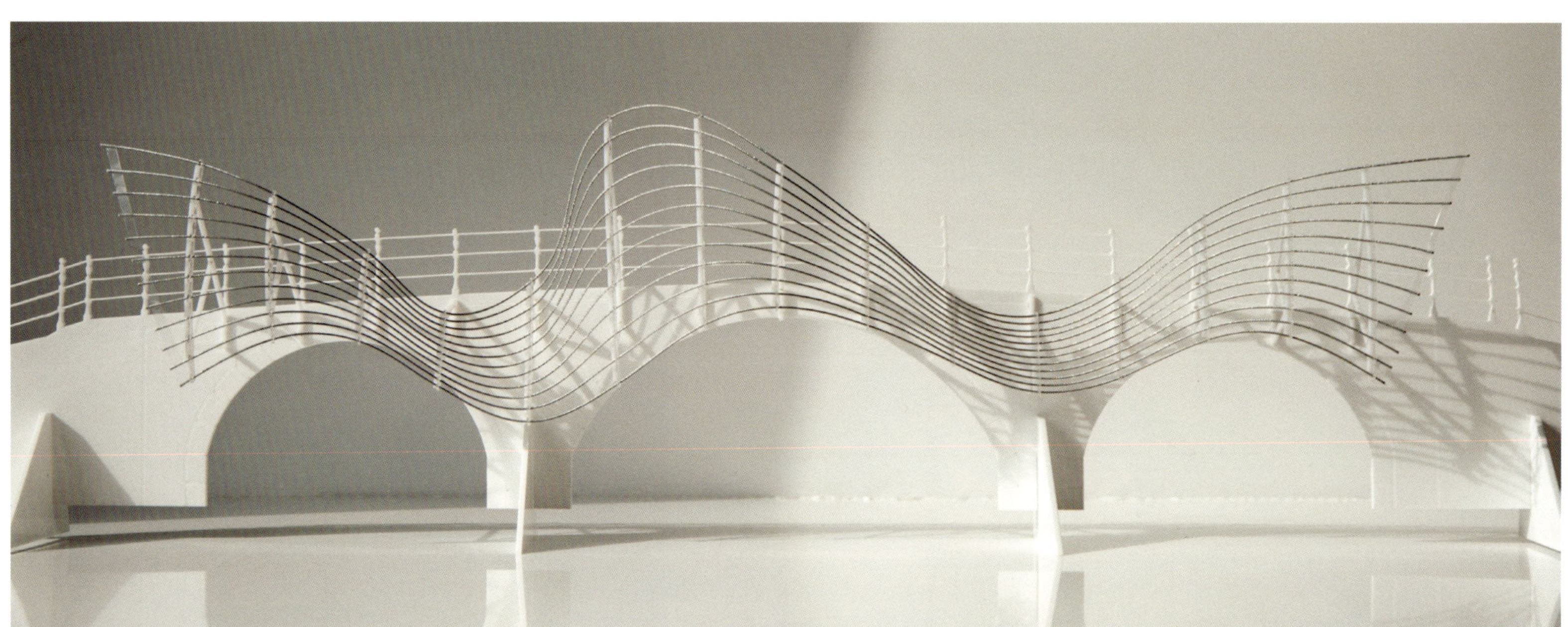

TESTING THE LIGHTING SCENARIOS

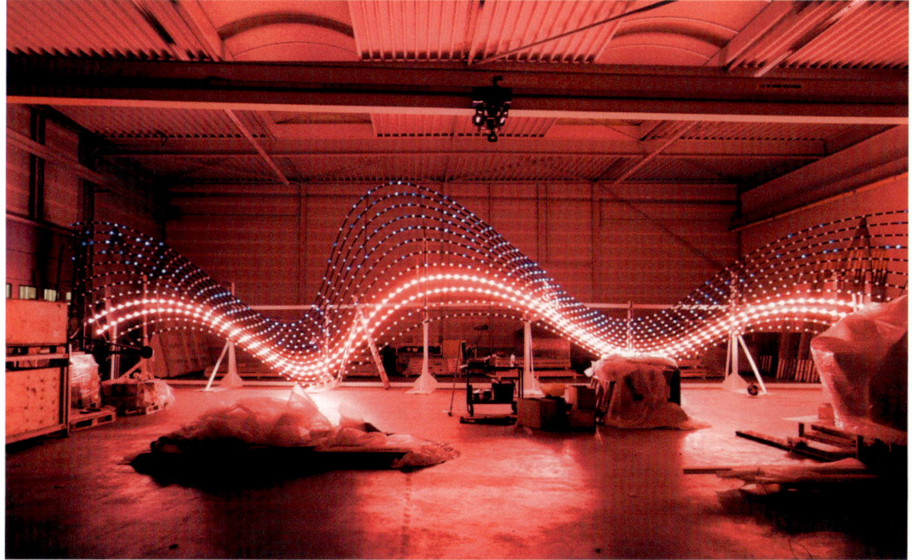

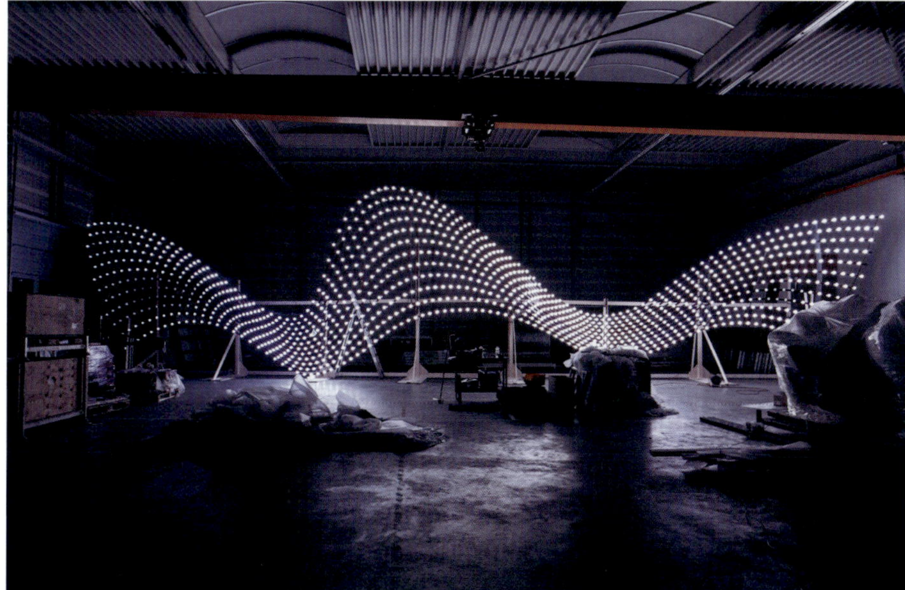

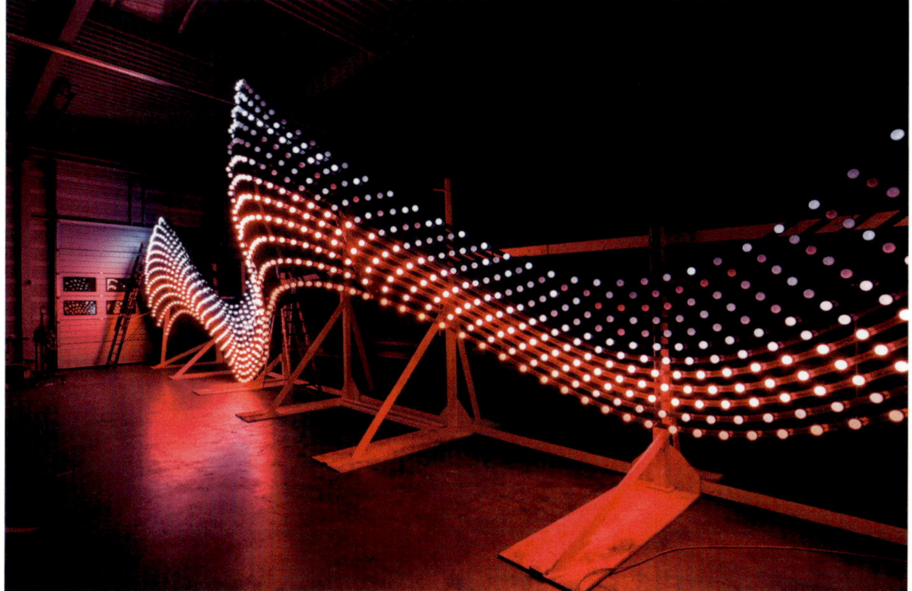

Specifics

/\/\/\/\ The design of the installation was inspired by the current lighting used on many of the old Amsterdam bridges: a simple light bulb. The key aspect of the project was the interactive intervention, meaning the illumination of the bulbs were linked to movement on the bridge itself, and below it on the water. Any movements detected by three motion sensors (two directed towards the road at the extremities and one directed toward the boats moving underneath) would trigger the bulbs. The system incorporated four different animations, either triggered by the data from the sensors or used during 'sleeping modus' when there was no traffic.

/\/\/\/\ In order to define the shape of the structure, the team worked with laser-cut scale models until the 3D organic shape was optimised. The 3D files were then sent to an engineering company who optimised the structure in relation to resistance to exterior forces, such as wind.

/\/\/\/\ Another aspect to consider was the fact that it was of utmost importance not to damage the monumental bridge in any way. All procedures needed to pass regulations of the city council; for example, the structure was effectively anchored under the street surface to the 'non-monumental' concrete re-enforcement core.

/\/\/\/\ During installation, it was very challenging to mount the artwork without disrupting the continuous traffic on the water. Special cranes were used to place the installation on the bridge itself.

CONSTRUCTION VISUAL

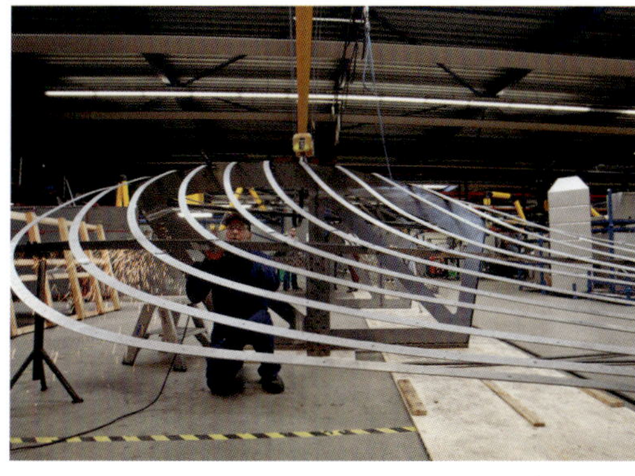

Photos Thomas Fasting

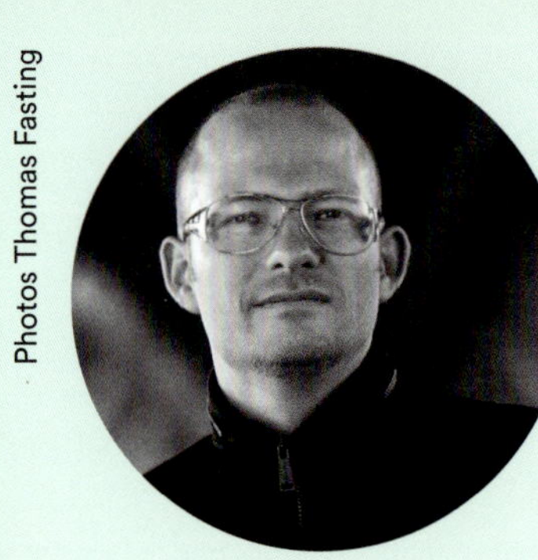

Tjep.

Tjep. is an Amsterdam-based team of misfit-fitting designers and non-designers, challenging themselves to question the world and make unexpected connections. Founded by Frank Tjepkema in 2001, the studio has garnered a reputation for delivering outstanding creative work across a broad field of expertise that includes interior, architectural, product, furniture, visual and jewellery design. Spanning both the commercial and cultural sectors with a diverse portfolio of acclaimed work, Tjep. fuses together peerless design with accomplished expertise in the latest production techniques. The studio's hallmarks are rich designs that make unexpected connections, playing with people's senses of the intuitive and rational.

TJEP.COM

Clockwork Snow

When **2011**
Where **Milan, Italy**
Client **Independent Ideas**

Tjep. was invited amongst six other international artists and designers to create a Christmas window Italy's most famous department store, La Rinascente on Piazza Duomo in Milan. Mechanical gears seem to drop from heaven like icy snowflakes, crystallising into a magical machine frozen in time. The gears represent the rational world, but here they transcend the purely utilitarian to become a romantic, haunting dream.

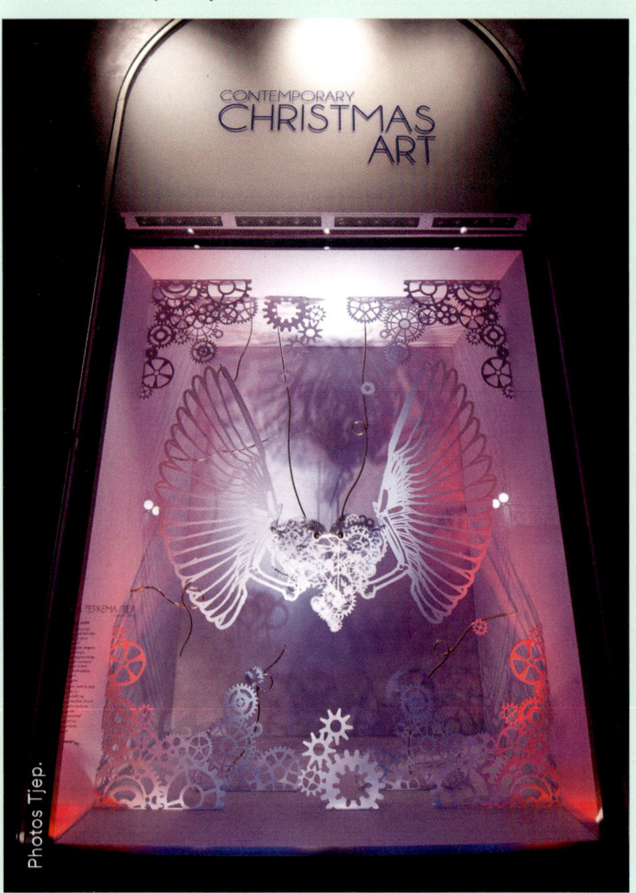

Photos Tjep.

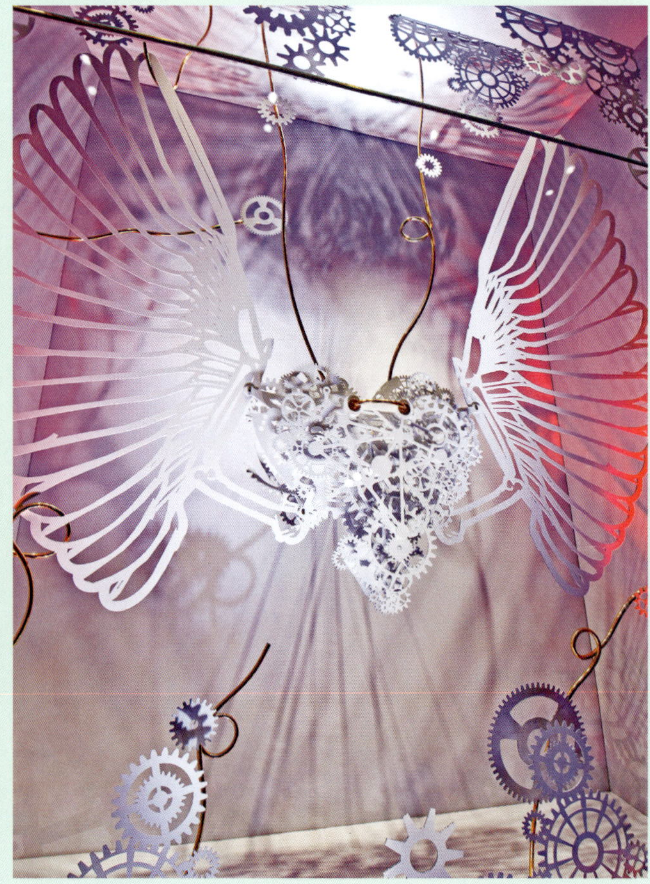

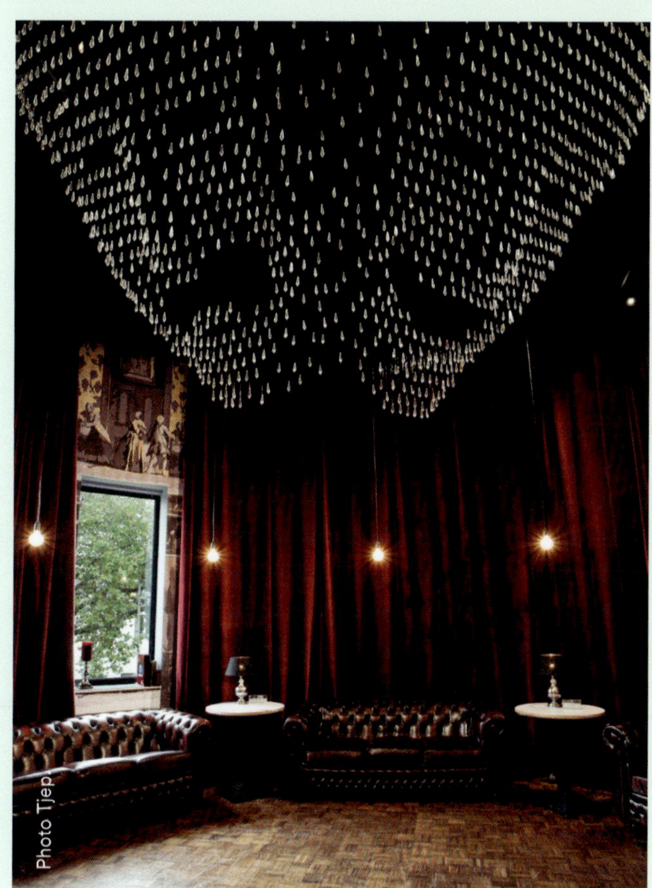

Photo Tjep.

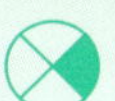 Zaantheater Foyer

When **2014**
Where **Zaandam, the Netherlands**
Client **Zaantheater**

Tjep. redesigned the actor's foyer in the Zaantheater. The concept sought to give the modern building stronger connections to the grand heritage of theatre. To this end, the team developed an approach that presents both traditional and modern features. A key aspect is the impressive chandelier that forms a grand visage from the great masks of 'tragedy' and 'comedy'. The chandelier uses 1200 Swarovski crystals to create a truly modern homage to the tradition of stage and screen. Along the walls, custom-made wallpaper harks back to the golden age of theatre with images from 17th century, whilst luxurious leather Chesterfields flank the diameter of the room, with deep red 'stage' curtains.

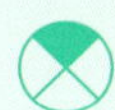 Dutch Tax Museum

When **2012**
Where **Rotterdam, the Netherlands**
Client **Belastingdienst**

Tjep. was responsible for creating a permanent exhibition for the Dutch Tax Museum in Rotterdam. Charged with redeveloping the interior design, exhibition and visitor experience, the team built a rich, interactive journey on the theme of trust and fairness, taking visitors on a historical tour of the Dutch tax system. The novel visitor-centric exhibition includes several unique Tjep. installation, such as the Mechanical Tax Heart, Magic Ceiling and Tenth Penny Theatre.

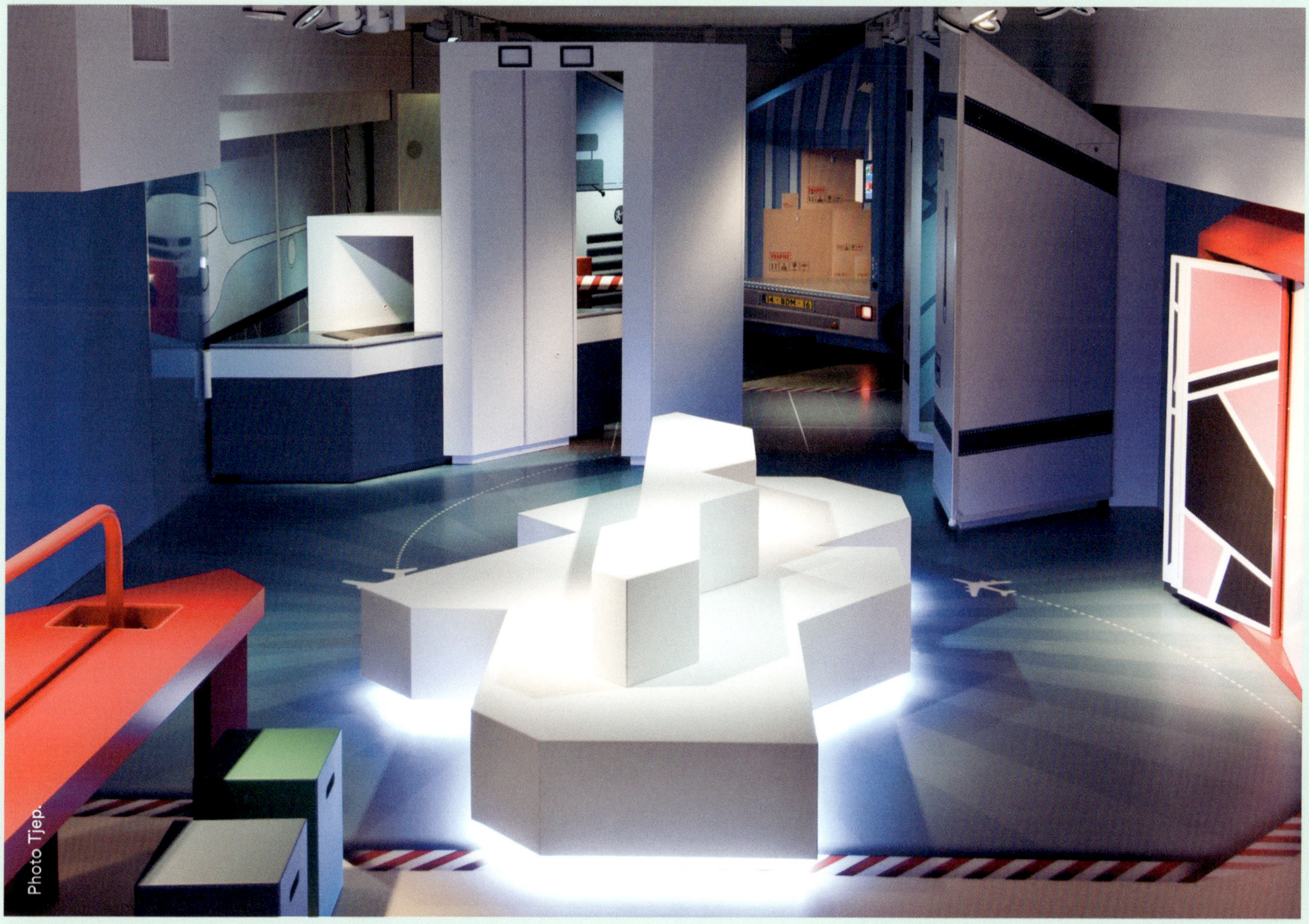

Photo Tjep.

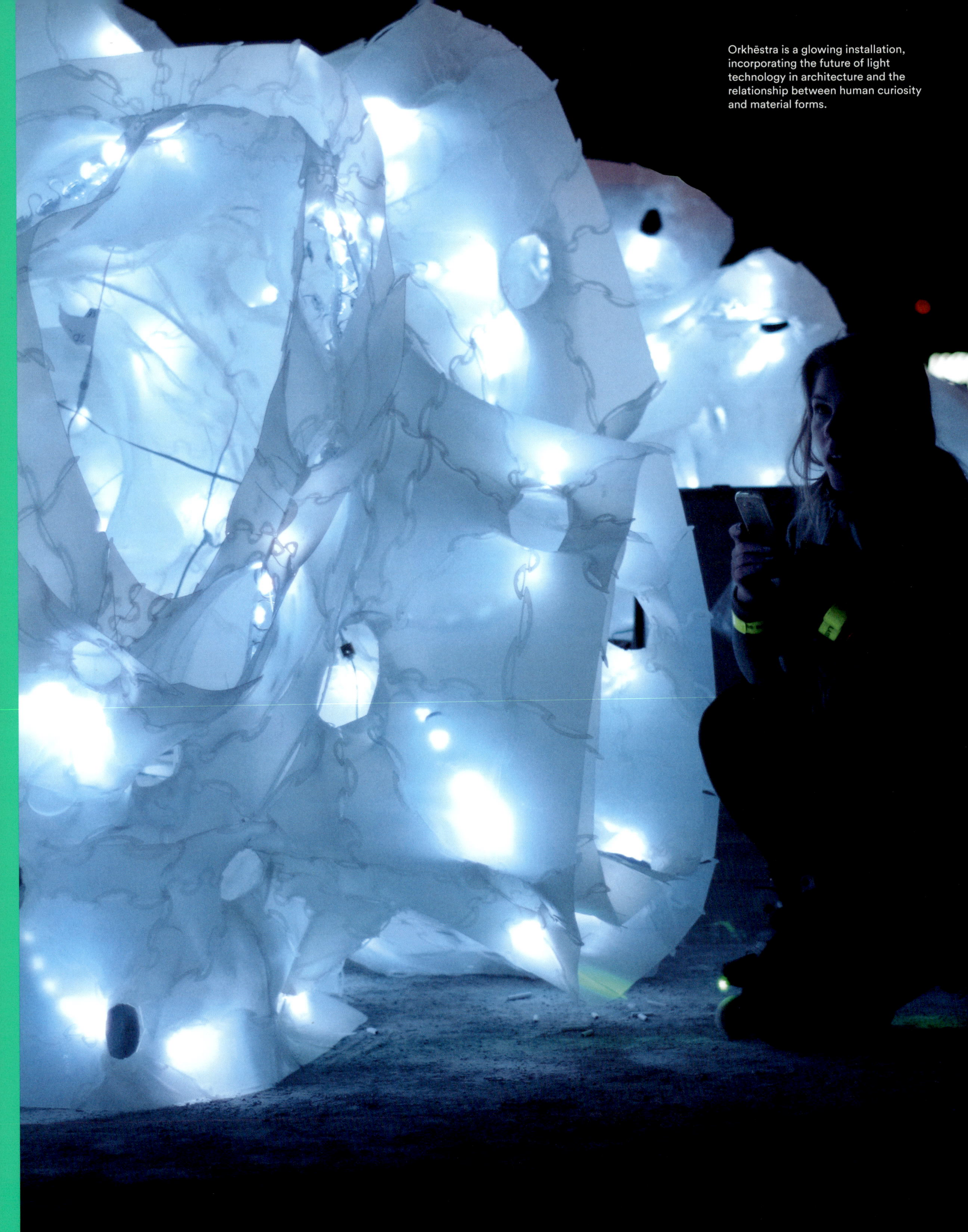

Orkhēstra is a glowing installation, incorporating the future of light technology in architecture and the relationship between human curiosity and material forms.

Orkhēstra

Looking to push the boundaries of architecture and light design in order to create a complex responsive environment, **Media Architecture Institute** joined Städelschule Architecture Class in 'orchestrating the depth of light' for Luminale 2014 in Frankfurt. The interactive installation – Orkhēstra – used media technology to animate architectural surfaces with light, triggered by visitors' camera flashes.

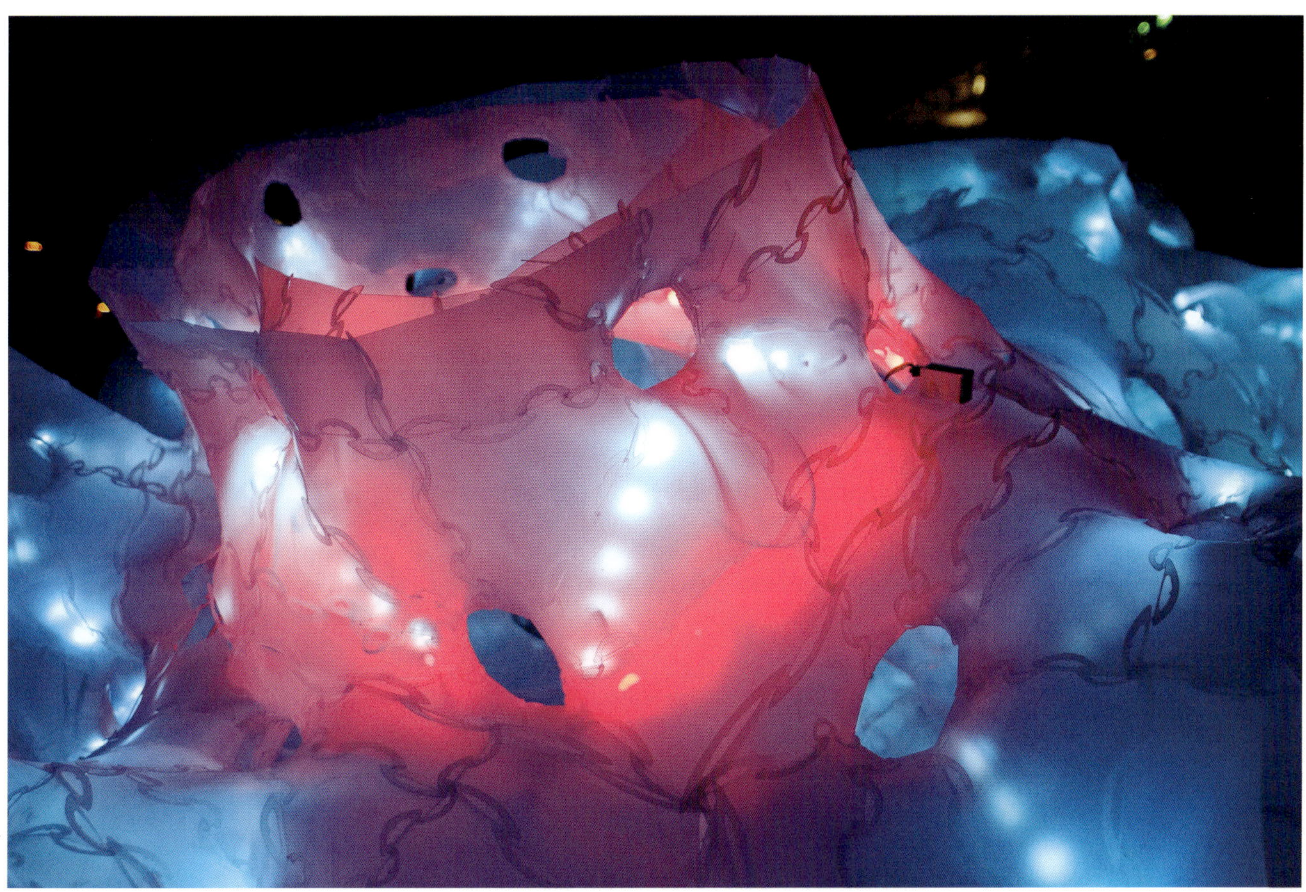

Photos Wolfgang Leeb

DELIVERING AN ORGANIC FORM THAT CHALLENGES THE USE OF ORTHOGONAL SURFACES FOR BUILDING COMPONENTS AND LED PLACEMENT

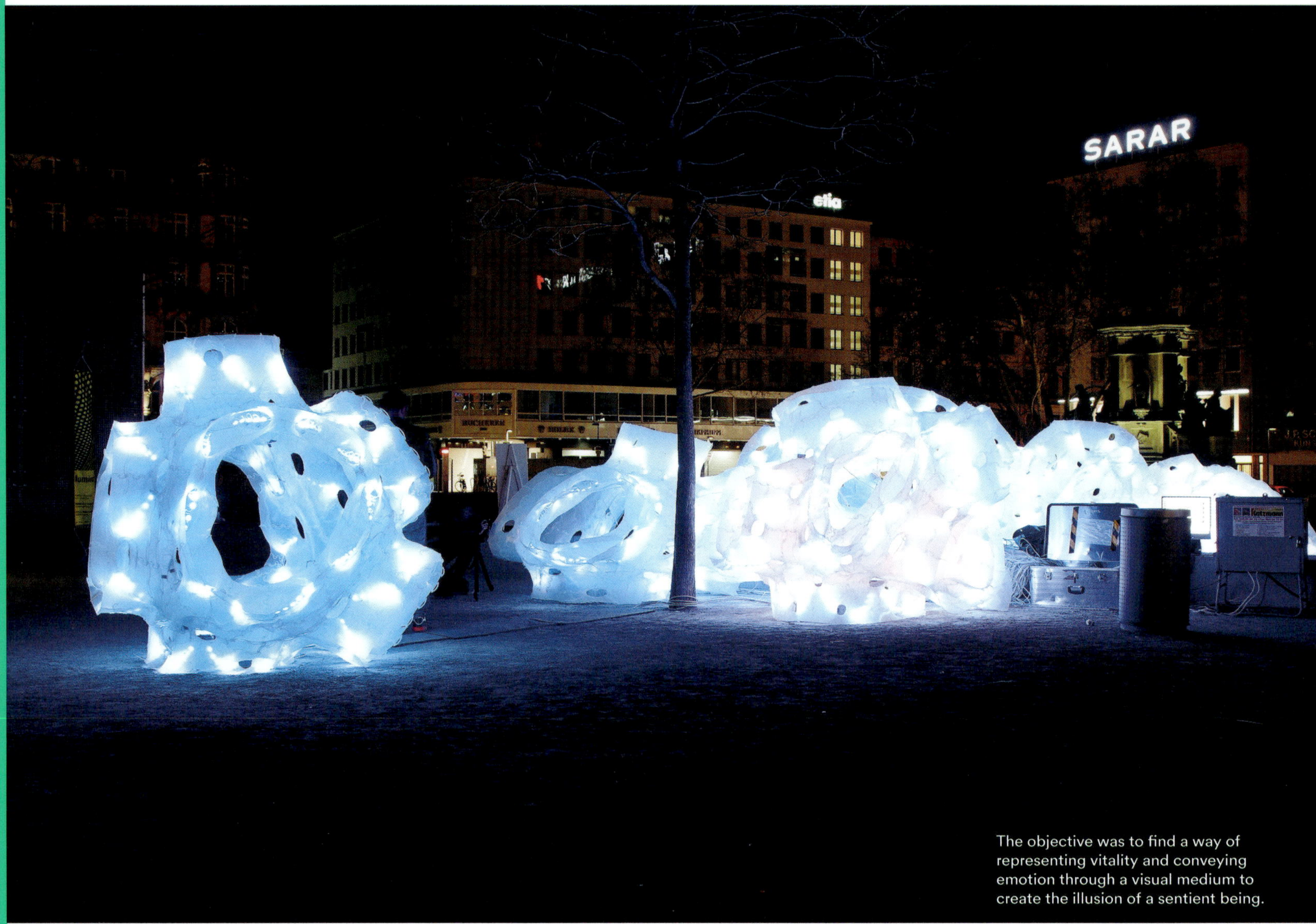

The objective was to find a way of representing vitality and conveying emotion through a visual medium to create the illusion of a sentient being.

A luminous, sculptural object popped-up during Luminale 2014 to bring together a playful interaction of state-of-the-art design expertise and media and interaction technology. The Orkhēstra installation was created by the Media Architecture Institute (MAI) in answer to the question: how can the current developments in computational architecture, light technology and interaction design be pushed to redefine the boundaries of what is possible in designing responsive environments?

In collaboration with Städelschule Architecture Class (SAC) in Frankfurt, the University of New South Wales (UNSW) in Sydney, Ludwig Maximillian University (LMU) in Munich and MF LED manufacturers in Shenzhen, the team created an interactive installation that brings together leading design expertise with media and interaction technology at an international light festival.

Orkhēstra extends the perception of what a screen can be, by using media technology to animate complex architectural surfaces. Boasting an organic shape, the structure is made up of polypropylene sheets, laser-cut into unique pieces to enfold and unfold in diverse perforations to form a complex, curved surface. Within its double layer, the team could create LEDs that illuminated the inner and outer skins on two independent media screens.

Different content could be displayed on either side of the media screen, with the installation animated in a controlled manner through a custom-made Java-based software. Triggered by visitors' camera flashes, data was fed directly into the LED controllers, so that the sculpture's responsive behaviour was swift.

The interactive sculpture was installed in a historic, central square in Frankfurt. Its interactive system, triggered by camera flashes, resulted in a wave of blinking lights that started at the point where it met the screen. By celebrating this location and promoting interaction with its visitors, the animated installation bridged the square's past and future. —

Designers
MAI, SAC, UNSW, LMU, MF
Location
Frankfurt, Germany
Client
Luminale 2014
Consultant
Bollinger+Grohmann Ingenieure
Manufacturer
MF LED
Date
March 2014

PLAN

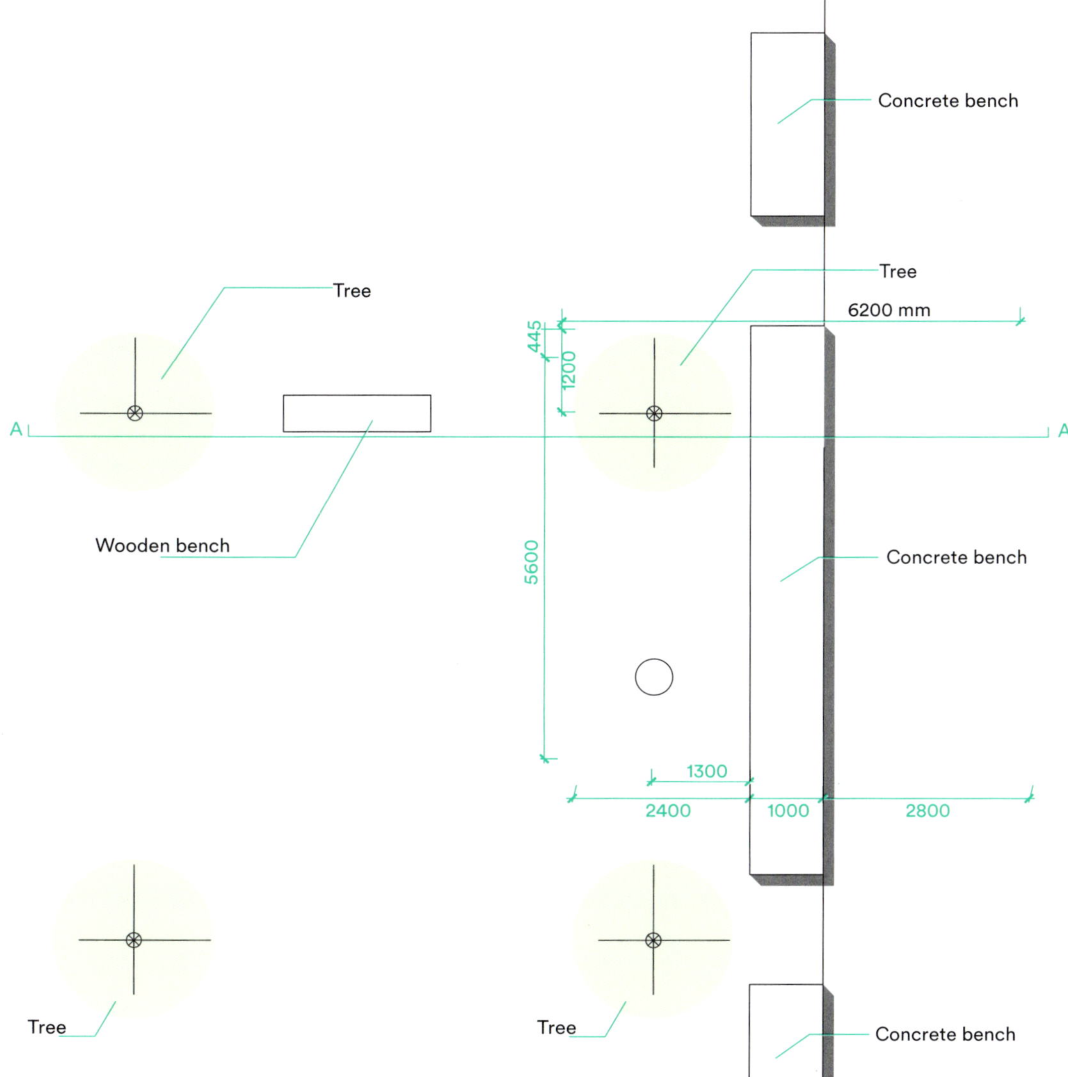

SECTION A

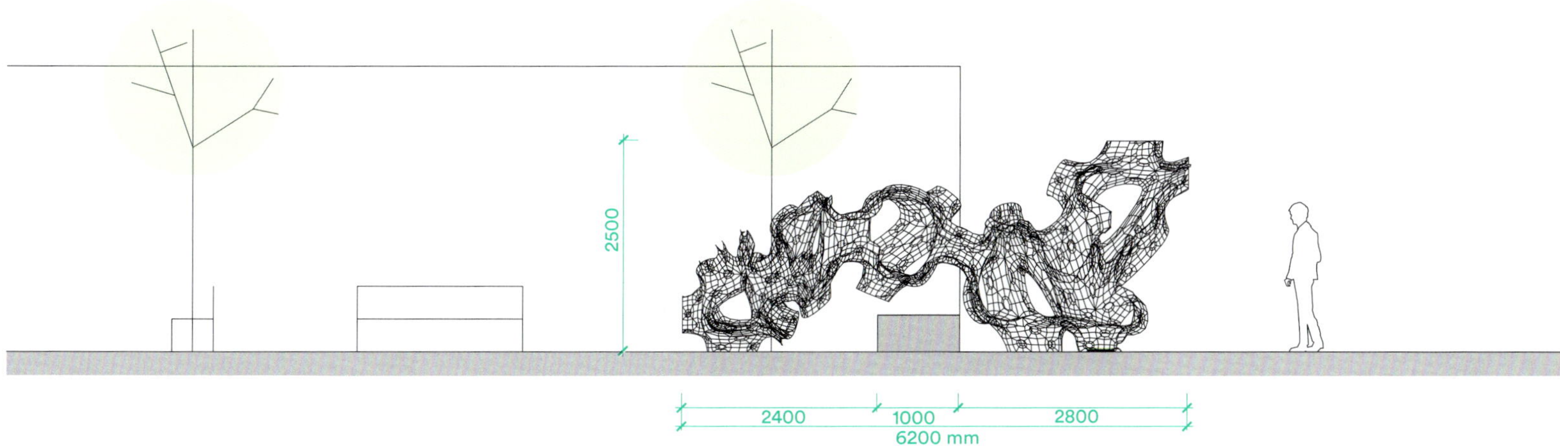

Drawings Gosha Mohammed

Specifics

The installation was developed using state-of-the-art computer modelling technology and delivers an organic form that challenges the conventional use of orthogonal surfaces for building components and LED placement.

The aim of pushing a media facade towards a complex, curved screen generated challenges that stepped beyond the fields of architecture and civil engineering. Cable management, LED positioning to enable a regular pixel pitch, suitable interactive content and ownership of interaction were but a few of the issues that arose with this question.

Different forms of interlocking mechanisms for the 'skin' assembly needed to be investigated, with a finger-joint system being found to be the most flexible in forming the organic structure.

By having a double-layer, the design team was able to hide all cables within the skin and also had the potential to create a double-layered screen, with LEDs illuminating the outer as well as the inner skin at the same time, yet independently.

Displaying content on the double curvature surface was not the only difficulty facing the design team when creating the interactive media software. The software also needed to control two different types of screen types, simultaneously.

Close collaboration with the LED manufacturer to develop the software ensured specific animation requirements. Where the 'outside' screens displayed a wave of blinking lights, the 'inside' screen supported a slow, pulsating ambient light. In order to do this, the 5500 pixel LEDs needed to be individually addressed and the input signal of the light sensors directly fed into the controller.

A timeframe of only 2 weeks meant that it was only possible for the team to install Orkhēstra with the help of SAC students.

Although the design team members were located across three continents in five different cities, they were able to overcome all challenges by acknowledging each participant's expertise and field of knowledge. All team members brought an openness to the demands and thinking of the other disciplines – the best way to master complex, interdisciplinary projects.

DRAWING OF THE SCULPTURE

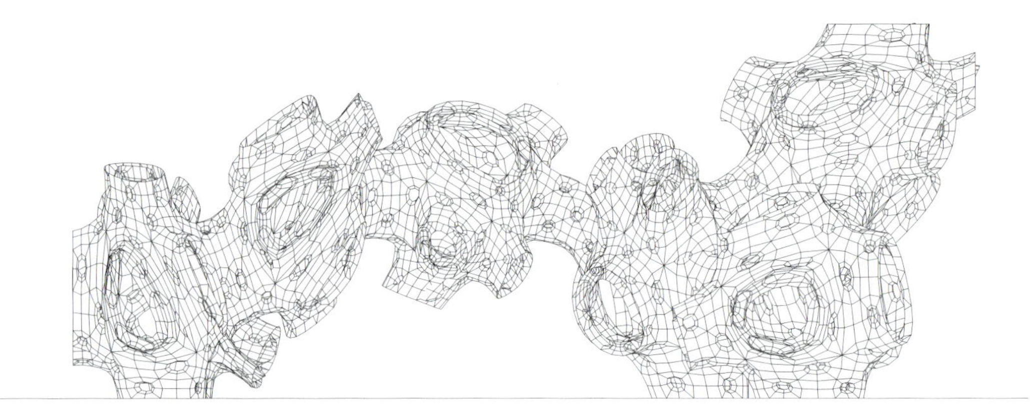

DETAIL OF INDIVIDUAL ELEMENT

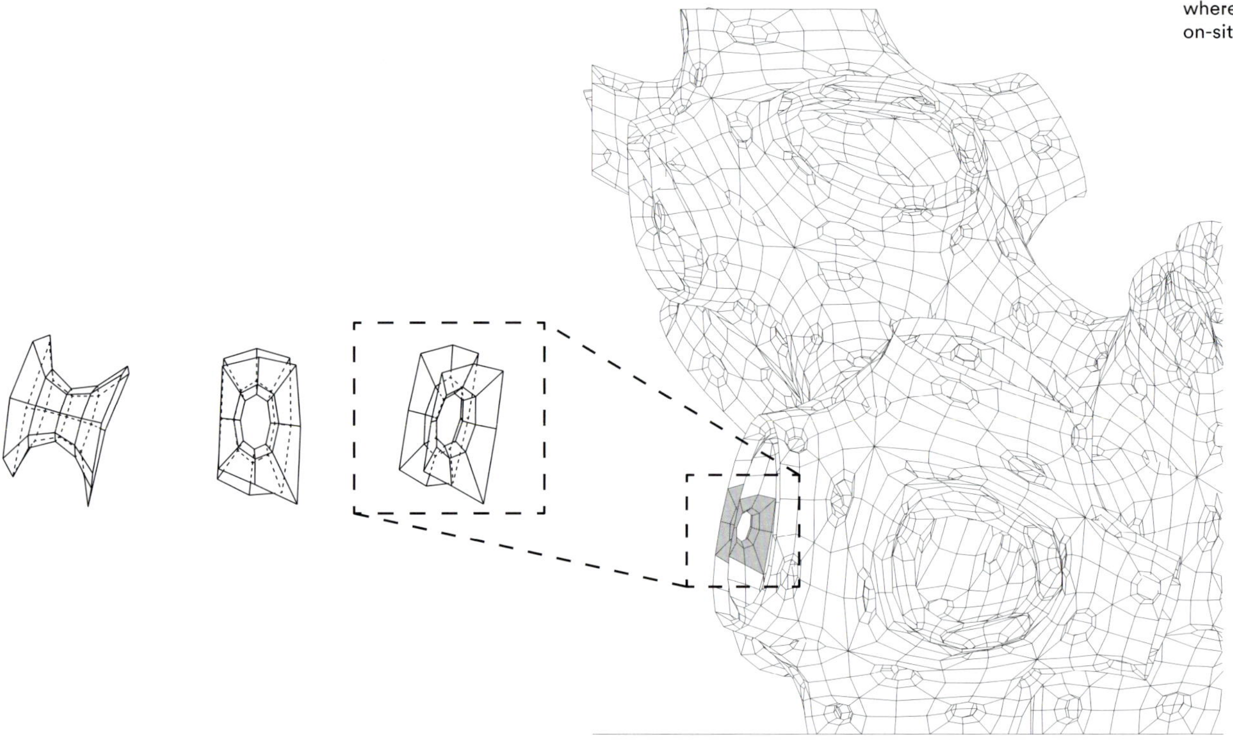

INTERLOCKING CONNECTION RESEARCH

Insertion

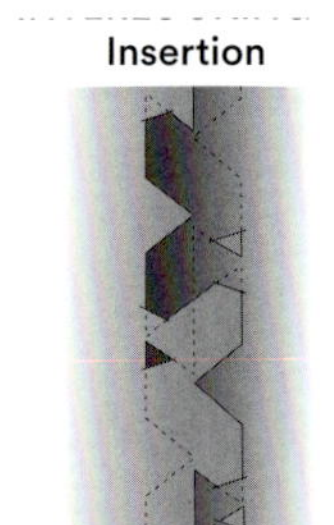

Finger joint

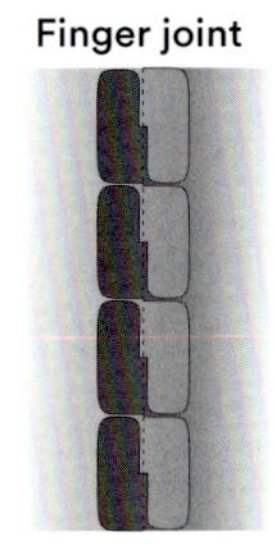

Dovetail

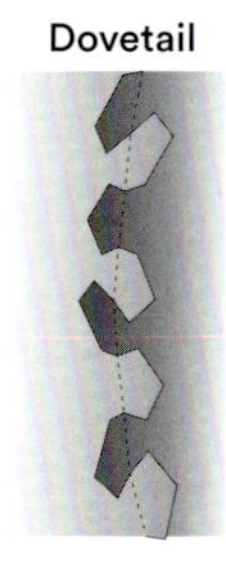

Bending

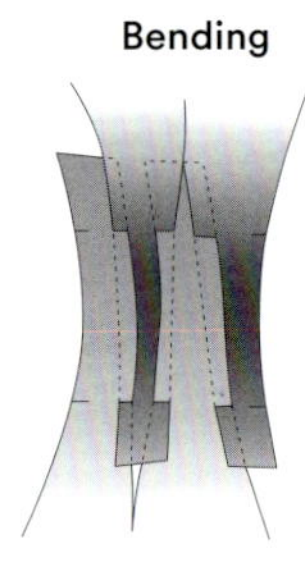

Set-up

Fabrication included laser-cutting of the polypropylene sheets to generate 7732 unique pieces, assembled with a finger-joint system to form a thick, double-layered outer skin. The elaborate shape is grounded on the designers' interest for parametric architecture and non-euclidean geometry.

Organic structure, with its double curvature surfaces reaching a height of 2.5 m, was made possible through the elastic-bending behaviour of the 1.2-mm-thick polypropylene sheets used in its construciton.

Lighting consisted of 5500 MF S25 LEDs with controls addressing each unit separately. Each S25 LED had three LEDs embedded in the pixel with capacity for displaying 16 million colours.

Software was customised for the installation. A Java-based system enabled fast responding interaction triggered by camera flashes, and created the required lighting effects to animate the complex depth of the architectural surfaces.

Assembly of the installation was made possible through a combined team effort where all participants worked off-site and on-site towards completion.

CONCEPT SKETCH

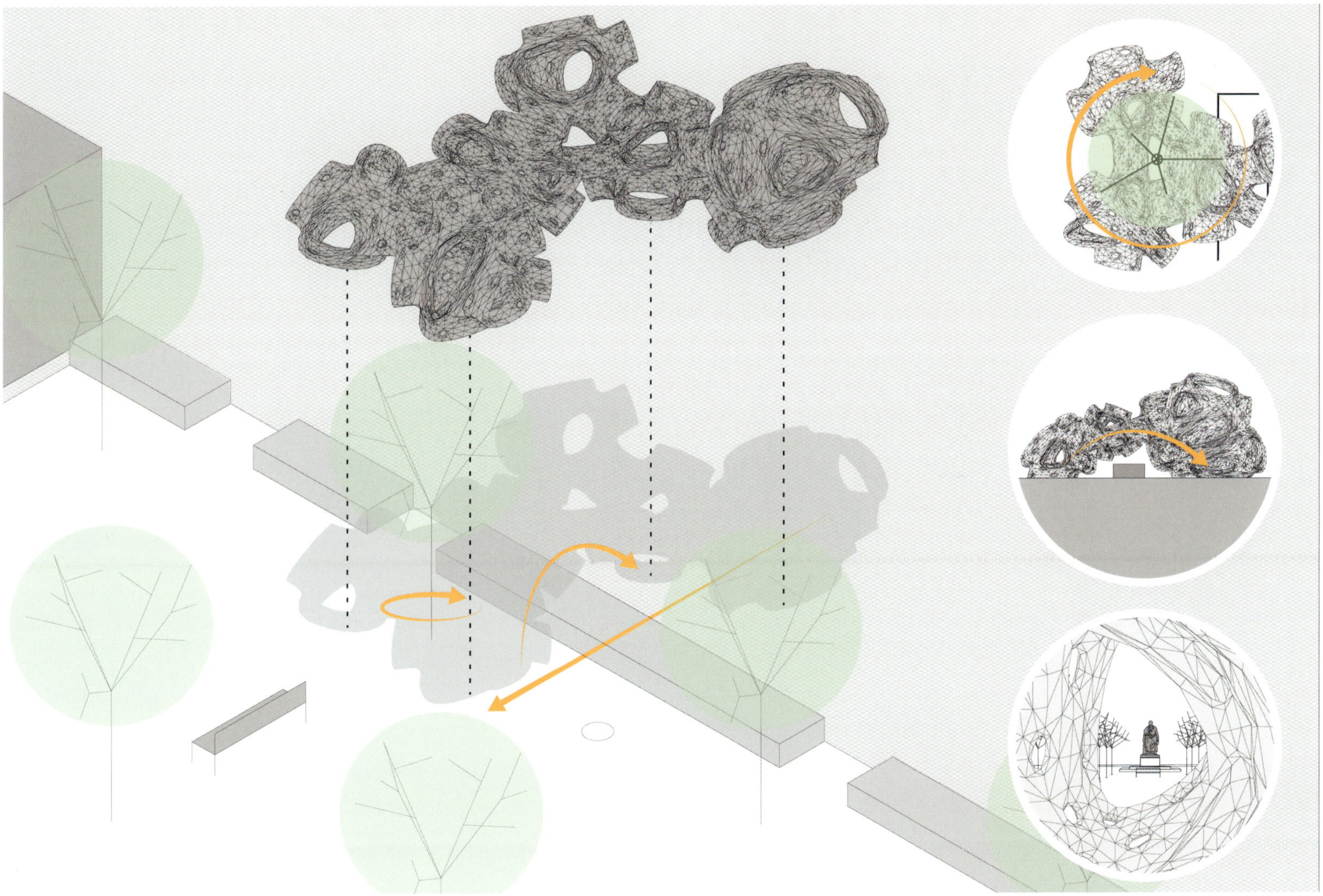

SOFTWARE INTERFACE/DETAILED VIEW

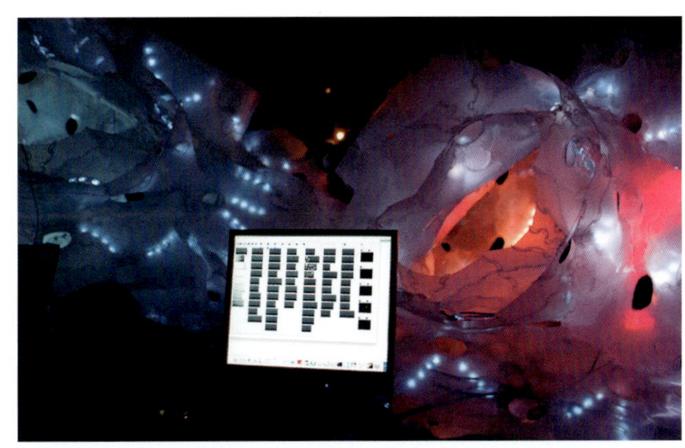

ORKHĒSTRA UTILISED MEDIA TECHNOLOGY TO ANIMATE ARCHITECTURAL SURFACES WITH LIGHT

RENDERING

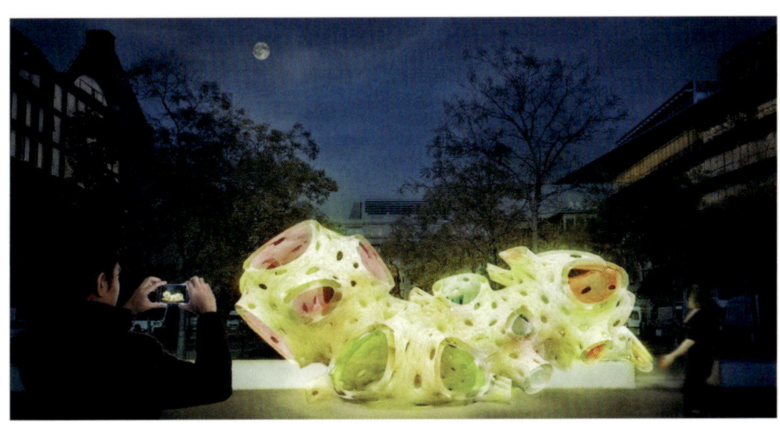

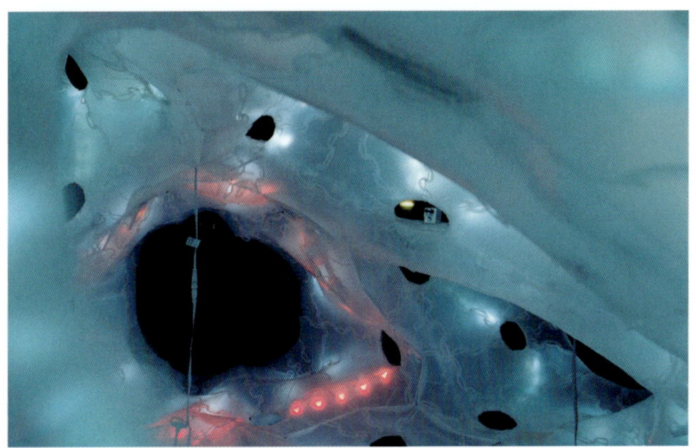

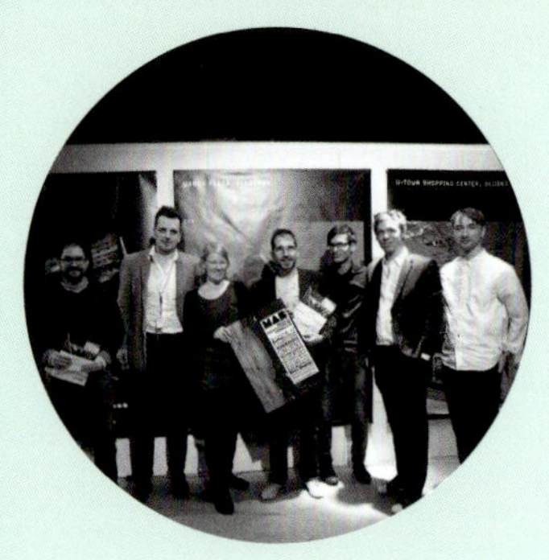

Media Architecture Institute

The Media Architecture Institute (MAI) complements the work of universities and research institutions with a flexible yet focused research activity, bringing together industry, education and academia. Founded in 2009 in Vienna, Austria the non-profit organisation expanded from its European base to a second office in Sydney, Australia in 2011. MAI regularly participates in international festivals, such as Vivid Festival Sydney and Luminale Frankfurt. In 2010, the team launched the Media Façades Summit and the Media Architecture Biennale, two important international events.

MEDIAARCHITECTURE.ORG

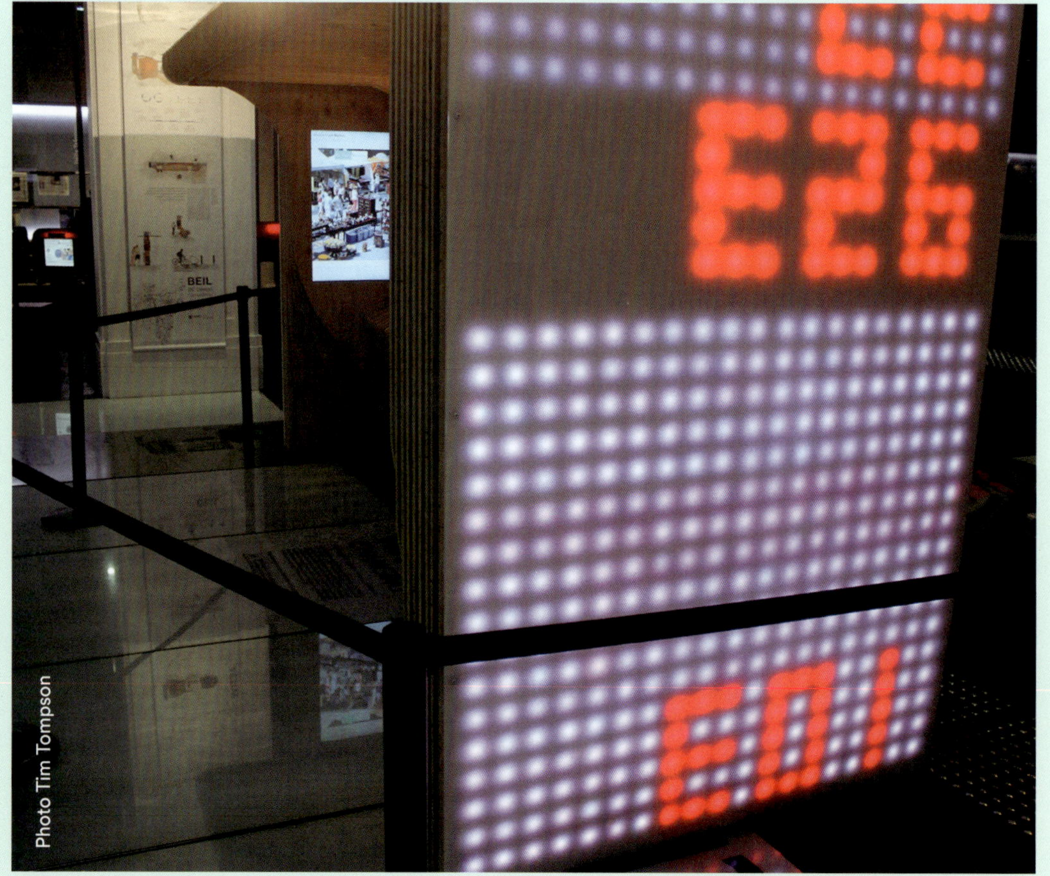

Photo Tim Tompson

Bus Stop of the Future

When **2014**
Where **Sydney, Australia**
Client **Sydney Design Week / Encircle Responsive Public Transport**

In this installation, LEDs and interactive technologies provided passengers with real-time information about timetable or routes. This mock-up developed with the Encircle Responsive Public Transport team equips bus stops with sensor and technology with the same computational capacity of a smartphone, playing a vital role in Smart City concepts.

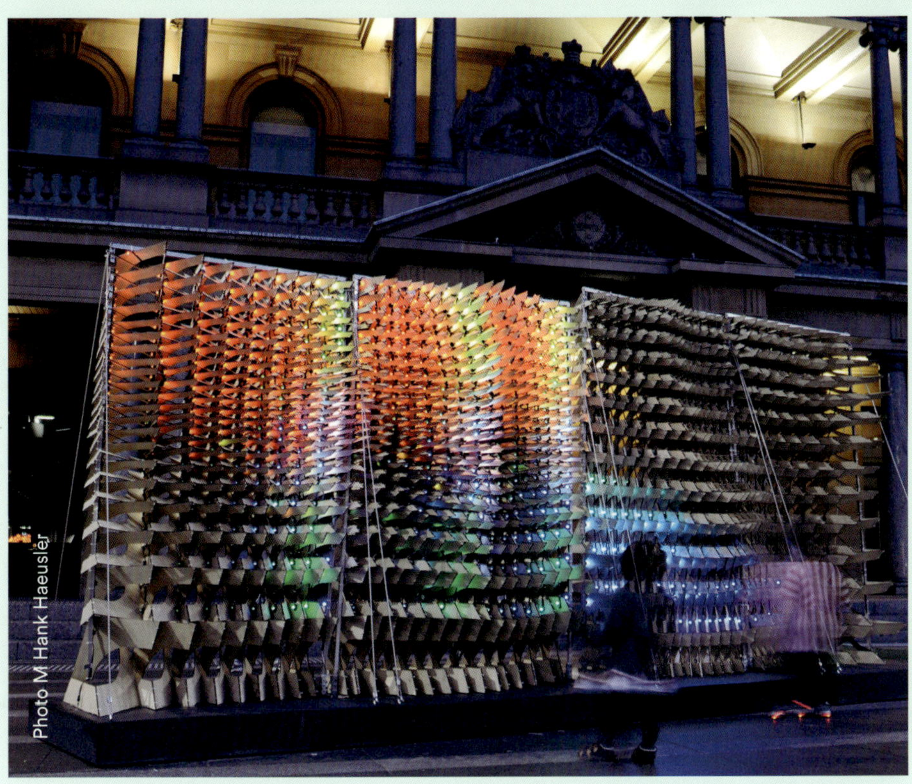

Photo M Hank Haeusler

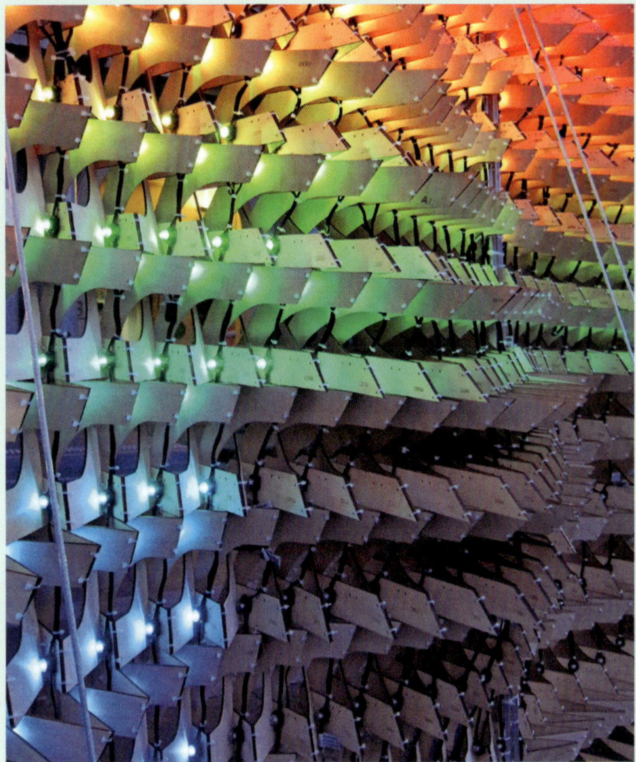

Hypersurface Architecture [Redux]

When **2012**
Where **Sydney, Australia**
Client **Sydney Architecture Festival / UNSW**

A total of 650 'digital bricks' of various sizes were installed in Sydney in 2012. controlled through parametric scripts in which MF S 18 LEDs activate plywood sheets, the system was designed and built by 40 computational design students at UNSW. This media wall installation explores the creative relation between 3D static space and 2D dynamic images.

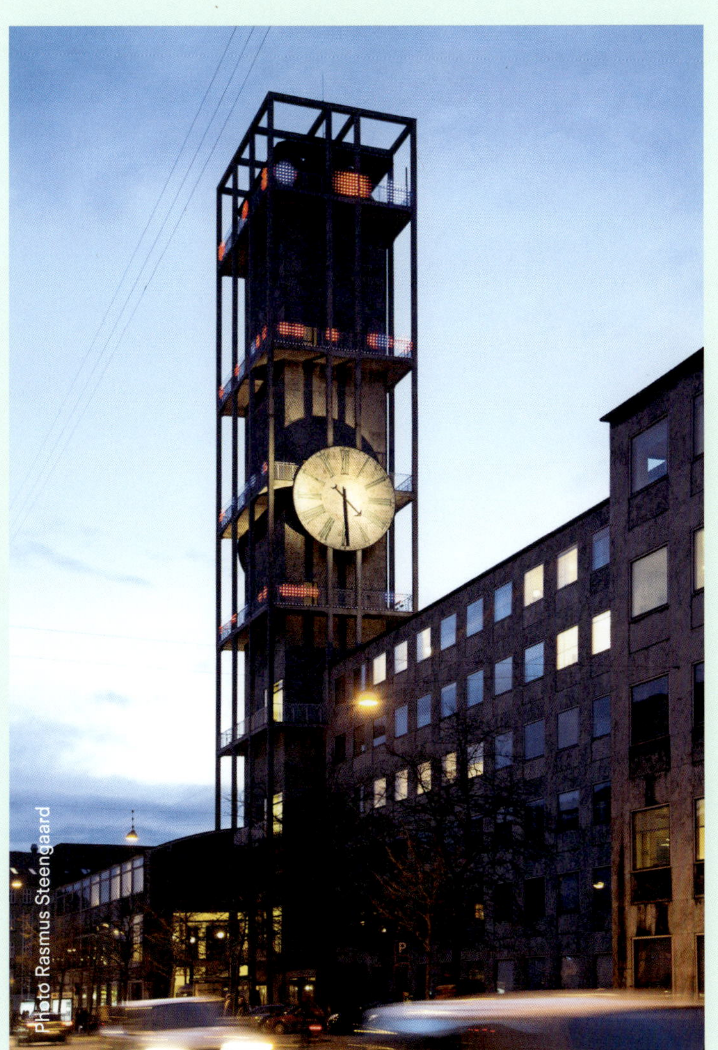

Photo Rasmus Steengaard

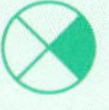

City Bug Report

When **2012**
Where **Aarhus, Denmark**
Client **Aarhus City Council / Aarhus University / CAVI**

During the Aarhus Media Architecture Biennale in 2012, this multimedia installation sat atop the city hall's tower. City Bug Report consisted of a system where Aarhus citizens could report problems occurring throughout the city via the internet. These problems, or 'bugs' were then displayed as a moving ball in a different stage, according to its timeframe and status.

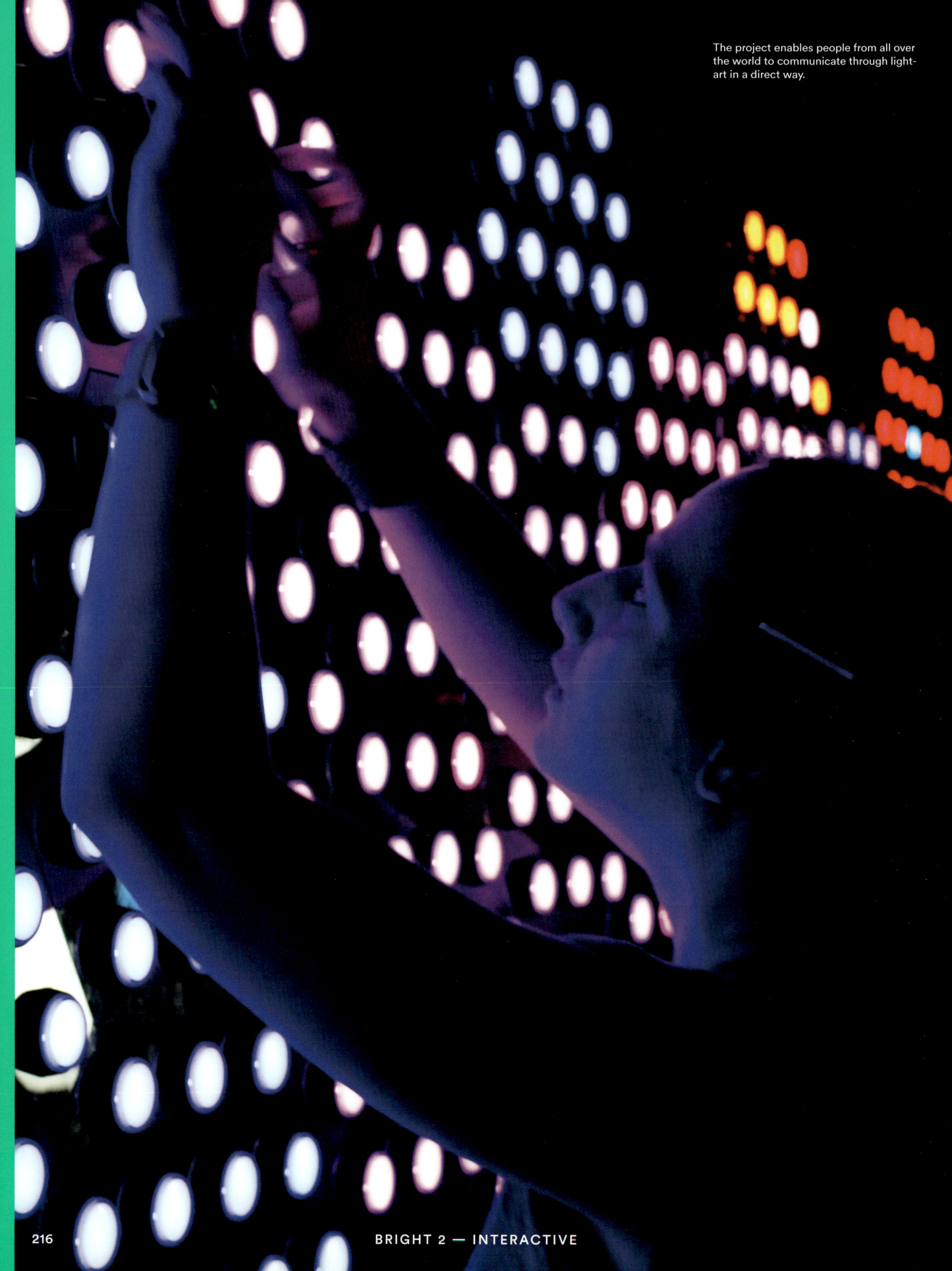

The project enables people from all over the world to communicate through light-art in a direct way.

Push Me!

An interactive light-art installation was initiated by Paul Klotz of **LED-Art** as a reflection upon the effect of how social media has been influencing our way of communication over a short period of time.

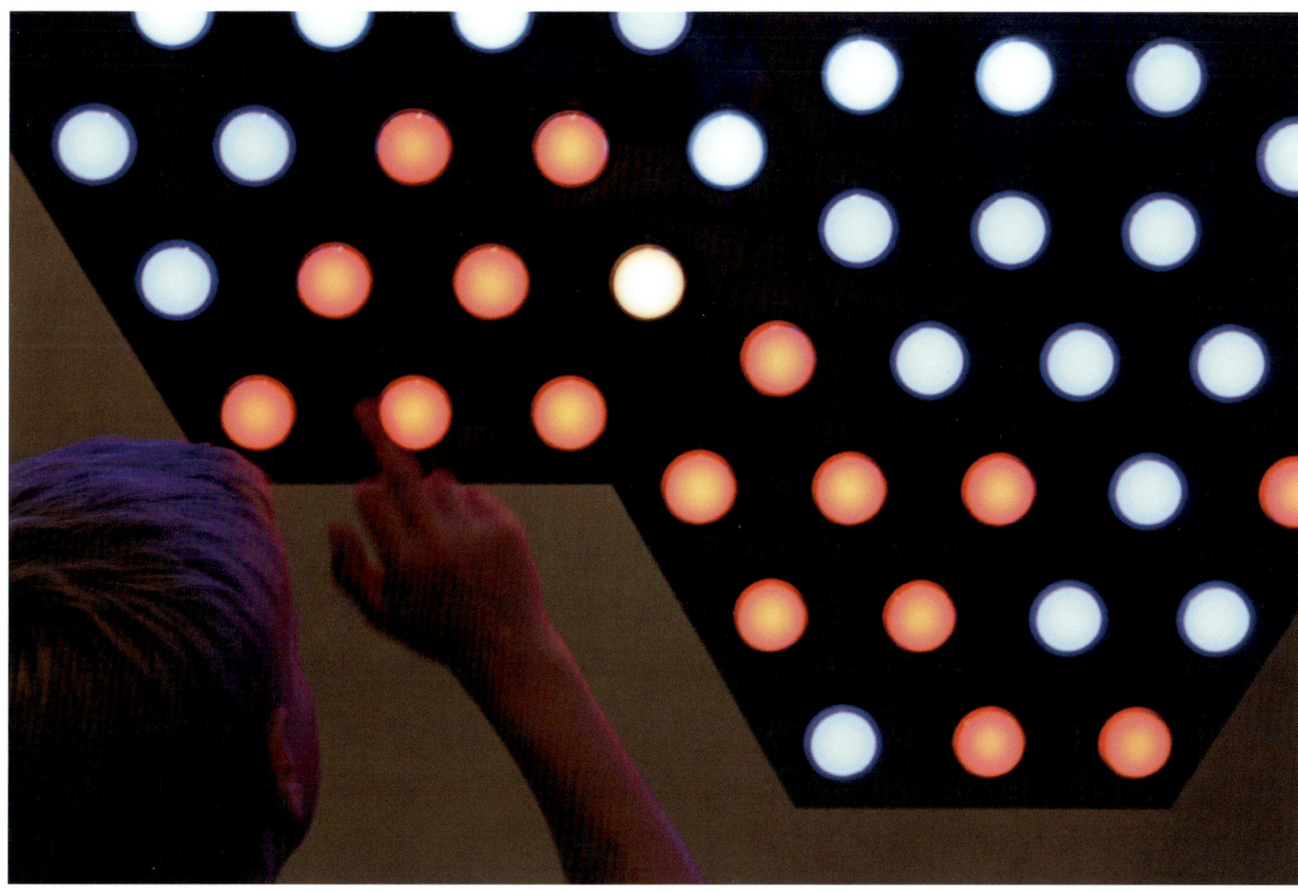

Photos Paul Klotz

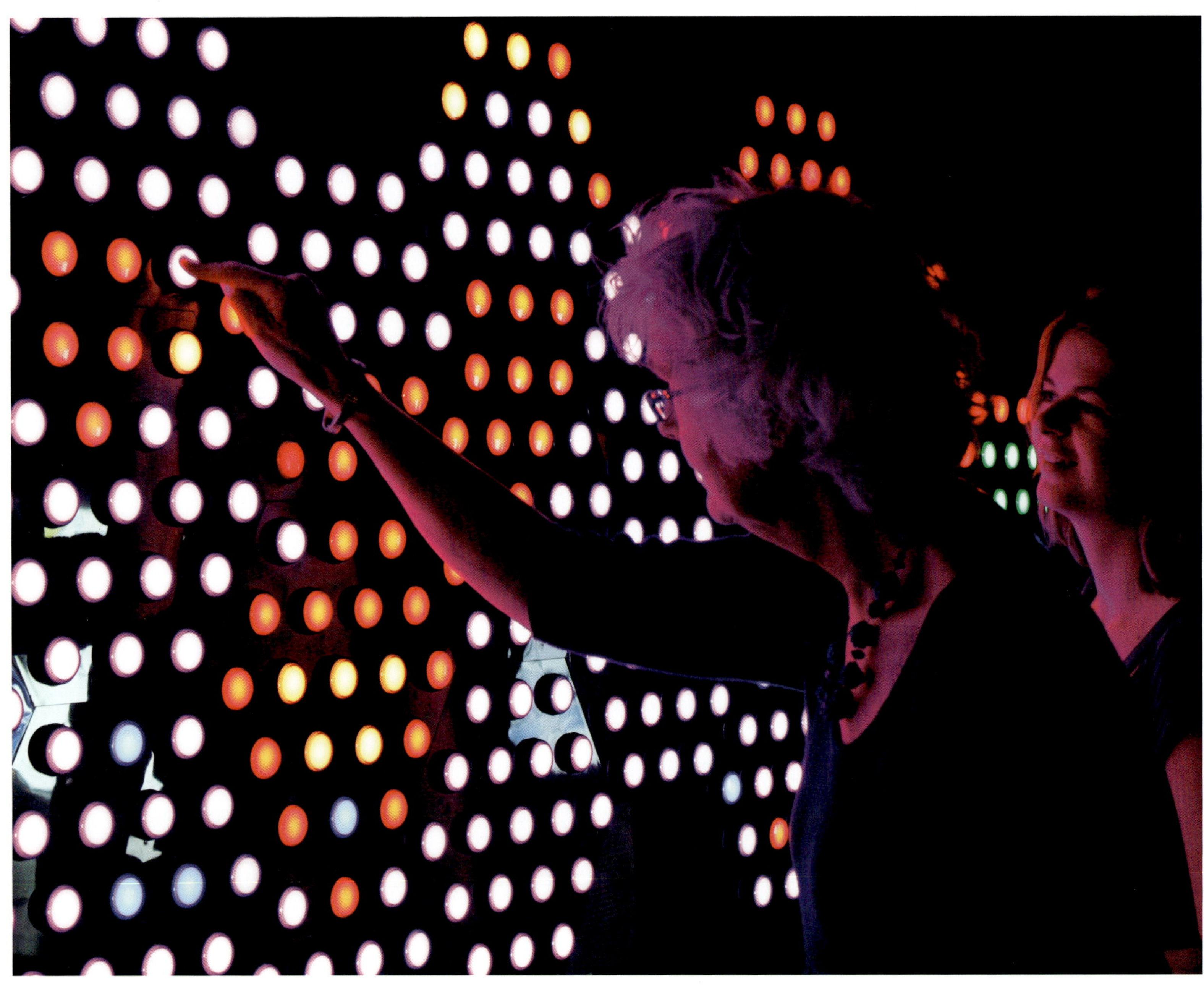

The designer's vision for the work aims to help users re-evaluate the significance of time and space and inter-human communication in an aesthetic way.

Push Me! is a light-art installation by Paul Klotz of Dutch creative studio LED-Art. The multi-colour, pixel-like illuminated work consists of a modular and tactile display that offers an experimental way to communicate. In this social media age, people decreasingly have one-on-one contact in the real world and instead are relying more on connections primarily via electronic means; Push Me! enables users to connect through art from different locations. The work therefore helps users to revaluate the significance of time and space and inter-human communication in an aesthetic way.

The set-up of the installation consists of a number of hexagon modules, each of which has 19 light-emitting circular buttons. These large light-pads change to a different colour every time they are pressed by a participant. Within the system, corresponding buttons are linked, but this only becomes clear when two people touch them simultaneously, which induces a special light effect to spread out across the entire set-up to indicate that such contact has been made. Two identical set-ups can be placed at two separate locations – even in different countries – to unite people by letting them create a connection with each other remotely.

The design of the installation required a tactile approach. The raised, spherical surface of the light buttons was selected because of the appealing nature – people are enticed to press them, as opposed to pressing a flat glass surface for example. The myriad of colours on the wall panel forms a constellation of lights that attracts passers-by to interact with the work. Push Me! can be considered as a light portal, allowing people to connect in a fun and accessible way – in effect, enabling them to create a new language. —

Designer
LED Art
Location
Nijmegen, the Netherlands
Client
Provincie Overijssel
Manufacturers
LED-Art (CNC milling/assembly), KITT Engineering (hardware and software engineering)
Date
November 2012

Set-up

Total area of the installation for illumination was 4.5 m^2.
Hexagonal tile framework is assembled with a total of 25 hexagonal tiles to form one large light panel, which has the possibility to be affixed to the wall or to be placed stand alone in free space utilising a custom-made aluminum framework. The installation is designed to incorporate any number of tiles so that it can be configured to create larger surfaces and different shapes.
Used light was Osram RGB LEDs, responsible for illuminating the buttons using a custom light guide. Each tile has 19 light buttons, with a total of 950 high-efficiency RGB LEDs incorporated in this piece.
Lighting control of each tile is over IP which makes the system very flexible. Control data may be distributed over standard RJ45 network cabling. Identification of the tile address can be set with DIP switches.
Specialist software to control the flow of data between physical and virtual installations at different sites was created by the team. In fact all tiles 'live' virtually on a server and are synchronised by a smart web service protocol.
Lightweight materials and a modular-construction were a vital part of the installation, as it needed to be easily transportable across the globe. This resulted in the modularity of the tiles, and the aluminium framework construction.
Development time from concept to completion of the installation, including all hardware and software, was about 6 months.

POSSIBLE TILE CONFIGURATIONS

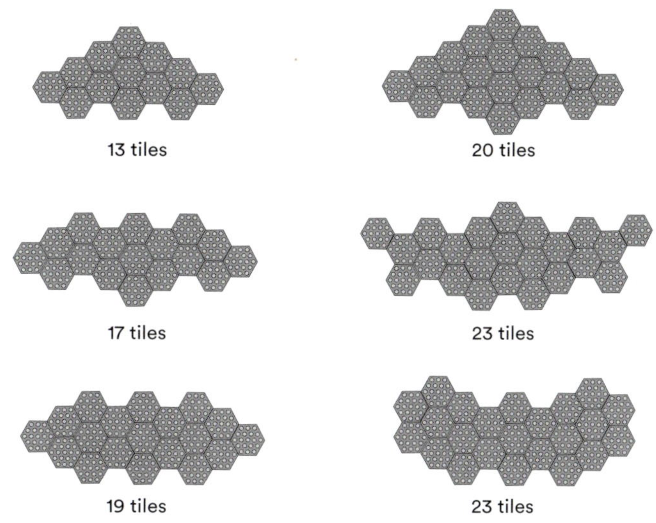

BUTTON/PANEL SECTION

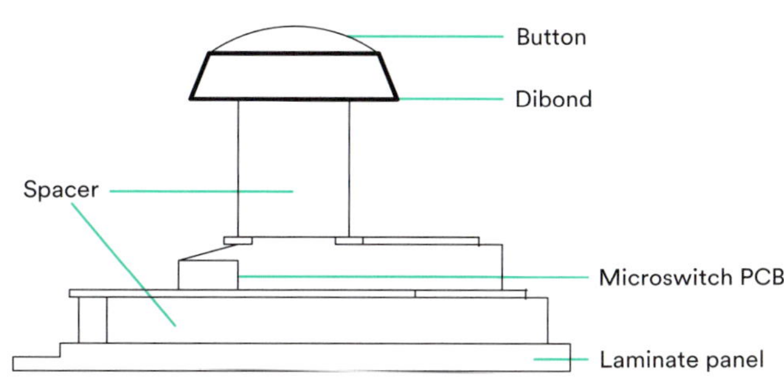

ELECTRICALS CLOSE-UP

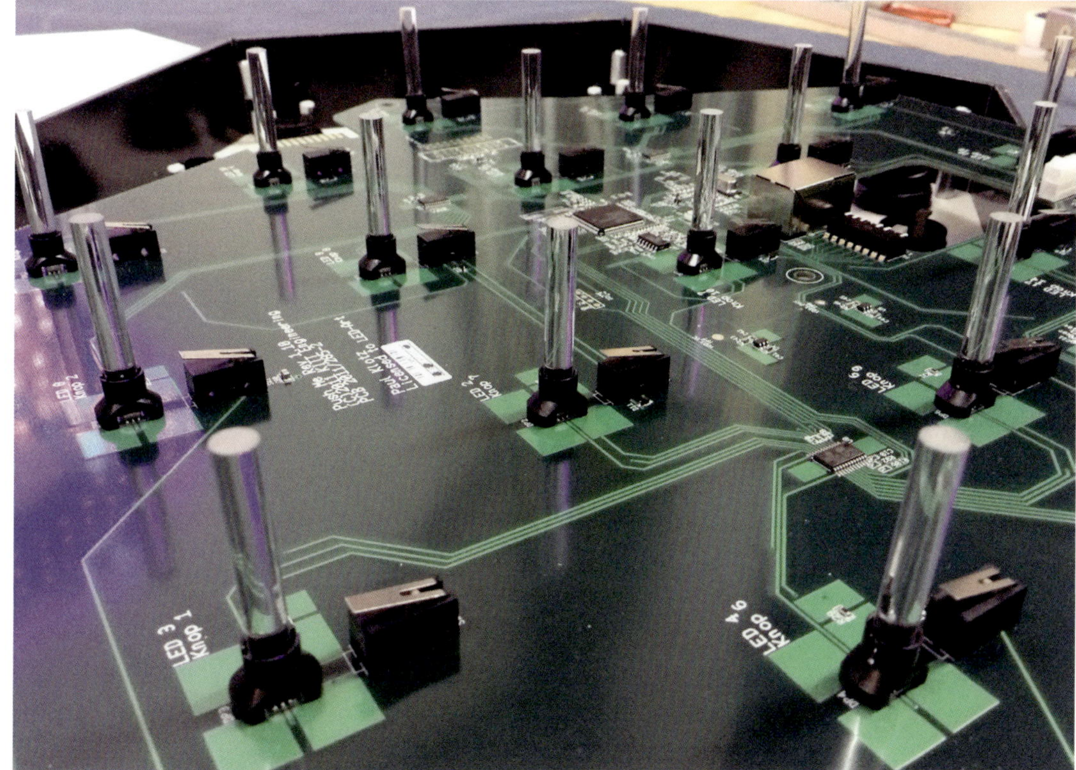

The behind-the-scenes circuitry on the reverse of the light panels reveals a network of connections, akin to the focus of the installation.

Specifics

This project was supported by Art E-Motion, granted by the Provincie Overijssel Further support was received from: AKKUH Hengelo, the Prins Bernhard Cultuurfonds, Stichting Stokroos, KITT Engineering, Smart XP-lab, Beeldbouw and the Videogilde.

Paul Klotz initiated the work as a reflection upon the effect of how social media has influenced our means of communication over a short period of time. One of the main criteria of the Provincie Overijssel was that the installation would be innovative, interact with the public and could be transported to different locations with ease.

The project enables people from all over the world to communicate through art in an accessible and direct way. One of the challenges in its design was to distribute all the data in a way that all colours would change instantly on both locations and give audio feedback at the same time for multiple buttons. This problem was solved in the software.

Another challenge was the distribution of light to the surface of the push buttons. The RGB LEDs on the PCB board had to be positioned about 30-mm higher, and a layout was required that would ensure there was no loss of light. Special light guides were laser-cut to distribute the light over the desirable distance so that the buttons were illuminated equally. A small, one button set-up with a single RGB LED was realised to test the optimal length of the light guides.

The large set-up is a constellation of 25 tiles with 19 illuminated buttons each. During shows, it is very interesting to see how people react on the open character of the installation and the communication when people on 'the other side' start pressing the buttons as well.

The hexagon-shaped modules were designed to create a pattern of light pixels that were divided evenly over a surface. The layout utilised no writing characters, so that users would be tempted to challenge other people to express themselves in a more figurative way.

During the conceptual stage of the project, a lot of effort was put into user research related to form and senses by testing several button designs and technologies, such as membrane switches, capacitive touch sensing and click-touch piezo technology. The old-school button was found most applicable in relation to the concept and the desired look-and-feel of the installation.

The lights were directly linked to each other by continuously updating their colour in real-time, related to the corresponding button on the twin set-up. Meaning that if the user alters button X to green by simply pressing it on set-up A, the corresponding button on set-up B is changing to green simultaneously. In this way, two living artworks co-exist at two places at the same time, enabling people on different locations to communicate and get involved in an open-ended play situation.

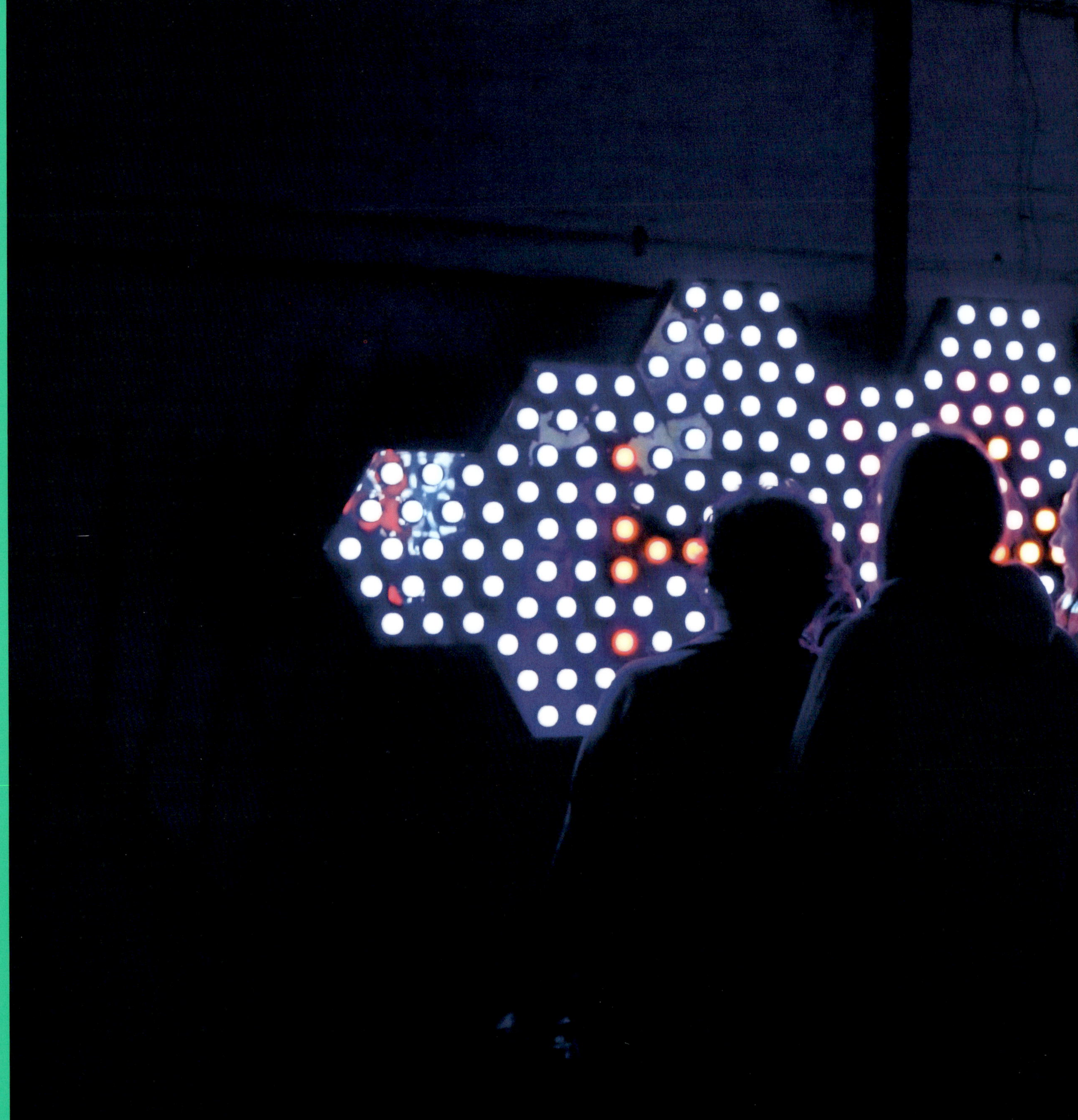

PUSH ME! IS LIKE A LIGHT PORTAL, ENABLING PEOPLE TO CONNECT IN A FUN AND ACCESSIBLE WAY

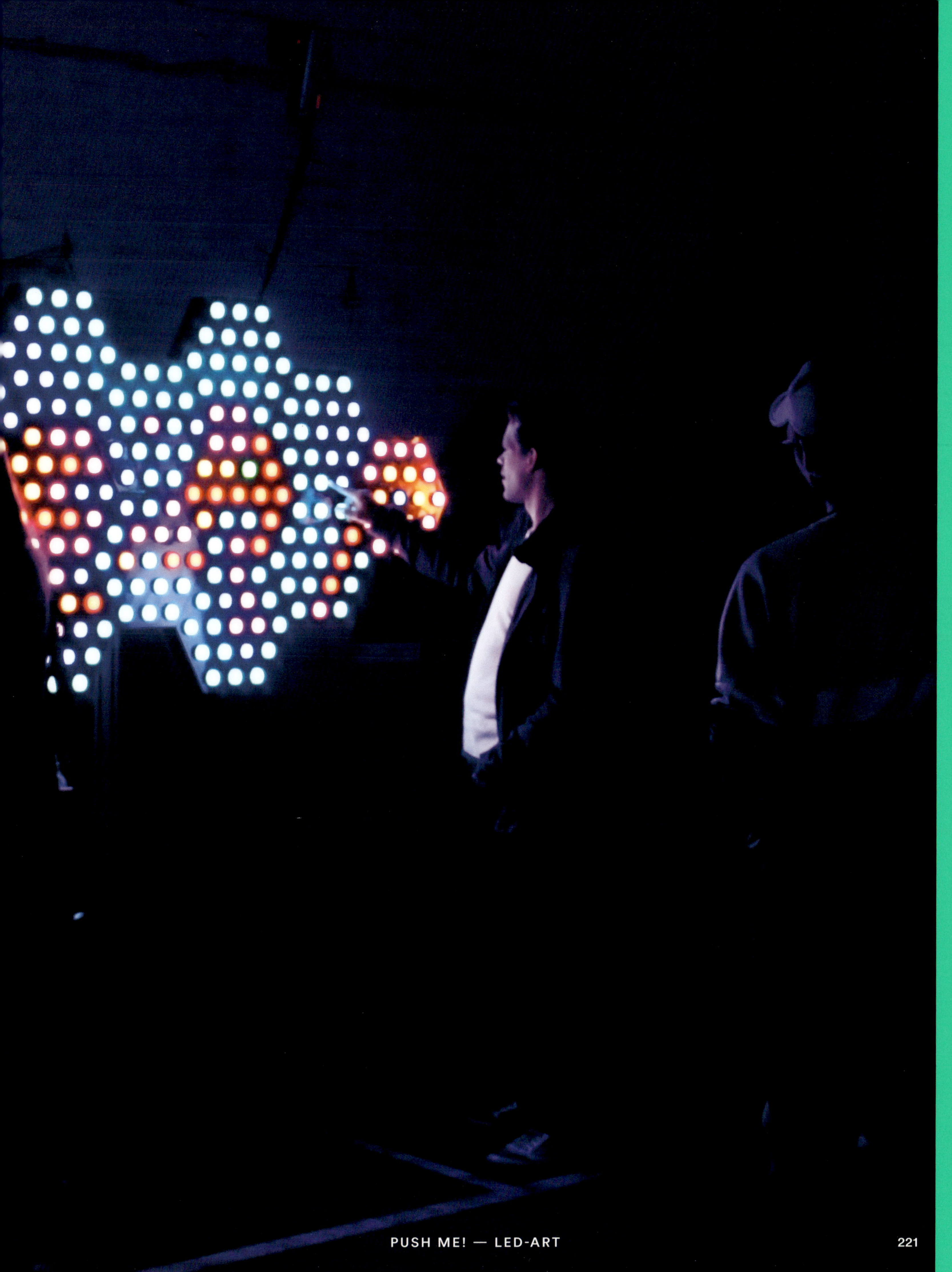

Photo Merel Brandon

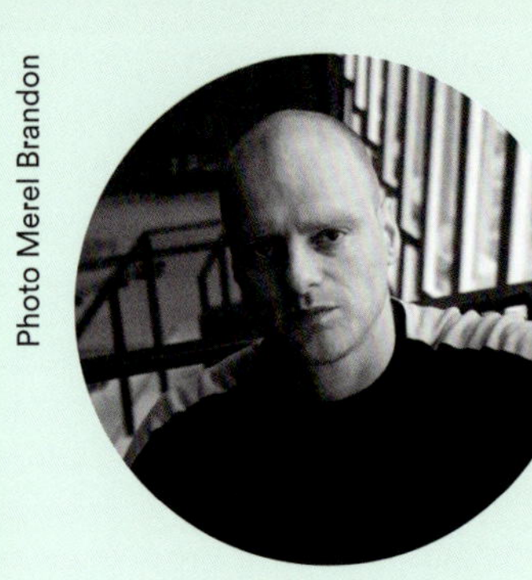

LED-Art

Founded by Paul Klotz, LED-Art creates interactive light art designed to enrich the consumer experience; both for public and private spaces. Klotz regards light as a variable that determines our state of mind. Visitor presence, environment and the desired identity or emotion are key elements in the designs. Klotz is currently doing a master's degree in industrial design at the Technical University of Eindhoven in the Netherlands. For the coming years, he envisions to fuse interactive art with parametric design and architecture in collaboration with other specialists.

LED-ART.NL

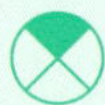

De Bellenblazer

When **2011**
Where **Enschede, the Netherlands**
Client **Prismare**

After an explosion in a fireworks factory, the suburb of Roombeek (near Enschede) was rebuilt with innovative and unique architecture. To mark this, Paul Klotz in collaboration with D Boekhout designed and developed De Bellenblazer – a 9-m-tall landmark in the centre of the neighbourhood. The interactive sculpture functions as a meeting point and at the same time a visual guide for visitors to the Prismare theatre building. The changing colours of the huge 'soap bubbles' are triggered by the moment of people walking nearby, as well as the local weather and wind force.

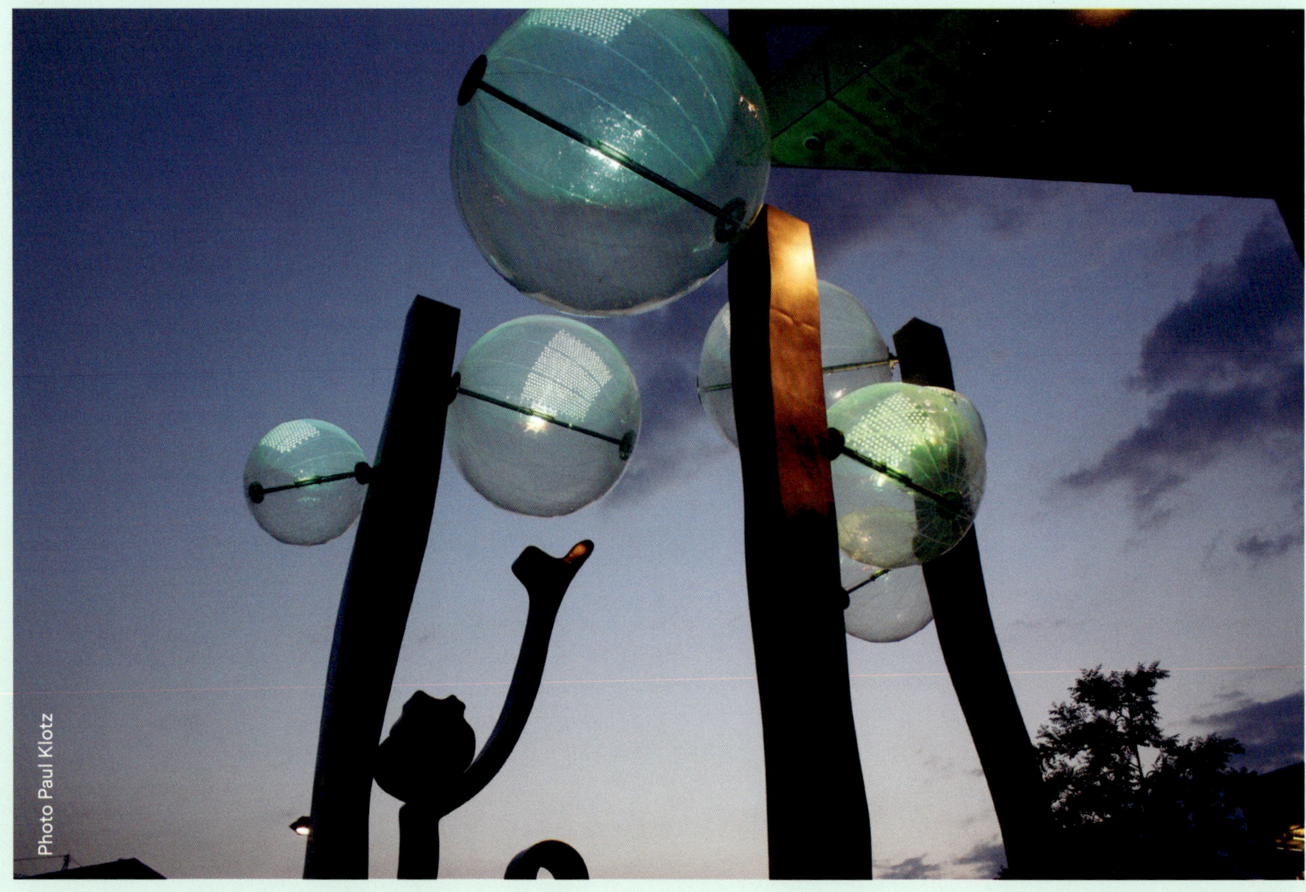

Photo Paul Klotz

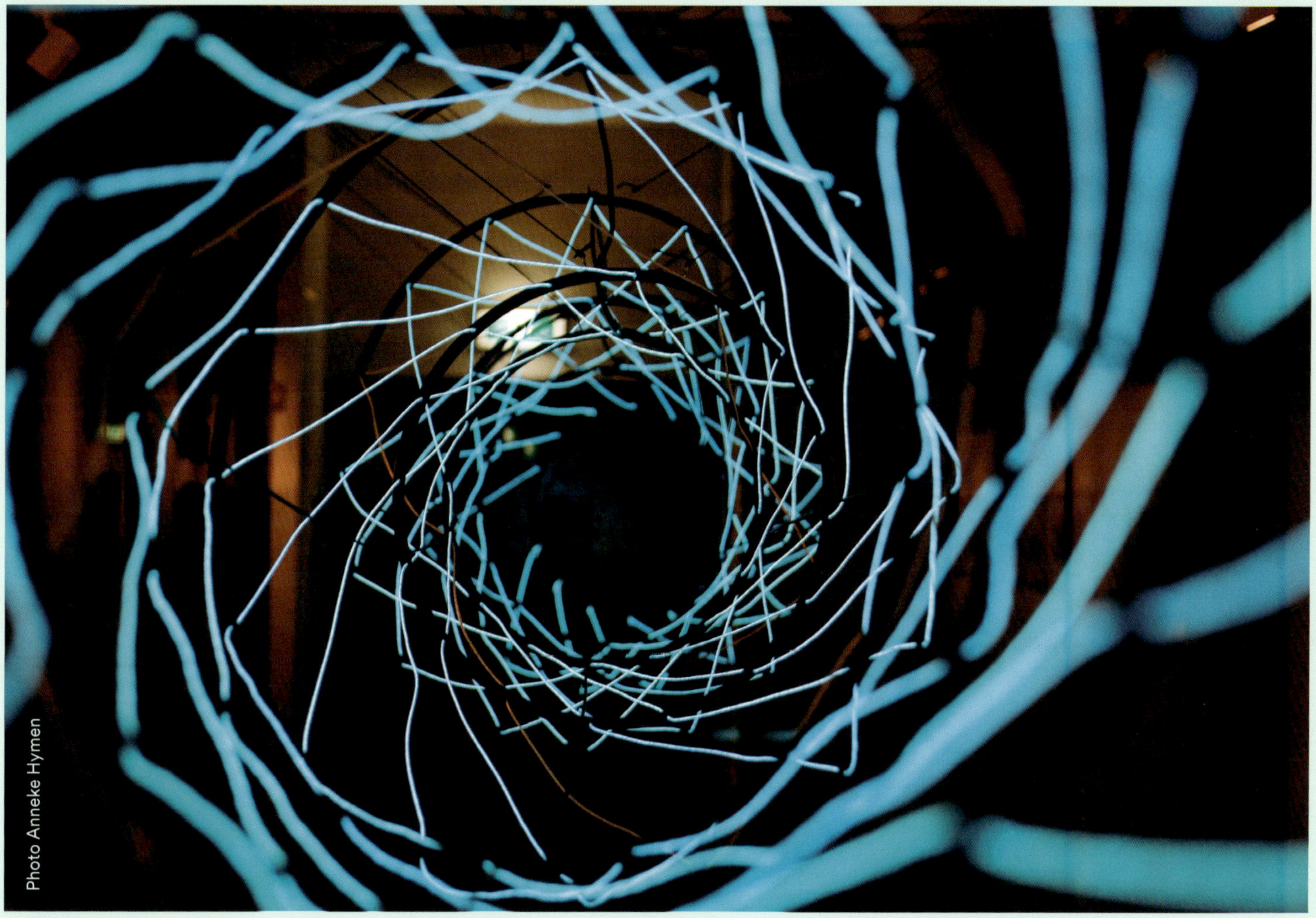

Photo Anneke Hymen

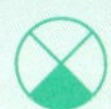

Tunnel Vision 2.0

When **2014**
Where **Nijmegen, the Netherlands**
Client **n/a**

Tunnel Vision 2.0 is an interactive light and sound installation. The shape of the work is based upon an abstraction of the 100-Hz buzz-tone produced by electrical generators. Users can interact with this work through a 'brain interface' which alters the animation of the light-emitting wires and their amplified sounds.

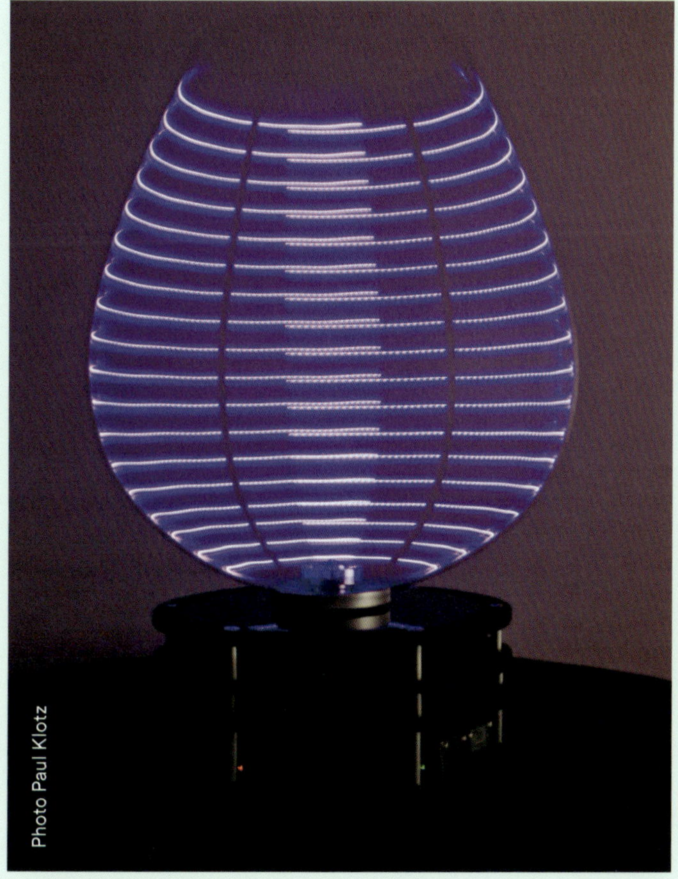

Photo Paul Klotz

Tech Flower

When **2013**
Where **Nijmegen, the Netherlands**
Client **n/a**

Tech Flower is an experimental installation showing the emotions of people at a certain location or event. A special algorithm is used that analyses Twitter data to reveal the emotions through animations of light, forming a tulip-like shape. This installation was created by Paul Klotz in collaboration with Leander Verstraelen and Frank Henselmans (Avans Hogescholen).

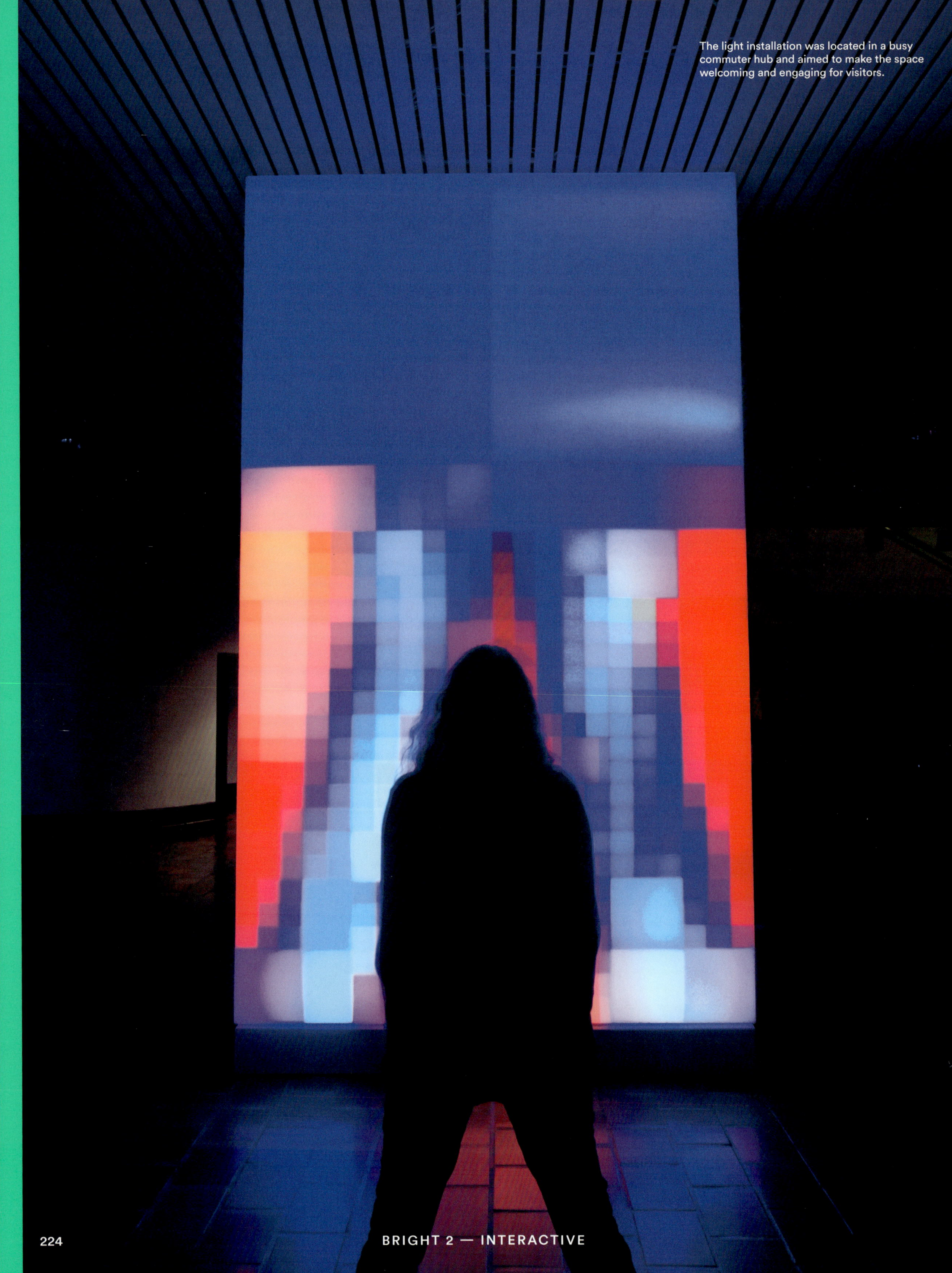

The light installation was located in a busy commuter hub and aimed to make the space welcoming and engaging for visitors.

Reflect

For the lobby of a Florida government building, **Ivan Toth Depeña** created an interactive light installation that engages visitors as it mirrors their image and movements in real-time. The illuminated piece elaborates on the idea of time and memory in space by vibrantly replaying the patterns created by participants' interactions.

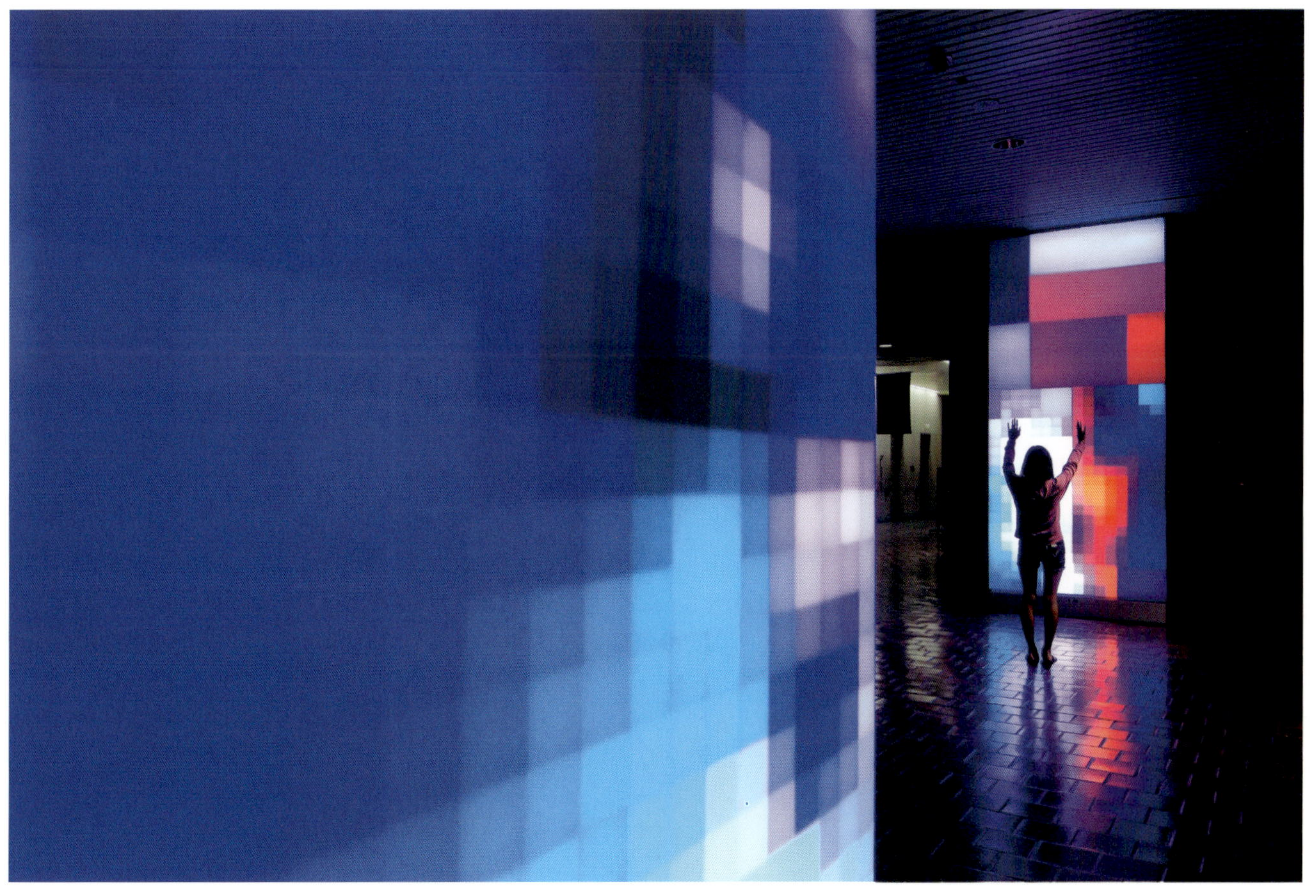

Photos Ivan Toth Depeña

An interactive light installation has animated the lobby of the Stephen Clark Government Center in Miami, Florida since 2012. As the name portrays, Reflect by Ivan Toth Depeña looks to mirror the dynamism of the lobby space by welcoming and engaging its visitors. The role of the installation is to elaborate on the idea of time and memory in space, engaging participants in new ways each day.

The chosen location is a hub for commuters, being an important stop on the city's MetroRail network. Depeña incorporated this notion of collective circulation by developing a system that uses input from motion sensors and infrared cameras to provoke individual reflection, interactivity and high-tech playfulness. Commissioned by Miami-Dade Art in Public Places, the vibrant artwork consists of a network of custom-designed LED columns spread throughout the lobby. A camera-tracking system reflects and records the commuters' movements, sends the information to a computer where custom-made software processes and abstracts the participants' mirrored images and sends these abstractions back to the light panels in real-time. The resulting image creates an ethereal replication that encourages visitors to play and explore the installation's responsiveness.

While the reflection of the real-time interactions animate the space during the day, at times of low traffic streams or at night, the installation's 'memory' plays back the patterns and interactions previously created. This display of the past – one of the key features of the piece's conceptual infrastructure – adds life to the lobby at the same time as it explores the notions of time and memory in space. —

Designer
Ivan Toth Depeña
Location
Miami, United States
Client
Miami-Dade Art in Public Places
Collaborator(s)/consultant
Focus Lighting, Dan Henry Design
Manufacturer
Philips
Date
November 2012

The colourful installation engaged passers-by as it explored the notions of time and memory in space.

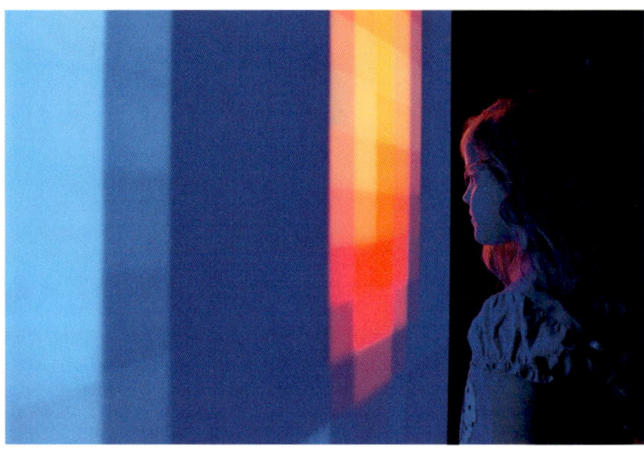

VIBRANT ILLUMINATIONS REPLAY THE PATTERNS CREATED BY PARTICIPANTS' INTERACTIONS

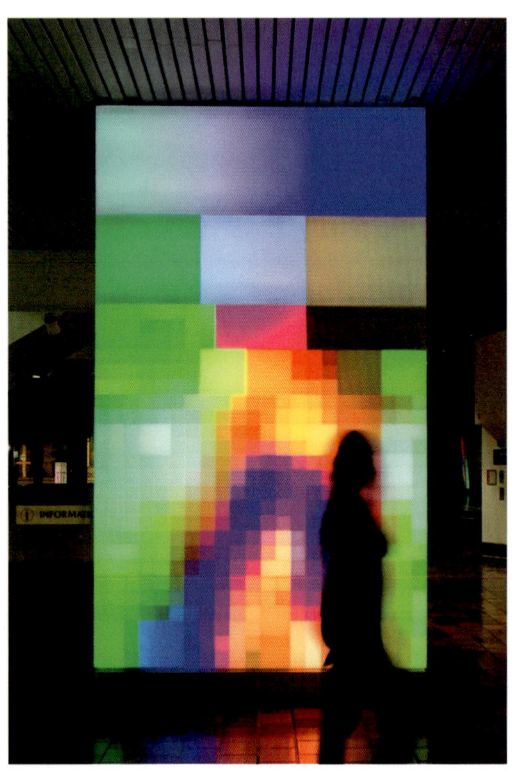

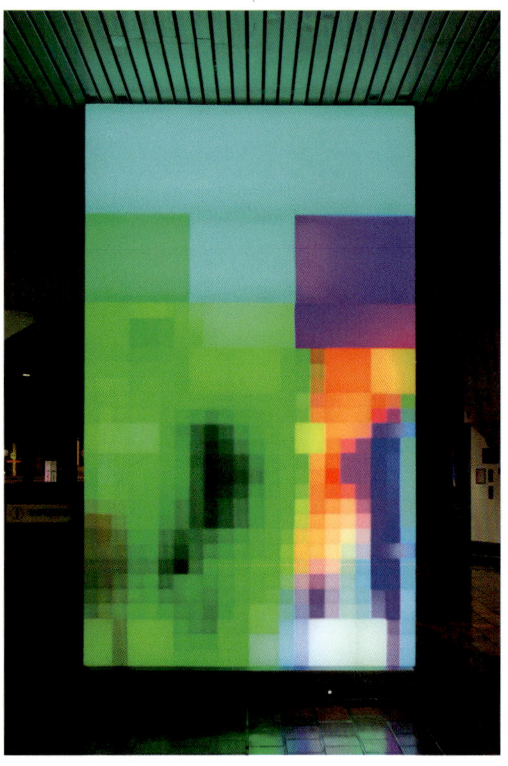

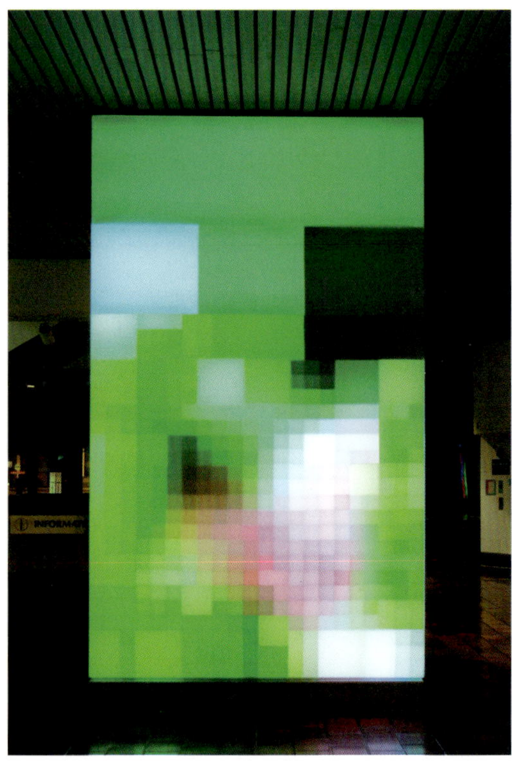

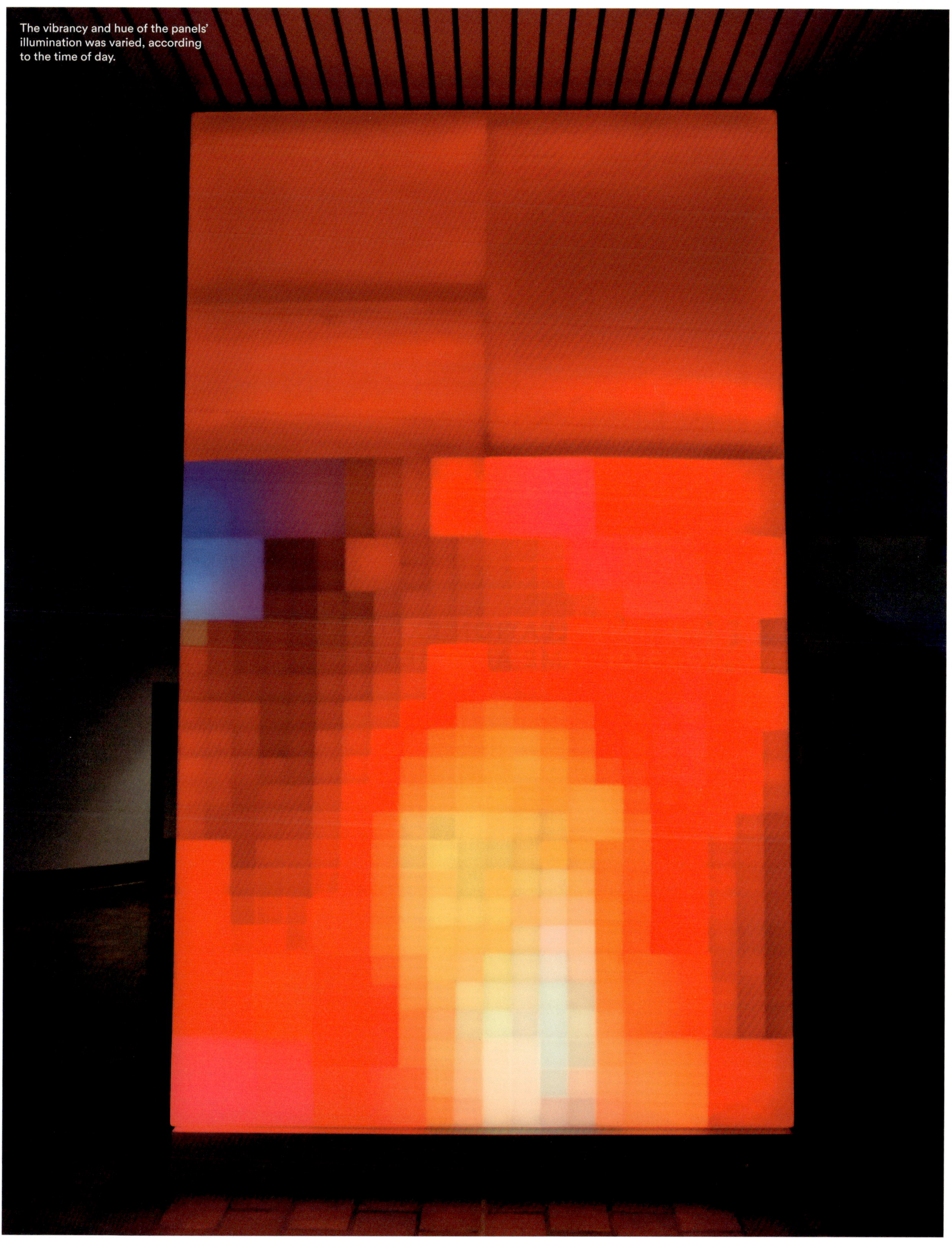

The vibrancy and hue of the panels' illumination was varied, according to the time of day.

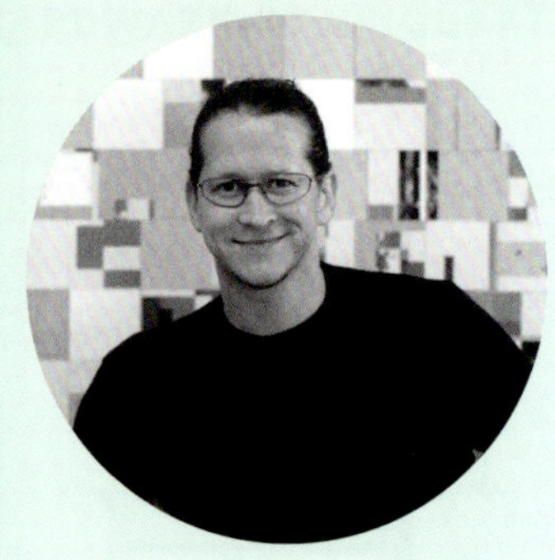

Ivan Toth Depeña

Living and working between Charlotte, NC and Miami, FL, award-winning American artist and architect Ivan Toth Depeña looks to bring together various creative disciplines in 'seamless harmony'. Painting, photography, light, video and installations are but some of the studio's creative outputs, which at times are fused together, illustrating the artist's quest for the surprise in serendipity. One of Depeña's recent works, a collaborative series of augmented reality public art projects throughout South Florida, was awarded the Knight Arts Challenge Miami Grant in early 2014.

IVANDEPENA.COM

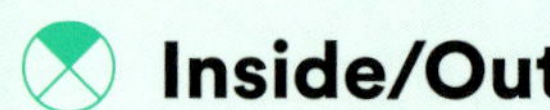

Inside/Out

When **2015**
Where **Albuquerque, New Mexico**
Client **New Mexico Arts and Department of Cultural Affairs**

Inside/Out uses the idea of architecture being a living organism injected with life as it is inhabited. Using various sensors, the activity that goes on within the building is abstractly simulated with light on the building's exterior and glass skin. Therefore, the 'interior experiences' are used to illuminate the facade.

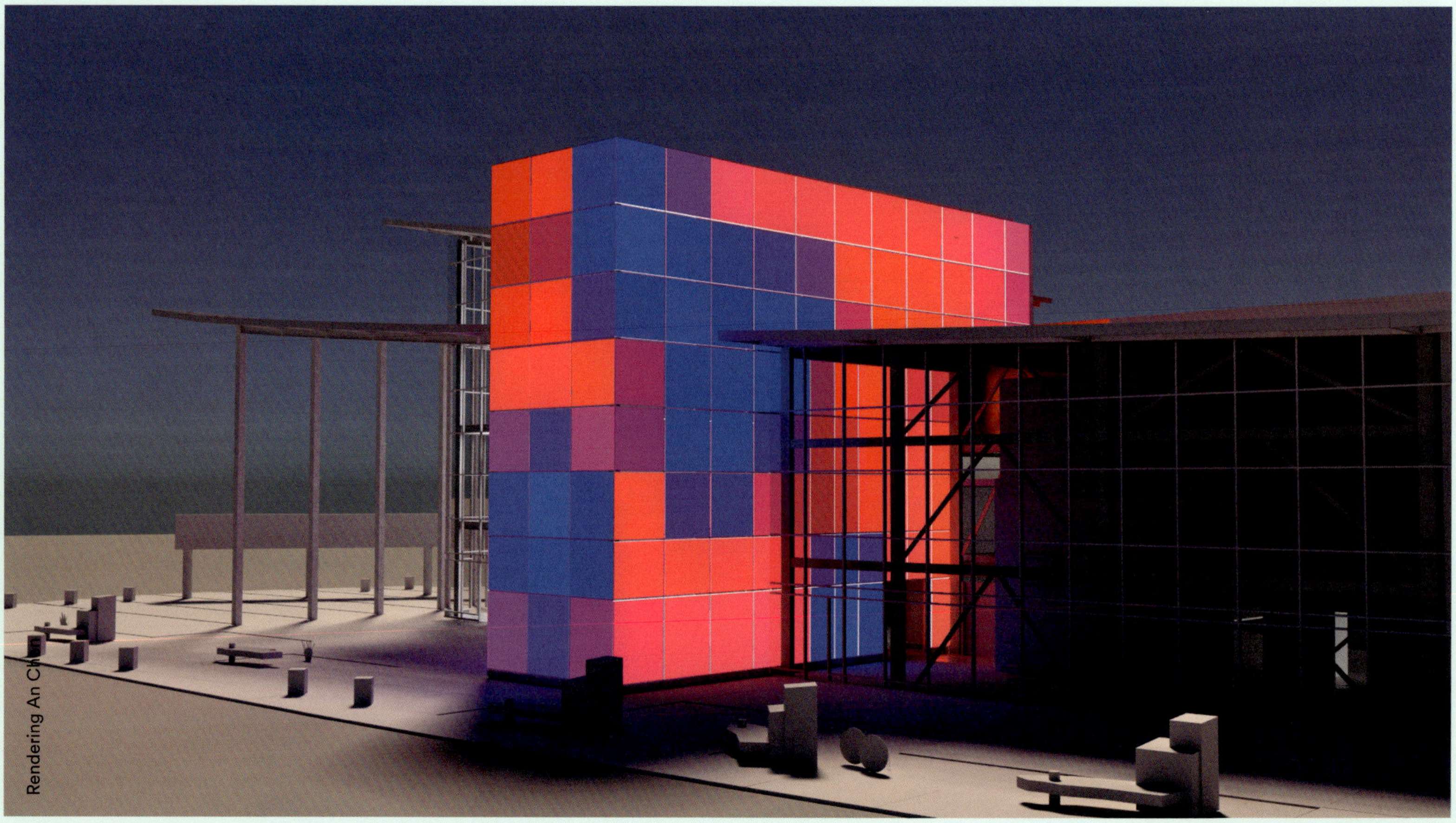

Rendering An Chen

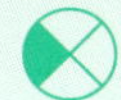

The Walkers

When **2013**
Where **New York, United States**
Client **Reed Krakoff**

This is a responsive light installation commissioned for Reed Krakoff's Madison Avenue flagship storefront. The vibrant display breaks away from traditional visual merchandising and becomes a work of public art that responds to pedestrians. The site-specific design features a series of coloured fluorescent tubes in custom-designed light boxes that respond to the physical presence of passers-by through the implementation of motion/camera tracking.

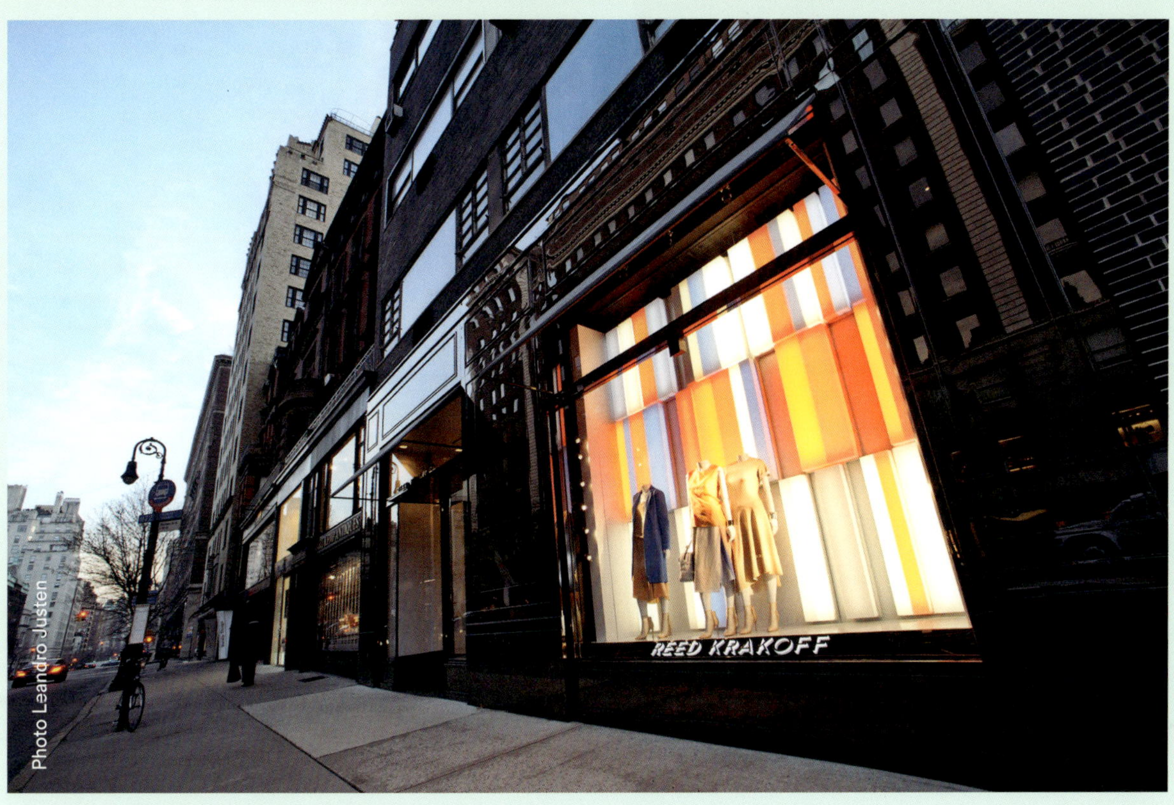

Photo Leandro Justen

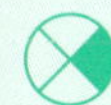

Color Field

When **2014**
Where **Denver, United States**
Client **RTD FasTracks and City of Lakewood, Colorado**

This installation casts patterns of colour across more than 30 m of ramps and walkways, interpreting nature's inherent chaos and representing it as an evolving, kaleidoscopic abstraction. Made of steel, laminated coloured glass and concrete poles that are lit by LEDs at night, the 'tree-like' structures were designed and created algorithmically.

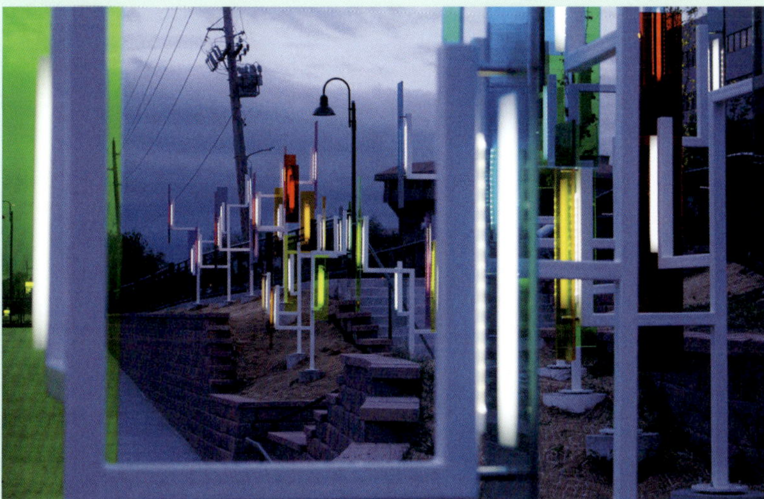

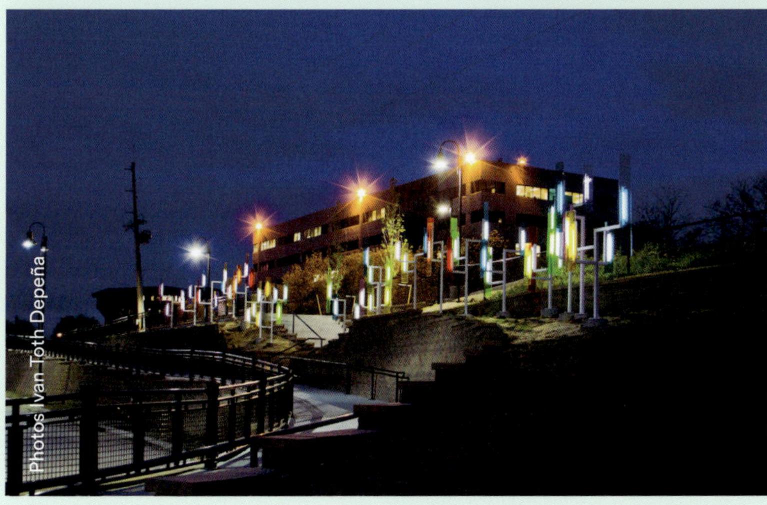

Photos Ivan Toth Depeña

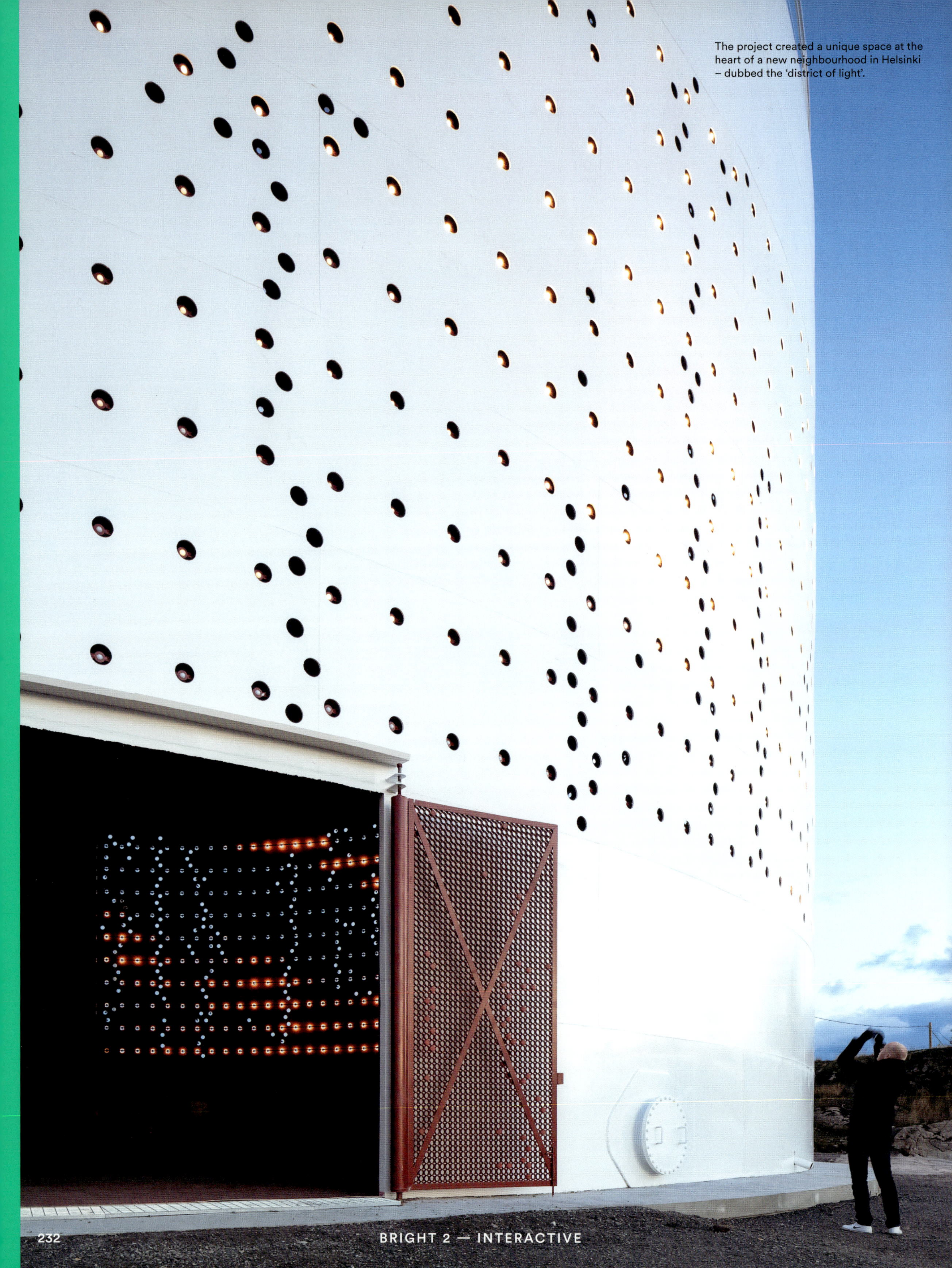

The project created a unique space at the heart of a new neighbourhood in Helsinki – dubbed the 'district of light'.

Silo 468

An interactive light-art installation in a disused oil silo was the **Lighting Design Collective**'s answer to urban redevelopment in Helsinki. The industrial landmark shone like a beacon thanks to its illuminated perforations, creating a visual experience from both outside and inside the unique civic space.

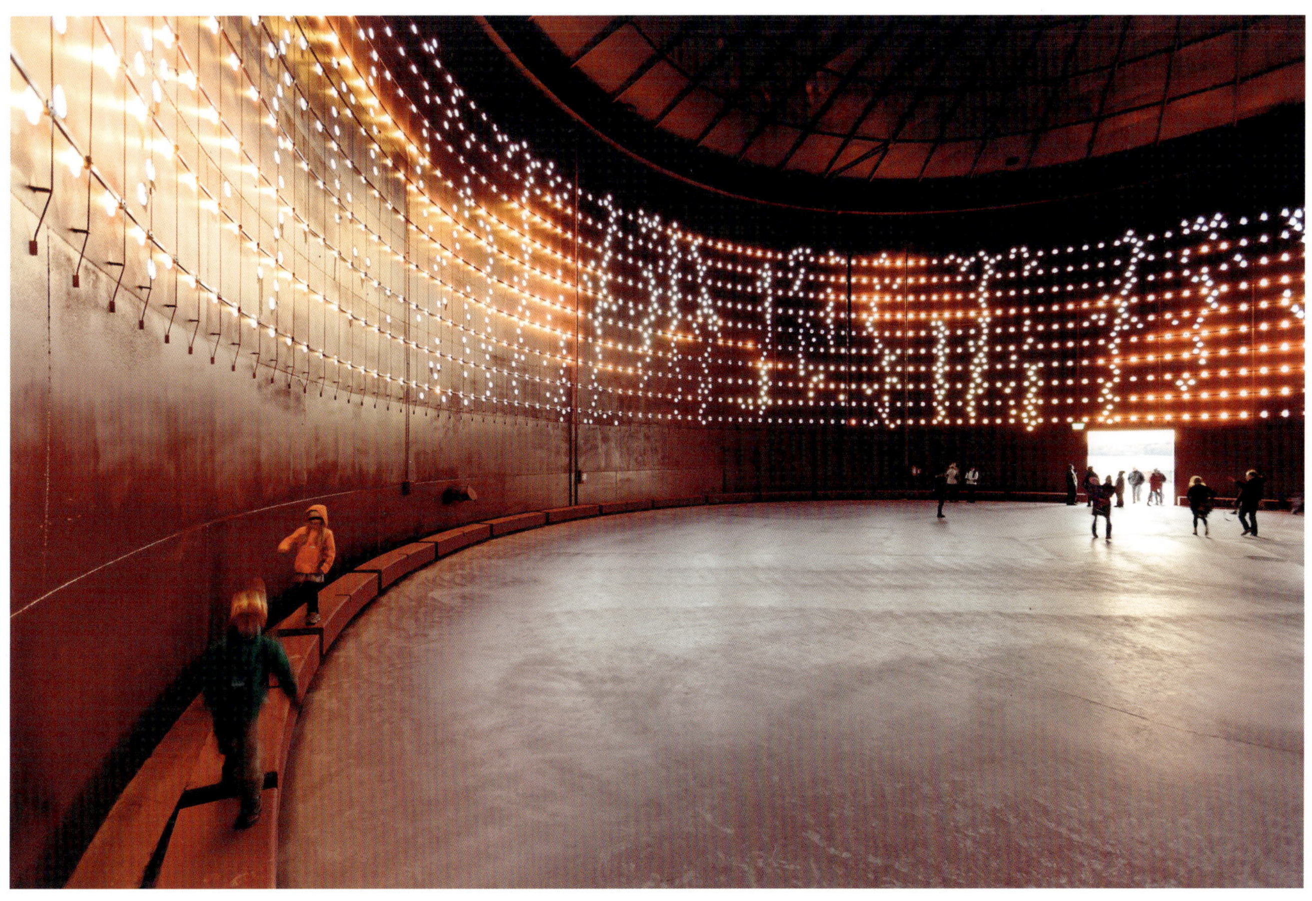

Photos Hannu Iso-oja, Tuomas Uusheimo

Located by the sea facing central Helsinki, a disused oil silo was converted into a mesmerising illuminated art installation by the team at Lighting Design Collective. The coastal site, its natural light and prevailing winds that are often present – along with the movement of the light on the water – were inspiration for the lighting concept. In addition, the walls of Silo 468 were perforated with 2012 holes, referring to the fact that the city was a World Design Capital during the year the silo was transformed.

The project signified the start of a major urban redevelopment for the City of Helsinki. It functioned to draw focus towards the unknown district, creating a destination landmark in the process. Through the use of natural and artificial light, it created a unique civic space at the heart of the new neighbourhood – dubbed the 'district of light'. During the early years of the area's development, the silo would shine like a beacon across the water, with its walls emanating with what seems (from a distance) pinpricks of light. Yet, it is when visitors get up-close – and in fact step inside the 1000 m^2 space – that the interactive lighting comes into its own.

The team developed bespoke software that refreshes every 5 minutes responding to various parameters derived from the wind. Over the course of one day, the dynamic light patterns changes in a fluid manner, natural in feel and never repeating. The interior is painted deep red (the colour corresponding to the former use of the silo) and comes into its own as the area gets populated. Daylight seeps through the pattern derived from original rust deposits on the walls. The north facing wall has no added perforations. In total, 450 steel mirrors are fitted behind the holes and move with the wind. In the daylight, the silo appears to glimmer and sparkle like the surface of water. At night, the warm white LED grid reflects light indirectly via the rust-red walls within the space. The moving patterns read as halos racing across the walls. At 02.30, timed to coincide with the last ferry departing back to the city, the lights are switched off. —

Designer
Lighting Design Collective
Location
Helsinki, Finland
Client
City of Helsinki
Collaborator(s)/consultant
Pöyry Finland, Granlund Oy, SunEffects
Manufacturers
Traxon Technologies, Meyer
When
October 2012

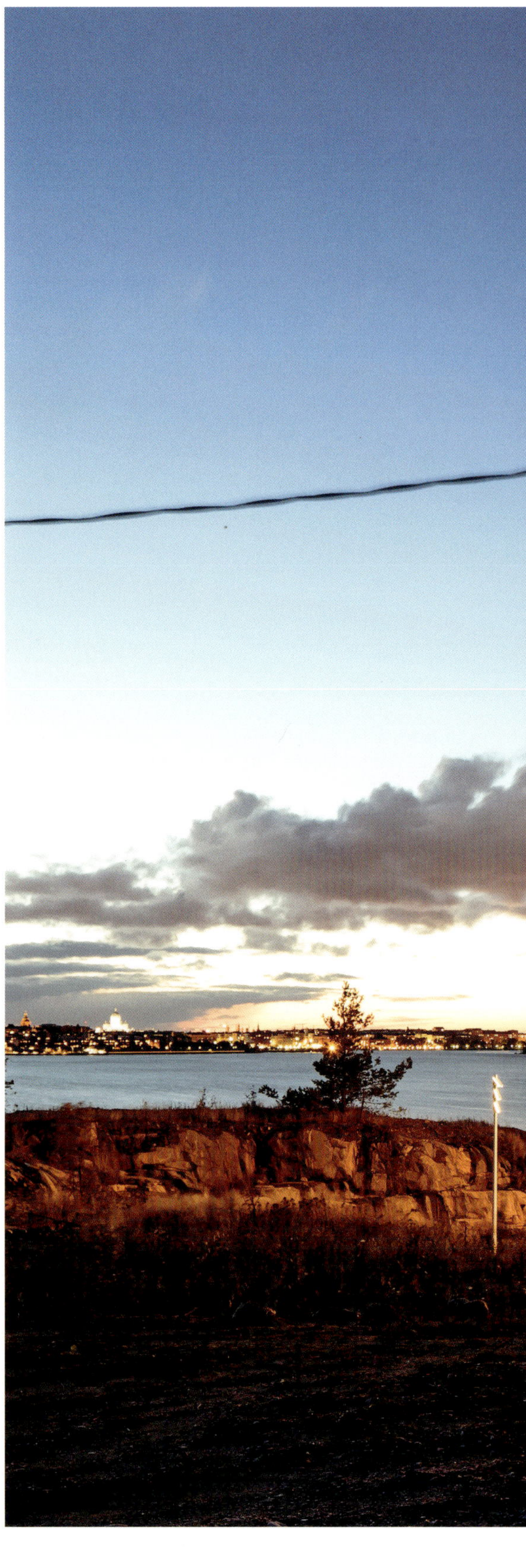

The interactive illumination patterns are fluid, natural in feel and never repeat.

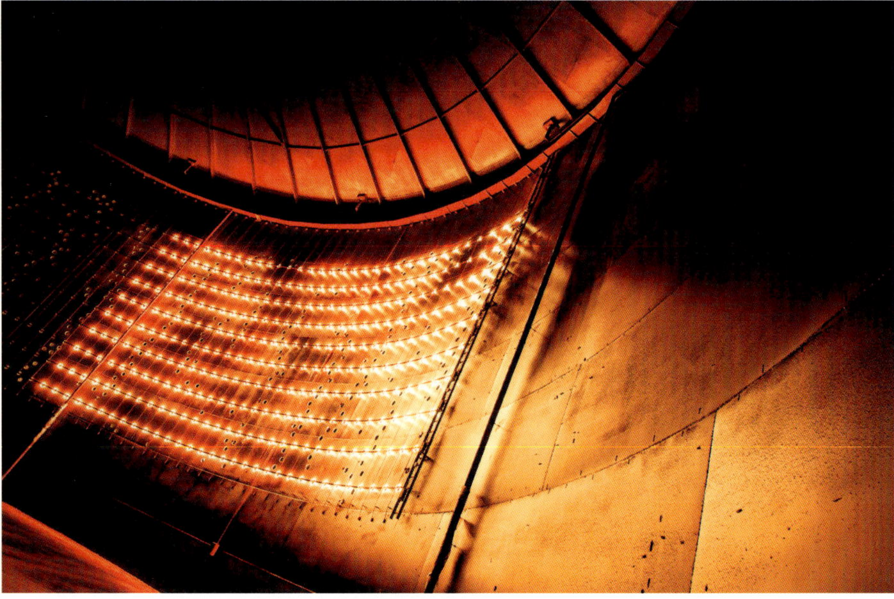

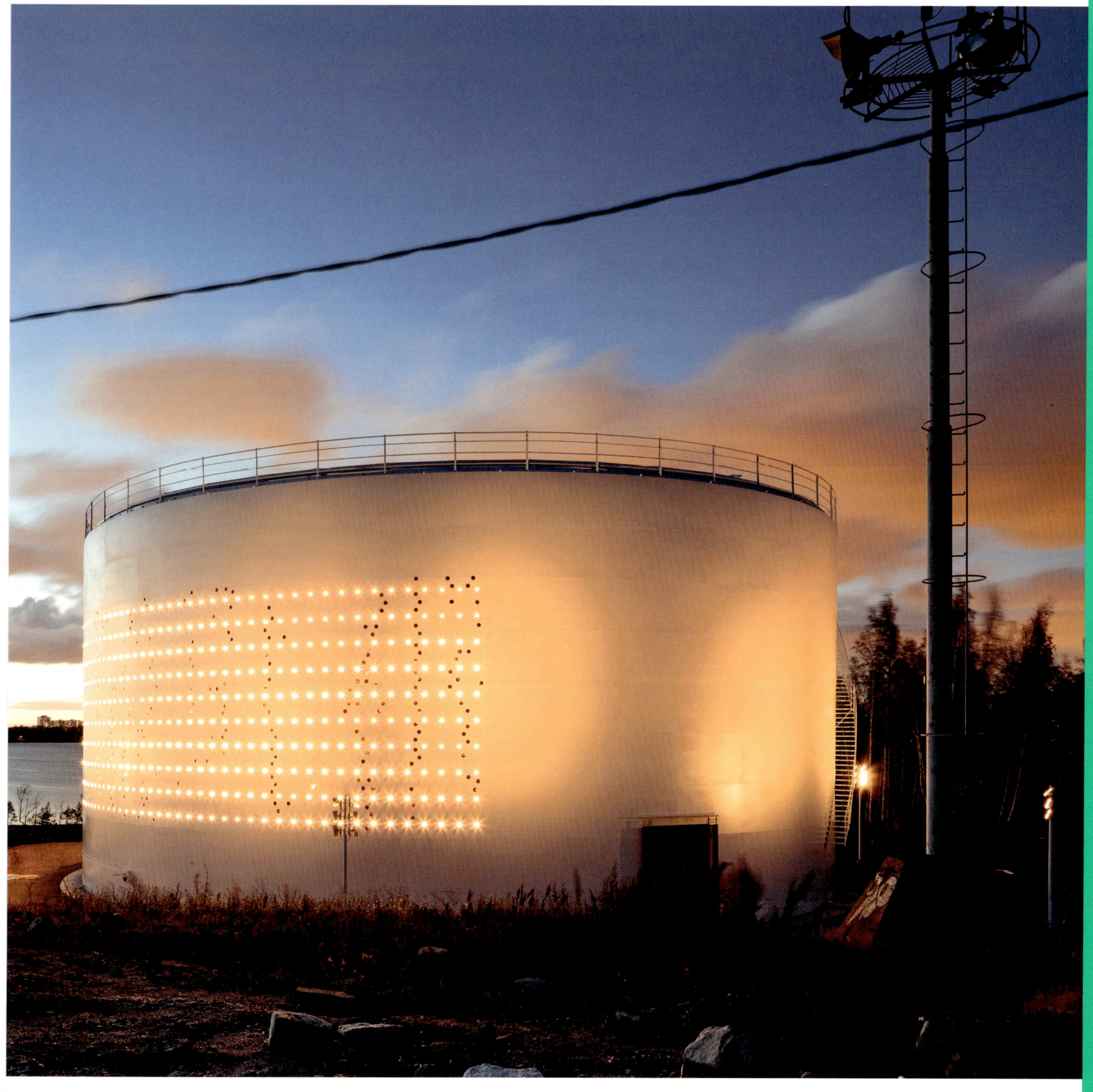

THE WARM WHITE LED GRID REFLECTS LIGHT INDIRECTLY VIA THE RUST-RED WALLS WITHIN THE SPACE

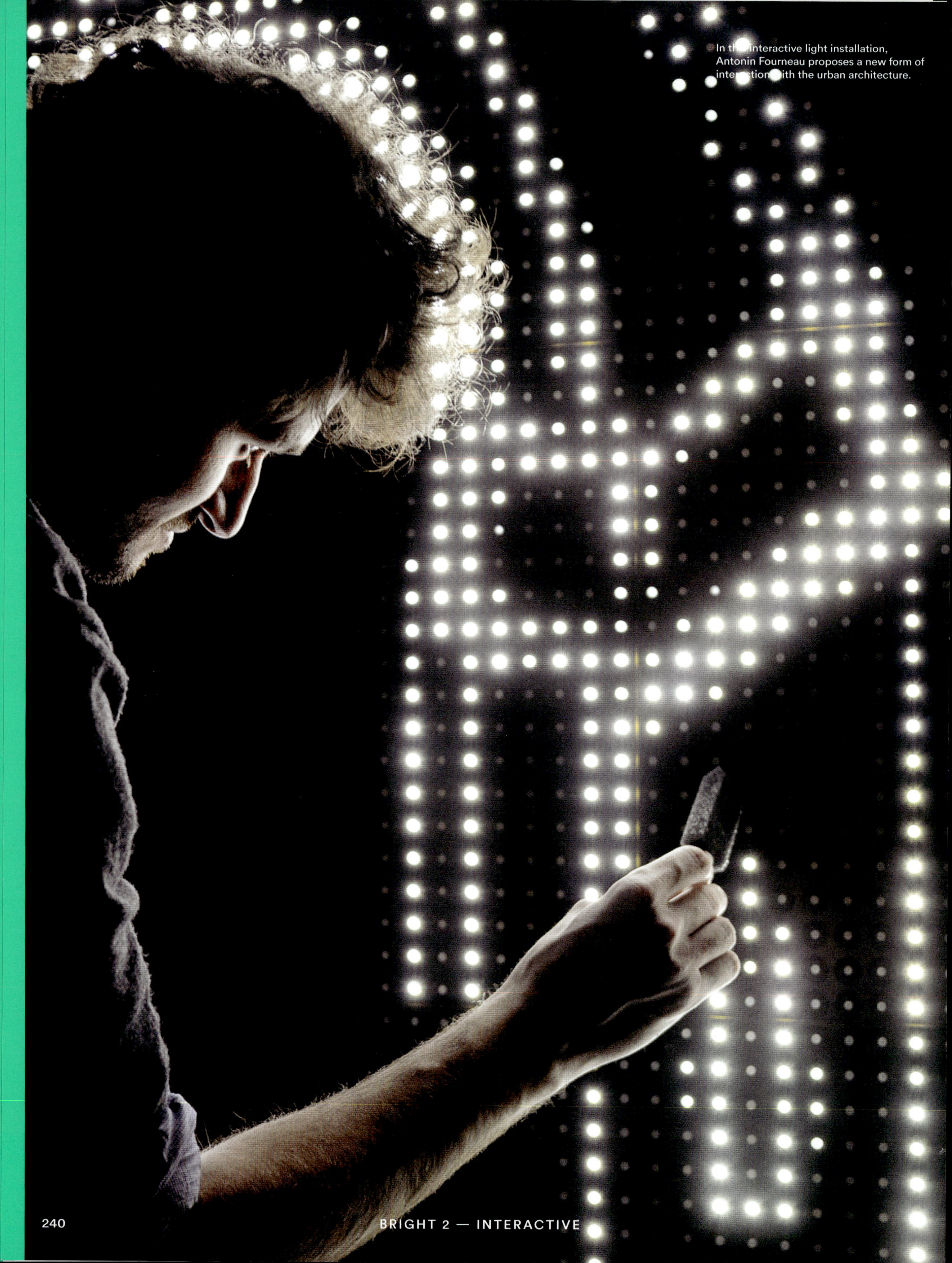

In this interactive light installation, Antonin Fourneau proposes a new form of interaction with the urban architecture.

Water Light Graffiti

Antonin Fourneau conceived and created Water Light Graffiti, an installation that celebrates the matter that we are all made up of and embraces the possibilities that lie within nature.

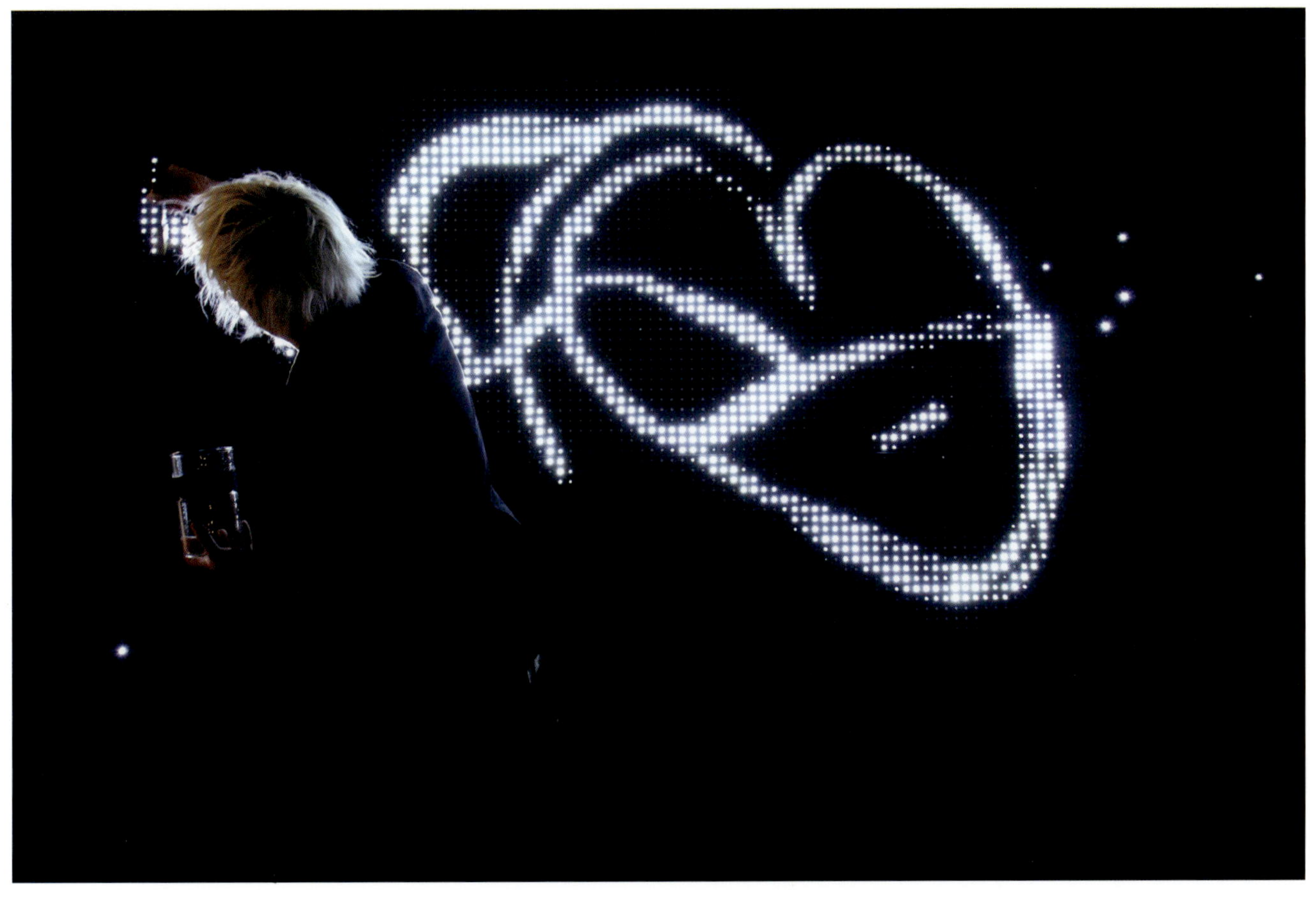

Photos Quentin Chevrie, FotoFilip

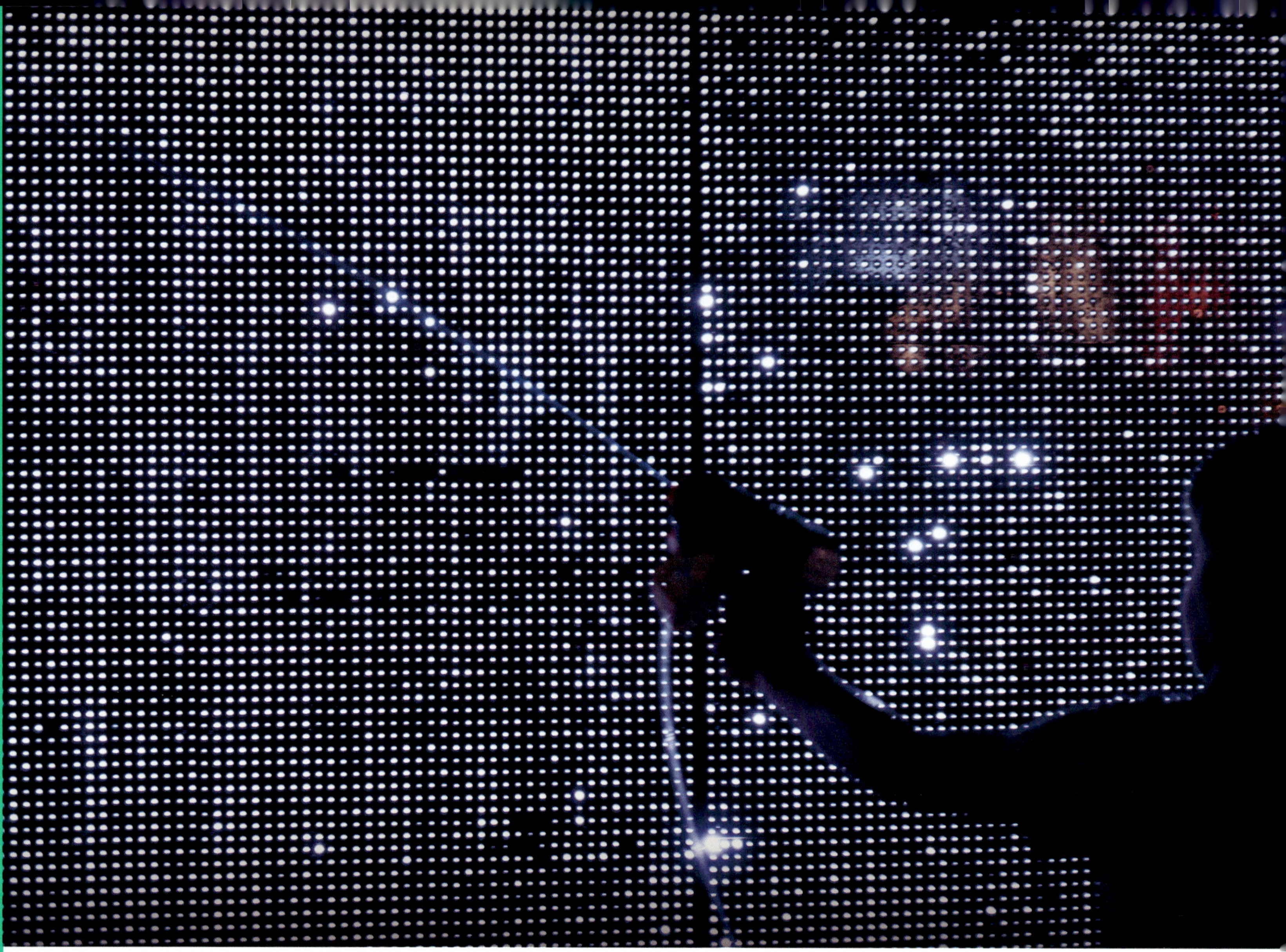

The quantity of water that is deposited around the LED increases the brightness.

CREATING ILLUMINATED GRAFFITI WITH JUST A WATER ATOMISER, SUPPORTING EPHEMERAL AND MORE ECO-FRIENDLY TAGGING

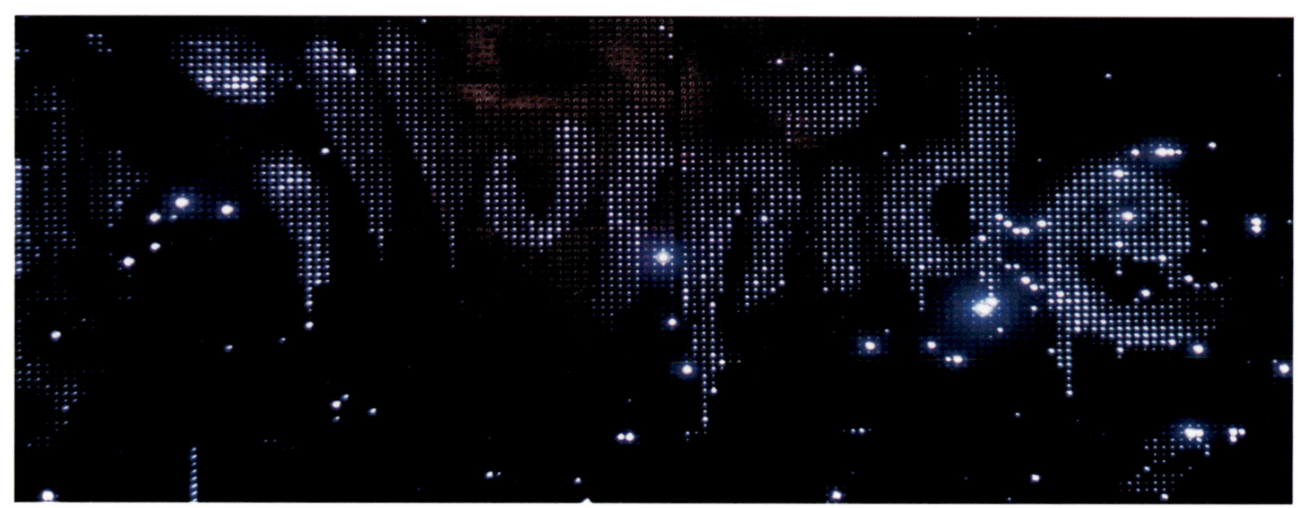

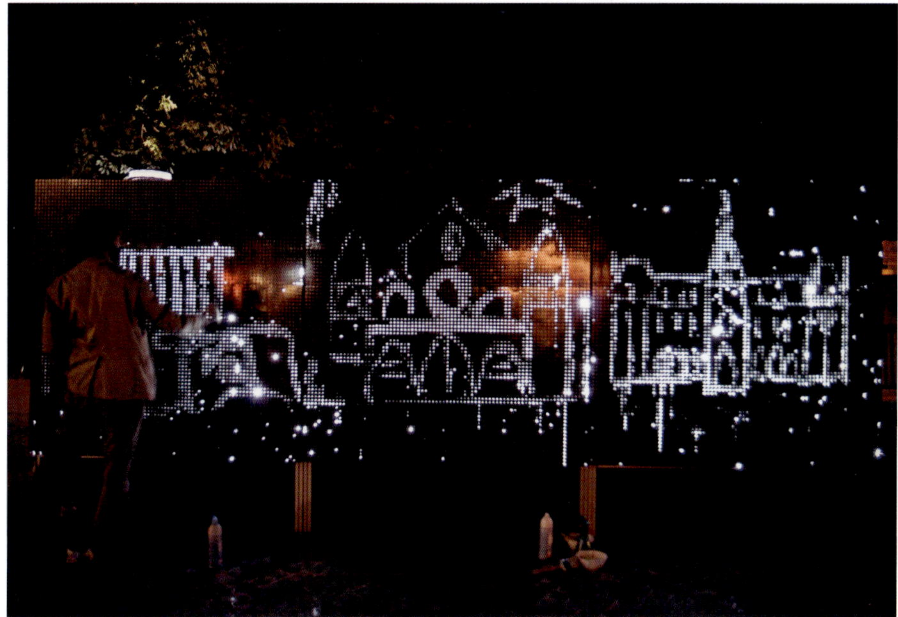

Designer
Antonin Fourneau
Location
Poitiers, France
Client
Poitiers Cultural Services
Collaborator(s)/ consultant
Guillaume Stagnaro, Jordan McRae
Producer
Art2M
Date
July 2012

When a drop of water is trapped in the tracks, it creates a connection that allows the electricity to go through (like a bridge) and lights the closest LED.

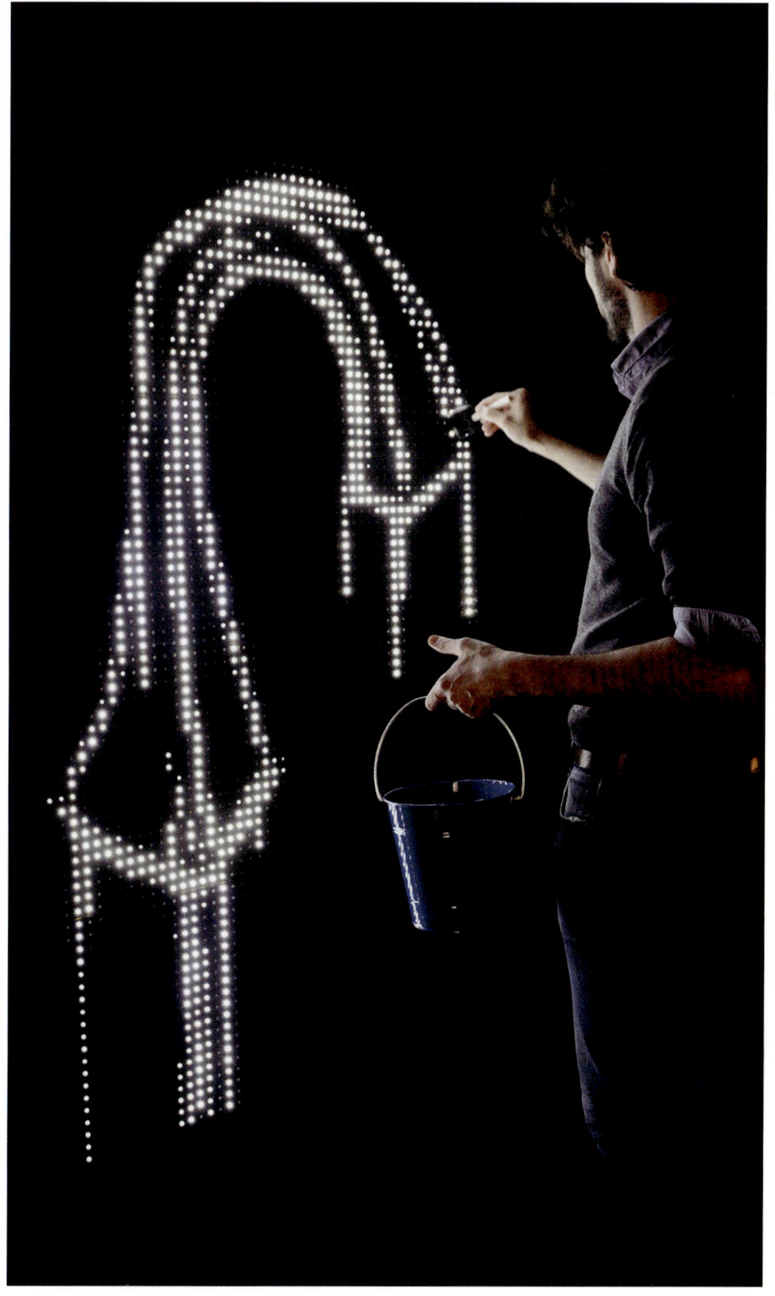

Inspired by the fluidity and possibility of water, Antonin Fourneau conceived and created an illuminated installation that celebrates the matter that we are all made up of and embraces the possibilities that lie within nature. Water Light Graffiti is a project of a material creation composed of several thousand LEDs, illuminated by the contact of water.

By touching the edges of an LED, the water creates an electrical bridge and allows current to pass through the circuit, which provides power to the LED below the surface. The quantity of water that is deposited around the LED increases the brightness: the wetter it is, the brighter the installation becomes. Based on this simple and clever idea, the aim of Water Light Graffiti is to propose a new smart material to draw or write ephemeral light messages. This project allows users to create graffiti with just a water atomiser. The entire ethos of the project supports ephemeral and more eco-friendly tagging or sketching – a non-permanent marker on our environments. Using water, which has no form and no colour, to draw light is completely magical for the public, and the older users even return to youth with this process.

This is a proposal for a new form of interaction with the urban architecture, by offering multiple possibilities. By combining nature and technology, users can play and experiment with the climate or the evaporation speed for instance. After experiencing the installation, people ask the designer all the time, how does it work? He doesn't give away his secrets though – instead he just relishes in the public's reaction, as he comments, 'I am always astonished by the public's first reaction, I love to see the expression in their faces as they watch how the water triggers the LED lights. It's very interesting to see how people are completely mesmerised, whatever their age, and how it is a pleasure for them to be part of this interactive art installation.' —

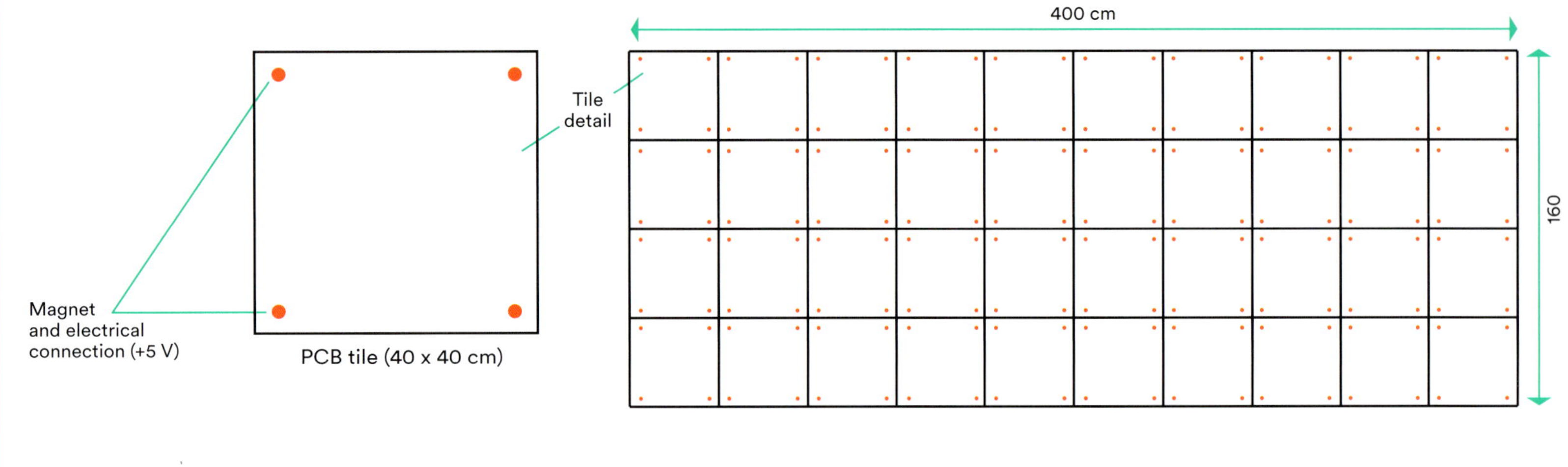

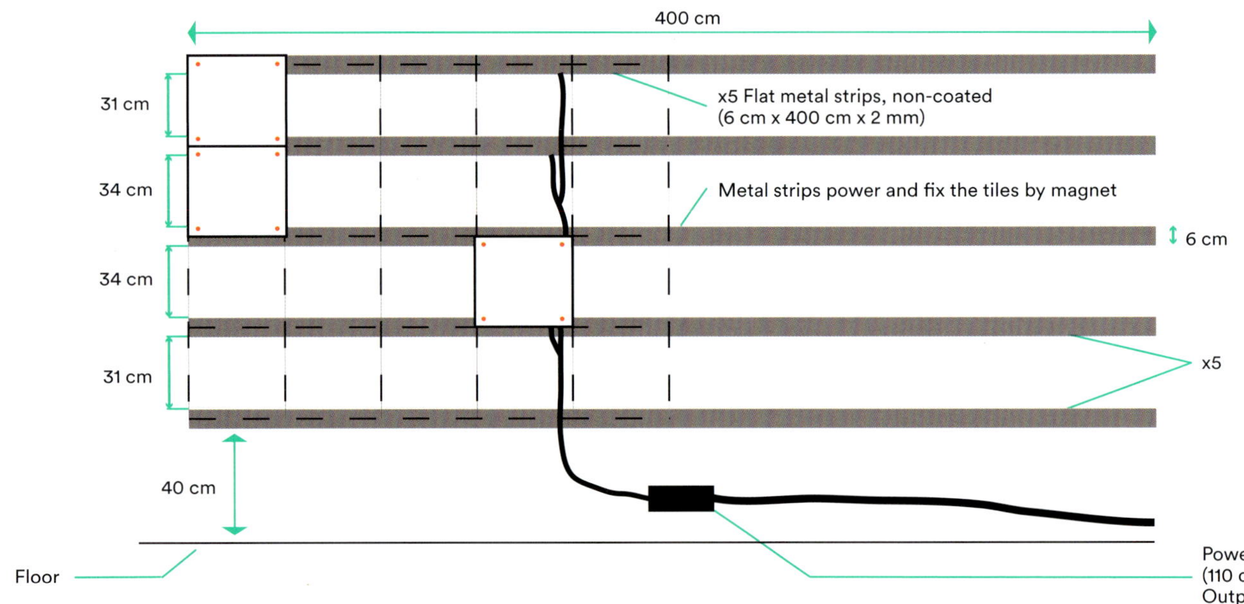

Set-up

Total surface area of the installation for illumination was approximately 40 m² (each panel is 40 x 40 cm, with a total of 260 panels in this particular set-up).
Component parts included mixed media, LEDs, waterproofed PCB (printed circuit board), magnets, electrical hardware and metal framework.
Tiled framework was assembled to create the installation in a modular form. Each tile is 0.40 x 0.40 m, and has 400 white LEDS incorporated. There are a total of 40 panels for a 4 x 1.6 m wall, which are affixed utilising a magnetised system.
Used light was with white LEDs that can be used in daylight and are strong enough to produce the same effect during the night. Power supply was 5 V/50 amp for every 40 tiles. The installation does not consume a vast amount of electricity as the wall is rarely fully lit during events; most of the time it averages 30 amp per show.
Use of water on the wall is made possible as each LED is surrounded by a flower-shaped double track. This shape increases the number of possible contacts with water. When a drop of water is trapped in those tracks, it creates a connection that allows the electricity to go through (like a bridge) and lights the closest LED.
Lightweight materials and a modular-construction were a vital part of the installation, as it needed to be easy to transport across the globe; a wall of 4 x 1.6 m can fit in three flight cases (50 x 50 x 35 cm) weighing 30 kg each.
Construction time has been drastically reduced since the very first prototype and now the set-up of an average-sized wall is just 4 hours.

WATER CREATES AN ELECTRICAL BRIDGE AND ALLOWS CURRENT TO PASS THROUGH THE CIRCUIT, WHICH ILLUMINATES THE LED

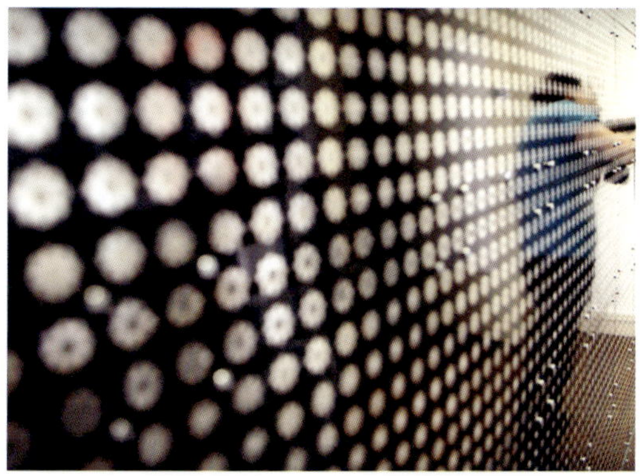

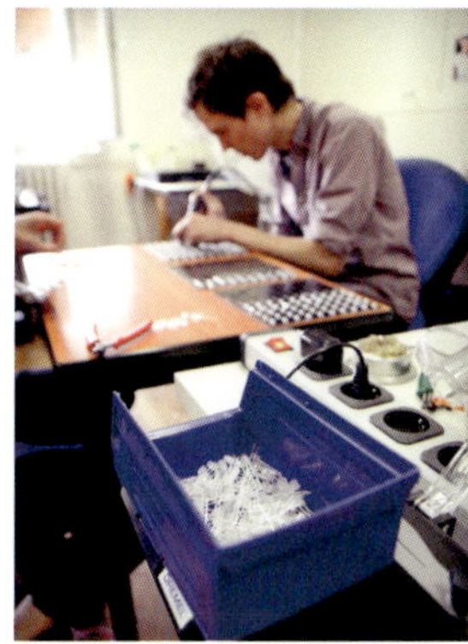

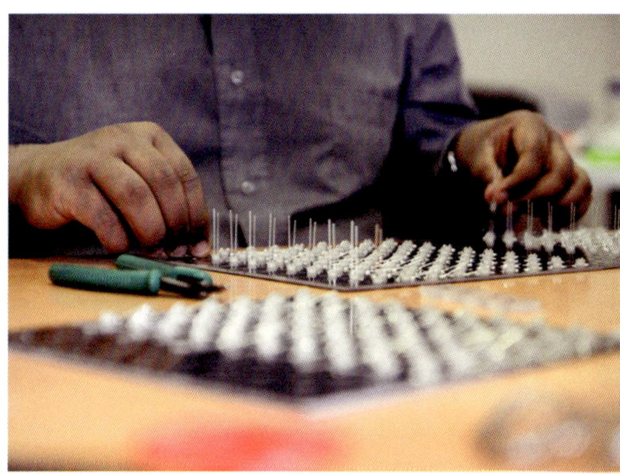

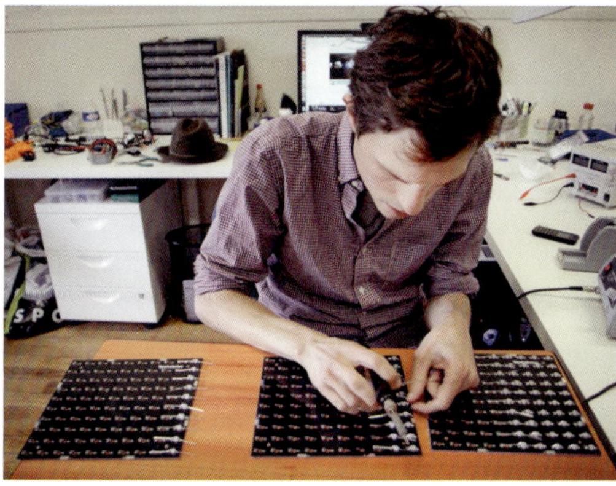

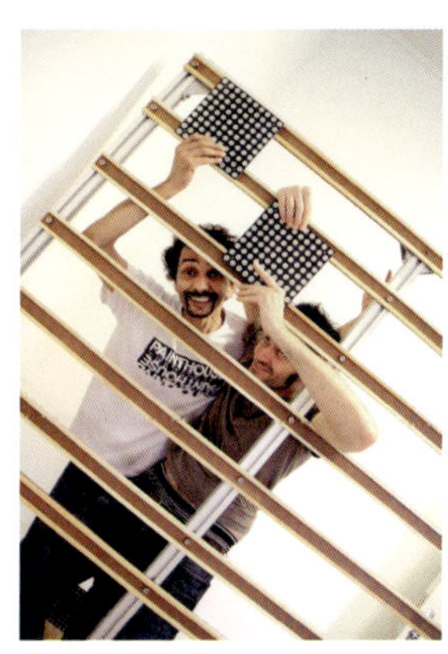

Each LED is surrounded by a flower-shaped double track, which increases the number of possible contacts with water.

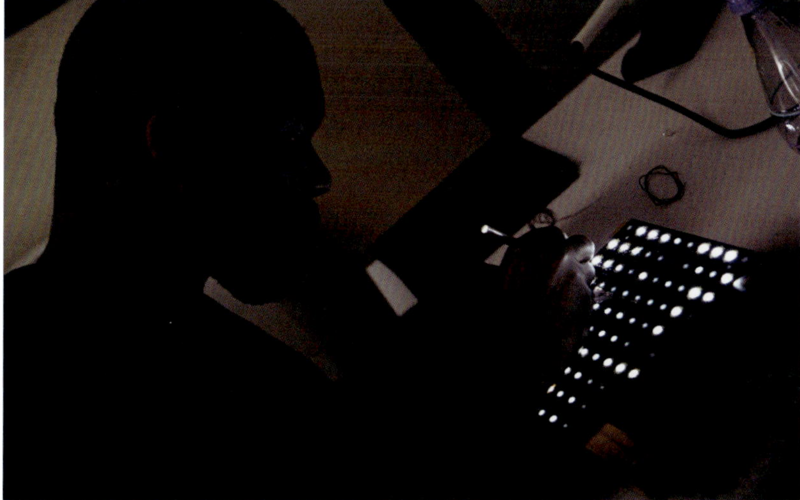

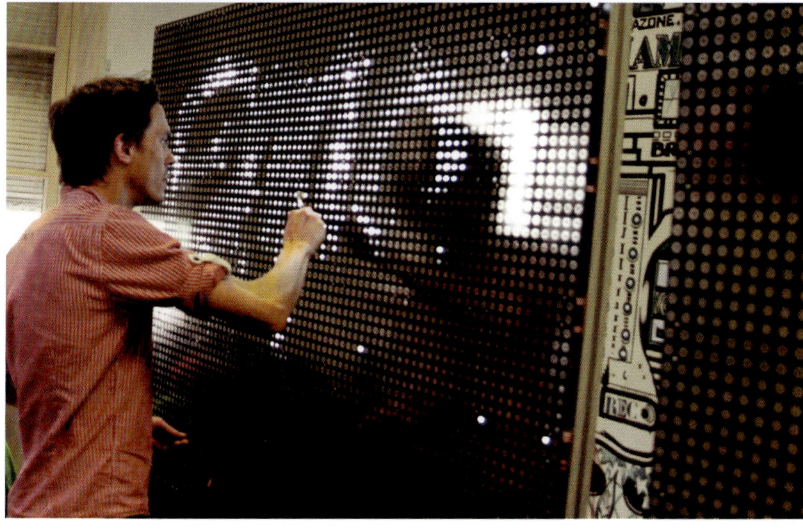

Specifics

Water Light Graffiti's patented system was first presented to the public in Poitiers, France in 2012 for a local art event. The Cultural Services of the city of Poitiers have been a great support for Antonin Fourneau. The installation has since appeared at numerous events.

The very first tests and prototypes by Antonin Fourneau were handmade and would measure 5 to 10 cm. The scale of the final project was also a challenge in itself: to have a project that is inspired by graffiti meant having a project that could grow to the size that graffiti works usually occupy, i.e. the actual size of a wall in a city.

Art2M supported the project during the prototyping phase by providing space and equipment. Thanks to the efforts of the team, the time of production in the manufacture of the light panels was drastically reduced. The manual production in the workshop began by taking up to 4 hours per tile to place all the components and has now progressed to industrial-scale production, where all components are placed by machines in the same timescale for a 4 m^2 wall.

Development of the product has allowed the use of materials that have increased quality. The team works with French companies who have a great understanding of this particular project.

An important requirement was the installation needed to be easily-transportable. The first version was not so easy to carry around, as the entire wall was one piece; the tiles were smaller (20 x 20 cm) and screwed to a wood structure. Imagining a new system that would allow the system to be built and dismantled easily and to put all the tiles in flight cases was a great challenge. The solution was a magnetised system: each tile is now fixed on metal strips with four magnets. The strips are powered with a power supply and work as a structure as well. This system also allows the installation to be very flat on a wall: properly installed, it is only 2-cm thick.

Another challenge was to make a material that is able to remain outdoors for a very long period and withstand the elements. Future versions of this installation will involve a system that can be built into concrete walls. Further developments include a prototyped 'coloured-light' installation, a version that is reactive to light, and a video game platform using a water gun which is now in the development stages.

Ultimately, the biggest challenge often is to defend simple ideas.

Photo Manuel Braun

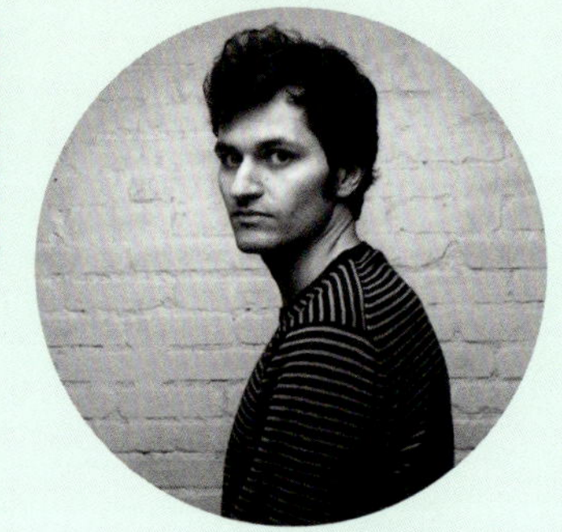

Antonin Fourneau

Antonin Fourneau is an artist or designer or inventor – and usually, all three all at once. An electronic enthusiast, he likes to play Dr Frankenstein with all these technologies. He has a portfolio of projects that has at its source one aspect: interaction. Working with technology and developing it in new ways to inspire audience participation is the main objective for the installations of his work, which can be found around the globe, including France, Abu Dhabi, the Netherlands, the United Kingdom and the United States. In collaboration with agency Art2M, Water Light Graffiti has encountered great success.

ATOCORP.FREE.FR
WATERLIGHTGRAFFITI.COM

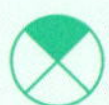

Sponge Game

When **2011**
Where **Lille, France**
Client **Kids Festival 'Fête Animation'**

This installation called for people to interact in a unique way – with a wet sponge – during a game that is normally played alone using a Nintendo Game controller. The sponge acted as a device that uses water to trigger an action in the Tetris game. Each person is a function of the game and the group must organise and collaborate to create a kind-of performance between people who do not know together, but who are in a proximity to have contact.

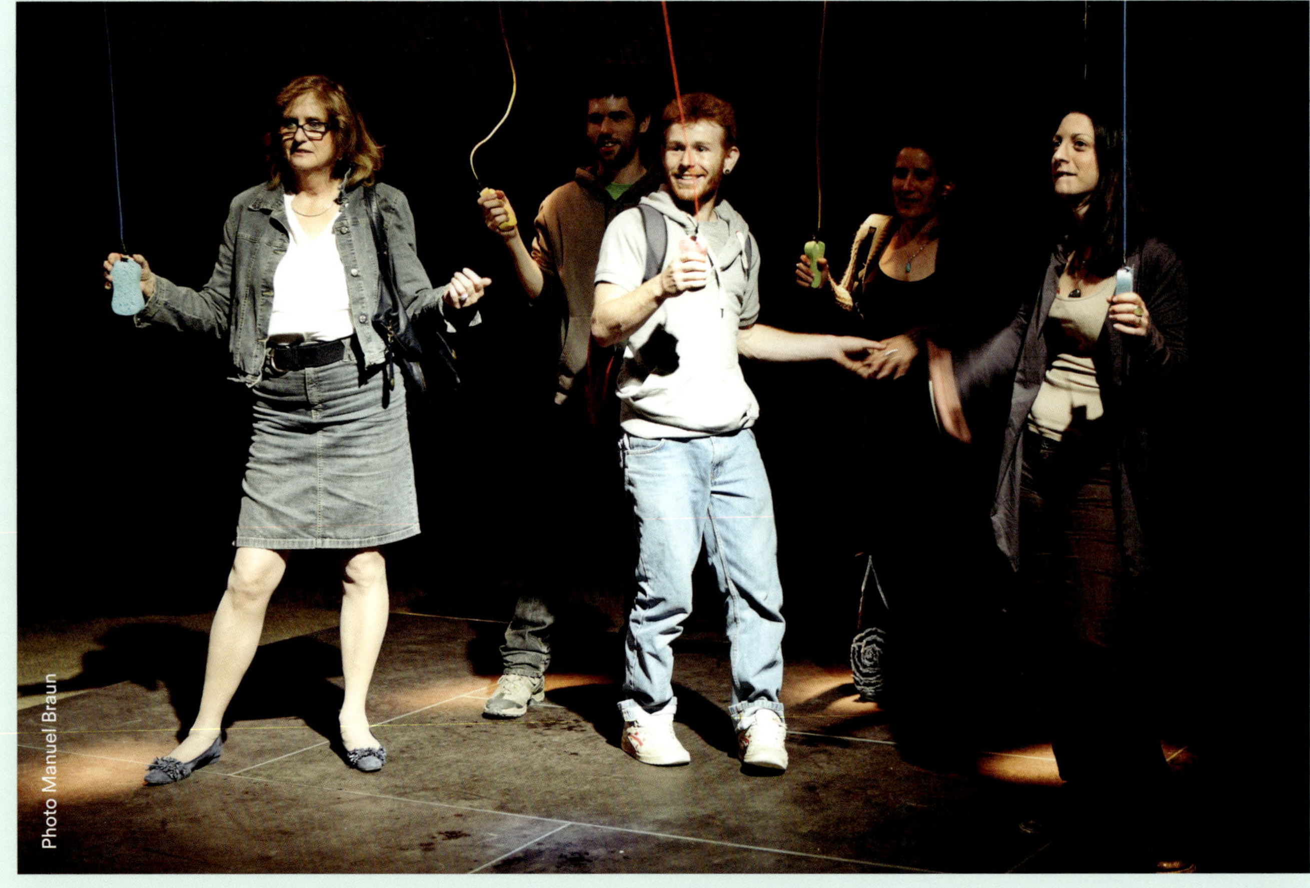

Photo Manuel Braun

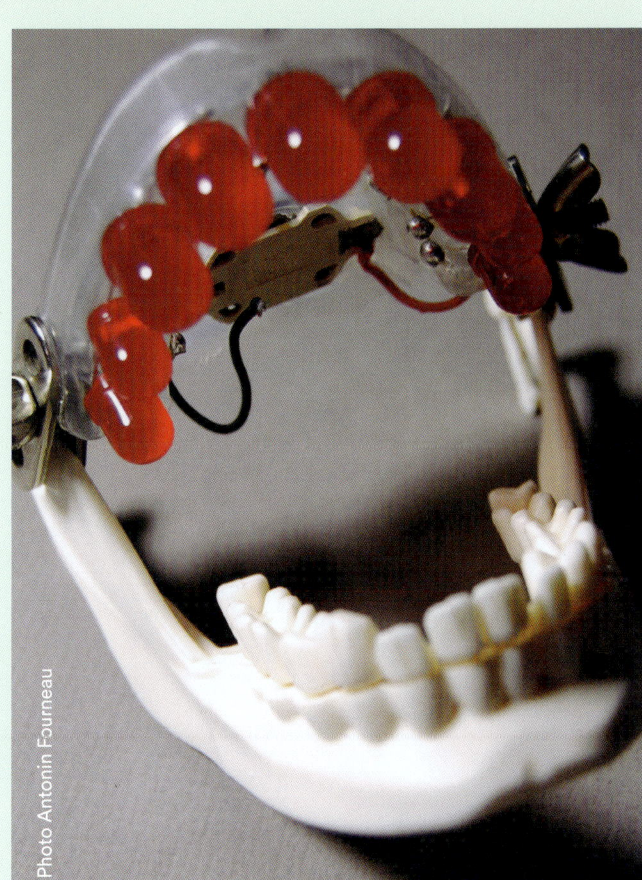
Photo Antonin Fourneau

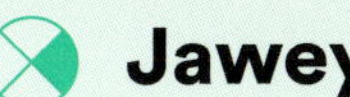

Jawey

When **2009**
Where **Paris, France**
Client **Gallery Duplex**

Jawey is a prosthesis for communicating by way of a light message. Inspired by Joey (the mechanical boy) of the book *The Empty Fortress* and the French rap star with the same name, this project plays the encounter between two worlds. The jaw looks very much like a dental prosthesis, but with the teeth replaced by LEDs that are triggered by water (saliva). Constructed of metal, silicone and resin, every time the LEDs come into contact with the tongue, they are illuminated.

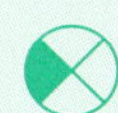

Eniarof

When **2013**
Where **Aix en Provence, France**
Client **Marseille-Provence 2013, European Capital of Culture**

Eniarof revisits the fun fairs and carnivals of the past century, drawing from old and emerging popular culture. Playing with the imagination, this innovative installation creates new forms of interaction for the public with friendly curioso and artistic creations that include karaoke, video games, art installations and performances. Eniarof is also an opportunity to develop new projects with different workshops organised before each event, with the carnival being the culmination of weeks of intense preparation.

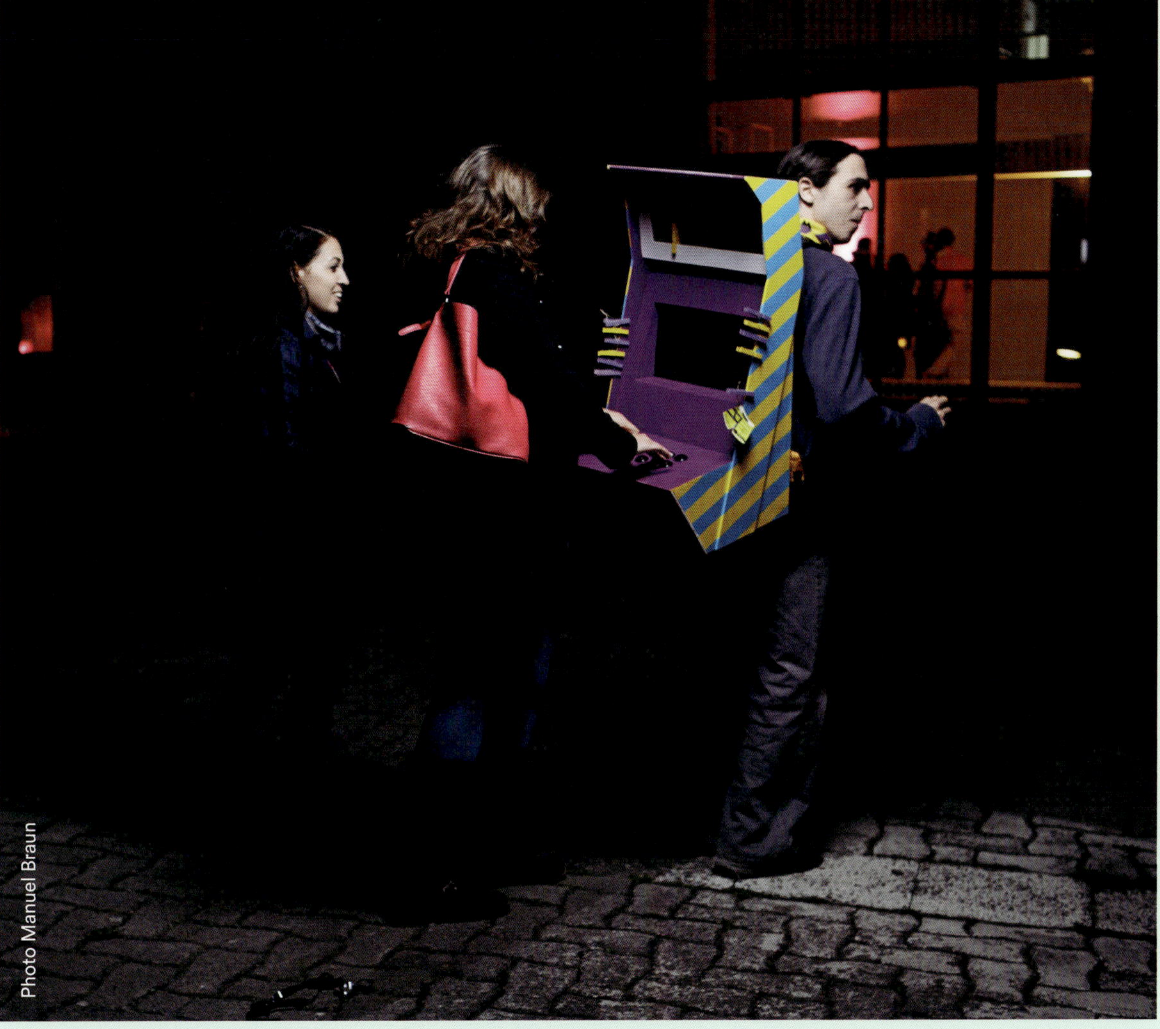
Photo Manuel Braun

Static

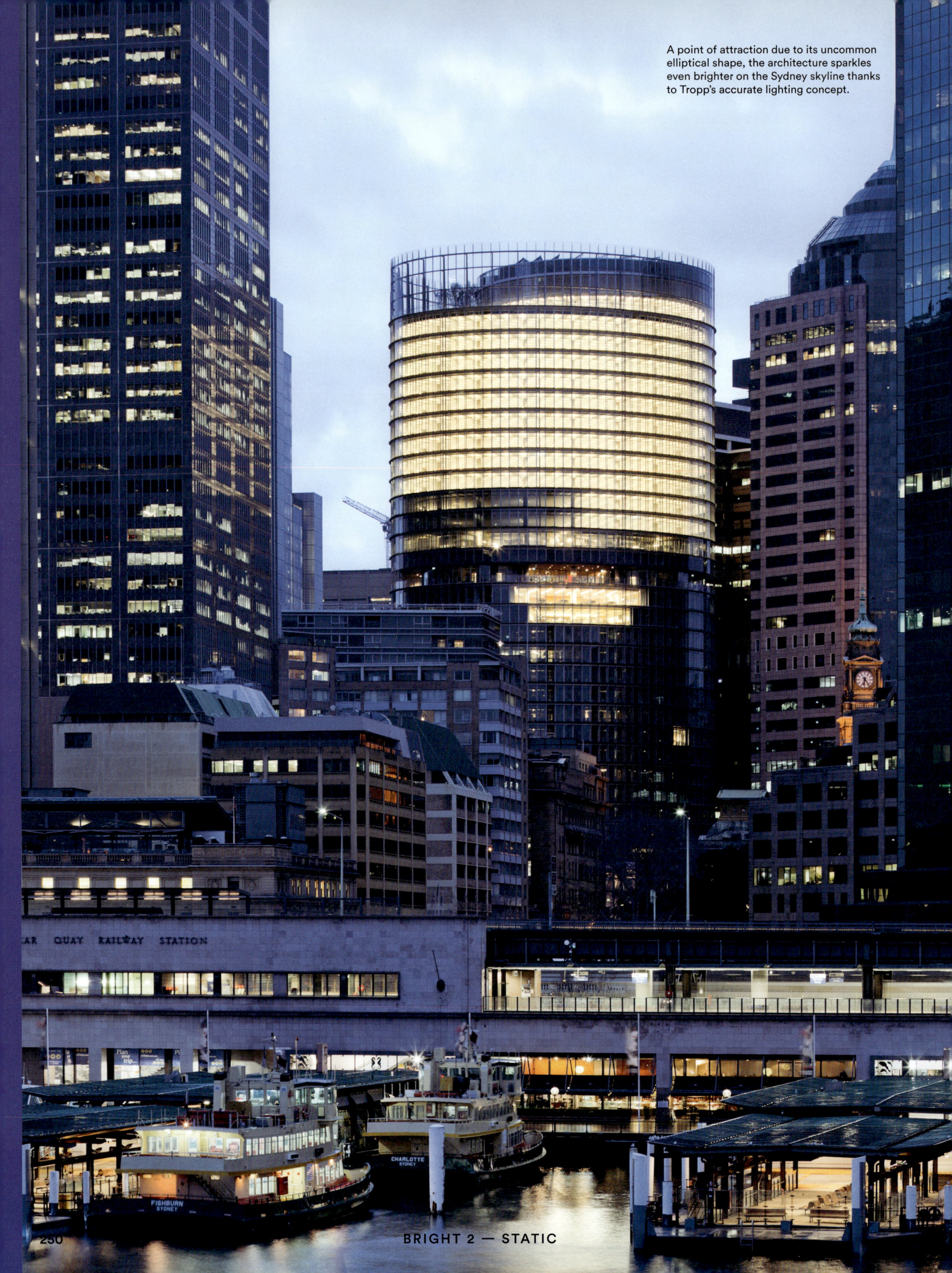

A point of attraction due to its uncommon elliptical shape, the architecture sparkles even brighter on the Sydney skyline thanks to Tropp's accurate lighting concept.

1 Bligh Sydney

When commissioned to illuminate an office tower in Sydney, Australia, **Tropp Lighting Design** opted for an approach to enhance the exceptional architecture. Under the hours of darkness, the lighting concept emphasises the unique characteristics of the building, in particular its elliptical contours, whilst ensuring it is not overly intrusive.

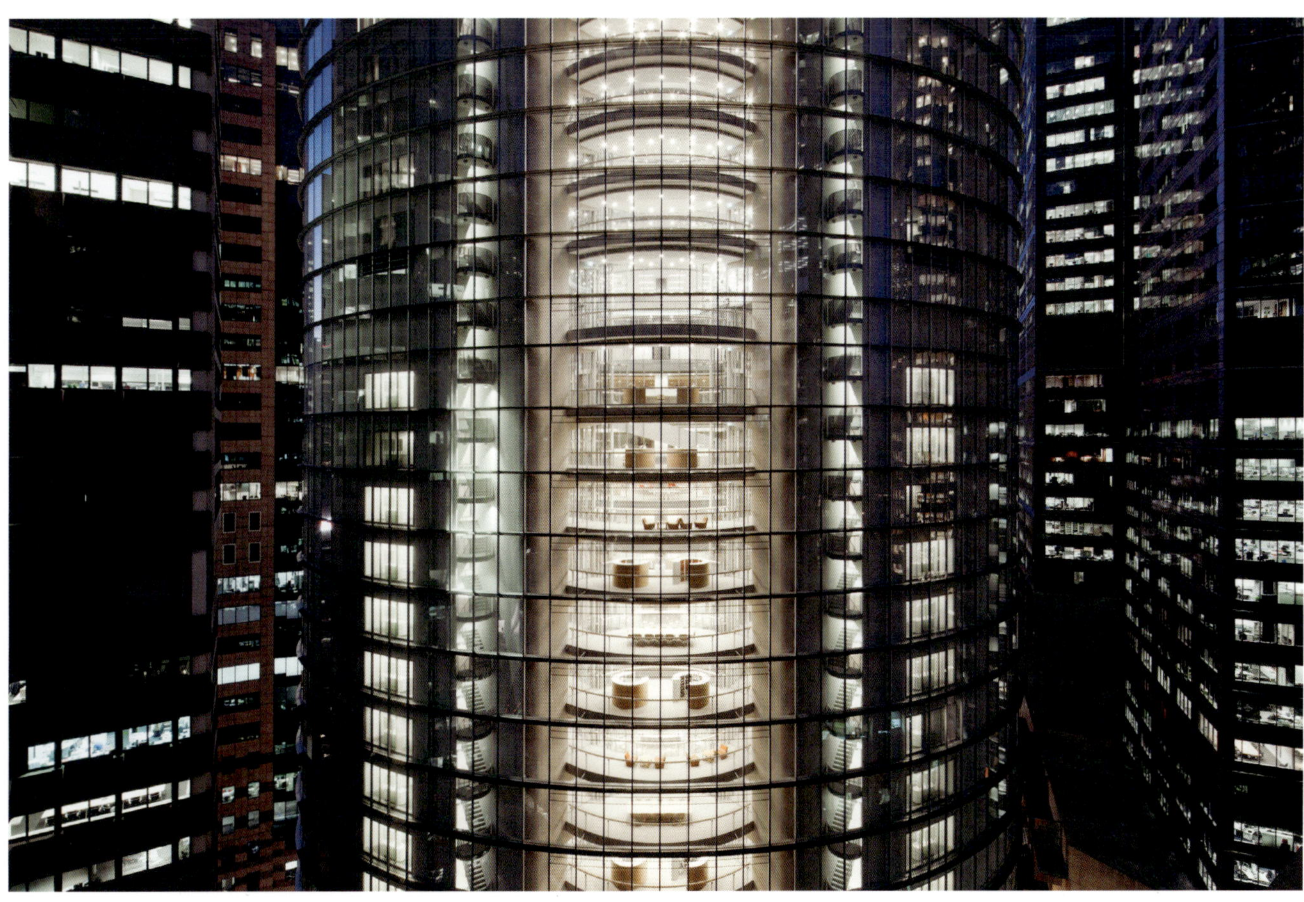

Photos HG Esch

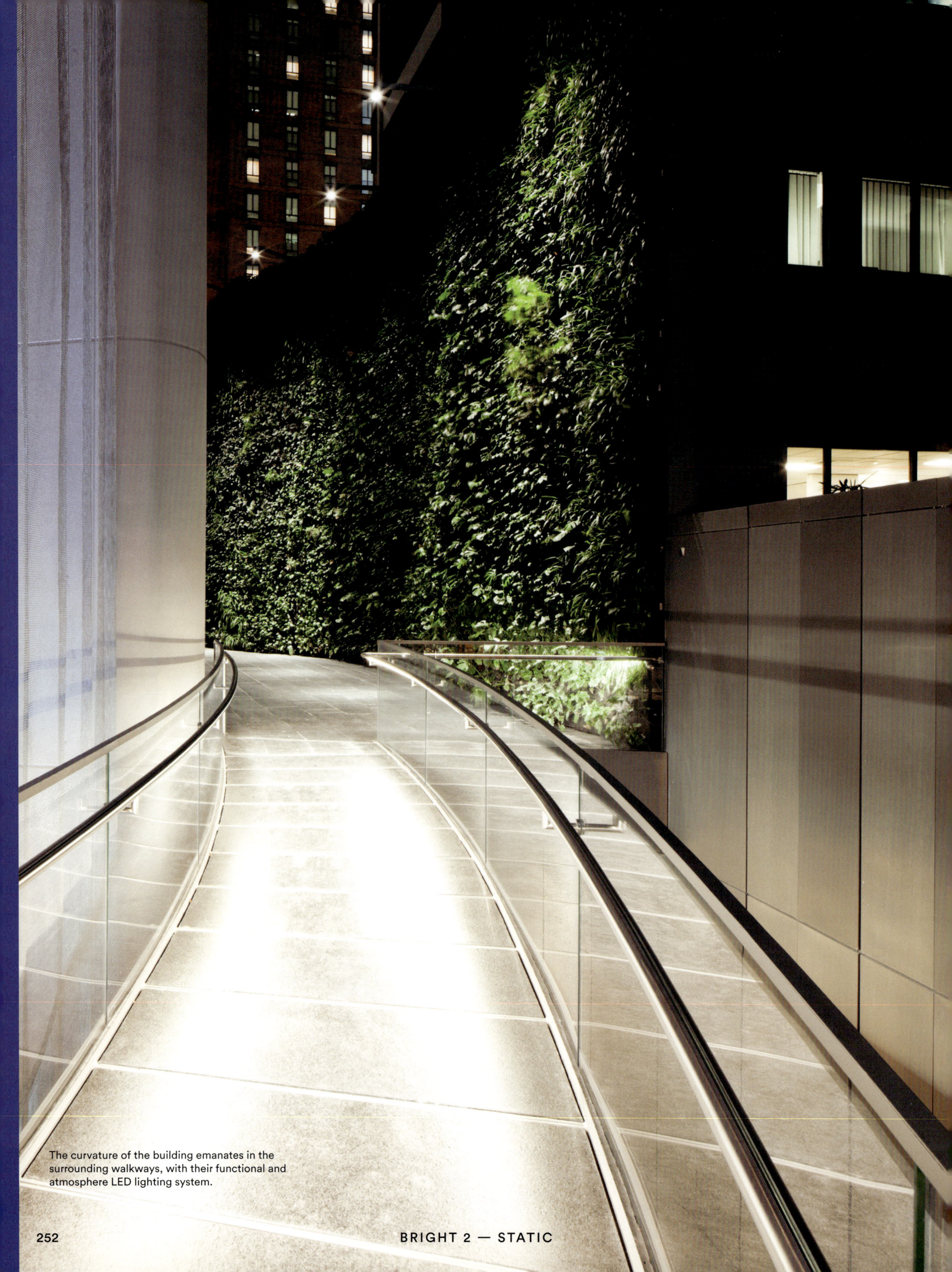

The curvature of the building emanates in the surrounding walkways, with their functional and atmosphere LED lighting system.

The lighting of the cascading plant wall provides the plaza with a fresh, luminous air.

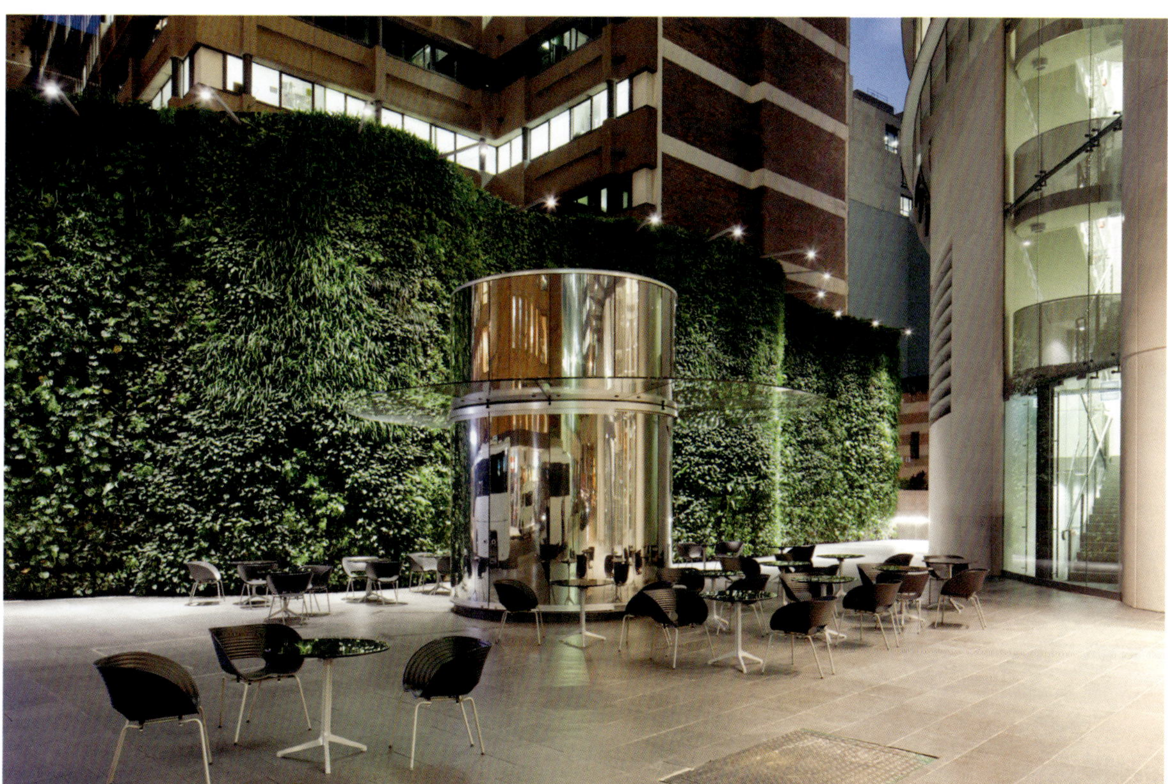

Designer
Tropp Lighting Design
Location
Sydney, Australia
Client
Dexus Property Group
Collaborator(s)/consultant
ingenhoven architects, Architectus, DS-Plan, Arup, Enstruct
Manufacturers
Various
Date
October 2011

A sparkling addition to the Sydney skyline was the 1 Bligh office tower, which opened its doors in the autumn of 2011. Rising to a height of 139 m, overlooking the city's picturesque harbour, the high-rise complex stands out as a solitaire at night-time in the midst of its rather cluttered surroundings. This is thanks to a lighting concept by the German studio Tropp Lighting Design, which accurately reinforces the architectural aspects of the building (designed by ingenhoven architects and Architectus).

The lighting concept aimed at a homogenously-illuminated facade, emphasising the unusual elliptical shape of the construction – ensuring it attracts attention, yet is not overly intrusive. This is realised by utilising suspended luminaires, which are alternated with perimeter lighting at the envelope of the glazed atrium that climbs the full height of the building. Colour temperature, light distribution and intensity of all lamps across the building's exterior is constant, creating a uniform appearance. Studio founder Clemens Tropp takes advantage of this consistency to highlight focal points by interrupting the building's 'coherent inner glow'. For instance, the transfer floor at mid-height is lit to appear as a break between the uniform office areas, as is the top floor and roof terrace.

The architects integrated a public plaza at street level, turning 1 Bligh into a natural point of convergence, a hub of activity and interaction. Therefore, the Tropp team needed to ensure that the entrance lighting was carefully designed to create a welcoming atmosphere, allowing visitors a smooth transition of illumination from exterior to interior, where they would encounter the impressive glass atrium. The plaza's open-air cafes have a cascading planted feature wall, illuminated to enhance the intense colour of the greenery and provide a fresh, luminous effect through an interplay between light and shadow. Moreover, discreet LED lighting fixtures were integrated into the handrail that lines the walkway, which acts as functional as well as atmospheric lighting. —

COLOUR TEMPERATURE, LIGHT DISTRIBUTION AND INTENSITY CREATE A UNIFORM APPEARANCE

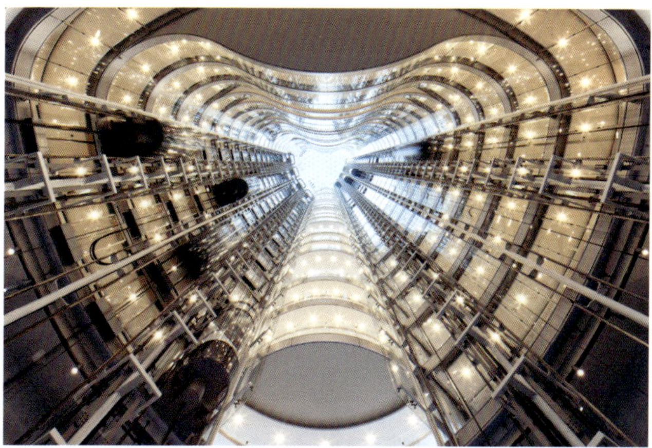

Tropp Lighting Design was responsible for the special lighting and master plan, with Arup Electrical in charge of the standard lighting.

Photo Frieder Blickle

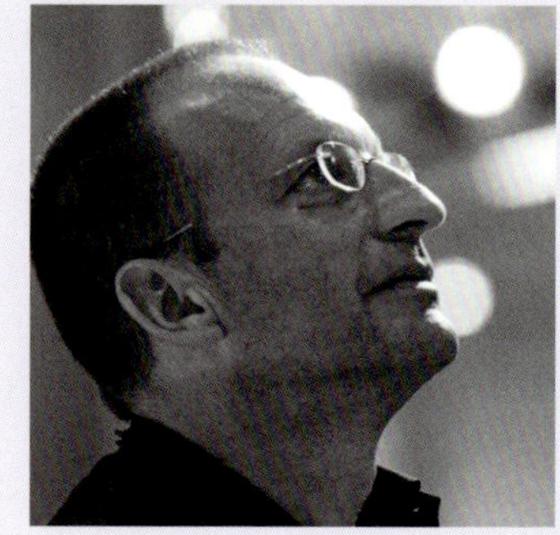

Tropp Lighting Design

German studio Tropp Lighting Design, founded in 2004 by electrical engineer Clemens Tropp, currently has seven staff members, among whom are key designers Daniel Meyer, Yvonne Weiss and Johannes Hoellerl. The team develops lighting design projects for numerous spaces, working on a number of architectural projects, from buildings, infrastructure and temporary constructions to outdoor and urban areas. Always striving to reveal a place's identity through its lighting, the studio balances high design standards with attention to economic, technical and environmental factors.

TROPP-LIGHTING.COM

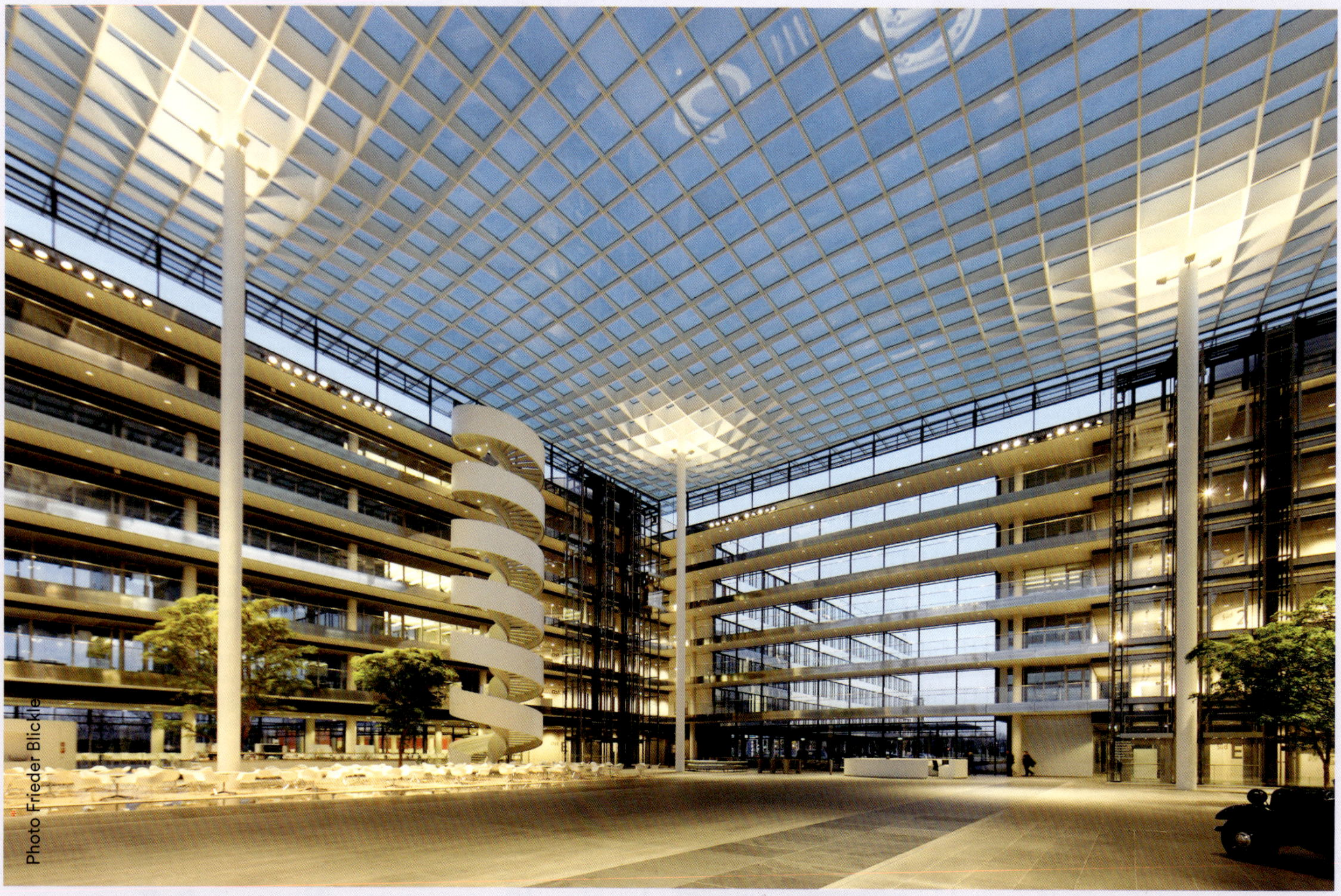

Photo Frieder Blickle

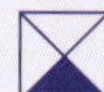

HDI-Gerling

When **2012**
Where **Hanover, Germany**
Client **HDI-Gerling Versicherung**

Sophisticated and aesthetically demanding solutions were developed for the lighting of the central atrium of the headquarters of the HDI-Gerling insurance group in Hanover, Germany. The open, transparent atrium is the heart of the six-storey office building (designed by ingenhoven architects), which offers modern, ecological and flexible workspaces.

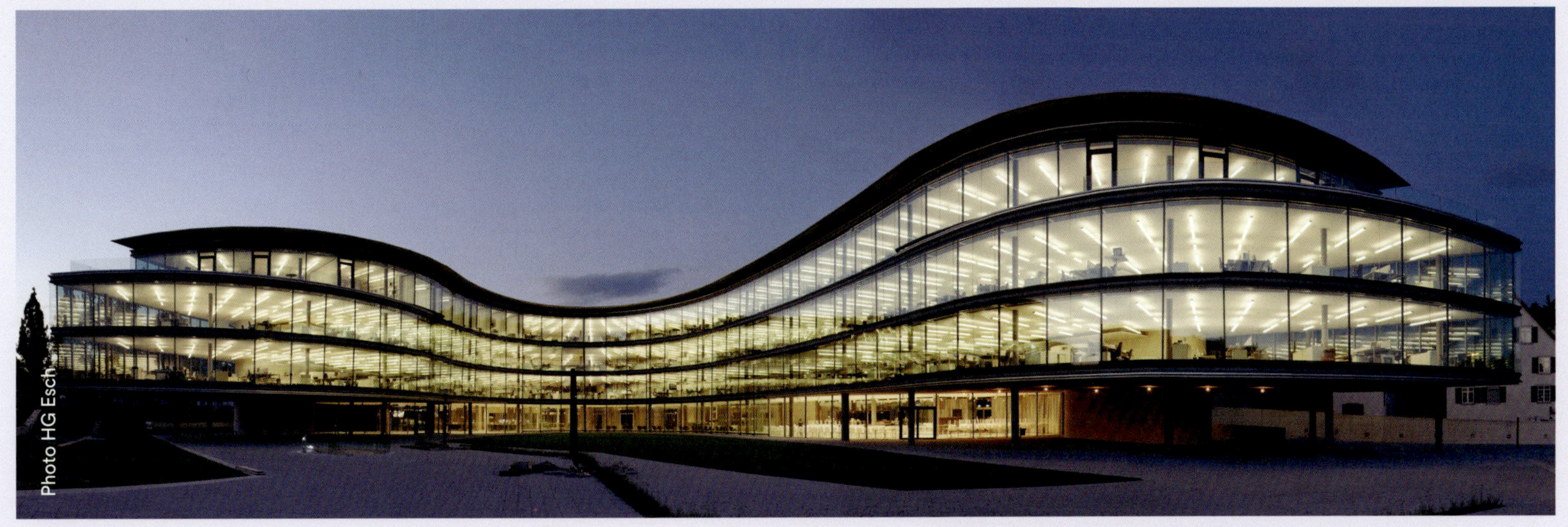
Photo HG Esch

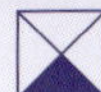

Swarovski Headquarters

When **2011**
Where **Zurich, Switzerland**
Client **Swarovski Immobilien AG**

The lighting of Swarovski Group's headquarters at Lake Zurich was realised by Tropp Lighting Design in 2011. The glass-fronted building (designed by ingenhoven architects) was accurately illuminated so that the architecture's swan-like curved shapes were emphasised – sparkling at night, like the glittering crystals the brand is known for.

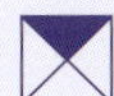

Hotel Lanserhof Tegernsee

When **2014**
Where **Tegernsee, Germany**
Client **Lanserhof Marienstein GmbH**

Tropp Lighting Design developed individual solutions for public areas, private rooms and treatment areas of the Hotel Lanserhof Tegernsee. Based in the midst of the German Alps, the health resort (designed by ingenhoven architects) aims to combine health, entertainment and hospitality by creating rooms that mimic the purity and tranquility of a cloister, which is enhanced by the lighting design.

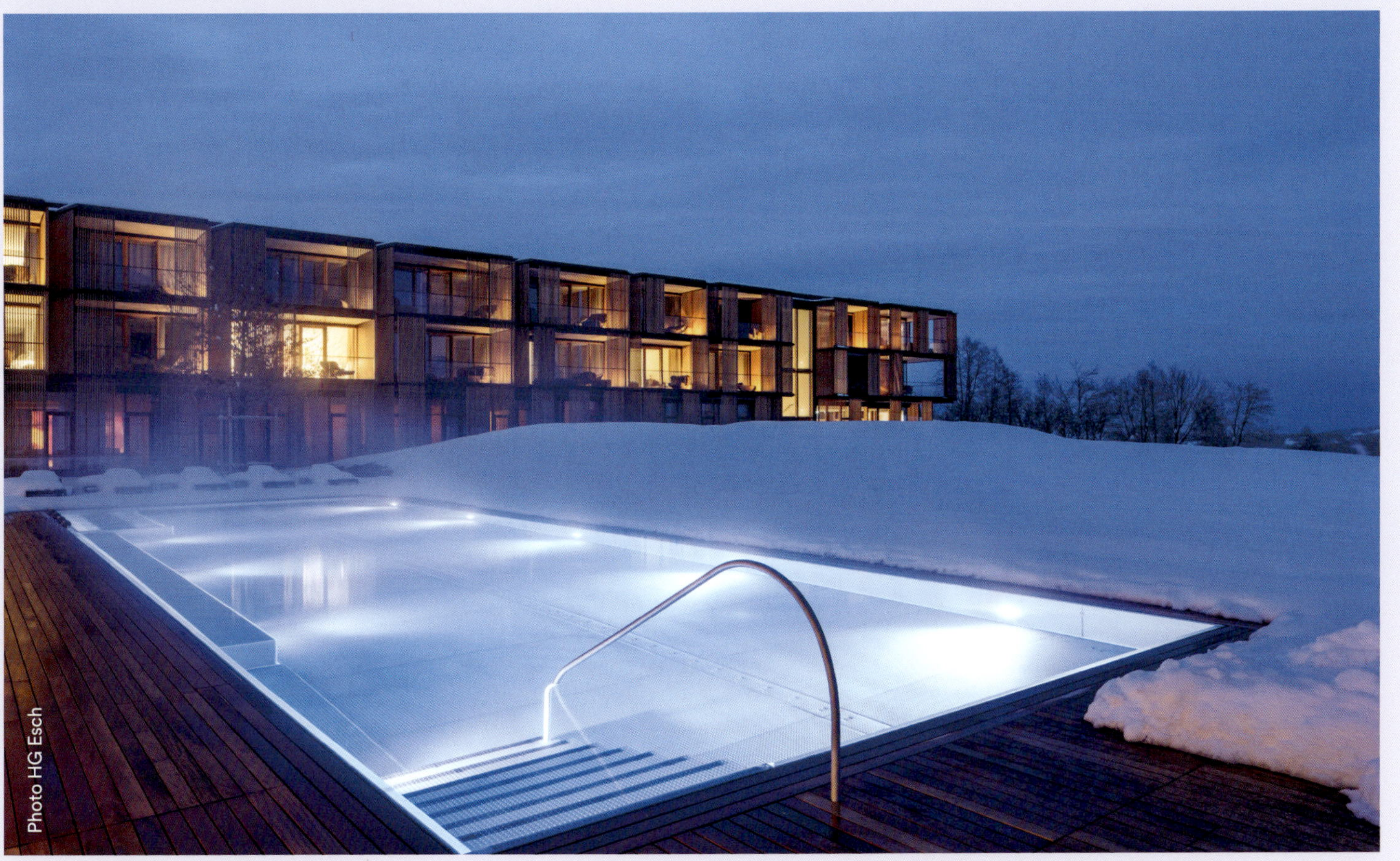
Photo HG Esch

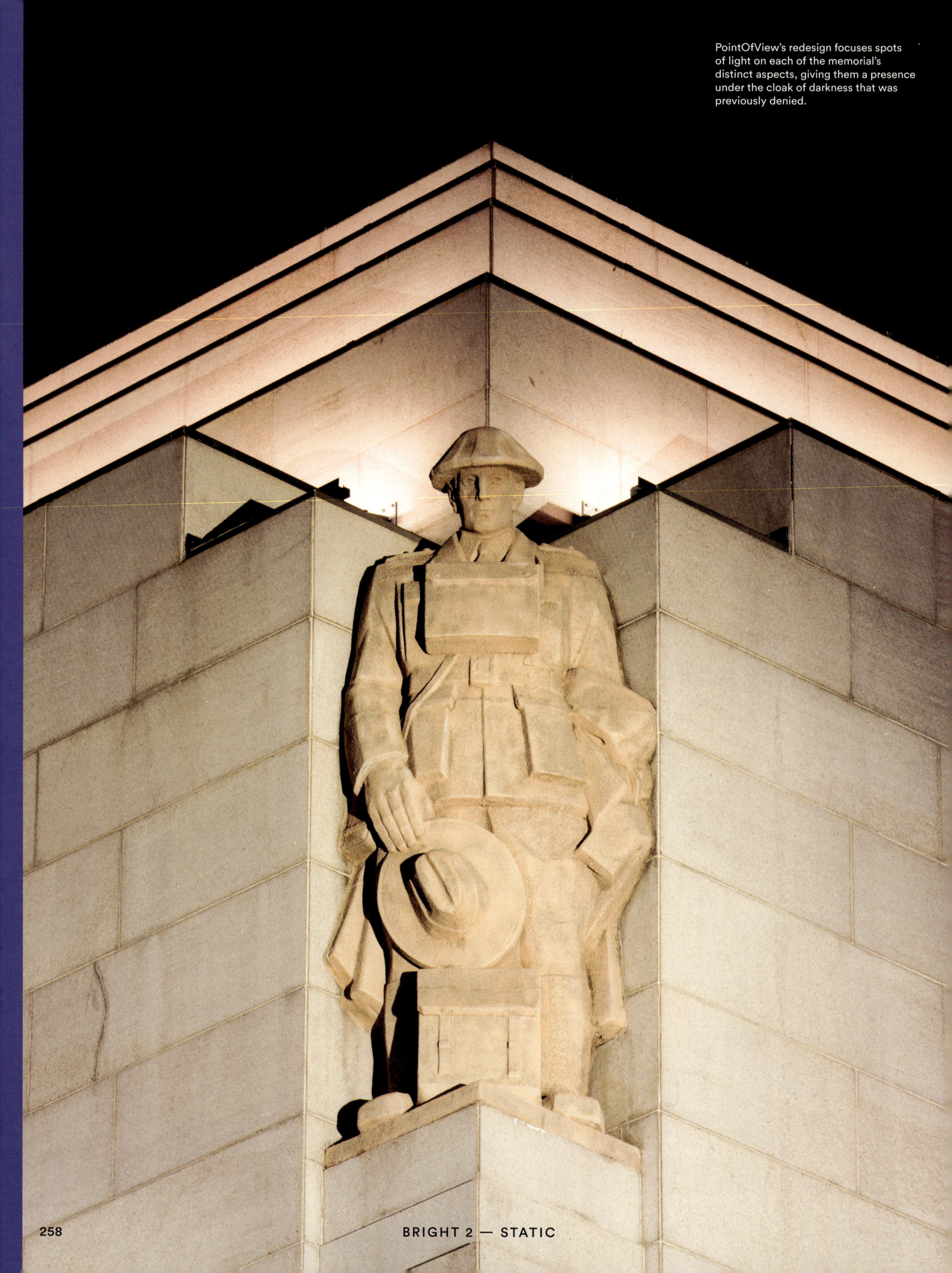

PointOfView's redesign focuses spots of light on each of the memorial's distinct aspects, giving them a presence under the cloak of darkness that was previously denied.

ANZAC Memorial

The lighting of the ANZAC Memorial in Sydney's Hyde Park was given a contemplative redesign by **PointOfView**, the result of a thorough study of the building's history and original design, and a careful adaptation of leading lighting technologies to prolong the presence of the site under the hours of darkness, without taking away from its familiar atmosphere.

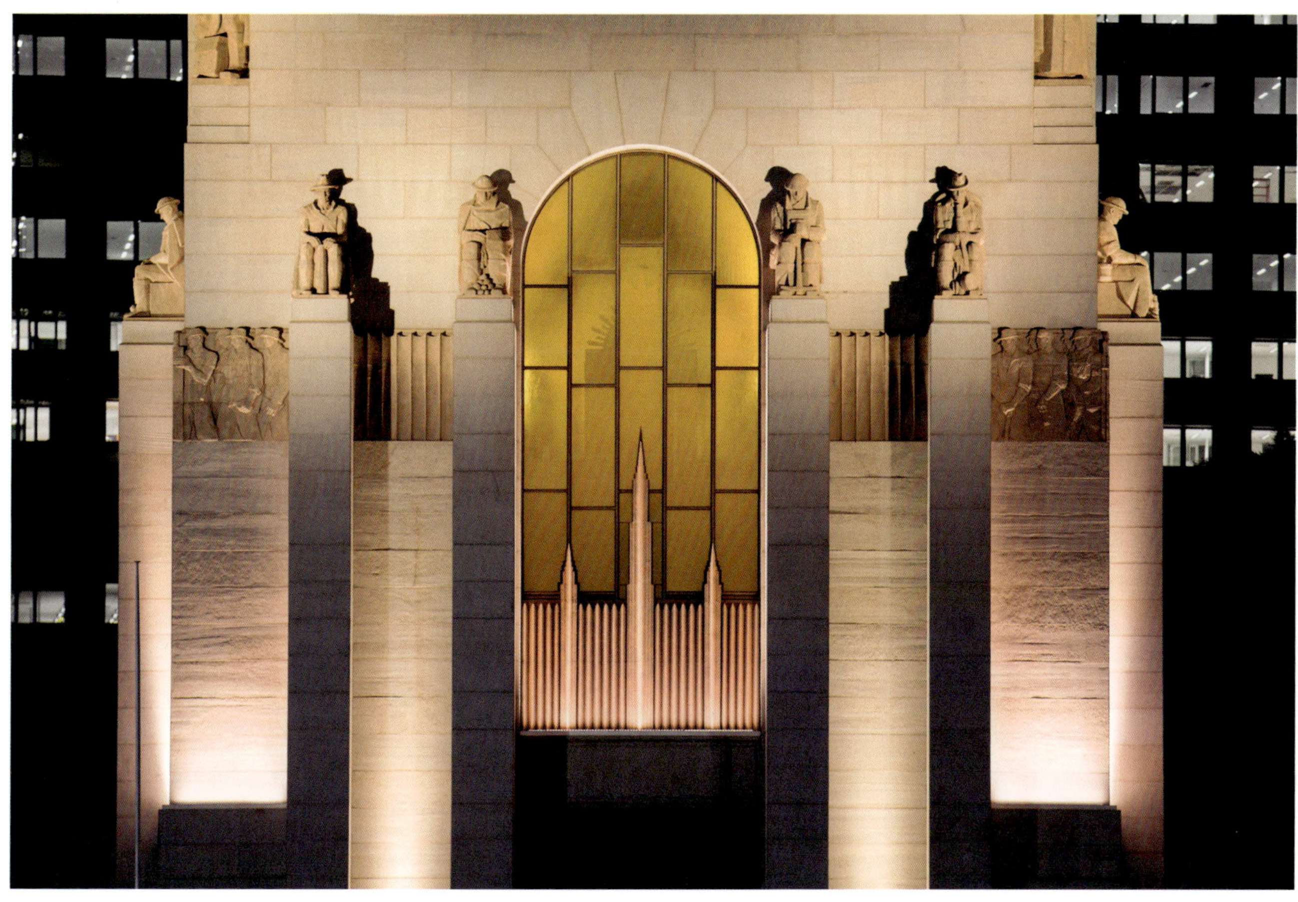

Photos Brent Winstone

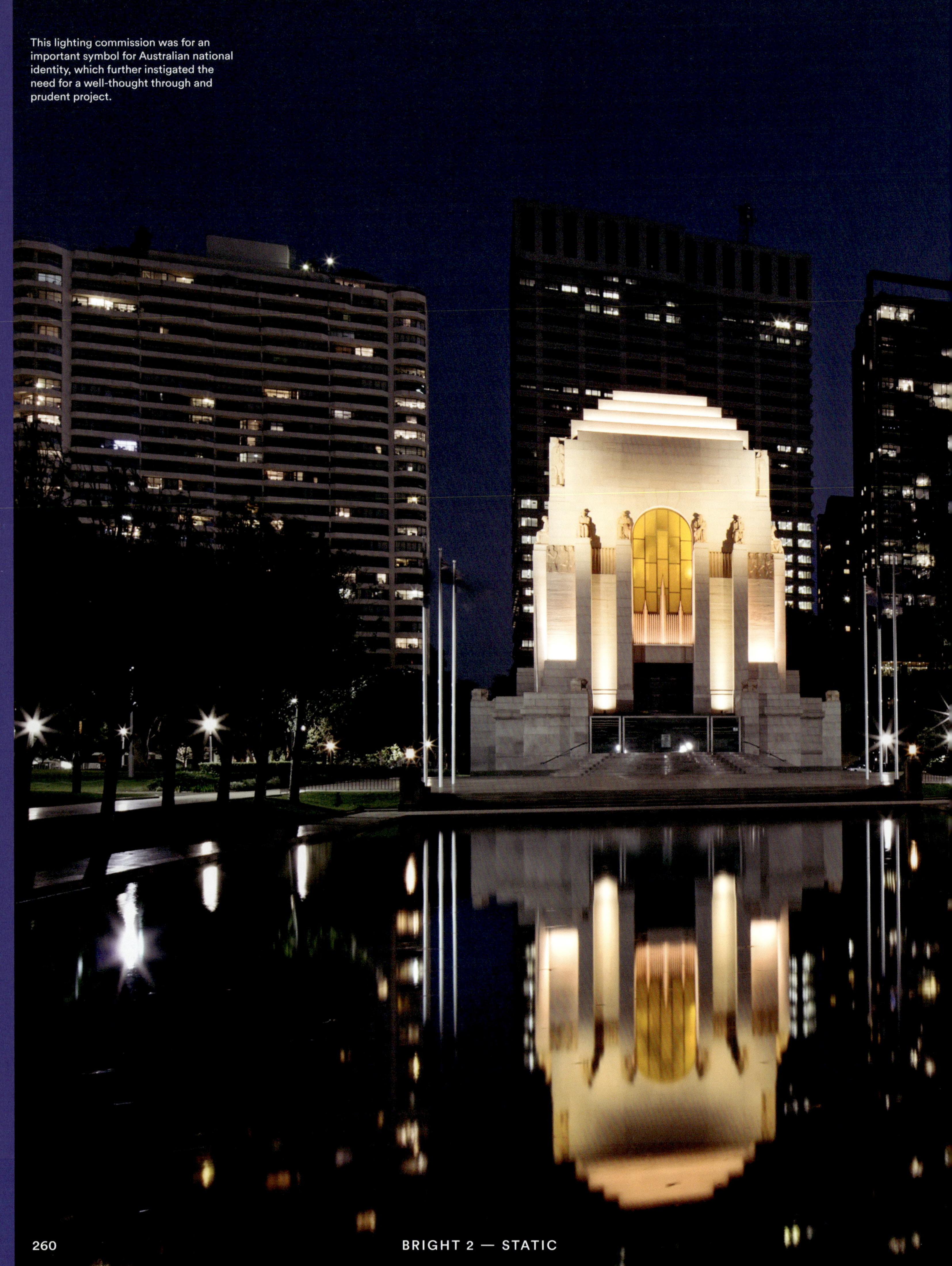

This lighting commission was for an important symbol for Australian national identity, which further instigated the need for a well-thought through and prudent project.

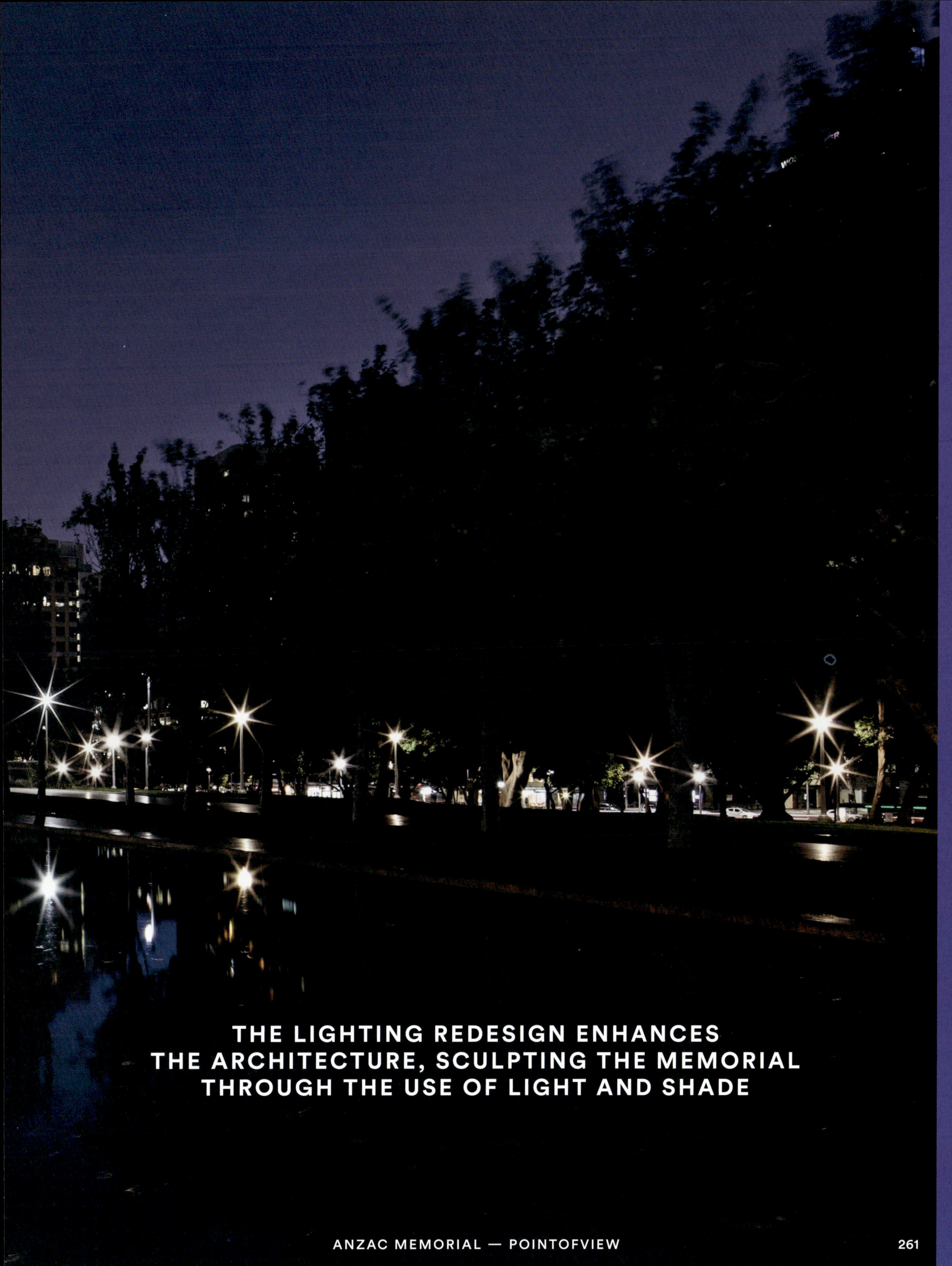

THE LIGHTING REDESIGN ENHANCES THE ARCHITECTURE, SCULPTING THE MEMORIAL THROUGH THE USE OF LIGHT AND SHADE

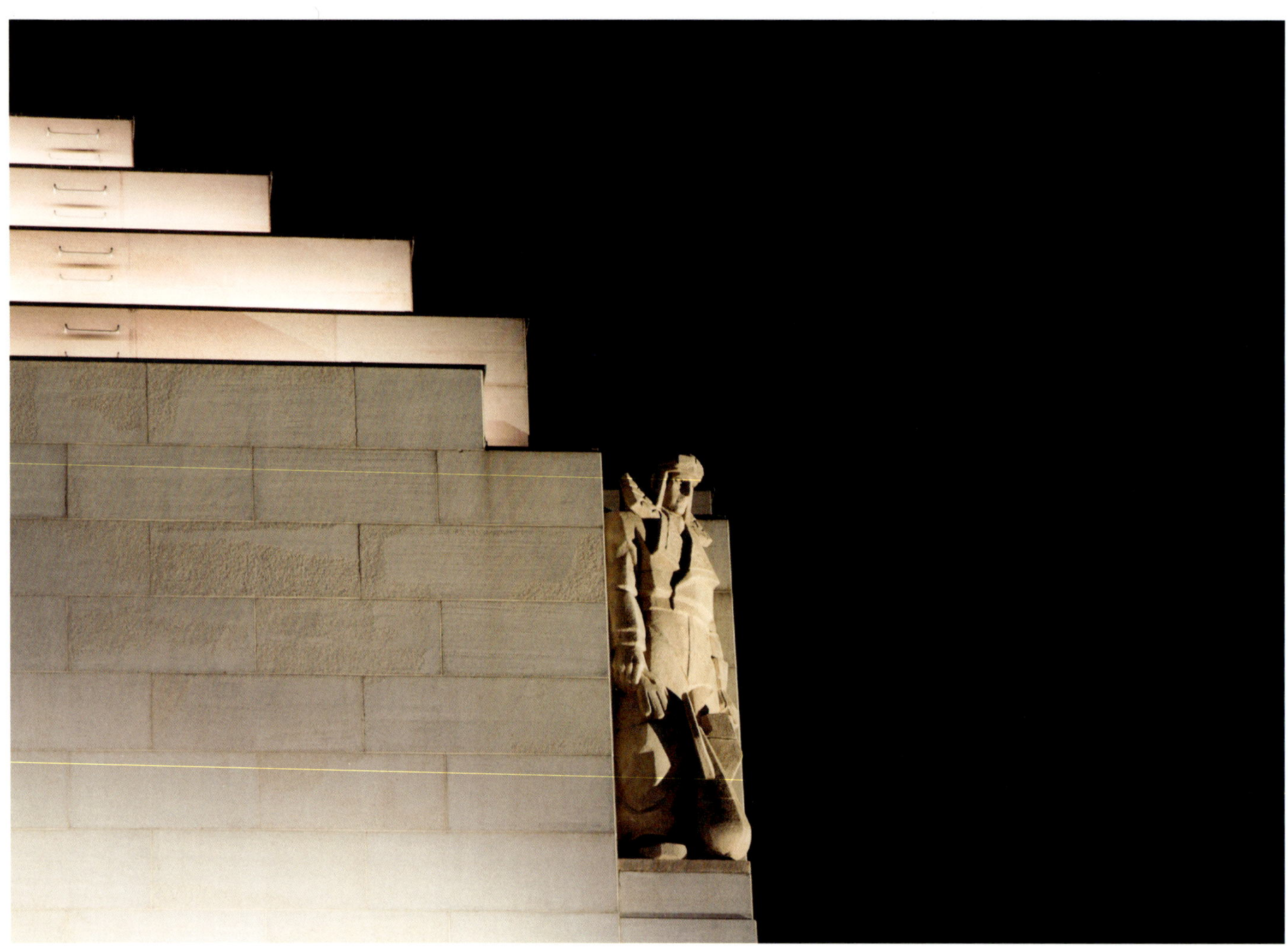

The redesign of the lighting for such a historic monument meant that each element needed to be carefully considered.

LED SPOTS SCULPT THE MONUMENT BY PLAYING WITH LIGHT AND SHADOW TO FOCUS ON KEY ELEMENTS

'Lighting the souls of the unforgotten' was the motto that Australian studio PointOfView used to describe this project – an accurate portrayal for the commission to illuminate the state memorial for soldiers who served their country in war. As you approach the steps leading up to the ANZAC Memorial, located at the southern end of Sydney's Hyde Park, visitors recognise this is the entrance to something very special. The importance of this place in the hearts and minds of the nation is significant, as it was to the team of designers commissioned this task.

The memorial site is made up of various aspects. The building was designed by Bruce Dellit in 1930, a leading proponent of Art Deco style. Four large sculptures adorn each corner of the building, representing the four seasons, while 16 smaller, seated figures represent the 'Arts of Peace and War'. Much like the cast granite sculptures sit in harmony with the building's architecture, the aim for the lighting redesign was to complement the memorial without changing its atmosphere.

For the past 80 years or so, this was a monument that remained hidden at night in its quiet corner of the park. A change was needed and PointOfView was brought on board to address this. Originally briefed over 4 years earlier to redesign the lighting for the memorial, plans were subsequently put on hold, to be revived for the ANZAC Centenary in 2014. 'Sometimes it seems that setbacks happen for the right reasons,' comments the studio's principal Mark Elliott. He continues, 'In a way, this project is a living example of how far lighting has come in such a short period of time, enabling us to use recent advances in lighting technologies in the beam control of spotlights.' The new design uses focused LED on each of the statues using existing floodlight pole locations which means that, now under the cloak of darkness, they have a presence that was previously denied them. The outcome is an enhancement of its composition by sculpting with light and shade.

The LED spots sculpt the monument by playing with light and shadow to focus on key elements, like statues. The small-scale linear LED fixtures available today allowed the designers to anchor the monument by casting light up to the corners of the building, which would have previously been impossible with more traditional lighting technologies. —

Designer
PointOfView
Location
Sydney, Australia
Client
New South Wales State Government
Collaborator(s)/consultant
NSW Government architects office, Stowe Australia (electrical contractor)
Manufacturers
We-ef Lighting, KKDC, Filix
Date
April 2014

Set-up

Used light was in the form of linear LEDs and LED spots.
Illumination of the facade and the monument's historic statues utilised eight existing poles onto which adjustable spotlights were fixed, used in combination with small-scale linear LEDs to cast light to the corners of the building.
Lighting control utilised circuits for all new lighting via an astronomical time-clock with manual override on/off switch plate.
Power cabling reticulation was coordinated with the existing services of the heritage building and all cables were concealed from view.
Testing and mock-ups drove the equipment selection, ensuring that the various suppliers provided colour renderings that matched each other, so that the stonework was consistently rendered.
Delivery time for the lighting in this project was 6 weeks from concept to completion, which was testament to great teamwork between designer and contractors with a supportive client.

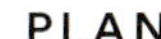

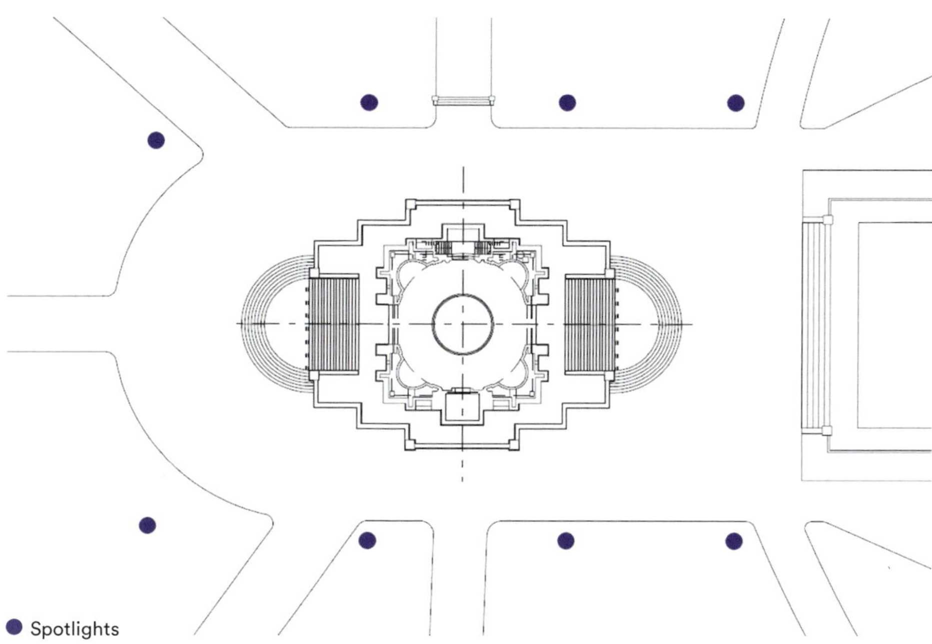

ELEVATION

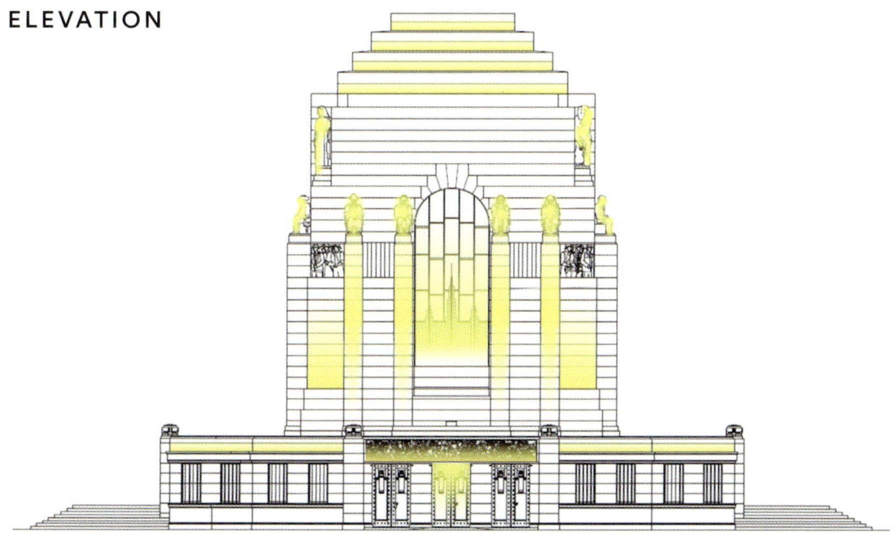

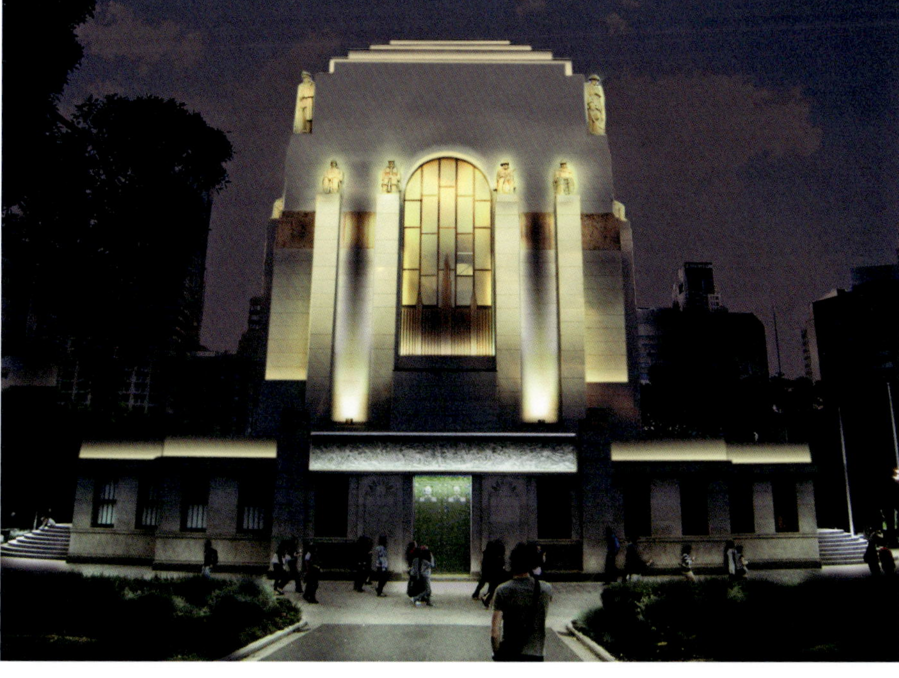

Specifics

Using visible fixtures to light each element of the war memorial without imposing on the historic aspects was challenging. On top of this, PointOfView found it all the more taxing to install degrees of feeling, raw emotion, subtlety and reverence into the project.

After the first plan to redesign the memorial's lighting was put on hold, the new brief required a short delivery time to be finalised for the ANZAC Centenary. This short timescale created its own challenges, which were met by the team – a testament to the talent and dedication of the designers to complete this accurate and contemplative design in only 6 weeks.

It was important to make use of artificial light and shadow on the building, enabling the designers to sculpt the architecture by focusing only on key elements, such as the buttresses and the ziggurat atop the memorial. The availability of LED fixtures of such a small-scale means it was possible to integrate lighting to anchor the corners of the building, something that would not have been possible at the time of the original briefing.

The miniature lense technology around the used LED light sources meant that accurate beam control and limited spill light ensured the onlooking residences did not suffer from any light pollution. With such a prominent architectural statement in the city, any such light fittings had to be completely concealed, so as not to impose on the architecture.

The internal lighting was used as part of the external lighting treatment. Within the memorial is a domed ceiling with 120,000 golden stars covering the ceiling in honour the men and women from New South Wales who enlisted for service in World War I. This dome was up-lit from within the existing wall sconces and the reflected light illuminates the glazing.

The extended lamp life of the LED equipment means that this monument will remain a beacon within Hyde Park and testament to the brave souls it represents for many years to come.

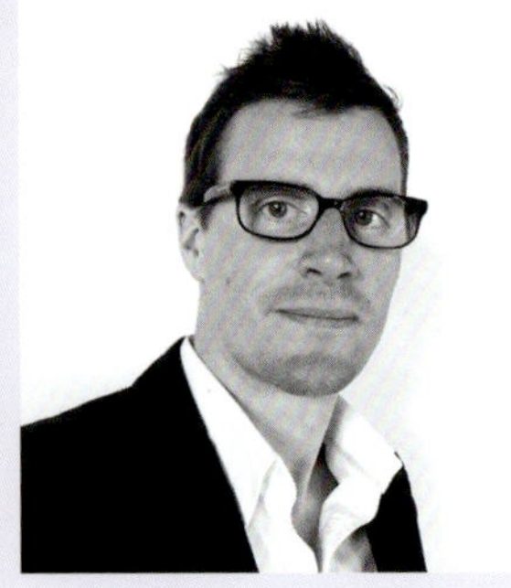

Point of View

PointOfView is headed-up by directors Mark Elliott (pictured) and Bernie Tan-Hayes in Sydney and Melbourne, respectively. Since it was founded in 2003, the award-winning consultancy has grown to become a worldwide respected firm, with over 700 projects realised across the globe covering a variety of sectors, from civic to retail with a focus on hospitality and lifestyle. The design consultancy specialises in lighting design, audio visual and theatre systems design, embracing innovation, individuality, proactivity and collaboration. The studio's projects strive to integrate and enhance the architectural environment, rather than impose upon it.

POV.COM.AU

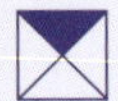

Hotel Hotel

When **2014**
Where **Canberra, Australia**
Client **Molonglo Group**

No two rooms are alike at Hotel Hotel in NewActon, the cultural centre of Canberra. With architecture by Fender Katselidis, interior design by March Studio and styling by Don Cameron, the hotel is made up of 68 rooms and 31 apartments featuring elegant furniture, natural materials, art and *'objets trouvé'*. Of special significance is the main staircase, constructed entirely from salvaged timber. All these aspects were highlighted in the PointOfView lighting scheme for the hotel.

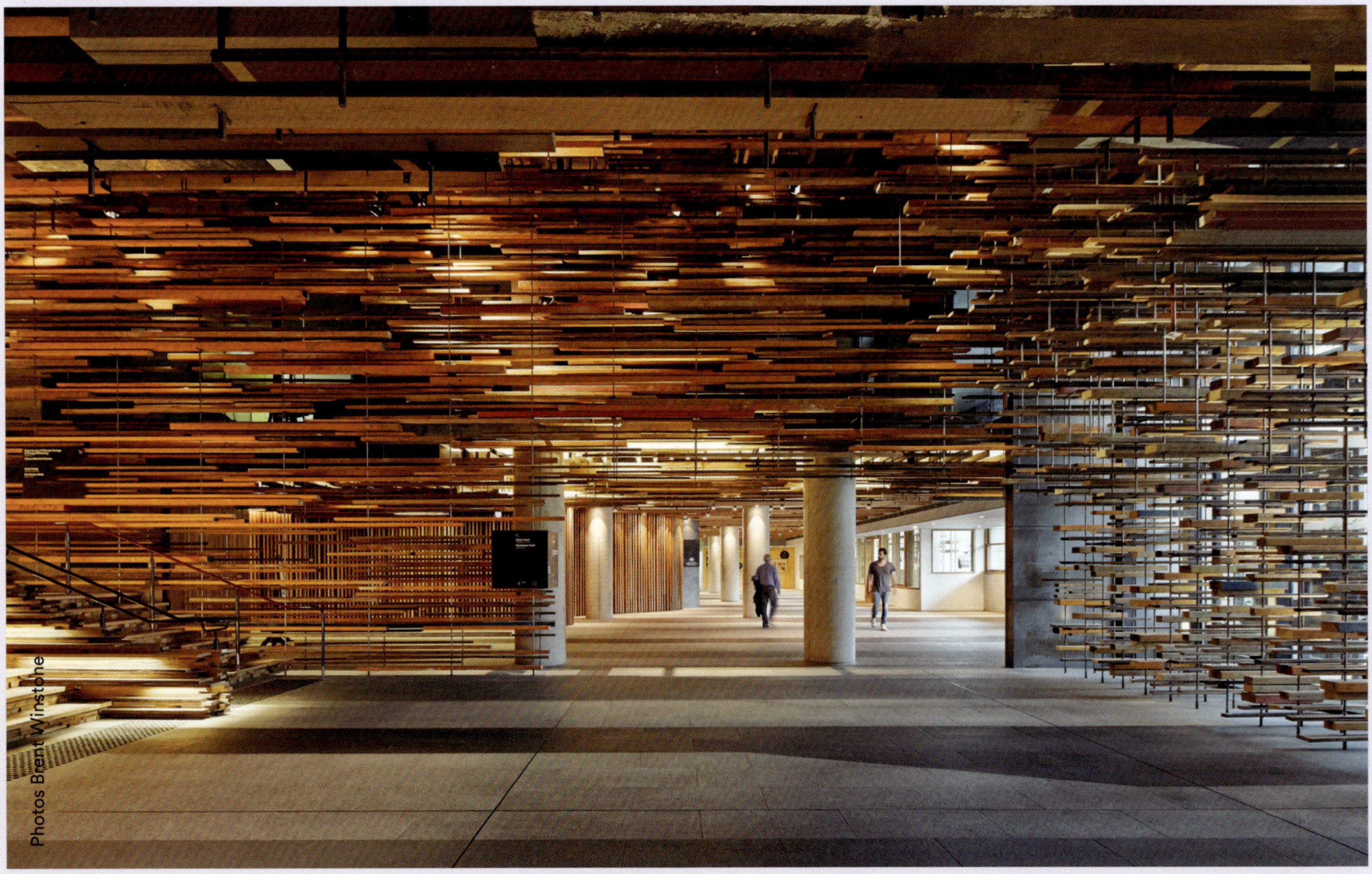

Photos Brent Winstone

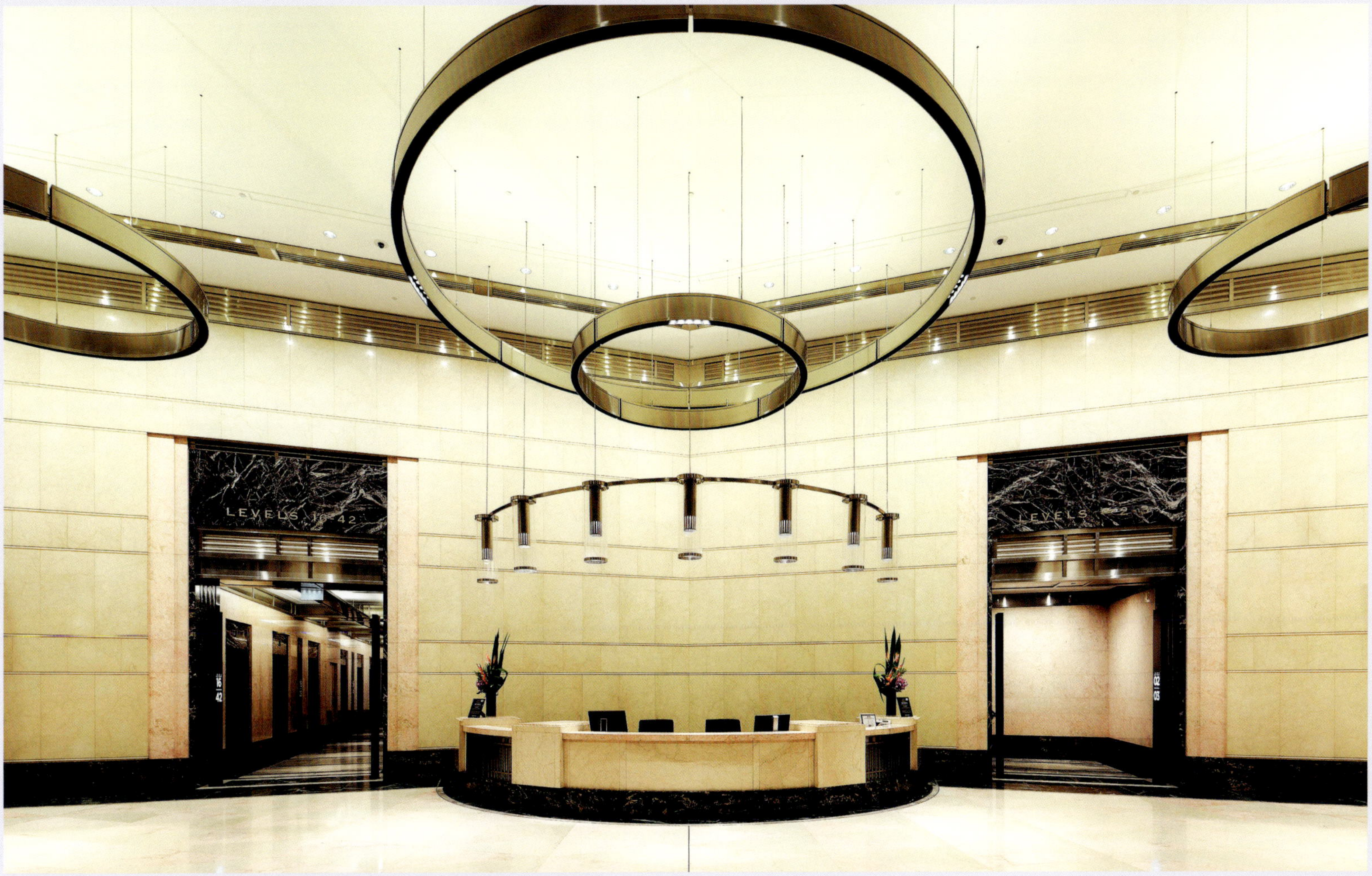

Chifley Tower Lobby

When **2013**
Where **Sydney, Australia**
Client **GIC**

The focus for PointOfView was the lobby of Chifley Tower – located on one of Sydney's most expensive sites. It was bought for $306 million in 1988, and continues to be admired today by the legal eagles, bankers, bondsmen, traders, developers and entrepreneurs who pass through its imposing entrance day after day. The lighting of the lobby also catches their attention, thanks to the PointOfView team. A design signature of POV is the custom lighting element which is showcased in the lobby. With its stainless steel rings and pendants decorating the space like an illuminated sculpture, the piece pays homage to the stone-clad art deco building.

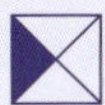

Conservatory Brassiere

When **2012**
Where **Melbourne, Australia**
Client **Crown Resorts**

Serving food every day of the year, the Conservatory Brassiere is the iconic all-day dining restaurant of the Crown Towers Hotel in Melbourne. Blainey North and Associates provided a stylish and dramatic interior, which cleverly disguises the true scale of this high turnover restaurant, and the PointOfView team complemented this with an accurate and appropriate lighting scheme.

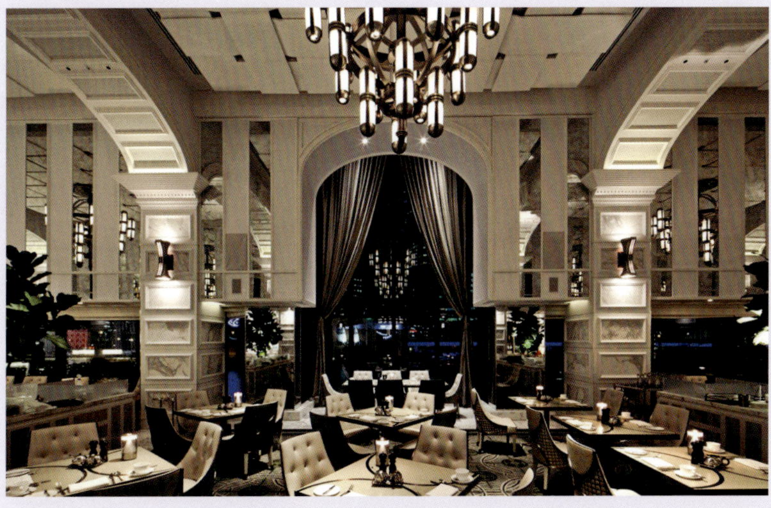

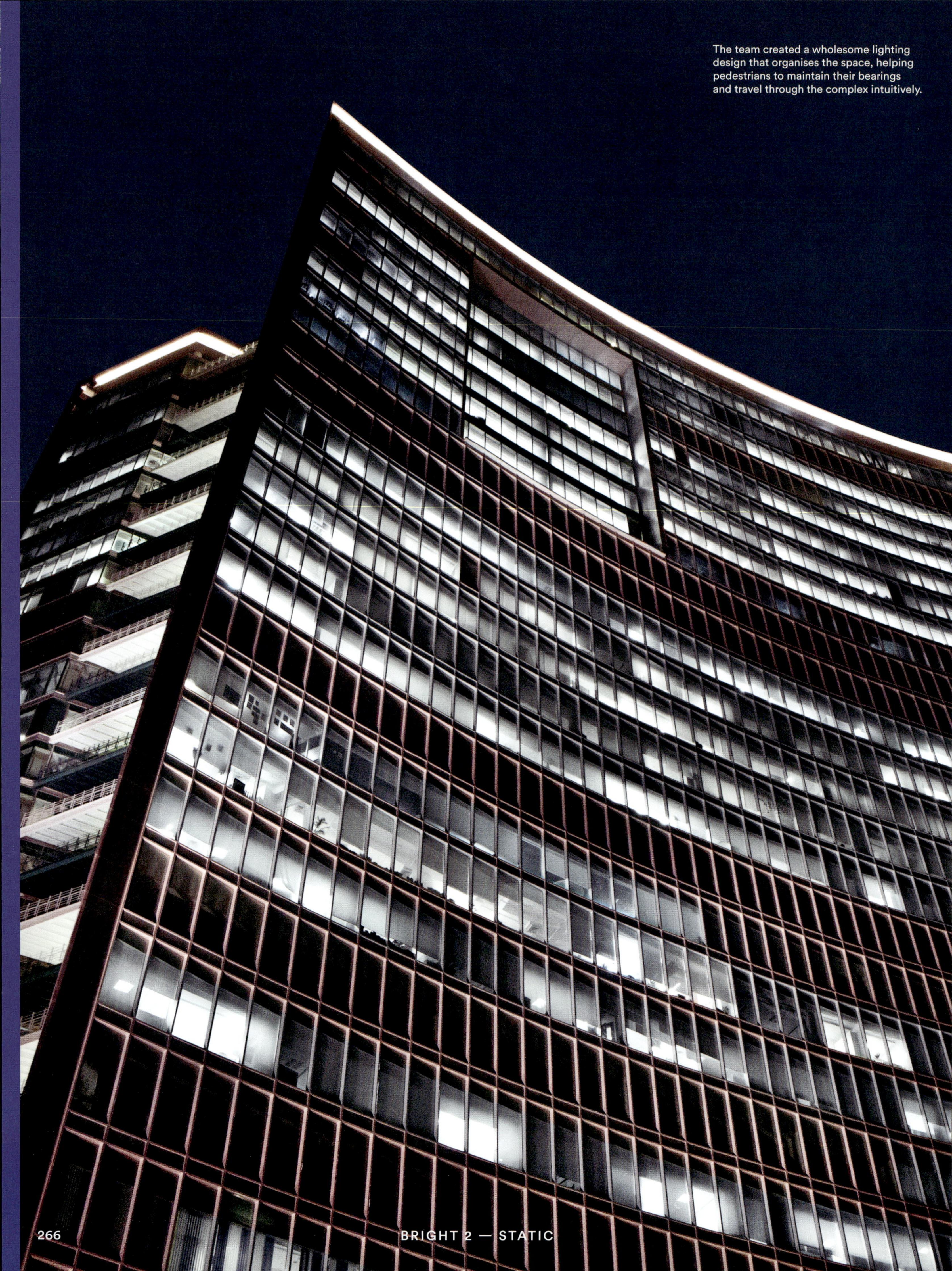

The team created a wholesome lighting design that organises the space, helping pedestrians to maintain their bearings and travel through the complex intuitively.

Brigade Gateway Complex

AWA Lighting Designers sought to create a paradigm shift for the lighting of public spaces and mixed-use environments when designing the Brigade Gateway Complex, an integrated lifestyle enclave located in Bangalore, India. The project's broad scope enabled the team to create diverse solutions and compelling compositions with landscape and facade elements.

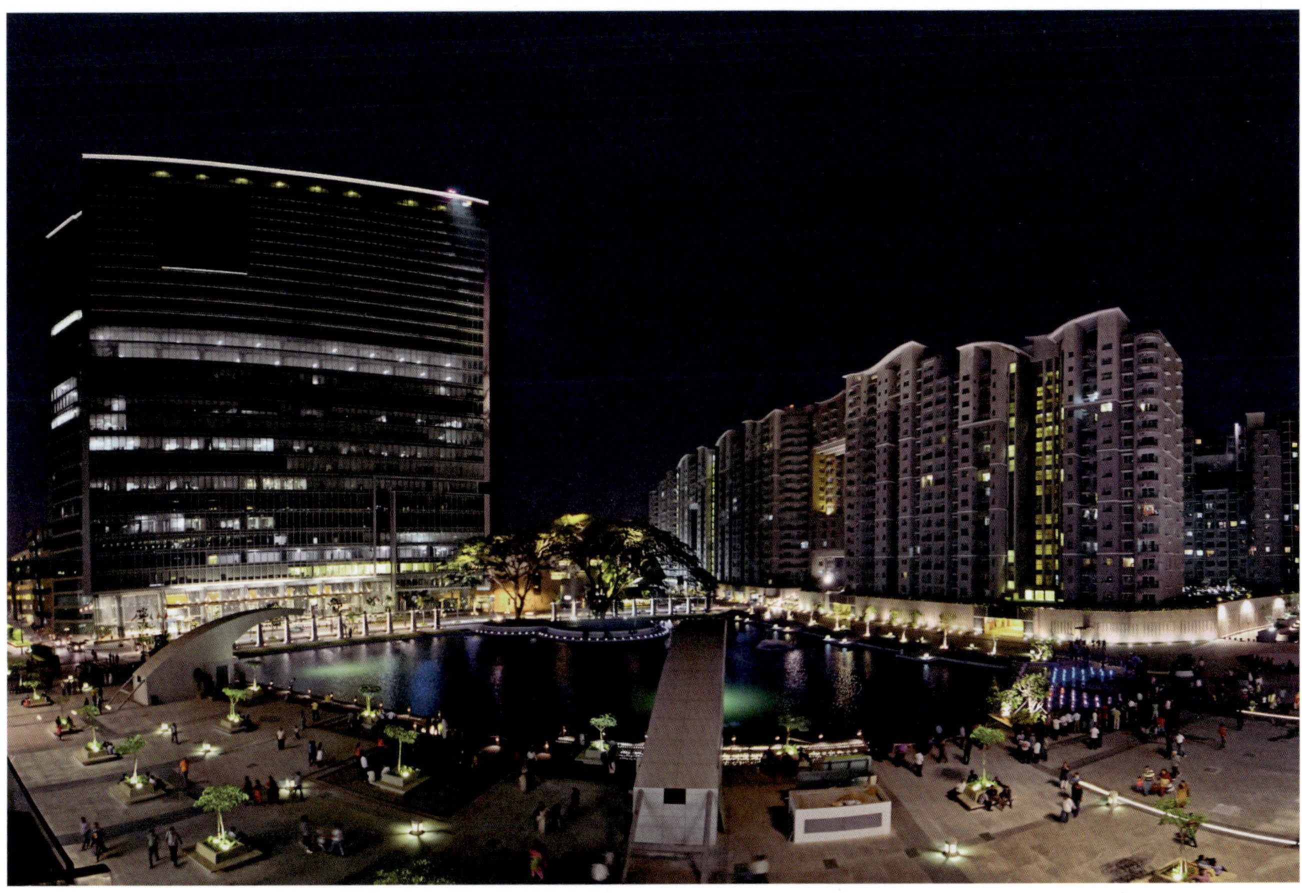

Photos AWA, Brigade Group, Michael Foley

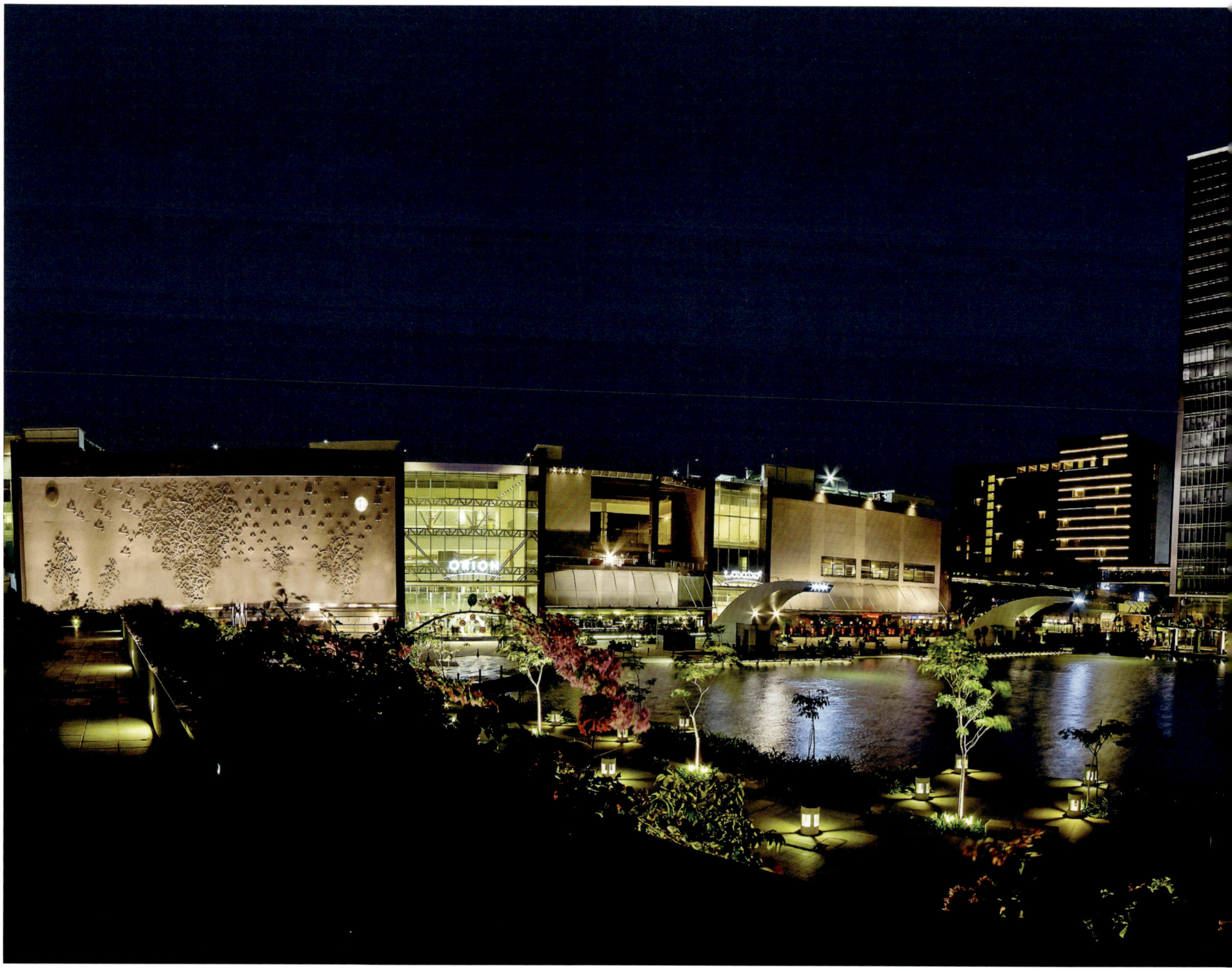

Moonlighting from the top of the World Trade Center dramatically lights the rain tree in front; this 200-year-old tree inspired the dappling foliage pattern on the floor and surrounding pathways.

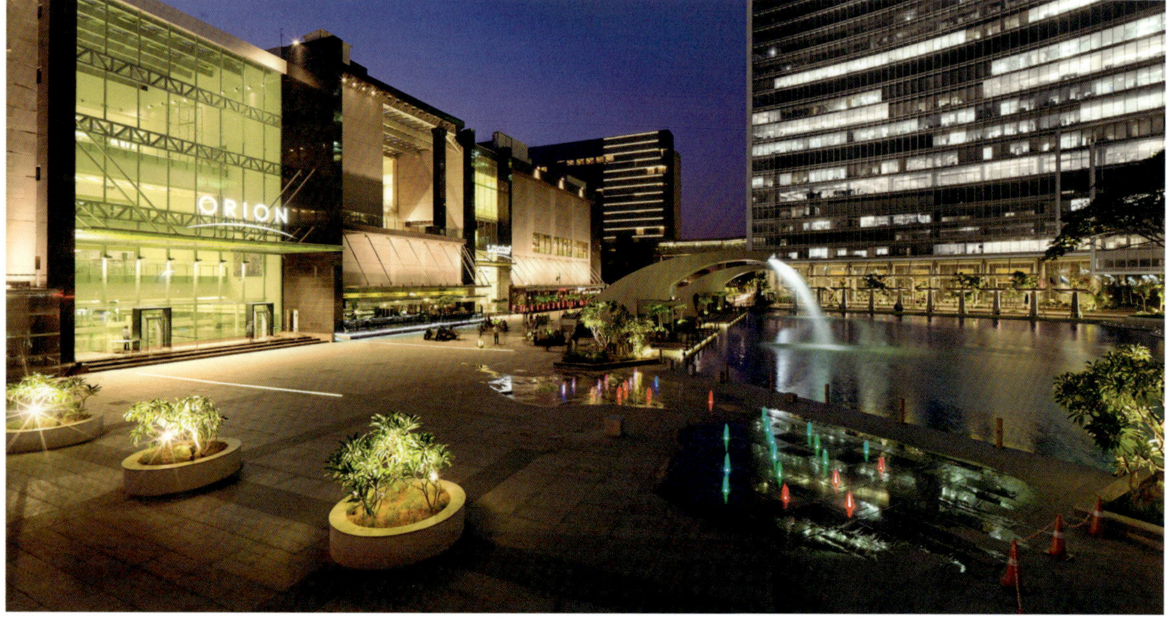

Designer
AWA Lighting Designers
Location
Bangalore, India
Client
Brigade Group
Collaborator(s)/ consultant
HOK, Michael Foley Design, Venkataramanan Associates
Manufacturers
Aldabra, Neo-Ray, Lightolier, Osram, HK Lighting, ERCO, Reggiani, Trilux, Wibre, Ligman
Date
July 2011

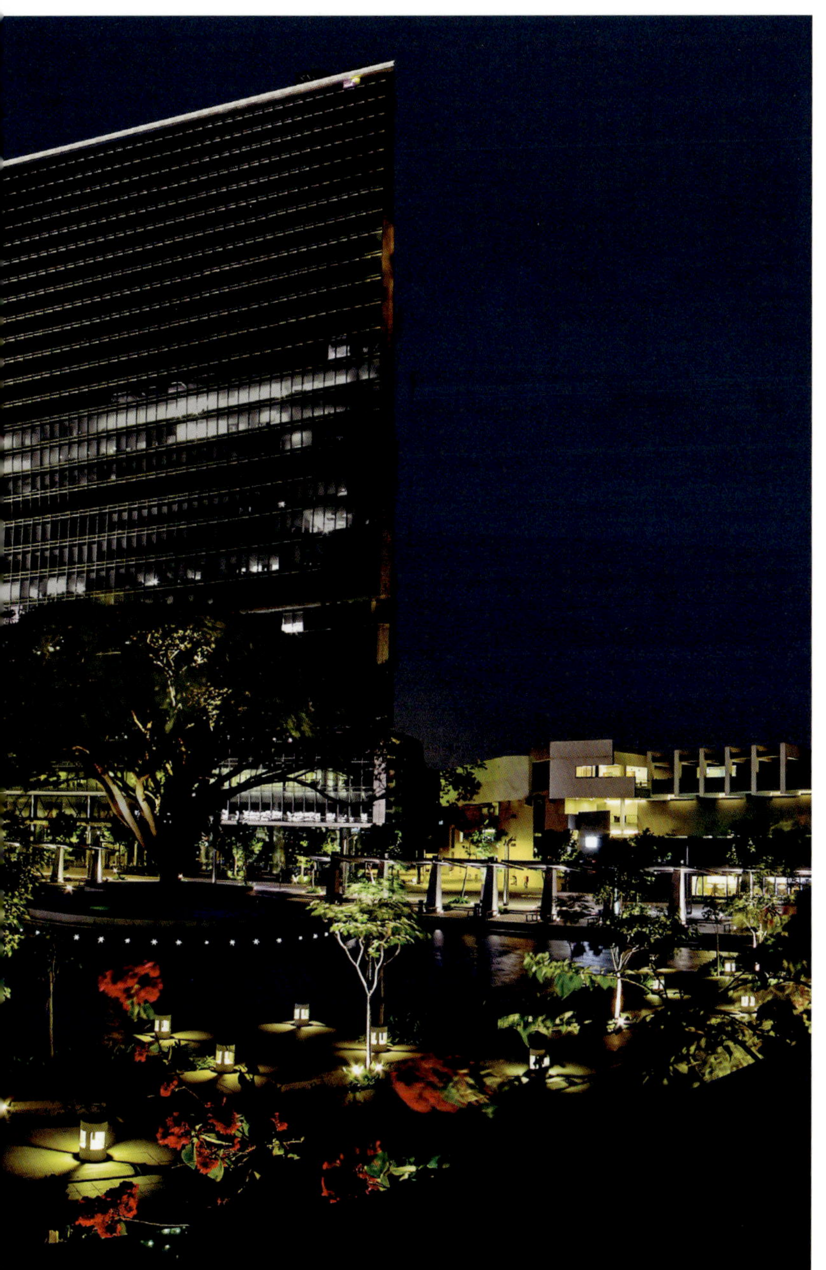

FACADES AND EXTERIOR SURFACES ARE LIT TO ENABLE PERCEPTUAL CLARITY

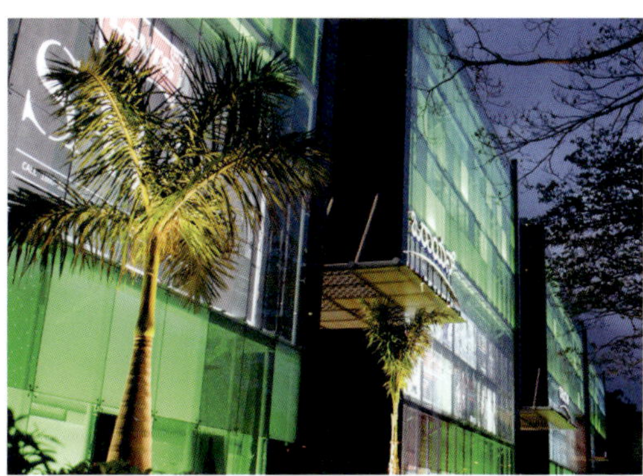

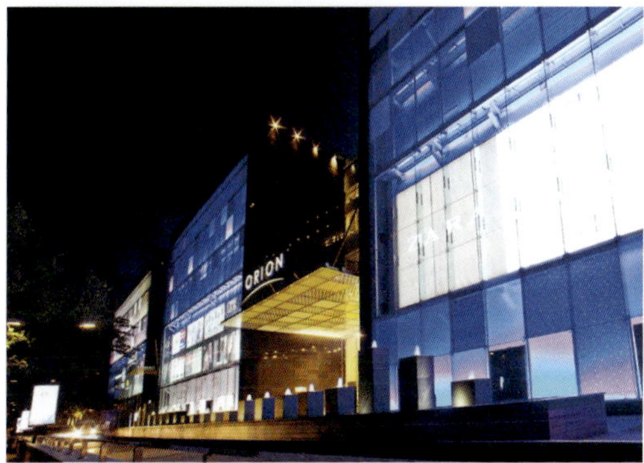

The mall's five-storey facade uses dynamic, colour changing features that attract attention from afar.

The Brigade Gateway is a vast lifestyle complex in Bangalore, which integrates residential and commercial facilities positioned around an artificial lake, as well as a school, hotel, hospital and shopping mall. Brooklyn-based studio AWA was commissioned to illuminate the exterior areas, including the main facade and landscaping elements. The complex was built according to the vision of architecture studio HOK and includes the first privately-owned World Trade Center in India.

Seeking to 'create a paradigm shift in the lit environment', AWA used the architectural focal points to set up a visual hierarchy. Facades and exterior surfaces were lit to enable perceptual clarity, with their illumination simultaneously spilling over the pedestrian pathways, functioning as supplementary street lighting. Moreover, linear progression of light fixtures in the exterior takes visitors to the complex's main destinations. The design team's intention was to sculpt the trees, water and pathways with illumination, avoiding harsh 'area' lighting.

The materiality of the WTC facade is highlighted using narrow beam projectors that allow the light to sweep across the facade, reinforcing the building's soaring height. Additionally, a radiant 'light crown' developed by the designers sits atop the building, creating a visual anchor for the city of Bangalore. The AWA team ensured accurate integration of other illuminated visual attractions, such as a public art and sculpture for the shopping mall's facade entitled the Tree of Life. The latter boasts an illuminated moon, with a lighting aspect synchronised with the lunar cycle. There is also a central dramatic theatre with a water feature, that is able to change the mood of the public space surrounding it through a lighting control system. Above all, AWA looked to produce a lighting programme that would function as a catalyst for social interactions. —

STREET LIGHT DESIGN

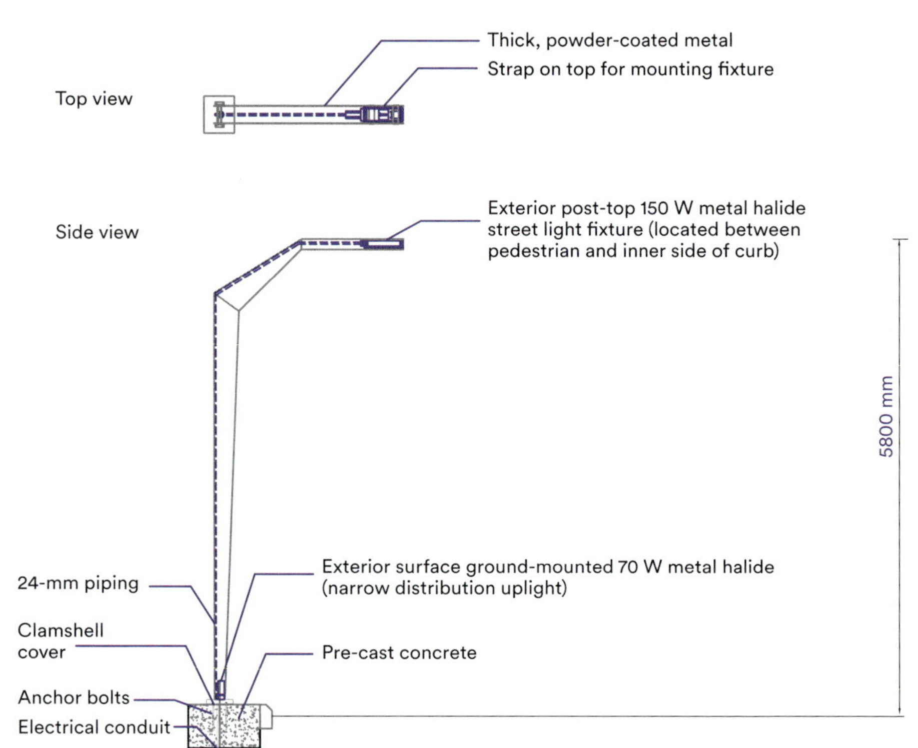

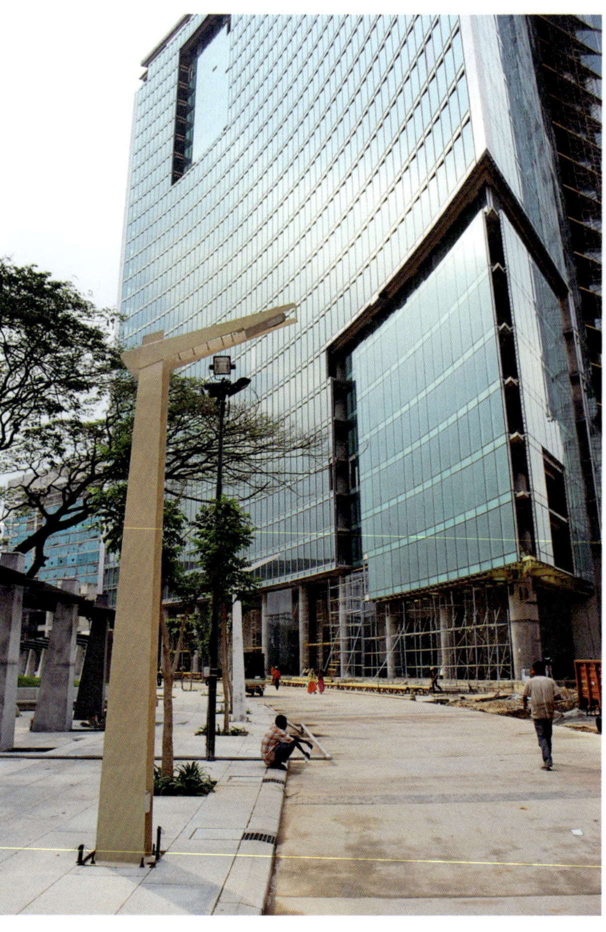

AWA SOUGHT TO CREATE A PARADIGM SHIFT IN THE LIT ENVIRONMENT

BOLLARD DESIGN

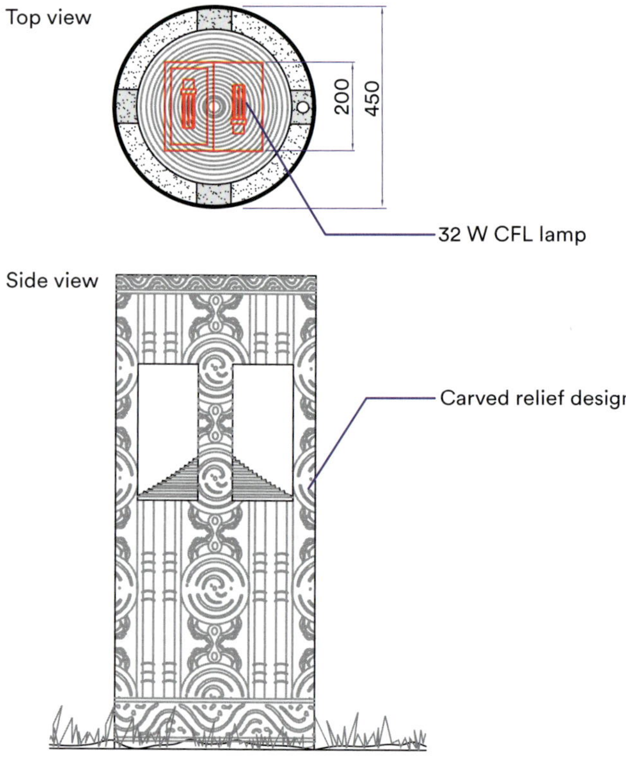

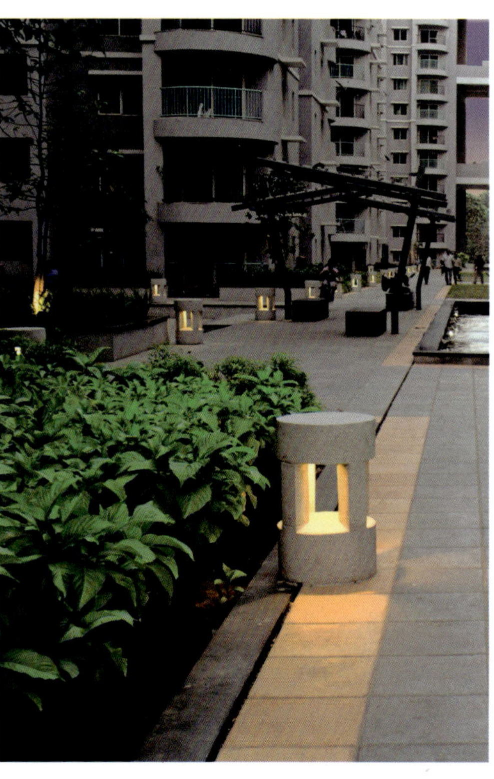

Set-up

Exterior illumination included pendant, recessed and surface mounted fixtures, with dimmable controls.
Luminaire types were numerous, including 26 W CFL granite bollard lights and 150 W metal halide street lights (with three solar panel options). Recessed linear white LEDs (5 W/RMT) were used on the vertical mullion of the facade, as well as the boardwalk. Exterior surface walls utilised 70 and 150 W metal halides, 28 W T5 fluorescent linear uplights and 4000 W Xenon narrow distribution lights, amongst others.
Lighting locations were decided upon according to user requirements: primary focal cluster, secondary focal cluster, primary focal point, secondary focal points, and collection points.
Pedestrian pathways utilised custom-made bollard fixtures made by local manufacturers from locally-sourced granite. Bollard lighting was used to avoid overhead lighting, in order to sculpt accurate way-finding.
All fixtures were custom-designed to coexist in harmony with the surrounding architecture.
Total area of the site is nearly 162,000 m^2.
Time of construction was 3 years.

RENDERED SITE PLAN

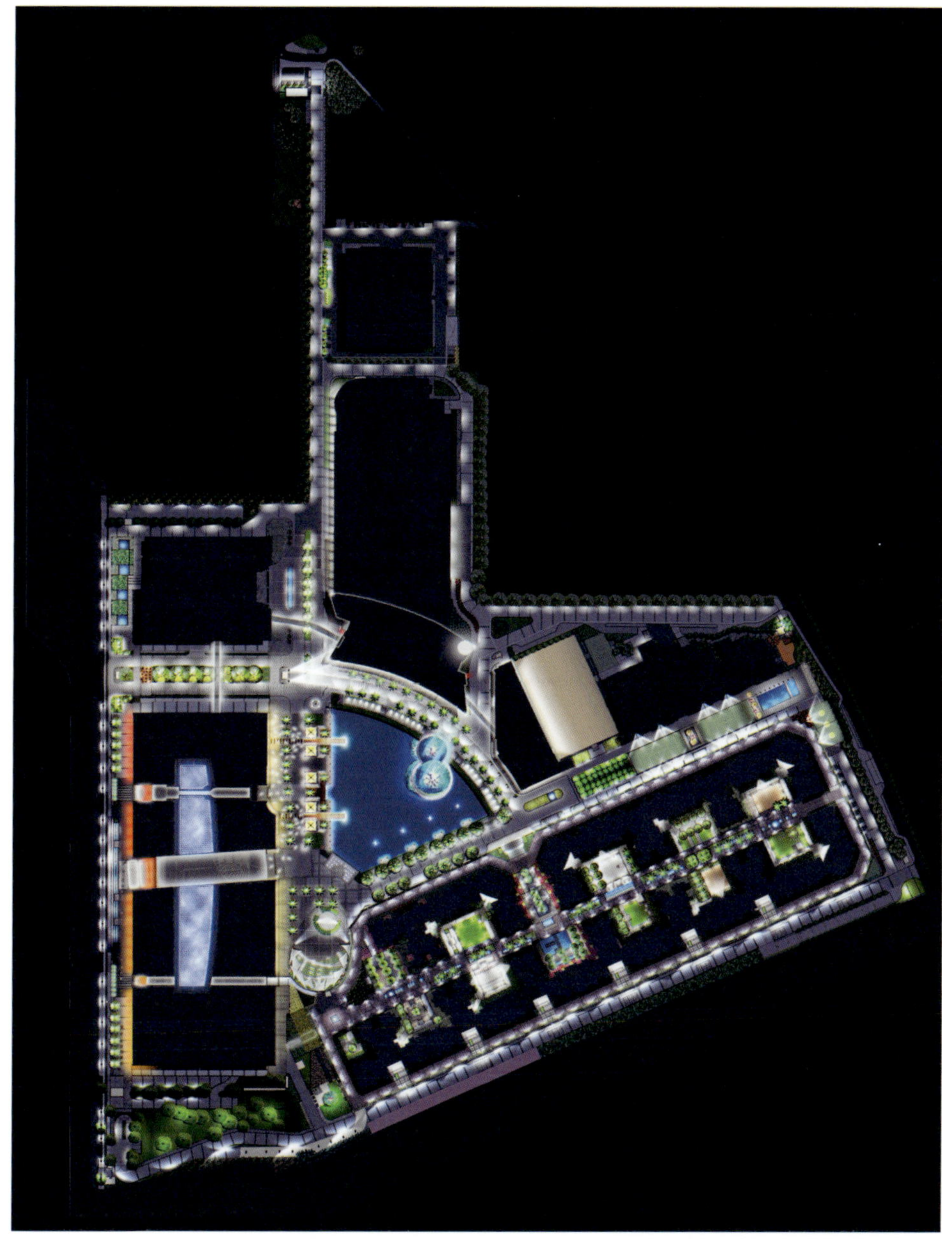

FOCAL POINT DIAGRAM

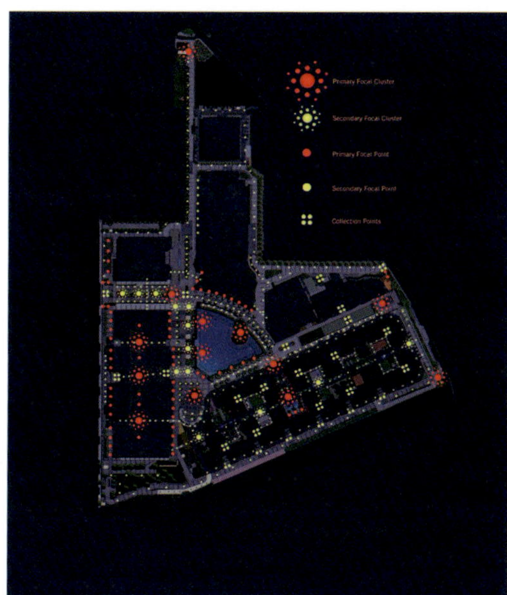

CIRCULATION DIAGRAM

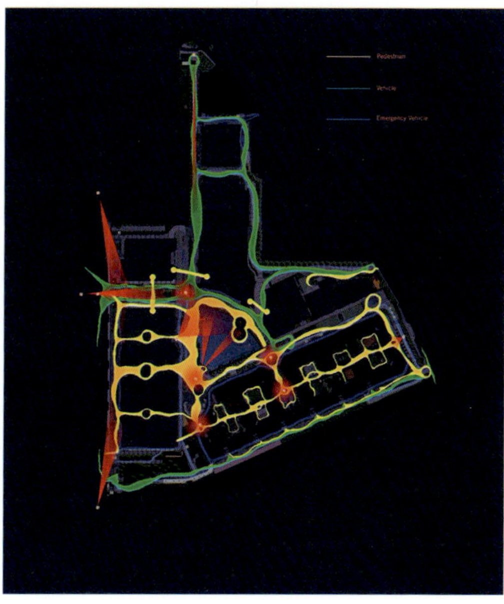

Specifics

/\/\/\ The lighting design solution would create a visual hierarchy by picking points of interest and appropriately accentuating different areas, providing areas of focus and way-finding tools within the overall visual experience. The fixtures were chosen such they were not only appropriate at night, but also tie-in to other landscape elements and street furniture during the day.

/\/\/\ The team decided to make a key feature of the most prominent building in the complex. The characteristics of the building were studied, following detailed design discussions with the architects. By using points of light to trace the WTC's edges from the ground-up, its soaring perspective was enhanced. Linear lights accentuate its crown.

/\/\/\ It was challenging to ensure accurate illumination of the residential wings and courtyard at night-time. The solution was a 'moon lit' design strategy which would create an adequate sense of safety and security. The entire podium is kept free of light poles so as to lead the user through without any visual clutter.

/\/\/\ The mall's five-storey facade required a different approach, utilising dynamic scenarios. The client asked to get the feel of 'Times Square' to the all-glass mall front facade, with the design approach incorporating a frit on the glass material so that locally-available light fixtures would make it as dramatic. Electronic media screens and a glowing display wall can change colour at the flick of a switch. The facade's lighting also illuminates the adjacent pedestrian pathway.

/\/\/\ A main focus for the design was the implementation of spaces for social interaction. Sculptural elements were considered vital in this respect, creating canopies and reversed arches to formulate a fun and inviting lighting atmosphere for pedestrians.

/\/\/\ Lighting the WTC facade from a biased location was carried out in order to accentuate the location of the mullions on the facade. The architectural bias detail meant that specific lighting treatment was required to highlight the mullions. The lights were all located on top of the security cabin, thereby not encumbering the landscape with extra poles.

/\/\/\ The inspiration for the facade lighting strategy came from observing the light from the setting sun as it hit the building, and so the creation of a similar light condition was aimed for.

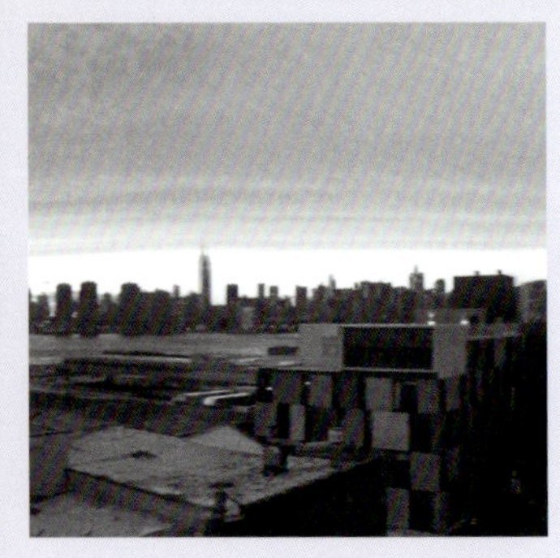

AWA Lighting Designers

International architectural lighting design firm AWA was founded in 2002 by Abhay Wadhwa. The Brooklyn-based studio designs and implements lighting solutions 'that evoke a sense of place instead of mere space' for infrastructural, hospitality, commercial and residential projects. With a team of over 20, including director R Sitaraman and associates Justin Moench and R Rajasimman, the firm utilises creative and technical methods, looking to develop conscious lighting design, remaining aware of the people who live, work, communicate and interact in the structures they illuminate.

AWALIGHTINGDESIGNERS.COM

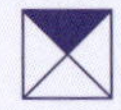

Cyber Hub

When **2013**
Where **Gurgaon, India**
Client **DLF**

For the Cyber Hub in Gurgaon, AWA wanted to create a lighting design that reinforced an 'architectural project that augments the sense of community and civic pride'. To do this, the firm created clear visual orientation, nodes connected through a lighting tapestry to make up a visually exciting and crisp environment at night.

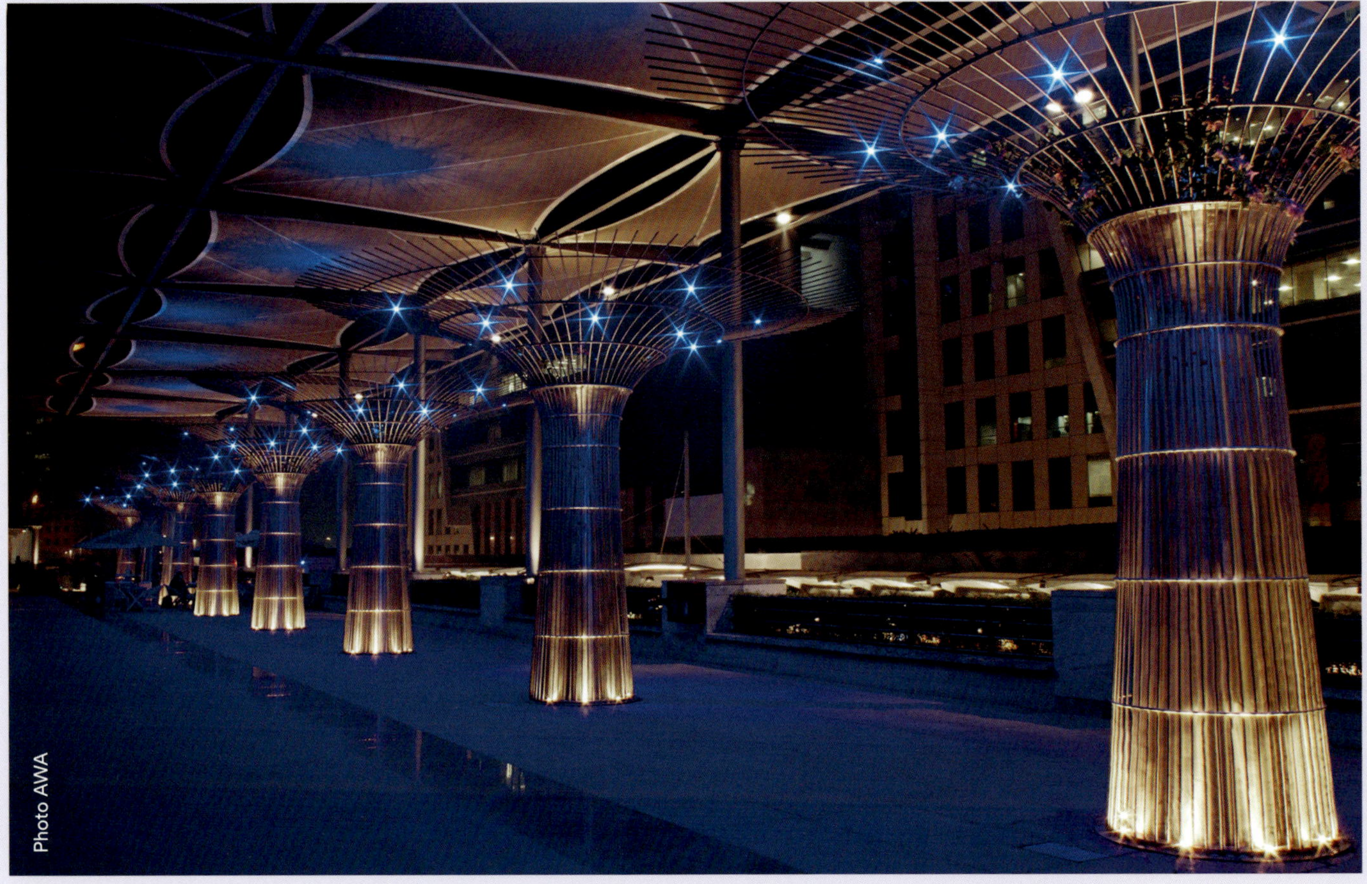

Photo AWA

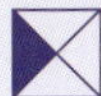

Singapore Chancery

When **2014**
Where **New York City, United States**
Client **HOK**

For the Singapore Chancery in New York City, AWA wanted to create a design that provided a layered lighting response to create a versatile facade so the client could adjust the visual appearance of the building at night. AWA designed the facade lighting to create a visual identification from in front of the building across the street, as well as creating an immediate impression from blocks away.

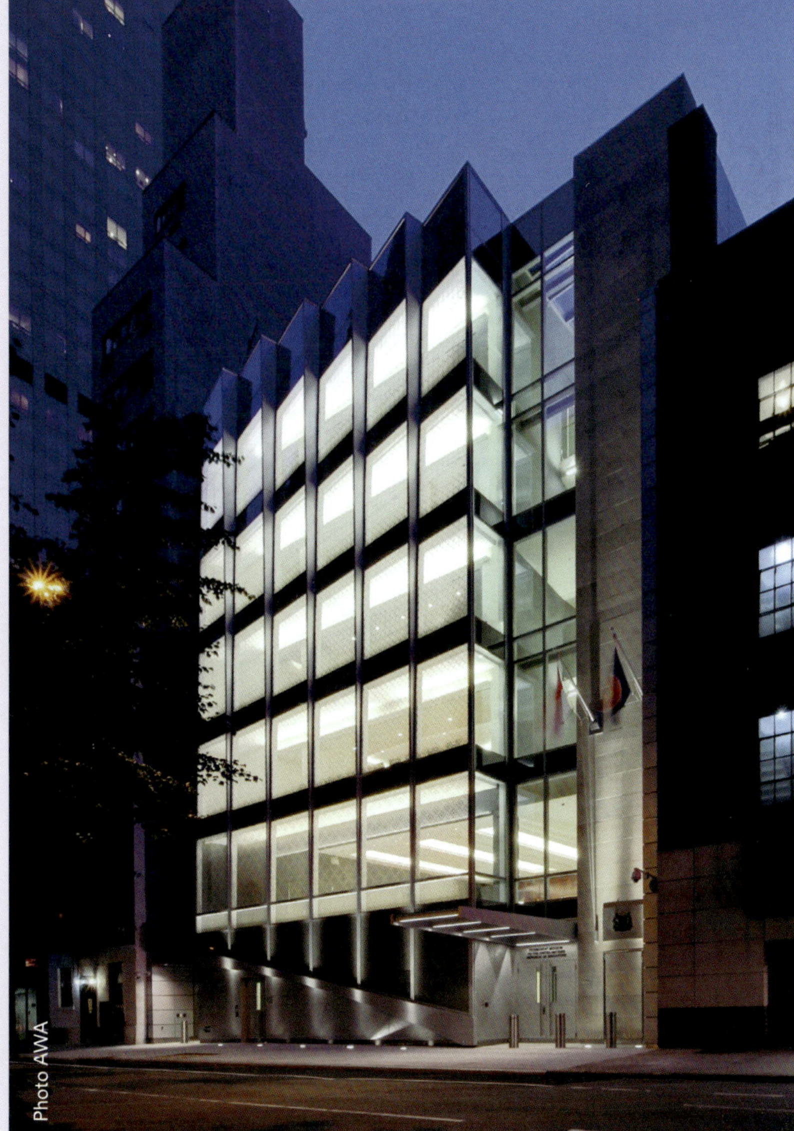
Photo AWA

First International Financial Center & TCG Financial Centre

When **2011**
Where **Mumbai, India**
Clients **Kohn Pederson Fox (FIFC), Raja Aederi (TCG)**

While focusing on meeting LEED standards, AWA developed a coherent lighting design for the exteriors of the First International Financial Center (left) and TCG Financial Centre (right), located in adjacent buildings in Mumbai. The lighting design highlights the 'floating tube' along each facade, by working with the architects (Kohn Pederson Fox for FIFC and Raja Aederi for TCG) to create a custom detail in the exterior glass mullion system to create the appearance of a continuous line of light between both buildings. The lighting design enhanced the salient design features to create an urban landmark.

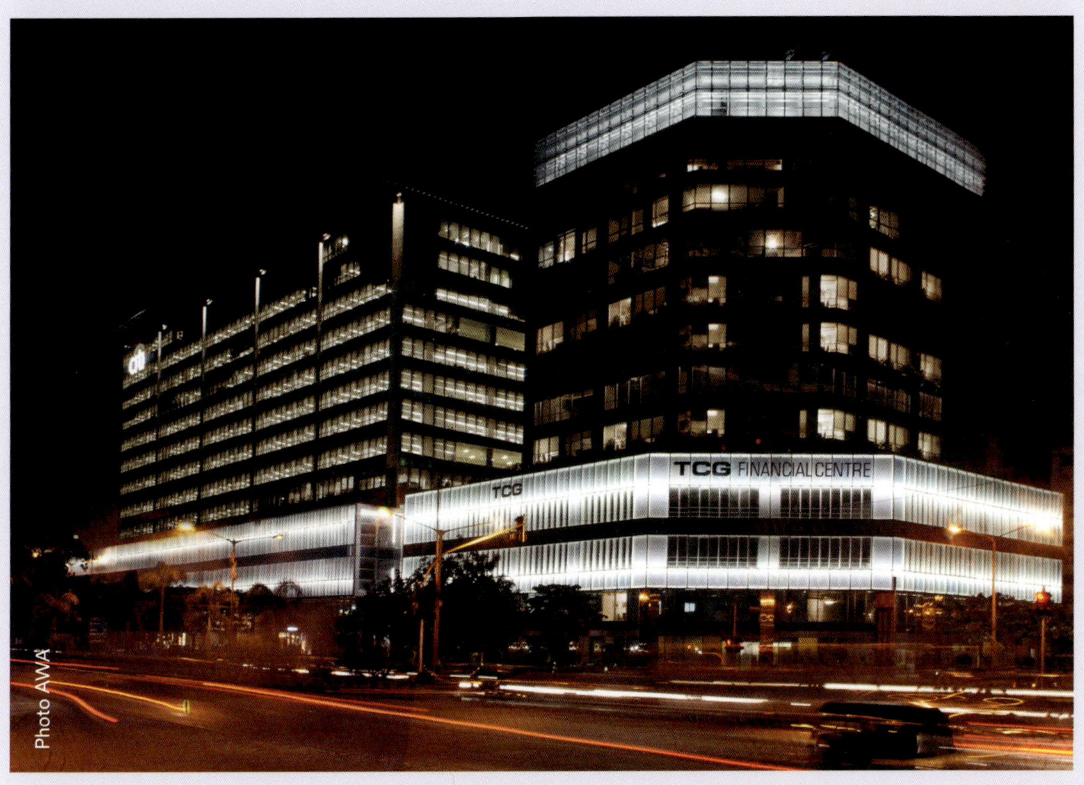

Photo AWA

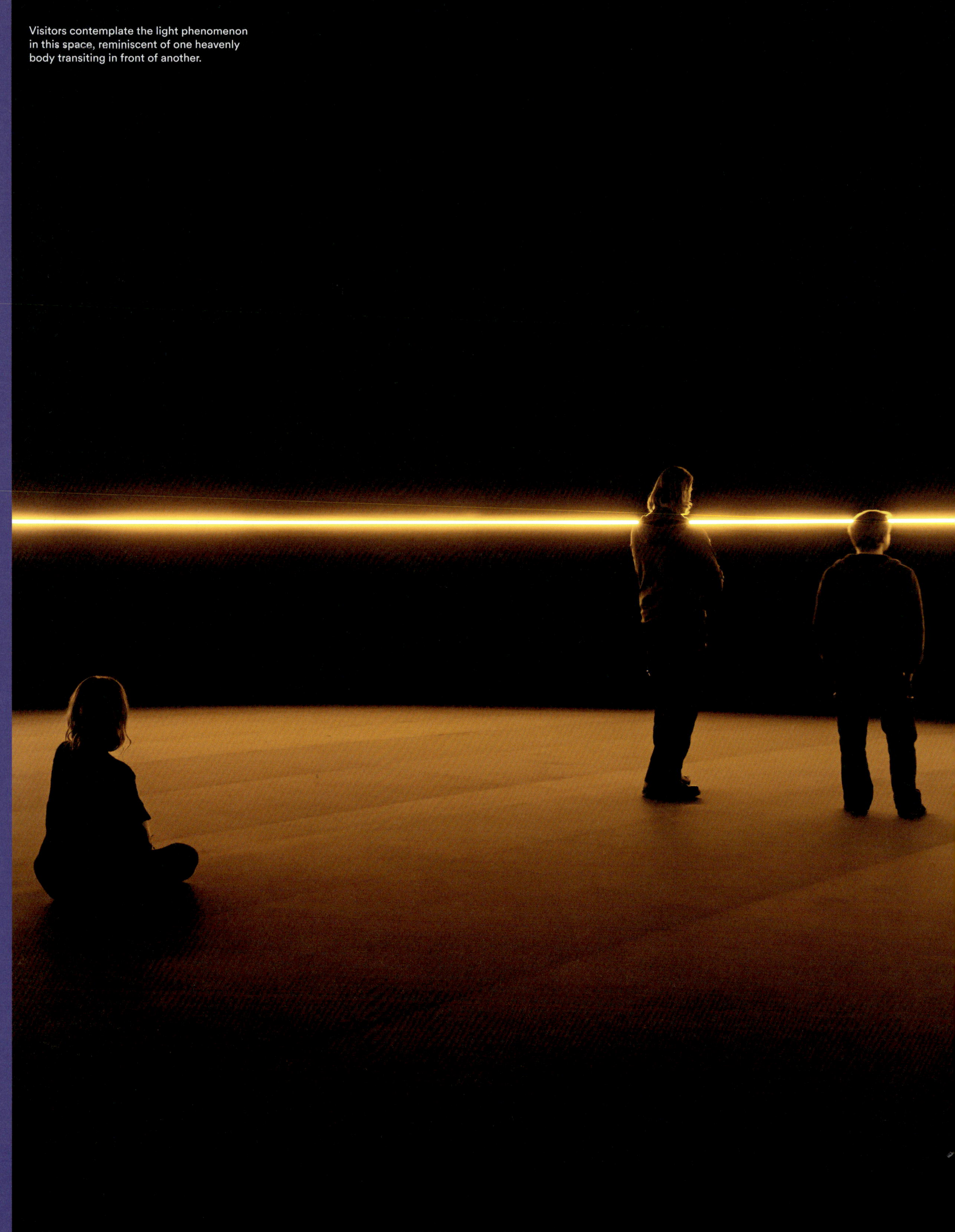

Visitors contemplate the light phenomenon in this space, reminiscent of one heavenly body transiting in front of another.

Contact

Olafur Eliasson's illuminated cosmos created within the Fondation Louis Vuitton in Paris explores relationships between self, space and universe. It envelops visitors in a choreography of moving light and shadows, seemingly transporting them into the darkness of outer space.

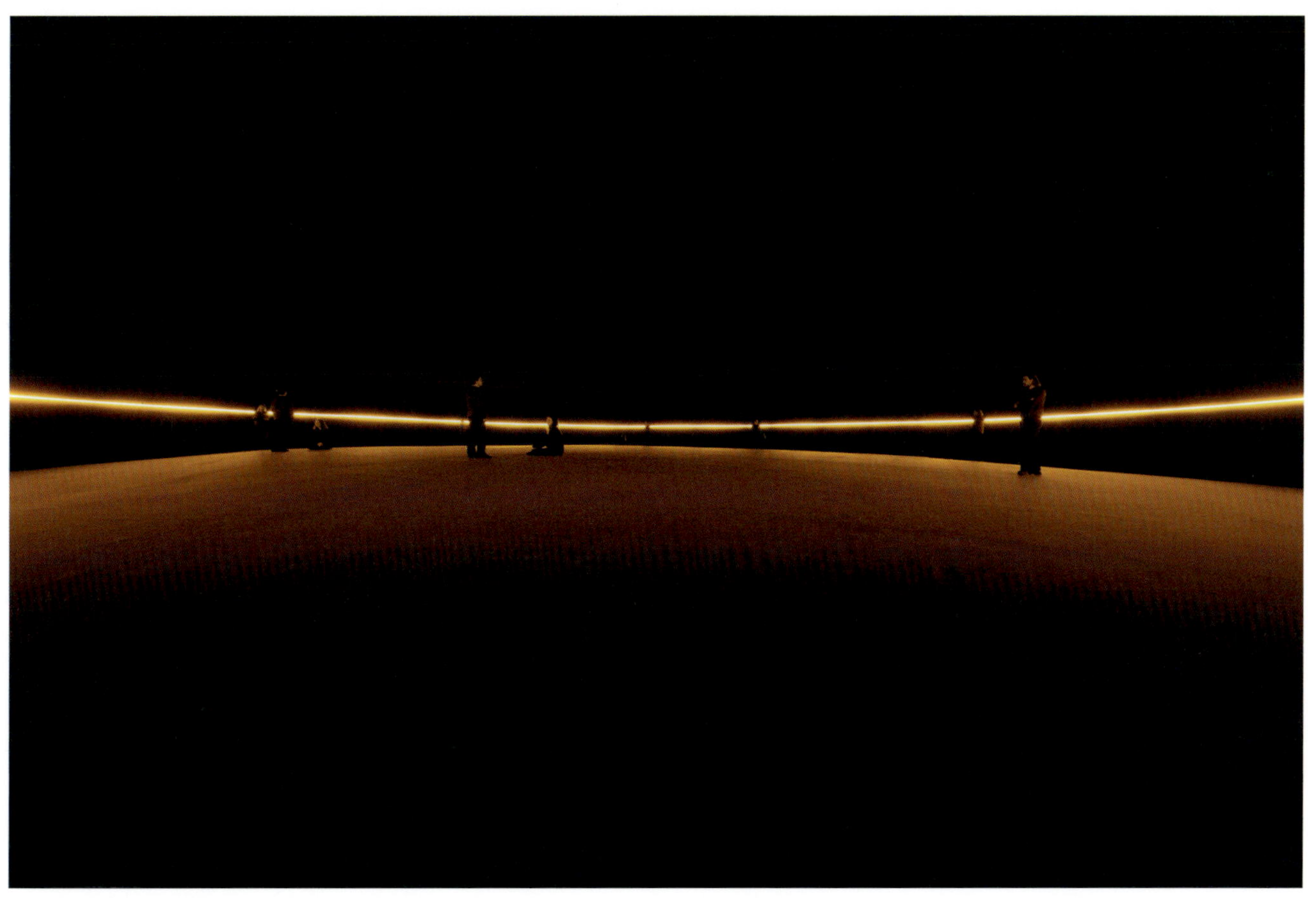

Photos Iwan Baan

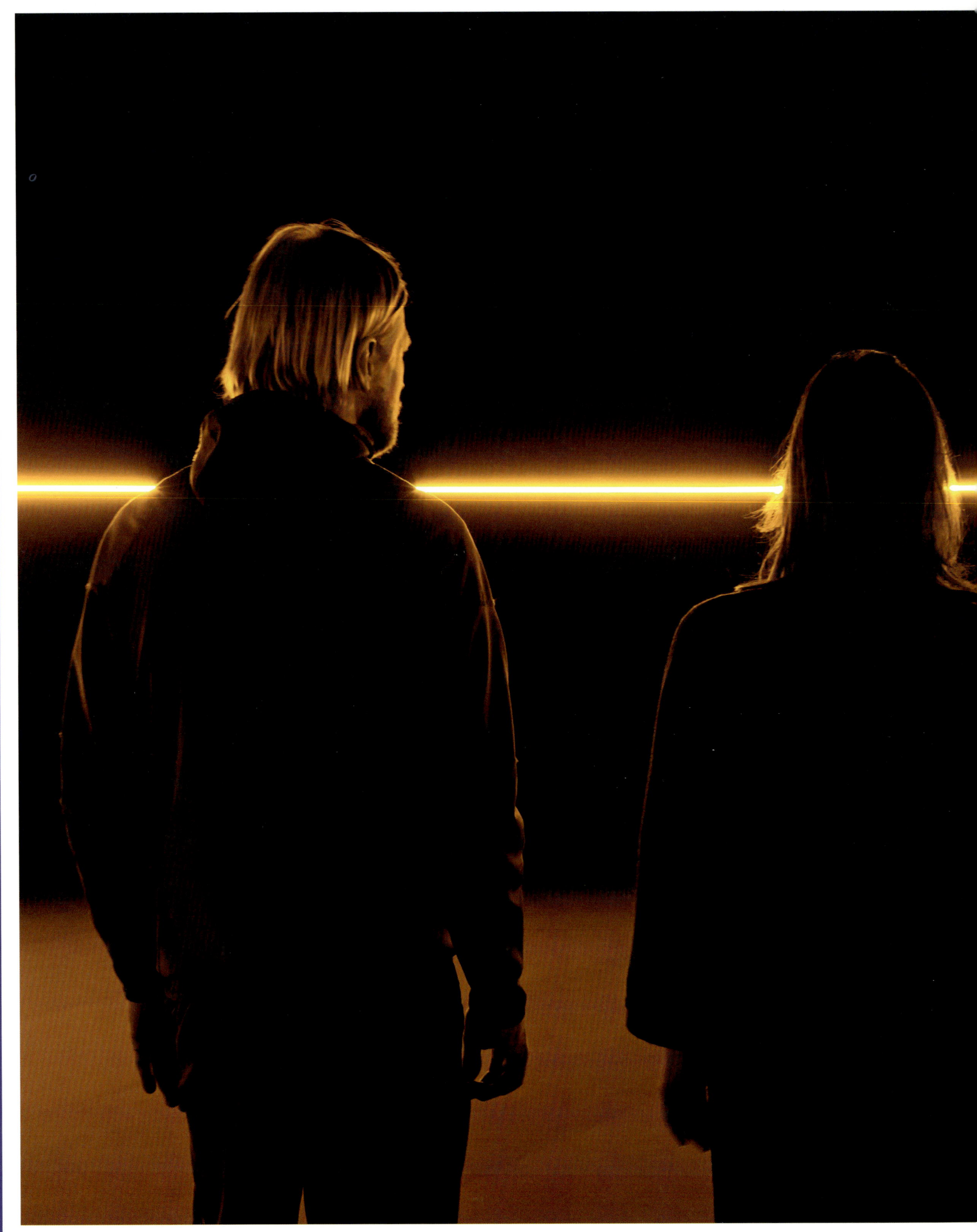

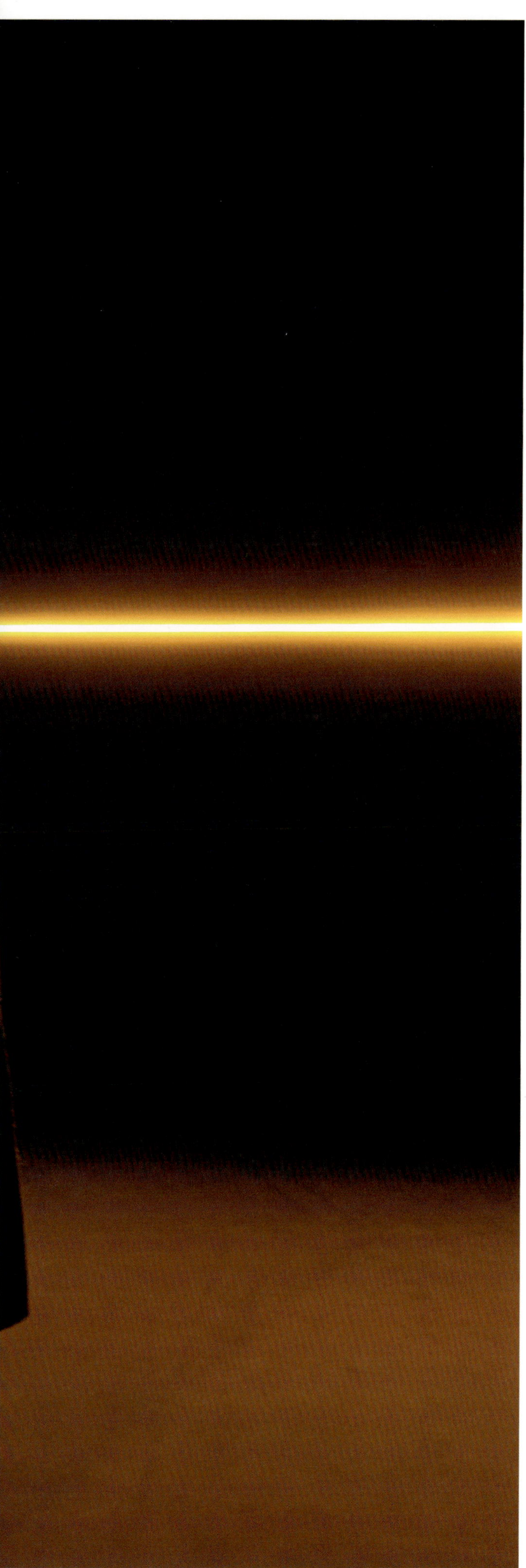

Designer
Olafur Eliasson
Location
Paris, France
Client
Fondation Louis Vuitton
Collaborator(s)/consultant
n/a
Manufacturer
n/a
Date
December 2014

VISITORS ARE PLUNGED INTO DARKNESS AND ENVELOPED IN A CHOREOGRAPHY OF MOVING LIGHT

Fondation Louis Vuitton launched an expansive exhibition by Olafur Eliasson in December 2014 and, in doing so, Eliasson became the first contemporary artist to exhibit at the Fondation's new building. Entitled *Contact*, the work explores 'the relations between self, space and universe' by creating a cosmos within the Fondation. The internationally-renowned Danish–Icelandic artist explains the work 'addresses that which lies at the edge of our senses and knowledge, of our imagination and our expectations. It is about the horizon that divides, for each of us, the known from the unknown.'

Tapping into the visitors' capacity for empathy, the artist strives to activate their participation, implicating them in a complex, multi-sensorial experience. As such, entering the vast space is akin to having been plunged into the darkness of outer space. The art installation envelops visitors in a choreography of moving light and shadows, and the sloping floor of the gallery space makes it feel as if they were traversing the top of a distant planet whilst witnessing a solar eclipse up-close.

On the roof of the building, an apparatus tracks the sun and, at certain hours of the day, directs light rays onto a multifaceted, geometric sculpture suspended within the building. A number of smaller optical devices distributed throughout specially-built passageways continue the artist's on-going investigations into the mechanisms of perception and the construction of space. Suzanne Pagé, the chief curator of the exhibition, comments, 'The constant oscillation between shadow/light, presence/absence and affirmation/doubt causes us to question our visual perceptions and, in consequence, our convictions. To this end, the route through the exhibition is derived from the geometry of the circle and founded upon the underlying principle of circularity.' —

Pitch black except for a single luminous horizon line, viewers may feel like they have become intergalactic explorers.

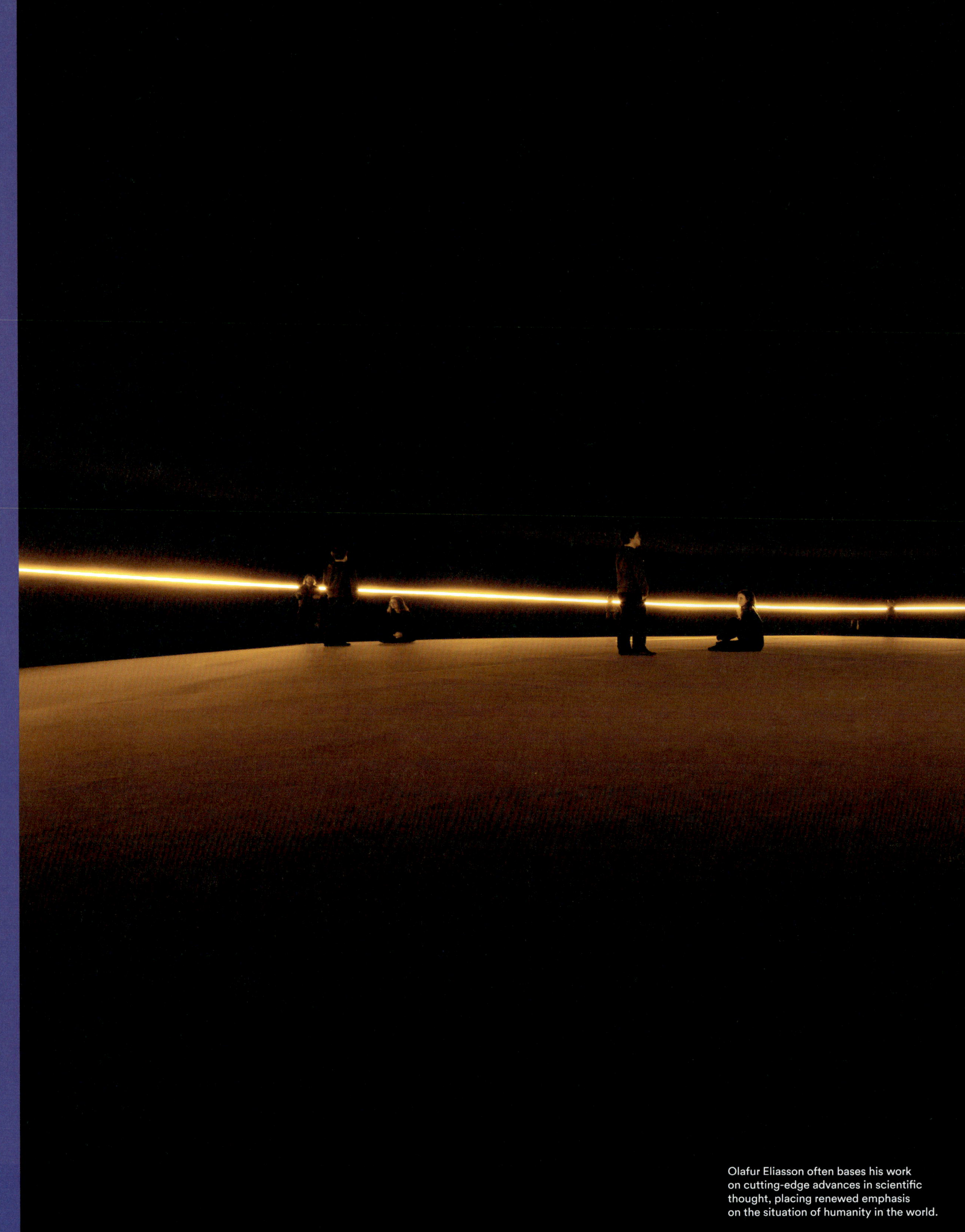

Olafur Eliasson often bases his work on cutting-edge advances in scientific thought, placing renewed emphasis on the situation of humanity in the world.

Specifics

ᐱᐱᐱ *Contact* is made up of numerous installations, including the featured other-worldly spatial experience of the same name. In total there are five works of art displayed as part of the exhibition, the journey through which is a coherent whole. It is about everything being interconnected, offering a sequence of events that should not be seen separately. These are not autonomous works of art, disconnected from the overall narrative; some might not even be works of art at all if they are taken in isolation.

ᐱᐱᐱ The show revolves around two large-scale installations, as well as transitional passages punctuated by three glass spheres – optical devices opening onto the outside, and thereby incorporating the exterior into the interior space.

ᐱᐱᐱ Eliasson wished to encourage visitors to think of themselves as if they were asteroids; to feel themselves floating through the space, meeting the artworks, meeting other visitors, and seeing them fly by. He wanted visitors to enjoy this temporal sequence and to try to take ownership of it by gradually slowing it down or speeding it up. 'It is about feeling your own presence, taking charts of your own trajectory, your own orbit, in your new 'asteroid' self.'

ᐱᐱᐱ To realise this work, the artist specifically worked within the context of Frank Gehry's strong architecture. 'While Gehry follows highly complex principles, I've chosen to use simple geometry, eventually reducing my intervention to basic shapes that provide the exhibition with a 'geometric metronome'. I've also created a number of spherical windows that allow glimpses of Gehry's building.'

Set-up

Total volume of the installation for illumination was 2555 m^3 (5.4 x 5.2 x 91 m).
Component parts included stainless steel, LED lights, coloured glass and mirror.
Used light was a continuous LED strip that created the impression of glowing, single horizon line.
Framework of the installation was in the form of LED strip lights around the circumference of the gallery space.
Optical devices included a multifaceted, geometric device to direct light rays at certain time of day, as well as smaller optical devices.
Construction of the installation was carried out on-site during the set up of the immersive, multi-room exhibition at the Frank Gehry-designed gallery.

PLAN

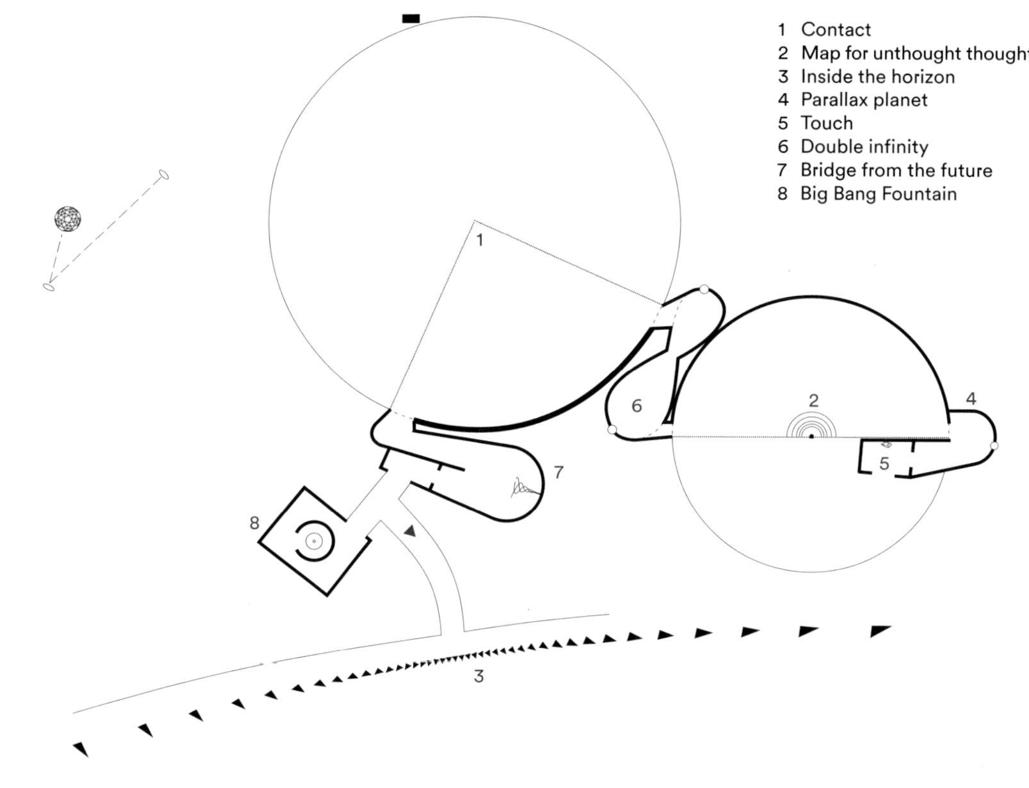

CONCEPT SKETCHES

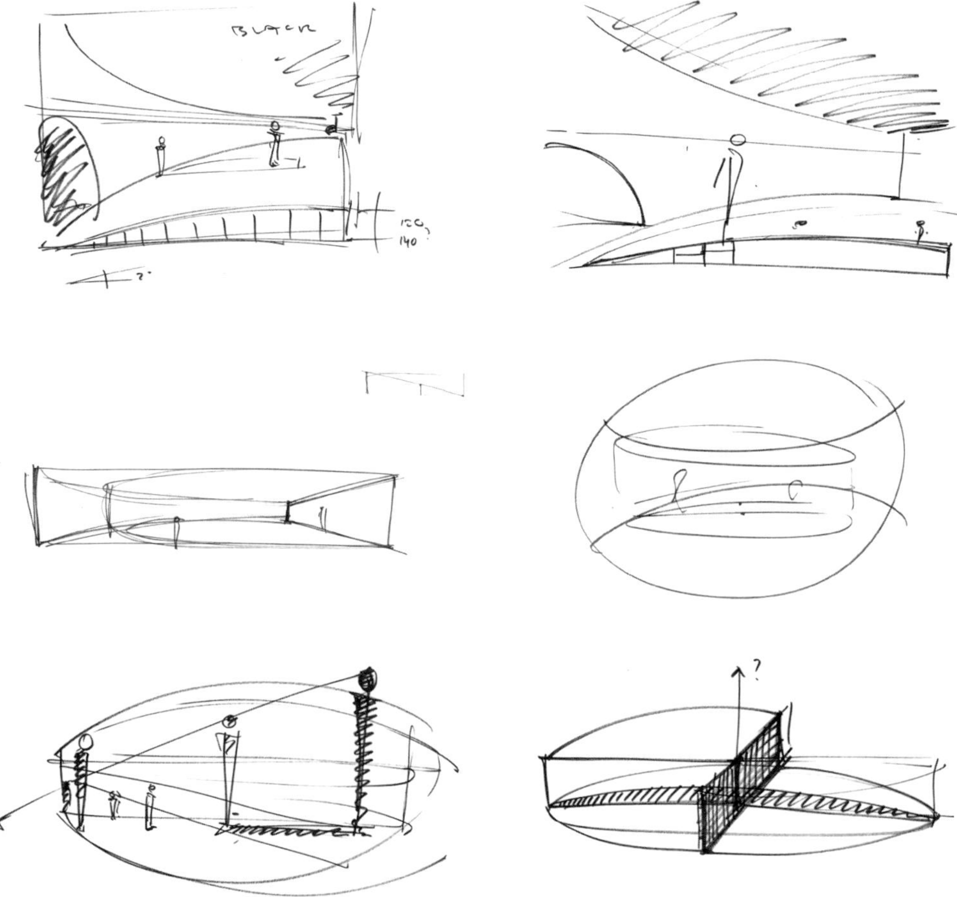

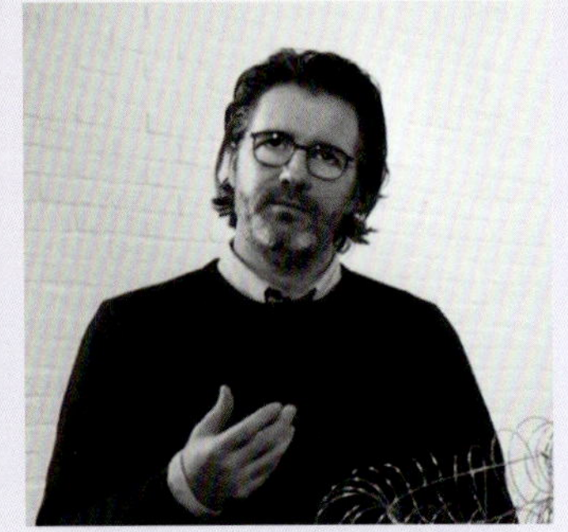

Olafur Eliasson

Olafur Eliasson's art is driven by his interests in perception, movement, embodied experience and feelings of self. Eliasson strives to make the concerns of art relevant to society at large. Art, for him, is a crucial means for turning thinking into doing in the world. Eliasson's diverse works – in sculpture, painting, photography, film and installations – have been exhibited widely throughout the world. Not limited to the confines of the museum and gallery, his practice engages the broader public sphere through architectural projects and interventions in civic space.

OLAFURELIASSON.NET

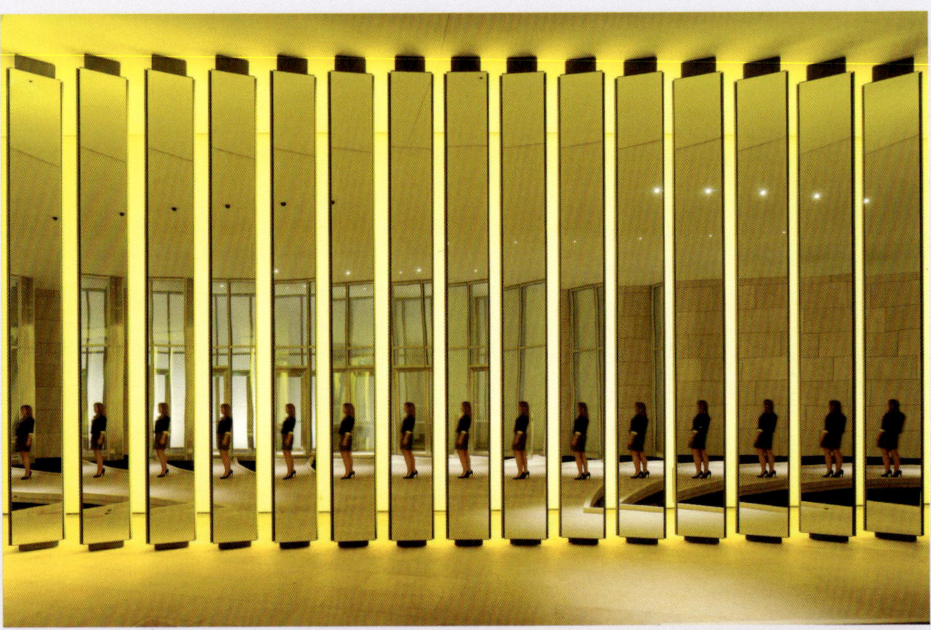

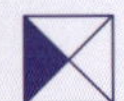

Inside the Horizon

When **2014**
Where **Paris, France**
Client **Fondation Louis Vuitton**

This site-specific installation saw the addition 43 triangular columns to the colonnade opposite the museum building. Two sides of each column are clad in mirrors, while the third is made of yellow glass tiles and illuminated from within. Running the full length of the colonnade, the work presents a vibrant play of light, shadows and reflections, and offers constantly changing perspectives. The work includes a unique sound composition by Samuli Kosminen and Olafur Eliasson, which emphasises the vibrant interplay of daylight, yellow light, shadows and reflections that offers constantly changing perspectives of the Fondation's architecture.

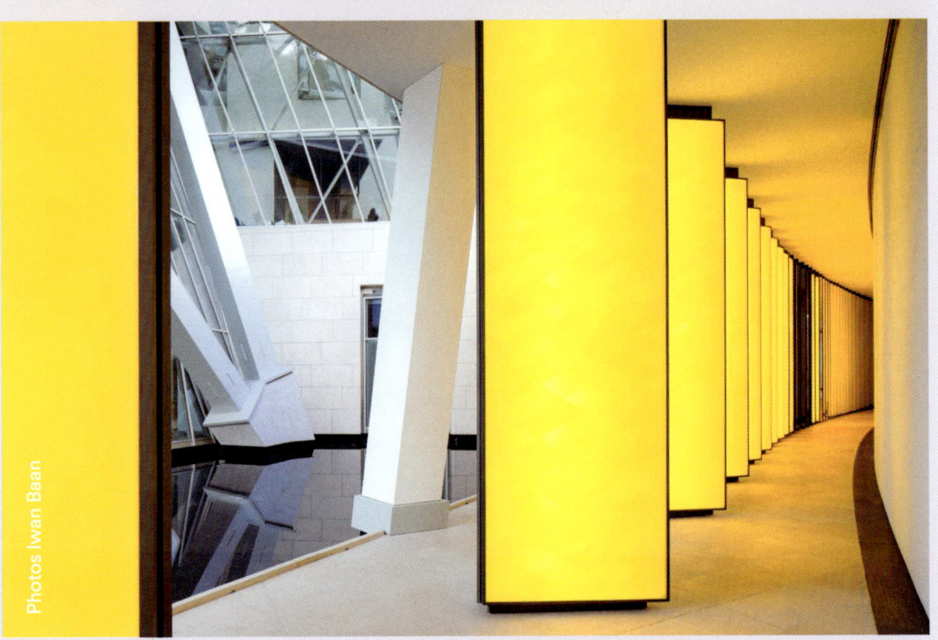

Photos Iwan Baan

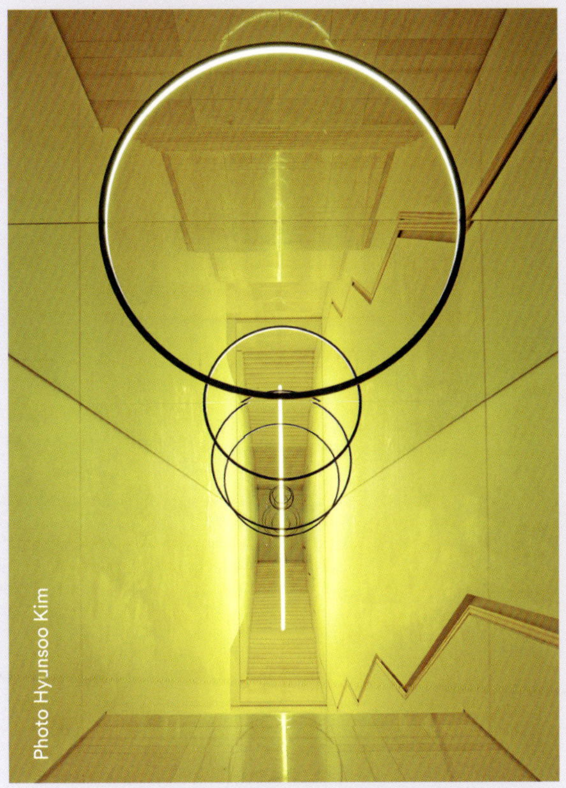

Photo Hyunsoo Kim

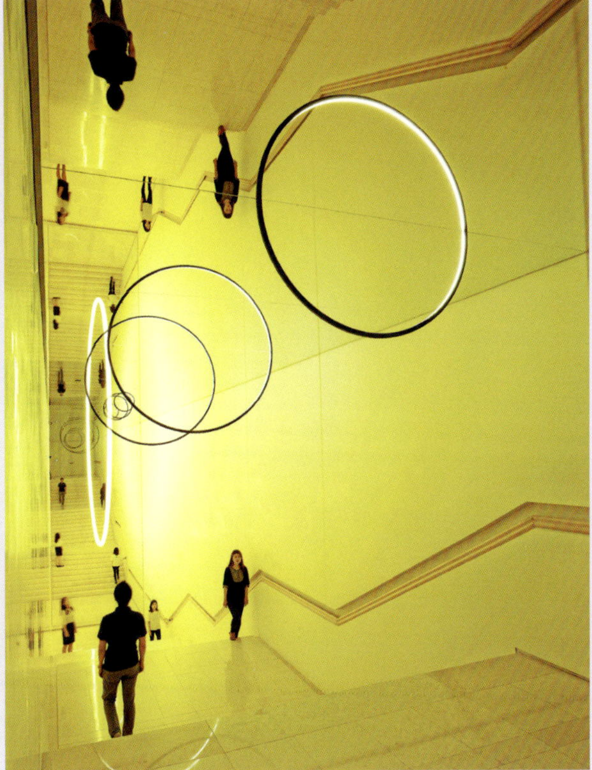

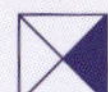

Gravity stairs

When **2014**
Where **Seoul, South Korea**
Client **Leeum, Samsung Museum of Art**

This installation shines on visitors as they traverse a stairway with an illuminated schematic model of the solar system positioned overhead, beneath a mirrored ceiling. The work is a play on perception thanks to the reflection of the particular arrangements of mirrors and lit ring segments in the ceiling mirrors. Semi-circular LED tubes look like complete spheres, thus replicating a glowing intergalactic display. The segments appear as full rings and a large, quarter ring attached to a wall-length mirror at the bottom of the stairs represents the sun.

Map for unthought thoughts

When **2014**
Where **Paris, France**
Client **Fondation Louis Vuitton**

This large semi-circular installation consists of five layers of criss-crossing bars that together form a pattern, based on five-fold symmetry, a moving light source and a mirrored wall. Moving about the space, visitors are able to map out distances with their bodies and movements. Viewers are at the centre of the piece – their shadow glides along a semicircle that is extended into a full circumference by a mirror. This shadow, shifting in scale, seems to orbit like an asteroid.

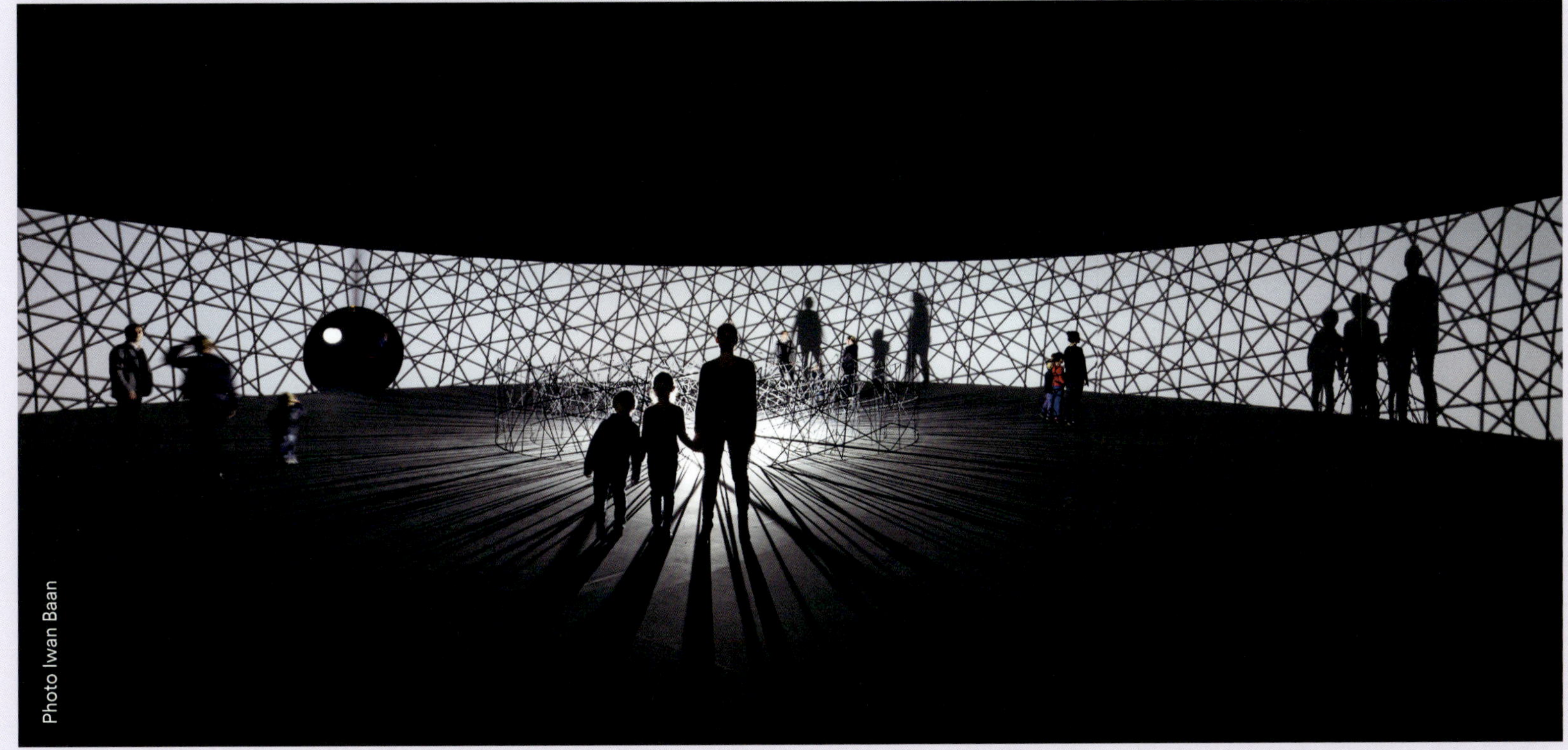

Photo Iwan Baan

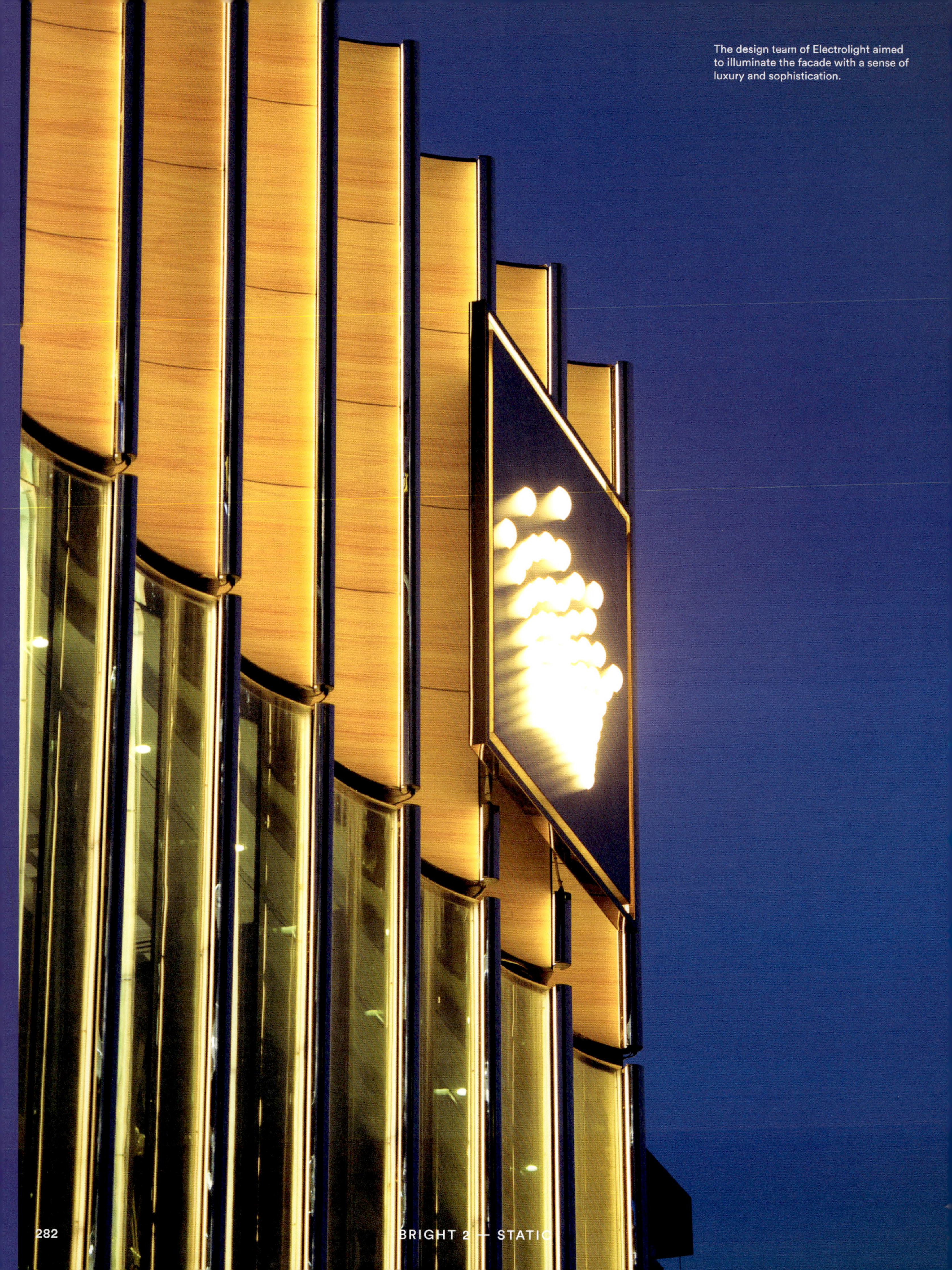

The design team of Electrolight aimed to illuminate the facade with a sense of luxury and sophistication.

Crown Towers Facade

Electrolight's specialist lighting design for the facade of the Crown Towers hotel in Melbourne boasts a smart composition of lighting elements which emphasises the building's sense of luxury and sophistication. The project showcases the studio's interest for the power of light to transform spaces and atmospheres.

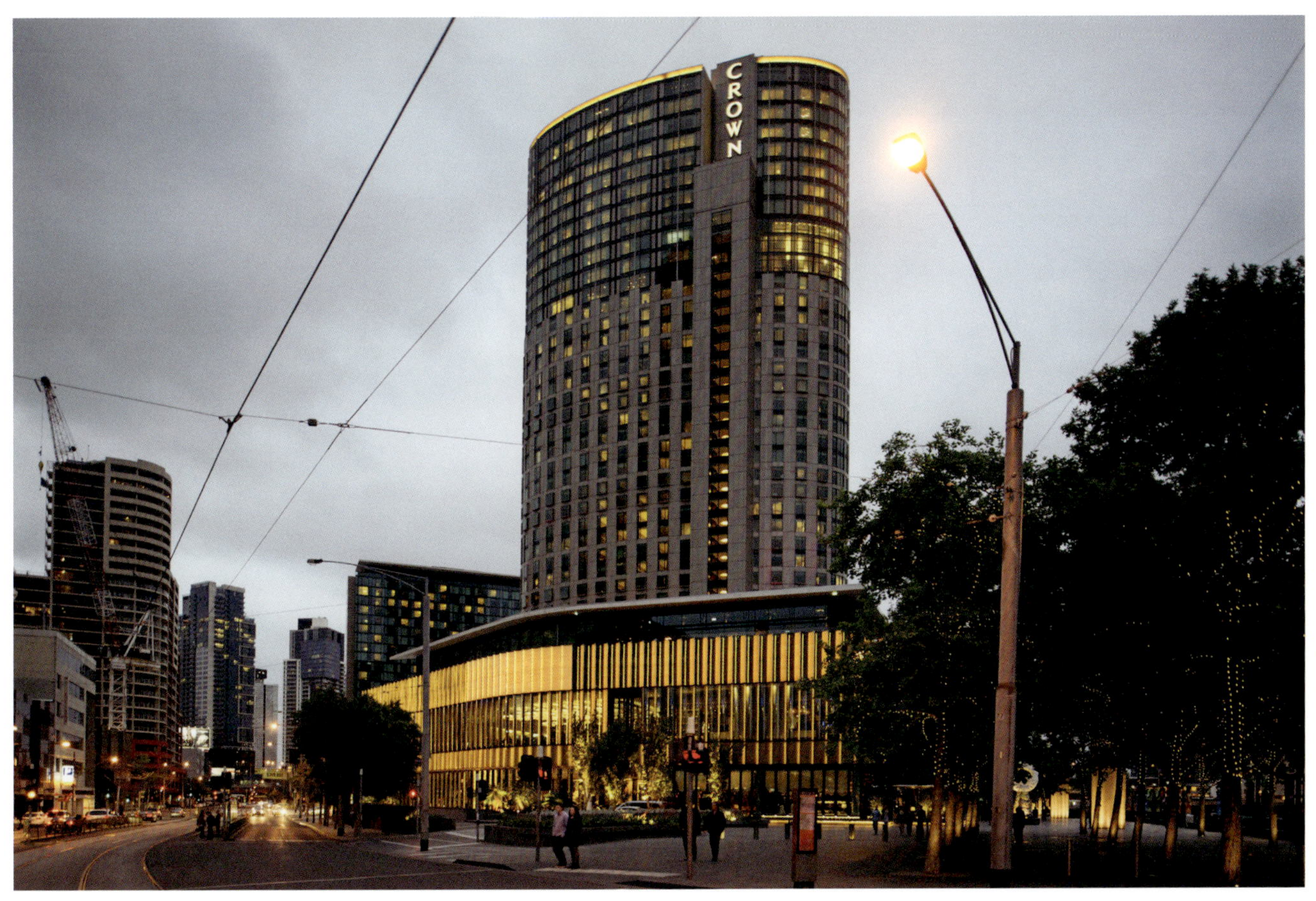

Photos Matt Irwin

Draping the facade in a lavish veil of lit glass was made possible thanks to cold cathode technology.

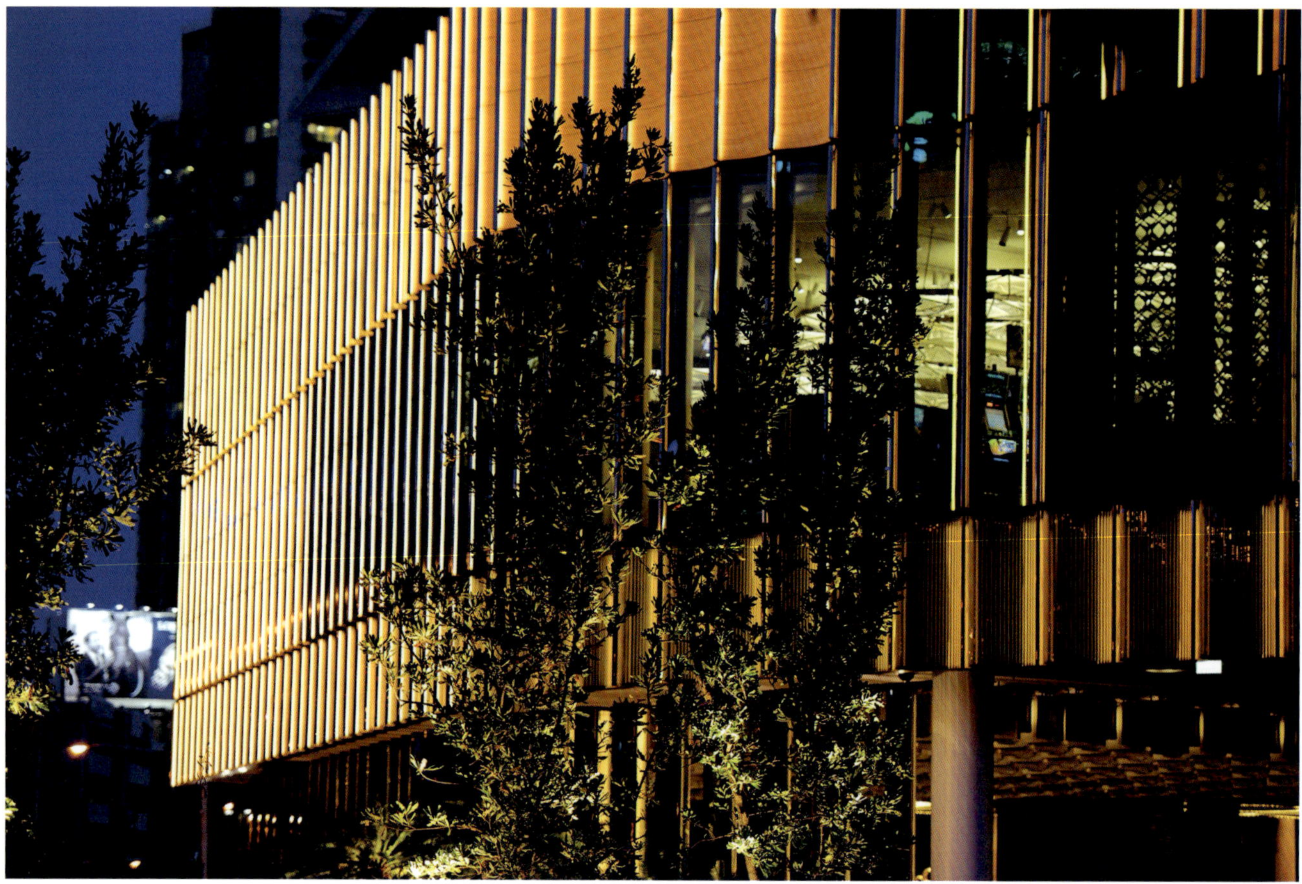

AN OPTIMAL INTERPLAY BETWEEN LIGHT AND SURFACE THAT WAS ALIGNED WITH NEARBY LIGHT SOURCES

The Crown Entertainment Complex in Melbourne, Australia is a vast hospitality space that integrates its own casino, hotels, function rooms, restaurants and shopping areas. In 2010, when its eastern entry underwent a major renovation, Electrolight was commissioned to help revitalise the 'Crown experience'. The studio was responsible for creating specialist lighting design for the new facade and the porte-cochère of the resort's most prestigious hotel, Crown Towers.

The dramatic new exterior – with its corrugated glass, sandstone and stainless steel – was emphasised with light. The design team selected cold cathode illumination, which reflected in each of the corrugations of the individual glass panels creating interesting effects. The concealed lighting system was seamlessly integrated into the facade's fabric, giving the impression that the building is draped by a lavish veil of lit glass. The properties of cold cathode allowed the designers to balance the facade's light temperature and align it to those of other light sources nearby, arriving at an optimal interplay between light and surface. Coupled with the thick glass panels, the lighting treatment adds depth and a gleam to the establishment, thus helping to lure customers inside.

The promise of luxury and sophistication continued at the hotel's porte-cochère, the entry point for all guests and visitors. Developed in close collaboration with Bates Smart Architects, the lighting scheme of the exterior area was an attempt to consolidate the entrance ceiling's illumination treatment into a whole, rather than applying separate gestures. This was achieved through the repetition of lighting elements. Alongside the entrance hall, a series of crystal-laden rings are suspended from the ceiling. Incorporating light, the rings 'create a spectacular entry statement,' as Electrolight puts it. —

Designer
Electrolight
Location
Melbourne, Australia
Client
Crown Resorts
Collaborator(s)/ consultant
Bates Smart Architects
Manufacturers
Coolon, We-ef, Endo, Beacon Neon
Date
March 2012

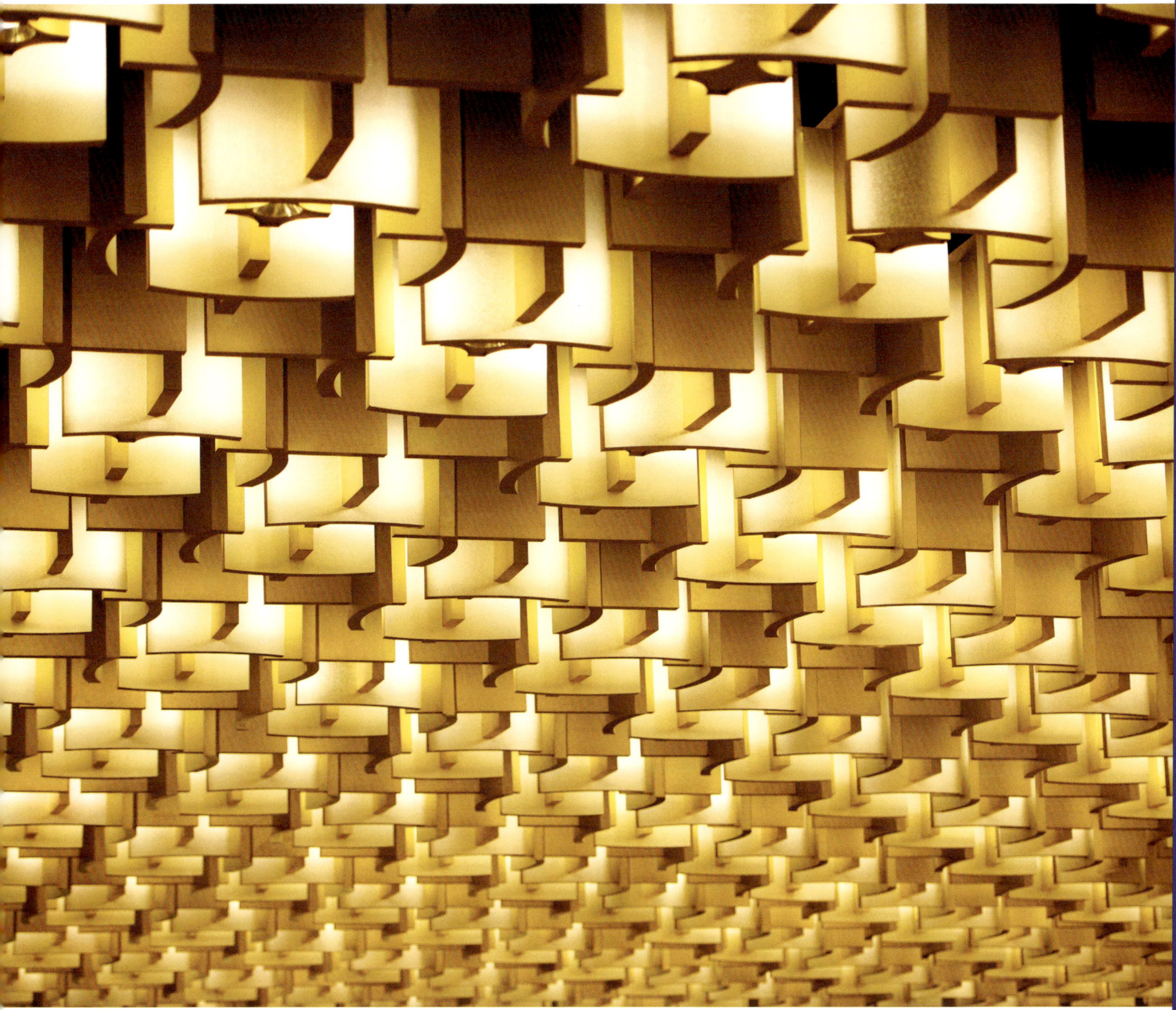

A coherent concept utilising complementary lighting gestures connected the exterior facade and the porte-cochère ceiling.

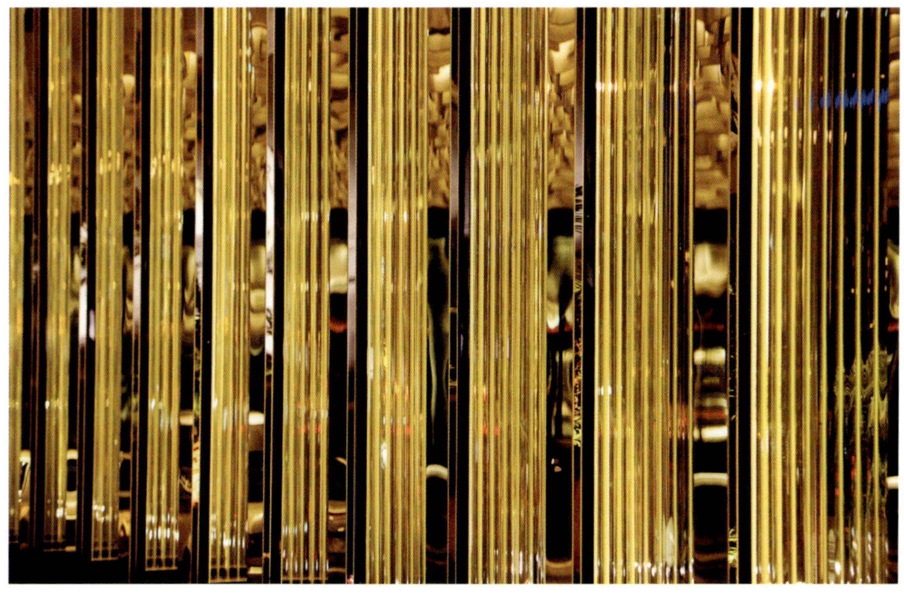

PLAN

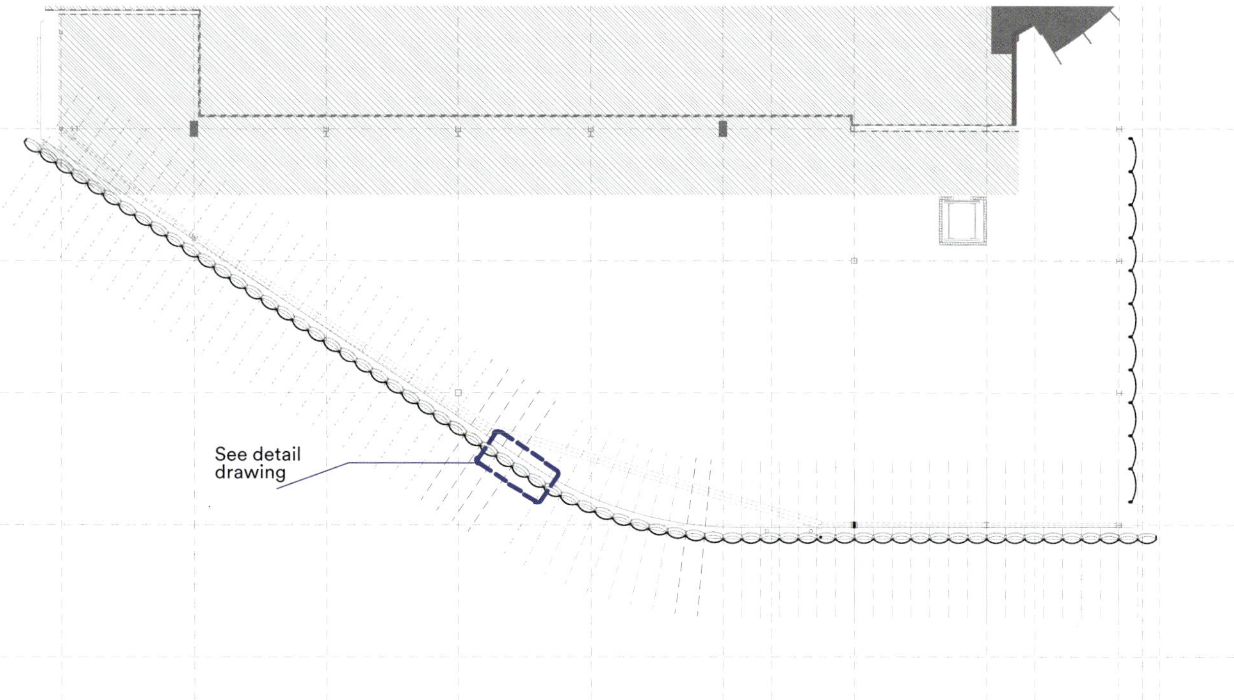

EAST FACADE ELEVATION

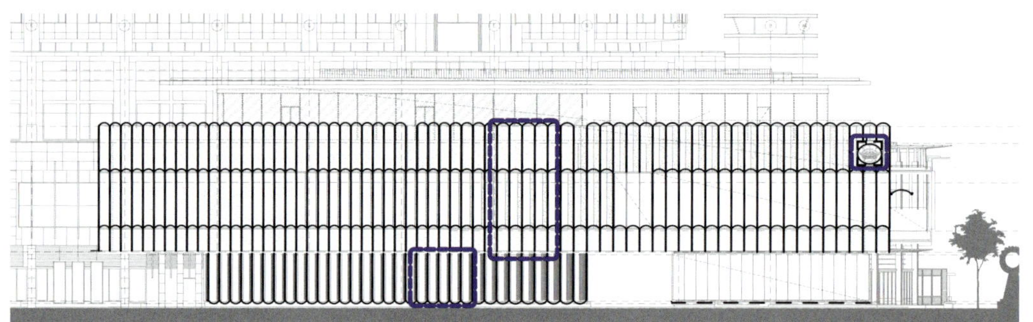

ILLUMINATED FACADE PANEL TYPES

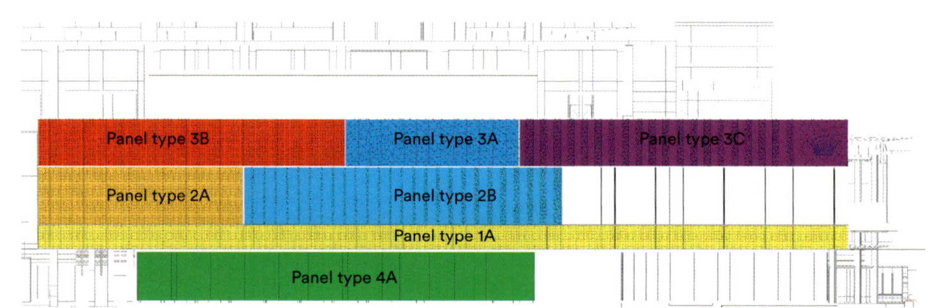

Set-up

Total surface area for the project was 1200 m^2.
Energy consumption per square metre of total energy was 11.89 W.
Used light was cold cathodes. This type of illumination was selected because of the long life properties and versatility of colour temperatures that was available. LED fixtures were used to highlight interior design and decoration features such as door handles, plants, artwork and columns.
Luminaires (cold cathode) of various lengths were supplied with 90 mA, 8 kV dimmable transformers (2 x runs per transformer, each concealed in an accessible location by the architect).
Neon current of the luminaires fixed to the corrugated glass was to be 100 mA.
Control boxes that indicated multiple circuits had luminaires evenly divided between the circuits.
Facade panels had cold cathode integrated in the mullion capping to either reflect in the corrugated glass or wash across the stone surface.
Crown sign was illuminated with a circuit of up to 10 V.
Construction time was 12 months.

EAST FACADE DETAIL

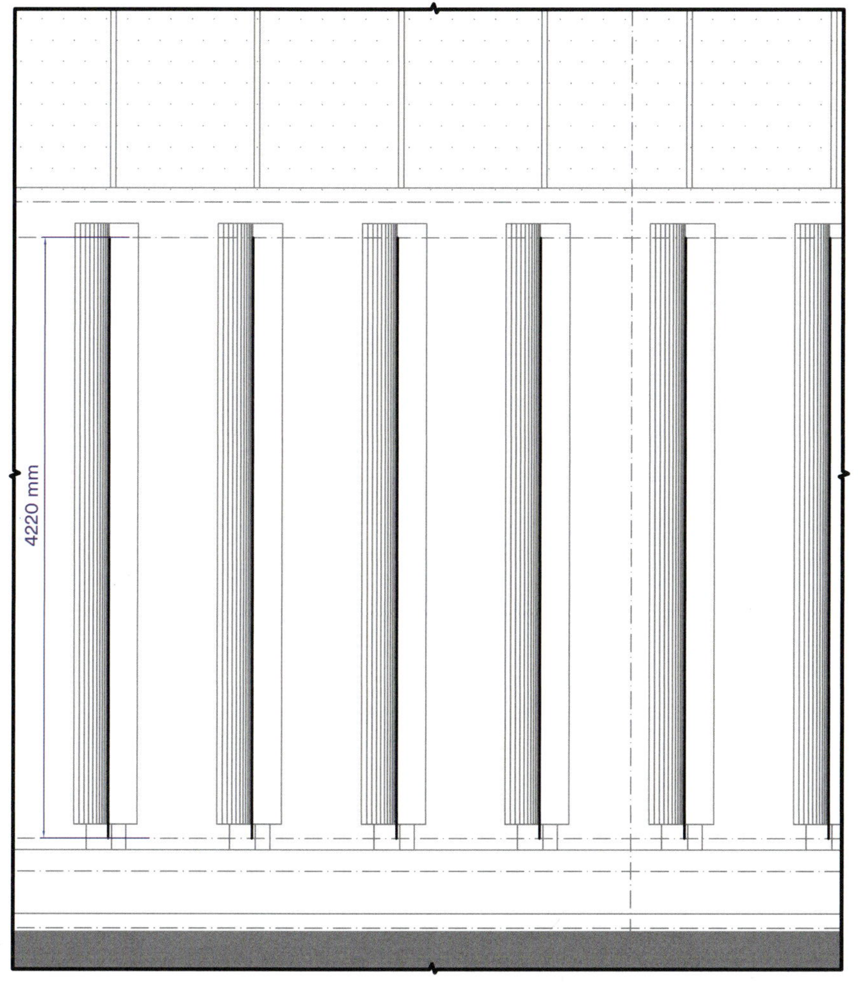

MULLION NEON LIGHT DETAIL (TYPE 1A)

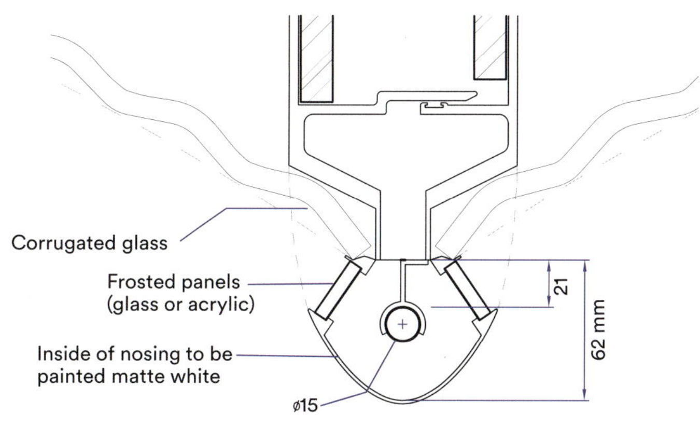

THE CONCEALED LIGHTING SYSTEM SEAMLESSLY INTEGRATED INTO THE FACADE'S FABRIC

FACADE SKETCHES

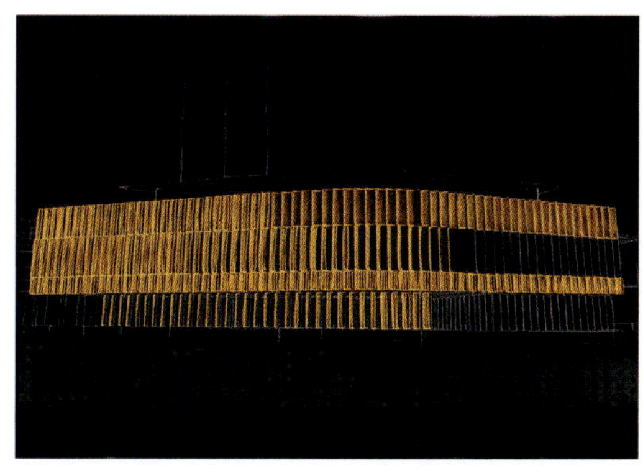

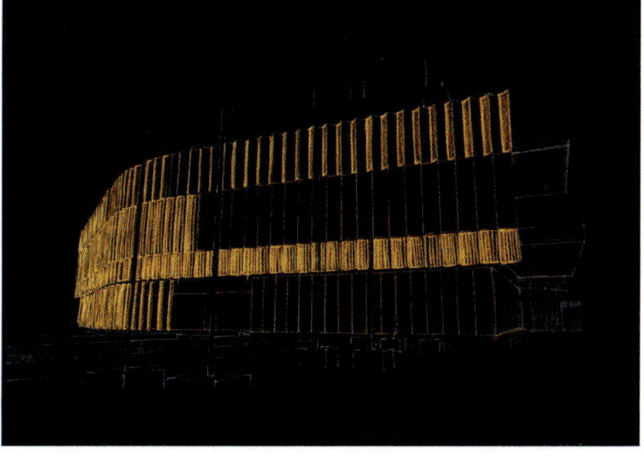

CONCEPT SKETCH

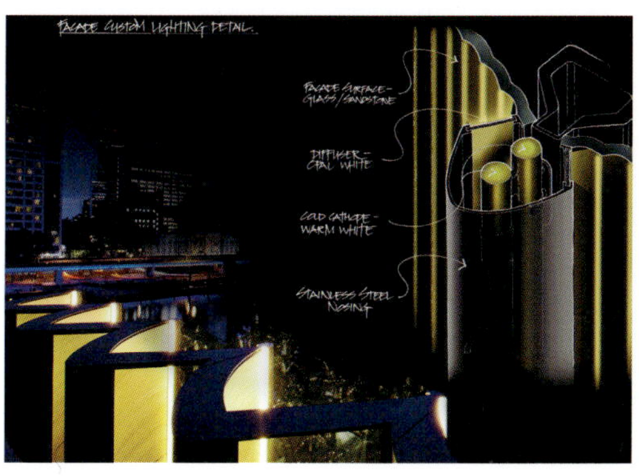

PROTOTYPE

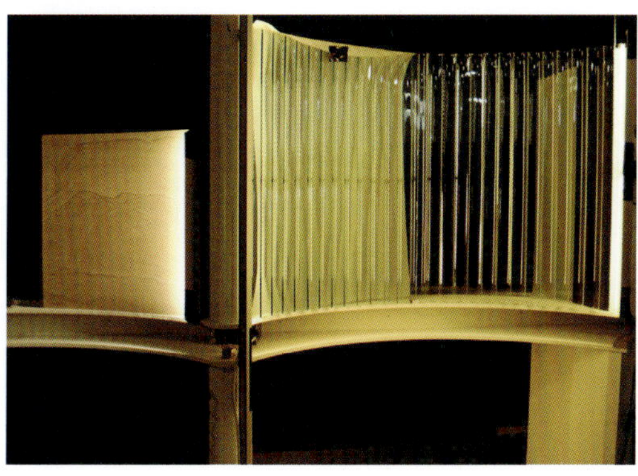

Specifics

ᴧᴧᴧ Lighting the glass facade and creating a unifying effect that spanned the entire frontage of the hotel was one of the biggest challenges. The conceptual idea behind the facade was a perfume bottle – beautifully crafted, elegant, sophisticated and luxurious. Key words used to describe the design intent were luxury, sparkle, glow and warmth. The lighting needed to reflect this.

ᴧᴧᴧ Extensive prototyping and testing was required in order to refine lighting arrangement and materials, in order to arrive to an optimal interplay between light and surface.

ᴧᴧᴧ For the hotel's porte-cochère, since the area is accessed by pedestrians and drivers alike, the lighting system needed to be designed to accommodate a wide variety of ambient lighting conditions, allowing drivers to safely adapt their vision as they enter from a sunlit street, for instance.

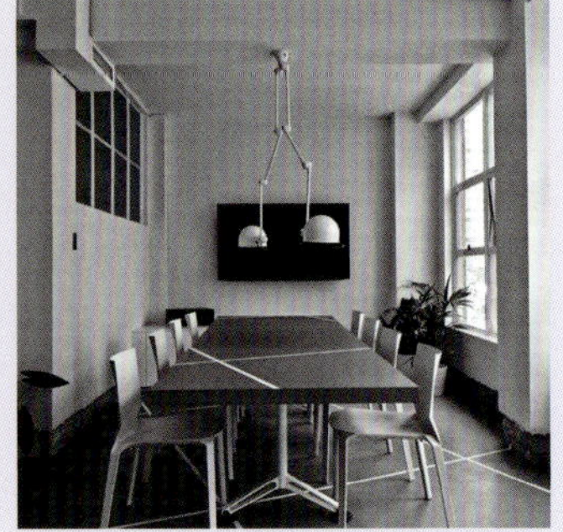

Electrolight

Electrolight is an independent lighting design consultancy developing a myriad of projects throughout the built environment. Founded in 2004 by Paul Beale, the award-winning firm has offices in Melbourne, Sydney and London, employing an experienced team of professionals who focus on interior and exterior lighting alike. The studio's passion for 'light and its ability to enhance and transform' and prestigious international reputation draws clients from sectors as diverse as architecture, engineering, visual art and computer modelling, as well as theatre, interior and industrial design.

ELECTROLIGHT.COM

171 Collins Street

When **2013**
Where **Melbourne, Australia**
Client **Charter Hall and Cbus Property**

Electrolight's specialist lighting design for the ground floor common areas of the premium office building at 171 Collins Street emphasises the internal architecture and reinforces a sense of luxury and sophistication. A custom-designed architectural recess with a frosted glass diffuser and internal LED strip at the base of the stone walls creates a glowing line that visually defines the perimeter of the spaces. Metal halide lamps are directed at the travertine stone walls of the entry, atrium and ground floor lobby so that the space is lit with reflected light infused with the warm hue of the stone. Environmentally sustainable principles were central to the lighting design.

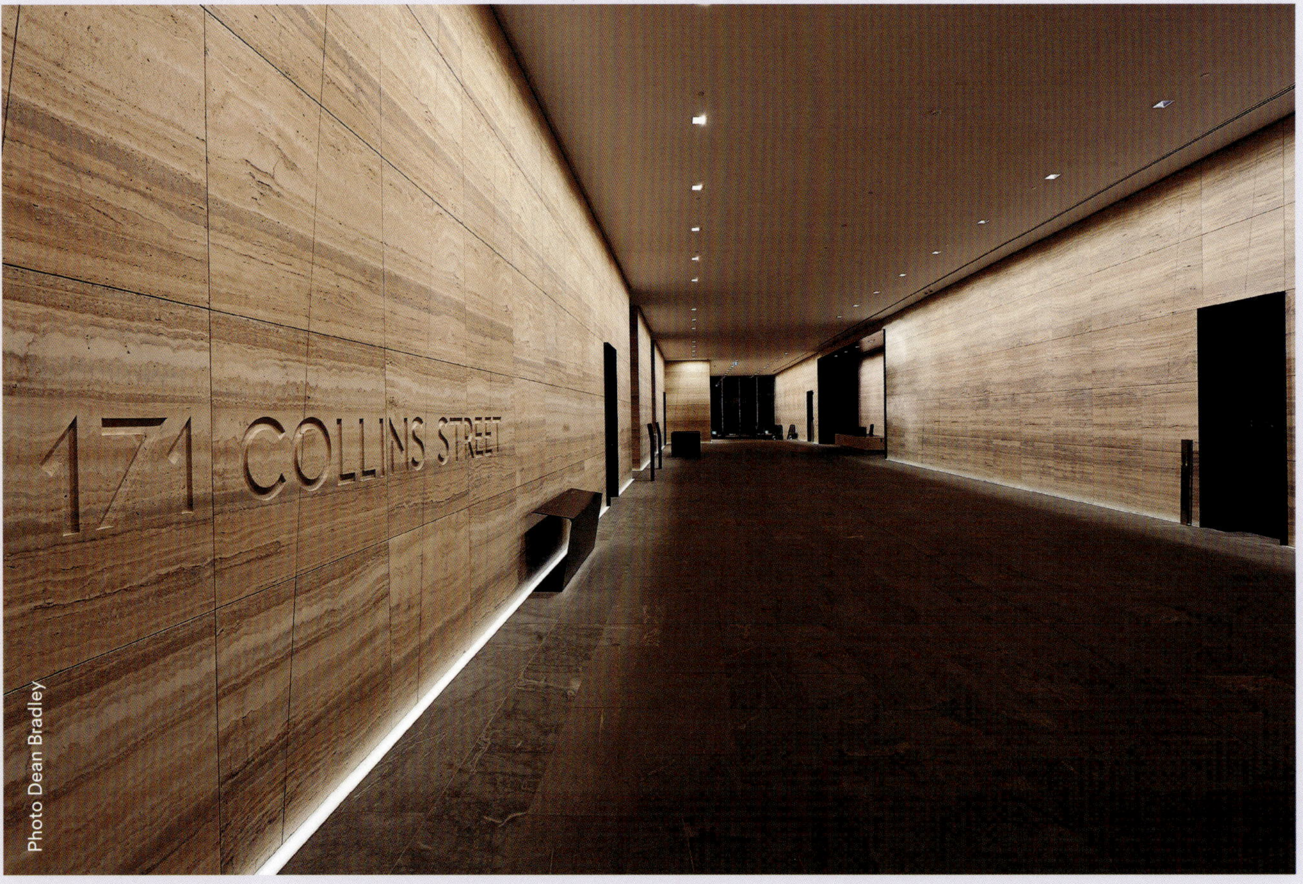

Photo Dean Bradley

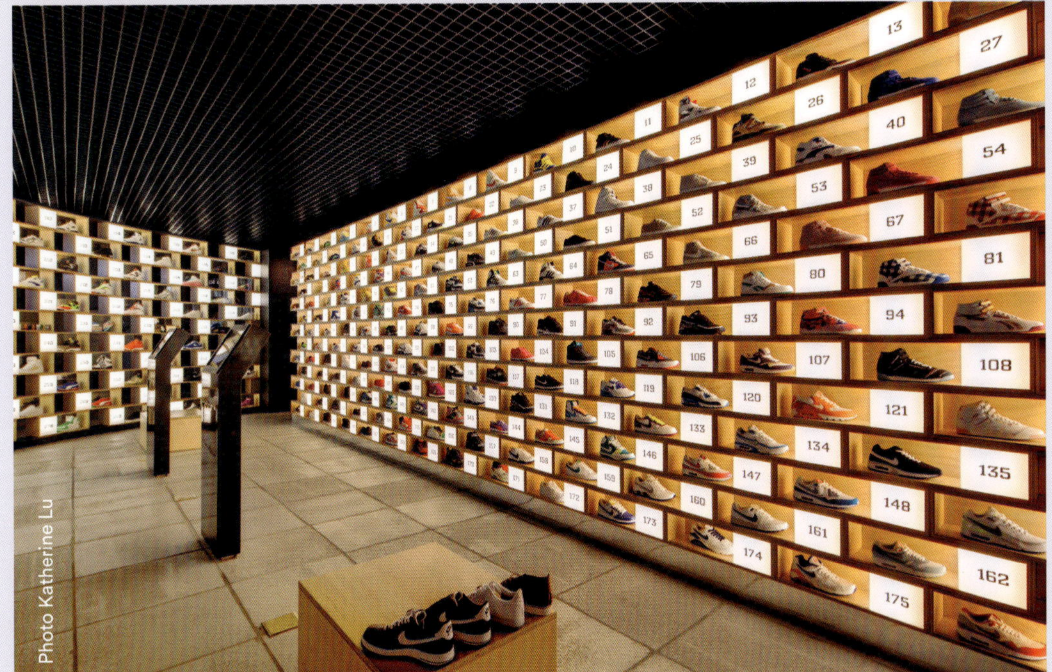
Photo Katherine Lu

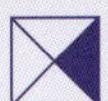

Sneakerology

When **2012**
Where **Sydney, Australia**
Client **Facet Studio**

A custom-made lighting system was paired with a store design by Facet Studio for Sneakerology. This boutique in central Sydney, selling exclusive and limited-edition footwear, boasts an innovative, vibrant and unique retail experience. The concept utilises a 'shoe-box' display system, with built-in light boxes presenting the footwear as items in a museum-like installation which can be navigated through touch screens.

Central Dandenong

When **2011**
Where **Dandenong, Australia**
Client **Places Victoria**

The safety, comfort and requirements of pedestrians have been the driving force behind Electrolight's lighting design for the revitalisation of Lonsdale and Langhorne Streets in Central Dandenong. A white LED and feature lighting scheme enhances the prestige of the site and renews Dandenong's community sense of civic pride by unifying the urban fabric and setting a benchmark in environmental performance for exterior lighting in Australia.

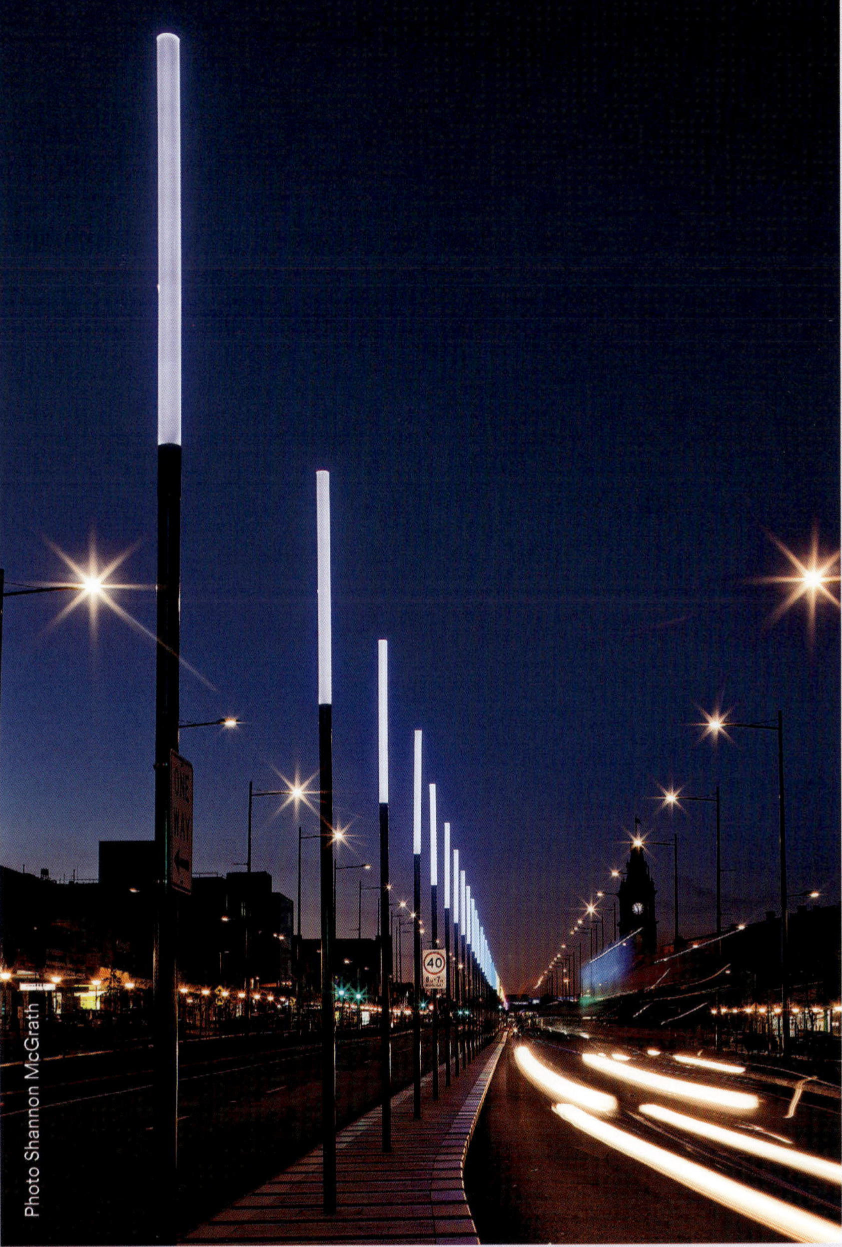
Photo Shannon McGrath

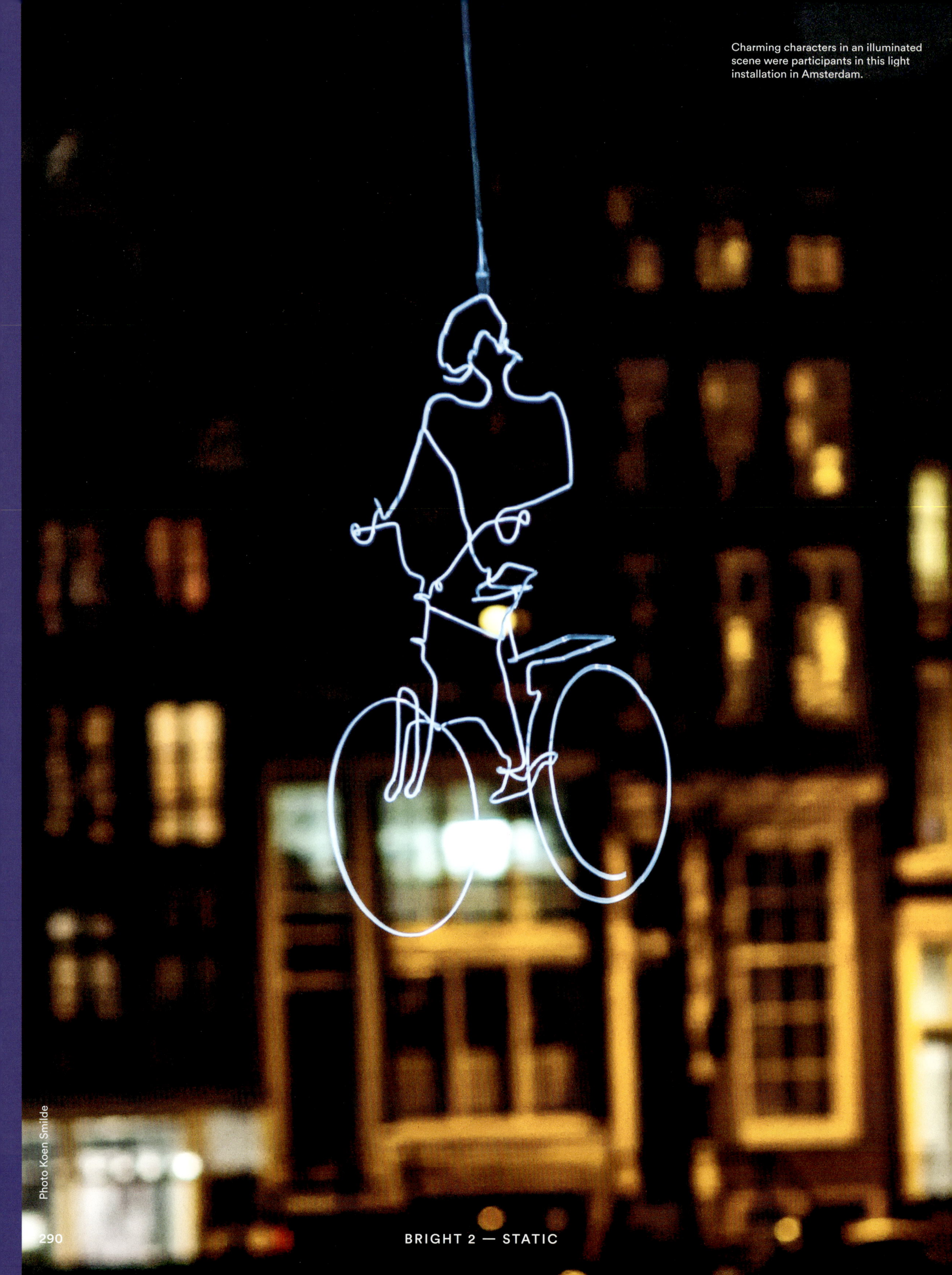

Charming characters in an illuminated scene were participants in this light installation in Amsterdam.

Photo Koen Smilde

Drawn in Light

Dutch artist and designer **Ralf Westerhof** created a spinning 3D-sculpture in light over a city canal, in response to the Amsterdam Light Festival's commission to brighten up the cold winter nights.

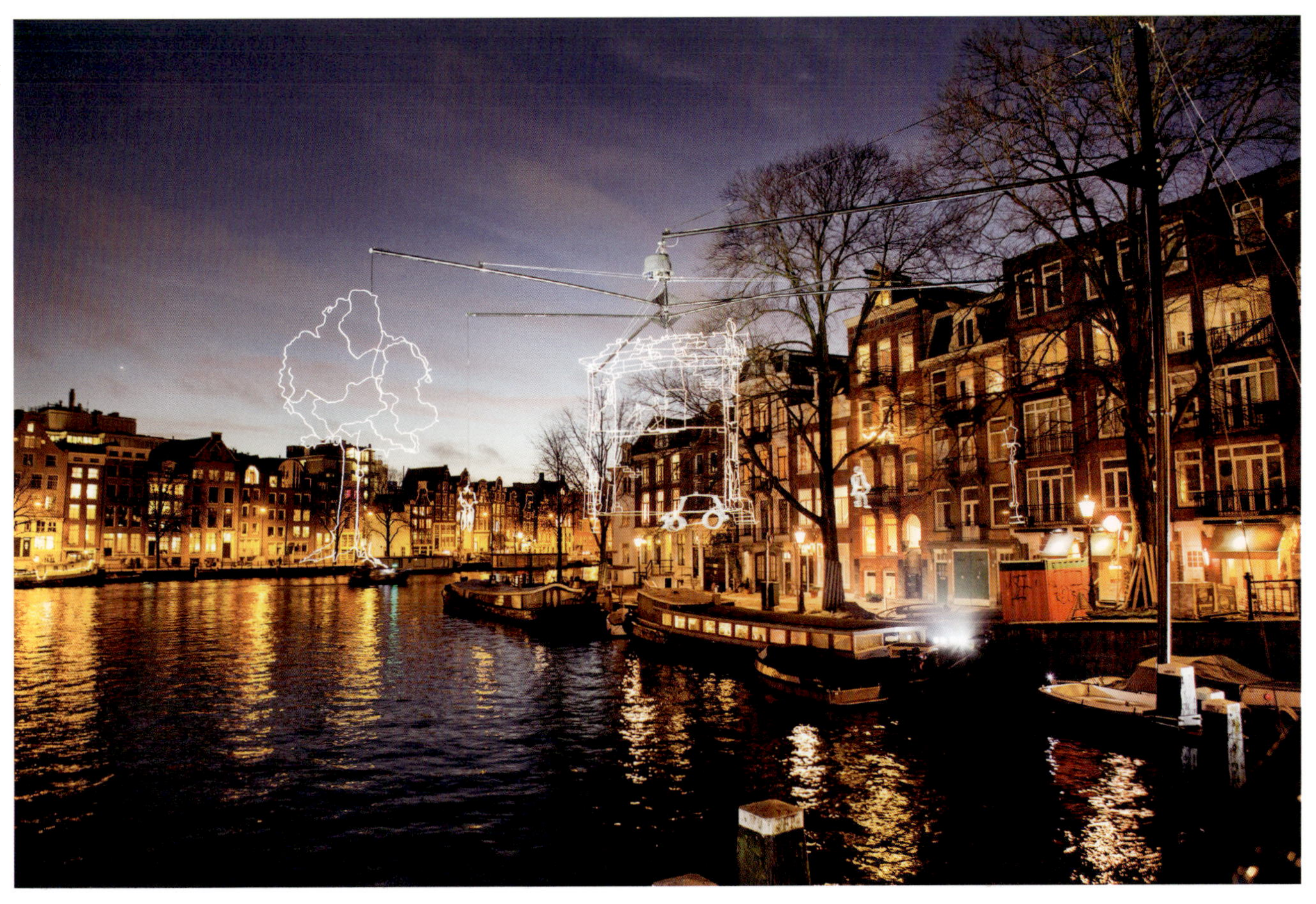

Photos Janus van den Eijnden

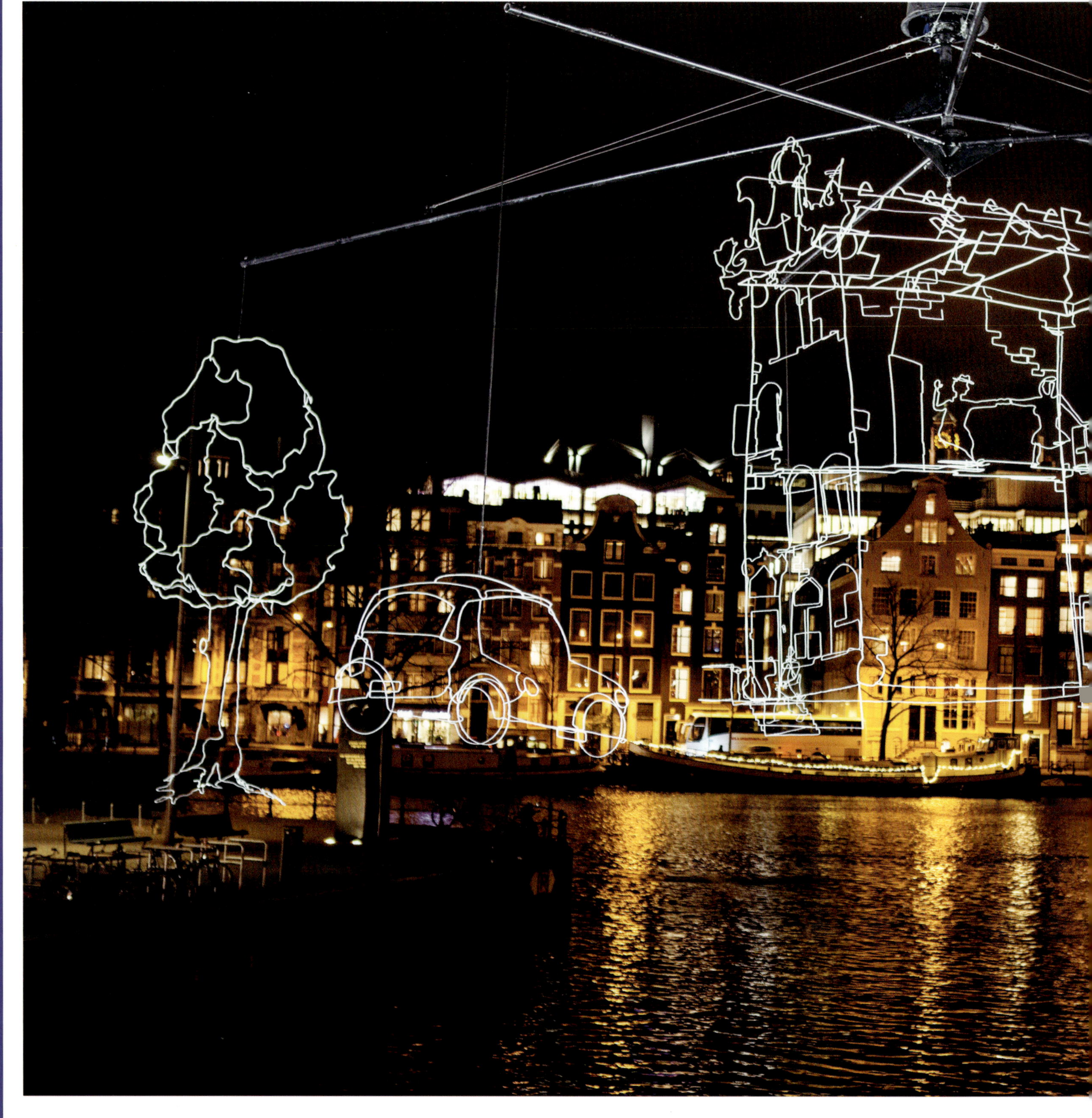

THE EYE-CATCHING CITYSCAPE AS AN EVER-CHANGING DRAWING MADE OF LIGHT

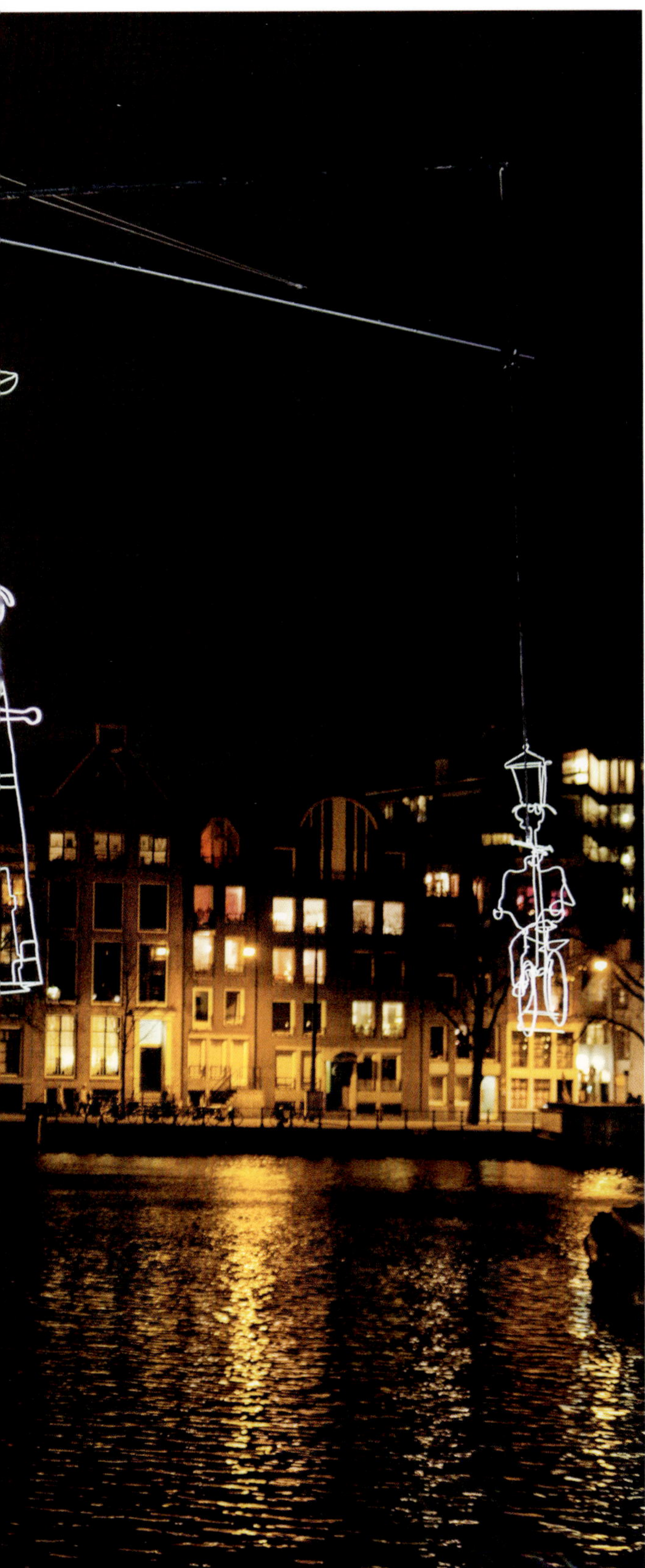

The installation sees a 3D sculpture spinning in light over the city canal.

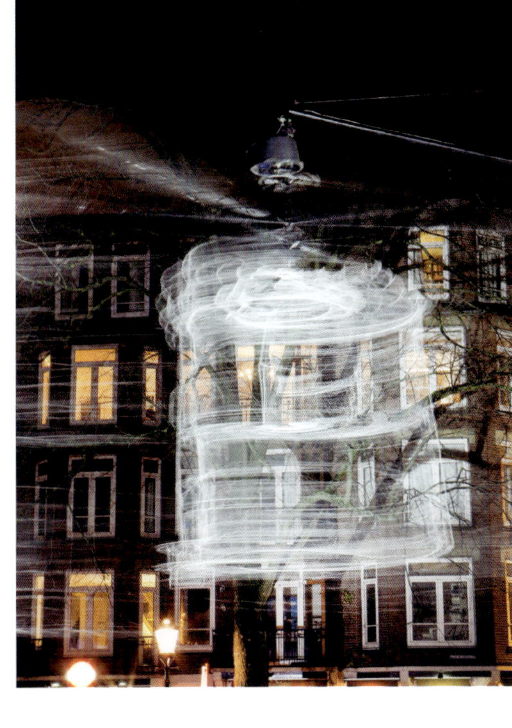

Designer
Ralf Westerhof
Location
Amsterdam, the Netherlands
Client
Amsterdam Light Festival
Collaborator(s)/consultant
Het Knutselparadijs
Manufacturer
Ralf Westerhof
Date
December 2013

Amsterdam was the location – and inspiration – for an eye-catching light installation that lit up a canal in the Dutch capital over the course of 2 months during a wintery holiday season. The Drawn in Light project by Ralf Westerhof shone bright as a beacon over the course of several foggy days and welcomed the New Year in with optimism. With figures containing minimal lines, the installation consists mainly of empty space, but still it feels very solid. The structure relies greatly on the exact placement of detailed parts in a spatial relation. Against the dark night sky, the bright three-dimensional line drawing of a typical Amsterdam-style canal building hung, slowly spinning in the air as elements of its surroundings gently orbitted around it.

This eye-catching cityscape was an ever-changing drawing made of light – developed as a progression of Westerhof's indoor works in which he also worked with light. In those works, the black lines of the 3D drawings cast shadows on their white surroundings, while for the Drawn in Light structure there are no surroundings to cast shadows on. In this work, all the attention goes to the lines of the drawing itself.

Crafted out of stainless steel rods that were hand-bent and welded, the mobile takes on the shape of a house, replete with a nearby tree and passing cyclist. Once the entire structure had been powder-coated white and suspended aloft over the water, LED spotlights provided accurate illumination. The constituent features of the construction included three layers that rotated independently, spinning on an axel propelled by an electric motor. As there is nothing else around to reflect light, the lines are brightly visible in the dark sky. In this work, all the attention goes to the lines of the drawing itself. In principle, thanks to the various light sources, the whole structure is lit all the time. Interestingly, when one light source illuminates a different spot of the structure, then it gives the viewers an extra perspective of the same image – still three dimensional and coherent, but slightly different. The experience for the spectator is different depending on the viewing angle, the light source placement and what part of the moving sculpture is facing the observer. —

PLAN

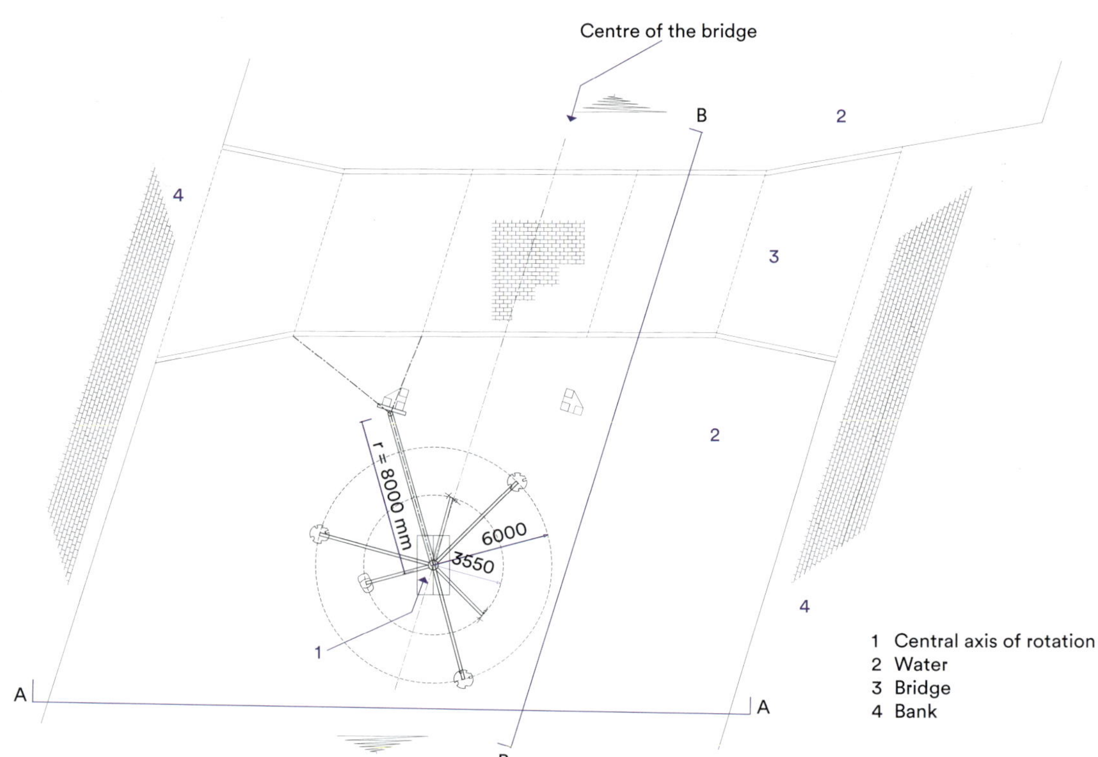

SCALE MODEL

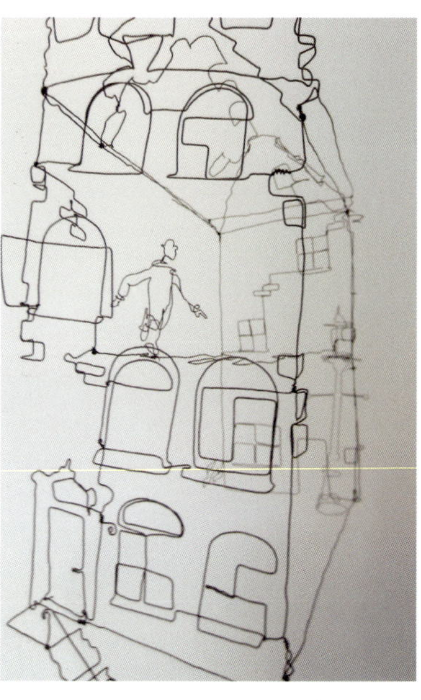

ENHANCED BY LED ILLUMINATION, THE SCULPTURE HAS AN INTRIGUING THREE-DIMENSIONAL FORM

SECTION A

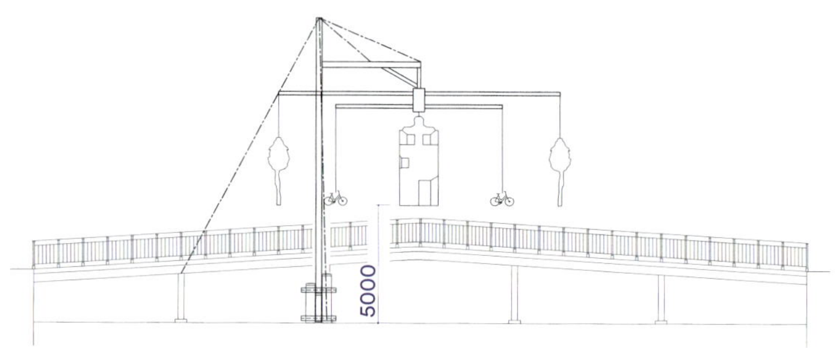

SECTION B

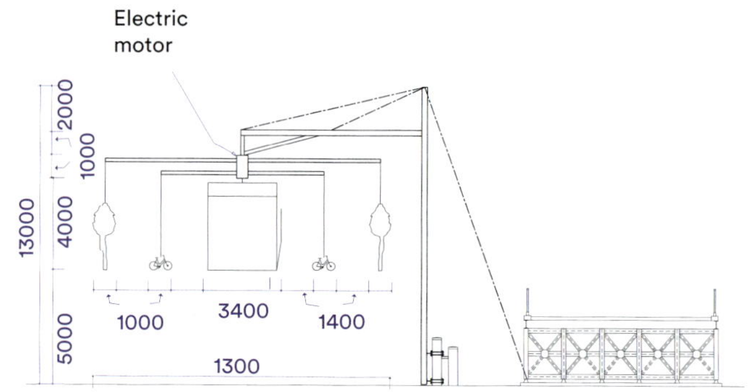

Set-up

16 LED PAR spotlights (56 Watt) were used to illuminate the moving structure, located on different sides (both sides of the canal and on the bridge).
Sculpture size was 49 m² in total, with the individual elements hanging from a frame that was 13-m high, with a horizontal pole of 8 m in length. The arms of the largest moving horizontal beams were 6 m. The dimensions of the largest component was the canal house at 3.8 x 2.2 x 3 m, with the tree being 3.8-m high and each streetlamp 1.5-m high.
Stainless steel wire, with a bright white powder coating (RAL 9016, Traffic White), was used to shape the cityscape. To bend the 10-mm wire for the roof required a pneumatic device, as well as cutting and TIG welding. The rest of the house was made of 6- and 4-mm wire, with the other elements made of 3-mm wire.
An electric motor was utilised so that the mobile moved around the axel at a slow and steady speed, with gear wheels and chains ensuring the different layers were propelled at different speeds and directions.
Construction time in total was 40 days: from sketch, research and preparation (21 days) to bending and welding (16 days), stainless-steal powder coating (1 day) and on-site build-up (1 day).

The characters in the streetscape were created by bending stainless steel into shape.

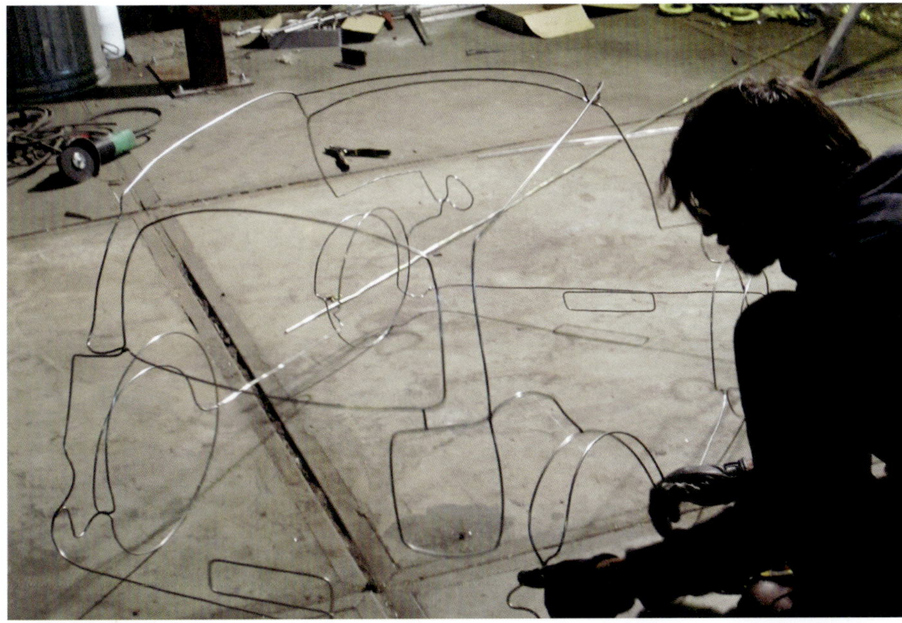

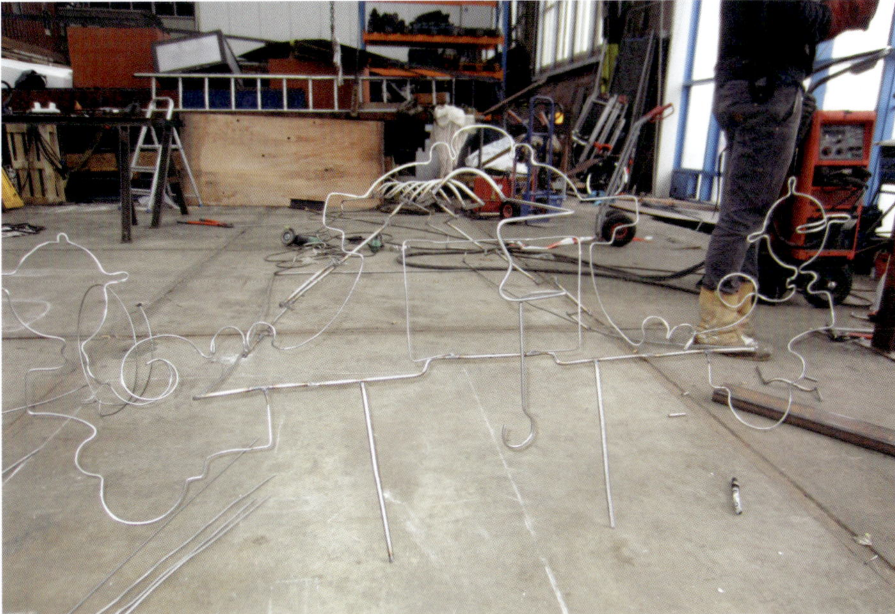

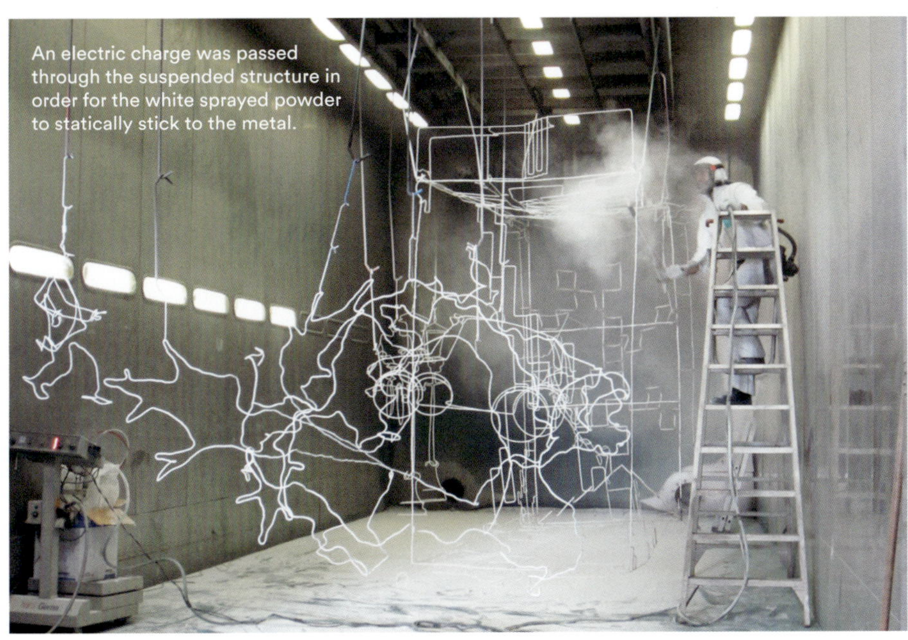

An electric charge was passed through the suspended structure in order for the white sprayed powder to statically stick to the metal.

Specifics

The initial scale model was made of 0.5-mm wire, and only 1-m wide. The challenge was to upscale the design and translate it into a large structure; initially aiming for it to be 6-m wide, the artist was challenged by the client to upscale even more, which led the final installation to be 13-m wide.

For the illumination, originally it was planned for the structure to be coverd with retro-reflective paint, similar to what is used on traffic signs. However, tests revealed that the reflection was only visible if the viewer would stand near a light source, as the light is not spread but bounces back to the source. This meant that the lines would not be visible from many viewing points. Further tests revealed the best way to make the whole sculpture visible was to make it bright, with white powder coating, and use 16 LED spots.

Another challenge was bending the stainless steel to make the required shapes. Thicker wire meant a stronger structure, but it still needed to be bendable enough to shape something. The canal house was the biggest challenge, as it had to be designed to have enough structural integrity to hang from one point for two windy winter months.

The elements are mostly empty space, with minimal lines that have very little wind resistance. Nevertheless, the structure needed to be prevented from tangling up in the wind. Entanglement would mean the work would self-destruct as the motor would continue spinning the layers in opposite directions. The elements therefore needed to be connected in a stiff way to the mobile. This means that the mobile moves as a whole in the wind, and the parts will keep a safe distance from each other even if the wind is strong and tilts the mobile. The only place that allows for this movement is the ring that connects the mobile with the frame it hangs from.

The mobile hangs from an axel that spins slowly by an electric motor. The house spins fastest, then the car, bike and pedestrian layer a bit slower in the opposite direction and the last layer, closest to the audience and with the longest distance to cover, goes even slower in the same direction as the house. This all means that the image that the audience sees is ever changing and they are unlikely to ever see exactly the same setting twice.

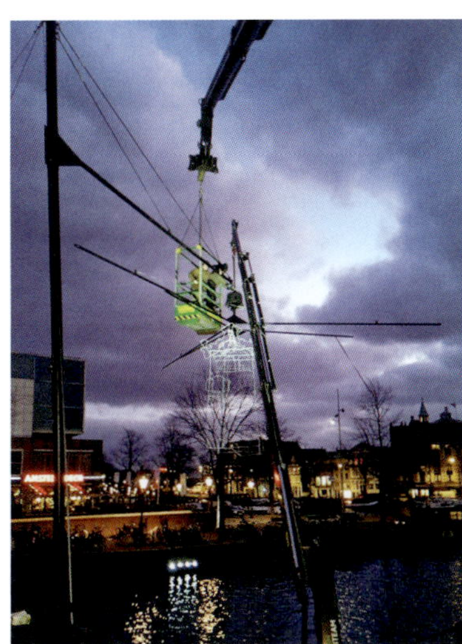

Using a Manitou telescopic handler and a boat wth a crane, the team worked hard over one (long) day to postition the sculpture over the canal.

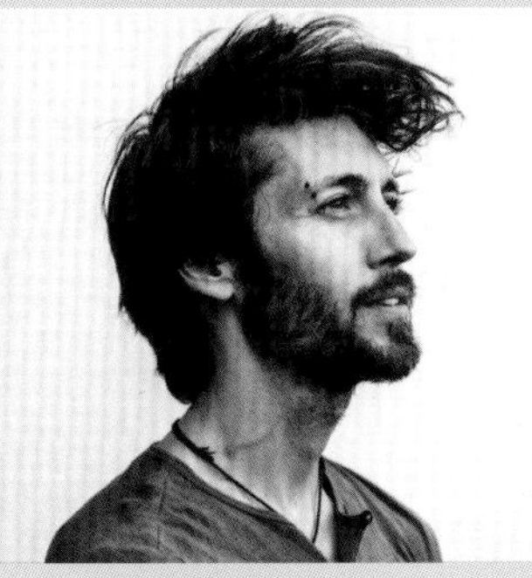

Photo Dim Balsem

Ralf Westerhof

Ralf Westerhof is a sculptor best known for his 3D drawings in wire. The minimal use of line in his often moving works requires the viewer to actively synthesise the whole image from the given parts – the opposite of our dominant, analytic way of dealing with the world. This internal switch in the mind leads to a temporary change in perception, which is also the major theme in his other works. His installations are accessible to all as they rely on common abilities to perceive and create instead of knowledge.

RALFWESTERHOF.NL

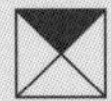

Perceptions

When **2012**
Where **Rotterdam, the Netherlands**
Client **n/a**

Thin metal wire, hand bent and hanging from thin threads on a mobile to keep their shape. The three more than life-size figures pivot on their own axis and each other, like planets. A projector lamp casts sharp shadows on the walls that show the figures from another perspective than that of the viewer. These shadows also take part in the new relations that are constantly created between the figures. The minimal lines of each moving figure need constant interpretation, as there are three figures and as many shadows that all need interpreting. Temporarily, the viewers' state of mind in relation to their outside world is not analysing but co-creating.

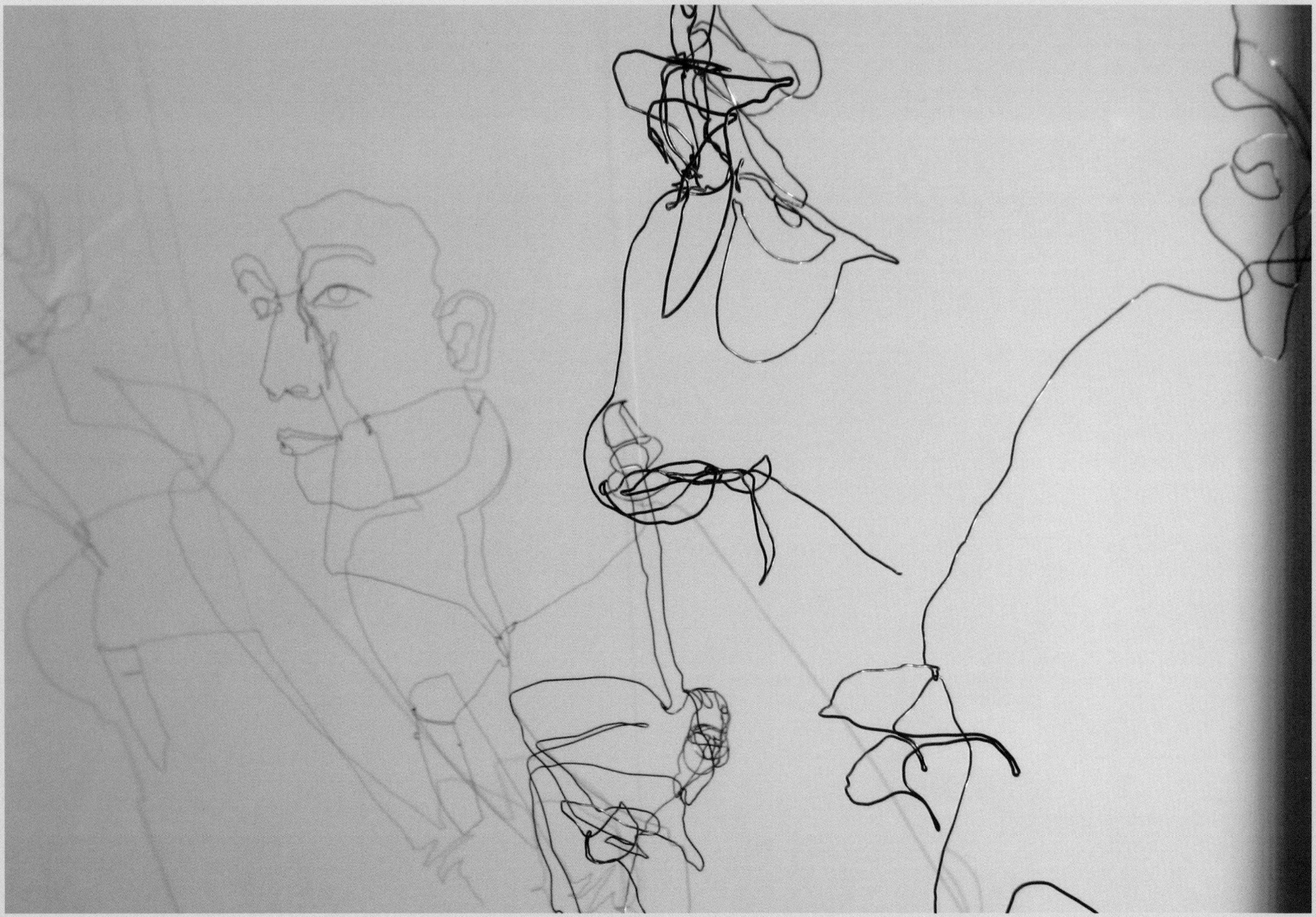

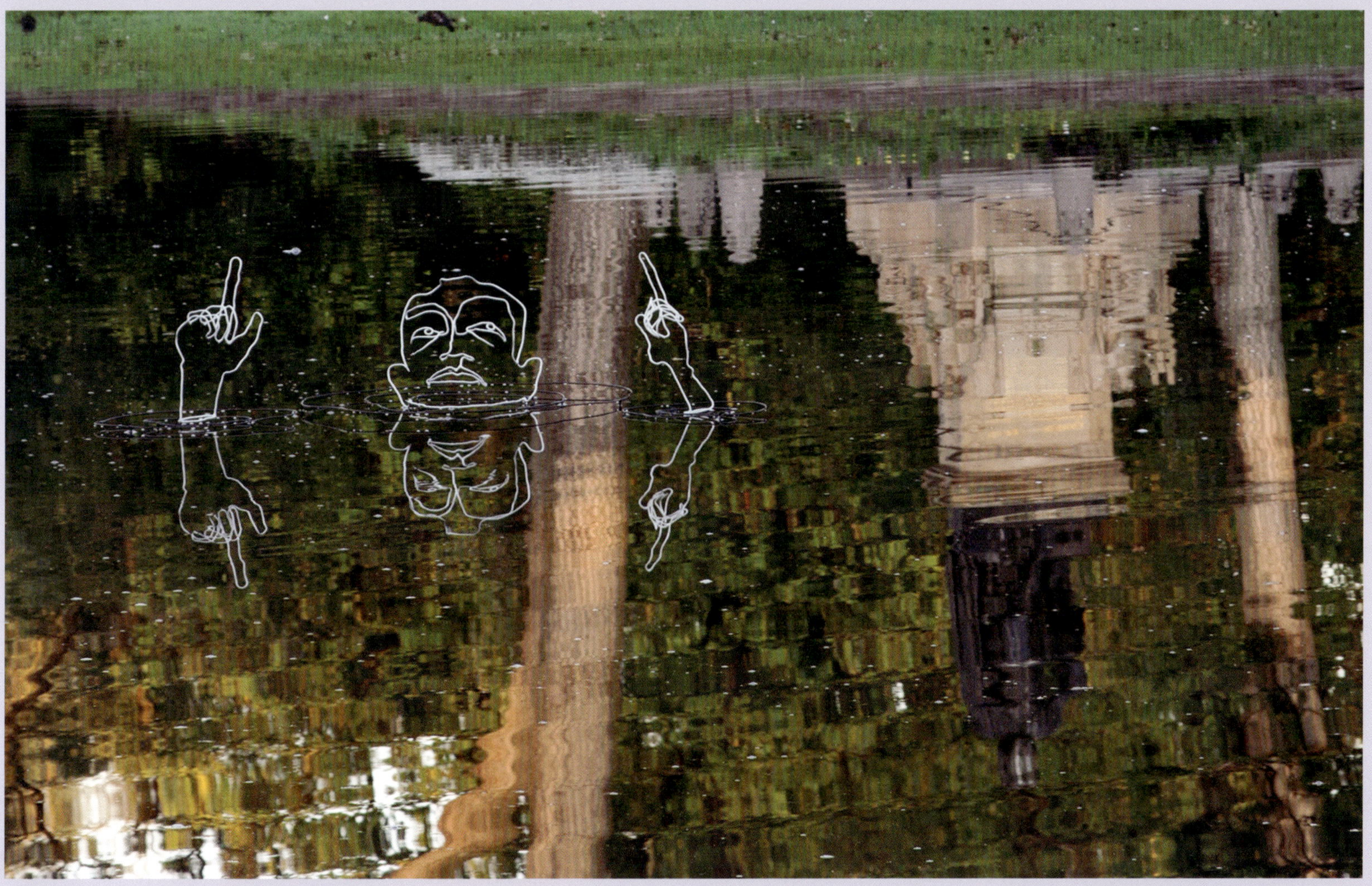

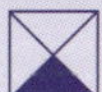

The Water Treader

When **2011**
Where **Amsterdam, the Netherlands**
Client **n/a**

This project was a reaction to the enormous budget cuts on culture in the Netherlands and it depicts a man keeping his head above the water. The first action of this 'guerrilla art project' was its sudden appearance in Amsterdam's famous Vondelpark. It was placed in the dark night by the artist wearing a wetsuit. The next floating sculpture (measuring 150 x 90 × 60 cm) appeared in 2012 and it could be placed and moved easier. This work toured the country, first guerrilla style and later legally.

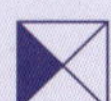

Portraits

When **2013**
Where **Leipzig, Germany**
Client **n/a**

Portraits is an ongoing series that started in 2011 in the Netherlands. The works are still ongoing and, in 2013, there was an installation in Germany. The portraits are slightly larger than life size. They hang from one point which makes them spin in the air currents of the space. Because they are a bit tilted, their emotional states seem to change slightly as they spin. The shadow that is cast also adds another perspective and a slightly different emotional state. This gives the faces a human like complexity, as if one can see through the superficial mask and see the real person underneath.

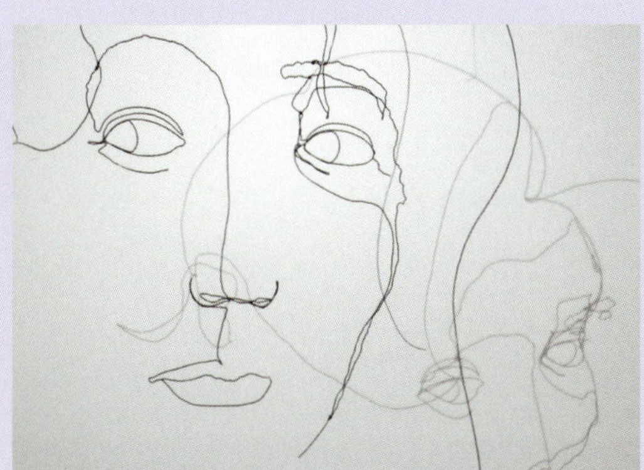

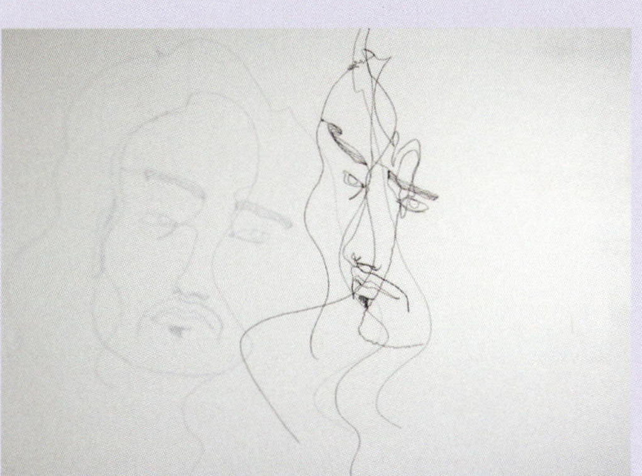

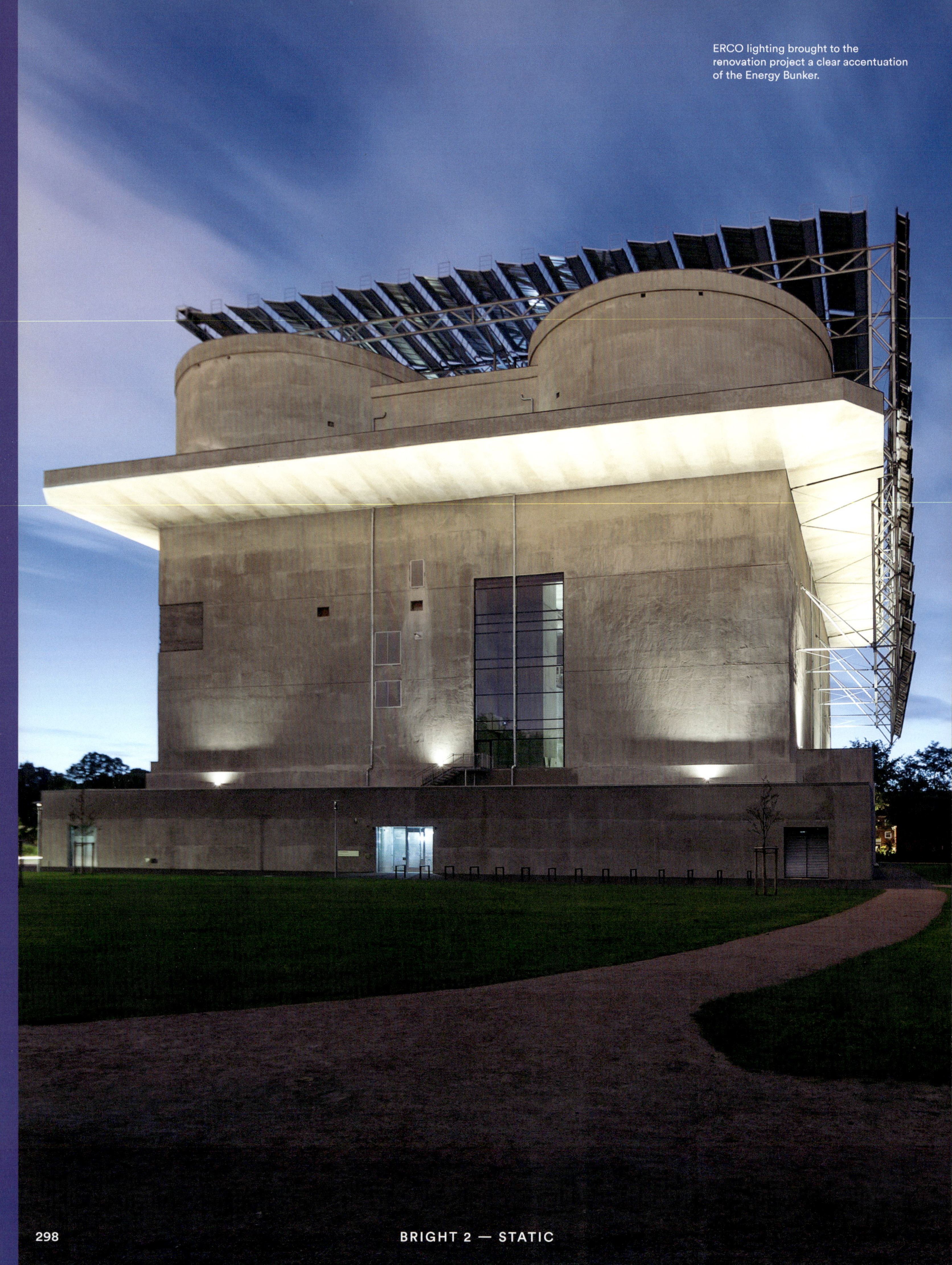

ERCO lighting brought to the renovation project a clear accentuation of the Energy Bunker.

Energy Bunker

Involved in the development of a lighting solution for the restoration project of a former World War II bunker in Hamburg was the lighting specialist **ERCO**, using the firm's own LED technology to answer to a programme that looked to replace the oppressive ideologies associated with such structures with a hopeful message about the region's sustainable future.

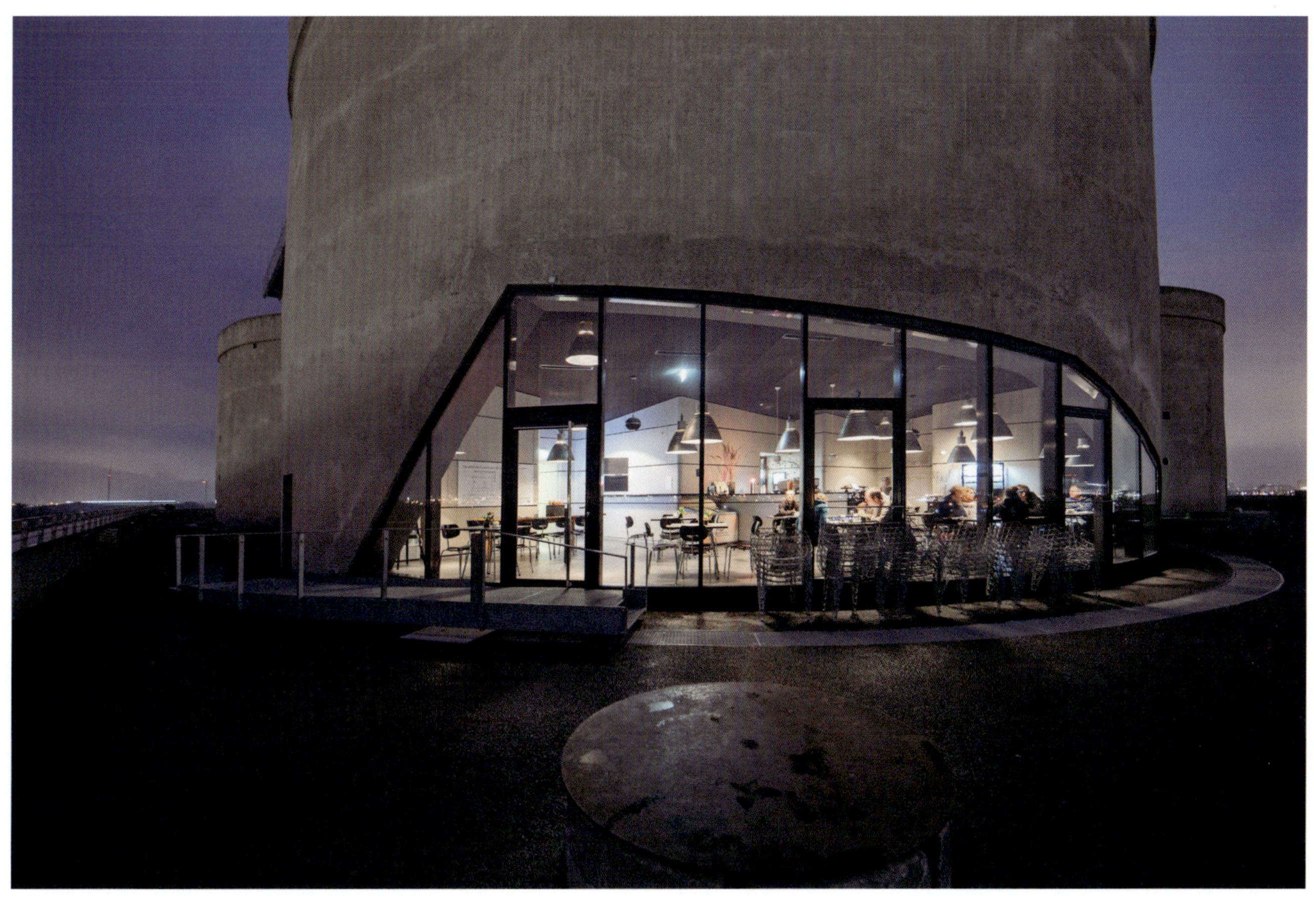

Photos Frieder Blickle

In the restoration of a colossal and somewhat oppressive building in the Wilhelmsburg district of Hamburg, ERCO was brought in to implement an accurate architectural lighting solution to accentuate the dramatic lines. A former air raid and anti-aircraft bunker built in 1943 to provide shelter for nearly 30,000 Hamburg residents during World War II, the structure was converted into a renewable energy plant and visitor centre in 2013. Commissioned by International Building Exhibition IBA Hamburg, the project was a collaboration between ERCO and the architectural firm HHS.

The building's 42-m-high concrete shell and load-bearing structure was redesigned to provide public and technical areas. Despite its many interventions, the bunker's original outline was not disturbed. At night, efficient lighting solutions mean that the building's architectural character is retained and emphasised, while the somber, oppressive history previously associated with it gave way to the cheerful promise of a sustainable future. A cafe and museum have been integrated in the upper levels, with one of the bunker's four turrets transformed into an exhibition area. The cafe spills out onto the cantilevered ledge that encircles the building, giving customers a 360-degree panorama from the open terrace.

Using a simple yet clever approach for the lighting design, the circumferential cantilever slab (at around 30 m above the site) was illuminated uniformly to differentiate the building's mass and transform the cantilever slab into a seemingly suspended element. Illumination of the top level was achieved largely through reflection, emphasising the form of the sculptural section whilst making its original purpose less intimidating. Luminaires were fitted to produce an optimum long-distance effect using discreet lighting from below, with the whole lighting concept responding sensitively to the historic heritage of the war monument. —

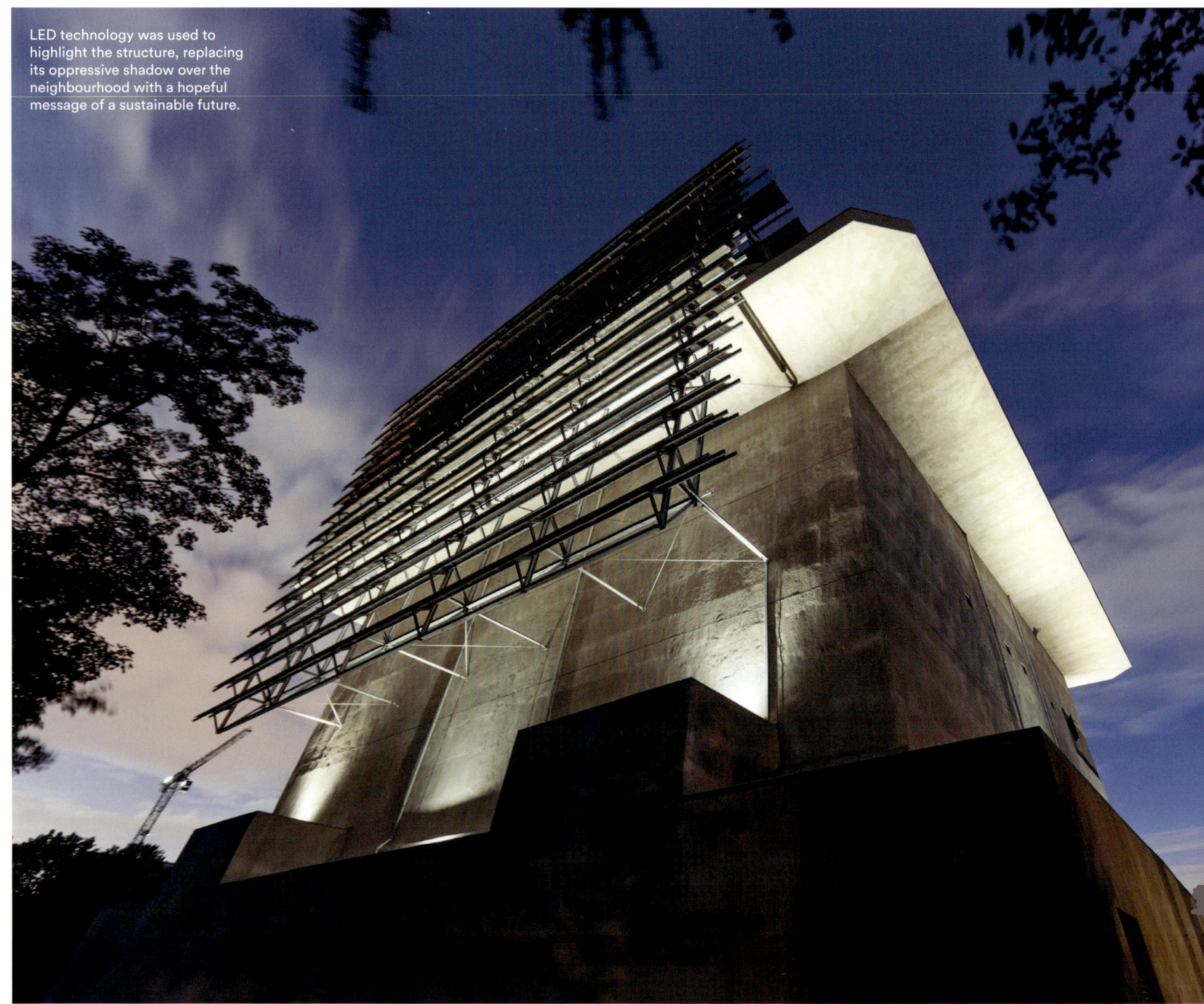

LED technology was used to highlight the structure, replacing its oppressive shadow over the neighbourhood with a hopeful message of a sustainable future.

Dramatic lighting and simple, yet clever solutions were implemented across all areas of the Energy Bunker in Hamburg.

Designer
ERCO
Location
Hamburg, Germany
Client
IBA Hamburg
Collaborator(s)/consultant
HHS Planer+Architekten
Manufacturer
ERCO
Date
April 2013

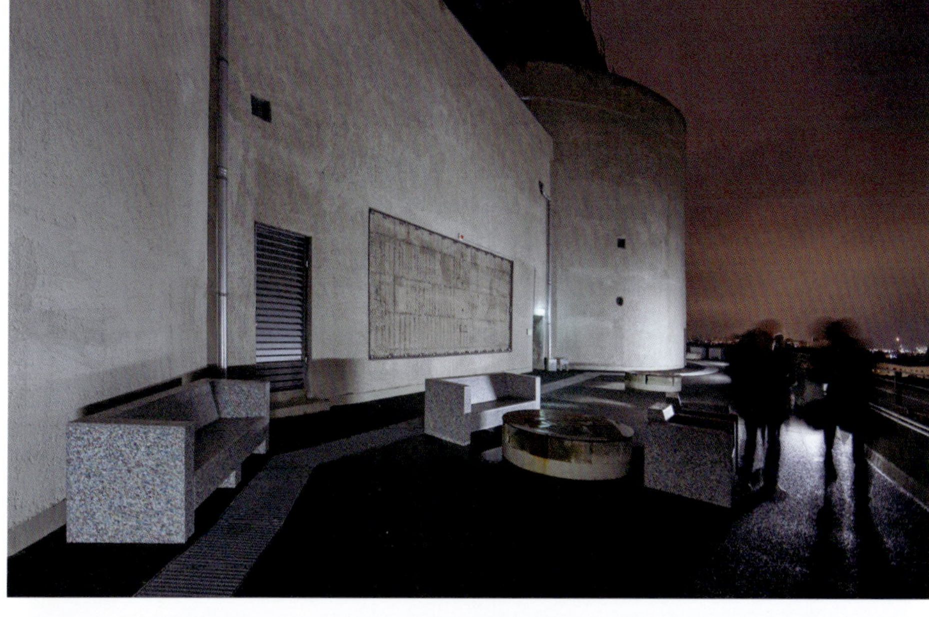

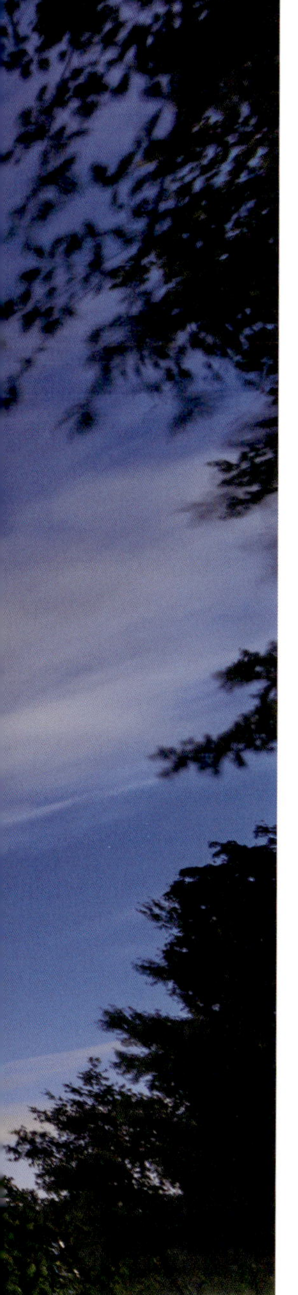

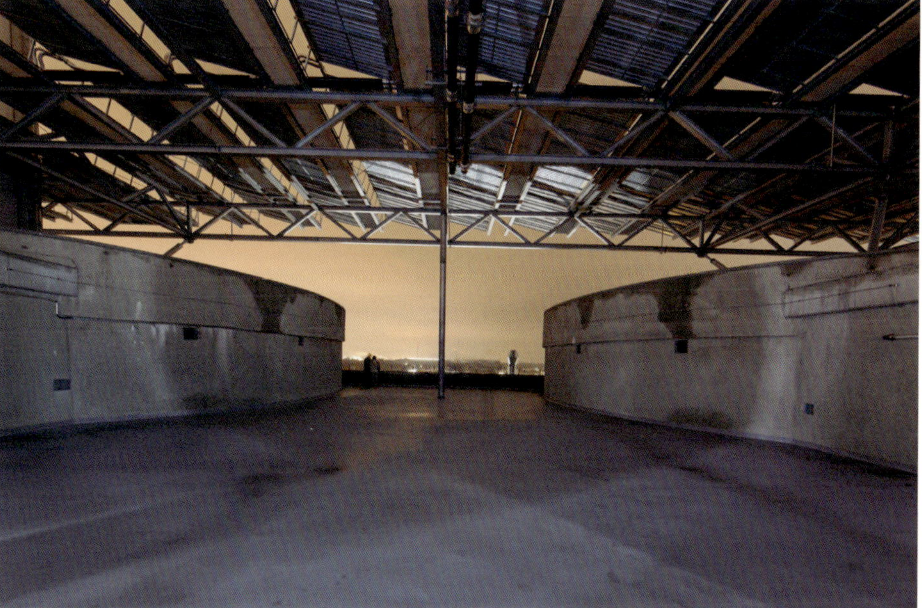

CLEVER LIGHTING SOLUTIONS TRANSFORM THE CANTILEVER SLAB INTO A SEEMINGLY SUSPENDED ELEMENT

PLAN

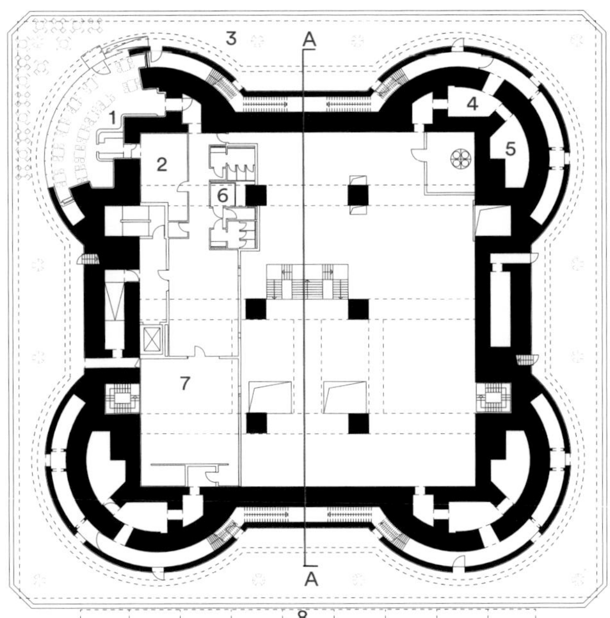

1 Cafe/bar and hospitality space
2 Kitchen
3 Cantilever slab/viewing balcony
4 Former gas lock
5 Former munition store
6 Lavatories
7 Roasting facility
8 Solar casing/steel structure

SECTION

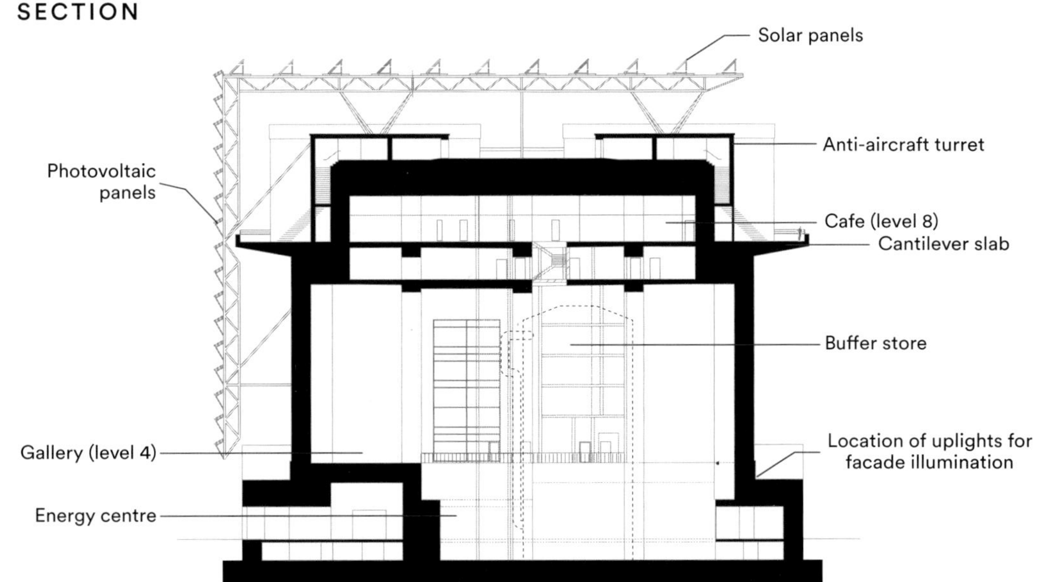

Set-up

Floodlights used for accent lighting of the concrete pillars are ERCO Grasshoppers (LED 18 W, 2250 lumen, 4000 K). The oval flood light distribution produces wide-beam illumination, avoiding unnecessary spill light. Three stronger floodlights (48 W LED) on each side of the building are also used to transform the cantilever slab.
Illumination of the vertical wall surfaces is achieved through ERCO Powercast luminaires (LED 48 W, 6000 lumen, 4000 K), which produces a uniform spread of light.
Spotlights used to illuminate the wall seen from the exterior are two ERCO Beamer Spots (LED 48 W, 6000 lumen, 4000 K).
Construction time for the entire renovation project was 3 years.

CONCEPT VISUAL

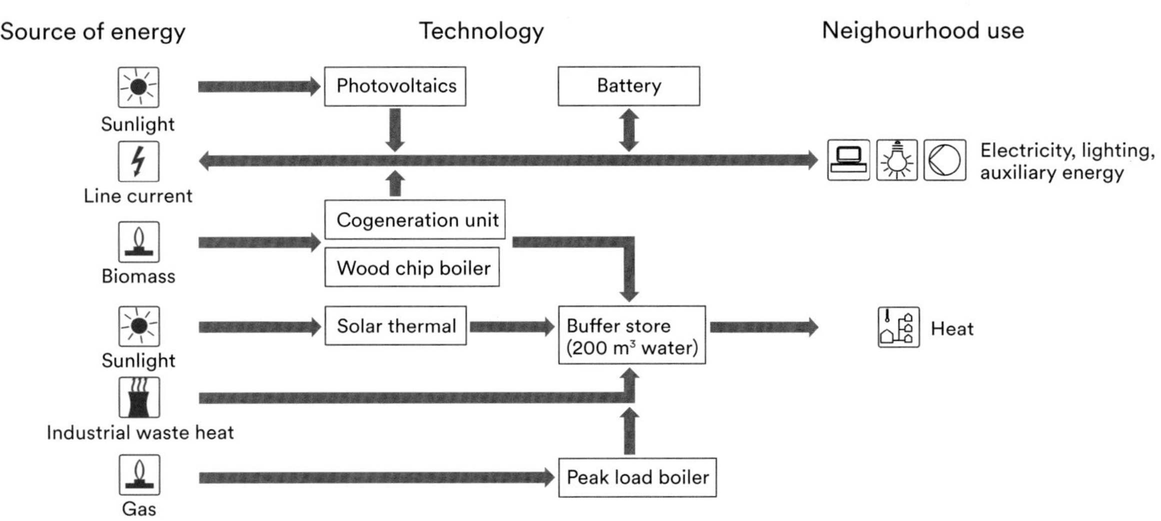

RENDERINGS

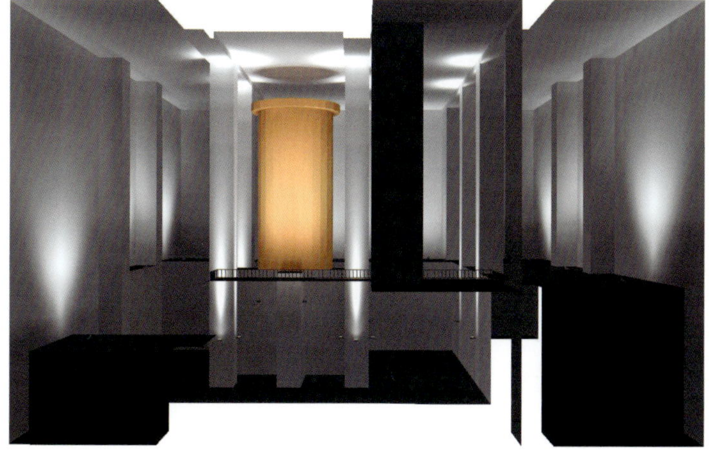

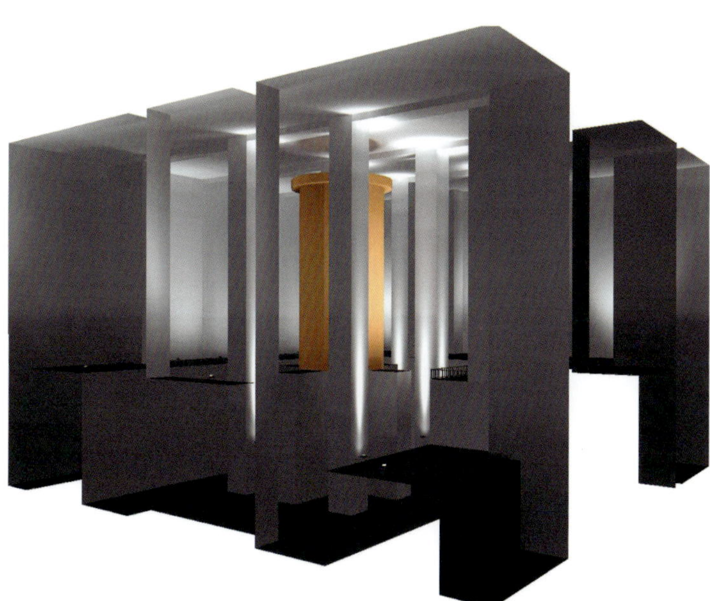

ACCURATE ARCHITECTURAL LIGHTING ACCENTUATES THE DRAMATIC LINES

Specifics

The biggest challenge for the lighting design of Energy Bunker was to eliminate through the use of light the oppressive ideology associated with the structure. Erected in 1943, the former flak bunker had defined the night sky of the Wilhelmsburg district as a dark mass for over 60 years, presenting itself as an imposing and intimidating structure through its size and monumentality alone.

The illumination of the radially arranged ammunition chambers at the top level of the building accentuates the rigour of the sculptural section, making its original purpose less intimidating. One of the four former turrets, now accessible to visitors, was left in its original condition.

Luminaires and discreet lighting respond to the war monument's historic heritage but refrain from staging the building as an artwork, underlining instead its current importance for the neighbourhood.

The interior lighting of the 25-m-high technology centre and its rough-textured, historic wall surfaces looked to create an impression of depth as well as emphasise the plain character of its reinforced concrete frame.

By using lighting tools with different brightness levels and light distributions park visitors are able to perceive its spatial depth and structure from outside through the large (measuring 7 x 17 m) opening. The chosen light colour emphasises the plain character of the cold-looking reinforced concrete frame.

The cantilever slab that surrounds the bunker's walls at nearly 30-m-high seems to divide the building in two parts. This division is emphasised at night through uniform illumination of the entire cantilever, transforming it into a seemingly suspended element and manifesting the symbolic nature of the this beacon project for the client, to lower the carbon footprint in an effort to combat climate change.

At the heart of the structure is a 200 m^3 water reservoir, which acts as a large heat buffer at the centre of a local heating network. A multiplicity of energy sources, including heat from a biomass thermal power plant, a wood burning unit, a solar thermal system on the bunker's roof and waste heat from a nearby industry plant feed the reservoir. Moreover, the thermal power station and the photovoltaic panels that cover the building's south facade feed into the public green energy grid.

Photo Studio Steve

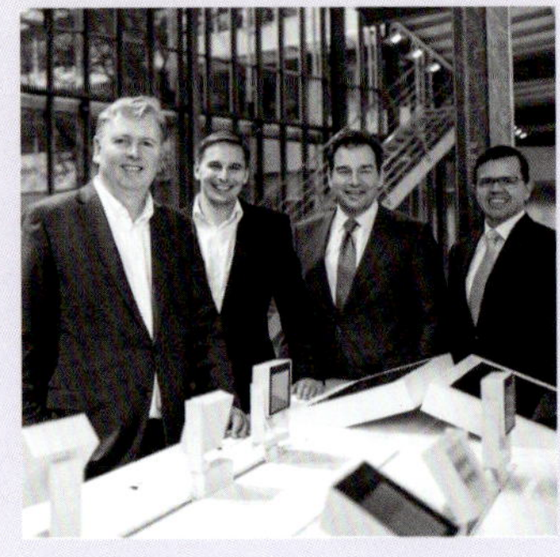

ERCO

Founded in 1934, the ERCO Light Factory is a leading international specialist in architectural lighting using LED technology. The company operates globally, with over 60 subsidiaries, branches and agencies in more than 40 countries worldwide. ERCO looks at digital light as the fourth dimension of architecture and develops, designs and produces digital luminaires with a focus on photometrics, electronics and design. The company works closely with architects, lighting designers and engineers to provide lighting tools and solutions in the following fields: work, shop, culture, community, hospitality, living, public and contemplation.

ERCO.COM

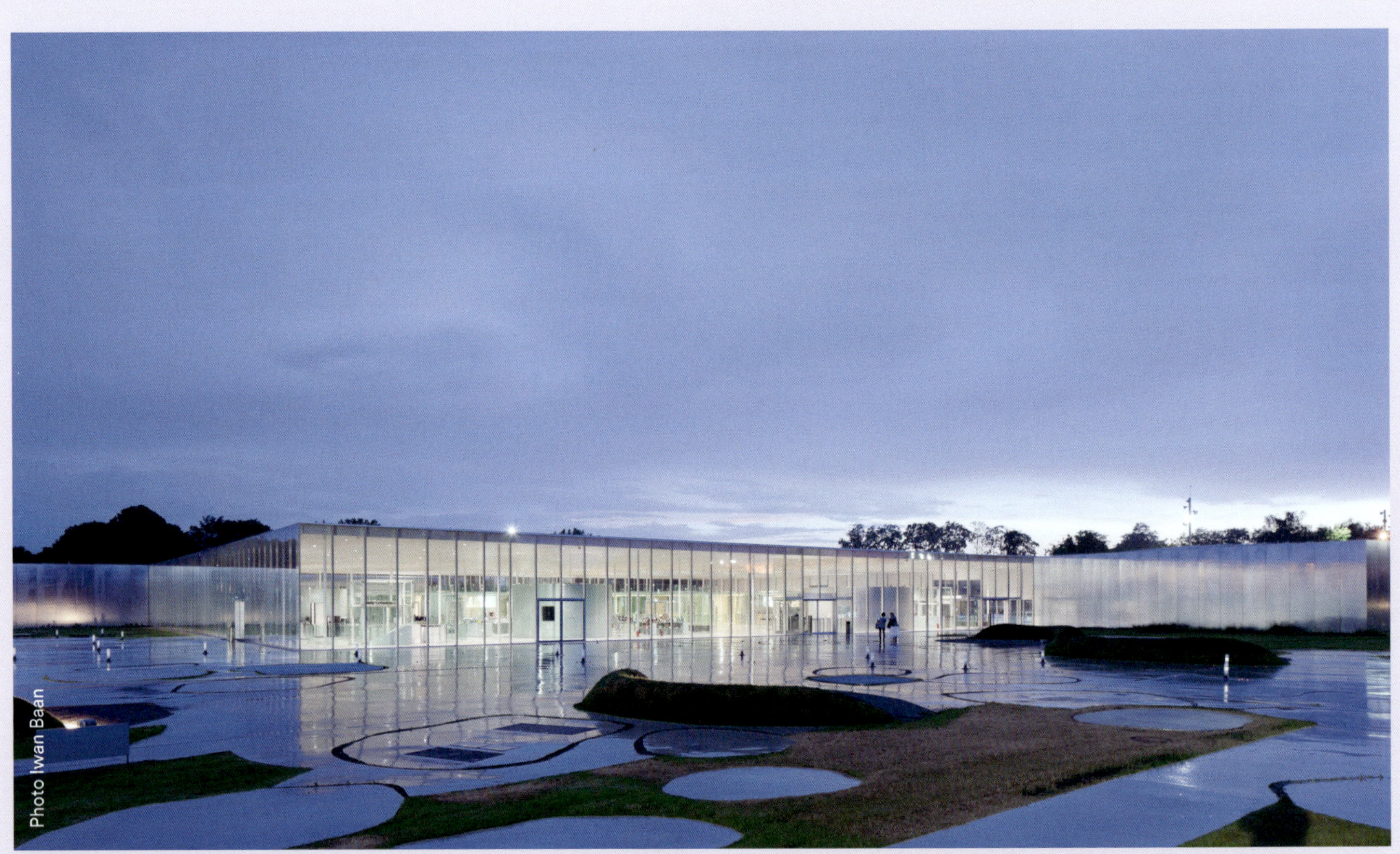

Photo Iwan Baan

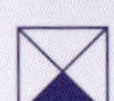

Louvre Lens

When **2012**
Where **Lens Cedex, France**
Client **Louvre Lens**

Designed by Tokyo-based architecture studio Sanaa and Arup's lighting designers, the Louvre Lens is a milestone in the world's museum landscape. This branch of the Paris Louvre is minimal and compact, unlike its older sibling. ERCO provided the lighting for the space, which contrasts with the Pas-de-Calais' history as a coal and heavy industry region.

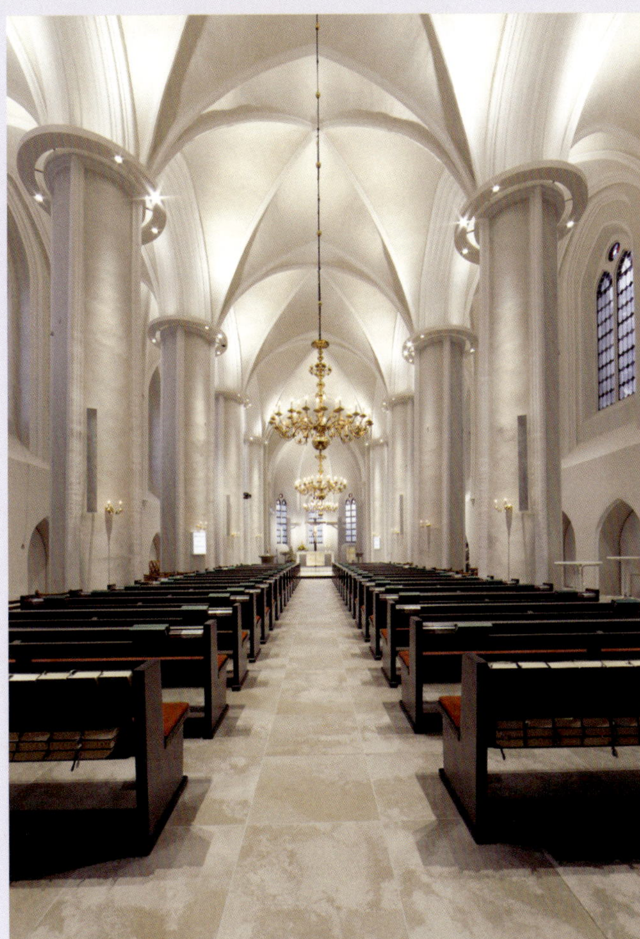

Harsefeld Church

When **2014**
Where **Harsefeld, Germany**
Client **Harsefeld Parish**

For the new lighting concept for the Harsefeld Church, Lucent developed a solution using ERCO LED lighting technology to bring back full glory to the historic structure. At a height of nearly 8 m, custom-made luminaire rings discreetly frame the nave pillars. A particularly challenging project though it was, ERCO was able to integrate the lighting tools into the space as unobtrusively as possible.

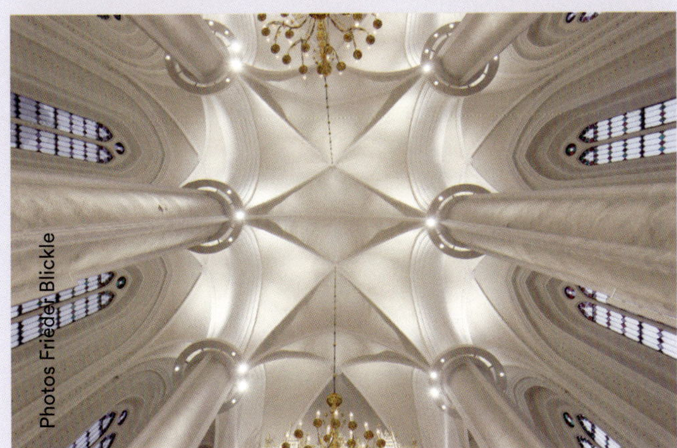

Photos Frieder Blickle

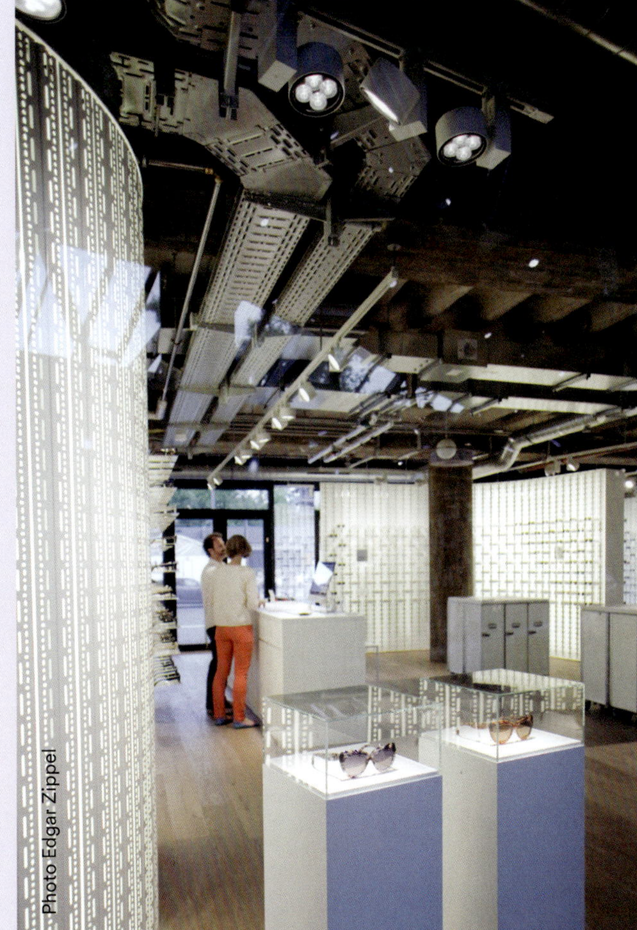

Photo Edgar Zippel

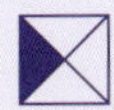

Mykita Store

When **2014**
Where **Berlin, Germany**
Client **Mykita**

ERCO LED spotlights were used to illuminate the eyewear store's key design element, the Mykita Wall. This is a 31-m-long backlit display structure, as well as the sales counter, and incorporates the metal trolleys where the glasses are stored. Moreover, diffused light on the support walls adds depth to the bold and minimalist design.

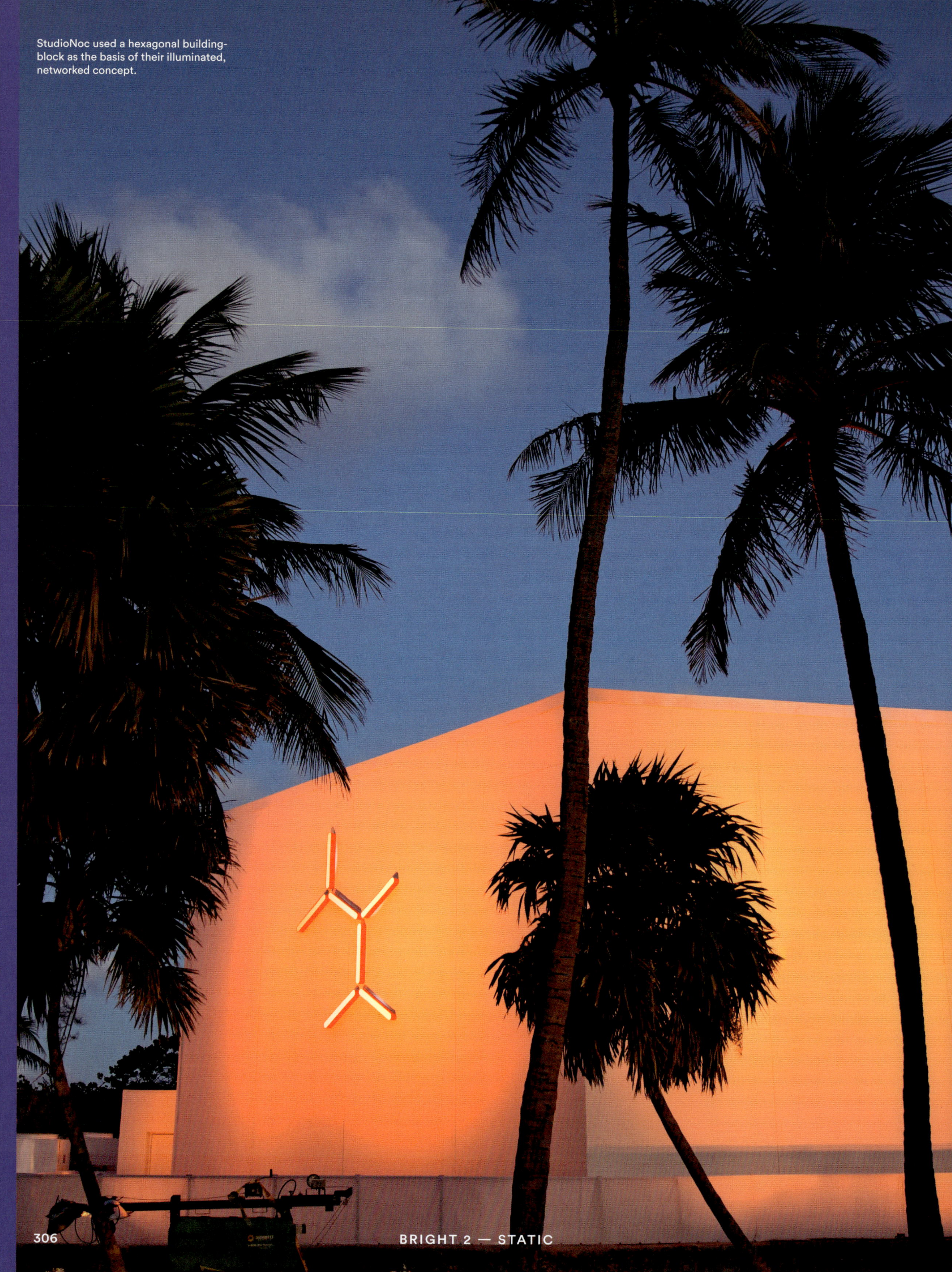

StudioNoc used a hexagonal building-block as the basis of their illuminated, networked concept.

Ericsson Nest

For Ericsson's Networked Society Forum event in Miami, **StudioNoc** created an illuminated beacon as a landmark location – a blank canvas to be filled with ideas and solutions, with a concept of increasing the strength of light on the facade as time progressed.

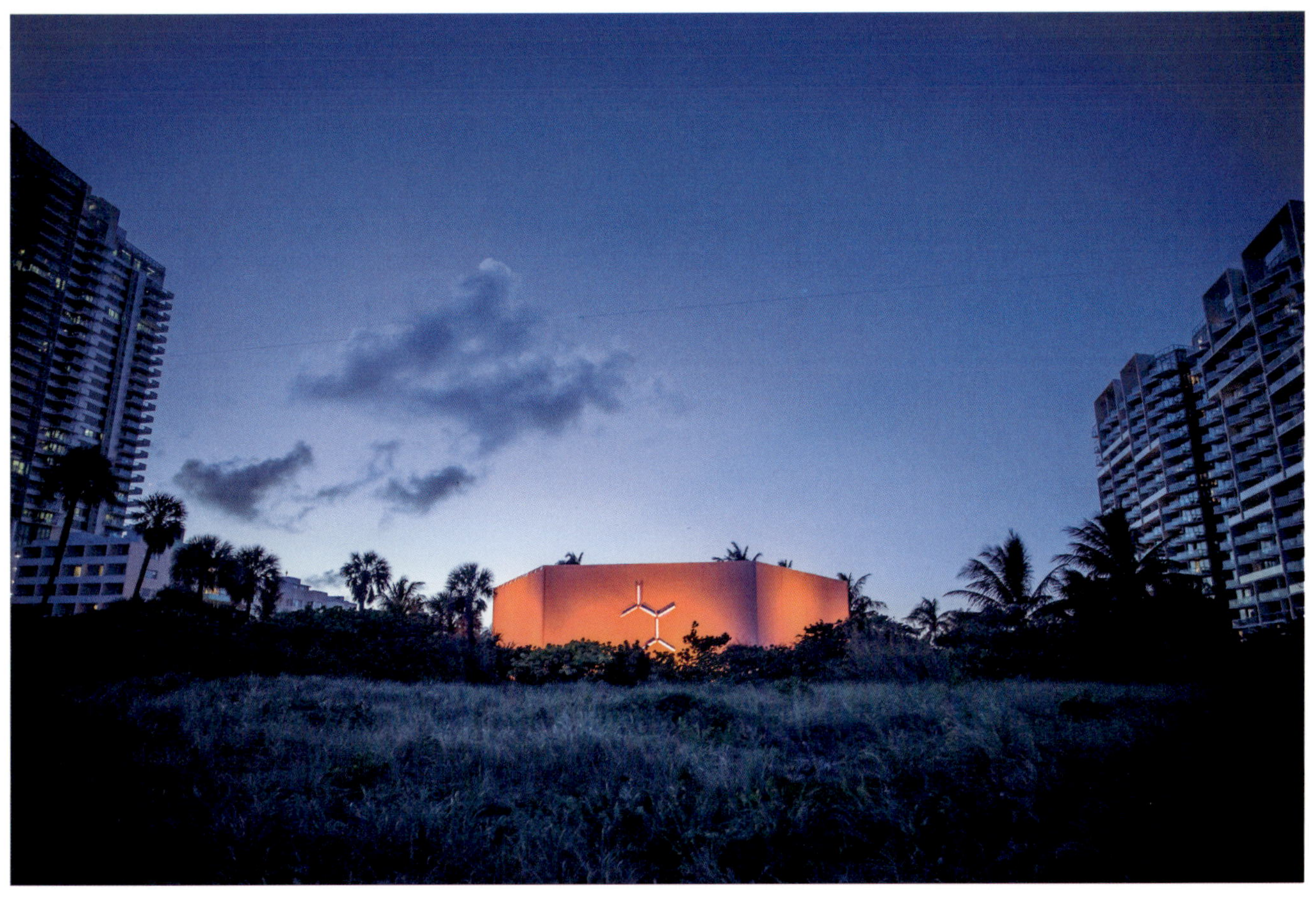

Photo Robert Chamarro

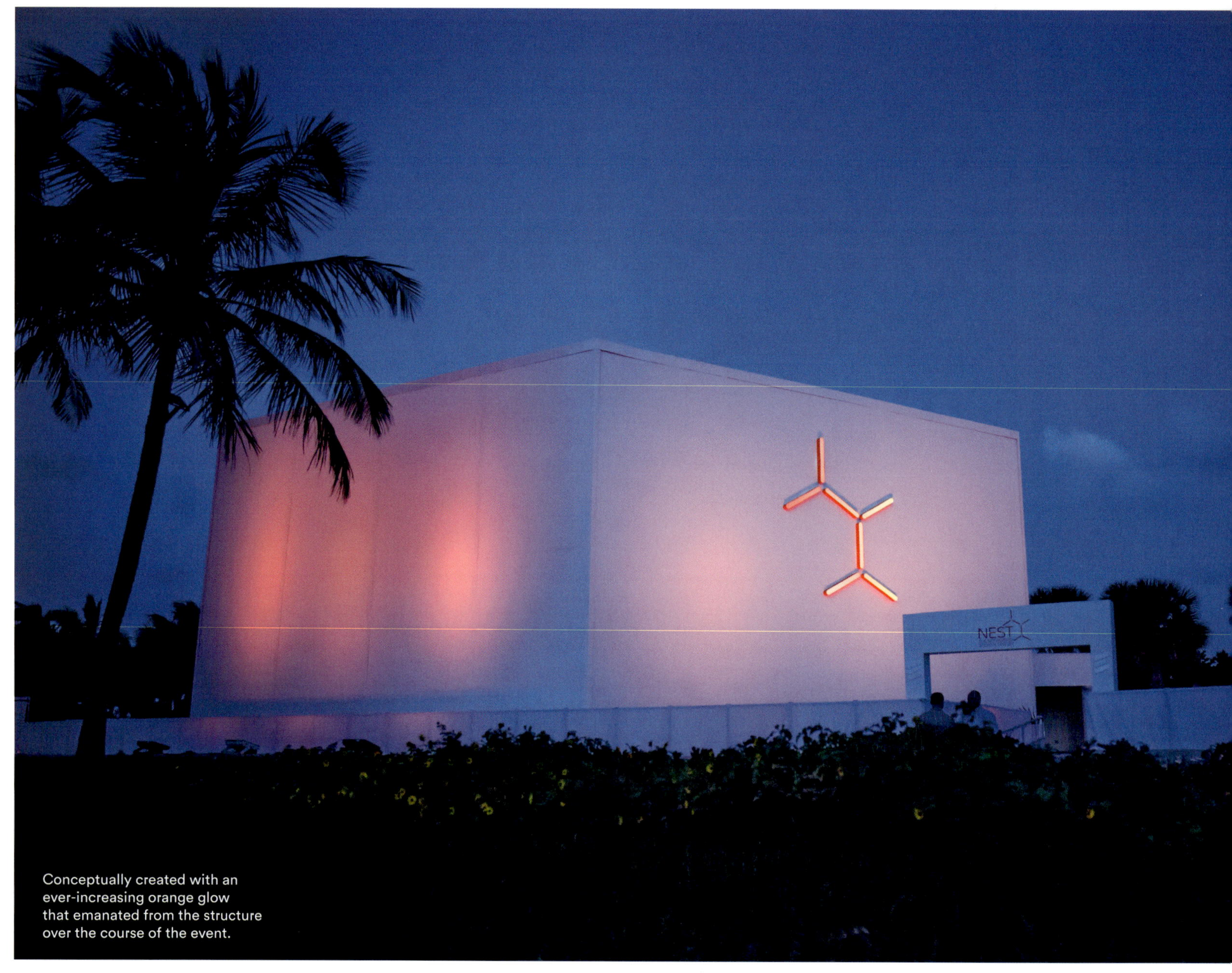

Conceptually created with an ever-increasing orange glow that emanated from the structure over the course of the event.

How could the information and communications technologies industry contribute to making the world a better place? And how could Ericsson utilise its influence and knowledge to enable answering that question and meet the needs of the industry? These were the questions that pre-empted the formation of the first Networked Society Forum – Nest – which took place in Hong Kong in 2011. StudioNoc was commissioned to create a destination for that inaugural event, and so the Swedish team was also asked to develop an engaging environment in 2013, when the next forum took place in Miami.

No stronger symbol holds the networked form than one of nature's most essential shapes: the hexagon. With its strong and ever-connecting character in line with the brand values of Ericsson, it was a natural progression that this should form the basis for the project's concept development. Comments StudioNoc's Jonas Pinzke, 'To reset the minds of the participants and truly spark the conversation, we created a unique meeting venue in the shape of a hexagon with an amphitheatre-style seating inside.' The six-sided form shaped the whole project and became the trademark notation throughout all touch points and communication with the participants – from the logo, invitation, graphics, badges and furniture.

Illumination of the architecture was a key priority in order to make this a shining beacon at the Miami beachfront location. A bespoke structure in the form of a 445 m^2 hexagon was erected with the monumental height of 9.5 m. Meeting one of the locality's requirements – to withstand hurricane-force winds – the whole structure was primarily designed with flexibility in mind, with the ability to be deployed virtually anywhere. The white facade symbolises a blank canvas, ready to be filled with new ideas and solutions made by the participants. As the event progressed and innovative ideas were created at the forum, the conceptual idea was to slowly increase the strength of illumination of the logo and the facade, as a dynamic interaction between interior and exterior. —

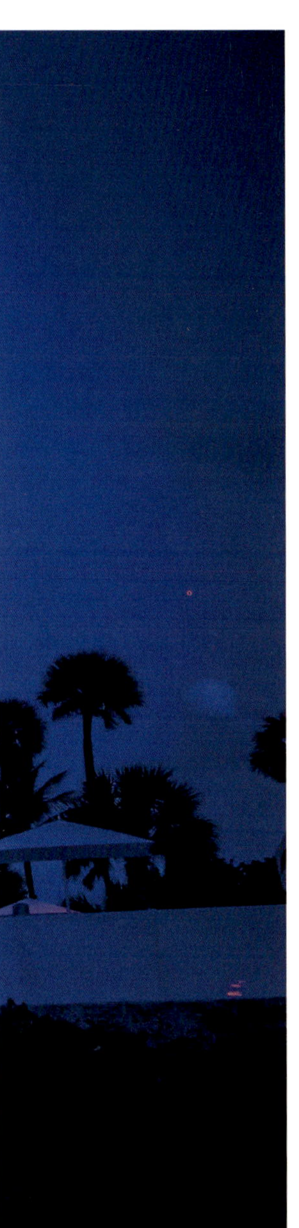

Designer
StudioNoc
Location
Miami, United States
Client
Ericsson
Collaborator(s)/
consultant
Revolve
Manufacturer
Event Star
Date
November 2013

STUDIONOC CREATED AN ILLUMINATED BEACON AS A LANDMARK LOCATION

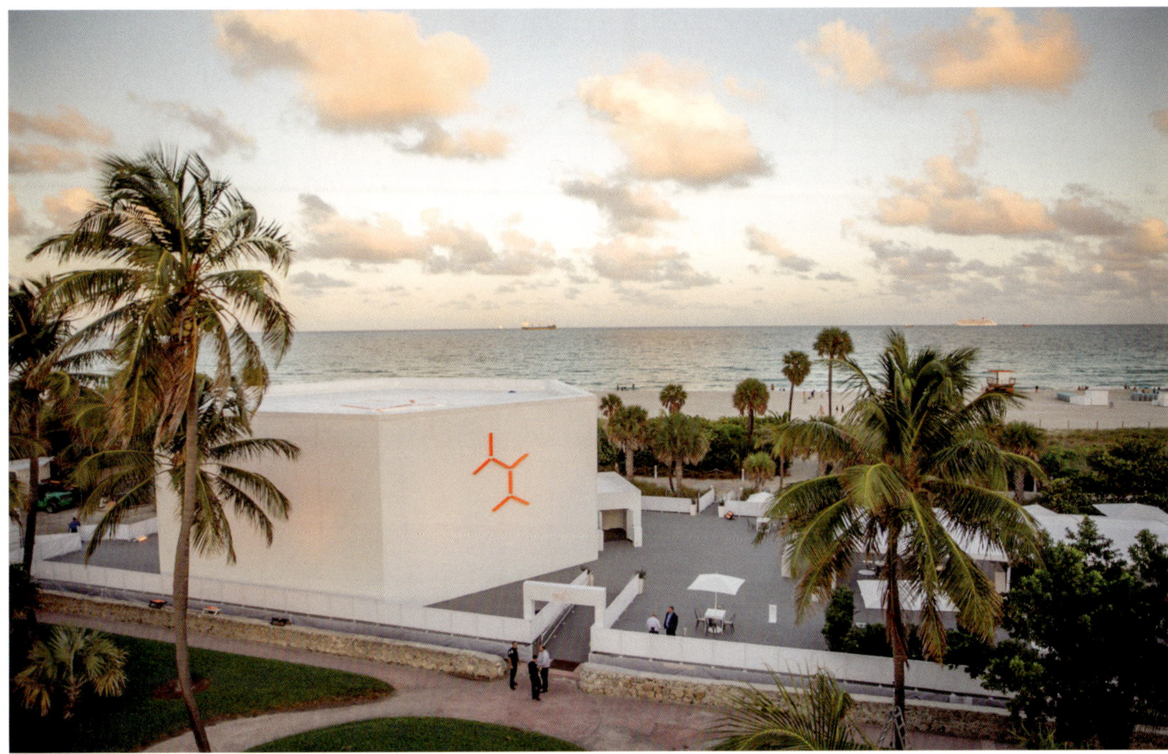

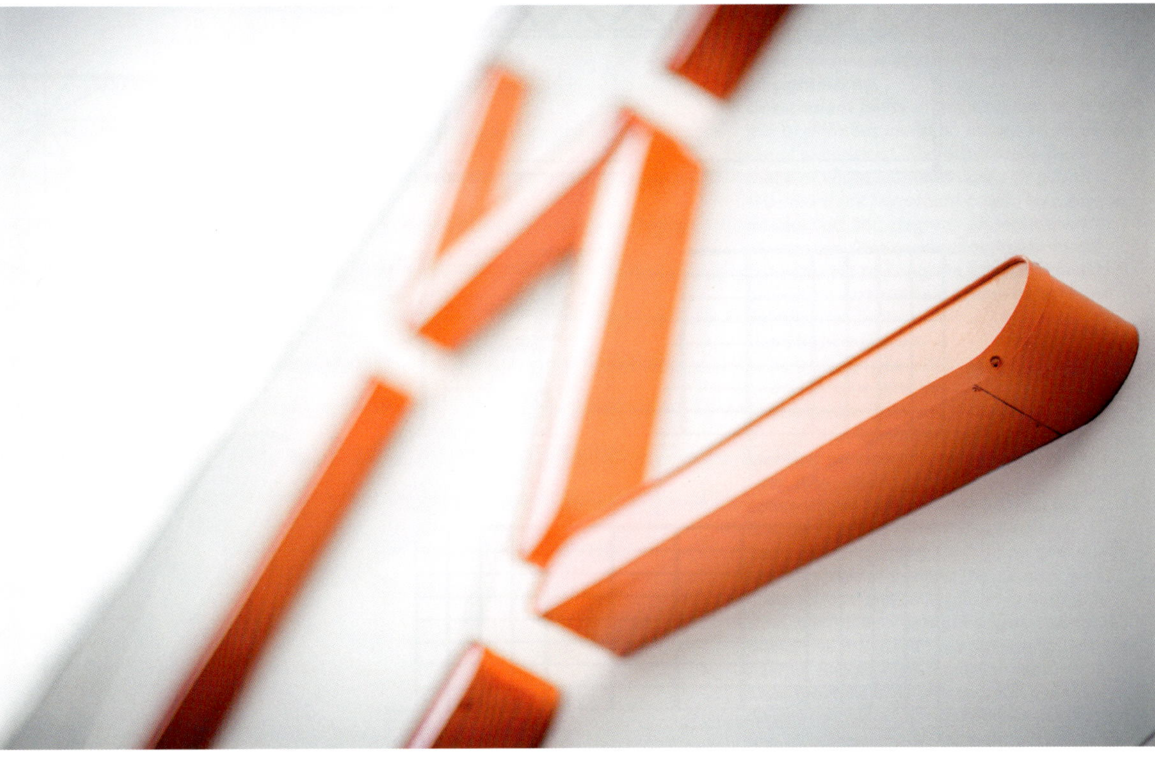

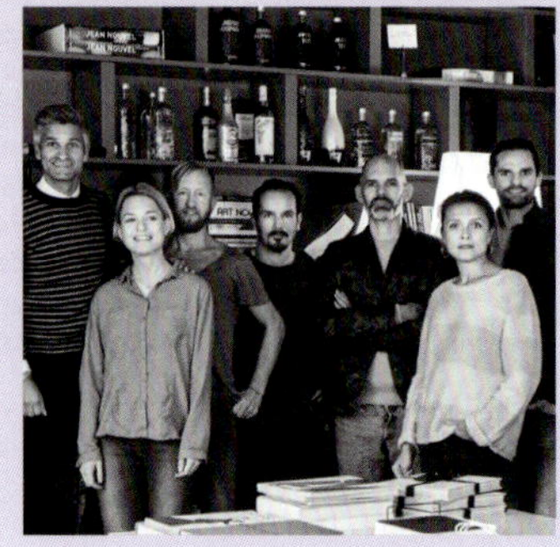

StudioNoc

StudioNoc is based in Stockholm, Sweden and was founded in 1994. The mission of the team of creative artisans is to deliver immersive experiences through 'no ordinary concepts'. With a common passion for transforming brands by infusing them with contemporary culture, the studio produces content based on live experiences – with focus on digital amplification – and creates exceptional moments, worth experiencing and sharing.

STUDIONOC.COM

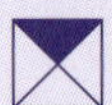

The Dollhouse

When **2012**
Where **Stockholm, Sweden**
Client **H&M Home**

The brief was to create a unique brand experience for 300 international journalists, in order to give them an experience that truly engaged all the senses. The challenge was to communicate six totally different interior concepts in an integrated, holistic and playful way. StudioNoc decided to approach the project by enabling the guests to realise a childhood dream – to enter a fantasy world for a day – and, thus, created a 'dollhouse' (in a traditional theatre in central Stockholm), with an open front to allow guests to get an overview of the collection as a whole.

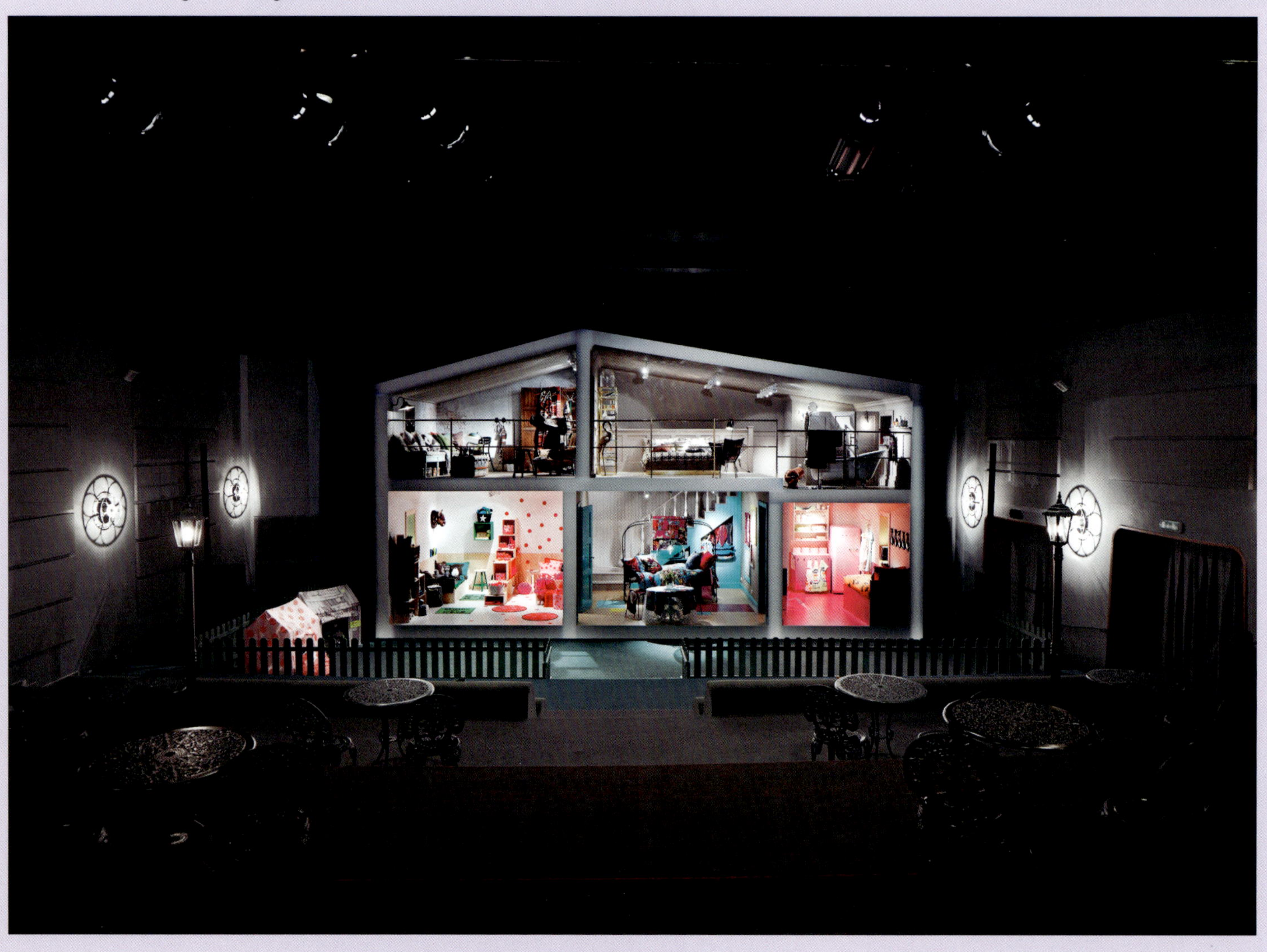

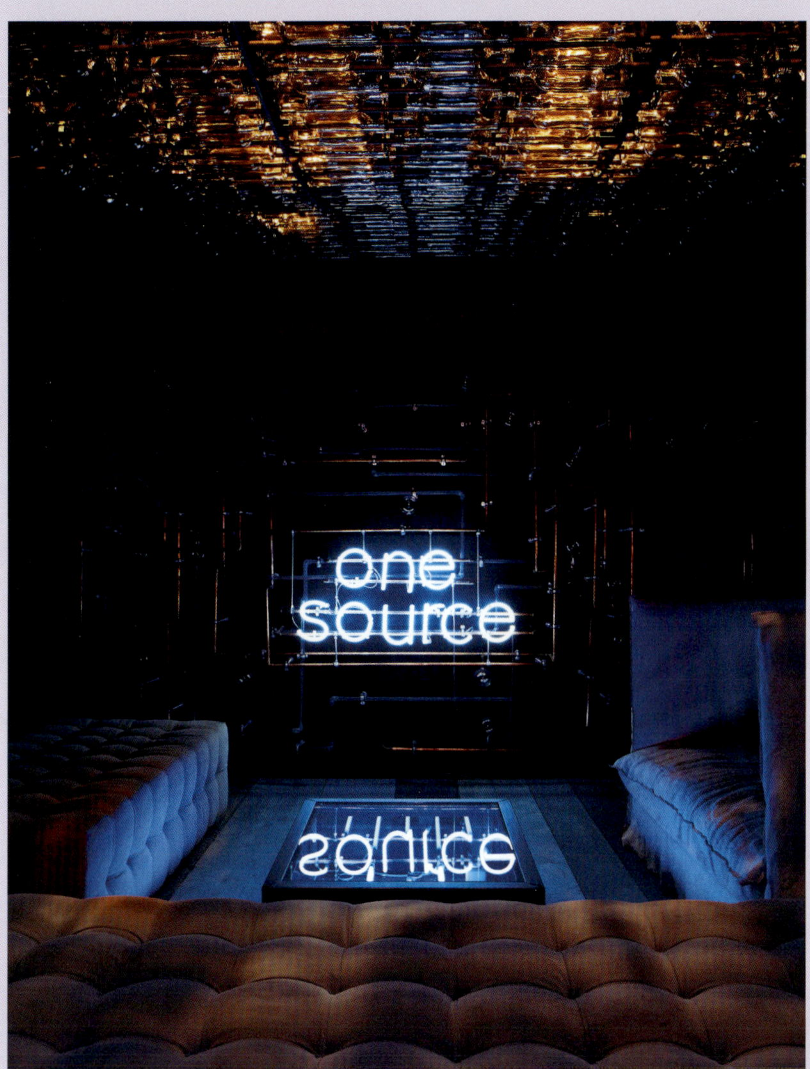

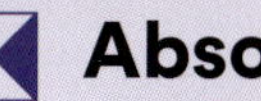

Absolut Atelier

When **2011**
Where **Stockholm, Sweden**
Client **Absolut**

Absolut Vodka is a world-renowned brand for which StudioNoc created a close, authentic and personal presentation space into which guests could be invited. The conceptual foundation was to allow them to then browse freely through the room settings, which each clearly conveyed the values of the drinks company. An old bank, in an 800 m² three-storey building, was turned into a brand home where Absolut Vodka could showcase itself through fashion, interior design and art.

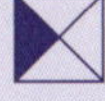

Grand Cuisine

When **2012**
Where **Milan, Italy**
Client **Electrlux**

When Electrolux launched a high-tech, premium product line – Grand Cuisine – at EuroCucina in Milan in 2012, StudioNoc was responsible for delivering a unique user experience. In order to present the technology to the stakeholders, the team created an immersive brand experience inside a room (15 × 6 m) with completely black surfaces. Specially-invited guests got the chance to see a presentation of the product line through an interplay between the host, films and spotlighted products – highlighted in the wall system with LED lights which were illuminated in sequence.

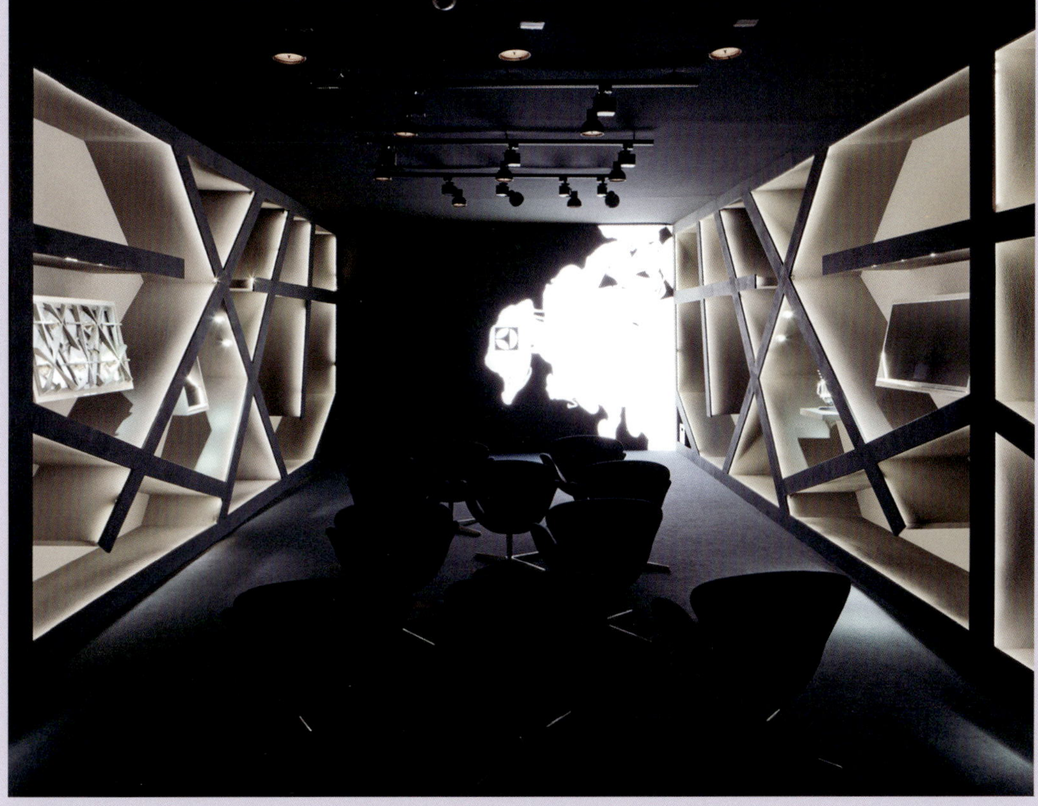

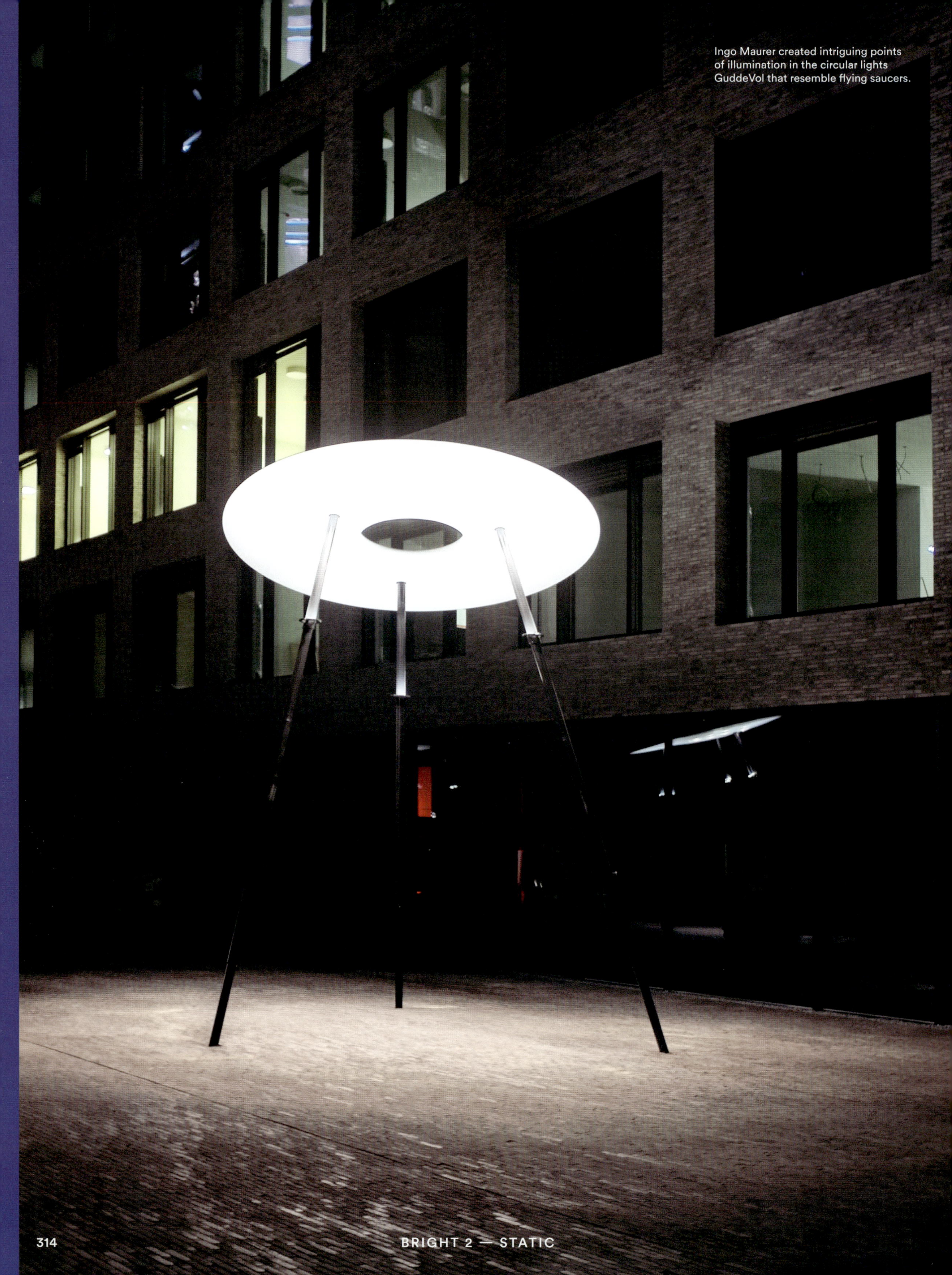

Ingo Maurer created intriguing points of illumination in the circular lights GuddeVol that resemble flying saucers.

GuddeVol

Imposing and grand street lights designed by **Ingo Maurer** and his team illuminate the pedestrian plaza alongside an area that was once home to Luxembourg's largest ironworks. The substantial GuddeVol lights are more akin to intriguing sculptures, which also have a UFO-quality about them – thanks to their seemingly hovering, illuminated crown and three-legged form.

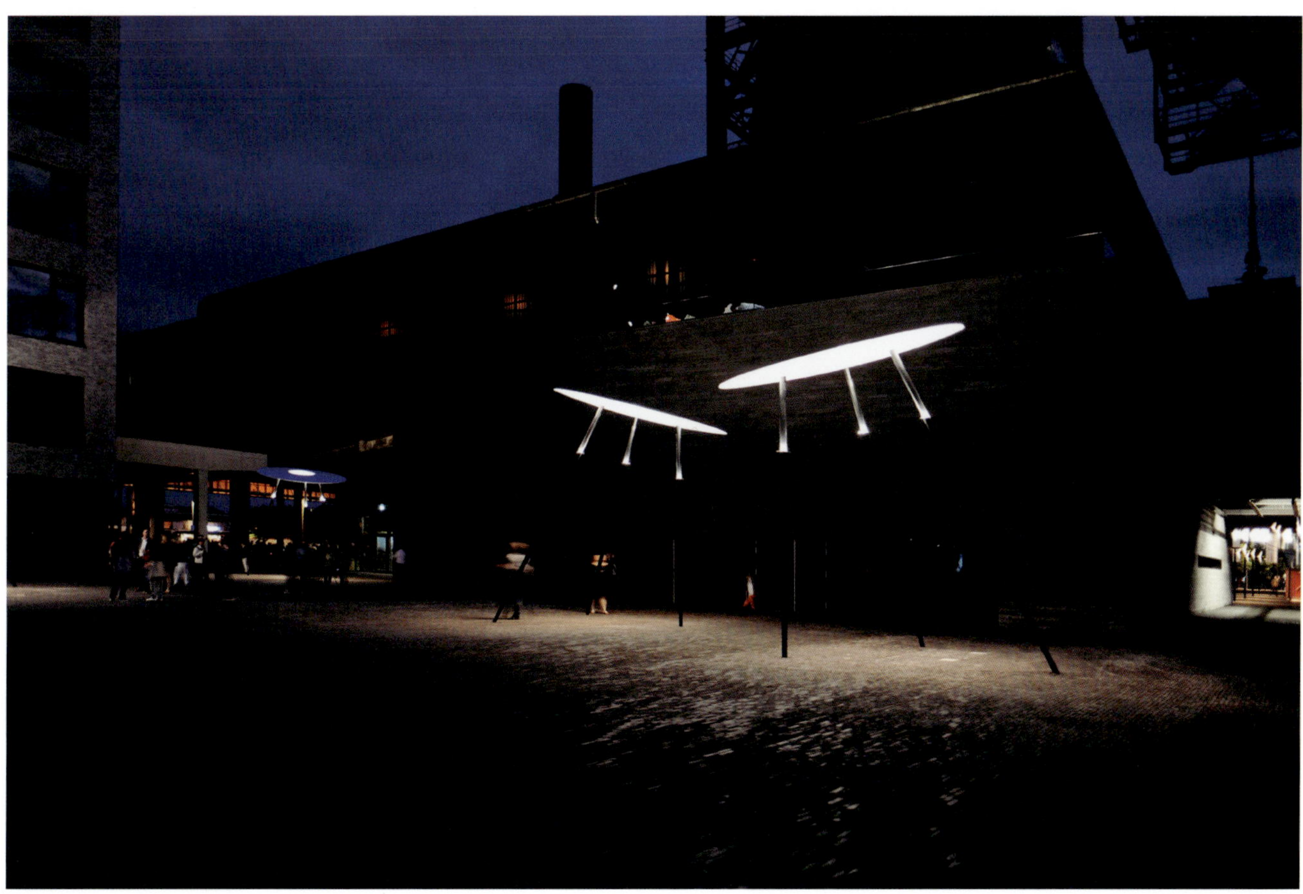

Photos Hagen Szczech

Until the 1990s, the Belval neighbourhood of Esch-sur-Alzette in Luxembourg boasted the country's biggest ironworks. The myriad of colossal, intriguing structures were then abandoned and left in decay until a conservation project was initiated to transform the dilapidated district into a new scientific and cultural centre. Commissioned to illuminate the redevelopment, it was in this context that Ingo Maurer's first ever street lights came about.

The Munich-based lighting company was asked to envision a solution for lighting the public square in the vicinity of the ironworks and blast furnaces. The outcome was the five GuddeVol luminaires, which resemble UFOs hovering nearly 5 m above ground. More akin to sculptures than to mere street lamps, each GuddeVol consists of a 4.2-m-diameter disc sitting at a slight angle on top of three matte-black poles. Curved openings on the underside of the disc are equipped with recessed LEDs, with uplights fixed around each pole, so that the discs themselves are transformed into radiating circles of light, in contrast to the dark surfaces of the furnaces looming nearby. One of the key features of this lighting system is that the circular discs light up the ground with a uniform brightness, yet avoiding any unwanted upward scattering of light. It was this feature, as well as the unique shape, that was very important in the early concept stages. From the outset, it was the designer's intention to create an independent profile for the lights that was not informed by the nearby industrial setting, nor pick up any of its characteristics.

The launch of the street lighting system, as well as illumination of the adjacent conservation area, was marked by a special inauguration ceremony and concert. Amidst the hovering circles of light, visitors actively engaged with the installation. Children used the bright area between the three legs of each GuddeVol as a stage for spontaneous dances, most certainly inspired by the celebratory atmosphere. —

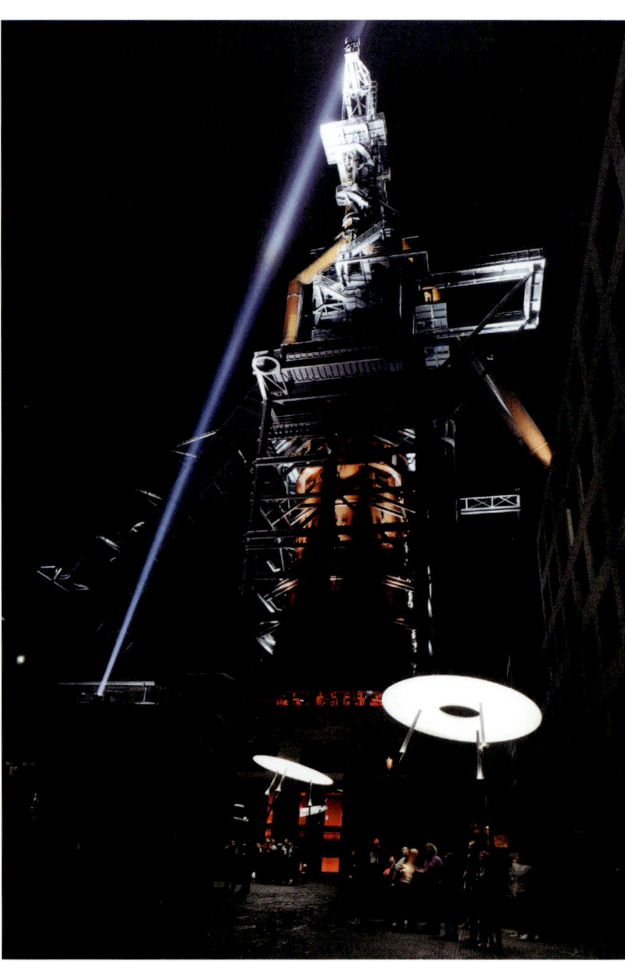

The illuminated circular discs contrast with the dark structures of the furnaces that loom in the distance.

Designer
Ingo Maurer
Location
Esch-sur-Alzette, Luxembourg
Client
Fonds Belval
Collaborator(s)/consultant
n/a
Manufacturer
Livebau Solutions
Date
July 2014

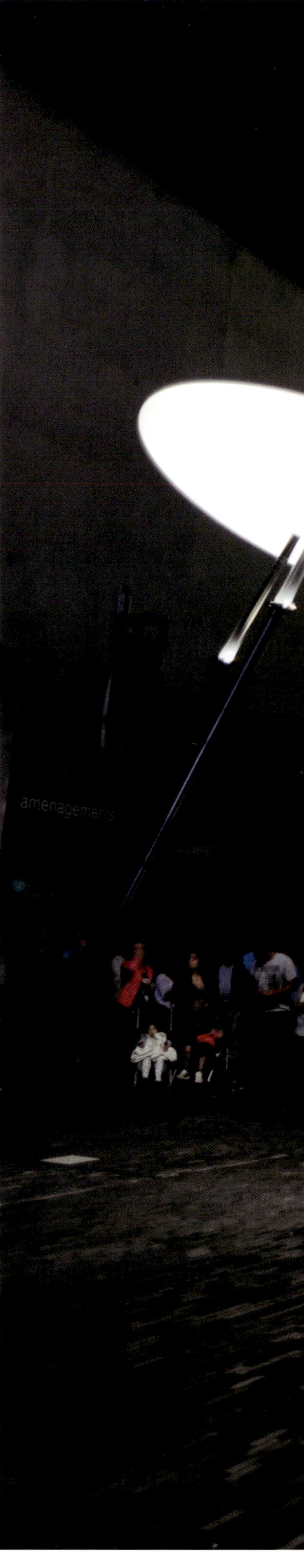

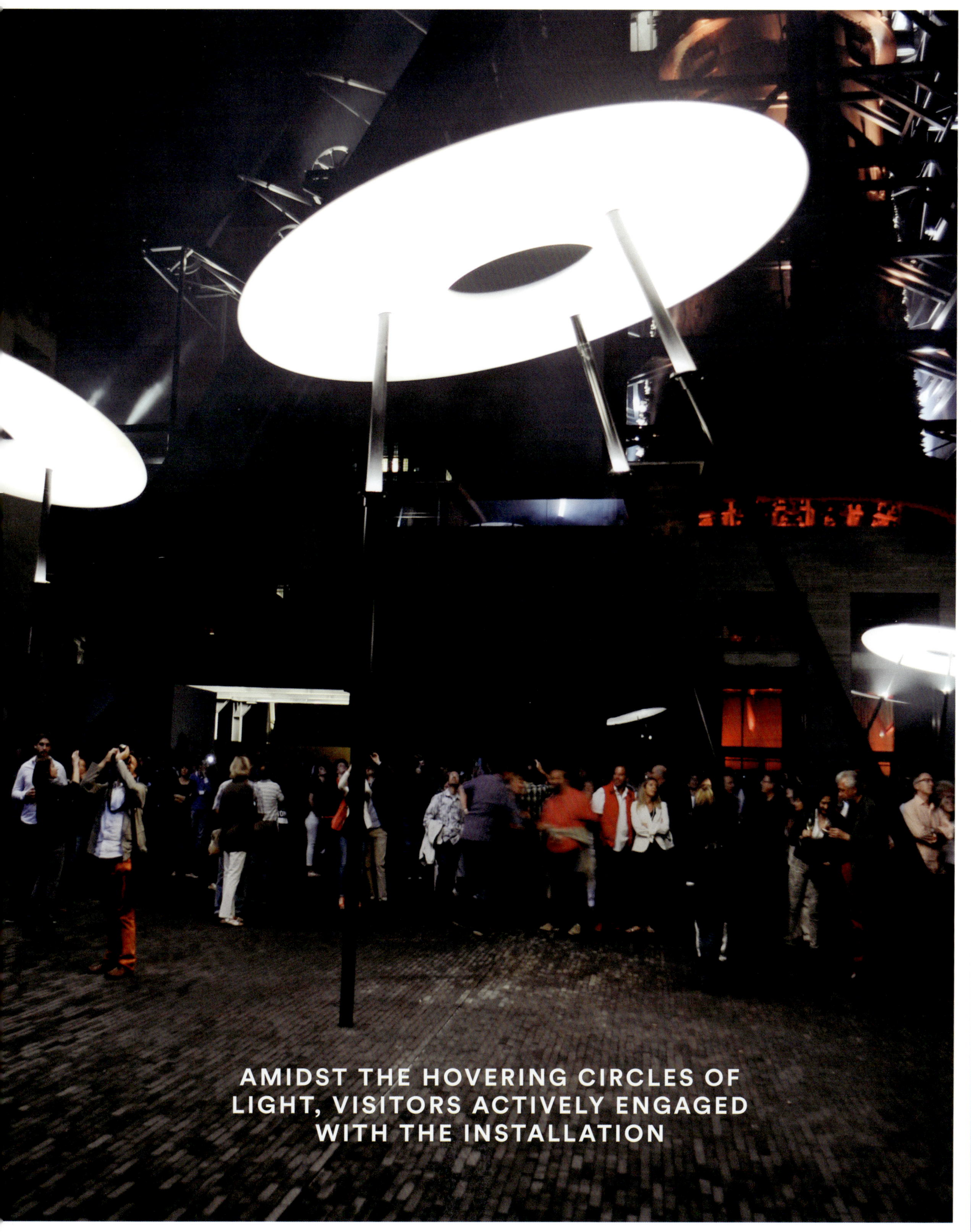

AMIDST THE HOVERING CIRCLES OF LIGHT, VISITORS ACTIVELY ENGAGED WITH THE INSTALLATION

PLAN

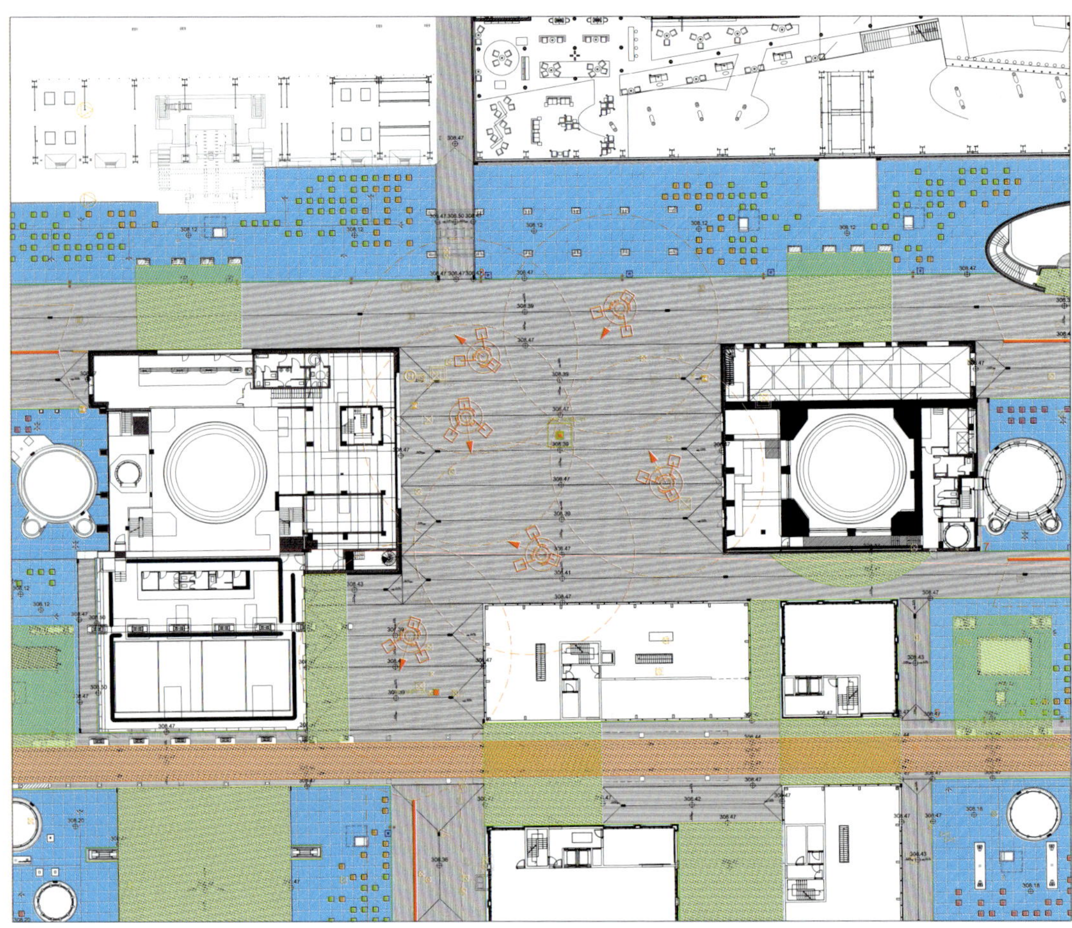

Set-up

Lighting specifications

- Downlights: 12 LED Citizen 40 W, totalling 480 W, 3600 K.
- Uplights: three LED Citizen 27 W, 3600 K.

Control systems used were DALI and DMX. Each lighting element can be remotely and separately controlled. Moreover, the combination of control systems allows for modulation and synchronisation with the street and the Blast Furnaces lighting.
Total power per object is 541 W.
Total weight per object is 1500 kg.
Construction of the lights utilised circular disks made out of glass fibre, with a steel structure within, which was necessary for stability.

ENVISIONING THE LIGHTING FOR A PUBLIC SQUARE IN THE VICINITY OF THE FURNACES

TECHNICAL DRAWINGS

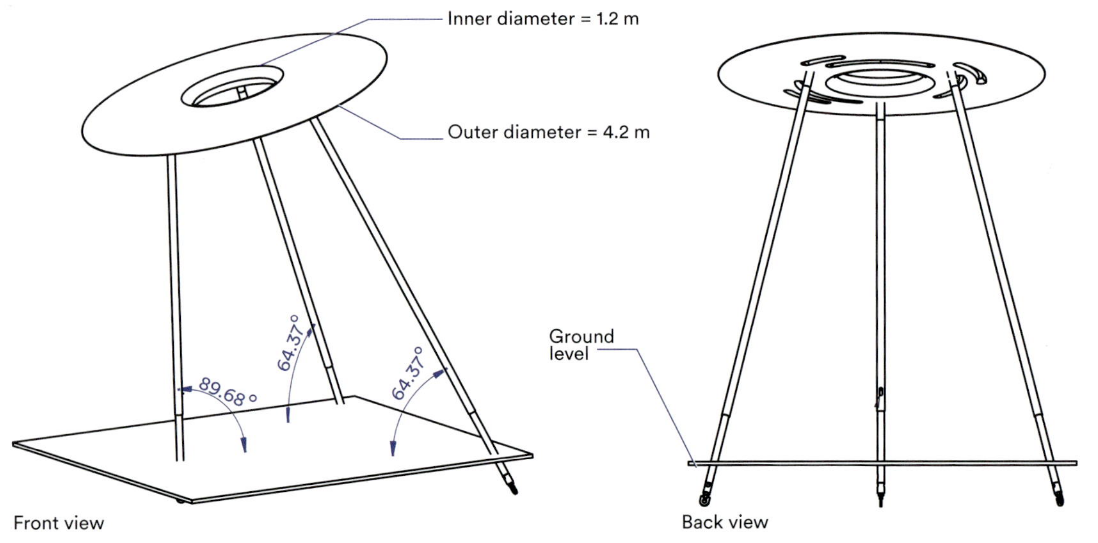

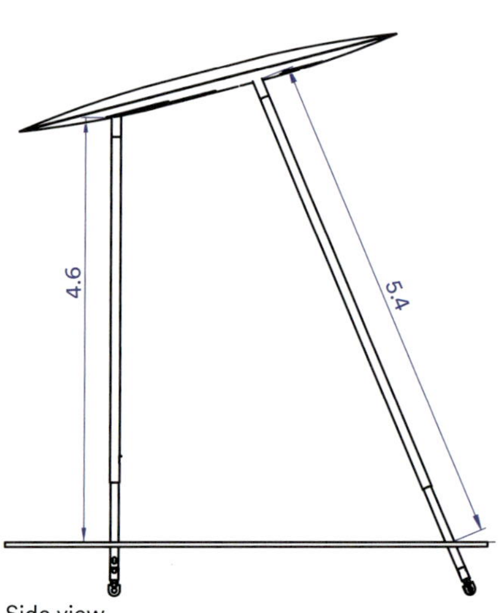

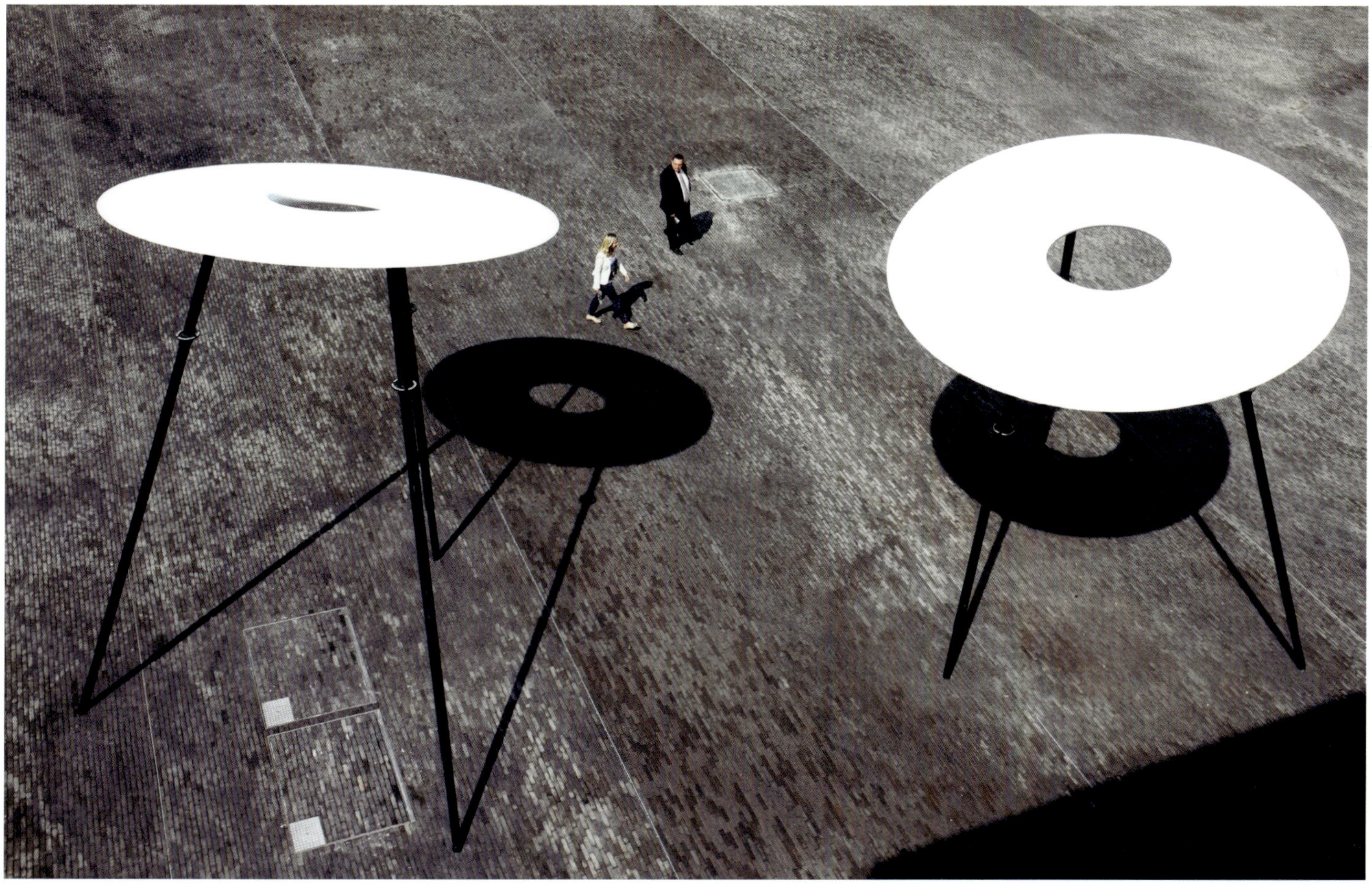

The lights are like sculptures that have a UFO-quality about them, with their illuminated crown and three-legged form.

RENDERINGS

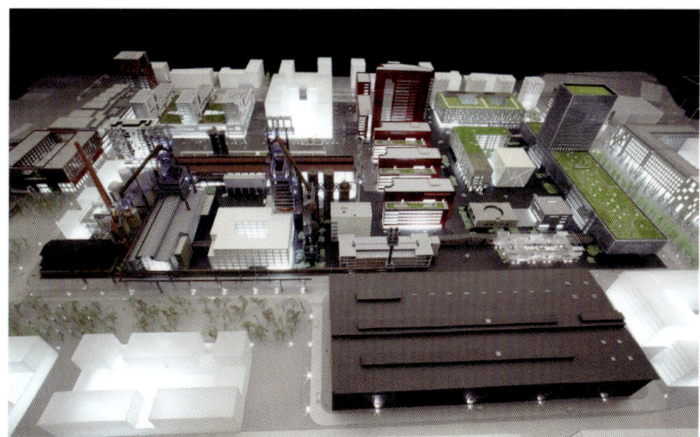

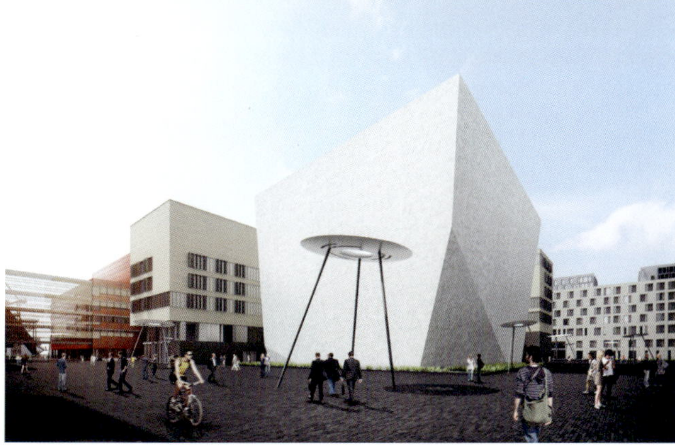

Renderings Fonds Belval

Specifics

In the design phase it was decided to give the primary circular aspect of the lights a slanting inclination. This was primarily carried out in order to add a dynamic element. It is from this aspect also that the luminaires get their name: vaguely reminiscent of a flying saucer, Ingo Maurer named his design GuddeVol, which translates to 'Have a good flight'.

The GuddeVol lamps were constructed in the firm's production grounds in Bavaria and, due to the large dimensions, the constituent parts had to be transported by a special truck to the final location.

On-site installation counted on custom-designed tools, as well as a coordinated team of steel craftsman and electricians working precisely during the set-up phase. With a tiny final tolerance of 5 mm, the set-up had to be very exact.

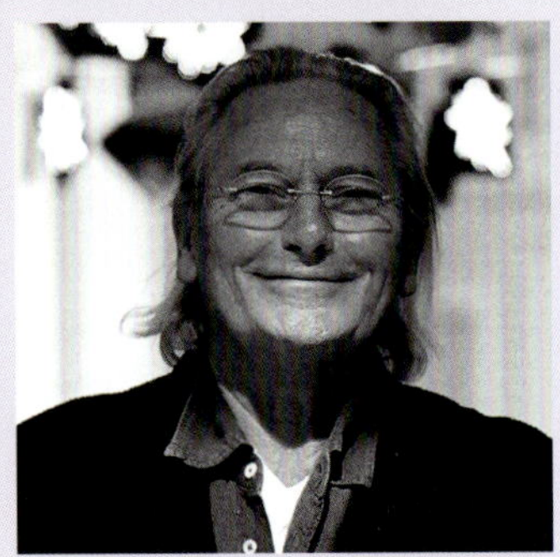

Ingo Maurer

In 1966, Ingo Maurer began designing lamps and lighting systems. His company has since then manufactured and distributed his designs worldwide. Usually produced in low-volume, some of his best known serial products include Maurer's very first design, Bulb from 1966, the 1984 low-voltage system YaYaHo and the winged bulb Lucellino, from 1992. In the 1990s, the Munich-based firm – which now counts a work force of over 60 people – began to move beyond individual lamps to design unique objects and lighting for interior and exterior spaces.

INGO-MAURER.COM

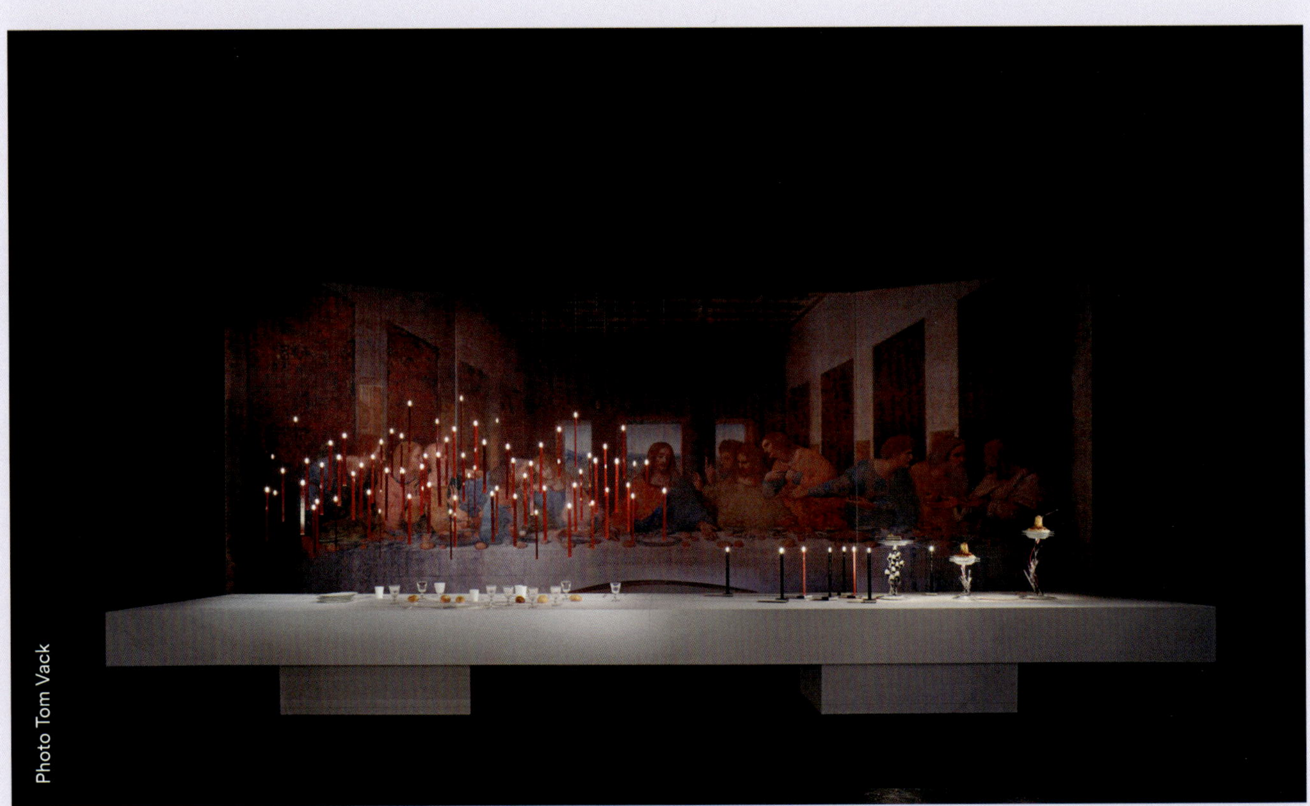

Photo Tom Vack

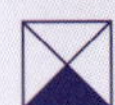

Flying Flames

When **2013**
Where **Milan, Italy**
Client **Ingo Maurer**

For an exhibition in Spazio Krizia in Milan, Ingo Maurer set up a Flying Flame chandelier over a white custom-made table. The composition was completed with a large print of Da Vinci's *Last Supper* and looked to draw attention to the uncanny chandelier. In collaboration with Moritz Waldemeyer, Flying Flames has a flexible system made up of LED 'candles' which are now part of Ingo Maurer's product collection.

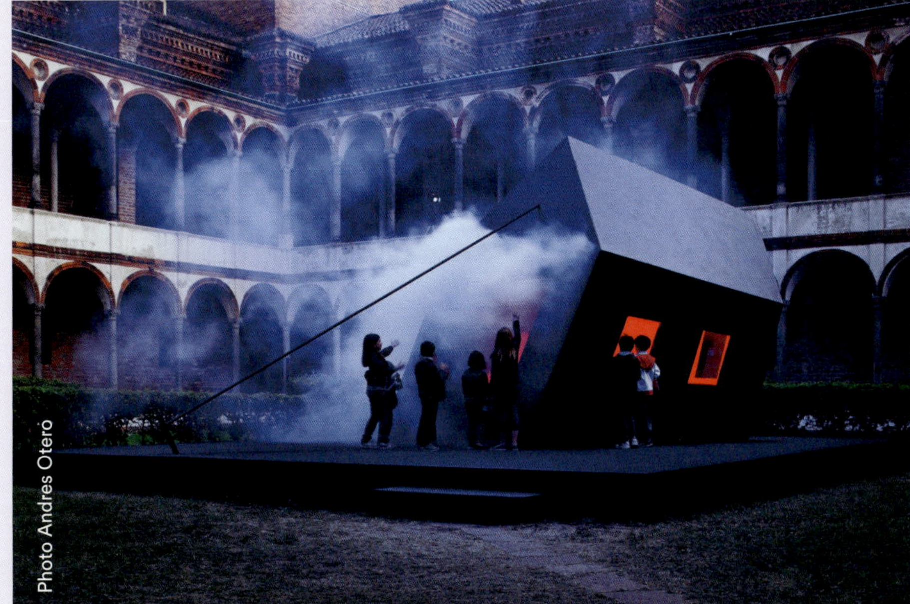
Photo Andres Otero

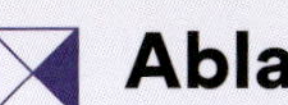

Ablaze

When **2011**
Where **Milan, Italy**
Client ***Interni***

A collaboration between Ingo Maurer and Axel Schmid, Ablaze was a temporary installation for *Interni* magazine's 2011 Mutant Architecture exhibition. The design team's original concept was to show the gradual transformation of a house burning down in slow-motion. The installation takes an abstract approach to this theme, showing the emotional power of fire and light.

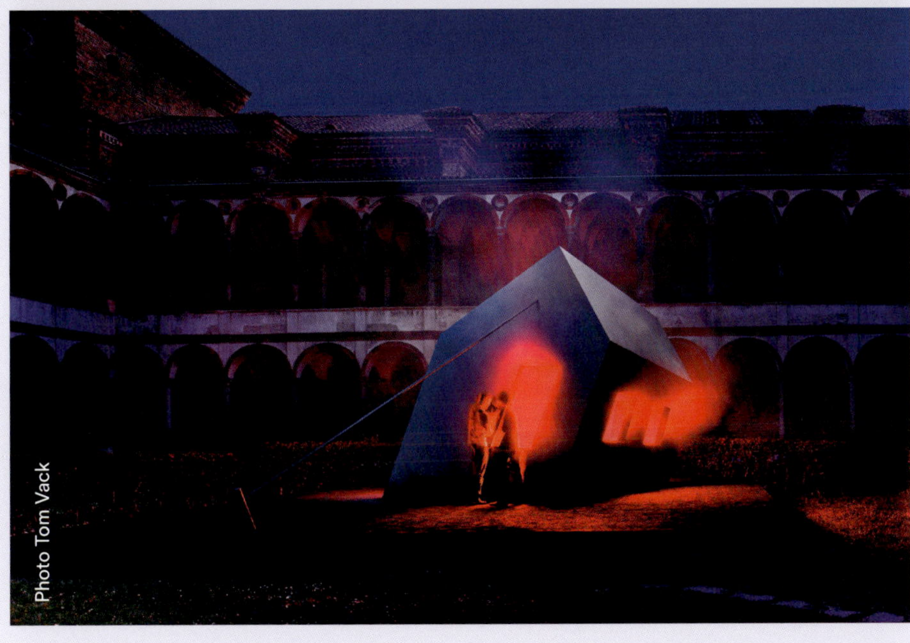
Photo Tom Vack

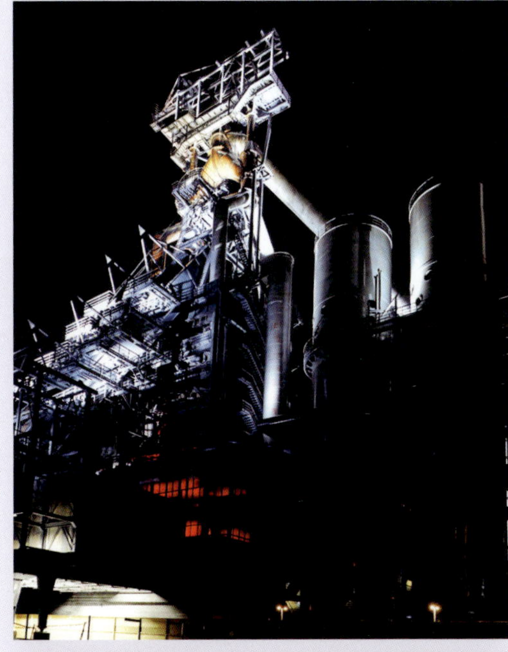

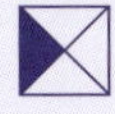

Blast Furnaces

When **2014**
Where **Esch-sur-Alzette, Luxembourg**
Client **Fonds Belval**

Part of the conservation project for the disused blast furnaces of Luxembourg's biggest ironworks, Ingo Maurer's lighting design was implemented to emphasise the spirit and role of the nearly 100-m-high blast furnaces in Esch-sur-Alzette. To this end, white lights were used to create contrast between bright and dark areas, in a lighting concept inspired by Fritz Lang and Sergej Eisenstein's black-and-white films.

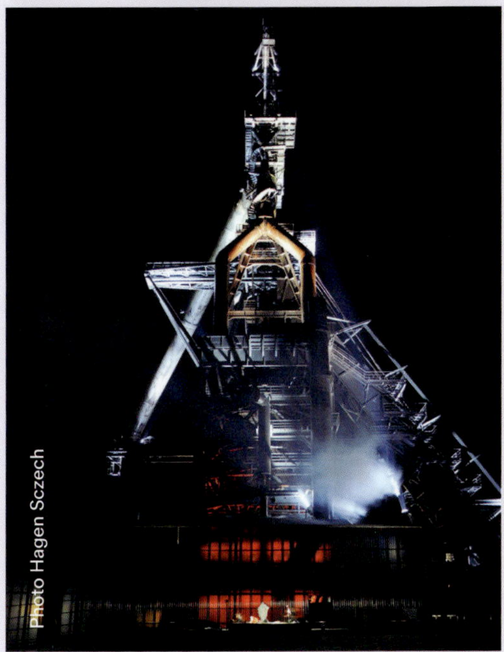
Photo Hagen Szczech

Ross Lovegrove created large-scale lights utilising a Barrisol skin wrapped around metallic loops to form a Mobius-type volumetric object.

Infinite Loop

The Infinite Loop installation by **Ross Lovegrove** consisted of a collection of three-dimensional lights in a highly volumetric form, presented in perfect synergy between Barrisol's innovative stretch material, LED lighting and an aluminium framework set in a futuristic setting.

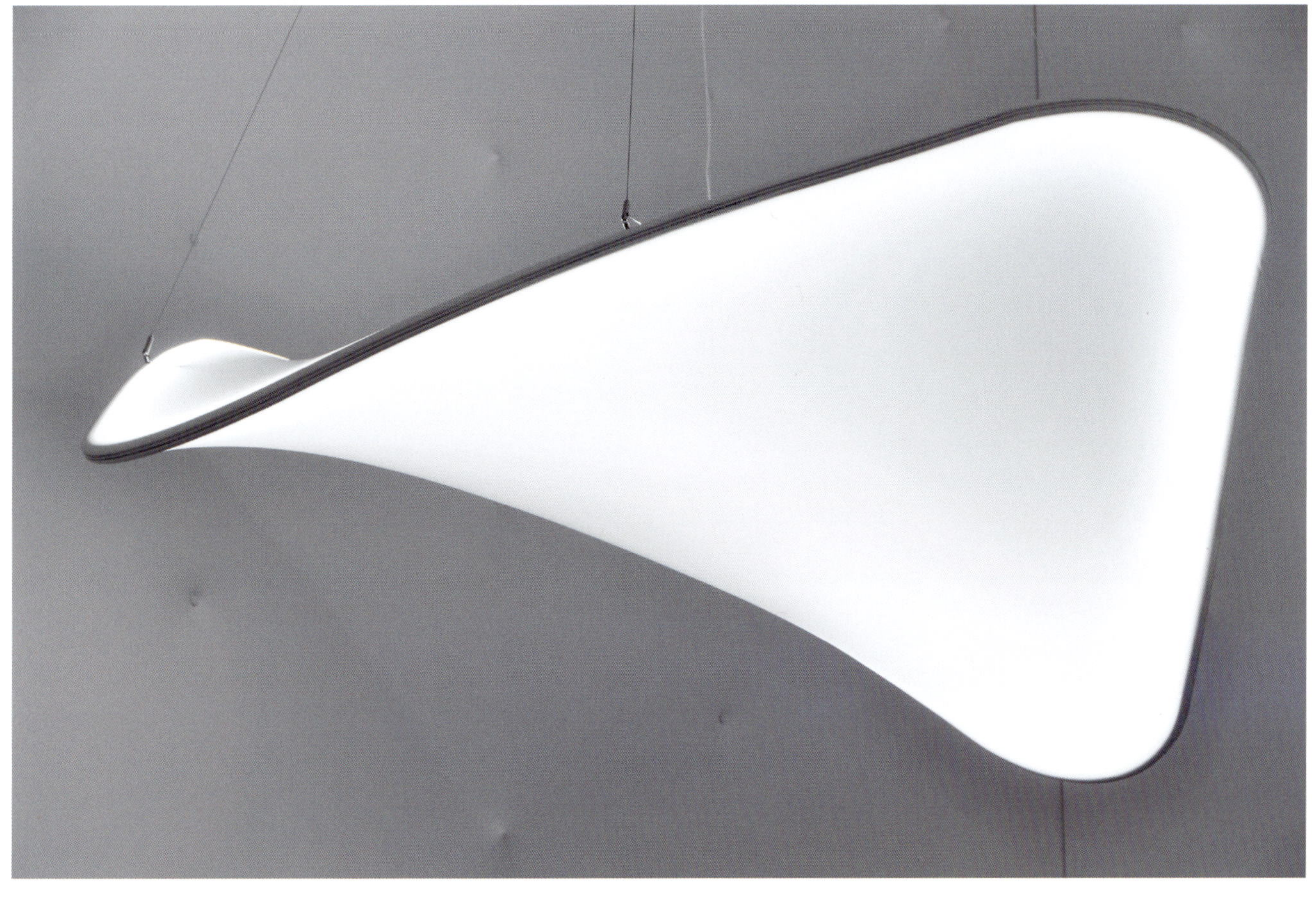

Photos Christoph Hermann, Michael Maheux, John Ross

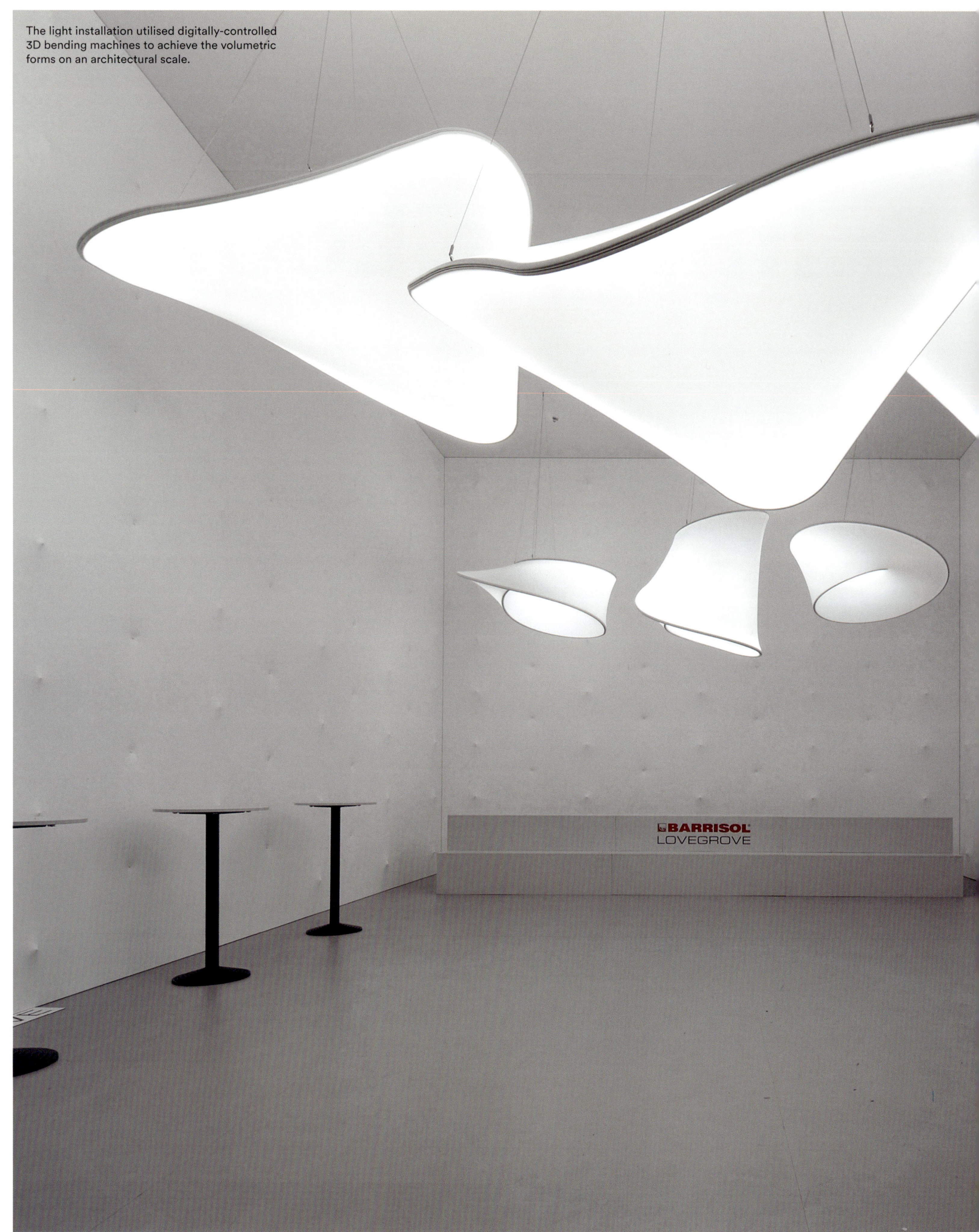

The light installation utilised digitally-controlled 3D bending machines to achieve the volumetric forms on an architectural scale.

Designer
Ross Lovegrove
Location
Kortrijk, Belgium
Client
Barrisol
Collaborator(s)/consultant
Barrisol
Manufacturer
Barrisol
Date
October 2014

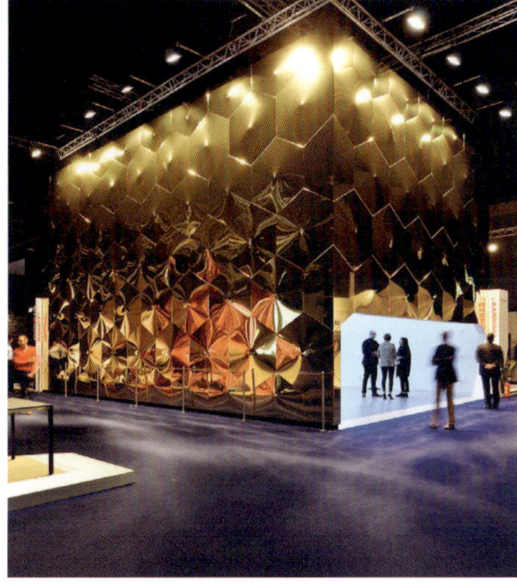

The 7-m-high pavilion was covered in a golden, mirrored material, contrasting with the serenity of the room's white interior.

At the 2014 Biennale Interieur in Belgium, designer Ross Lovegrove collaborated with Barrisol to present his new lighting collection for the French company. Highlighting the brand's stretch-form material, the illuminated Infinite Loop installation was presented in a distinctive golden volume. Within the bustle of the interior design fair, the glistening Barrisol pavilion enticed visitors inside to discovered an all-white, futuristic space where Lovegrove's three-dimensional lights took pride of place.

The installation itself included large pendants that balanced aesthetic form with subtle lighting function. The pieces in the collection are highly volumetric, created from a perfect synergy between the stretch material, LED lighting and an aluminium framework. The room's interior took on a serene-white tactile form using material with acoustic dampening properties for the wall surfaces. The presentation of the Infinite Loop lights illustrate a unique application of the firm's products that have a range of acoustic dampening, mirroring and translucency properties. Lovegrove utilised Barrisol's advanced manufacturing technology, such as numerically-controlled cutting and digitally-controlled 3D bending machines to achieve a high level of precision for the components of the installation.

The large-scale luminaires are constructed from a metal framework covered with the stretch material. The aluminium frame is built out of a single continuous metal loop, which holds the LED strip light source. The surrounding Barrisol skin forms a Mobius-type volumetric object. Three different types of loops, ranging from horizontal to vertical structures, have potential for customisation. This makes them adaptable for a wide variety of architectural spaces, including hotel lobbies, halls and high-rise atriums. —

RENDERINGS

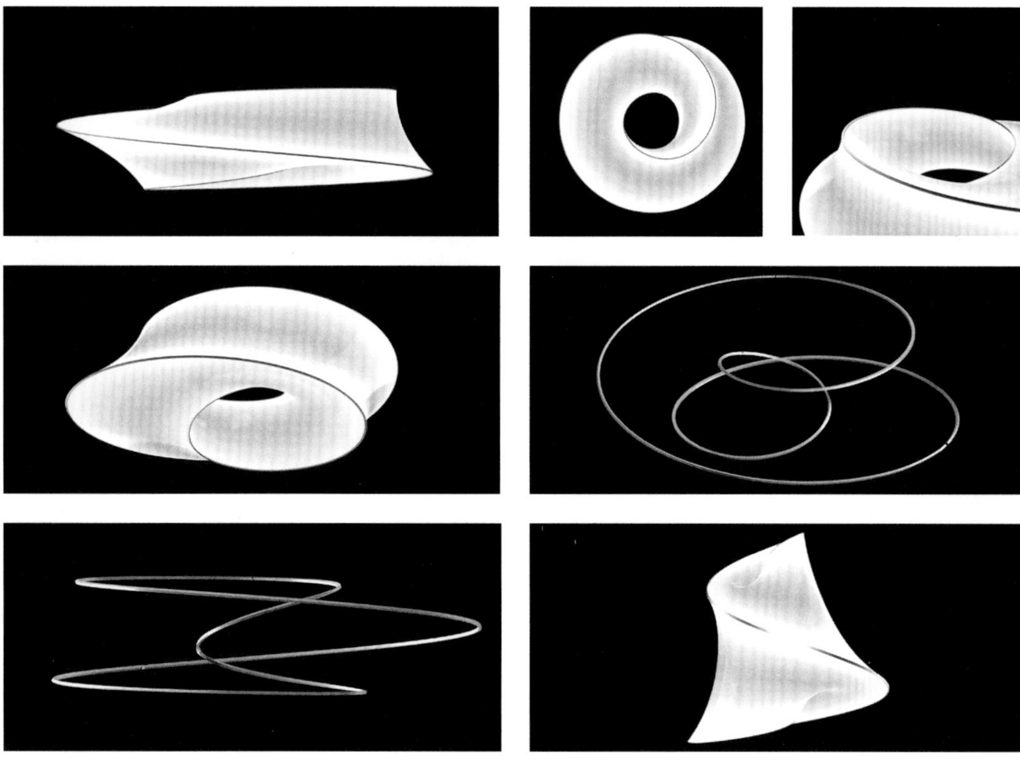

COMPUTATIONAL DIAGRAMS

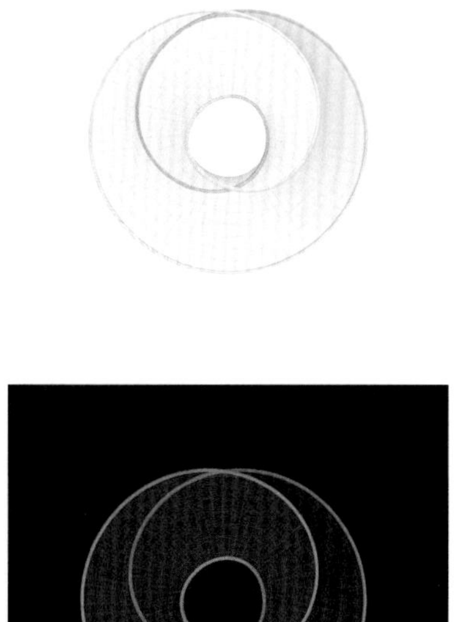

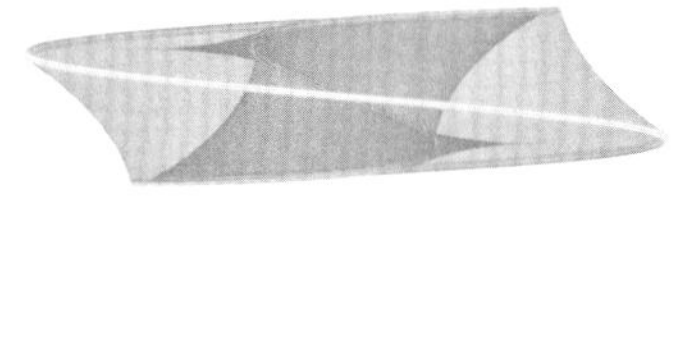

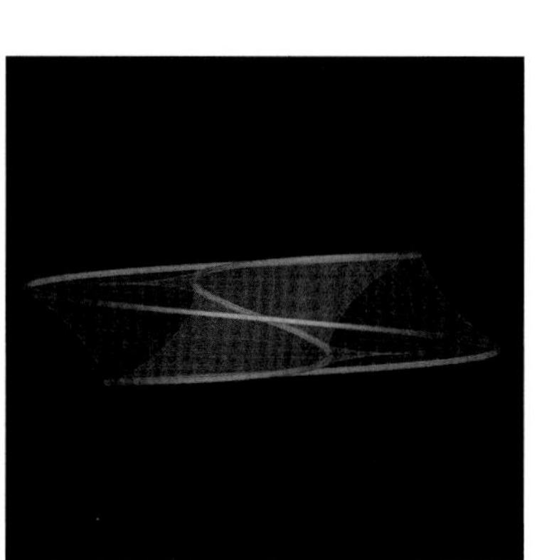

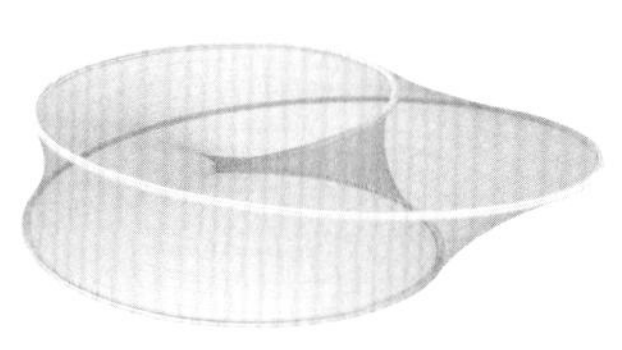

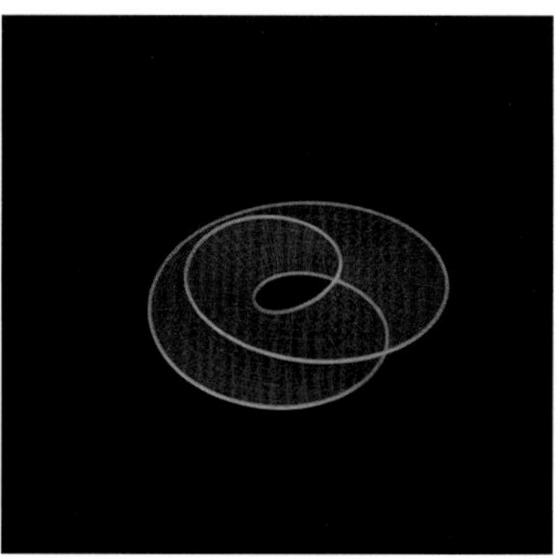

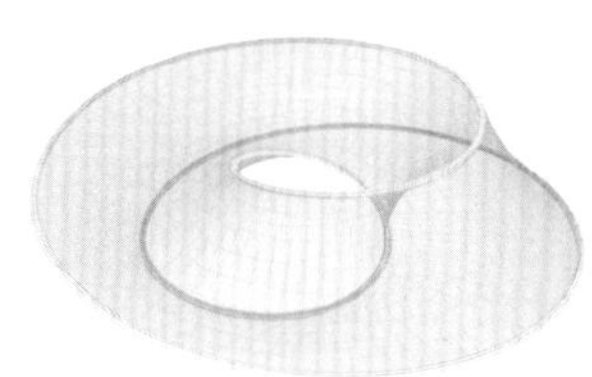

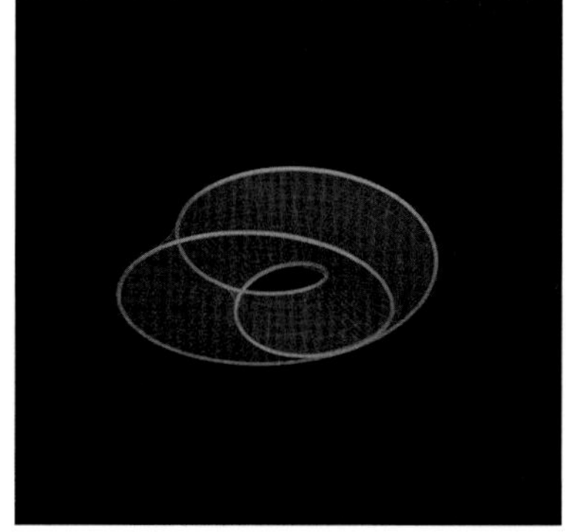

Specifics

ΛΛΛΛ By intensively studying the stretch behaviour of the Barrisol material, the team found that it had an ability to take the form of minimal surfaces, as they are known in mathematical terms. This means that while the sheet itself is flat, when it is cut out with precision and mounted in a highly-controlled way, it can create 3D volumetric shapes. Parametric modelling programs (digital systems that generate geometries based upon an initial input of data) were then used to determine the form of the lights.

ΛΛΛΛ Developing and fine-tuning the production process was complex and time consuming. Beneficially creating various versions of form and scale can now be achieved fairly easily allowing customisation and unique low-batch productions for clients.

ΛΛΛΛ Both Manta and Infinite Loop are based on the same core principles forming a perfect synergy between the stretch material, LED lighting and metal framework. However, while the Infinite Loop light is a highly volumetric and three-dimensional, the Manta light consists of various Barrisol foils with different intensities of stretch to provide a highly-layered acoustic dampening and uniform lighting. This slim framework forms a rhomboid, which allows for many different large modular arrangements.

ΛΛΛΛ Various new forms and smaller-scale versions are currently in development. Furthermore, plans are underway for further developing the production process allowing custom, low-batch versions, creating a bridge between one of a kind installation work and conventional mass-production products.

ΛΛΛΛ Parametric tools were created allowing to digitally simulate the stretch behaviour of the Barrisol material. While first concentrating on the aesthetical output, further investigations on the amount of stretching and limits of the material were also conducted.

ΛΛΛΛ The biggest challenge involved the investigations to overcome misalignment between the digital simulations and the actual material behaviour. In order to achieve the same output, the production tolerances had to be extremely low. In addition, the Barrisol team's new techniques and tools were utilised to ensure homogenous and wrinkle-free lights.

CONSTRUCTION VISUALS

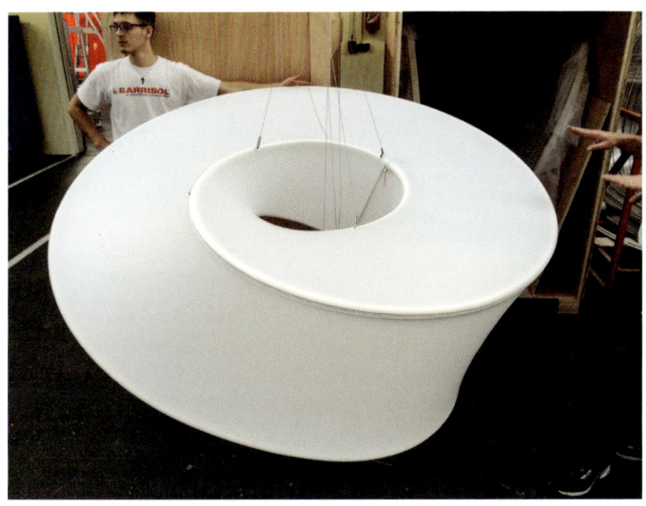

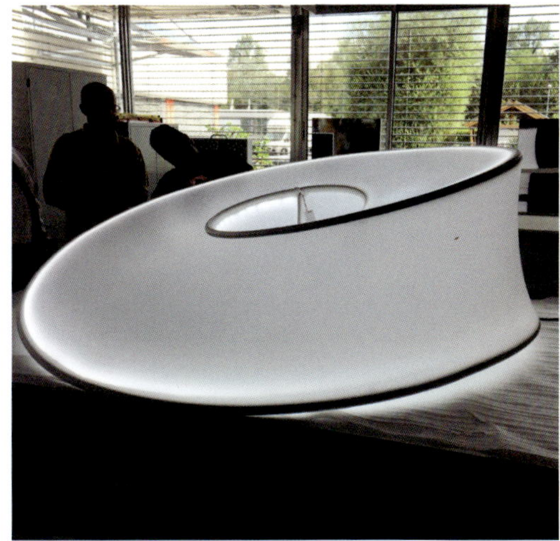

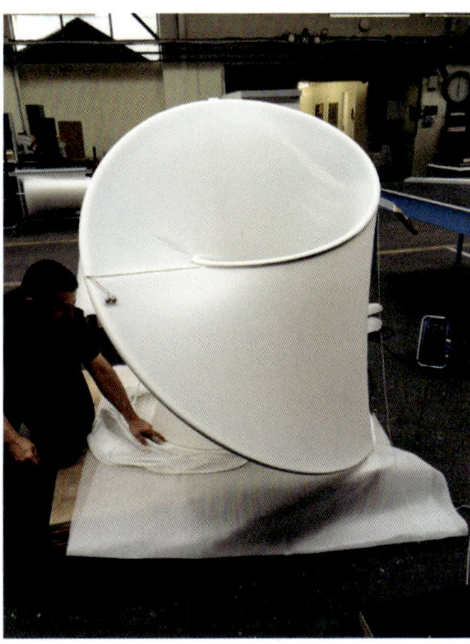

INFINITE LOOP LIGHTS ILLUSTRATE A UNIQUE APPLICATION OF BARRISOL'S PRODUCTS

ON-SITE INSTALLATION

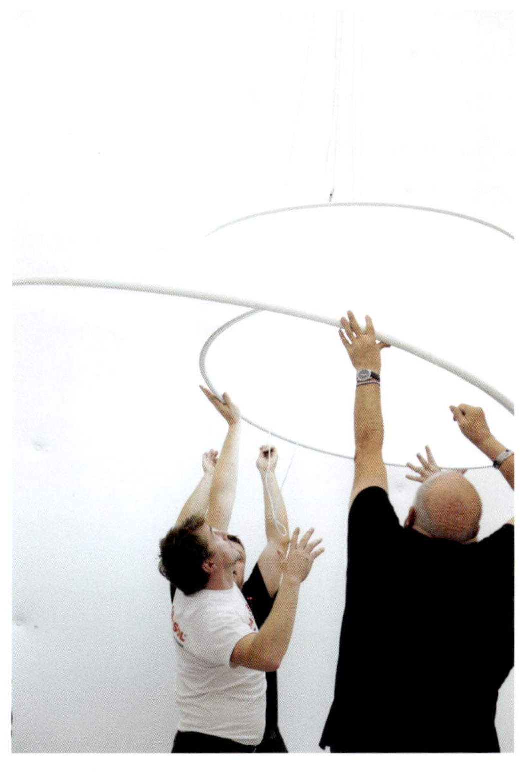

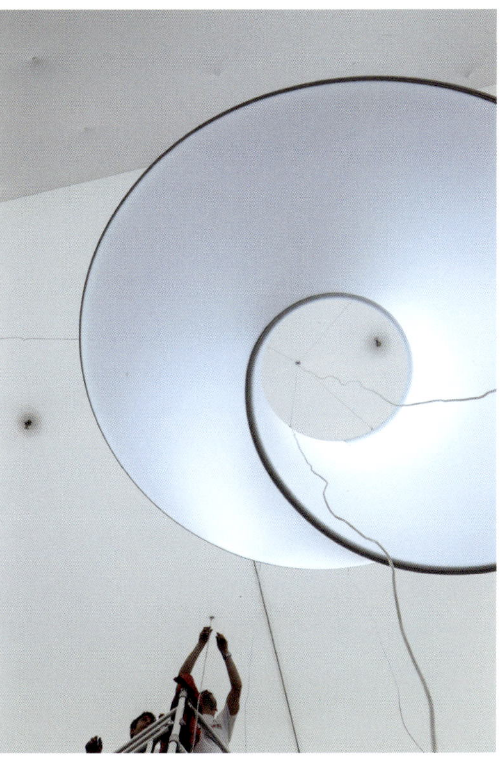

Set-up

Total volume of the installation for illumination was 639 m³ (7.8 x 7.0 x 11.7 m).
Component parts included LED strip light, diffusive reflective Barrisol stretch foil, bent aluminium, custom lighting sequence, electrical hardware and metal framework.
Used light was LED strip lights positioned along the bent aluminium rails.
Installation included a total of six light sculptures which each incorporated LED strips and bent aluminium: 2 x Infinite Loop 1 (with 11.4-m aluminium extrusion); 1 x Infinte Loop 2 (with 9.3-m aluminium extrusion); and Manta light (with 7.1-m aluminium extrusion).
Lighting control could be individually controlled for all internal LEDs, as well as the overall internal light of the pavilion's interior. The set-up included 5-minute-sequences of various lighting scenarios showcasing different ambient modes.
Metal framework of the custom-built pavilion was constructed with an internal metal structure, clad with Barrisol stretch foil. This internal metal framework has been built to provide accurate fixing points for the light sculptures.
Construction time in total was 14 days; arrangement of the LED–aluminium tubes were carried out on-site.

Photo John Ross

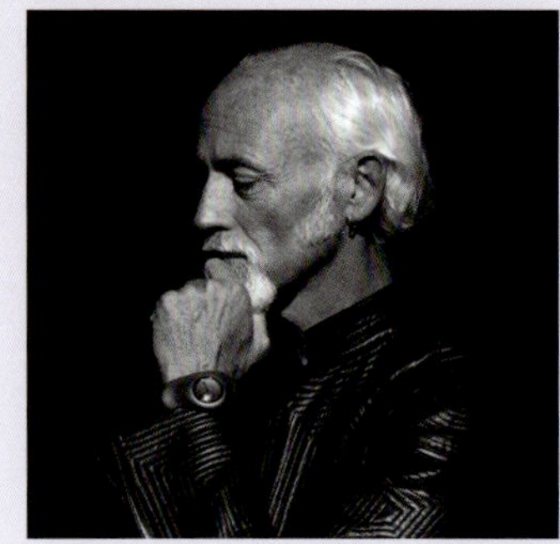

Ross Lovegrove

Ross Lovegrove is a designer and visionary whose work is considered to be at the very apex of stimulating a profound change in the physicality of our three-dimensional world. Inspired by the logic and beauty of nature, his design possess is a trinity between technology, materials science and intelligent organic form, creating what many industrial leaders see as the new aesthetic expression for the 21st century. There is always embedded a deeply human and resourceful approach in his designs, which project an optimism, and innovative vitality in everything he touches, from cameras to cars to trains, aviation and architecture.

ROSSLOVEGROVE.COM

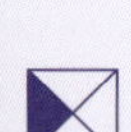

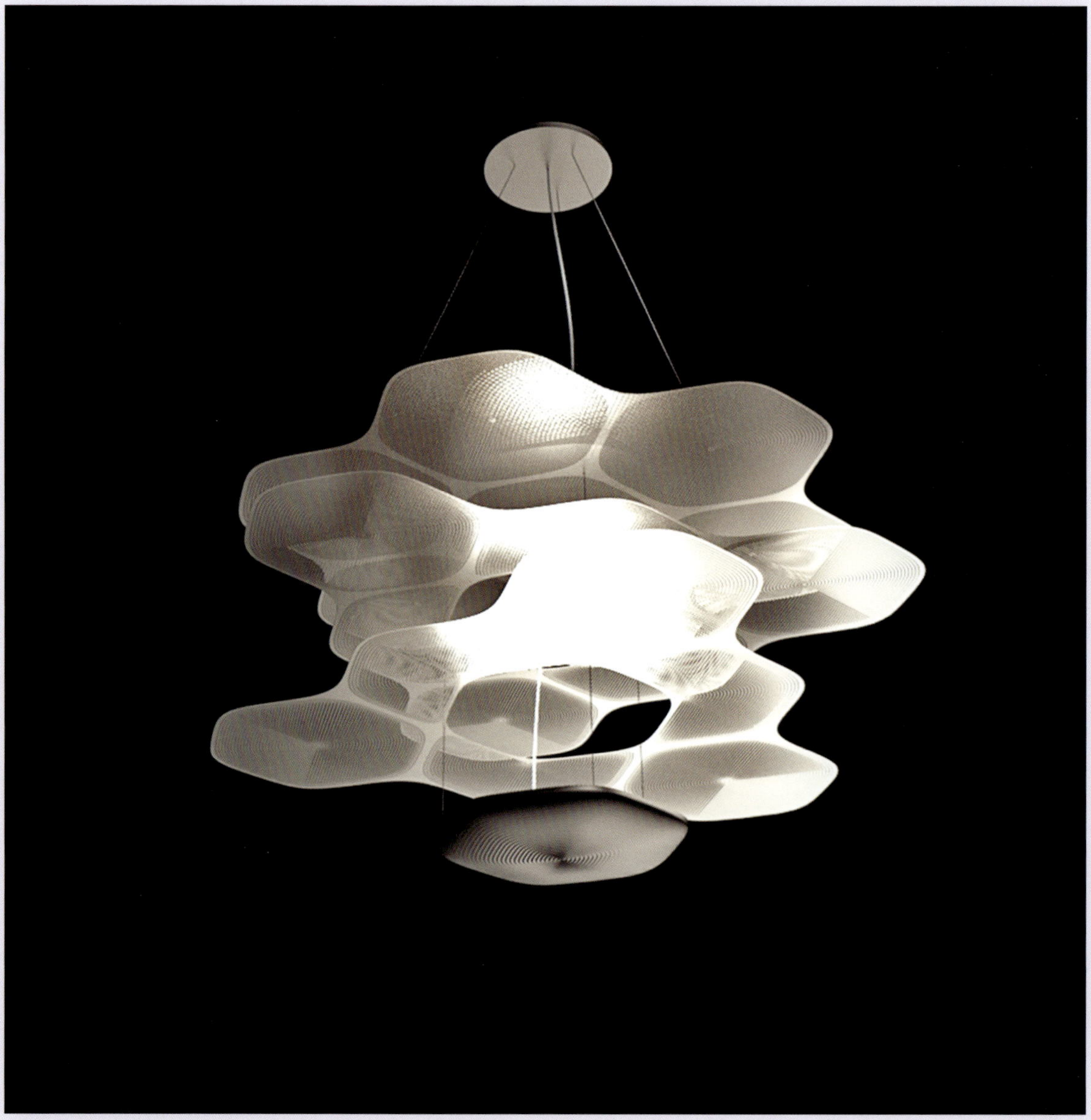

Space Cloud

When **2014**
Where **London, United Kingdom**
Client **Artemide**

This highly-modular suspension lighting system has a great potential for creating large density installations. It can be configured in vertical or horizontal layers, as light-capturing surfaces allow both absorption and reflection through each of the sheets. Spatially designed in a hexagonal matrix, the lightweight construction is made from 100% aluminium that has been anodised so that the colour/surface treatment is an integral part of its physicality, as in space technology. Its concept was inspired by NASA's photos from space of the Earth.

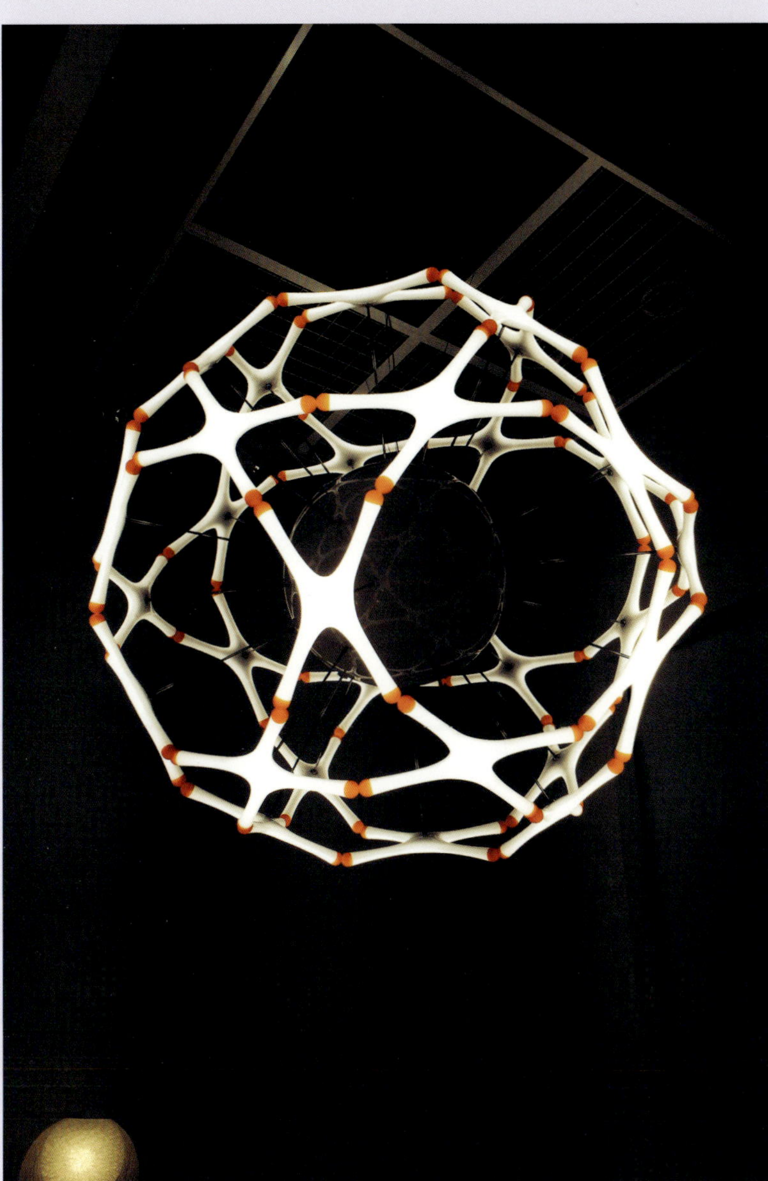

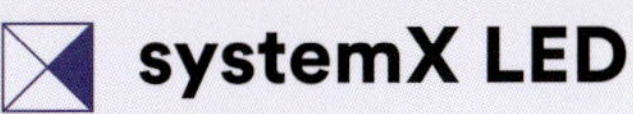

systemX LED

When **2014**
Where **Tokyo, Japan**
Client **Yamagiwa**

The basic application of systemX LED involves interconnections among modules, varied connectors and wires to form a variety of configurations, from grid surface, parabolic pattern and circle layers to elaborate forms, such as a globe. Designed by Ross Lovegrove to give an overall effect suggesting seamlessness between the luminaire itself and the light it casts upon the ceiling and surrounding environment. Each module can be dimmed individually through a programming system.

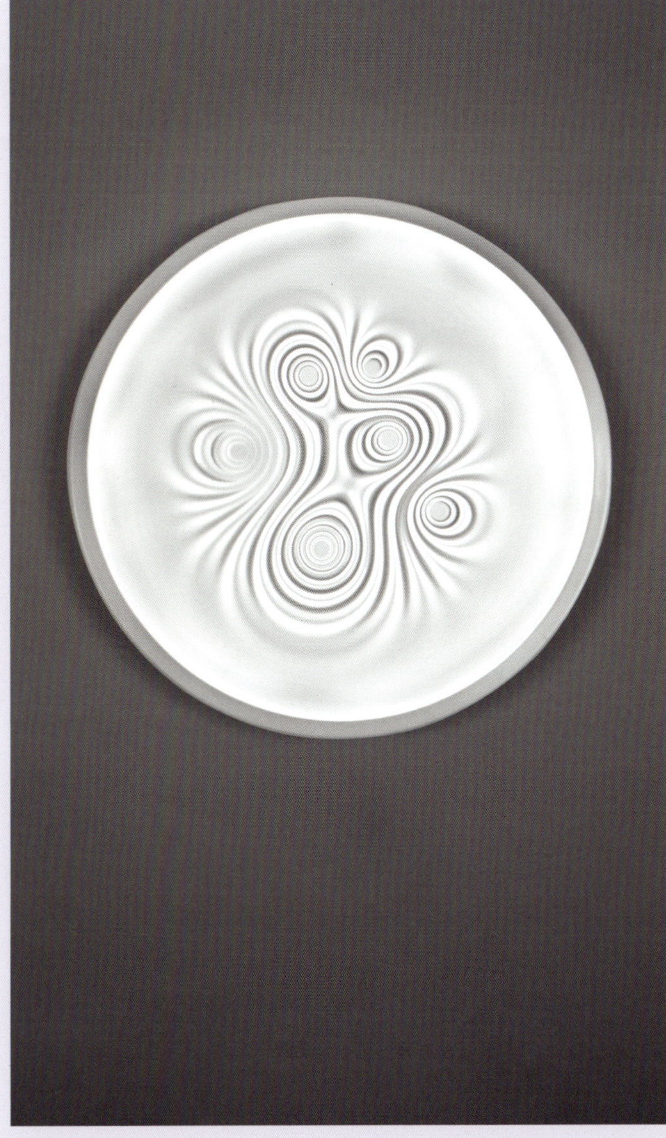

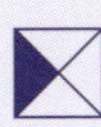

Nebula

When **2012**
Where **London, United Kingdom**
Client **Artemide**

Generated from advanced parametric scripting where the vortices act as light-capturing forms, Nebula emits a glowing light from a ring of LEDs located in its circumference. It includes a single aluminium sheet which articulates an ornamental pattern to reflect and diffuse light. Surface implementations are raised towards the centre to accentuate the organic forms as it captures the surrounding light. The projects reflects Ross Lovegrove's interest and research in generative designs further into digital baroque.

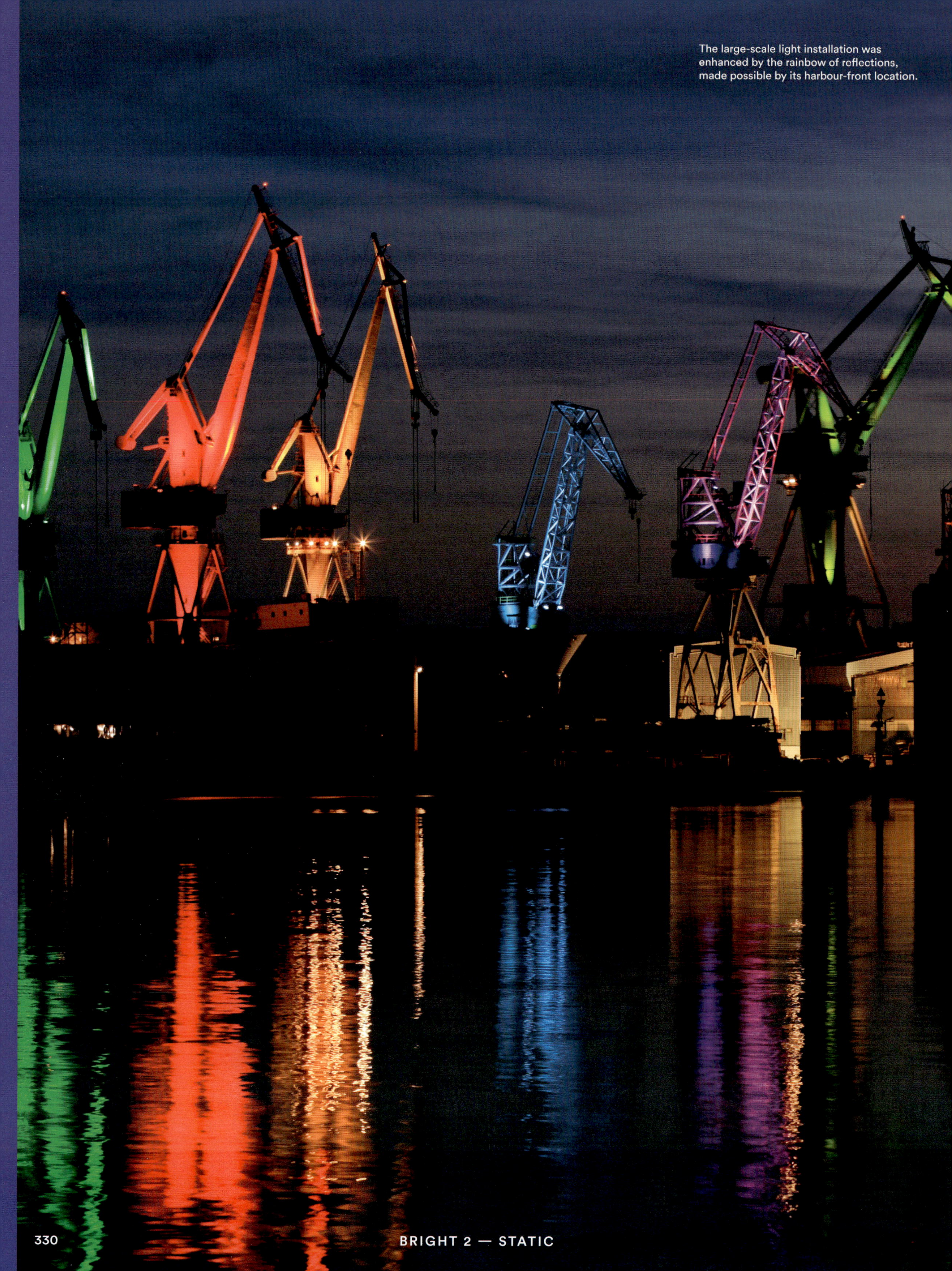
The large-scale light installation was enhanced by the rainbow of reflections, made possible by its harbour-front location.

Lighting Giants

Rising high above Uljanik, a 200-year-old shipyard in Croatia, eight majestic cranes bathed in vibrant colours illuminates the skyline of Pula. The Lighting Giants installation, designed and devised by the internationally-renowned lighting design practice **Skira**, includes a blend of technology and history, functioning as a monument to the city's past.

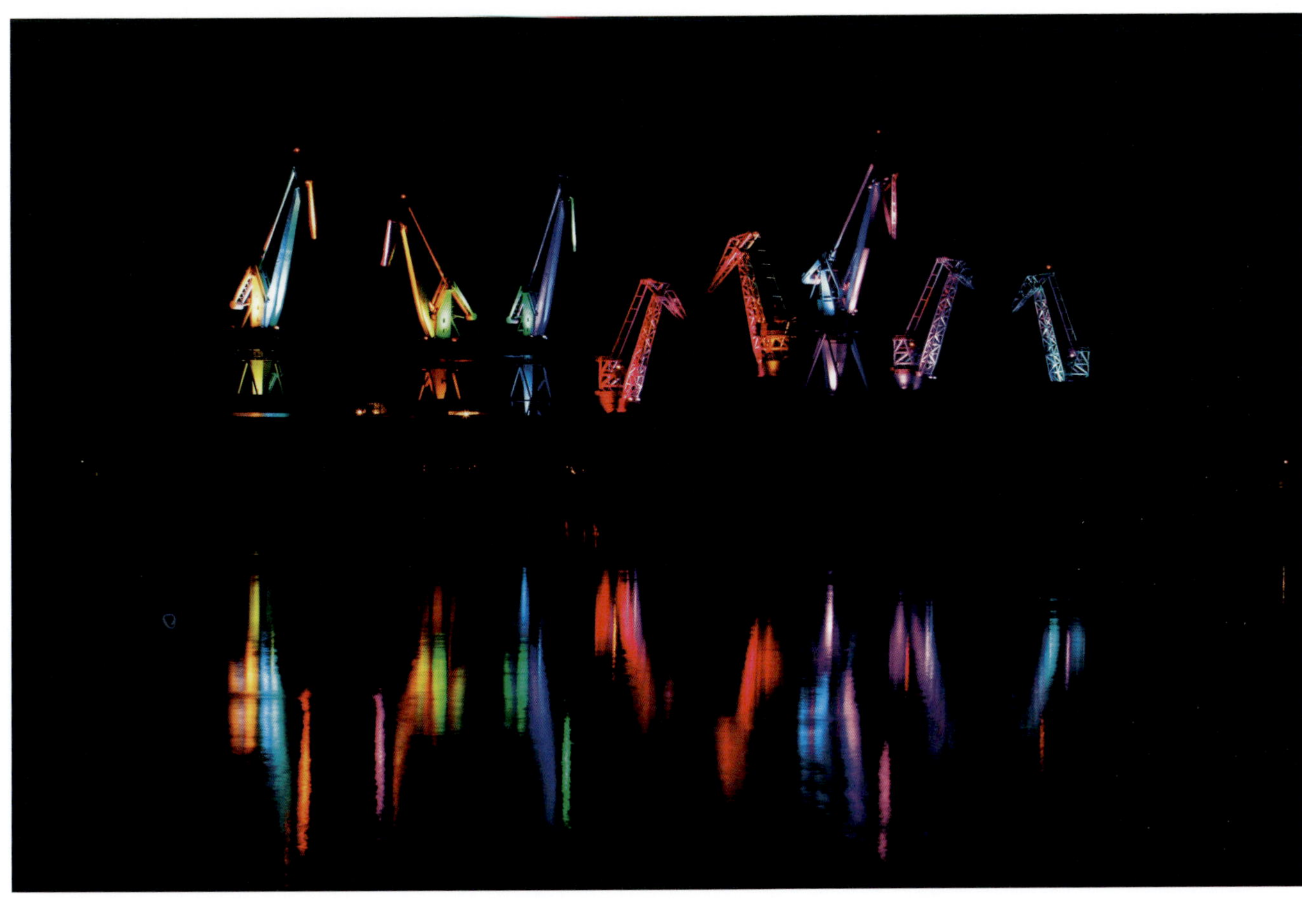

Photos Goran Šebelić

THE LIGHTING GIANTS BURST INTO LIFE
WITH A VIBRANT RAINBOW OF COLOUR

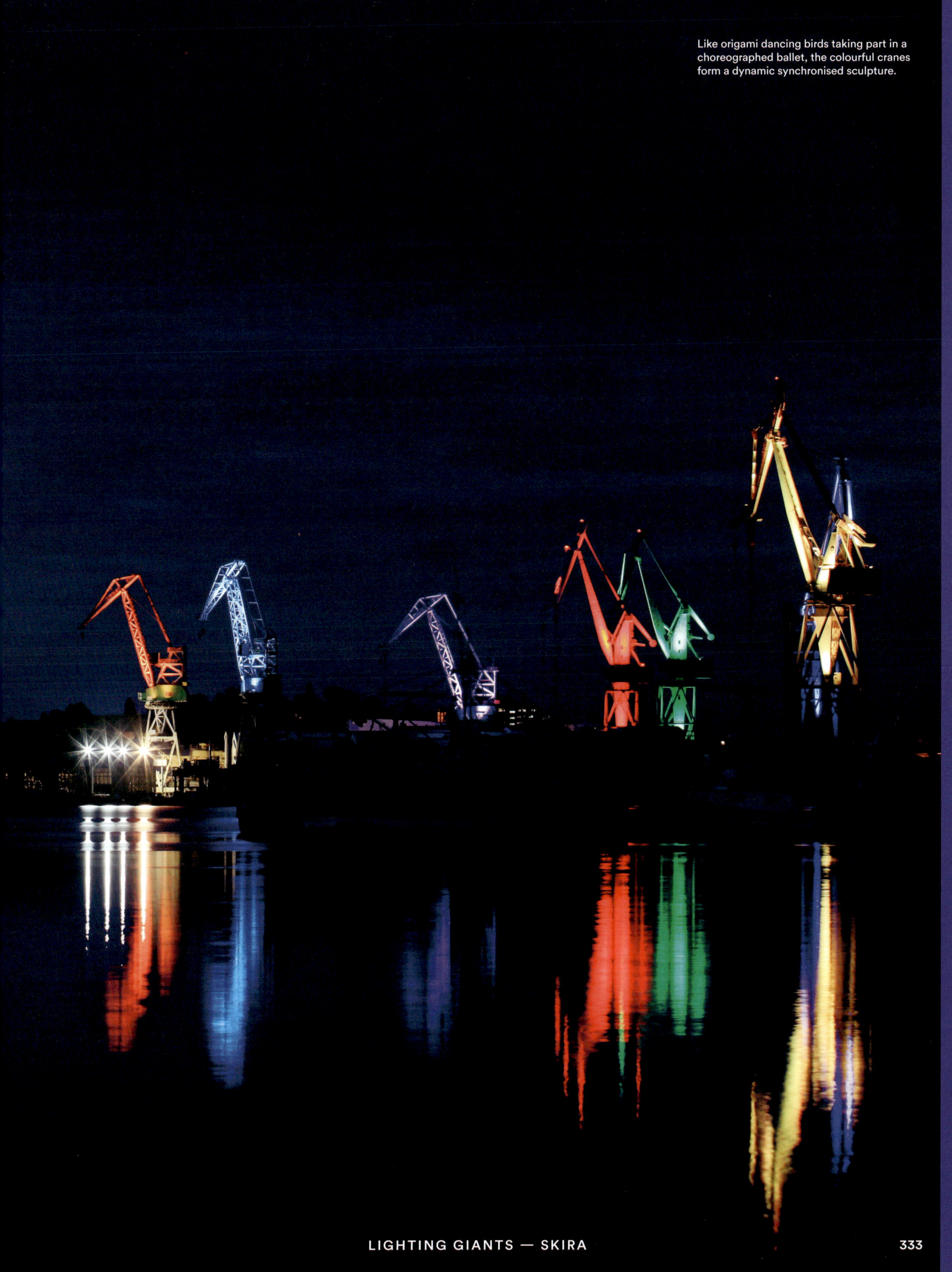

Like origami dancing birds taking part in a choreographed ballet, the colourful cranes form a dynamic synchronised sculpture.

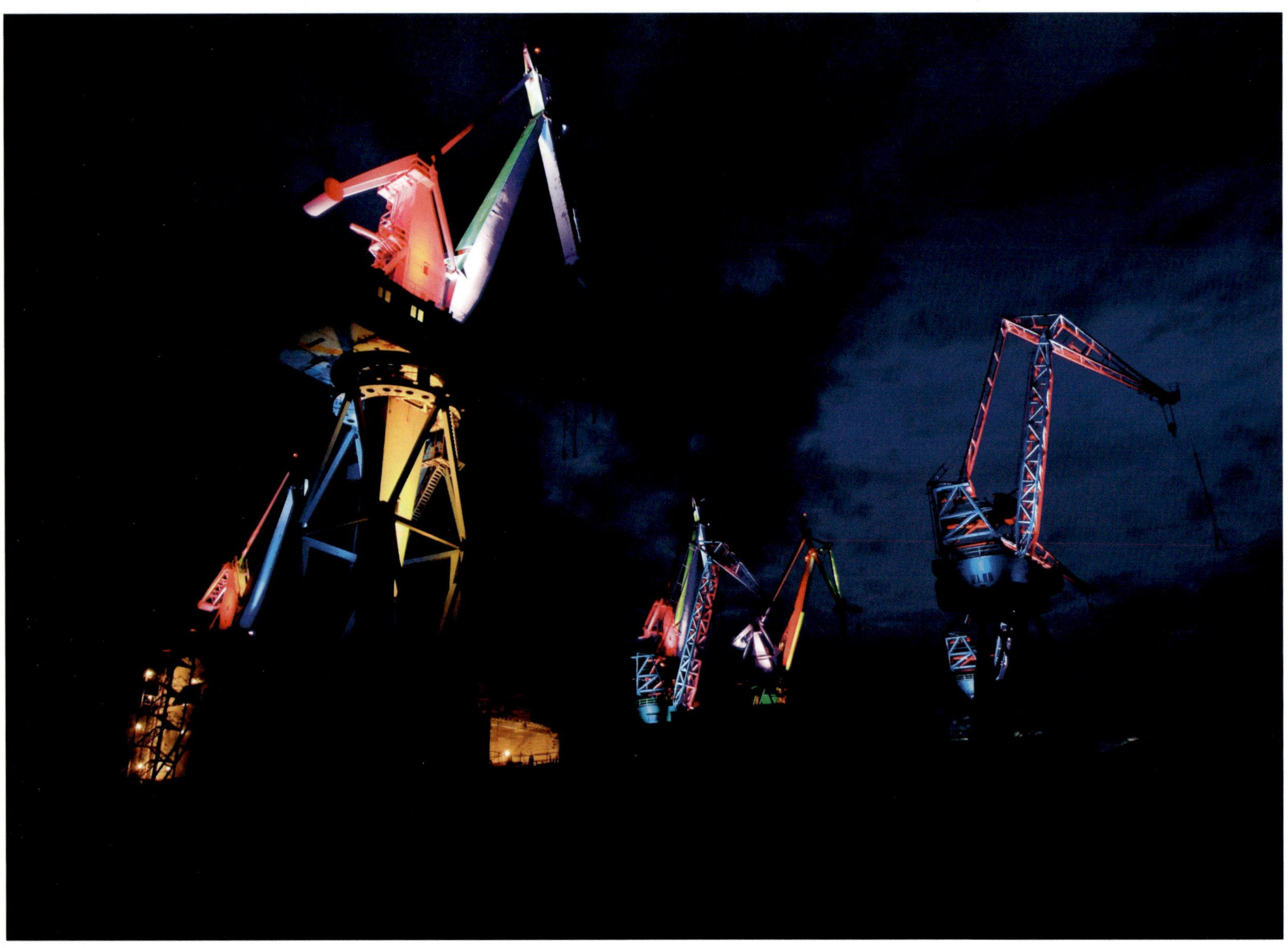

The metallic giants are dressed for different illumination scenarios, suitable for numerous occasions and celebrations.

MOVING IN UNISON, THE COLOSSAL CRANES RESEMBLE A COLOURFULLY-CHOREOGRAPHED DANCE TROUPE

Designer
Skira
Location
Pula, Croatia
Client
Tourist Board of Pula
Collaborator(s)/consultant
Philips, Uljanik Shipyards
Manufacturer
Philips
Date
May 2014

One of the world's oldest working shipyards can be found in the Mediterranean city of Pula, Croatia. Built in 1856, Uljanik was on the verge of being relocated when Pula-born lighting designer Dean Skira devised a plan for highlighting the industrial shipyard cranes that dominate the skyline of the city's bay. The Lighting Giants' cranes were first illuminated during Visualia, the festival of lights in collaboration with the Tourist Board of Pula, with an aim to make the shipyard the focal point of the city once again.

The eight gigantic cranes are metallic monuments, undertaking their own 'gentle dance of steel' as they help to create towering commercial ships. Still functioning during the day in the active shipyard, it is under the hours of darkness that the Lighting Giants take to the stage in their full glory. The vibrant sculptures required illumination that was suitable for such large-scale installations, and fitting for the job were the 73 LED spotlight fixtures that could be programmed in 16,000 ways to accommodate different lighting, adding a vertical axis to the night landscape with various static illumination scenarios.

The cranes are at once functional and monumental, making this a unique project that is constantly changing as the colossal structures move. For its inauguration, the Lighting Giants burst into life with a rainbow of colour. The luminous finale in front of thousands of visitors at the Pula seafront, saw the cranes moving in unison to an atmospheric musical score, like a colourfully-choreographed dance troupe. Still to this day, they continue to animate the skyline in the night sky, illuminated every hour during the evenings. —

CRANE DRAWING, WITH LIGHTING POSITIONS

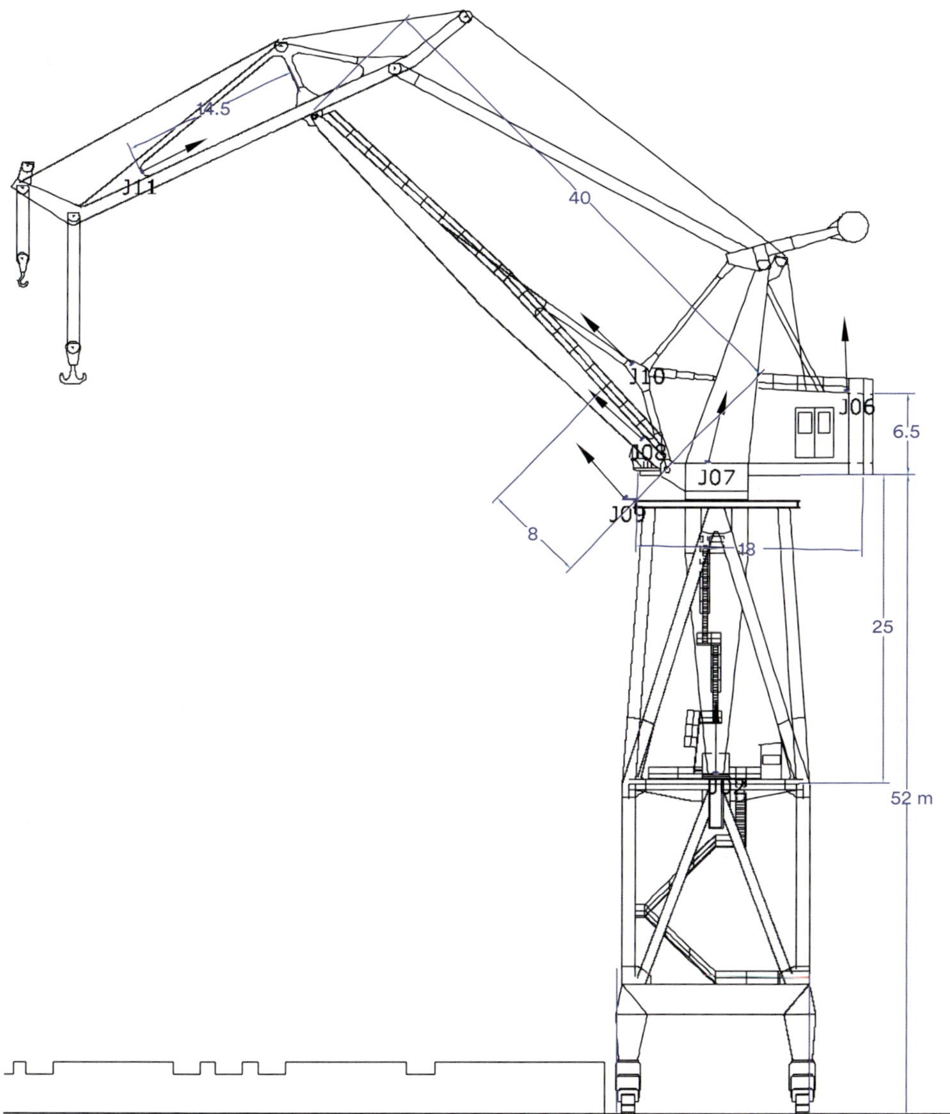

Specifics

This project was originally conceived by Dean Skira in 1998, yet it was only accepted and adopted by the Tourist Board of Pula more than 15 years later. In May 2014 it finally became a reality, thanks to the support of local bodies, industry and several private companies. The Croatian Ministry of Tourism selected Lighting Giants among the 85 development projects from the 'Innovative Tourism 2013' programme, awarding it a grant of 40,000 EUR. Apart from this purpose-allotted incentive, the project was fully-privately funded.

One major challenge was the fact that the main focus of this project were the colossal cranes, which are functioning daily in the active shipyard. With the support of the shipbuilders, the disruption to daily activities in the set-up phase for lighting this installation was avoided.

Numerous tests were carried out relating to the various colourways for illuminating the cranes, and also the accurate intensity of light required to ensure good visibility of the installation from a great distance. The main aim of the project was to link the shipyard to the city, so it was important that the colourful visuals could be appreciated from all angles.

Each crane had a minimum of seven spotlights positioned on various levels and at various locations across its structure. Focusing and positioning was carried out whilst the cranes were in a fixed ('parked') position.

The software and programming also presented the designers with a challenge: a non-wired solution was needed for the lighting control system (so as not to interfere with the daily functioning of the cranes). The final project included a sophisticated remote control system of lighting and scenography.

A computational program was used to predict the spotlight position. Focusing and fine-positioning was carried out manually for each spot, with special three-axis carriers designed to enable fine-tuning.

Set-up

Luminaires included 73 Philips LED spotlights (ColorReach Powercore gen2 fixtures), weighing 40 kg each. Each of these lighting elements are specifically designed for use in large-scale installations.
Programming of the Philips spotlights meant that a total capacity of 16,000 different variations of colour and intensity was made possible, due to each unit consisting of 64 LED chips. At least half of the fixtures were individually addressable and controllable.
Eight cranes on the harbour front were selected to be active participants in the spectacle. The location of each crane was an important factor when considering the possible lighting scenarios for the inauguration ceremony.
Blinds were used to prevent unnecessary light dispersion and pollution, but also to focus precisely the beams of light at the correct angles and in the desired directions.
Power for the illumination of the cranes was provided through separate channels placed inside the tracks (which allow the cranes to move) and then wired throughout their structure. The electronic equipment consisted of an iPlayer control system in order run the software program for the lighting scenarios via a wireless signal.
Construction time was 7 months.

CRANES WITHOUT LIGHTING

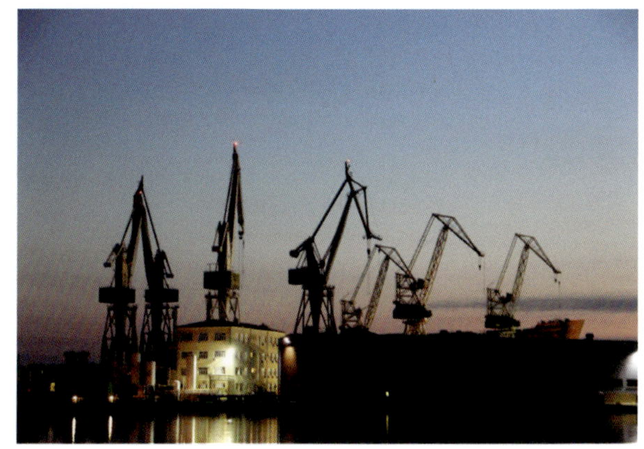

TESTING & LIGHTING SET-UP

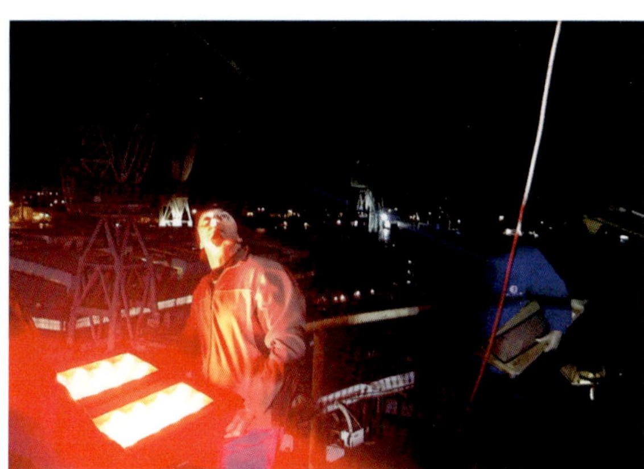

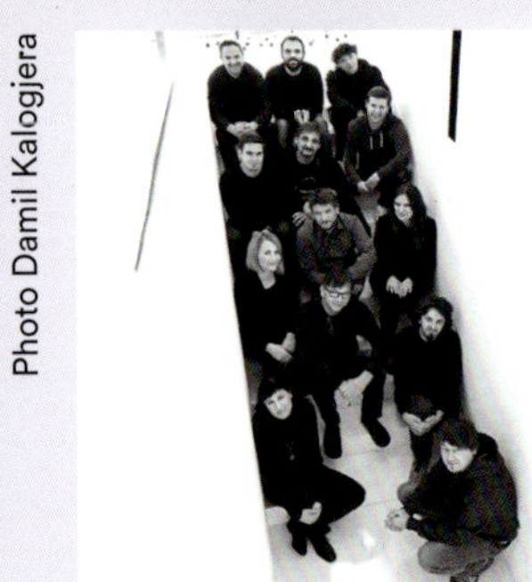

Photo Damil Kalogjera

Skira

Skira is an award-winning lighting design practice based in Croatia. Founded in 1990 by Dean Skira, the firm has currently a team of 15 people working on architectural lighting design projects, urban master planning, product innovation and high-end integration systems. The team works together with clients, architects and designers to create lighting design that isn't simply utilitarian, but also provides sustainable and easily-manageable solutions.

SKIRA.HR

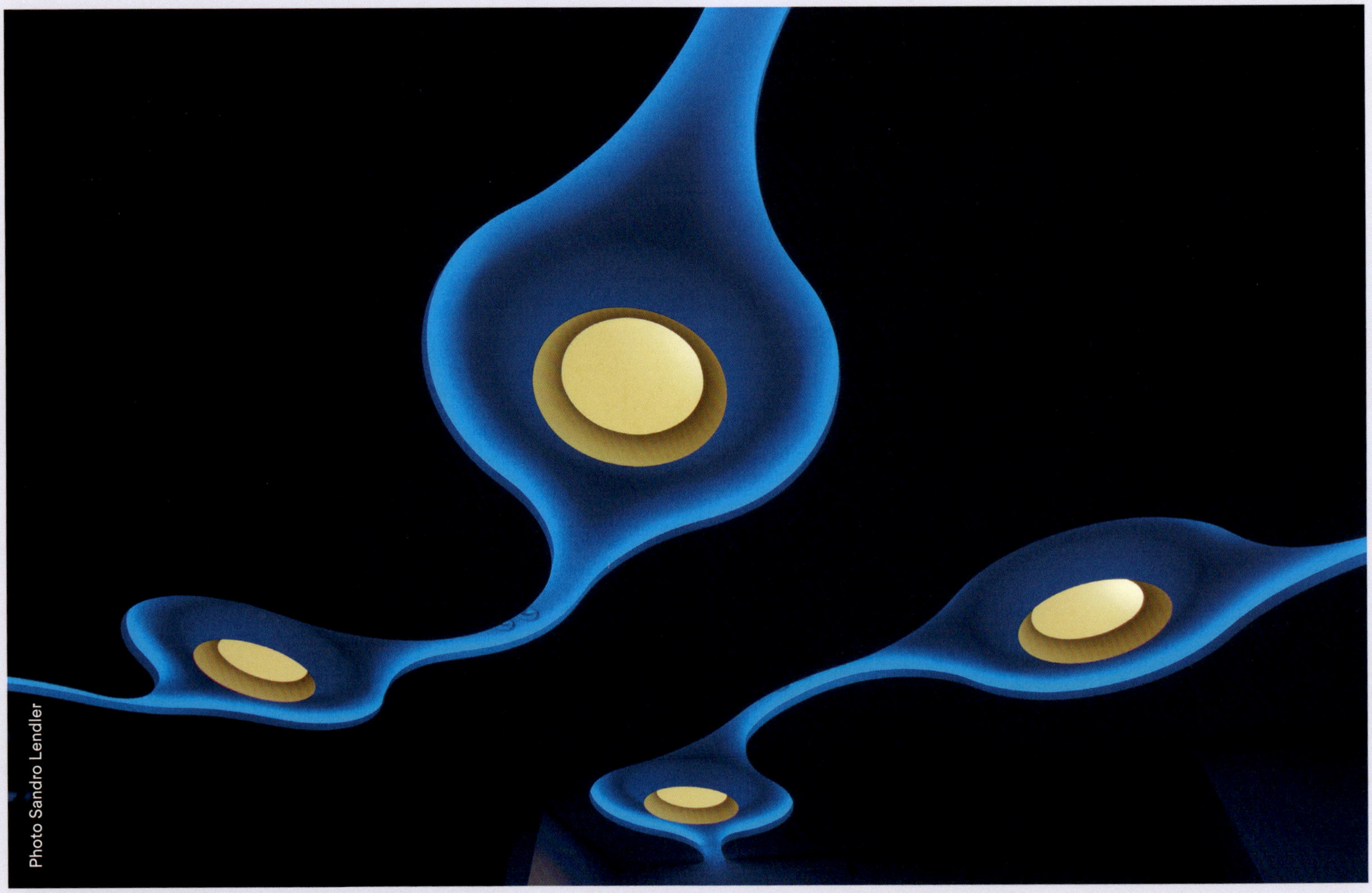

Photo Sandro Lendler

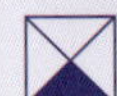

Novamed

When **2012**
Where **Zagreb, Croatia**
Client **Polyclinic Novamed**

Fun, dynamic, colourful and inviting – all words used to describe the atmosphere that Skira looked to create for the lighting in a clinic in Zagreb. Ever changing scenes of light that look to portray images of human blood cells animate the ceilings, reducing the formality usually associated with medical institutions. The project received a special citation IALD Award for the most successful translation of a visual theme into light.

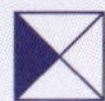

Hotel Bellevue

When **2014**
Where **Mali Lošinj, Croatia**
Client **Jadranka Hotels**

A luxurious five-star destination, Hotel Bellevue features the latest advances in lighting technology and management systems. Sitting in one of the most beautiful Croatian bays, the hotel boasts an elegant visual experience for guests, including sculptural installations, and integrated lighting that emphasises the architecture's details while leaving room for the play of shadows and darkness.

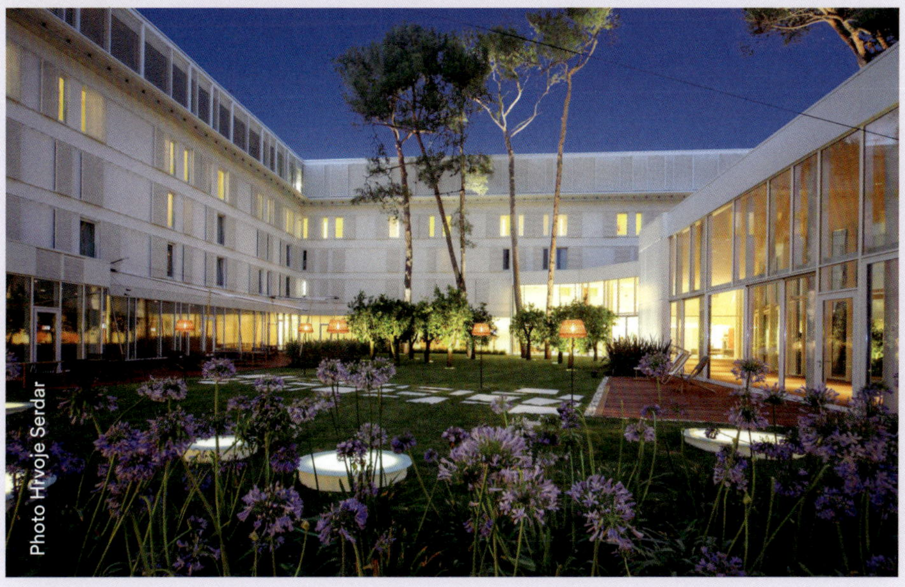
Photo Hrvoje Serdar

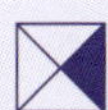

House of Light

When **2014**
Where **Pula, Croatia**
Client **Lumenart**

In House of Light, Skira's headquarters, 'architecture accompanies light and light enhances architecture,' illustrating one of the firm's design mottos. Taking into account the different seasons, the building's white facades are tilted in many directions in order to maximise daylight intake.
A conspicuous, white form during the day, at night the building becomes the stage for playful lighting scenarios.

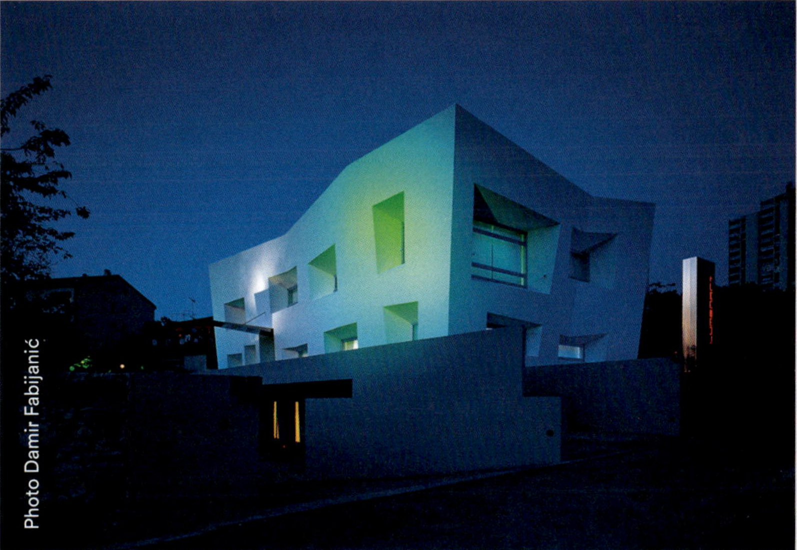
Photo Damir Fabijanić

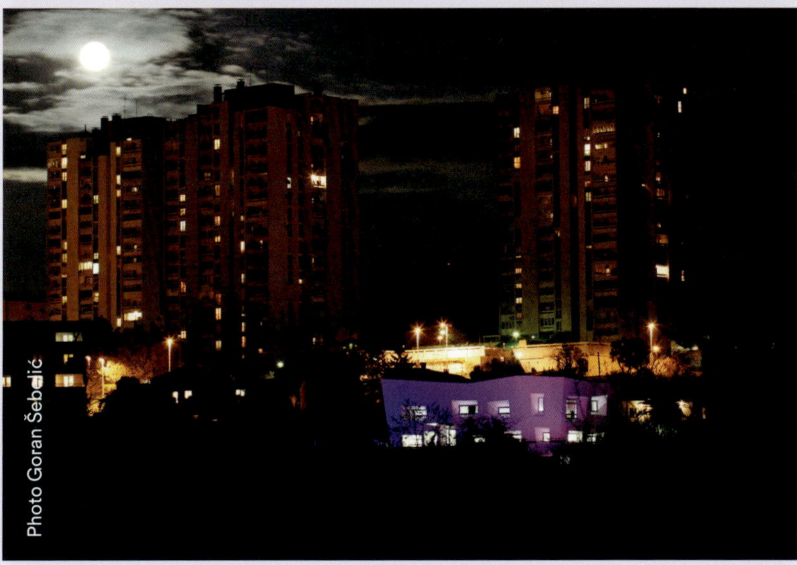
Photo Goran Šebelić

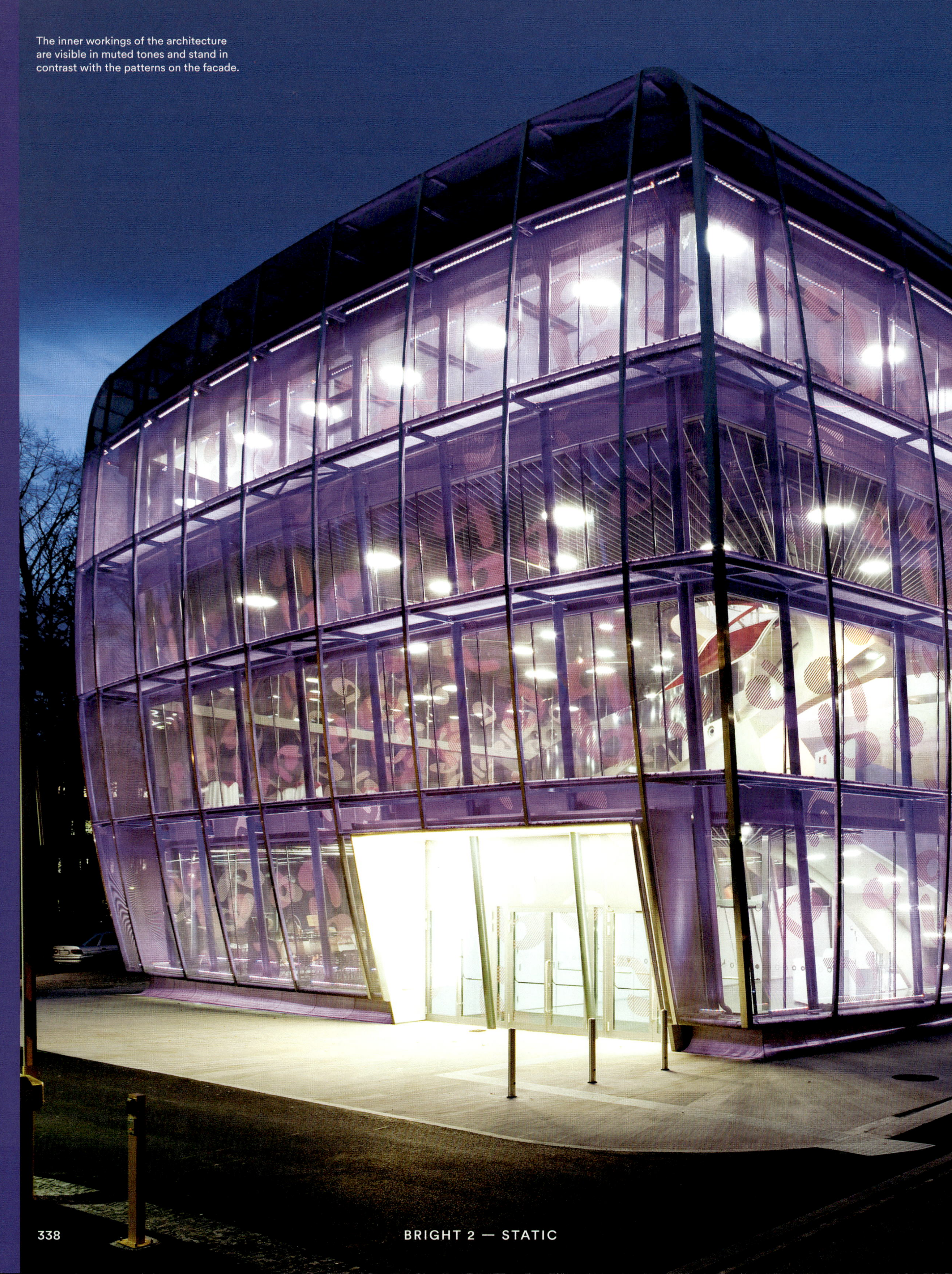

The inner workings of the architecture are visible in muted tones and stand in contrast with the patterns on the facade.

MUMUTH

The eye-catching contours of a music theatre in Graz are emphasised thanks to the lighting products of **XAL**, ensuring that every illuminated aspect is like a connection to the sounds emanating from within, leading to the whole space feeling connected and alive.

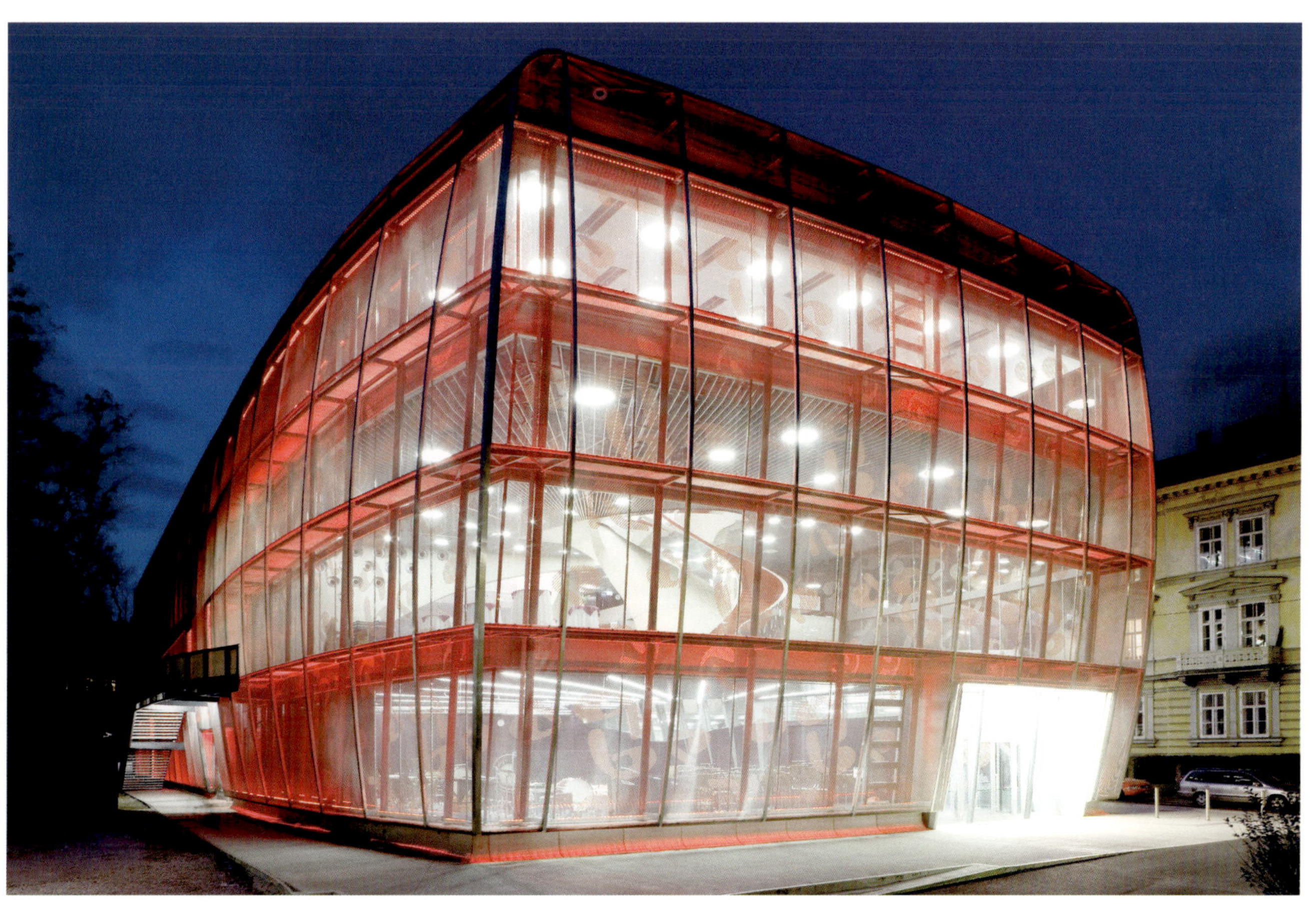

Photos Paul Ott

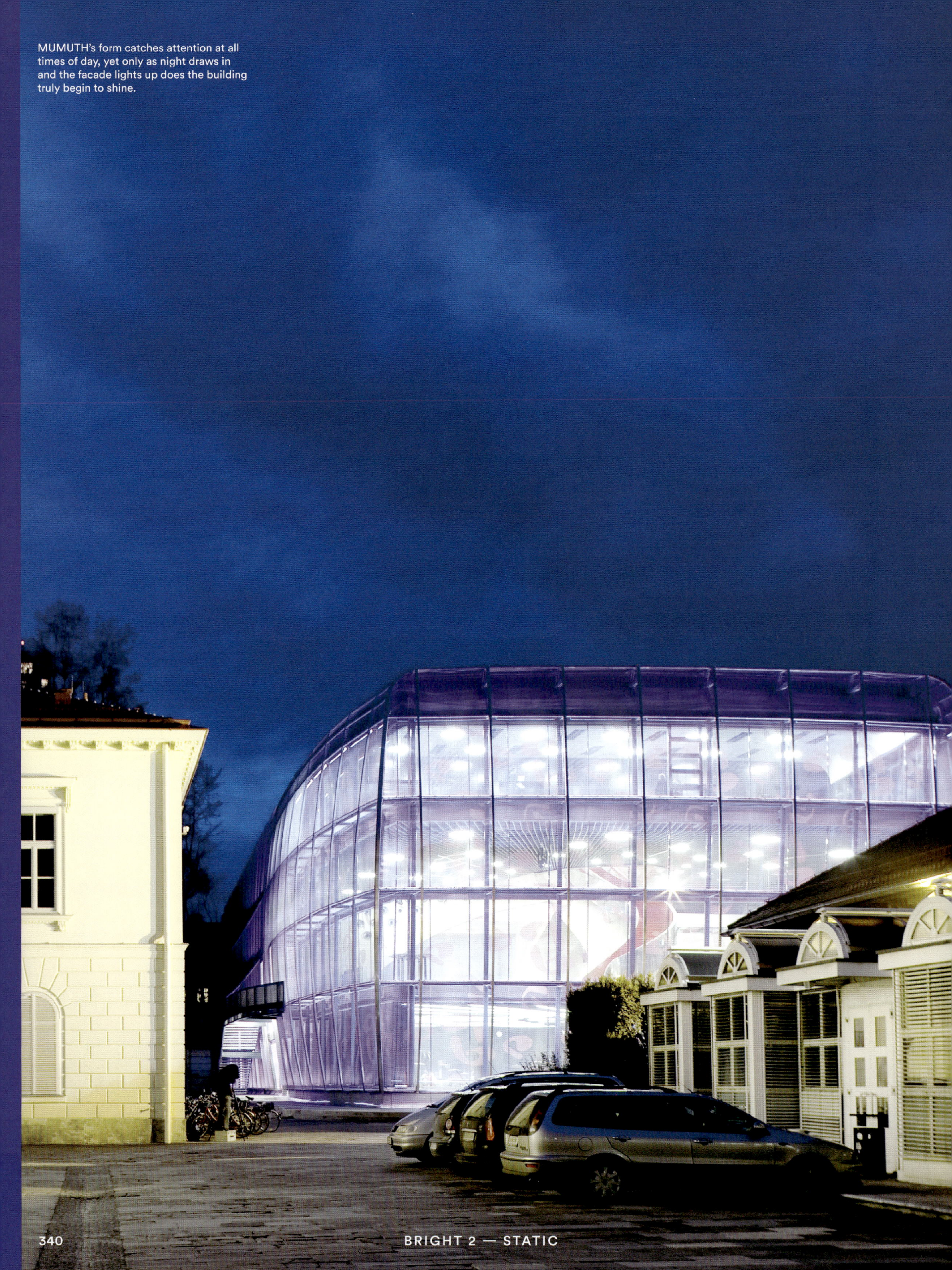

MUMUTH's form catches attention at all times of day, yet only as night draws in and the facade lights up does the building truly begin to shine.

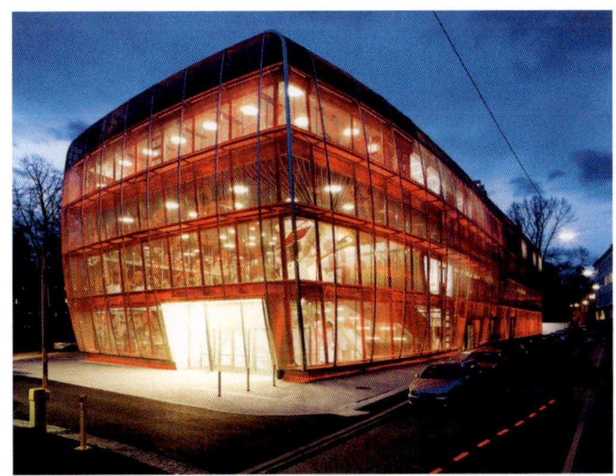

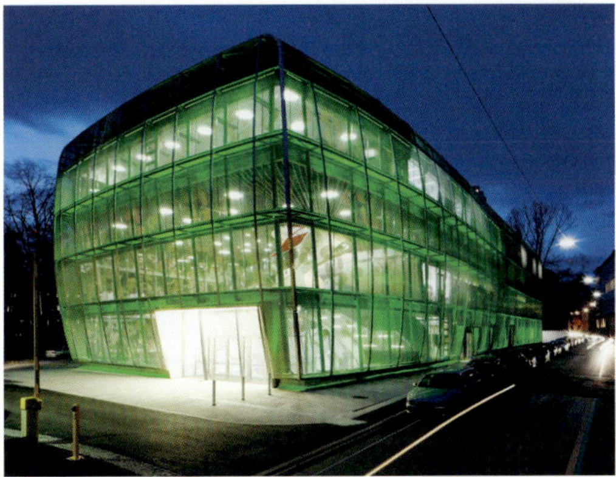

Illumination of the facade linked the architecture with the music, giving it a specific rhythm.

The topic of 'acoustic space and its dramatic effect' may have been adequately researched by everyone from Le Corbusier to Xenakis, but for the architects of UNStudio it nevertheless still holds a vast amount of fascination and influence, as shown by the company's design of the 'Haus für Musik und Musiktheater' (MUMUTH for short) at the University of Music and Performing Arts in Graz, Austria.

The architecture of the centre – in which students not only experience artistic education at the highest level, but can also test the emergence of their future professional life with public appearances under professional conditions – aims to express that it is a place where music 'lives'. The form of the building itself appears to be bulging, representing the life-force of music, resulting in the structure itself appearing to breathe. In the interior, the free-flowing space of the foyer is made possible by a spiralling constructive element – the 'Twist'. Everything revolves around this massive concrete construction, including the illumination. Artistically welding the various levels of the building is the lighting concept, with a centrality reinforced by the interplay between linear and round luminaires. Additionally, the roof window with its dark wooden slats is accurately lit, so that it makes the spiral rotation appear like a fan.

The concept for the facade lighting sought to give a restrained appearance during the day yet become increasingly transparent as night-time draws in, thus revealing insights into the goings-on within the building. This approach has been implemented with a combination of facade-located Stila RGB luminaires and an outer web of expanded stainless steel metal. XAL outdoor lights were arranged with special brackets between the glass and the metallic web, so that the light is reflected and gives the impression that the web is coloured. —

REPRESENTING THE LIFE-FORCE OF MUSIC, THE BUILDING APPEARS TO BREATHE

Designer
XAL
Location
Graz, Austria
Client
University of Music and Performing Arts
Collaborator(s)/consultant
UNStudio
Engineers
Klauss Elektro (lighting design, planning), Siemens Bacon (implementation)
Date
March 2009

The 'Twist' is a load-bearing spiral element that links the public space of the foyer with the music rooms above.

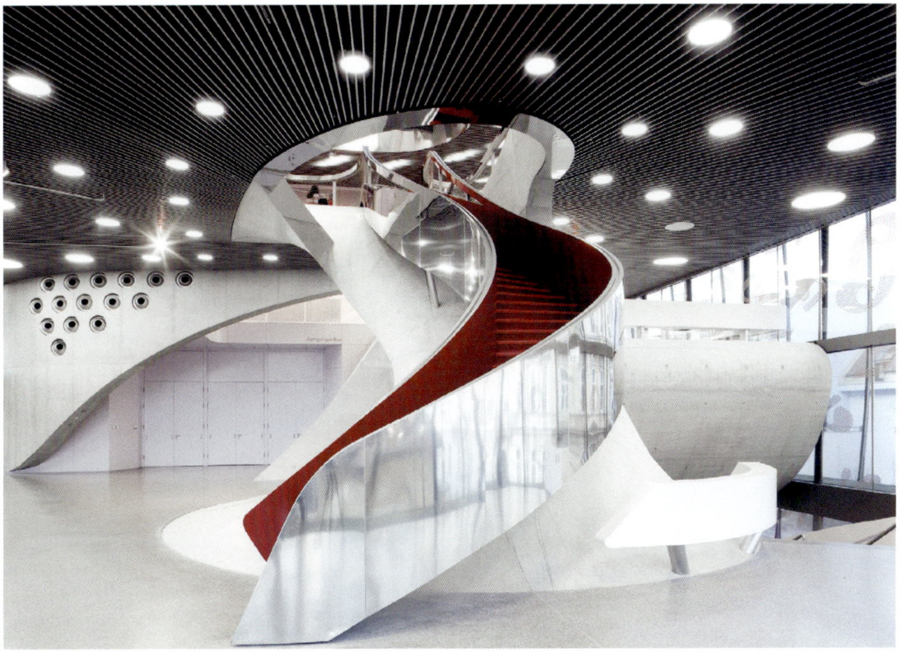

PLAN

1 Foyer
2 The 'Twist'
3 Large theatre
4 Office space
5 Lavatories

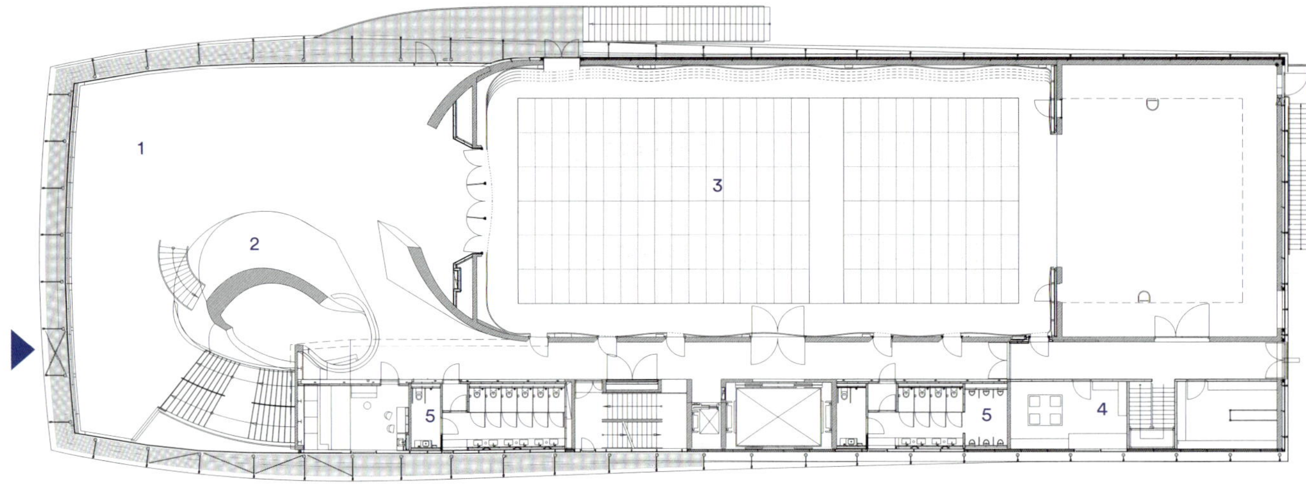

SECTION A

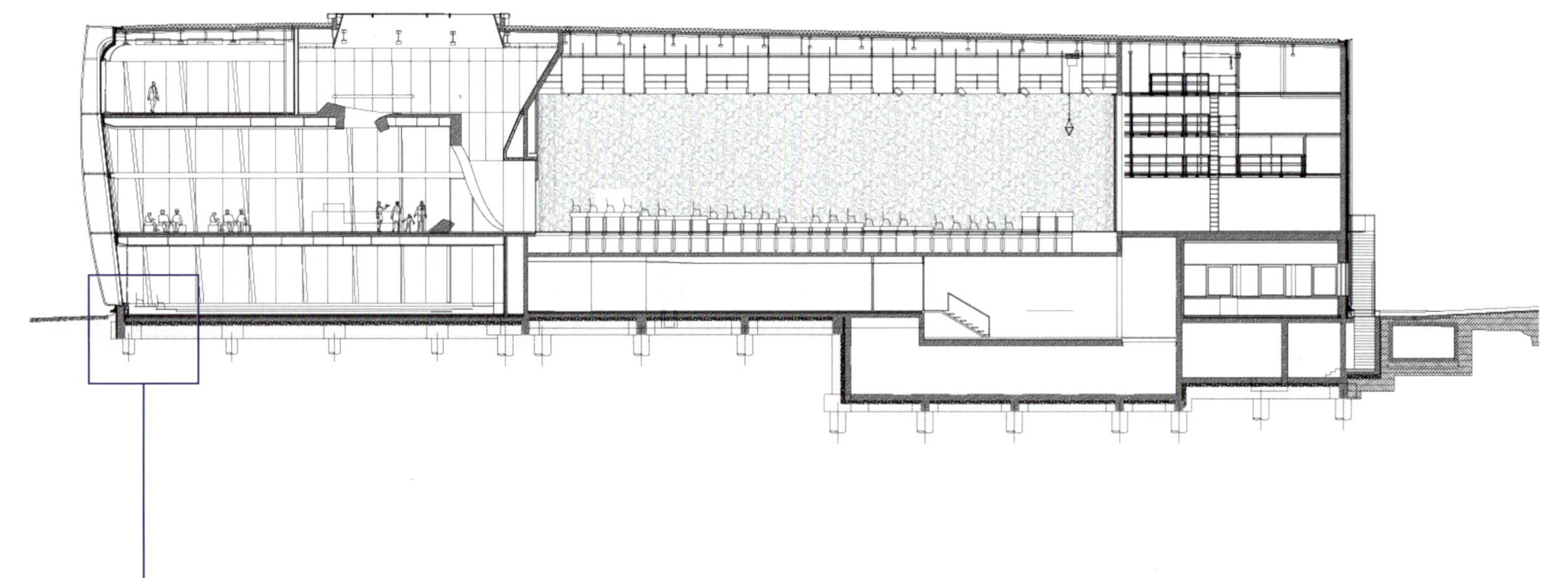

FACADE DETAIL

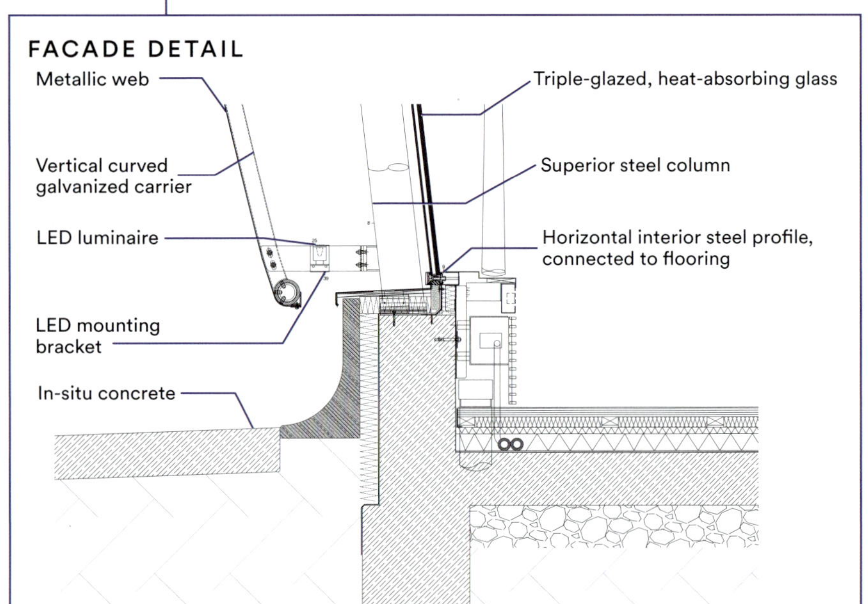

SECTION B

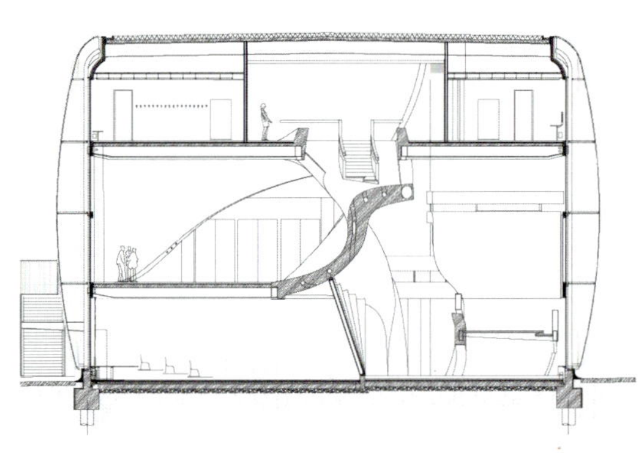

SITE PLAN

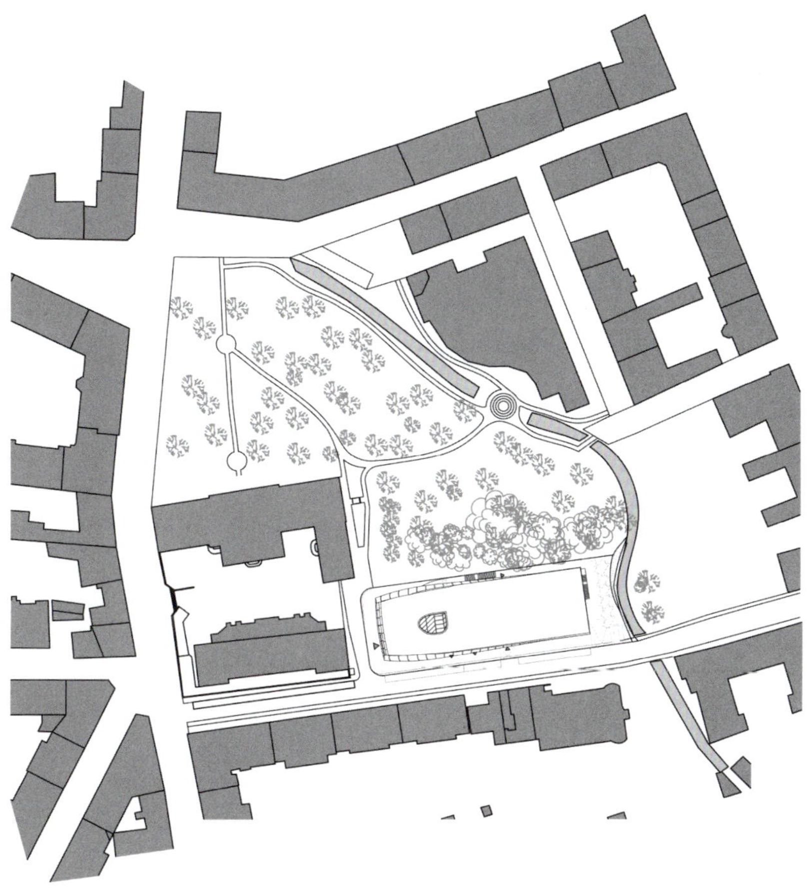

SPECIAL BRACKETS BETWEEN THE GLASS AND THE METALLIC WEB RESULT IN REFLECTED LIGHT

CONSTRUCTION

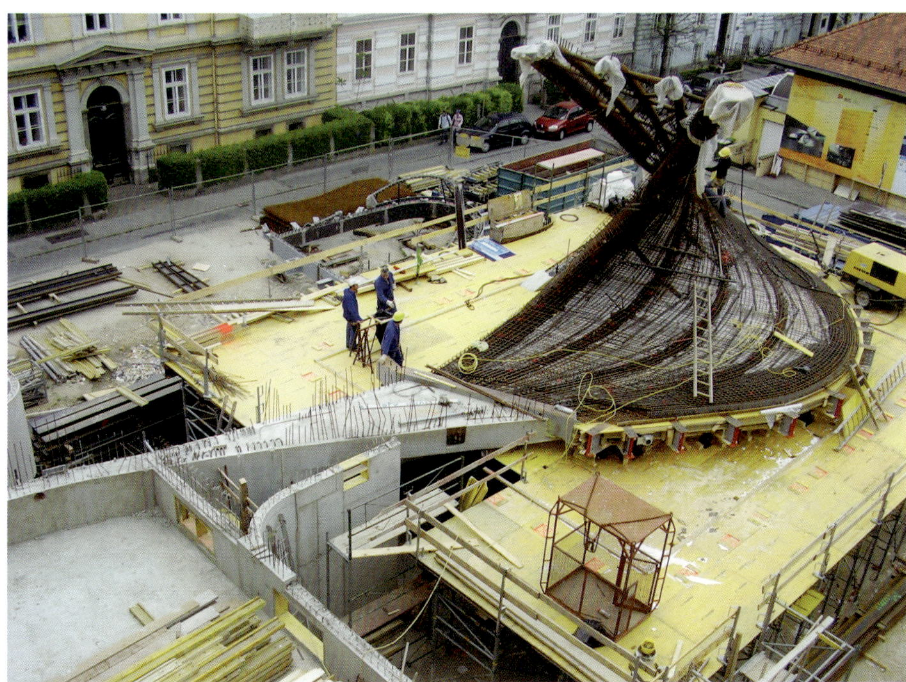

The spiralling concrete construction was the first architectural feature to be completed.

Specifics

In overall terms, the implementation was a major challenge right from the very start (first plans 2006/2007). Back then, XAL had already established its reputation and kudos as a pioneer in technology and was ahead of its time. This leading expertise in light planning combined with the latest generation of LEDs has enabled the development of fully optimised illumination at the cutting edge of technology.

A special feature of the optical system is the special lenses of the Stila RGB lights. These offer two advantages due to their elliptical light beam: anti-glare for pedestrians, which is due to the 6-degree depth of the light beam angle and means that people passing by outside are not dazzled at all; and early colour mixing, which is related to the 24-degree width of the light beam angle, which means the different colours mix even close to the ground and thus provide a uniformly illuminated appearance of the entire facade.

Another challenge was the differences in temperature which facade lighting in Central Europe is subject to. The relatively large temperature variations directly at the facade are more extreme than usually observed in outside areas, but they are brought under control by thermal pressure valves. This ensures the sealing of the lights is guaranteed.

The building's exterior was a blank canvas, generating the opportunity to return to the theme of music in a new way. Repetition generates an aggregate with densifications, intensifications and intervals, so a repetitive pattern was utilised in the form of a metallic web which became a glittering mesh at night. Accurate lighting with XAL luminaires between the glass and the metallic web gives the impression that the web is coloured.

Set-up

Used light totalled 250 linear metres of XAL Stila RGB luminaires to surround the entire building, at 33 W/m. The interior luminaires used for the the ceiling were XAL Vela Round.
Component parts included mixed media, LEDs, diffusive reflective material, custom lighting sequence, electrical hardware and metal framework.
Illumination monitoring included 1725 separately controllable lighting addresses.
Lighting console incorporated a DMX controller designed at the request of the client, so that students can operate the lighting scenes themselves as part of their training. They can therefore control the desired lighting independently, according to the event, via control consoles.
Circuit boards totalled a number of 575, which each used 15 LEDs/15 lenses. The total number of LEDs/lenses was 8625.
Facade mounting brackets for the LEDs consisted of round tubes from Edelstahl39, with the XAL outdoor lights fixed using a strap and round stainless-steel pipes.
Project duration took approximately 5 years in planning stages, with light implementation on-site being completed within 6 months.

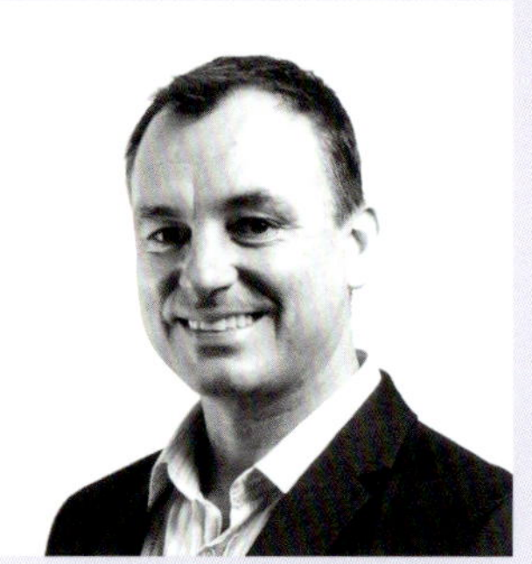

XAL

Austrian lighting firm XAL was founded in 1989. Over the years, the focus has remained constant with an objective for the company being to develop quality lighting solutions. Extraordinary in terms of technology and design, these solutions can be perfectly integrated into all kinds of indoor and outdoor contexts. Research and development is just as important as first-class knowledge in terms of production and logistics. These competencies enable the firm to respond flexibly to customer requirements, as well as realising specially-tailored solutions to meet the tightest of schedules, according to Dr Christian Schraml, managing director of XAL (pictured).

XAL.COM

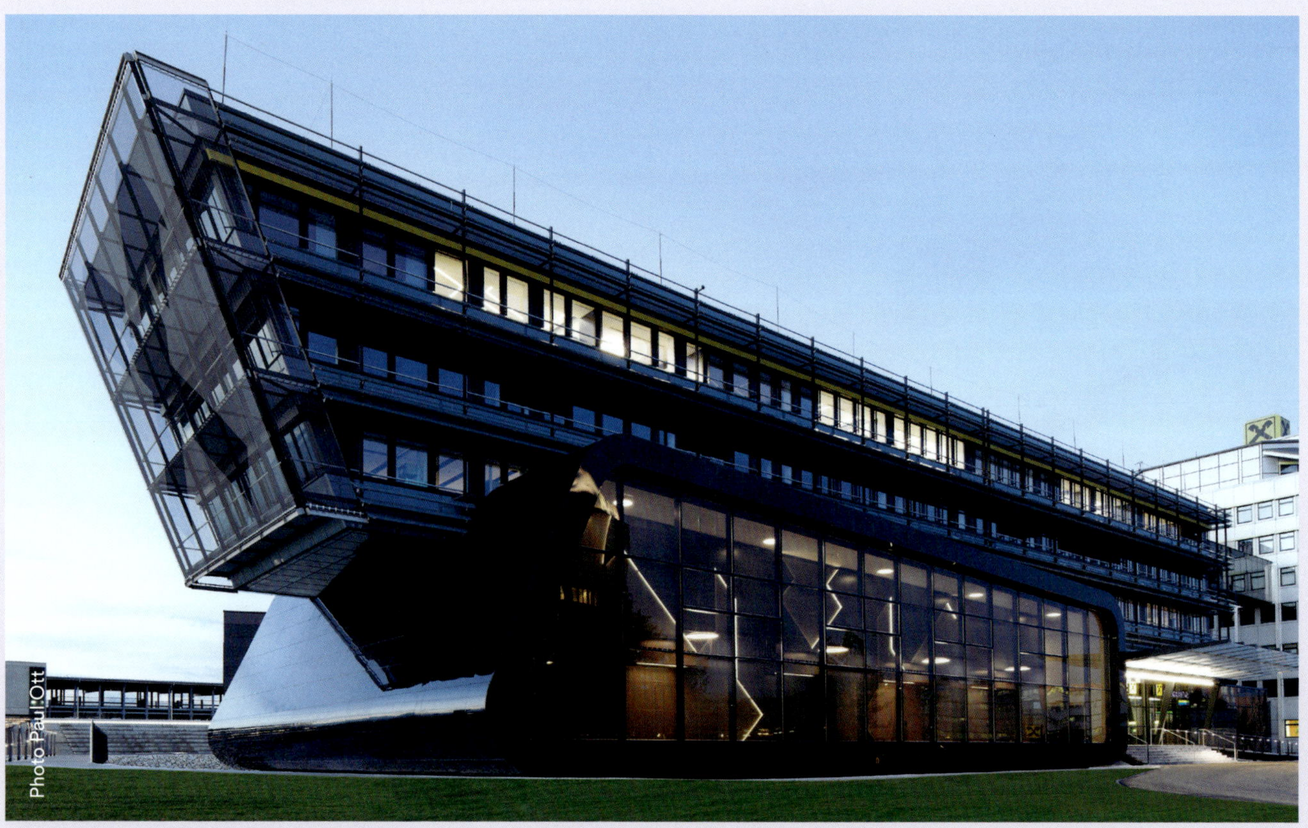
Photo Paul Ott

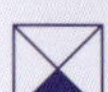

RLB Headquarters

When **2013**
Where **Graz, Austria**
Client **Raiffeisen Landes Bank**

Bringing into one building the entire breadth of business areas of the banking company RLB was the task of architects at Strohecker ZT. The building has multifunctional uses, with a lighting concept to match which was designed by Klauss Elektro-Anlagen to reflect the diversity of areas of activity. Stretched light fixtures – bands of lighting with rounded edges and large-format lighting surfaces – perfectly fit the image of a modern office complex. The use of free-standing luminaires for employee workspaces provides the possibility of individualised lighting as needed.

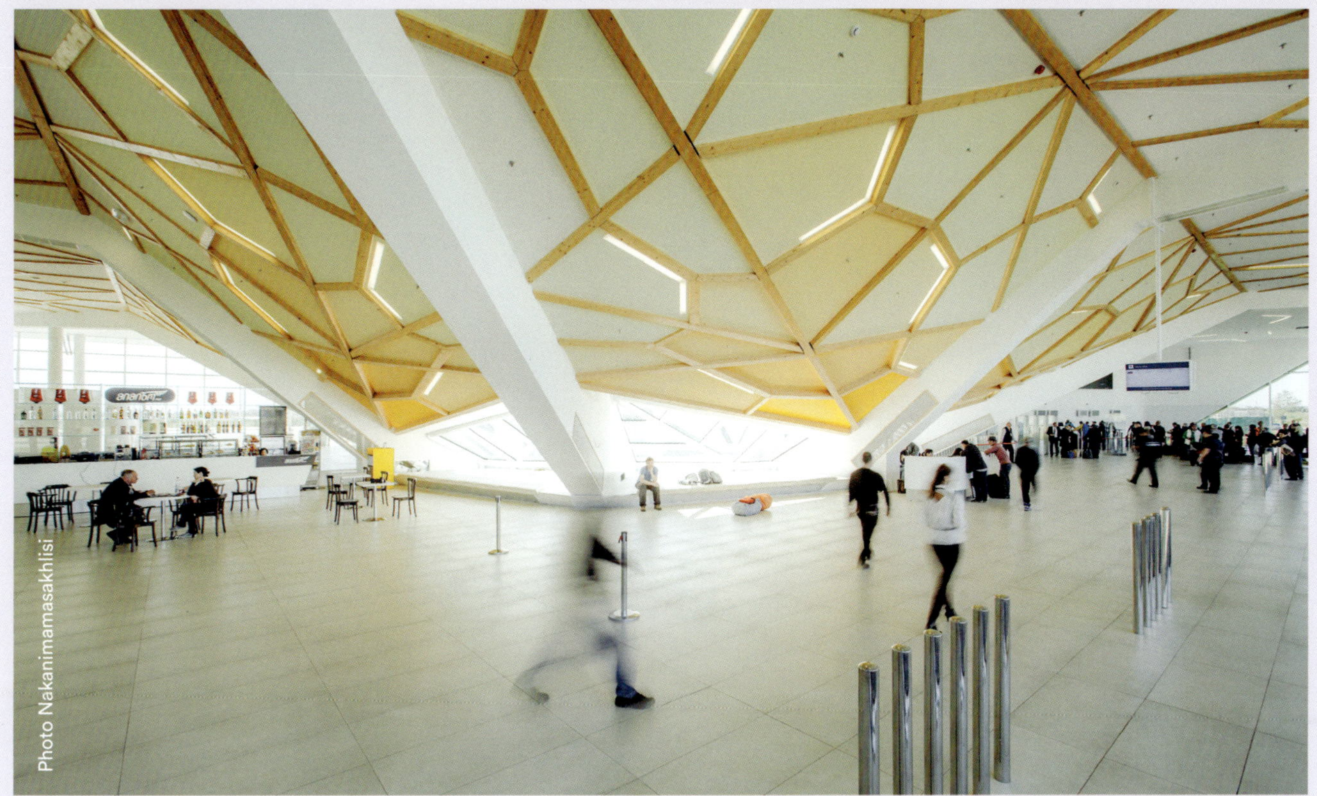
Photo Nakanimamasakhlisi

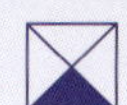

Kutaisi International Airport

When **2013**
Where **Kutaisi, Georgia**
Client **United Airports of Georgia**

UNStudio's airport design integrated both the historic landscape and the local vernacular style into the concept. Architect Ben von Berkel sought to place the social and rich experiential aspects of travelling at the centre of focus again. The lighting concept of Primo Exposures integrated 650 m of special XAL recessed lights, positioned in selected corners of the cell structure. Over the suspended ceiling, these spread out from the terminal hall to the outdoor areas, extending onto the concrete overhangs in front of the glazed facade.

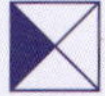

Humanic Shoe Store

When **2013**
Where **Munich, Germany**
Client **Humanic**

In this retail space by Design und Bauabteilung Leder & Schuh, each area is staged like a scene from its own film with the focus at all times on the shoes. The lighting design by Vedder Lichtmanagement created contrast in the white rear-wall area using XAL illumination. From cool floodlighting and fiery spots on the goods and the red chairs, to the under-stairs aspect with large floor luminaires forming a spatial bracket. The light setting throughout is as reserved as it is effective: materials and colours come alive using light clusters and light intensity, thereby giving the rooms depth and excitement.

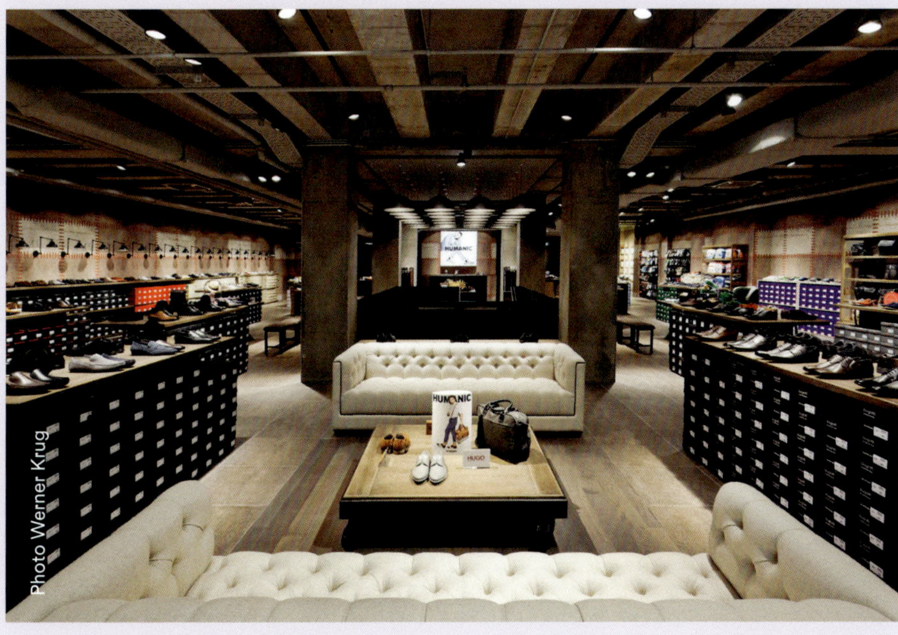
Photo Werner Krug

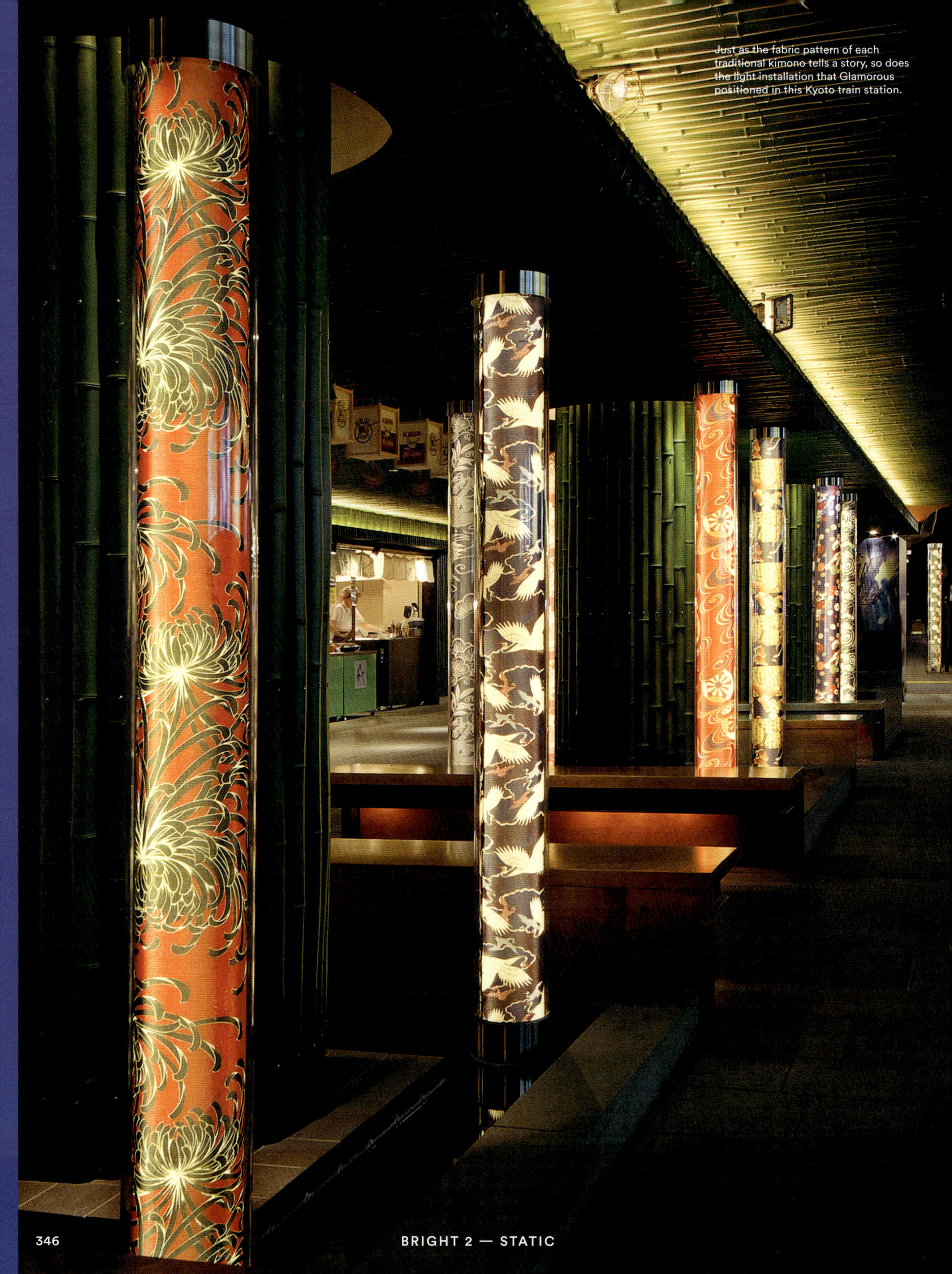

Just as the fabric pattern of each traditional kimono tells a story, so does the light installation that Glamorous positioned in this Kyoto train station.

Randen Arashiyama Station

By creating an 'illuminated kimono forest' – columns clad in glowing Japanese textile – in the station and village area around the only tram-line in Kyoto, **Glamorous** sought to maximise the visual and social impact around the station by lighting the traditional material and its translucent weave.

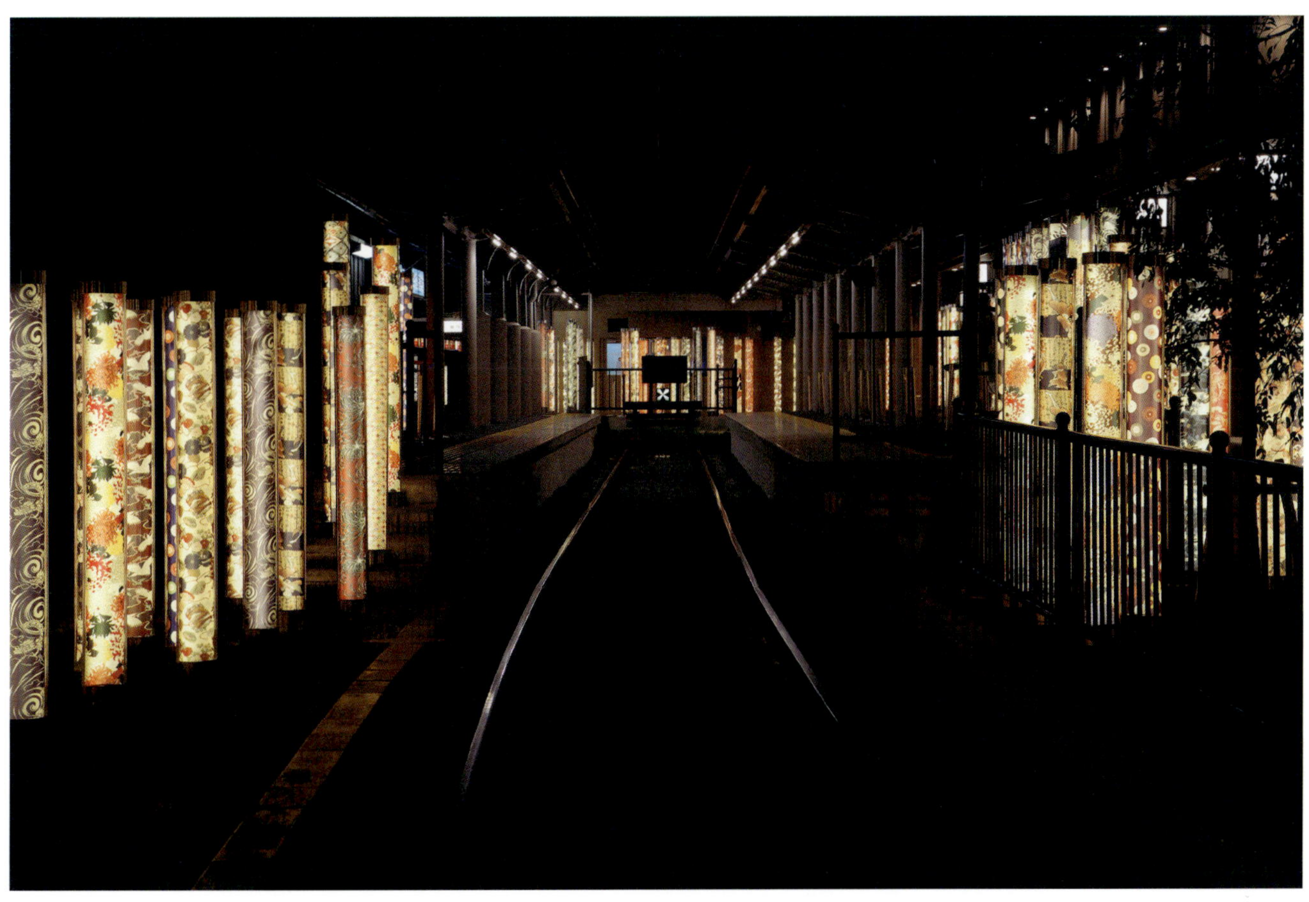

Photos Seiryo Yamada

Randen Arashiyama station in a western outpost of Kyoto is a popular destination for visitors to the many surrounding world cultural heritage sites, temples and gardens. It is here that the design office Glamorous was tasked with the job of creating a platform that would maximise visual and social impact of the area. The studio renovated the station 10 years previously with innovative and ecological materials as well as indirect lighting, which attracted more people and helped the local economy. After the success of the project, Glamorous was commissioned to design the platform area to activate the community further.

Yasumichi Morita's team obliged with a glowing installation of 600 light poles, a contemporary re-interpretation of the bamboo forests for which the area is famous. The poles, which are over 2-m-high, are wrapped in another local speciality, *yuzen* kimono fabrics. These textiles use a special paste-resist dye process resulting in a translucent weave, which Glamorous exploited for this project. The designers chose 16 patterns from the hundreds of *yuzen* options available and had them dyed in contemporary colours. There are two colour versions for each pattern, bringing the total number of pole variations to 32. The custom fabrics were then coated in acrylic to protect them from the elements. When lit by LED lights, the colours and patterns sing out.

Weaving their way over approximately 3000 m^2 of station ground, the poles carve out new areas of public space in what was formerly a quite featureless landscape. Vibrant yet soft, the coloured lights provide a welcome change from the harsh fluorescent lighting all too often seen in public transit spaces. This is one enchanting 'illuminated kimono forest' where commuters are encouraged to linger. —

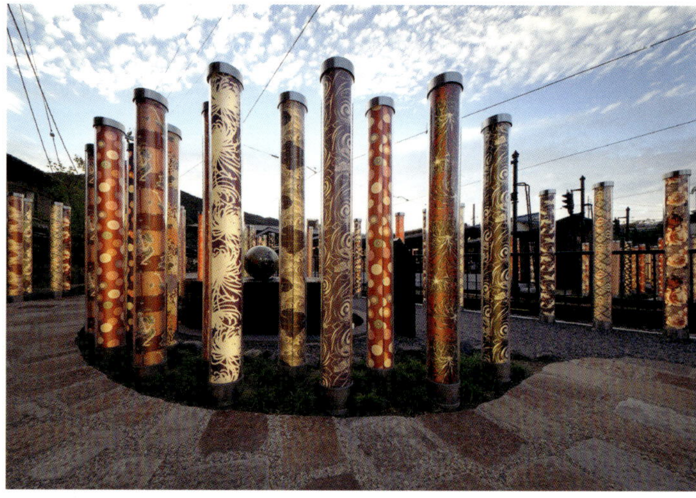

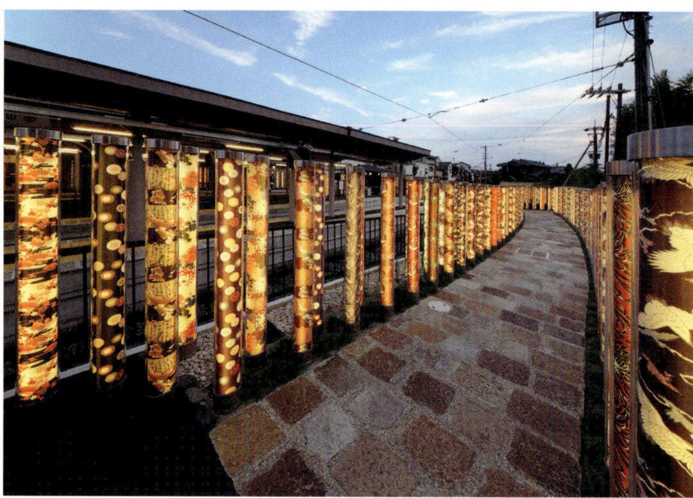

This lighting of the kimono forest is a cordial gesture for both locals and visitors alike, creating a visual, social and emotional impact for all passengers who pass through the station.

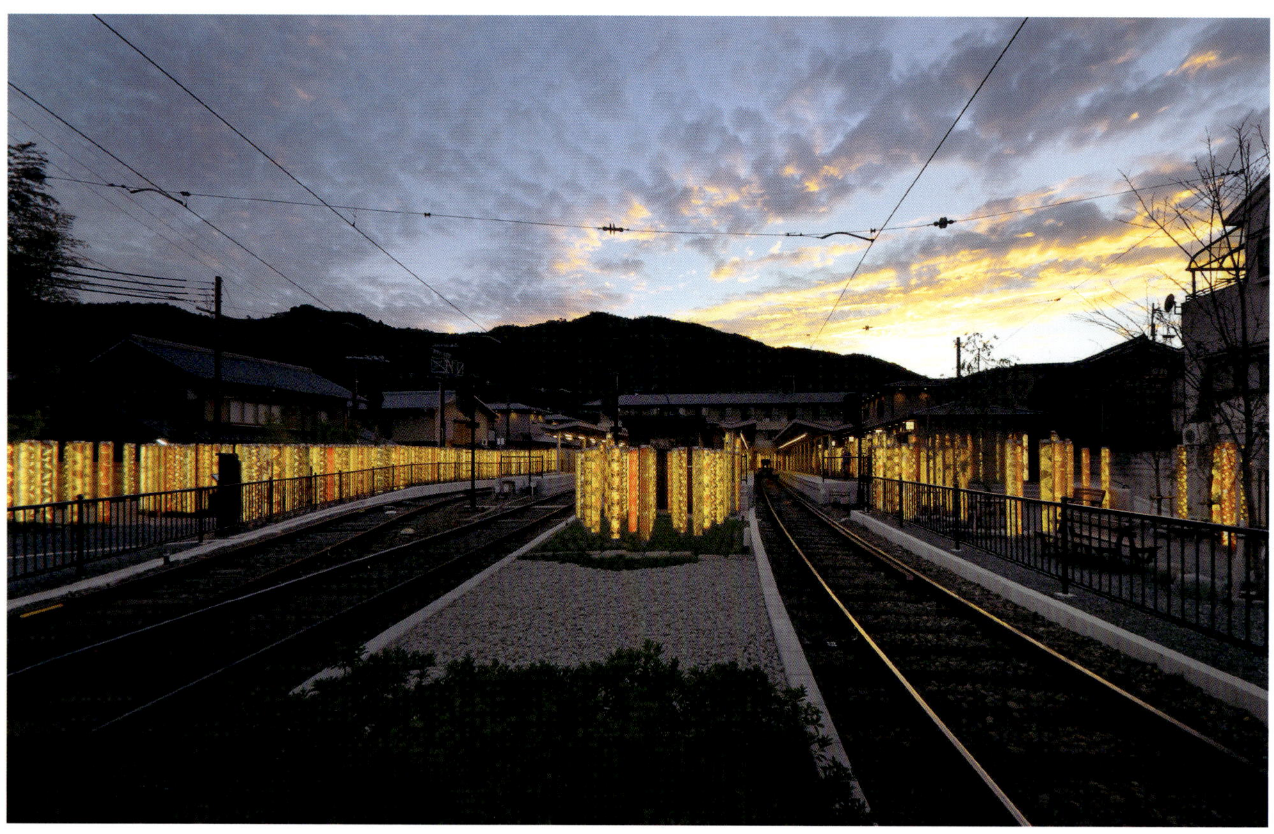

Designer
Glamorous
Location
Kyoto, Japan
Client
Keifuku Electric Railroad
Collaborator(s)/consultant
Heian Kensetsu Kogyo, Hitoshi Takamura
Textile manufacturer
Pagong
Date
July 2013

PLAN

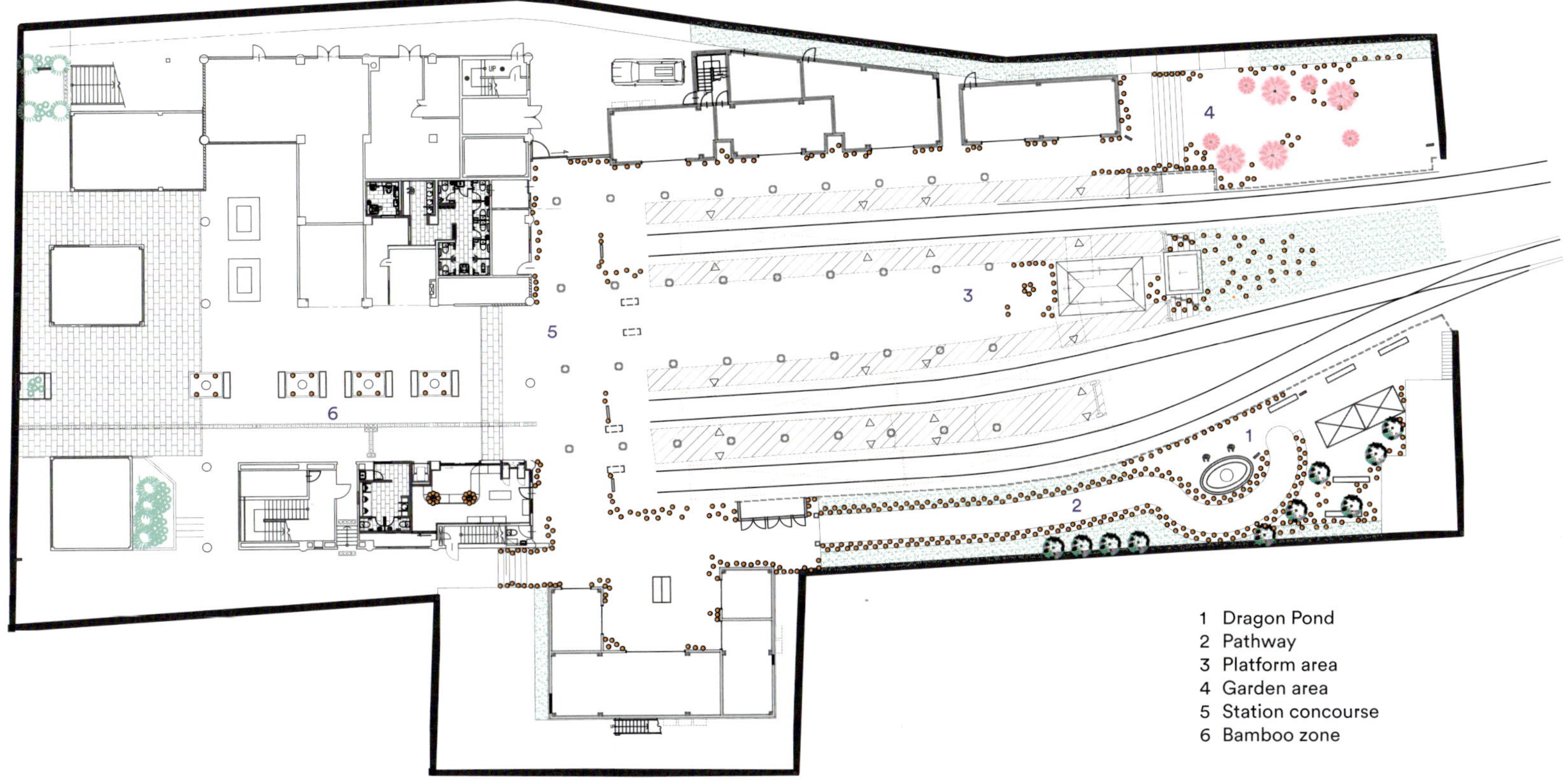

LIGHT POLE DRAWINGS

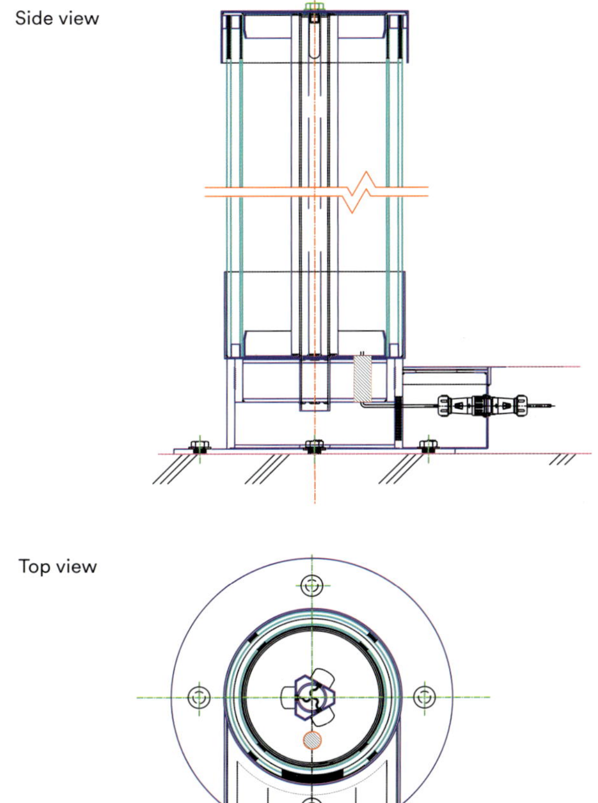

Set-up

Installation area measured approx. 3000 m² of station ground.
Light poles were each of 20 cm in diameter and 210 cm in height. A total of 600 poles were constructed to simulate a bamboo forest, illuminated with LEDs to create a different expression and ambience at night time for passengers through the station.
Light source included a quantity three LEDs that were fixed within each column to evenly illuminate the fabric. These can be individually-controlled and can produce a maximum of 39 W light output per pole.
Fabric was the local *yuzen* kimono cloth. A total 376.8 m² of material was utilised in 16 different patterns, each of which were available in two colour-ways. As the installation was outside, exposed to harsh weather at times, the fabrics were treated and applied with an acrylic finish in order to protect them from the elements.
Construction time was a total of 6 months, from dyeing the fabrics to the final installation on-site in the station.

Specifics

Every *yuzen* pattern tells a story; this was important for the design team in communicating their own story within the installation. Linking the strong historical background was one aspect the team was keen to incorporate for the users of the station.

In line with the cultural heritage aspects of the surrounding area, one major challenge was to adhere to the strict regulations in regards to signatures, colours, heights, etc. in order to preserve the scenic beauty. To clear all these regulatory issues step-by-step took 3 years.

Another important aspect was to consider were the everyday commuters who regularly use the railway station. A priority was to maintain the function of the location and ensure that the installation did not obstruct the passengers but enhance and add something new to the experience of travelling.

The ideal positioning and number of the light poles was a major consideration in the research phase of the project. The journey begins at the 'Dragon Pond' – symbolising the guardian of the station – and winds its way towards the platform areas, with columns and shelters lined with bamboo to add to the enchanted forest concept.

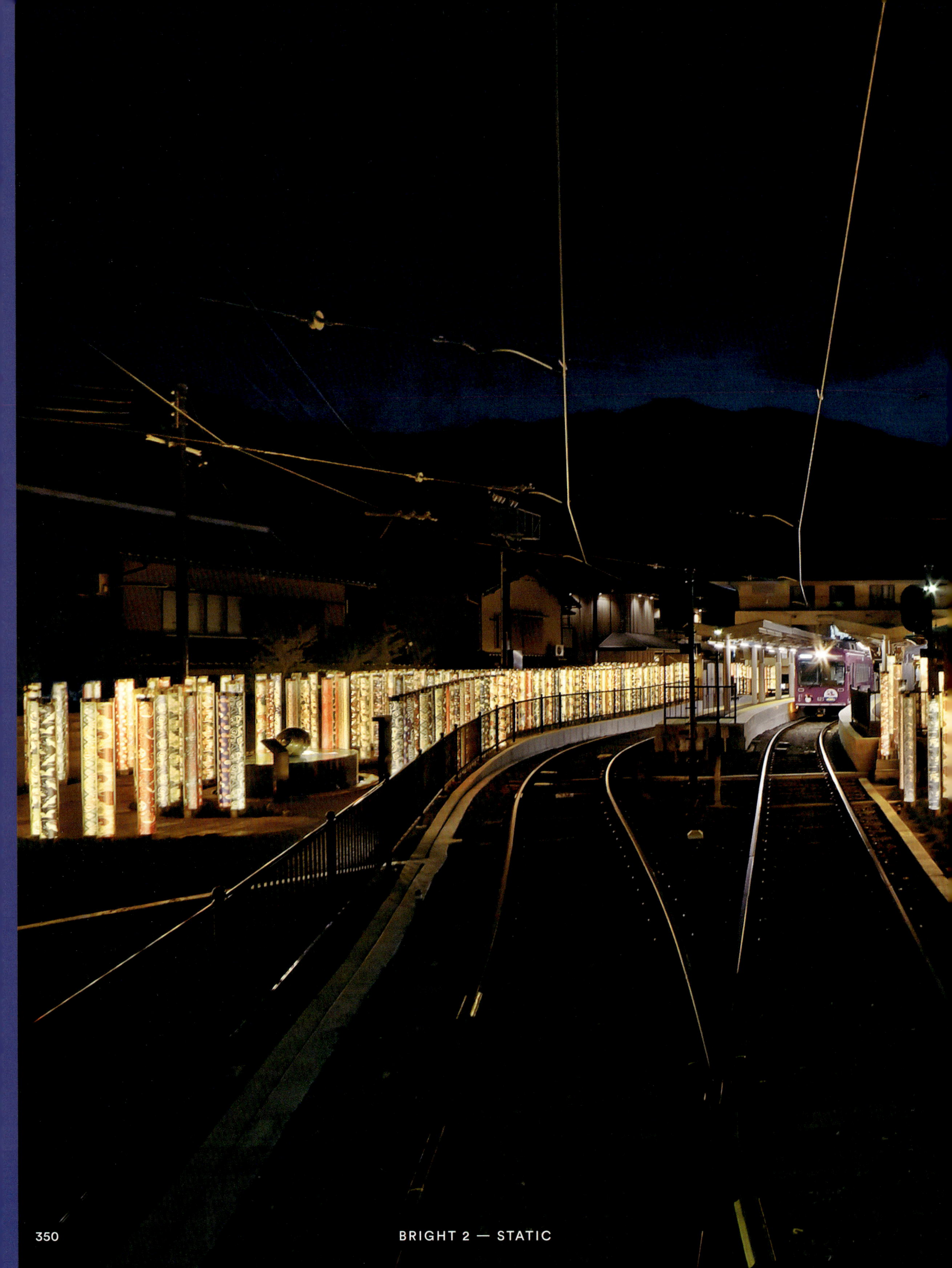

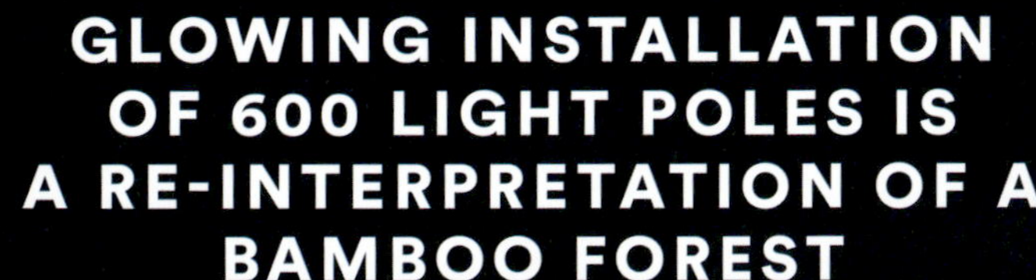

GLOWING INSTALLATION OF 600 LIGHT POLES IS A RE-INTERPRETATION OF A BAMBOO FOREST

Photo | Susa

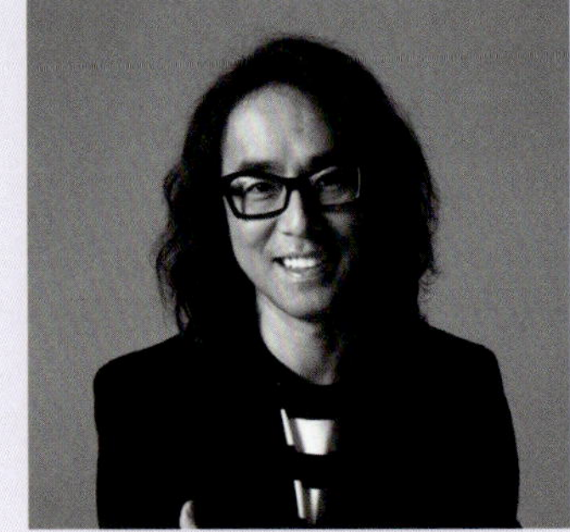

Glamorous

Yasumichi Morita founded the interior-design firm Glamorous in 2000. Rather than relying on traditional design processes, Morita's team uses the notion of glamour to spur on the excitement desired in all the interiors that the studio completes. Having established a reputation in Japan over a number of years, Glamorous is expanding its interest in global hospitality design, as well as graphic and product design.

GLAMOROUS.CO.JP

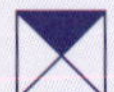

Sunluxe

When **2014**
Where **Macau, China**
Client **Sunluxe**

Sunluxe is a high-end brand with accessories, jewellery and watches available all over the world. When commissioned to realise a new store's interior design, Glamorous applied a luxurious concept. Taking the shape of an existing column located at the centre of the shop as the design motif, huge column-shaped chandeliers were added to both sides of the space to give a rhythmic balance. Taking advantage of the high ceiling and indirect lighting, a chic and glittering palette of bronze and gold colours complete the space.

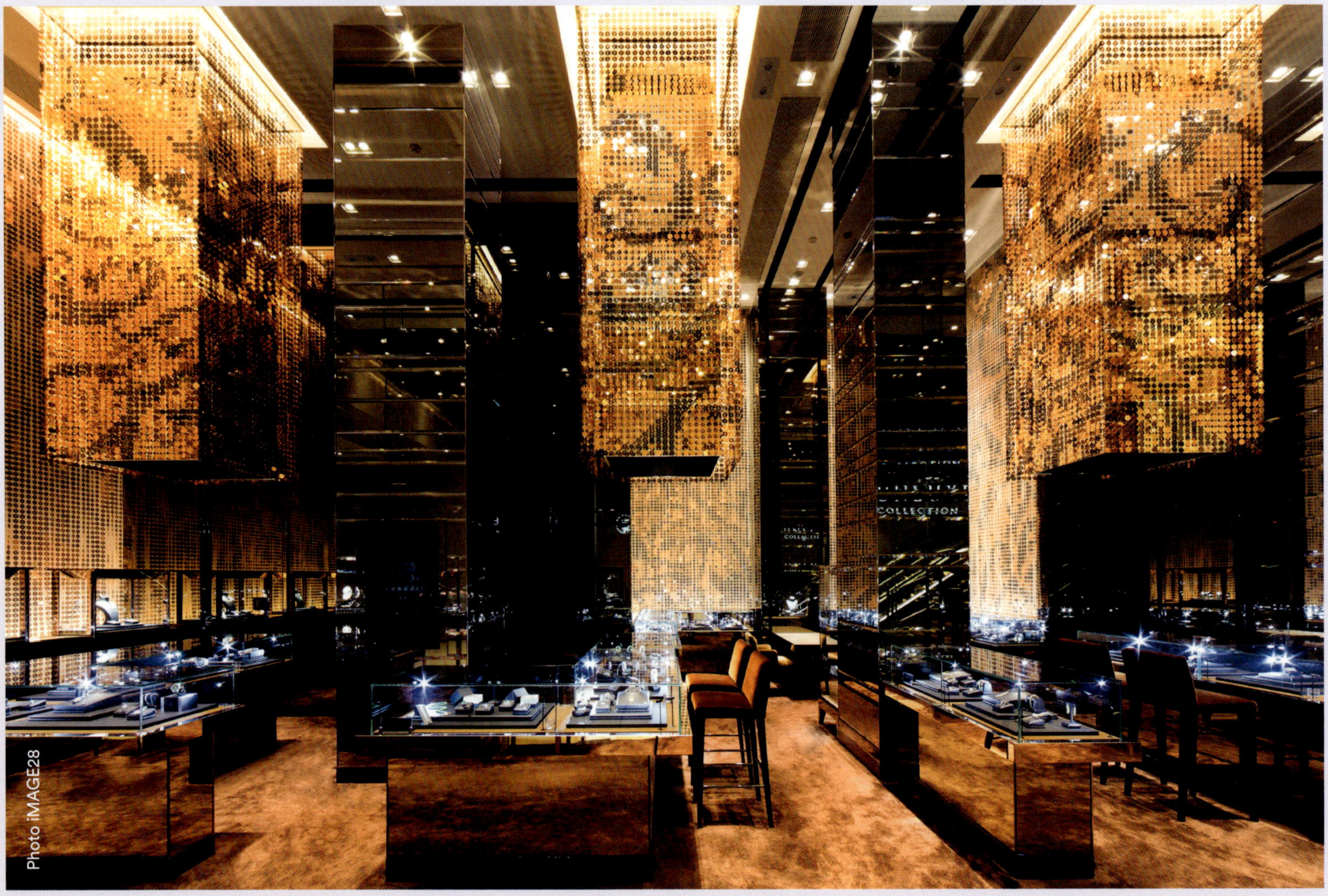

Photo iMAGE28

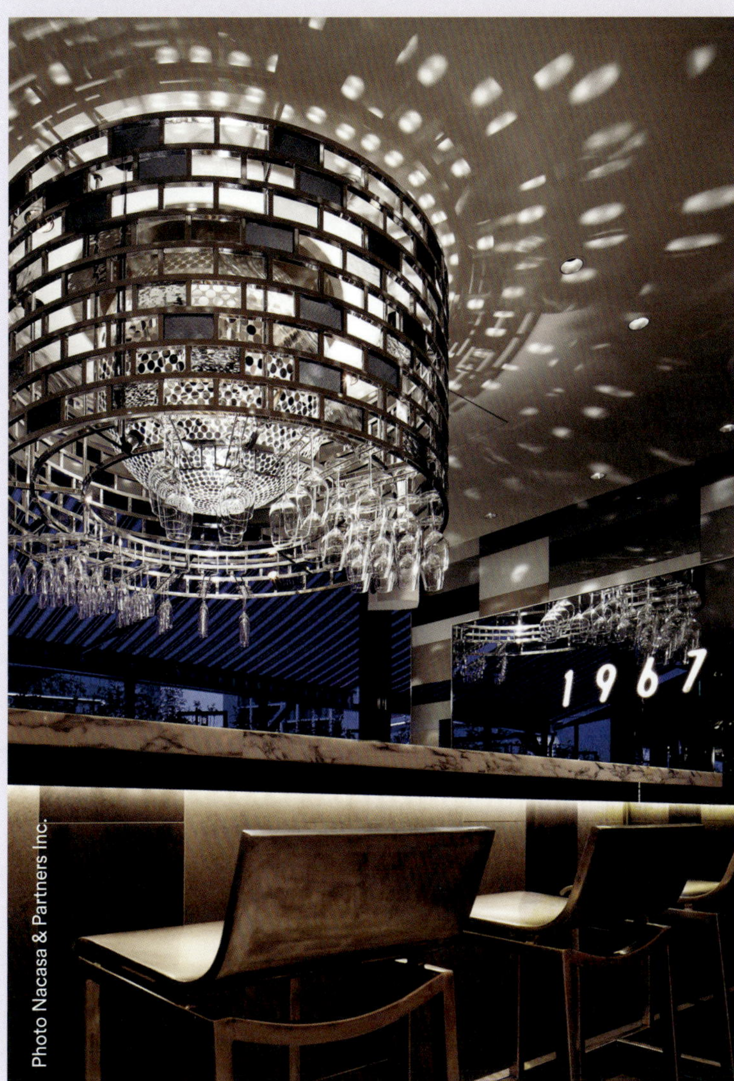

Photo Nacasa & Partners Inc.

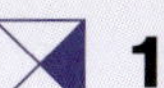

1967

When **2013**
Where **Tokyo, Japan**
Client **Diamond Dining**

The bar and lounge 1967 was named after three guys – the producer, the director and the principal designer, Yasumichi Morita of Glamorous – who all worked on this project and who were all born in that year. The stylish lounge is a luxury escape from the surrounding busy streets of the Roppongi neighbourhood in Tokyo. An eye-catching aspect above the marble-topped countertop in the bar is the contemporary chandelier.

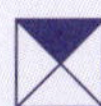

D

Where **2012**
Where **Tokyo, Japan**
Client **Naoki Ito Office**

When bar D opened in the busy area of Roppongi in central Tokyo, it gave its clientele an impression of sublime calmness – as if the time has stopped in this secluded space. The back wall is lined not in velvet drapes but with illuminated chain curtains, giving a golden glow to the small space. Bespoke lamps adorning the dark wood countertop radiate a diffuse glow, efficient to alleviate the spatial tension and create an intimate atmosphere between guests and bartender.

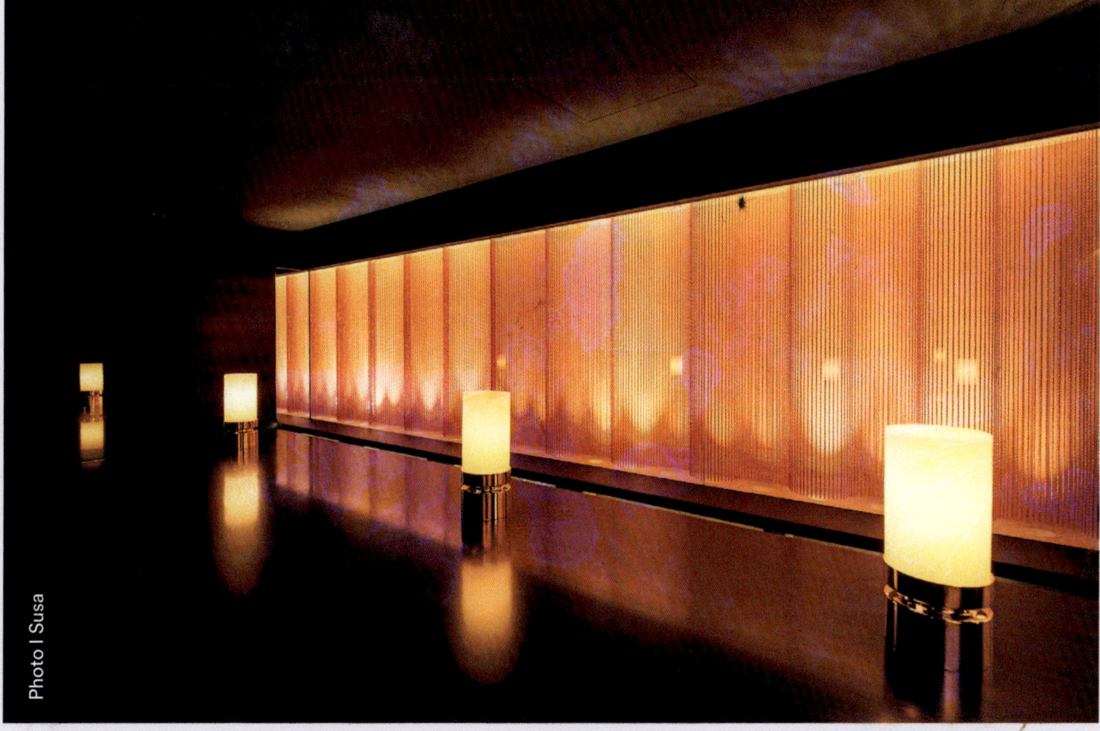

Photo I Susa

Studio Roosegaarde's tribute to Vincent van Gogh is an illuminated bike path that glows in the dark as it winds its way through the Dutch countryside.

Photos Daan Roosegaarde

Van Gogh Bike Path

An illuminated bike path that glows in the dark – much like the artwork that inspired it, *The Starry Night* – has been created in the Dutch countryside by **Studio Roosegaarde** as a fitting tribute to Vincent van Gogh.

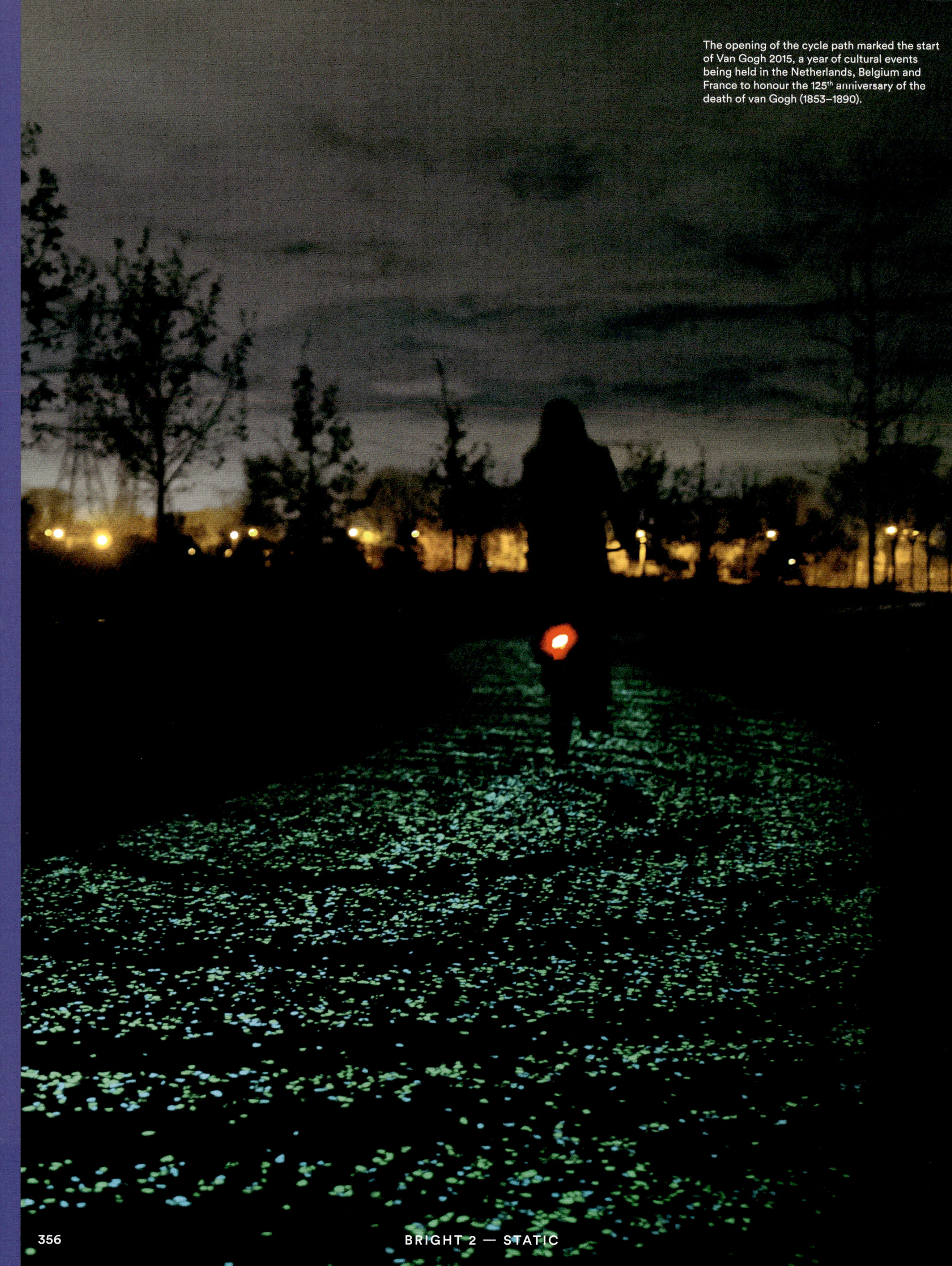

The opening of the cycle path marked the start of Van Gogh 2015, a year of cultural events being held in the Netherlands, Belgium and France to honour the 125th anniversary of the death of van Gogh (1853–1890).

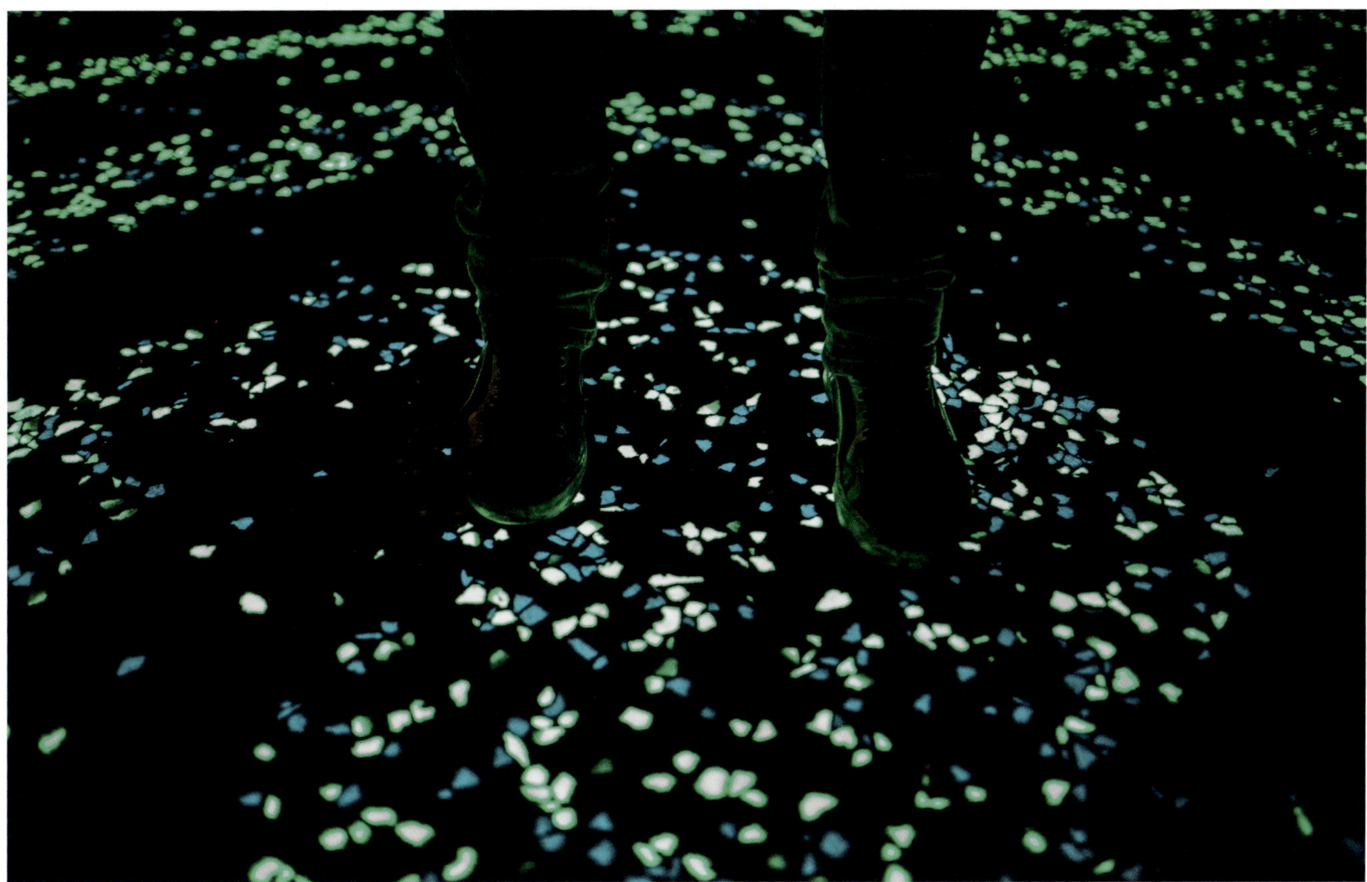

The community of Eindhoven made available not only a financial contribution but also the ground which served as a testing site for developing the cycle path.

Designer
Studio Roosegaarde
Location
Nuenen, the Netherlands
Client
Province of Brabant, City of Eindhoven and the Van Gogh Foundation
Collaborator(s)/consultant
Heijmans
Manufacturer
Heijmans
Date
November 2014

TWINKLING STONES EMBEDDED INTO THE PATH CREATE A PLAY ON LIGHT AND POETRY

The illuminated Van Gogh-Roosegaarde cycle path combines innovation and design with cultural heritage and tourism. Studio Roosegaarde was commissioned, by the body in change of a programme of events in 2015 to commemorate the 125th anniversary of Vincent van Gogh's death, to create a fitting tribute to the famous Dutch artist. The installation in the countryside near Eindhoven is a pathway illuminated by thousands of twinkling stones, that feature glow-in-the-dark technology and solar-powered LED lights – inspired by Van Gogh's work *The Starry Night*.

The 1-km-long section forms part of a cycle route in the southern Dutch province of Brabant, which begins at the city of Nuenen – where Van Gogh lived in 1883 – and links two watermills that feature in his masterpieces. After dark, visitors are amazed by a design that incorporates light and colour. The project involves contemporary design and technical innovations, where the path's surface 'charges' during the day and then emanates a radiant glow at night-time. Daan Roosegaarde teamed up with construction-services business Heijmans, an engineering firm which develops roads that are sustainable utilising smart technologies with concepts that combine light, energy and signage that react according to the present traffic situation.

Special innovative technology is utilised, illuminated by thousands of twinkling stones embedded into the path's surface. This creates a play on light and poetry, as Roosegaarde explains, 'I wanted to create a place that people will experience in a special way; technical combined with experience – that's what techno-poetry means to me.' —

THE PATH'S SURFACE 'CHARGES' DURING THE DAY AND EMANATES A RADIANT GLOW AT NIGHT

DAYTIME BIKE ROUTE

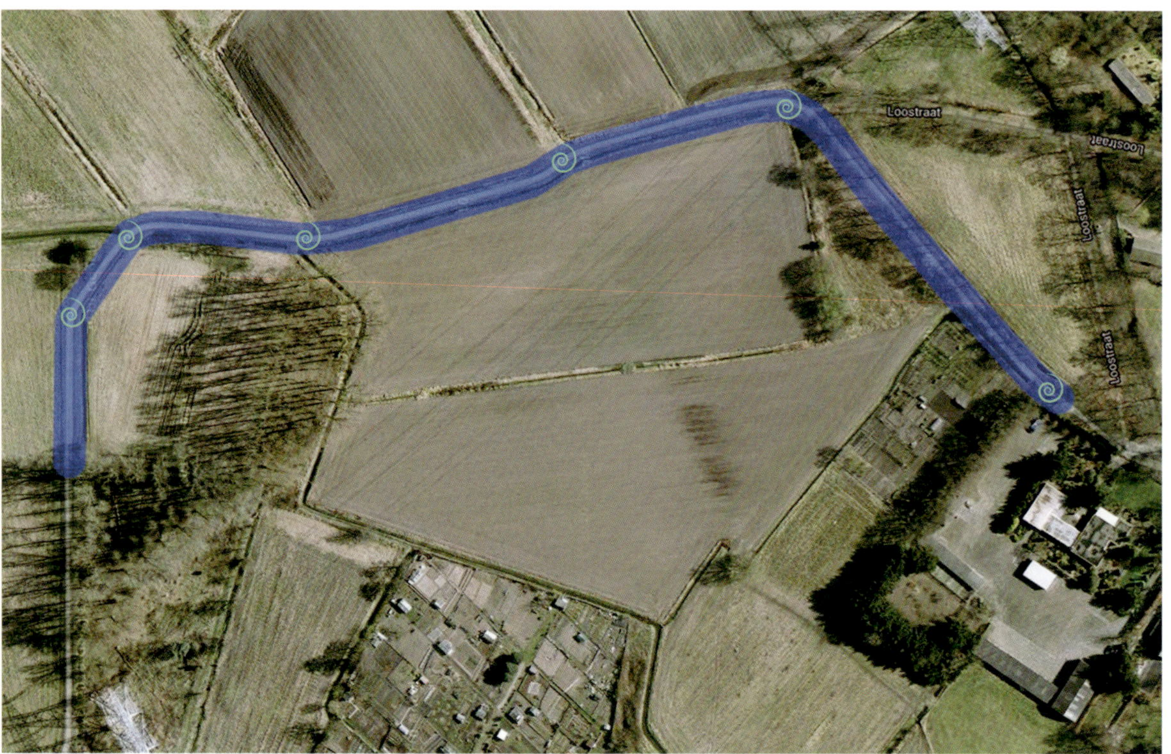

NIGHT-TIME BIKE ROUTE

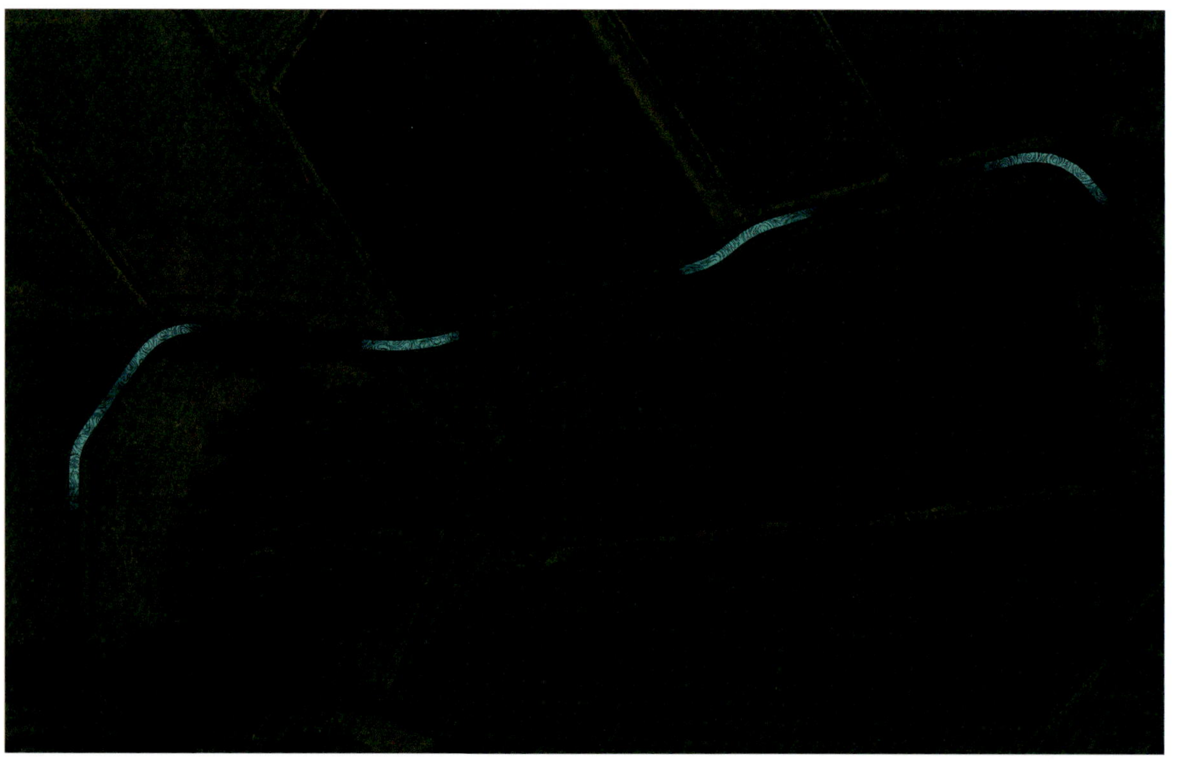

The 1-km-long section forms part of a cycle route in the Dutch province of Brabant.

Set-up

Total length of the installation for illumination was 600 m, which formed part of the Brabant cycle route (which was in total, 335 km).
Constituents of the pathway construction include a new layer of asphalt, scattered with thousands of luminous stones.
Used light came from the innovative technology that enables the pathway's stones to charge during the day and emit light in the evening.
Construction time for the path along the 600-m route was a total of 3 weeks. In the lead up to this, a technical team which was made up of representatives from Heijmans and Studio Roosegaarde, along with Daan Roosegaarde, worked on the path for 10 months.

Specifics

/\/\/\ The collaboration between Roosegaarde and Heijmans is a true example of innovative industries. The design and interactivity from Studio Roosegaarde and the craftsmanship of Heijmans were fused into one common goal: innovation of the Dutch landscape.

/\/\/\ Heijmans and Studio Roosegaarde translated a 600-m-long section of the 335-km-long Van Gogh cycle route into a contemporary design. The bicycle path section has been given a new layer of asphalt, scattered with thousands of luminous stones. What has resulted is the most innovative and artistic bicycle path in the Netherlands. The 600-m-long path runs past the place where Vincent van Gogh lived from 1883 to 1885. The bicycle path, designed by Daan Roosegaarde, consists of thousands of tiny, luminous stones and is inspired by the famous Van Gogh painting.

/\/\/\ Innovative technology enables them to charge during the day and emit light in the evening. The luminous stones are formed in a pattern inspired by the famous painting *The Starry Night* (1889). The painting depicts a night scene with yellow stars above a small, hilly town. People experience a ride across the bicycle path as if they are cycling through Van Gogh's very own masterpiece.

/\/\/\ During the development phase, the team worked with a priority that the bicycle path lighting should be as subtle as possible to ensure minimal intrusion on the habitat of animals. To this purpose, the intensity of the light-giving stones has been adapted. By incorporating lighting into the bicycle path itself, additional street lighting is unnecessary.

/\/\/\ Together with Studio Roosegaarde, Heijmans is giving consideration to the Dutch landscape in a new way. Light, energy and interaction play an important role in this process. This innovative cycle path and the development of Glowing Lines are examples of how innovation and design go hand in hand with culture and recreation.

DETAIL OF THE TWINKLING PATH

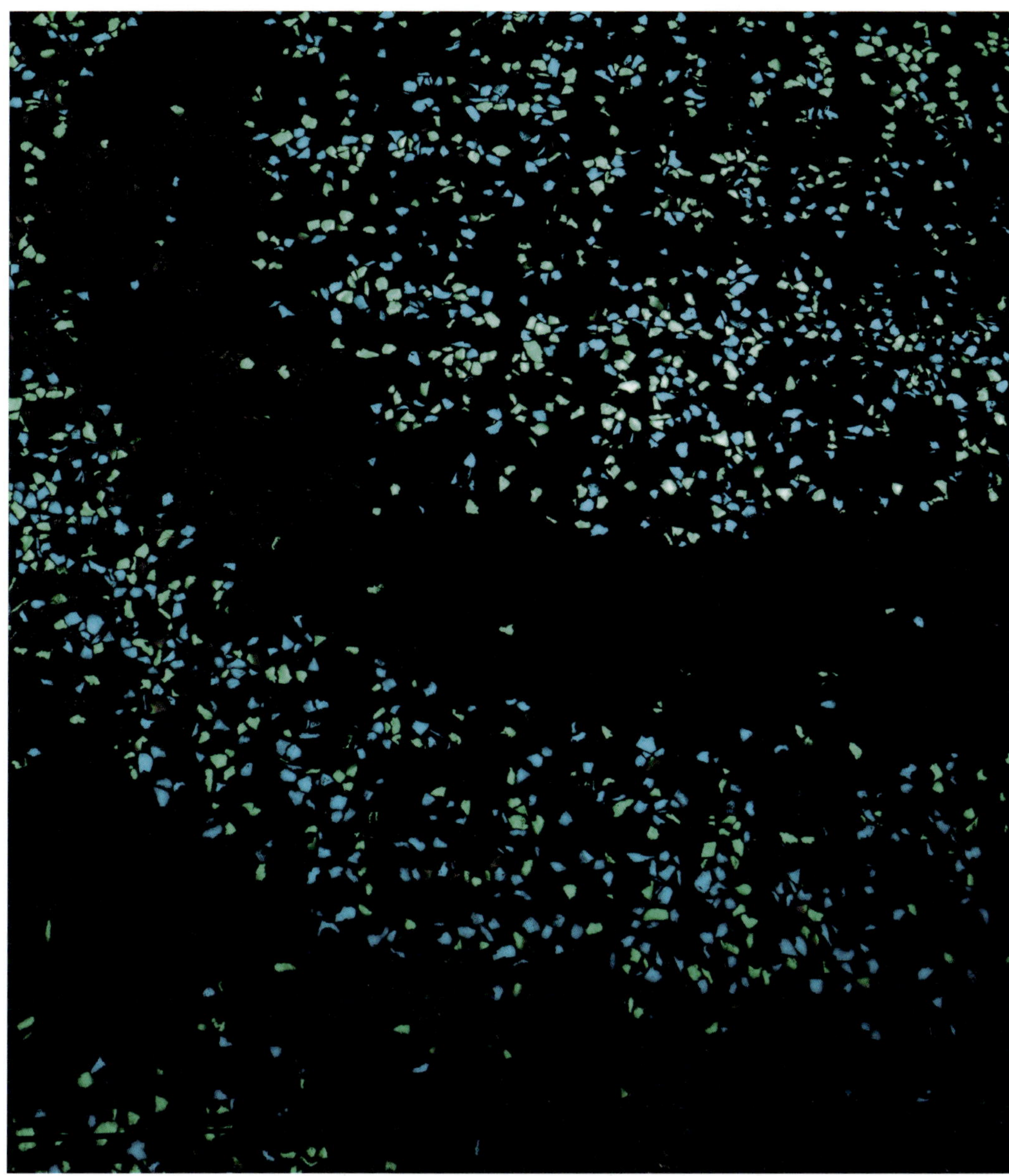

LAUNCH EVENT

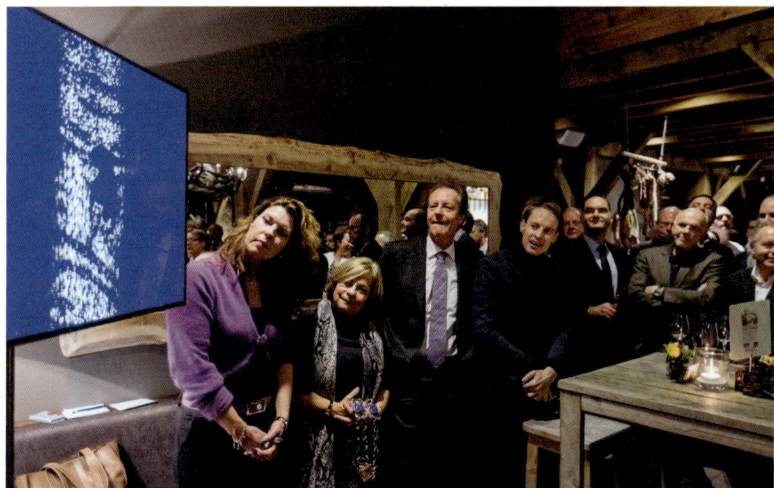

Daan Roosegaarde discussing the design at the launch of the cycle path, marking the start of the Van Gogh celebrations.

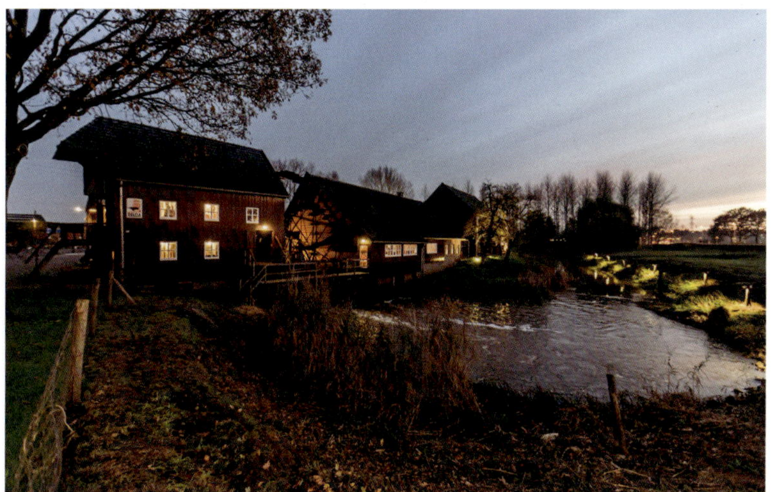

The path begins at the city of Nuenen, where Van Gogh lived in 1883, and links two watermills that feature in his masterpieces.

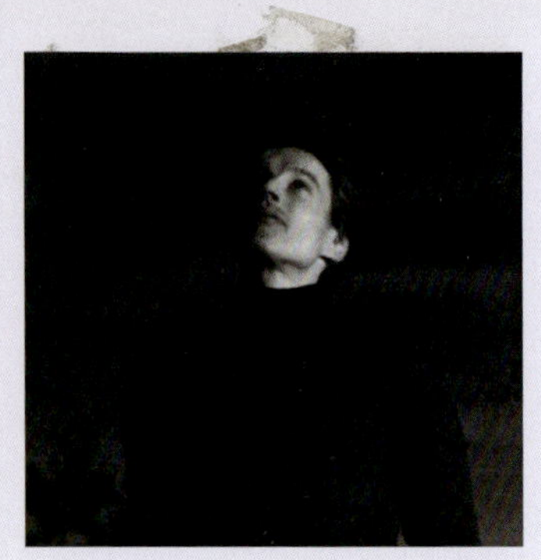

Studio Roosegaarde

In a world shifting between the analogue and the digital, Studio Roosegaarde is the social design lab of artist Daan Roosegaarde with his team of designers and engineers. The studio creates interactive designs that explore the dynamic relation between people, technology and space. The studio develops its own innovations and is internationally known for interactive projects, such as Dune, Intimacy and Smart Highway. Roosegaarde has been the focus of exhibitions at Rijksmuseum Amsterdam, Tate Modern and Victoria & Albert Museum in London and National Museum in Tokyo, and the winner of numerous international innovation awards.

STUDIOROOSEGAARDE.NET

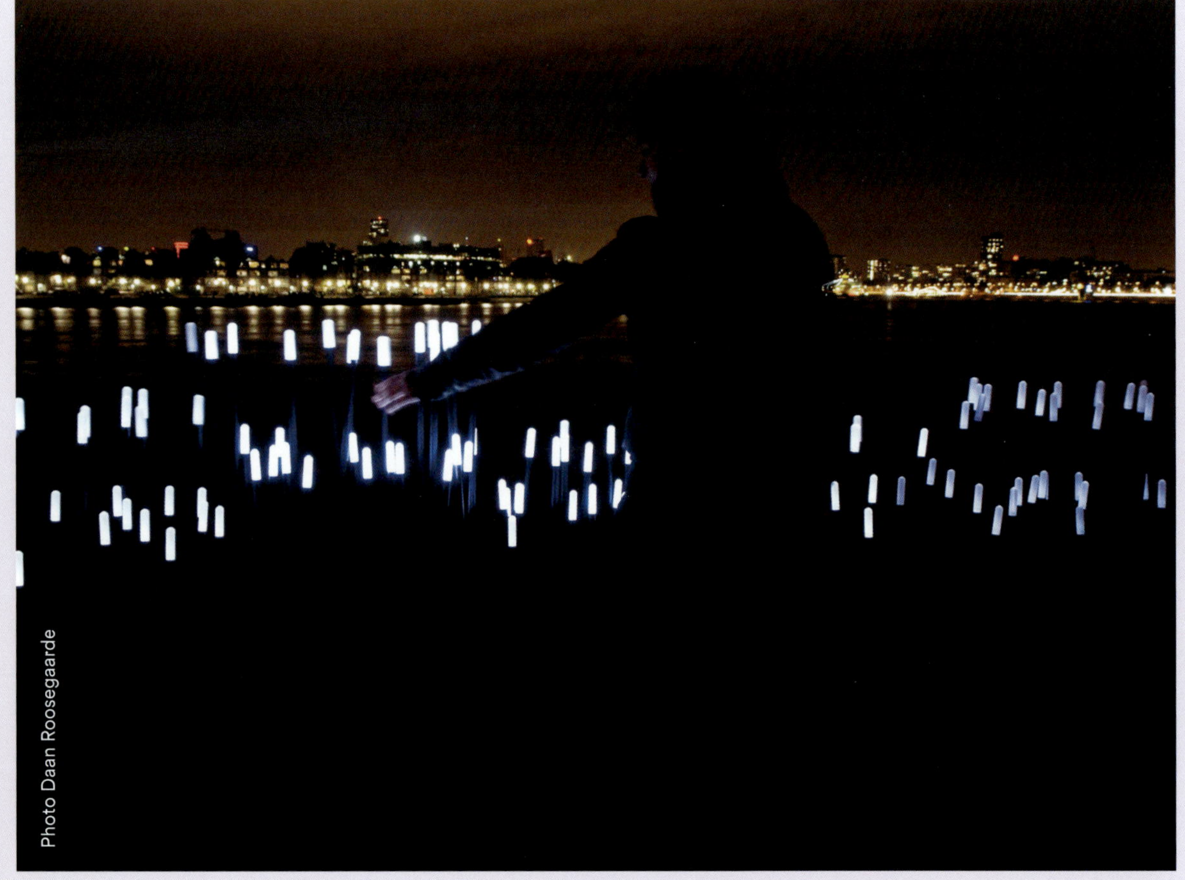

Photo Daan Roosegaarde

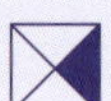

Dune 4.2

When **2011**
Where **Rotterdam, the Netherlands**
Client **City of Rotterdam**

Dune is a public interactive landscape that interacts with human behaviour. This hybrid of nature and technology is composed of large amounts of fibres that brighten according to the sounds and motion of passing visitors. The 60-m permanent installation of Dune 4.2, situated alongside the Maas River in Rotterdam, utilises fewer than 60 W of energy. Within this setting, Rotterdam citizens enjoy their daily 'walk of light'.

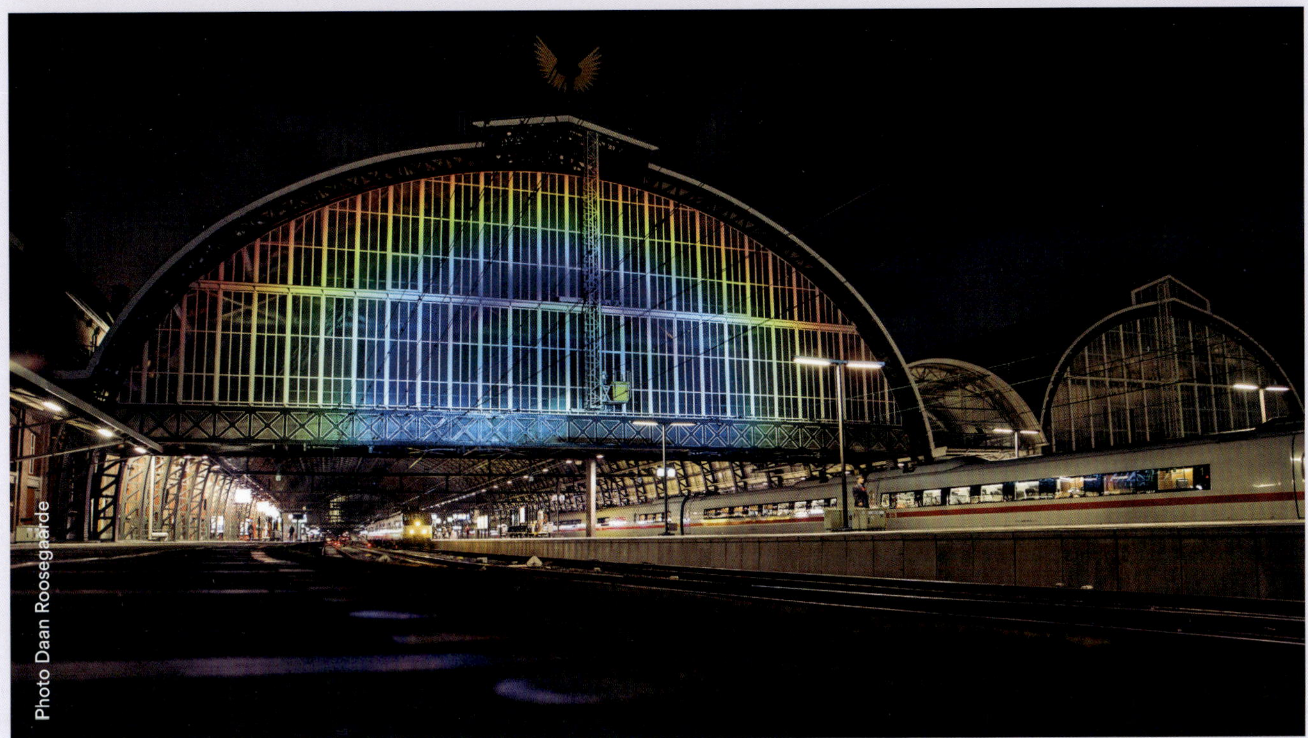

Photo Daan Roosegaarde

Rainbow Station

When **2014**
Where **Amsterdam, the Netherlands**
Client **Self-commissioned**

Rainbow Station connects the historic Amsterdam Central Station by using astronomy science to create a site specific 45 × 25 m rainbow of light. Working with astronomers of the University of Leiden, the artist has unravelled light efficiently into a spectrum of colours via liquid-crystal technology. The artwork marks the celebration of the 125th anniversary of the railway station, and the start of the UNESCO International Year of Light 2015. Rainbow Station can be seen every day for a brief moment within one hour after sunset.

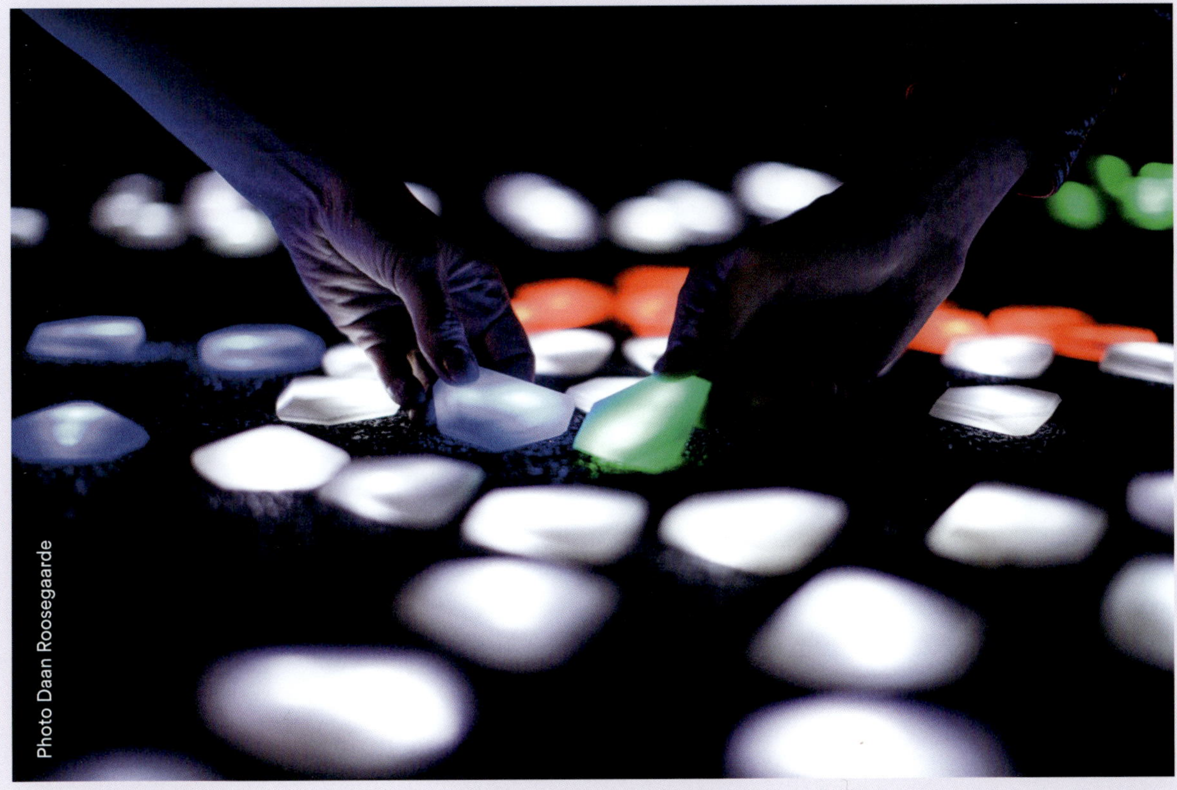

Photo Daan Roosegaarde

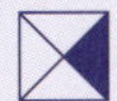

Crystal

When **2013**
Where **Eindhoven, the Netherlands**
Client **City of Eindhoven**

This installation is made up of hundreds of crystals of light, which brighten when touched. People can play with them and share their stories of light. Each 'crystal' contains LEDs that are wirelessly charged via a magnetic surface beneath. Once visitors start adding, moving or sharing them, the basic breathing of the constituent parts change. It is this lighting behaviour of the crystals as they move from 'excited' to 'bored' that keeps visitors curious.

Index

Designers

ACT Lighting Design

p.110

Avenue Wielemans Ceuppens 45/1
1190 Brussels
Belgium
+32 2 340 60 30
info@actlightingdesign.com
actlightingdesign.com

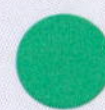

Antonin Fourneau

p.240

13, rue des Ecluses Saint-Martin
75010 Paris
France
+33 77 4478 61
info@waterlightgraffiti.com
waterlightgraffiti.com

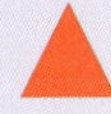

Arup

p.086

Naritaweg 118, Beta Building
1043 CA Amsterdam
the Netherlands
+31 203058500
amsterdam@arup.com
arup.com

AWA Lighting Designers

p.266

61 Greenpoint Avenue Suite 603
Brooklyn, NY 11222
United States
+1 212 473 9797
newyork@awalightingdesigners.com
awalightingdesigners.com

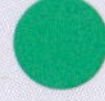

Christine Lyschik

p.168

Christian-Pless-Str. 11–13
63069 Offenbach
Germany
+49 177 2789 899
mail@christinelyschik.de
christinelyschik.de

Cinimod Studio

p.038

Unit 108, Canalot Studios
222 Kensal Road
London W10 5BN
United Kingdom
+44 208 969 3960
enquiries@cinimodstudio.com
cinimodstudio.com

Creative Machines

p.136

3113 East Columbia Street
Tucson, AZ 85714
United States
+1 520 294 0939
info@creativemachines.com
creativemachines.com

Daily tous les jours

p.144

5445 De Gaspé, Suite 420
Montreal, QC H2T 3B2
Canada
+1 514 296 6772
hello@dailytlj.com
dailytouslesjours.com

Daydreamers Design

p.070

PO Box 9089
Hong Kong Central
Hong Kong
+852 6431 7855
daydreamers.d3@gmail.com
bit.ly/daydreamersdesign

Electrolight

p.282

414 Lonsdale Street, Level 5
Melbourne 3000
Australia
+61 3 9670 2694
info@electrolight.com
electrolight.com

ERCO

p.298

Brockhauser Weg 80–82
58507 Luedenscheid
Germany
+49 2351 551 0
info@erco.com
erco.com

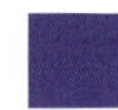

Glamorous

p.346

2F, 2-7-25 Motoazabu, Minatoku
106-0046 Tokyo
Japan
+81 3 5475 1037
info@glamorous.co.jp
glamorous.co.jp

Glasbau Hahn

p.118

Hanauer Landstrasse 211
60314 Frankfurt
Germany
+49 69 94417 0
info@glasbau-hahn.de
glasbau-hahn.de

Grimanesa Amorós Studio

p.006

54 N Moore St, 5th floor
New York, NY 10013
United States
+1 212 941 9787
studio@grimanesaamoros.com
grimanesaamoros.com

Har Hollands Lichtarchitect

p.014

Maasstraat 23
5626 BB Eindhoven
the Netherlands
+31 40 2621111
har@hollands.info
hollands.info

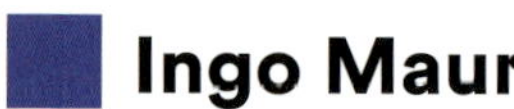

Ingo Maurer
p.314
Kaiserstrasse 47
80801 Munich
Germany
+49 89 381 606 0
info@ingo-maurer.com
ingo-maurer.com

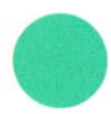

Ivan Toth Depeña
p.224
Charlotte, NC
and Miami, FL
United States
ivandepena.com

Jan Plecháč and Henry Wielgus
p.046
Cimburkova 6
Prague 13000
Czech Republic
+42 732729784
info@janandhenry.com
janandhenry.com

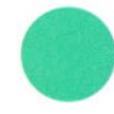

LED-Art
p.216
Claustrum 13
6515 GH Nijmegen
the Netherlands
+31 641525072
info@led-art.nl
led-art.nl

lichtundsoehne
p.118
Warschauer Strasse 60
10243 Berlin
Germany
+49 163 27 07 98 3
info@lichtundsoehne.de
lichtundsoehne.de

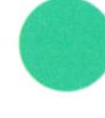

Lighting Design Collective
p.232
Carrera de San Jeronimo 16, 2D
28014 Madrid
Spain
+34 911885851
info@ldcol.com
ldcol.com

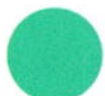

Like Architects
p.184
Praça Coronel Pacheco, 2
4050 453 Porto
Portugal
+351 220 123 744
info@likearchitects.net
likearchitects.net

Loop.pH
p.094
231 Church Street, Units 2 & 4
Stoke Newington
London N16 9HP
United Kingdom
+44 207 812 9188
info@loop.ph
loop.ph

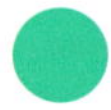

Matthijs Munnik
p.152
Amperestraat 95a
2563 ZS The Hague
the Netherlands
+31 6 48641770
matthijs.munnik@gmail.com
matthijsmunnik.nl

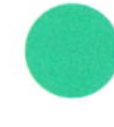

Media Architecture Institute
p.208
1502/355 Kent Street
Sydney 2000
Australia
+61 417 117281
institute@mediaarchitecture.org
mediaarchitecture.org

Moment Factory
p.062
6250 Hutchison, Suite 200
Montreal, QC H2V 4C5
Canada
+1 514 843 8433
info@momementfactory.com
momentfactory.com

NE-AR
p.118
Dreikönigsstrasse 35
60594 Frankfurt
Germany
+49 69 173 209 850
info@ne-ar.com
ne-ar.com

Olafur Eliasson
p.274
Christinenstr. 18/19, Haus 2
10119 Berlin
Germany
studio@olafureliasson.net
olafureliasson.net

Ombrages
p.030
1015 Avenue Wilfrid-Pelletier
Quebec, QC G1W 0C5
Canada
+1 418 780 2220
info@ombrages.com
ombrages.com

Omni Pictures
p.102
Leeds
United Kingdom
studio@omnipictures.com
omnipictures.com

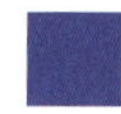

PointOfView
p.258
207 Clarence Street, Level 3
Sydney 2000
Australia
+61 2 9818 6355
sydney@pov.com.au
pov.com.au

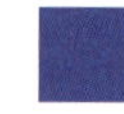

Ralf Westerhof
p.290
Minahassastraat 61
1094 RV Amsterdam
the Netherlands
+31 6 41505660
studio@ralfwesterhof.nl
ralfwesterhof.nl

realities:united
p.022
Falckensteinstrasse 48
10997 Berlin
Germany
+49 30 20646630
info@realu.de
realities-united.de

Ross Lovegrove

p.322

21 Powis Mews
London W11 1JN
United Kingdom
+44 20 7229 7104
general@rosslovegrove.com
rosslovegrove.com

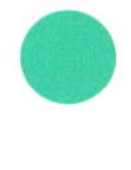

Singapore University of Technology and Design (SUTD)

p.176

20 Dover Drive
138682 Singapore
Singapore
+65 6303 6675
enquiry@sutd.edu.sg
sutd.edu.sg

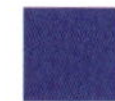

Skira

p.330

Veruda 60B
52100 Pula
Croatia
+385 52 535 940
info@skira.hr
skira.hr

SmartLight

p.078

Jules Verneweg 21-09
5015 BE Tilburg
the Netherlands
+31 13 544 975 8
info@lightsolutions.nl
lightsolutions.nl

Speirs + Major

p.054

8 Shepherdess Walk, Third Floor
London N1 7LB
United Kingdom
+44 20 7067 4700
info@speirsandmajor.com
speirsandmajor.com

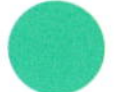

Studio Drift

p.160

Asterweg 20, B1
1031 HN Amsterdam
the Netherlands
+31 20 840 6993
info@studiodrift.com
studiodrift.com

studioheyhey

p.118

Niddastrasse 64
60329 Frankfurt
Germany
+49 69 56 99 85 72
info@studioheyhey.com
studioheyhey.com

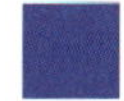

StudioNoc

p.306

Kungsgatan 5
111 43 Stockholm
Sweden
+46 8 141111
hi@studionoc.com
studionoc.com

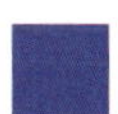

Studio Roosegaarde

p.354

Coenecoop 620
2741 PV Waddinxveen
the Netherlands
+31 182 769213
mail@studioroosegaarde.net
studioroosegaarde.net

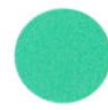

Tjep.

p.200

Veembroederhof 204
1019 HC Amsterdam
the Netherlands
+31 20 362 42 96
goodnews@tjep.com
tjep.com

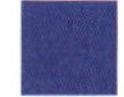

Tropp Lighting Design

p.250

Marienplatz 5
82362 Weilheim
Germany
+49 881 92486 90
info@tropp-lighting.com
tropp-lighting.com

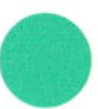

Unstable

p.128

Reykjavik
Iceland
+354 661 8166
info@unstablespace.com
unstablespace.com

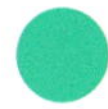

Venividimultiplex

p.192

Eerste Keucheniusstraat 8hs
1051 HR Amsterdam
the Netherlands
+31 6 2729 6447
contact@venividimultiplex.com
venividimultiplex.com

XAL

p.338

Auer-Welsbach-Gasse 36
8055 Graz
Austria
+43 316 3170 0
office@xal.com
xal.com

Projects

ARCHITECTURAL ILLUMINATION

LIGHT INSTALLATIONS

Credits

Bright 2
Architectural Illumination
and Light Installations

Publisher
Frame Publishers

Production
Carmel McNamara

Authors
Carmel McNamara and Ana Martins

Graphic Design
Mariëlle van Genderen and Federica Ricci

Prepress
Edward de Nijs

Trade distribution USA and Canada
Consortium Book Sales & Distribution, LLC.
34 Thirteenth Avenue NE, Suite 101,
Minneapolis, MN 55413-1007
United States
T +1 612 746 2600
T +1 800 283 3572 (orders)
F +1 612 746 2606

Trade distribution Benelux
Frame Publishers
Laan der Hesperiden 68
1076 DX Amsterdam
the Netherlands
distribution@frameweb.com
frameweb.com

Trade distribution rest of world
Thames & Hudson Ltd
181A High Holborn
London WC1V 7QX
United Kingdom
T +44 20 7845 5000
F +44 20 7845 5050

ISBN: 978-94-91727-41-2

The Koninklijke Bibliotheek lists this publication in the Nederlandse Bibliografie: detailed bibliographic information is available on the internet at http://picarta.pica.nl

Printed on acid-free paper produced from chlorine-free pulp. TCF ∞
Printed in the Netherlands

987654321